This book is part of the

ALLYN AND BACON SERIES IN ENGINEERING

Consulting Editor: FRANK KREITH
UNIVERSITY OF COLORADO

Dedicated to

D. C. SPRIESTERSBACH

A Dean of Deans

A division of Simon & Schuster
160 Gould Street
Needham Heights, Massachusetts 02194

Haug, Edward J.
Computer aided kinematics and dynamics of mechanical systems / Edward J. Haug.
p. cm.
Bibliography: p.
Includes index.
Contents: v. 1. Basic methods.
ISBN 0-205-11669-8 (v. 1)
1. Machinery, Kinematics of—Data processing. 2. Machinery, Dynamics of—Data processing. I. Title.
TJ175.H34 1989
621.8′11—dc 19 88-24111
CIP

Printed in the United States of America

10 9 8 7 6 5 4 3 2 1 93 92 91 90 89

COMPUTER AIDED KINEMATICS AND DYNAMICS OF MECHANICAL SYSTEMS

VOLUME I: BASIC METHODS

EDWARD J. HAUG
Center for Computer Aided Design
and
Department of Mechanical Engineering
College of Engineering
The University of Iowa

ALLYN AND BACON
Boston London Sydney Toronto

Contents

Preface

OBJECTIVES

The field of kinematics and dynamics of mechanical systems has progressed from a manual graphics art to a highly developed discipline in analytical geometry and dynamics. Large displacements and rotations that occur in the kinematic and dynamic performance of mechanical systems lead to nonlinear mathematical models that must be formulated and analyzed. The analytical complexity of nonlinear algebraic equations of kinematics and nonlinear differential equations of dynamics makes it impossible to obtain closed-form solutions in most applications. The literature in this field, therefore, contains a large number of specialized techniques and elegant analytical methods for analysis of special-purpose mechanisms and machines. This situation contrasts sharply with the literature on finite element structural analysis and electronic circuit analysis, both of which are much more systematically developed and oriented toward both computer formulation and solution of their governing equations.

This text presents basic methods for the analysis of the kinematics and dynamics of planar and spatial systems, using a Cartesian coordinate approach that is applicable to broad classes of mechanisms and machines. The approach taken places emphasis on systematic methods that are used in computer formulation and solution of the governing equations of kinematics and dynamics, thus removing the burden of routine derivation and computation from the engineer. Basic theoretical considerations that must be understood by the engineer in preparing models for computer simulation and in interpreting results, to assure that reality has indeed been modeled, are emphasized throughout the text. The objective is to prepare the reader to use a large-scale computer code that implements tedious computation, but requires its user to make modeling judgments and to critically evaluate reasonableness of results.

The text is intended for use in junior or senior undergraduate and beginning graduate engineering instruction and for self-study by practicing engineers who have a batchelor's degree. Matrix methods for vector analysis and multivariable calculus are introduced and used throughout the text. Equations of kinematics are developed using only basic methods of vector analysis and calculus.

Equations of dynamics are derived, beginning with two of Newton's basic laws of motion for particles: (1) force equals mass times acceleration, and (2) reaction forces are equal in magnitude and opposite in direction. Numerical methods are presented in adequate detail for an understanding of all the calculations that are required for computer solution of the equations of kinematics and dynamics.

The focus of the text is on the kinematic and dynamic analysis of moderate- and large-scale mechanical systems that are encountered in diverse fields of application. Considerable emphasis is therefore placed on modeling mechanical systems and interpreting the results obtained from computer simulations. Examples are presented to illustrate the physical difficulties associated with redundant constraints, lock-up of mechanisms, and singular configurations that prevent a mechanism or machine from performing the intended function. Since poorly designed mechanisms and machines may fail physically, it is clear that the mathematics that represents their performance must get into trouble. The mathematical consequences of mechanism and machine physical failure are identified and illustrated to assist the reader in interpreting the numerical results and determining the physical implications of the performance of the system.

This text serves as the theoretical foundation for a large-scale computer code called the Dynamic Analysis and Design System (DADS), which is commercially available from Computer Aided Design Software Incorporated, P.O. Box 203, Oakdale, Iowa 52319. This computer code is available, at minimal cost, for use in academic instruction. Examples and exercises are presented in the text, using the DADS computer code to present formulations and results for a variety of applications, to interpret results, and to investigate design alternatives that may improve performance. While the reader can gain insights with analytical examples and exercises in the text, it is very difficult to develop a practical capability in kinematics and dynamics of machines without substantial experience with real mechanical systems. It is therefore recommended that the DADS computer code be used in conjunction with the text, so that the reader can create models and gain experience in actual kinematic and dynamic analysis.

ORGANIZATION OF TEXT

Chapter 1 discusses the scope of the kinematics and dynamics of mechanical systems and considerations that guide the development of computer-aided engineering methods in kinematics and dynamics.

Chapters 2 through 8 comprise Part One of the text, which is concerned with planar kinematics and dynamics of mechanical systems. Chapter 2 presents a self-contained development of the matrix algebra and multivariable calculus notation and methods that are used throughout the text. Cartesian coordinates that are used to position and orient bodies in the plane and a library of kinematic constraints between bodies are presented and analyzed in Chapter 3. Methods of controlling, or driving, the motion of mechanisms and for carrying out position, velocity, and acceleration analysis are presented and illustrated with analytical examples. Numerical methods for solving equations of kinematics are presented in Chapter 4. Matrix methods for solving linear equations and the Newton–Raphson method for solving nonlinear equations are developed and illustrated. Planar kinematic systems are modeled and analyzed in Chapter 5 to illustrate the use of methods developed to evaluate performance and consider the effects of design changes.

The equations of dynamics for planar systems are developed in Chapter 6 to include internal forces due to springs and dampers, calculation of reaction forces in joints and drivers, and the determination of equilibrium configurations. Numerical methods for solving the equations of dynamics and statics are presented in Chapter 7. A brief, but self-contained, development of numerical integration methods that are suitable for large-scale system dynamic analysis is presented. Dynamic analyses of realistic examples, including those whose kinematics are studied in Chapter 5, are presented in Chapter 8.

Chapters 9 through 12 comprise Part Two of the text, which treats the kinematics and dynamics of mechanical systems that move in three-dimensional space. Euler parameters are introduced for orientation of bodies in space in Chapter 9. A library of kinematic constraints is derived and the governing kinematic equations for determining position, velocity, and acceleration are developed and analyzed. Kinematic analyses of spatial systems are presented in Chapter 10. Dynamic equations for spatial systems are developed in Chapter 11, as a direct extension of the derivation for planar systems in Chapter 6. Finally, dynamic analyses of spatial systems are carried out in Chapter 12.

USE OF TEXT IN TEACHING AND SELF-STUDY

The contents of this text have been used in four different courses and course sequences at the University of Iowa during the past eight years. One option is to teach a one-semester course on planar kinematics and dynamics of mechanical systems, at the undergraduate level, based on Part One of the text. The second one-semester course taught, at a slightly higher level, reviews basic ideas of planar systems and progresses to Part Two of the text, plus Chapters 4 and 7, for the study of spatial kinematics and dynamics of mechanical systems. A third single-semester course has been offered in planar and spatial kinematics alone. Finally, the entire content of the text has been used as the basis for a two-semester sequence that is taught at the senior undergraduate and beginning graduate level on the kinematics and dynamics of both planar and spatial systems.

In virtually all course offerings that have used this text, it has been found that project assignments using the DADS code have been a valuable tool in permitting students to formulate realistic models and obtain numerical feedback to gain analysis experience. One major challenge in teaching kinematic and dynamic analysis of mechanical systems is to develop a physical understanding of mechanisms and machines and the techniques for modeling and interpreting results. While special-purpose computer programs can be written to implement the computational methods presented in the text, it is far more efficient to use an available and thoroughly debugged computer program such as DADS to permit students to develop facility in the modeling and analysis of realistic mechanical systems. Studies of the effect of design variations on the kinematic and dynamic performance of machines, as in the examples treated in Chapters 5, 8, 10, and 12, are remarkably effective in motivating students and in gaining insights into use of the tools of kinematics and dynamics of mechanical systems.

CONTENTS OF VOLUME II: ADVANCED AND INTERDISCIPLINARY METHODS

Advanced and interdisciplinary topics in computer aided kinematics and dynamics of mechanical systems will be treated in a forthcoming companion volume. The foundations for the second volume include the basic formulation and numerical methods contained in this text, fundamental concepts of dynamics that are contained in a related text, *Intermediate Dynamics* [35], and recently developed fundamental theory and numerical methods.

The contents of Volume II will include methods of computer graphics for model formulation and animated graphics for interpretation of computer predictions. Feedback control and hydraulic subsystems, which are integral to dynamic performance of modern mechanical systems, will be modeled using formulations similar to those presented in this text.

Advanced relative coordinate kinematic and dynamic formulations that offer potential for high-speed kinematic and dynamic simulation will be developed and computational methods that are suitable for their implementation will be presented. Emerging multiprocessor parallel computer methods that provide a unique opportunity for high-speed computation, to include real-time man-in-the-loop simulation, will be developed with relative coordinate formulations. This body of emerging technology represents a major extension of the basic methods presented in the current text.

Special problems and interdisciplinary considerations in modern machine dynamics will be developed, including friction, stiction, impact, and intermittent motion; analysis of singular kinematic configurations and dynamic stability; definition of boundaries of robot and manipulator workspaces; and analysis of flexible multibody dynamics. Methods developed in these areas will be applicable with the basic formulations presented in this text and with advanced relative coordinate formulations.

Advanced numerical methods for the analysis and solution of differential-algebraic equations of motion, using emerging manifold theoretic and differential geometric methods of mathematics and numerical analysis, will be developed.

These methods provide powerful new tools for solving the equations of motion of broad classes of mechanical systems. Finally, design sensitivity analysis and optimization methods that predict the effect of design variations on dynamic performance and that optimize designs will be presented and illustrated.

ACKNOWLEDGMENTS

The author extends thanks to the following reviewers, whose contributions have enriched the text:

Professor Robert A. Lucas
Lehigh University

Professor David Hutton
Washington State University

Professor H. J. Sommer III
Pennsylvania State University

Special thanks are due to many colleagues and students who have made major contributions to development of this text over the past decade, including K. E. Atkinson, D. S. Bae, R. R. Beck, R. S. Hwang, D. Y. Jo, (S. S. Kim)^2, O. K. Kwon, N. K. Mani, P. E. Nikravesh, T. Park, J. O. Song, R. A. Wehage, S. C. Wu, and W. S. Yoo.

Finally, the author is indebted to C. R. Geyer, R. L. Huff, J. L. Markham, C. K. Mills, and S. M. VanFosson for their patience and dedication in preparation of many drafts of the manuscript and illustrations.

E. J. H.

CHAPTER ONE

Elements of Computer-Aided Kinematics and Dynamics

The impact of the digital computer on all fields of science and engineering is already significant, to the point of becoming dominant in many disciplines. Well-developed computer software has already revolutionized the analysis of structures and electronic circuits. The situation, however, is quite different in the kinematics and dynamics of mechanical systems. While the potential for computational techniques in this field is at least as great as for structures and circuits, development has lagged behind. The objective of this text is to present basic methods for the computer formulation and solution of the equations of kinematics and dynamics of mechanical systems, to permit mechanical engineers to enjoy some of the benefits of modern computer methods that have served structural and electrical engineers for some time. The purpose of this chapter is to define the scope of the subject, to introduce prototype applications, and to outline methods that may be brought to bear for computer formulation and solution of the governing equations of kinematics and dynamics.

1.1 SCOPE OF MECHANICAL SYSTEM KINEMATICS AND DYNAMICS

For the purposes of this text, a *mechanical system* is defined as a collection of interconnected rigid bodies that can move relative to one another, consistent with joints that limit relative motion of pairs of bodies. The motion of a mechanical system may be prescribed by defining the time history of the position or relative position of some of its bodies. The motion of the system is then determined by algebraic kinematic relations or from differential equations of motion and externally applied forces, in which case the motion of the system is determined by laws of physics. Kinematics and dynamics of mechanical systems are characterized by large amplitude motion, which leads to geometric nonlinearity that is reflected in the algebraic equations of constraint and differential equations of motion.

Three basically different kinds of analysis are employed in the design of mechanical systems:

Kinematic analysis of a mechanical system concerns the motion of the system independent of forces that produce the motion. Typically, the time history of position or relative position of one or more bodies in the system is prescribed. Time histories of position, velocity, and acceleration of the remaining bodies are then determined by solving systems of nonlinear algebraic equations for position and linear algebraic equations for velocity and acceleration.

Dynamic analysis of a mechanical system concerns the motion of the system that is due to the action of applied forces. A special case of dynamic analysis is the determination of an equilibrium position of the system under the action of forces that are independent of time. The motion of the system, under the action of applied forces, is required to be consistent with kinematic relations that are imposed on the system by joints that connect bodies in the system. The equations of dynamics are differential equations or a combination of differential and algebraic equations.

Inverse dynamic analysis is a hybrid form of kinematic and dynamic analysis in which the time history of positions or relative positions of one or more bodies in the system is prescribed, leading to complete determination of position, velocity, and acceleration of the system from the equations of kinematics. The equations of motion of the system are then solved, with known position, velocity, and acceleration, as algebraic equations to determine the forces that are required to generate the prescribed motion.

An important consideration that serves to classify mechanical systems concerns the source of forces that act on the system. This is particularly important in modern mechanical systems in which some form of control is exerted. Force effects due to electrical and hydraulic feedback control subsystems play a crucial role in the dynamics of modern mechanical systems. The scope of mechanical system dynamics is, therefore, heavily dependent on the classes of force systems that act on the system.

The most elementary form of force that acts on a mechanical system is *gravitational force,* which is normally taken as constant and acting perpendicular to the surface of the earth. Other relatively simple forces that act on bodies in a system, due to interaction with their environment, include aerodynamic forces and friction and damping forces that act due to the relative motion of the components of the system. An important class of forces that act in a mechanical system is associated with *compliant elements,* such as coil springs, leaf springs, tires, shock absorbers, and a multitude of other deformable components that have reaction forces and moments associated with them. Forces due to compliant elements act between bodies in the system and are functions of their relative position and velocity.

To be more concrete regarding classes of mechanical system kinematic and

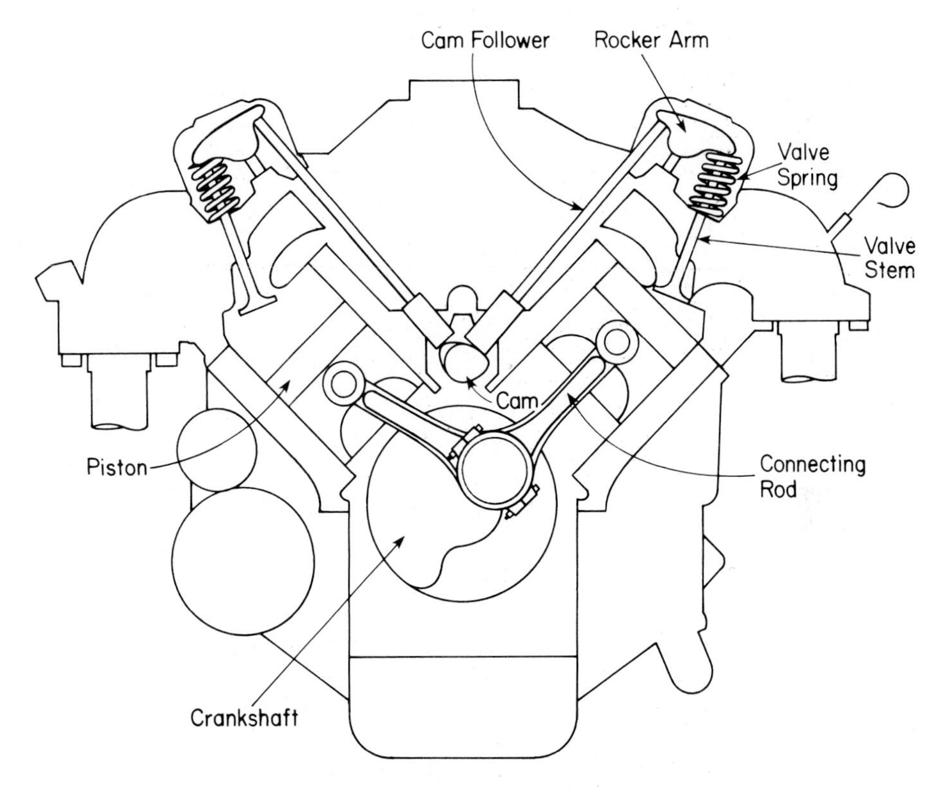

Figure 1.1.1 Cross section of a V-8 engine.

dynamic applications to be considered in the text, it is helpful to review a few typical engineering applications. The V-8 engine shown schematically in Fig. 1.1.1 contains many moving parts and illustrates a number of the most common mechanisms employed in machine design. The crankshaft of the engine rotates in lubricated bearings and contains eccentric rotational bearings with connecting rods, which are subsequently coupled through rotational bearings to translating pistons that move in combustion cylinders. The crankshaft–connecting rod–piston assembly comprises what is commonly called a *slider–crank mechanism,* which is used in this application and in many other machine components. The basic purpose of this mechanism is to transfer forces that are induced by combustion of fuel on the pistons into torques that act about the axis of rotation of the crankshaft, hence inducing the rotational motion that is used to propel a vehicle or to drive rotating machinery. Cams are typically used to induce the precisely timed motion of the cam–follower, which controls the position of the valve stem through a rocker arm, to open and close the intake and exhaust valves during engine operation. To close a valve and to maintain contact between the cam and follower, valve springs are used, as shown. While there are numerous

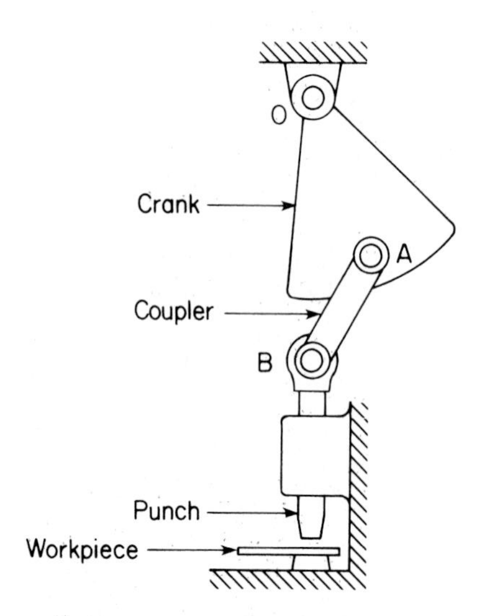

Figure 1.1.2 Punch mechanism.

other mechanisms within an engine, these basic components provide examples of typical machine elements.

A second application of the slider–crank mechanism is the punch mechanism shown schematically in Fig. 1.1.2. The crank is rotated about pivot point O through some angle of oscillation. The coupler, which is pivoted with the crank at point A and with the punch at point B, transmits load to the punch that translates in the machine housing, to provide reciprocating motion of the punch. Careful control of dimensions permits a modest torque that acts through a substantial driving angle of the crank to be transformed into a very large force that acts through a short range of motion of the punch, to deform or cut the workpiece.

A third example of a slider–crank mechanism is the fly-ball governer shown in Fig. 1.1.3. Relatively massive balls are attached to arms that are pivoted to a rotating shaft, so that they rotate with the shaft. Coupler arms are pivoted in the ball arms and a collar that is constrained to translate along the shaft. The ball arms, couplers, collar, and shaft form two slider–crank mechanisms that operate in spatial motion. This entire mechanism rotates with angular velocity ω of the shaft. As the shaft speeds up, centrifugal forces act on the balls to throw them out, causing the collar to move upward, hence increasing the distance s shown in Fig. 1.1.3.

The purpose of the fly-ball governor is to control the operating speed of an engine. A mechanism couples the position s of the collar to the fuel feed of an internal combustion engine that drives the shaft. The mechanism is designed so that, at the desired speed of the engine, centrifugal forces on the balls and gravitational and spring forces that act on the mechanism reach an equilibrium

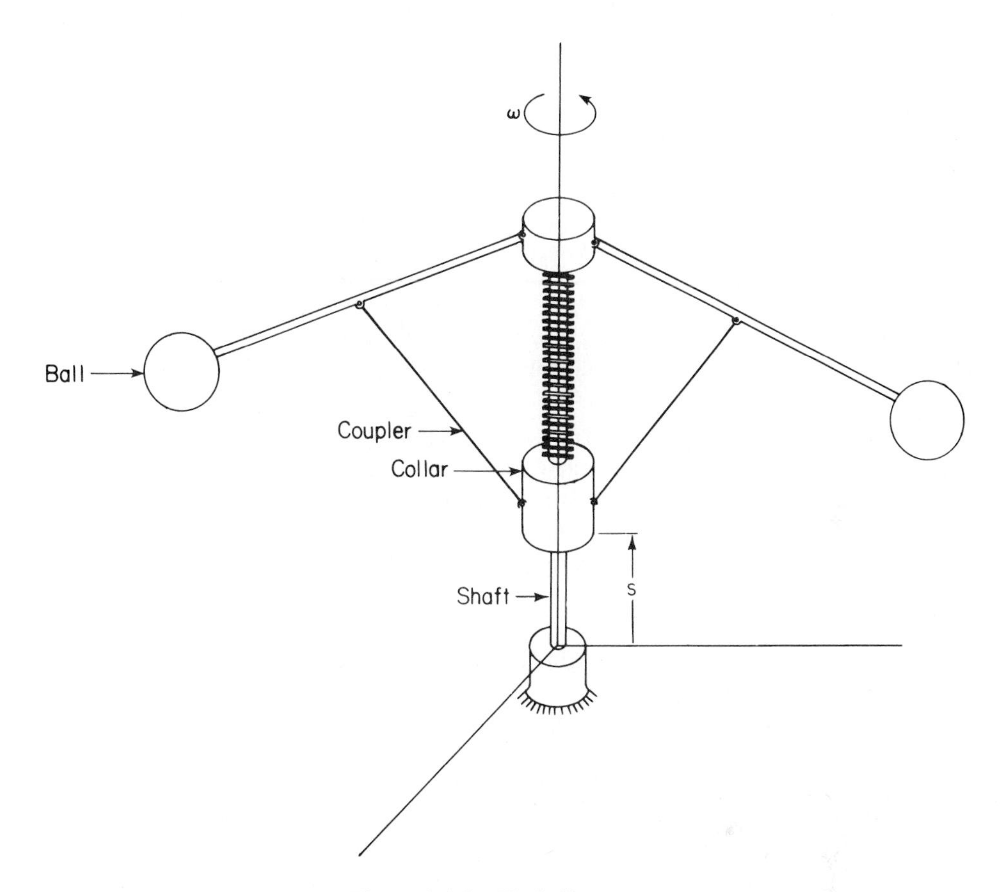

Figure 1.1.3 Fly-ball governor.

state with the collar at a given height. If an increased load is applied to the engine that reduces the angular velocity ω of the shaft (e.g., a vehicle encountering a hill or a lawn mower encountering tall grass), the balls will drop and the collar will move downward. The mechanism that couples the position of the collar with the fuel intake provides additional fuel, which in turn speeds the engine and causes centrifugal forces on the balls to increase, raising the balls toward their nominal height and reducing fuel feed to its nominal value. This is a typical application of a slider–crank mechanism.

Another mechanism that is commonly encountered in mechanical design is the *four-bar linkage,* such as that shown in the vehicle-suspension application of Fig. 1.1.4. The suspension linkage on each side of the vehicle is made up of upper and lower control arms that are pivoted in rotational joints in the frame and in the wheel assembly. This mechanism permits motion of the wheel assembly relative to the frame and transmission of road forces to the frame through a coil suspension spring and shock absorber, as indicated in Fig. 1.1.4.

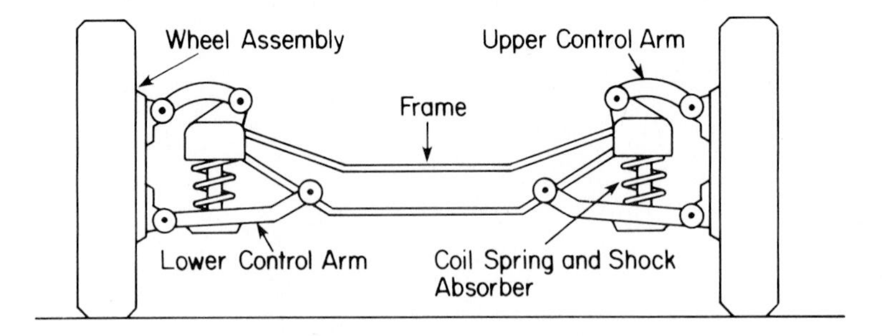

Figure 1.1.4 Suspension linkage.

The dimensions of the arms and attachments are carefully designed to cause the wheels to remain in as nearly a vertical position as possible during roll motion of the vehicle. Suspension springs and dampers are designed to provide vehicle stability and to transmit loads with small variation to the frame of the vehicle, even though extreme variations in force occur between the tire and road surface.

The windshield wiper mechanism shown in Fig. 1.1.5 is another application of four-bar linkages that transmits motor-driven rotation of the crank to the reciprocating motion of windshield wipers. The crank and left rocker arm are pivoted in the vehicle frame at points A and B. The crank coupler is pivoted in the crank at point C and in the left rocker arm at point D. The crank, crank coupler, left rocker, and frame of the vehicle constitute a four-bar linkage. Since the distance from B to D is greater than the distance from A to C, a full rotation of the crank causes only a partial rotation of the left rocker arm, leading to the

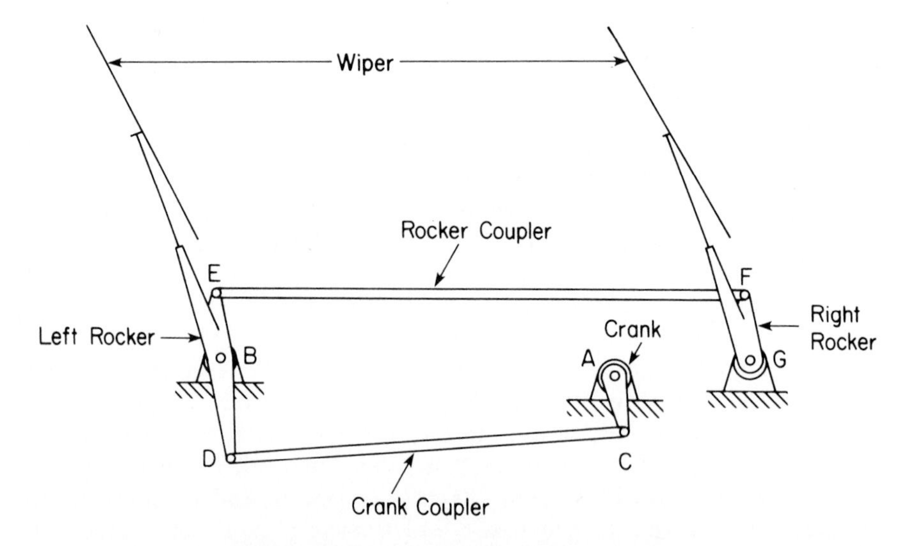

Figure 1.1.5 Windshield wiper mechanism.

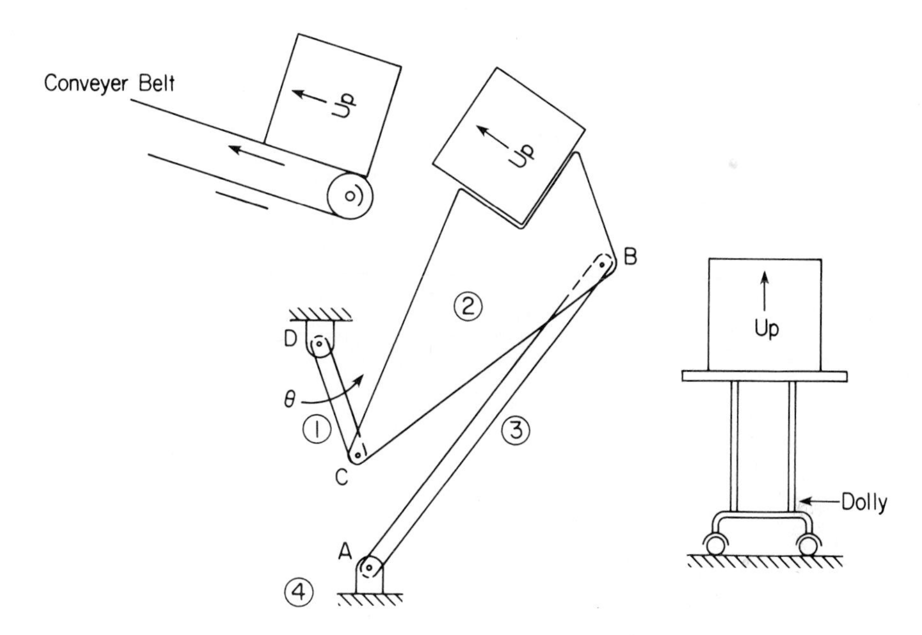

Figure 1.1.6 Material-handling mechanism.

desired reciprocating motion of the left windshield wiper. The dimensions of the various links are carefully selected to generate the desired range of motion. A second four-bar linkage is formed by the right rocker arm that is pivoted in the frame of the vehicle at point G and the rocker coupler that is pivoted in the left and right rocker arms at points E and F. This second linkage transmits reciprocating motion from the left rocker arm to the right rocker arm, hence driving the right windshield wiper.

Still another example of a four-bar linkage is the material-handling mechanism shown in Fig. 1.1.6. The crank (body 1) is pivoted in ground (body 4) at point D and with the material handler (body 2) at point C. The material handler is in turn connected to the follower arm (body 3) at point B and the follower arm is pivoted in ground at point A. The purpose of the mechanism is to permit counterclockwise rotation θ of the crank to lower the material-handling arm to a position that permits loading of cargo from a dolly located on the floor. Subsequent clockwise rotation of the crank raises the cargo so that it can be transmitted to a conveyer belt and moved to another station within a fabrication or storage facility. It is geometrically clear from the schematic diagram of Fig. 1.1.6 that the dimensions of components of the mechanism must be carefully selected so that the material handler is in the proper position and orientation for both pickup and deposit of the cargo onto the conveyer belt.

Gears, typified by the circular gear pair shown in Fig. 1.1.7, are commonly used in mechanical equipment to transmit rotation and torque at varying speeds and magnitudes, respectively, for both control of motion and transmission of

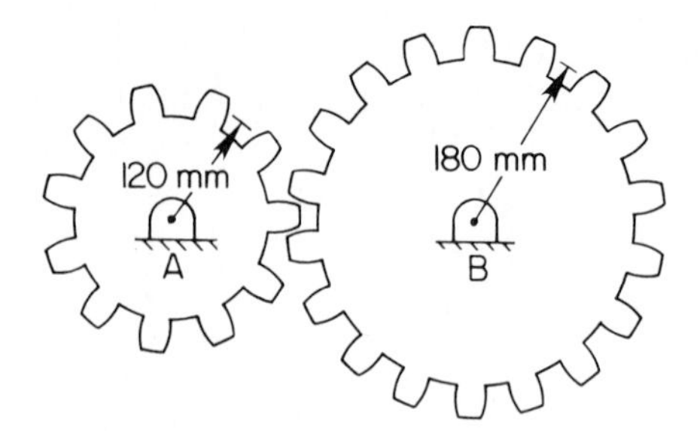

Figure 1.1.7 Gear pair.

power. While the details of the design of gear teeth are not treated in this text, it is presumed that the geometry of the gears is designed so that continuous contact is maintained at the gear pitch circles, shown in Fig. 1.1.7 with radii of 120 and 180 mm. If the smaller gear is driven, the larger gear follows. One full revolution of the larger gear requires 1.5 revolutions of the smaller gear. However, one unit of torque applied to the smaller gear is transmitted as 1.5 units of torque to the shaft of the larger gear. Gearing mechanisms, therefore, permit great latitude in the adjustment of speeds of shafts and torques transmitted.

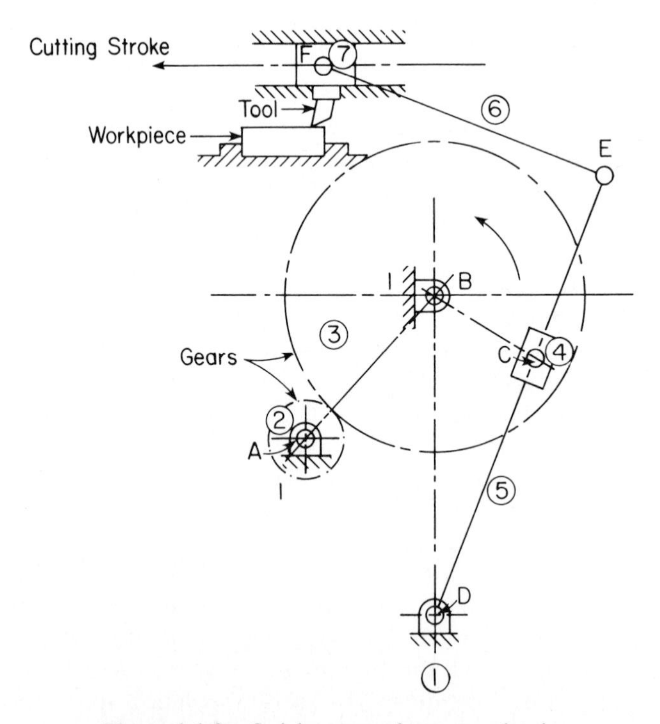

Figure 1.1.8 Quick-return shaper mechanism.

Compound mechanisms, such as the quick-return shaper mechanism shown schematically in Fig. 1.1.8, are made up from combinations of many of the basic kinematic couplings that have been encountered in the preceding examples. In this application, body 1 is designated as ground. Bodies 2 and 3 are gears of substantially different pitch diameter, with gear 2 pivoted in ground at point A and gear 3 pivoted in ground at point B. A motor drives the shaft of gear 1, resulting in smaller angular velocity of the larger gear. A coupler, body 4, is attached to the larger gear with a rotational joint at point C. A rocker arm, body 5, is pivoted in ground at point D and slides freely through the coupler of body 4. Thus, as body 3 rotates, body 5 undergoes an oscillating motion. Another coupler, body 6, is connected by rotational joints with body 5 at point E and with the slider, body 7, at point F. Body 7 translates relative to ground and carries a cutting tool that contacts a workpiece and removes material. By carefully selecting the dimensions of the components of the machine, the cutting tool can be made to move to the left (the cutting stroke) at relatively low speed and to return more quickly to the beginning of the cutting stroke.

The robot, or manipulator, of Fig. 1.1.9 is made up of nine bodies, including ground (body 1). The first degree of freedom is the rotation angle q_1 of the base (body 2) about a vertical axis fixed in ground. The second degree of freedom is the rotation q_2 of the pivot arm (body 3) about the horizontal axis fixed in body 2. The third degree of freedom is translation q_3 of the boom (body 4) in a guide that is fixed in body 3. The fourth degree of freedom is rotation q_4

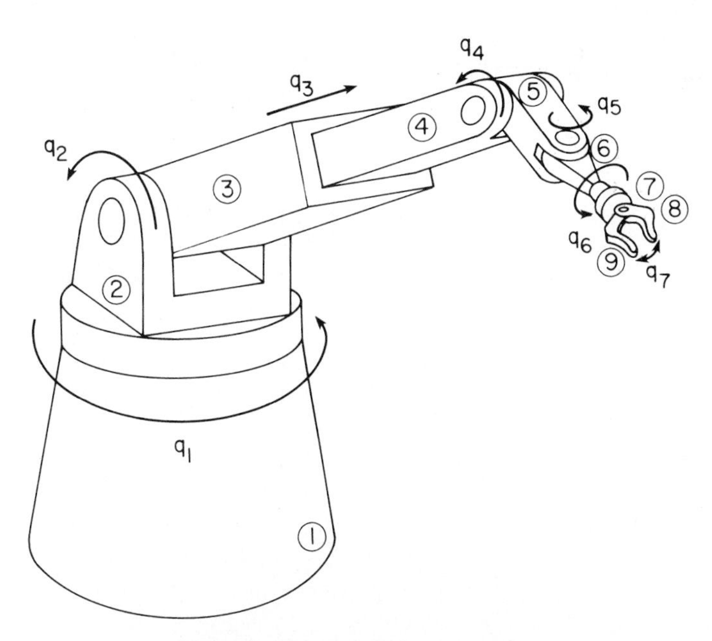

Figure 1.1.9 Robotic manipulator.

of the first wrist pivot (body 5) relative to body 4. The fifth degree of freedom is rotation q_5 of the second wrist pivot (body 6) relative to body 5. The sixth degree of freedom is rotation q_6 of the hand mechanism (body 7) relative to body 6. The final degree of freedom is relative rotation q_7 of the robot fingers (bodies 8 and 9). Such a mechanism permits the end-effector to grasp and manipulate workpieces.

The final compound mechanism example that illustrates the scope of study to follow is the vehicle of Fig. 1.1.10, whose suspension system is shown schematically in Figs. 1.1.11 and 1.1.12. This commonly employed high-performance vehicle suspension consists of a McPherson strut front suspension and a trailing arm rear suspension. Each front wheel assembly is attached to the chassis of the vehicle through a lower control arm and a telescoping strut assembly, as shown in Fig. 1.1.11. Concentric with the strut assembly are suspension spring and damping components. The spherical joints at the top and bottom of the strut assembly permit steering rotation of the wheel assembly about the strut. The more elementary rear suspension shown in Fig. 1.1.12 is simply a control arm that is pivoted in the chassis to permit the rear wheel assembly to move relative to the chassis. Spring and damping components attached between the rear control arm and chassis provide for support of the chassis and cushioning of extreme tire–road forces.

These examples represent typical machines that are encountered in mechanical system kinematic and dynamic analysis and design. The breadth of such

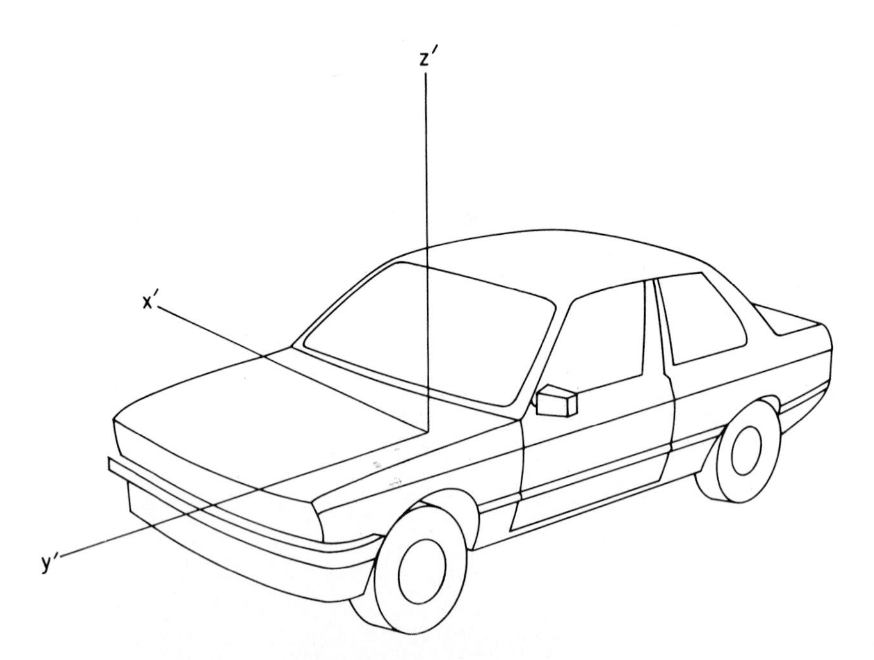

Figure 1.1.10 Automobile.

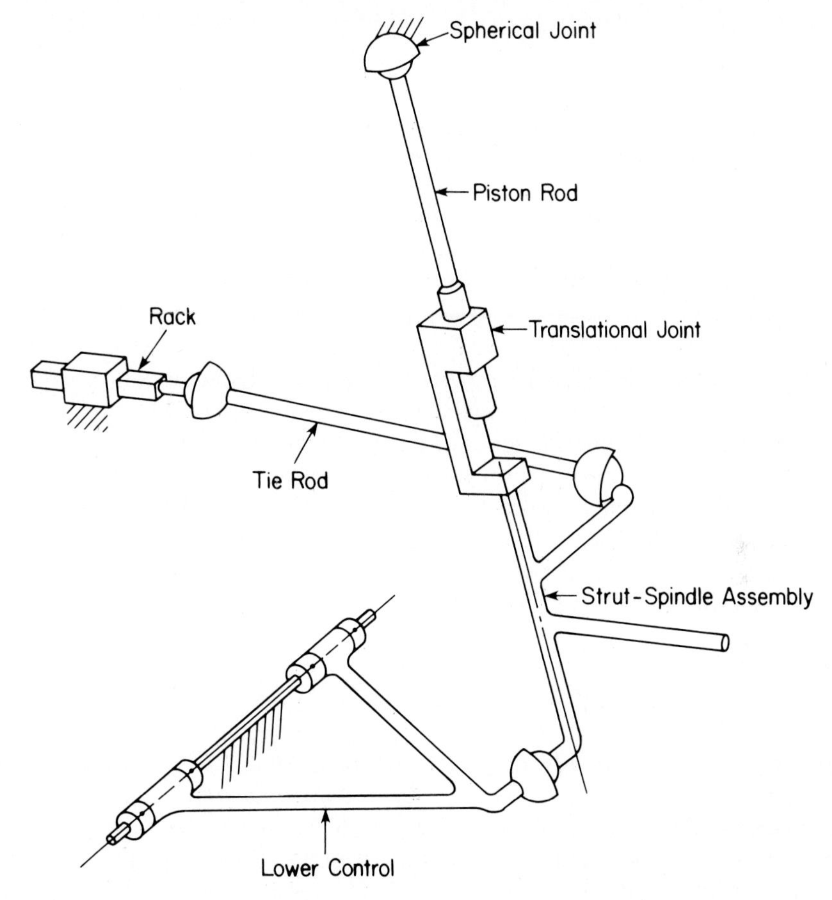

Figure 1.1.11 McPherson strut front suspension.

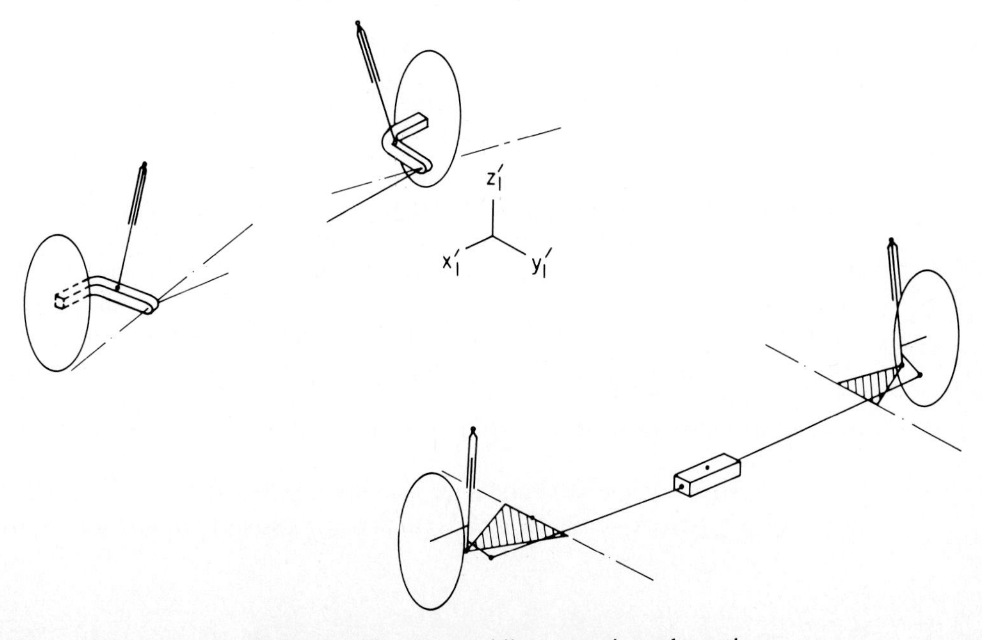

Figure 1.1.12 Automobile suspension schematic.

applications is extensive. While applications and environments differ greatly, many technical similarities permit the development of a uniform approach to computer-aided kinematic and dynamic analysis.

1.2 CONVENTIONAL METHODS OF KINEMATIC AND DYNAMIC ANALYSIS

Due to the nonlinear nature of large displacement kinematics, the mechanism designer has traditionally resorted to graphical techniques and physical models for the kinematic analysis of mechanical systems [1, 2]. As might be expected, such methods are limited in generality and rely heavily on the designer's intuition. For more contemporary treatments of mechanism and machine dynamics, References 3 through 6 may be consulted. The conventional approach to the dynamic analysis of mechanical systems is to use Lagrangian methods of formulating the system equations of motion in terms of a minimal set of variables that define absolute or relative position and orientation. Numerous texts on dynamics [4, 7–9, 35] provide the fundamentals that are needed for mechanical system dynamic analysis.

Mathematical models of kinematic and dynamic systems with several degrees of freedom have traditionally been characterized by "clever formulations" that take advantage of the properties of a specific system to obtain simplified forms of the equations of kinematics and dynamics. The ingenious selection of position and orientation variables can occasionally lead to a formulation with independent variables that allows manual derivation, but rarely analytical solution of the equations of motion. More often, analysis of systems with even three or four degrees of freedom leads to massive algebraic manipulation in constructing equations of motion. The "clever formulation" approach is, therefore, limited to relatively simple mechanical systems. Some extension has been achieved using symbolic computation [10], in which the computer is used to carry out differentiation and algebraic manipulation, creating terms that are required in the equations of motion.

After the governing equations of motion have been derived by manual or symbolic computation methods, the engineer or analyst is still faced with the problem of obtaining a solution of the differential equations and initial conditions. Since these equations are highly nonlinear, the prospect of obtaining closed-form solutions is remote, except in very simple cases. With the advent of digital computers, engineers began to use the computer and available numerical integration methods to solve their equations of motion. This, however, still involved a substantial amount of time and personnel for deriving equations of motion and writing ad hoc digital computer programs to carry out numerical integration.

In contrast to the traditional ad hoc approach that has been employed in mechanical system kinematics and dynamics, a massive literature has evolved in finite element structural analysis [11, 12] and the analysis of electronic circuits [13, 14]. Developments in these fields are characterized by the same technical approach. Rather than relying on "clever formulations," a "systematic approach" is taken and digital computers are used for both the formulation and solution of the governing equations. Through systematic formulation and selection of numerical techniques, user-oriented computer codes have been developed that are capable of simulating a broad range of structures and circuits. The overwhelming success of finite element structural and electronic circuit analysis computer codes suggests that such a formulation be adopted for mechanical system kinematics and dynamics.

1.3 OBJECTIVE OF COMPUTATIONAL KINEMATICS AND DYNAMICS

The objective of computational methods in kinematics and dynamics is to create a formulation and digital computer software that allow the engineer to (1) input data that define mechanical systems of interest and automatically formulate governing equations of kinematics and dynamics, (2) automatically solve nonlinear equations for kinematic and dynamic response, and (3) provide computer graphics output of results of simulations to communicate results to the designer or analyst. The essence of this objective is to make maximum use of digital computer power for rapid and accurate data manipulation and numerical computation, hence relieving the engineer of tedious and error-prone calculations that heretofore have been carried out manually or with ad hoc computer programs.

As suggested by advances in computer-aided finite element structural and electronic circuit analysis, a systematic approach to the formulation and solution of the equations of kinematics and dynamics of mechanical systems is required to implement computations in a user-oriented computer program. Great care must be taken to consider the numerous alternatives that are available in selecting a formulation and numerical methods to achieve this objective.

Several computer programs for kinematic and dynamic analysis were developed in the late 1960s and early 1970s [15–17] using relative coordinates between bodies. These programs are satisfactory for many applications. An alternative method of formulating system constraints and equations of motion, in terms of global Cartesian coordinates, was introduced in the late 1970s [18–20], bypassing topological analysis and making it easier for the user to supply constraints and forcing functions. This approach leads to a general-purpose computer program, with practically no limitation on the type of mechanism or machine that can be analyzed. The penalty, however, is a larger system of equations to be solved.

To be specific concerning some of the alternatives and trade-offs that exist in the field of computational kinematics and dynamics, an elementary example is

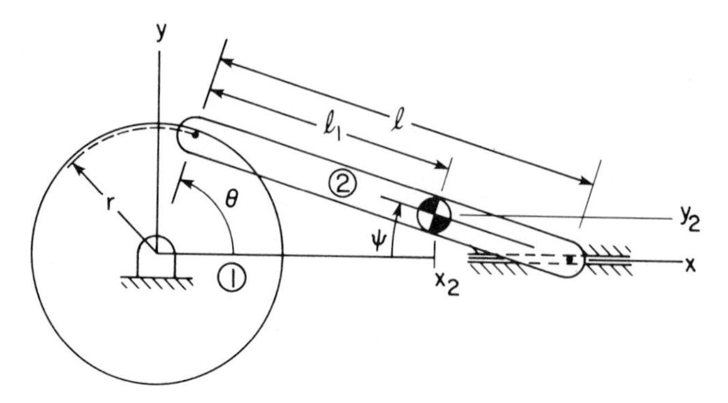

Figure 1.3.1 Elementary slider–crank model.

employed to discuss approaches to system dynamics. Consider a simplified model of the slider–crank mechanisms of Figs. 1.1.1 through 1.1.3 idealized to include the slider mass at the right end of the connecting rod (body 2), as shown in Fig. 1.3.1. The center of mass of the connecting rod has been adjusted to reflect the incorporation of the mass of the slider as a point mass at the right end, which must slide along the x axis. The center of mass of the crank is at its pivot with ground.

It is clear from simple trigonometry that once the crank angle θ (called an independent *Lagrangian generalized coordinate*) is fixed, as long as $\ell > r$, the angle ψ and positions x_2 and y_2 can be determined, with some analytical complexity. A single, highly nonlinear, and complicated second-order differential equation of motion in the independent variable θ can then be derived. It must, of course, be solved numerically.

A more systematic approach to deriving equations of motion for the simplified slider–crank of Fig. 1.3.1 is to first consider the bodies as being disconnected, as shown in Fig. 1.3.2. In this formulation, the angular orientation ϕ_1 of the crank and coordinates x_2 and y_2 of the center of mass of body 2 and its angular orientation ϕ_2 are taken as position and orientation coordinates called *Cartesian generalized coordinates*. To assemble the linkage, however, these four variables must satisfy three kinematic relations. Specifically, points A_1 and A_2 must concide in order to have a rotational joint between the crank and connecting rod, leading to two algebraic constraint equations. Similarly, for point B_2 on the connecting rod to slide along the x axis, it is necessary that its y coordinate be zero, leading to an additional constraint equation. These three algebraic equations comprise three constraints among the four generalized coordinates ϕ_1, x_2, y_2, and ϕ_2. Since these equations are nonlinear, a closed-form solution for three of the variables in terms of the remaining variable is difficult to obtain.

The Lagrange multiplier form of the equations of motion [4, 7–9, 35, or Chapter 6, herein] may be written for this system as four second-order differential equations and three algebraic equations of constraint, for four Cartesian

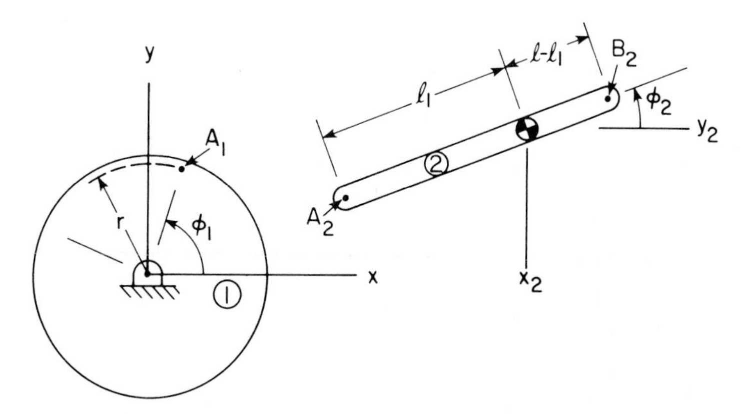

Figure 1.3.2 Cartesian coordinates for slider–crank.

generalized coordinates and three Lagrange multipliers. This is a mixed system of *differential–algebraic equations* that must be solved to determine the motion of the system. The form of this large number of equations is simple, however, permitting computer generation and solution.

The systematic Cartesian coordinate approach is adopted in this text, since it is well suited for both computer formulation and solution of the governing equations of kinematics and dynamics. The general-purpose Dynamic Analysis and Design System (DADS) computer code [27] has been developed, based on the methods presented in this text, and is available for use with the text as an instructional tool. To enhance the student's experience with realistic kinematic and dynamic applications, examples and exercises using the DADS computer code are presented in Chapters 5, 8, 10, and 12.

1.4 GUIDE TO THE TEXT CONTENTS

The text begins with Part One (Chapters 2 through 8) on planar kinematics and dynamics and associated numerical analysis and modeling methods. Spatial kinematic and dynamic analysis formulations are presented in Part Two (Chapters 9 through 12), and their use in the analysis of spatial mechanical systems is illustrated. This organization has been selected to permit the beginning reader to master basic concepts and to gain experience in the kinematic and dynamic analysis of planar systems for which the level of analytical complexity is minimal. Having developed confidence in basic methods of formulation and numerical analysis, the reader may proceed to the study of spatial systems, which is characterized by a higher degree of technical complexity, but follows the same basic approach and utilizes identical numerical methods. The reader with some experience in kinematics and dynamics who is primarily interested in spatial systems can proceed directly to Part Two of the text without any loss of continuity.

Part One of the text begins with Chapter 2, which presents planar vector analysis methods and matrix and multivariable calculus notations that are used throughout the text. Of key importance in Chapter 2 is the formulation of vector analysis techniques in terms of the matrix operations that are needed for digital computer implementation. Differentiation of vectors, multivariable differential calculus, and virtual displacement concepts that are needed in subsequent kinematic and dynamic analysis are also developed.

The planar Cartesian kinematics formulation is developed in Chapter 3. Standard constraints are formulated and governing equations for position, velocity, and acceleration analysis are derived and illustrated. Numerical methods for kinematic analysis (applicable to both planar and spatial systems) are presented in Chapter 4. Computational methods of assembling the required equations and matrices for kinematic analysis are discussed and numerical methods for solving linear and nonlinear equations are presented. Chapter 5 presents and illustrates methods for planar kinematic modeling and analysis. Modeling and analysis methods are illustrated using the DADS computer code. Technical difficulties that are encountered due to poorly formulated system models are illustrated to guide the reader in interpreting the results of numerical analysis and to demonstrate that it is easy to define a mechanical system that will not perform the intended kinematic functions.

The governing equations of planar system dynamics are derived in Chapter 6. This development is self-contained, beginning with Newton's laws of motion for a particle and developing the equations of motion of planar rigid bodies. The virtual work method is introduced for use throughout the text. The Lagrange multiplier form of equations of motion for constrained dynamic systems is developed. Methods for inverse dynamic and equilibrium analysis are presented. Numerical methods for solving equations of motion (applicable to both planar and spatial systems) are presented in Chapter 7. The process of formulating mixed differential–algebraic equations of motion is discussed, and algorithms that may be employed to reduce such systems to an integrable form are presented. Numerical integration methods are outlined, and their use in algorithms for numerical solution of the differential–algebraic equations of motion is presented. Chapter 8 presents methods for planar dynamic analysis. Examples used in Chapter 5 for kinematic analysis are employed to illustrate methods for the dynamic analysis of planar systems and to study the effect of design variations on dynamic performance.

Part Two of the text is devoted to the kinematics and dynamics of spatial systems. Chapter 9 develops the theory of spatial position and orientation of rigid bodies. Euler parameter orientation variables are defined and their properties are developed as required for analysis and applications. Spatial kinematic equations for constrained multibody systems are developed for a library of joints, and the computation of needed derivatives is presented. Spatial kinematic modeling and analysis methods are discussed in Chapter 10, where examples are analyzed using the DADS code, to illustrate modeling methods and their use in kinematic analysis. The importance of careful definition of kinematic joints is emphasized, and pitfalls that arise due to improper formulation of mechanical system models are illustrated.

Chapter 11 presents a self-contained derivation of the equations of motion of spatial multibody systems. The virtual work approach introduced in Chapter 6 is employed. Finally, methods of spatial dynamic modeling and analysis are presented in Chapter 12 and illustrated through the study of examples using the DADS code.

Part One

PLANAR SYSTEMS

Chapters 2 through 8 are devoted to the kinematics and dynamics of systems in which all bodies move in a plane or in parallel planes. Initial focus on this class of systems is motivated by the following considerations: (1) analytical representation of the position and orientation of bodies in a plane is much less complex than for bodies in space, (2) analytical and modeling concepts are most easily learned in the context of planar systems, and (3) once the engineer has a clear understanding of the concepts and modeling methods, the study of spatial systems involves only analytical and algebraic extensions. Consistent with the objective of restricting attention in Part I to planar systems, only planar vector analysis and planar position and orientation generalized coordinates are introduced and used. The matrix and differential calculus methods presented in Chapter 2 are, however, adequate for both planar and spatial applications.

The reader who is new to the field of the kinematics and dynamics of machines is encouraged to thoroughly master concepts in the planar setting. To develop a sound understanding and facility with mathematical and numerical analysis methods, the reader is encouraged to build on a physical foundation of mechanics and machine applications, in order to achieve a practical capability in the kinematics and dynamics of machines. The insidious presence of nonlinearity in virtually all aspects of the kinematics and dynamics of machines leads to intricacies in mathematical and numerical analysis that are not as easily solved as is the case in linear structural mechanics and associated finite element methods. Pathological forms of behavior, such as lock-up and branching of kinematic solutions, can best be understood and overcome by an engineer who develops both a firm mathematical foundation and a clear physical understanding of the behavior of mechanisms and machines. Examples and exercises are presented throughout the text to assist the engineer in developing mathematical skills and in relating them to the mechanics of machines. The reader is encouraged to study the examples thoroughly and to carry out independent analysis of mechanisms of his or her own choice.

CHAPTER TWO

Planar Vectors, Matrices, and Differential Calculus

Vector and matrix algebra form the mathematical foundation for kinematics and dynamics. Geometry of motion is at the heart of both the kinematics and dynamics of mechanical systems. Vector analysis is the time-honored tool for describing geometry. In its geometric form, however, vector algebra is not well suited to computer implementation. In this chapter, a systematic matrix formulation of planar vector algebra, referred to as the algebraic vector representation, is presented for use throughout Part One. This form of vector representation, in contrast to the more traditional geometric form, is easier to use for both formula manipulation and computer implementation. Multivariable differential calculus plays a key role in kinematic analysis for both writing and solving the equations of motion of mechanical systems. Basic ideas and notations of matrices and multivariable differential calculus are developed in this chapter for use throughout the text. Key formulas are summarized at the end of the chapter for easy reference.

2.1 GEOMETRIC VECTORS

The *geometric vector* $\vec{a}$ in Fig. 2.1.1, beginning at point A and ending at point B, is defined as the directed line segment from A to B. The *magnitude of a vector* $\vec{a}$ is its length and is denoted by a or $|\vec{a}|$.

Multiplication of a vector $\vec{a}$ *by a scalar* $\alpha > 0$ is defined as a vector in the same direction as $\vec{a}$, but having magnitude αa. Multiplication of a vector $\vec{a}$ by a scalar $\alpha < 0$ is the vector with magnitude $|\alpha| a$ and opposite direction of $\vec{a}$. The *negative of a vector* is obtained by multiplying the vector by -1. It is a vector with the same magnitude but opposite direction. A *unit vector,* that is, a vector having a magnitude of 1 unit, in the direction $\vec{a} \neq \vec{0}$ is $(1/a)\vec{a}$.

Two vectors $\vec{a}$ and $\vec{b}$ are added according to the *parallelogram rule,* as

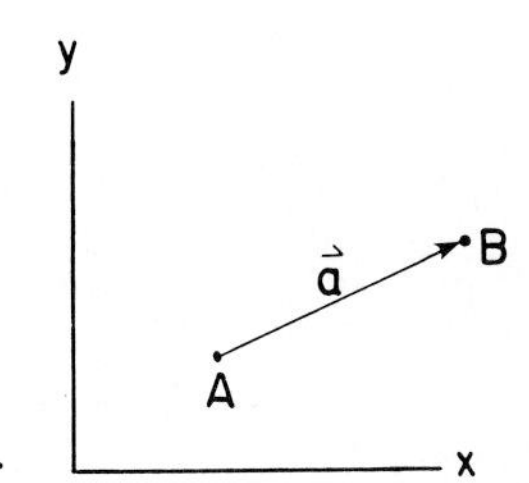

Figure 2.1.1 Vector from point A to point B.

shown in Fig. 2.1.2. In vector notation, the vector sum is written as

$$\vec{c} = \vec{a} + \vec{b} \tag{2.1.1}$$

Addition of vectors and multiplication of vectors by scalars obey the following rules [21]:

$$\begin{aligned} \vec{a} + \vec{b} &= \vec{b} + \vec{a} \\ (\alpha + \beta)\vec{a} &= \alpha\vec{a} + \beta\vec{a} \end{aligned} \tag{2.1.2}$$

where α and β are scalars.

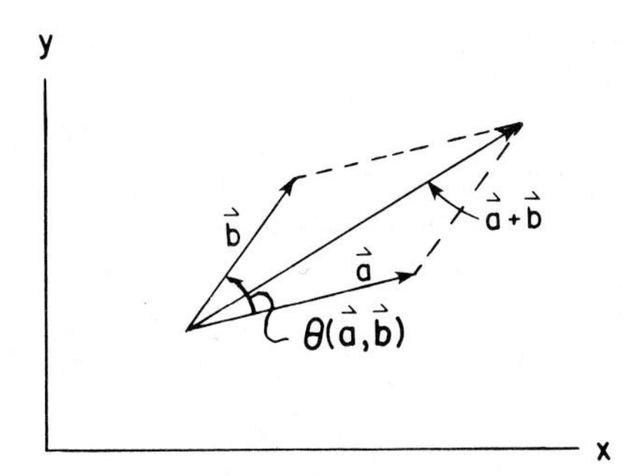

Figure 2.1.2 Addition of vectors.

Orthogonal reference frames are used extensively in representing vectors. Use in this text is limited to *right-hand orthogonal reference frames*; that is, the y axis is oriented $\pi/2$ rad counterclockwise from the x axis, as shown in Figs. 2.1.1 and 2.1.2. Such frames are called *Cartesian reference frames.*

A vector $\vec{a}$ can be resolved into components a_x and a_y along the x and y axes of a Cartesian reference frame, as shown in Fig. 2.1.3. These components are called the *Cartesian components of the vector.* Here, the *unit coordinate vectors* $\vec{i}$ and $\vec{j}$ are directed along the x and y coordinate axes, as shown in Fig. 2.1.3. In terms of the components of a vector and the unit coordinate vectors,

$$\vec{a} = a_x\vec{i} + a_y\vec{j} \tag{2.1.3}$$

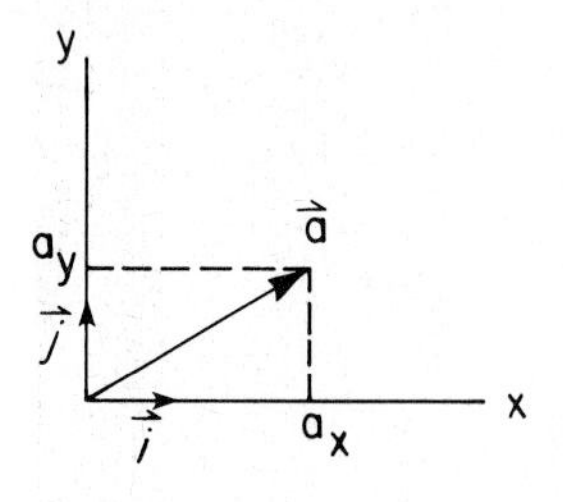

Figure 2.1.3 Components of a vector.

Addition of vectors $\vec{a}$ and $\vec{b}$ may now be expressed in terms of their components as

$$\begin{aligned}\vec{c} = \vec{a} + \vec{b} &= (a_x + b_x)\vec{i} + (a_y + b_y)\vec{j} \\ &\equiv c_x\vec{i} + c_y\vec{j}\end{aligned} \tag{2.1.4}$$

where $c_x = a_x + b_x$ and $c_y = a_y + b_y$ are Cartesian components of vector $\vec{c}$. Thus, addition of vectors occurs component by component. Using this idea, three vectors $\vec{a}$, $\vec{b}$, and $\vec{c}$ may be added, to verify (Prob. 2.1.1) that

$$(\vec{a} + \vec{b}) + \vec{c} = \vec{a} + (\vec{b} + \vec{c}) = \vec{a} + \vec{b} + \vec{c} \tag{2.1.5}$$

Example 2.1.1: Given vectors $\vec{a} = \vec{i} + 2\vec{j}$, $\vec{b} = -2\vec{i} + \vec{j}$, and $\vec{c} = \vec{j}$, their vector sum is

$$\begin{aligned}\vec{a} + \vec{b} + \vec{c} &= (\vec{i} + 2\vec{j}) + (-2\vec{i} + \vec{j}) + (\vec{j}) \\ &= -\vec{i} + 4\vec{j}\end{aligned}$$

Denote the angle from vector $\vec{a}$ to vector $\vec{b}$ by $\theta(\vec{a}, \vec{b})$, with counterclockwise positive, as shown in Fig. 2.1.2. The *scalar product* (or *dot product*) of two vectors $\vec{a}$ and $\vec{b}$ is defined as the product of the magnitudes of the vectors and the cosine of the angle between them; that is,

$$\vec{a} \cdot \vec{b} = ab \cos \theta(\vec{a}, \vec{b}) \tag{2.1.6}$$

Note that if the vectors are nonzero, that is, if $a \neq 0$ and $b \neq 0$, then their scalar product is zero only if $\cos \theta(\vec{a}, \vec{b}) = 0$. Two nonzero vectors are said to be *orthogonal vectors* if their scalar product is zero. Since $\theta(\vec{b}, \vec{a}) = 2\pi - \theta(\vec{a}, \vec{b})$ and $\cos(2\pi - \alpha) = \cos \alpha$, the order of terms appearing on the right side of Eq. 2.1.6 is immaterial; so

$$\vec{a} \cdot \vec{b} = \vec{b} \cdot \vec{a} \tag{2.1.7}$$

Example 2.1.2: Let $\vec{u}$ be a unit vector, as shown in Fig. 2.1.4. Geometrically, $\vec{u} \cdot \vec{a} = a \cos \theta(\vec{u}, \vec{a})$ is the projection of $\vec{a}$ onto the directed line segment defined by $\vec{u}$, as shown in Fig. 2.1.4. Note that this is a geometric property of the scalar product, independent of the reference frame.

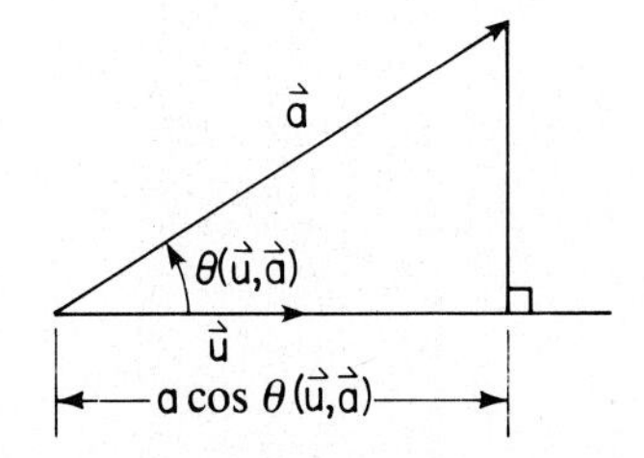

Figure 2.1.4 Projection of $\vec{a}$ onto $\vec{u}$.

Based on the definition of the scalar product, the following identities hold for the unit vectors $\vec{i}$ and $\vec{j}$:

$$\begin{aligned} \vec{i} \cdot \vec{j} &= \vec{j} \cdot \vec{i} = 0 \\ \vec{i} \cdot \vec{i} &= \vec{j} \cdot \vec{j} = 1 \end{aligned} \tag{2.1.8}$$

For any vector $\vec{a}$, since $\theta(\vec{a}, \vec{a}) = 0$,

$$\vec{a} \cdot \vec{a} = aa \cos 0 = a^2 \tag{2.1.9}$$

While not obvious on geometrical grounds, the scalar product satisfies the relation [21]

$$(\vec{a} + \vec{b}) \cdot \vec{c} = \vec{a} \cdot \vec{c} + \vec{b} \cdot \vec{c} \tag{2.1.10}$$

Using Eq. 2.1.10 and the identities of Eq. 2.1.8, a direct calculation yields

$$\vec{a} \cdot \vec{b} = a_x b_x + a_y b_y \tag{2.1.11}$$

From Eqs. 2.1.9 and 2.1.11, note that

$$a = \sqrt{\vec{a} \cdot \vec{a}} = \sqrt{a_x^2 + a_y^2} \tag{2.1.12}$$

It is often helpful to define a vector that is perpendicular to a given vector $\vec{a}$, has the same length as $\vec{a}$, and is oriented $\pi/2$ rad counterclockwise from $\vec{a}$. This vector is denoted as $\vec{a}^\perp$ and is written as

$$\vec{a}^\perp = -a_y \vec{i} + a_x \vec{j} \tag{2.1.13}$$

as shown in Fig. 2.1.5. To verify that $\vec{a}^\perp$ is indeed orthogonal to $\vec{a}$, note that

$$\vec{a}^\perp \cdot \vec{a} = -a_y a_x + a_x a_y = 0$$

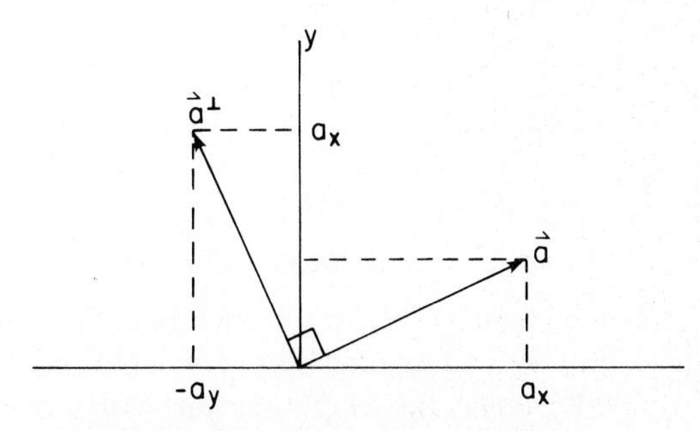

Figure 2.1.5 Vector perpendicular to $\vec{a}$.

Example 2.1.3: Let $0 \leq \text{Arccos}\,\alpha < \pi$ be the principal value of the arccos function. Since, if $-\pi < \phi < 0$, $\text{Arccos}(\cos\phi) = -\phi$, knowing $\cos\phi$ does not determine ϕ; that is, $\text{Arccos}(\cos\phi) = \mp\phi$. Thus, from Eq. 2.1.6 and Fig. 2.1.2,

$$\begin{aligned}\text{Arccos}(\cos\theta(\vec{a}, \vec{b})) &= \text{Arccos}\left(\frac{\vec{a}\cdot\vec{b}}{ab}\right)\\ &= \mp\theta(\vec{a}, \vec{b})\end{aligned} \tag{2.1.14}$$

To resolve the sign ambiguity, note first that the plus sign is selected in Eq. 2.1.14 if $\vec{b}$ is directed to the left of $\vec{a}$ and the minus sign is selected if $\vec{b}$ is to the right of $\vec{a}$. Since $\vec{a}^{\perp}$ is directed to the left of $\vec{a}$, as shown in Fig. 2.1.5, then $\vec{b}$ is to the left of $\vec{a}$ if $\vec{a}^{\perp}\cdot\vec{b} \geq 0$ and to the right otherwise. Defining

$$\text{sgn}\,x = \begin{cases} 1 & \text{if } x > 0 \\ 0 & \text{if } x = 0 \\ -1 & \text{if } x < 0 \end{cases}$$

then

$$\theta(\vec{a}, \vec{b}) = \text{sgn}(\vec{a}^{\perp}\cdot\vec{b}) \times \text{Arccos}\left(\frac{\vec{a}\cdot\vec{b}}{ab}\right) \tag{2.1.15}$$

Using Eqs. 2.1.11 through 2.1.13, this relation can be written in terms of the components of $\vec{a}$ and $\vec{b}$ (Prob. 2.1.2).

2.2 MATRIX ALGEBRA

Matrix notation [22] permits systematic representation of a system of equations. Matrix manipulation also allows for organized development, simplification, and solution of systems of equations.

A *matrix* is a rectangular array of numbers, taken here to be real. If it has m rows and n columns, the *dimension of the matrix* is said to be $m \times n$. A matrix is denoted by a boldface capital letter if m and n are greater than 1 and is written in the form

$$\mathbf{A} \equiv [a_{ij}] \equiv \begin{bmatrix} a_{11} & a_{12} & \cdots & a_{1n} \\ a_{21} & a_{22} & \cdots & a_{2n} \\ \vdots & \vdots & & \vdots \\ a_{m1} & a_{m2} & \cdots & a_{mn} \end{bmatrix} \tag{2.2.1}$$

where element a_{ij} is located in the ith row and jth column. The *transpose of a matrix* is formed by interchanging rows and columns and is designated by the superscript T. Thus, if a_{ij} is the i-j element of matrix $\mathbf{A}$, a_{ji} is the i-j element of its transpose $\mathbf{A}^T$.

A matrix with only one column is called a *column matrix* and is denoted by a boldface lowercase letter; for example, $\mathbf{a}$. A matrix with only one row is called a *row matrix* and is denoted by a boldface lowercase letter. An $m \times n$ matrix can be considered as being constructed of n column matrices $\mathbf{a}_j = [a_{1j}, \ldots, a_{mj}]^T$, $j = 1, \ldots, n$, or m row matrices $\mathbf{b}_i = [a_{i1}, \ldots, a_{in}]$, $i = 1, \ldots, m$; that is,

$$\mathbf{A} = [\mathbf{a}_1, \mathbf{a}_2, \ldots, \mathbf{a}_n] = \begin{bmatrix} \mathbf{b}_1 \\ \mathbf{b}_2 \\ \vdots \\ \mathbf{b}_m \end{bmatrix} \tag{2.2.2}$$

Example 2.2.1: The 2×3 matrix

$$\mathbf{A} = \begin{bmatrix} 1 & 2 & 0 \\ 2 & 3 & 1 \end{bmatrix}$$

may be thought of as made up of the three 2×1 column matrices

$$\mathbf{a}_1 = \begin{bmatrix} 1 \\ 2 \end{bmatrix}, \quad \mathbf{a}_2 = \begin{bmatrix} 2 \\ 3 \end{bmatrix}, \quad \mathbf{a}_3 = \begin{bmatrix} 0 \\ 1 \end{bmatrix}$$

or the two 1×3 row matrices

$$\mathbf{b}_1 = [1,\ 2,\ 0], \qquad \mathbf{b}_2 = [2,\ 3,\ 1]$$

That is,

$$\mathbf{A} = [\mathbf{a}_1,\ \mathbf{a}_2,\ \mathbf{a}_3] = \begin{bmatrix} \mathbf{b}_1 \\ \mathbf{b}_2 \end{bmatrix}$$

A *square matrix* has an equal number of rows and columns. A *diagonal matrix* is a square matrix with $a_{ij} = 0$ for $i \neq j$ and at least one nonzero diagonal term. An $n \times n$ diagonal matrix $\mathbf{A}$ can be represented as

$$\mathbf{A} \equiv \operatorname{diag}[a_{11},\ a_{22}, \ldots, a_{nn}] \tag{2.2.3}$$

The $n \times n$ *identity matrix*, denoted $\mathbf{I}$ or $\mathbf{I}_n$, is a diagonal matrix with $a_{ii} = 1$, $i = 1, \ldots, n$. A *zero matrix* of any dimension, designated as $\mathbf{0}$, has $a_{ij} = 0$, for all i and j.

If two matrices $\mathbf{A}$ and $\mathbf{B}$ have the same dimension, they are defined to be *equal matrices* if $a_{ij} = b_{ij}$, for all i and j; that is, entries in the same position are equal. The *sum of two matrices* $\mathbf{A}$ and $\mathbf{B}$ that have the same dimension is a matrix with the same dimension defined as

$$\mathbf{C} = \mathbf{A} + \mathbf{B} \tag{2.2.4}$$

where $c_{ij} = a_{ij} + b_{ij}$, for all i and j. That is, matrices with the same dimension add component by component. The *difference between two matrices* $\mathbf{A}$ and $\mathbf{B}$ of the same dimension is defined as

$$\mathbf{D} = \mathbf{A} - \mathbf{B}$$

where $d_{ij} = a_{ij} - b_{ij}$, for all i and j. If three matrices have the same dimension, then (Prob. 2.2.1)

$$(\mathbf{A} + \mathbf{B}) + \mathbf{C} = \mathbf{A} + (\mathbf{B} + \mathbf{C}) = \mathbf{A} + \mathbf{B} + \mathbf{C} \tag{2.2.5}$$

Similarly, for matrices **A** and **B** with the same dimension (Prob. 2.2.2),

$$\mathbf{A} + \mathbf{B} = \mathbf{B} + \mathbf{A} \tag{2.2.6}$$

Example 2.2.2: The sum of the 2×2 matrices

$$\mathbf{A} = \begin{bmatrix} 1 & 1 \\ 2 & 1 \end{bmatrix}, \quad \mathbf{B} = \begin{bmatrix} 0 & 2 \\ 1 & 1 \end{bmatrix}$$

is

$$\mathbf{C} = \mathbf{A} + \mathbf{B} = \begin{bmatrix} 1 & 3 \\ 3 & 2 \end{bmatrix}$$

and the difference is

$$\mathbf{D} = \mathbf{A} - \mathbf{B} = \begin{bmatrix} 1 & -1 \\ 1 & 0 \end{bmatrix}$$

Let **A** be an $m \times p$ matrix and **B** be a $p \times n$ matrix, written in the form

$$\mathbf{A} \equiv [a_{ij}] = \begin{bmatrix} \mathbf{d}_1 \\ \mathbf{d}_2 \\ \vdots \\ \mathbf{d}_m \end{bmatrix}, \quad \mathbf{B} \equiv [b_{ij}] = [\mathbf{b}_1, \mathbf{b}_2, \ldots, \mathbf{b}_n] \tag{2.2.7}$$

where $\mathbf{d}_i$, $i = 1, \ldots, m$, are rows of **A** with p elements and $\mathbf{b}_i$, $i = 1, \ldots, n$, are columns of **B** with p elements. The *matrix product* of **A** and **B** is defined as the $m \times n$ matrix

$$\mathbf{C} = \mathbf{AB} \tag{2.2.8}$$

where

$$c_{ij} = \sum_{k=1}^{p} a_{ik} b_{kj} \tag{2.2.9}$$

or, in terms of the rows of **A** and columns of **B**,

$$c_{ij} = \mathbf{d}_i \mathbf{b}_j \tag{2.2.10}$$

Example 2.2.3: The product of the 2×2 matrix $\mathbf{B}$ of Example 2.2.2 and the 2×3 matrix $\mathbf{A}$ of Example 2.2.1 is the 2×3 matrix

$$\mathbf{C} = \mathbf{BA} = \begin{bmatrix} 0 & 2 \\ 1 & 1 \end{bmatrix} \begin{bmatrix} 1 & 2 & 0 \\ 2 & 3 & 1 \end{bmatrix} = \begin{bmatrix} 4 & 6 & 2 \\ 3 & 5 & 1 \end{bmatrix}$$

It is important to note that the product of two matrices is defined only if the number of columns in the first matrix equals the number of rows in the second matrix. From the definition of matrix multiplication, in general,

$$\mathbf{AB} \neq \mathbf{BA} \tag{2.2.11}$$

To see that this is the case, the matrices of Example 2.2.2 yield

$$\mathbf{AB} = \begin{bmatrix} 1 & 1 \\ 2 & 1 \end{bmatrix} \begin{bmatrix} 0 & 2 \\ 1 & 1 \end{bmatrix} = \begin{bmatrix} 1 & 3 \\ 1 & 5 \end{bmatrix} \neq \begin{bmatrix} 4 & 2 \\ 3 & 2 \end{bmatrix} = \begin{bmatrix} 0 & 2 \\ 1 & 1 \end{bmatrix} \begin{bmatrix} 1 & 1 \\ 2 & 1 \end{bmatrix} = \mathbf{BA}$$

In fact, the products $\mathbf{AB}$ and $\mathbf{BA}$ are defined only if both $\mathbf{A}$ and $\mathbf{B}$ are square and of equal dimension.

If $\mathbf{A}$ and $\mathbf{B}$ are $m \times p$ matrices and $\mathbf{C}$ is a $p \times n$ matrix [22],

$$(\mathbf{A} + \mathbf{B})\mathbf{C} = \mathbf{AC} + \mathbf{BC} \tag{2.2.12}$$

Similarly, if $\mathbf{A}$ is an $m \times p$ matrix, $\mathbf{B}$ is a $p \times q$ matrix, and $\mathbf{C}$ is a $q \times n$ matrix [22],

$$(\mathbf{AB})\mathbf{C} = \mathbf{A}(\mathbf{BC}) = \mathbf{ABC} \tag{2.2.13}$$

Multiplication of a matrix $\mathbf{A}$ *by a scalar* α is defined as

$$\mathbf{C} = \alpha\mathbf{A} \tag{2.2.14}$$

where $c_{ij} = \alpha a_{ij}$, for all i and j; that is, all terms in the matrix are multiplied by the same scalar.

If $a_{ij} = a_{ji}$, for all i and j, the square matrix $\mathbf{A} = [a_{ij}]$ is called a *symmetric matrix*; that is, $\mathbf{A} = \mathbf{A}^T$. If $a_{ij} = -a_{ji}$, for all i and j, $\mathbf{A}$ is called a *skew-symmetric matrix*; that is, $\mathbf{A} = -\mathbf{A}^T$. Note that in this case $a_{ii} = 0$, for all i.

The transpose of the sum of two matrices is the sum of their transposes [22]; that is,

$$(\mathbf{A} + \mathbf{B})^T = \mathbf{A}^T + \mathbf{B}^T \tag{2.2.15}$$

Also, if $\mathbf{A}$ is an $m \times p$ matrix and $\mathbf{B}$ is a $p \times n$ matrix, then [22]

$$(\mathbf{AB})^T = \mathbf{B}^T\mathbf{A}^T \tag{2.2.16}$$

A set of column matrices $\mathbf{a}_j$, $j = 1, \ldots, m$, is called *linearly dependent* if there are constants α_j, $j = 1, \ldots, m$, that are not all zero such that

$$\sum_{j=1}^{m} \alpha_j \mathbf{a}_j = \mathbf{0}$$

If a set of column matrices is not linearly dependent, it is called *linearly independent.* Equivalently, column matrices $\mathbf{a}_j$, $j = 1, \ldots, m$, are linearly independent if and only if

$$\sum_{j=1}^{m} \alpha_j \mathbf{a}_j = \mathbf{0}$$

implies that $\alpha_j = 0$, $j = 1, \ldots, m$.

Example 2.2.4: To see if the column matrices

$$\mathbf{b}_1 = \begin{bmatrix} 1 \\ 1 \\ 0 \end{bmatrix}, \quad \mathbf{b}_2 = \begin{bmatrix} 1 \\ 0 \\ 1 \end{bmatrix}, \quad \mathbf{b}_3 = \begin{bmatrix} 1 \\ 2 \\ -1 \end{bmatrix}$$

are linearly dependent, form

$$\sum_{j=1}^{3} \beta_j \mathbf{b}_j = \mathbf{0}$$

This can be viewed as a system of equations for β_1, β_2, and β_3:

$$\beta_1 + \beta_2 + \beta_3 = 0$$

$$\beta_1 + 2\beta_3 = 0$$

$$\beta_2 - \beta_3 = 0$$

Thus, $\beta_1 = -2\beta_3$ and $\beta_2 = \beta_3$ satisfy all three equations for any value of β_3; for example, $\beta_3 = 1$. Thus, $\mathbf{b}_1$, $\mathbf{b}_2$, and $\mathbf{b}_3$ are linearly dependent.

Consider the $p \times m$ matrix $\mathbf{A} = [\mathbf{a}_1, \ldots, \mathbf{a}_m]$. If a linear combination of the columns $\mathbf{a}_i$ of $\mathbf{A}$ is zero (see Prob. 2.2.5),

$$\mathbf{A}\boldsymbol{\alpha} = \sum_{j=1}^{m} \alpha_j \mathbf{a}_j = \mathbf{0} \tag{2.2.17}$$

for some $\boldsymbol{\alpha} = [\alpha_1, \alpha_2, \ldots, \alpha_m]^T \neq \mathbf{0}$, then the columns of $\mathbf{A}$ are linearly dependent. Otherwise, they are linearly independent. Rows $\mathbf{d}_i$ of $\mathbf{A}$ (see Eq. 2.2.7) are linearly dependent if (Prob. 2.2.5)

$$\boldsymbol{\beta}^T\mathbf{A} = \sum_{i=1}^{n} \beta_i \mathbf{d}_i = \mathbf{0} \tag{2.2.18}$$

for some $\boldsymbol{\beta} = [\beta_1, \ldots, \beta_n]^T \neq \mathbf{0}$. Otherwise, they are linearly independent.

Example 2.2.5: The matrix $\mathbf{B} = [\mathbf{b}_1, \mathbf{b}_2, \mathbf{b}_3]$, with columns given in Example 2.2.4, has linearly dependent columns, as shown in Example 2.2.4. To check for linear dependence of its rows, the product of Eq. 2.2.18 is

$$\boldsymbol{\beta}^T\mathbf{B} = [\beta_1,\ \beta_2,\ \beta_3]\begin{bmatrix} 1 & 1 & 1 \\ 1 & 0 & 2 \\ 0 & 1 & -1 \end{bmatrix} = [0,\ 0,\ 0] = \mathbf{0}$$

From the first two equations, $\beta_2 = -\beta_1 = \beta_3$. The third equation is then $\beta_1 - 2\beta_1 + \beta_1 = 0$. Thus, if $\beta_1 = 1$ and $\beta_2 = \beta_3 = -1$, the equation is satisfied; so the rows of $\mathbf{B}$ are linearly dependent.

The *row rank* (*column rank*) of a matrix is defined as the largest number of linearly independent rows (columns) in the matrix. The row and column ranks of any matrix are equal [22], hence defining the *rank* of the matrix. The rank of a matrix is also equal to the dimension of the largest square submatrix (obtained by deleting rows and columns) with nonzero determinant [22] (Prob. 2.2.6). A square matrix with linearly independent rows (columns) is said to have *full rank.*

When a square matrix does not have full rank, it is called a *singular matrix.* A square matrix $\mathbf{A}$ of full rank is called a *nonsingular matrix.* For such a matrix, there is an *inverse* [22] denoted by $\mathbf{A}^{-1}$, such that

$$\mathbf{A}\mathbf{A}^{-1} = \mathbf{A}^{-1}\mathbf{A} = \mathbf{I} \tag{2.2.19}$$

where $\mathbf{I}$ is the identity matrix. Using the definition of Eq. 2.2.19 and Eq. 2.2.16, it may be shown that [22] (Prob. 2.2.7)

$$(\mathbf{A}^{-1})^T = (\mathbf{A}^T)^{-1} \tag{2.2.20}$$

and that (Prob. 2.2.8)

$$(\mathbf{A}\mathbf{B})^{-1} = \mathbf{B}^{-1}\mathbf{A}^{-1} \tag{2.2.21}$$

A special nonsingular matrix that arises often in kinematics is called an *orthogonal matrix,* with the property that

$$\mathbf{A}^T\mathbf{A} = \mathbf{A}\mathbf{A}^T = \mathbf{I} \tag{2.2.22}$$

That is, from the definition of Eq. 2.2.19,

$$\mathbf{A}^{-1} = \mathbf{A}^T \tag{2.2.23}$$

Since constructing the inverse of a nonsingular matrix is time consuming, it is important to know when a matrix is orthogonal. In case a matrix is orthogonal, its inverse is easily constructed using Eq. 2.2.23.

Example 2.2.6: The 2×2 matrix

$$\mathbf{A} = \begin{bmatrix} \cos\phi & -\sin\phi \\ \sin\phi & \cos\phi \end{bmatrix}$$

is orthogonal, for any value of ϕ, since

$$\mathbf{A}^T\mathbf{A} = \begin{bmatrix} \cos^2\phi + \sin^2\phi & 0 \\ 0 & \cos^2\phi + \sin^2\phi \end{bmatrix} = \mathbf{I}$$

2.3 ALGEBRAIC VECTORS

Recall, from Eq. 2.1.3, that a vector $\vec{a}$ can be written in component form as

$$\vec{a} = a_x\vec{i} + a_y\vec{j} \tag{2.3.1}$$

The vector $\vec{a}$ is thus uniquely defined by its Cartesian components, which may be written using matrix notation as

$$\mathbf{a} = \begin{bmatrix} a_x \\ a_y \end{bmatrix} = [a_x, \ a_y]^T \tag{2.3.2}$$

This is the *algebraic representation of a vector* in the x-y Cartesian reference frame. An *algebraic vector* is defined as a column matrix. When an algebraic vector represents a geometric vector in the plane (i.e., it contains the x-y Cartesian components of the vector), the algebraic vector has two components.

If two vectors $\vec{a}$ and $\vec{b}$ are represented in algebraic form as $\mathbf{a} = [a_x, a_y]^T$ and $\mathbf{b} = [b_x, b_y]^T$, then their vector sum $\vec{c} = \vec{a} + \vec{b}$ is represented in algebraic form by (Prob. 2.3.1)

$$\mathbf{c} = \mathbf{a} + \mathbf{b} \tag{2.3.3}$$

Similarly, $\vec{a} = \vec{b}$ if and only if the components of the vectors are equal; that is, $\mathbf{a} = \mathbf{b}$. Multiplication of a vector $\vec{a}$ by a scalar α occurs component by component, so the geometric vector $\alpha\vec{a}$ is represented by the algebraic vector $\alpha\mathbf{a}$. A *zero algebraic vector,* denoted by $\mathbf{0}$, has all its components equal to zero.

Algebraic representation of vectors allows algebraic vectors with more than three components to be defined (i.e., algebraic vectors with higher dimension than 2). An algebraic vector with n components is called an *n vector.* For example, the algebraic vectors $\mathbf{a} = [a_x, a_y]^T$, $\mathbf{b} = [b_x, b_y]^T$, and $\mathbf{c} = [c_x, c_y]^T$ may be combined to form the 6-vector

$$\begin{aligned} \mathbf{d} &= [a_x, a_y, b_x, b_y, c_x, c_y]^T \\ &= [\mathbf{a}^T, \mathbf{b}^T, \mathbf{c}^T]^T \end{aligned}$$

Since there is a one-to-one correspondence between geometric vectors in a plane and 2×1 algebraic vectors formed from their components, no distinction other than notation will be made between them in the remainder of Part I.

The scalar product of two geometric vectors may be expressed in algebraic form, using the result of Eq. 2.1.11, as

$$\vec{a} \cdot \vec{b} = a_x b_x + a_y b_y = \mathbf{a}^T\mathbf{b} \tag{2.3.4}$$

The vector $\vec{a}^\perp$ that is orthogonal to $\vec{a}$, given by Eq. 2.1.13, is represented in algebraic form as

$$\mathbf{a}^\perp = \begin{bmatrix} -a_y \\ a_x \end{bmatrix} = \mathbf{R}\mathbf{a} \tag{2.3.5}$$

where **R** is the *orthogonal rotation matrix* (Prob. 2.3.2):

$$\mathbf{R} = \begin{bmatrix} 0 & -1 \\ 1 & 0 \end{bmatrix} \tag{2.3.6}$$

The matrix **R** is called orthogonal since, consistent with Eq. 2.2.22,

$$\mathbf{R}^T\mathbf{R} = \begin{bmatrix} 1 & 0 \\ 0 & 1 \end{bmatrix} = \mathbf{I}$$

Finally, by direct computation,

$$\mathbf{RR} = -\mathbf{I} \tag{2.3.7}$$

Example 2.3.1: Using the result of Example 2.1.3, specifically Eq. 2.1.15, and Prob. 2.3.2,

$$\begin{aligned}\theta(\mathbf{a}, \mathbf{a}^{\perp}) &= \operatorname{sgn}(\mathbf{a}^{\perp T}\mathbf{a}^{\perp})\operatorname{Arccos}\frac{\mathbf{a}^T\mathbf{a}^{\perp}}{aa} \\ &= \operatorname{Arccos}\frac{-a_x a_y + a_y a_x}{a^2} \\ &= \operatorname{Arccos} 0 = \frac{\pi}{2}\end{aligned}$$

which shows that the operation $\mathbf{Ra} = \mathbf{a}^{\perp}$ indeed rotates a vector **a** by an angle $\pi/2$ counterclockwise. Using this result and the result of Prob. 2.3.3, the result of successive applications of **R** to a vector **a** is shown in Fig. 2.3.1.

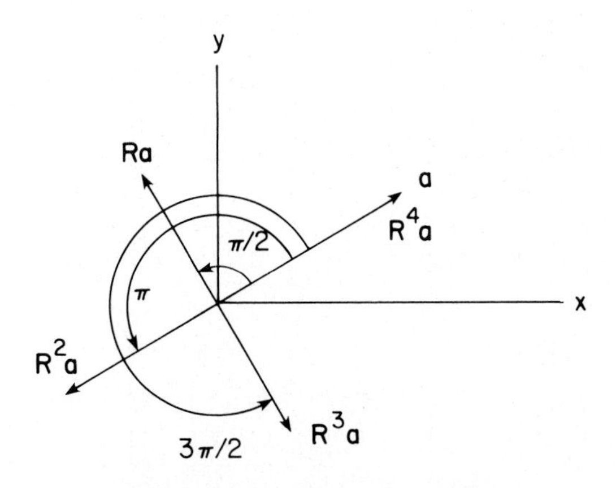

Figure 2.3.1 Rotation of a vector **a**.

2.4 TRANSFORMATION OF COORDINATES

It is shown in Section 2.3 that, in a fixed x-y Cartesian reference frame, geometric vectors are represented by algebraic vectors that contain their components. The components of a vector, however, are defined in a specific Cartesian reference frame. Consider a second Cartesian x'-y' frame, with the same origin as the x-y frame, with angle ϕ between the x and x' axes, counterclockwise positive, as shown in Fig. 2.4.1.

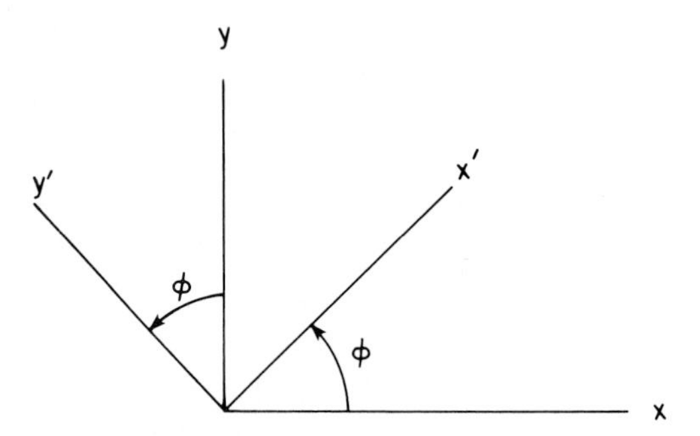

Figure 2.4.1 Two Cartesian reference frames.

A vector $\vec{s}$ in Fig. 2.4.2 can be represented by algebraic vectors

$$\begin{aligned} \mathbf{s} &= [s_x, s_y]^T \\ \mathbf{s}' &= [s_{x'}, s_{y'}]^T \end{aligned} \qquad \textbf{(2.4.1)}$$

in the x-y and x'-y' frames, respectively.

By elementary trigonometry,

$$\begin{aligned} s_x &= s_{x'} \cos\phi - s_{y'} \sin\phi \\ s_y &= s_{x'} \sin\phi + s_{y'} \cos\phi \end{aligned} \qquad \textbf{(2.4.2)}$$

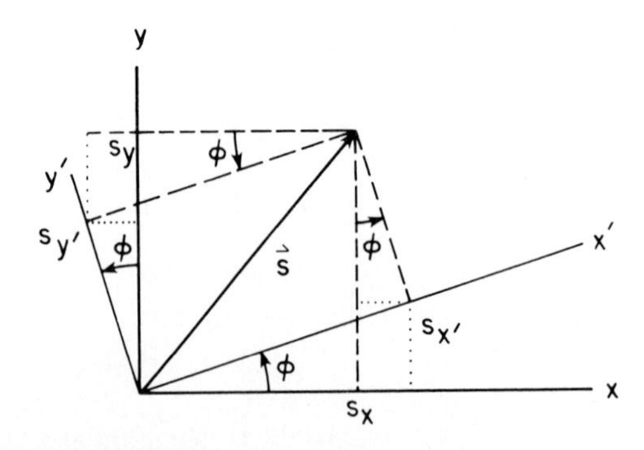

Figure 2.4.2 Vector $\vec{s}$ in two frames.

Thus, $\mathbf{s}$ and $\mathbf{s}'$ are related by the matrix transformation

$$\mathbf{s} = \mathbf{A}\mathbf{s}' \tag{2.4.3}$$

where $\mathbf{A}$ is the planar *rotation transformation matrix*

$$\mathbf{A} = \mathbf{A}(\phi) \equiv \begin{bmatrix} \cos\phi & -\sin\phi \\ \sin\phi & \cos\phi \end{bmatrix} \tag{2.4.4}$$

By direct expansion, note that

$$\mathbf{A}^T\mathbf{A} = \begin{bmatrix} \cos^2\phi + \sin^2\phi & -\cos\phi\sin\phi + \sin\phi\cos\phi \\ -\cos\phi\sin\phi + \sin\phi\cos\phi & \sin^2\phi + \cos^2\phi \end{bmatrix} = \mathbf{I} \tag{2.4.5}$$

Thus, $\mathbf{A}$ is orthogonal and, by Eq. 2.2.23,

$$\mathbf{A}^T = \mathbf{A}^{-1} \tag{2.4.6}$$

The inverse of the transformation of Eq. 2.4.3 is thus

$$\mathbf{s}' = \mathbf{A}^T\mathbf{s} \tag{2.4.7}$$

When the origins of the x-y and x'-y' frames do not coincide, as shown in Fig. 2.4.3, the foregoing analysis is applied between the x'-y' and translated x-y frames, as shown in Fig. 2.4.3. If the vector $\mathbf{s}'^P$ locates point P in the x'-y' frame, then in the translated x-y frame this vector is just $\mathbf{A}\mathbf{s}'^P$. Thus,

$$\begin{aligned} \mathbf{r}^P &= \mathbf{r} + \mathbf{s}^P \\ &= \mathbf{r} + \mathbf{A}\mathbf{s}'^P \end{aligned} \tag{2.4.8}$$

where $\mathbf{r}$ is the vector from the origin of the x-y frame to the origin of the x'-y' frame, as shown in Fig. 2.4.3.

Consider the pair of x_i'-y_i' and x_j'-y_j' frames shown in Fig. 2.4.4. An arbitrary vector $\mathbf{s}$ in the x-y frame has representations $\mathbf{s}_i'$ and $\mathbf{s}_j'$ in the x_i'-y_i' and x_j'-y_j' frames, respectively; that is,

$$\mathbf{s} = \mathbf{A}_i\mathbf{s}_i' = \mathbf{A}_j\mathbf{s}_j' \tag{2.4.9}$$

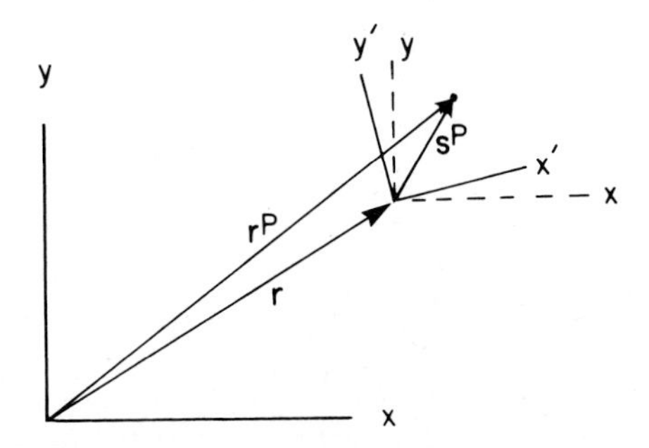

Figure 2.4.3 Translation and rotation of reference frames.

where $\mathbf{A}_i$ and $\mathbf{A}_j$ are transformation matrices from the x_i'-y_i' and x_j'-y_j' frames to the x-y frame, respectively. Since $\mathbf{A}_i$ and $\mathbf{A}_j$ are orthogonal, Eq. 2.4.9 yields

$$\mathbf{s}_i' = \mathbf{A}_i^T\mathbf{A}_j\mathbf{s}_j' \equiv \mathbf{A}_{ij}\mathbf{s}_j' \tag{2.4.10}$$

Since $\mathbf{s}_j'$ is an arbitrary vector,

$$\mathbf{A}_{ij} = \mathbf{A}_i^T\mathbf{A}_j \tag{2.4.11}$$

Thus, $\mathbf{A}_{ij}$ is the transformation matrix from the x_j'-y_j' frame to the x_i'-y_i' frame. A direct calculation shows that $\mathbf{A}_{ij}$ is an orthogonal matrix; that is,

$$\mathbf{A}_{ij}^T\mathbf{A}_{ij} = \mathbf{A}_j^T\mathbf{A}_i\mathbf{A}_i^T\mathbf{A}_j = \mathbf{A}_j^T\mathbf{A}_j = \mathbf{I}$$

Using the definition of $\mathbf{A}_i$ and $\mathbf{A}_j$ in Eq. 2.4.4 and standard trigonometric identities, Eq. 2.4.11 may be expanded to obtain

$$\mathbf{A}_{ij} = \begin{bmatrix} \cos(\phi_j - \phi_i) & -\sin(\phi_j - \phi_i) \\ \sin(\phi_j - \phi_i) & \cos(\phi_j - \phi_i) \end{bmatrix} = \mathbf{A}(\phi_j - \phi_i) \tag{2.4.12}$$

which is just the orthogonal rotation transformation matrix due to a rotation of the x_j'-y_j' frame by angle $\phi_j - \phi_i$, relative to the x_i'-y_i' frame. Since $\mathbf{A}_{ij}$ transforms vectors from the x_j'-y_j' frame to the x_i'-y_i' frame, this result could have been anticipated from Fig. 2.4.4.

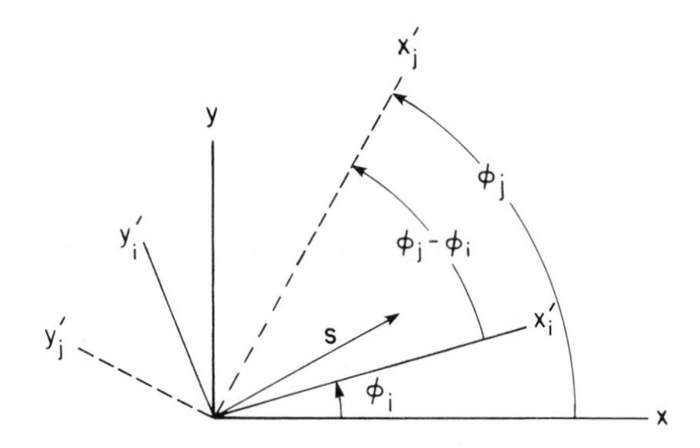

Figure 2.4.4 Three reference frames with coincident origins.

Example 2.4.1: If actuators control angles θ_1 and θ_2 in the positioning mechanism shown in Fig. 2.4.5, then $\phi_1 = \theta_1$ and $\phi_2 = \theta_1 + \theta_2$. Thus,

$$\mathbf{r}^Q = \mathbf{A}(\theta_1)\mathbf{s}_1'^Q$$

$$\mathbf{r}^P = \mathbf{r}^Q + \mathbf{A}(\theta_1 + \theta_2)\mathbf{s}_2'^{QP}$$

where $\mathbf{s}_2'^{QP}$ is the vector fixed in the x_2'-y_2' frame that locates point P relative to

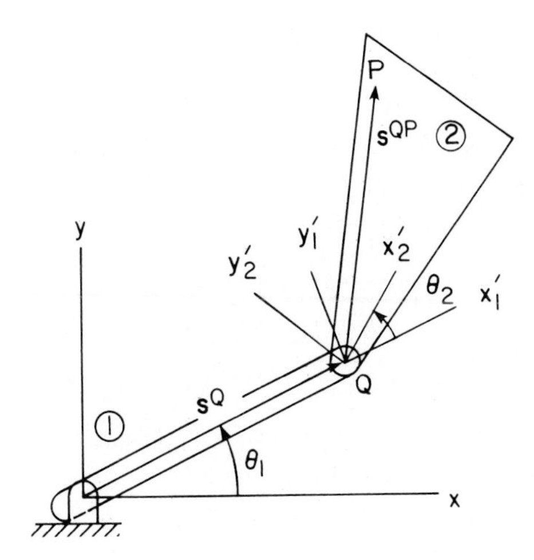

Figure 2.4.5 Two-body positioning mechanism.

point Q. Thus,

$$\mathbf{r}^P = \begin{bmatrix} \cos\theta_1 & -\sin\theta_1 \\ \sin\theta_1 & \cos\theta_1 \end{bmatrix} \mathbf{s}_1'^Q + \begin{bmatrix} \cos(\theta_1+\theta_2) & -\sin(\theta_1+\theta_2) \\ \sin(\theta_1+\theta_2) & \cos(\theta_1+\theta_2) \end{bmatrix} \mathbf{s}_2'^{QP} \quad \textbf{(2.4.13)}$$

Example 2.4.2: Using Eq. 2.4.12, where $\phi_2 - \phi_1 = \theta_2$ in Fig. 2.4.5,

$$\mathbf{A}_{12}(\theta_2) = \begin{bmatrix} \cos\theta_2 & -\sin\theta_2 \\ \sin\theta_2 & \cos\theta_2 \end{bmatrix}$$

Thus, the vector $\mathbf{s}^{QP}$ from point Q to point P in Fig. 2.4.5, represented in the x_1'-y_1' frame, is

$$\mathbf{s}_1'^{QP} = \mathbf{A}_{12}\mathbf{s}_2'^{QP}$$

In the x-y frame, this is

$$\mathbf{s}^{QP} = \mathbf{A}_1\mathbf{A}_{12}\mathbf{s}_2'^{QP}$$

where

$$\mathbf{A}_1(\theta_1) = \begin{bmatrix} \cos\theta_1 & -\sin\theta_1 \\ \sin\theta_1 & \cos\theta_1 \end{bmatrix}$$

Thus, since $\mathbf{s}_1'^Q$ and $\mathbf{s}_2'^{QP}$ are constant, $\mathbf{r}^P$ as a function of θ_1 and θ_2 is

$$\mathbf{r}^P = \mathbf{A}_1(\theta_1)\mathbf{s}_1'^Q + \mathbf{A}_1(\theta_1)\mathbf{A}_{12}(\theta_2)\mathbf{s}_2'^{QP} \quad \textbf{(2.4.14)}$$

Example 2.4.3: The slider–crank mechanism shown in Fig. 2.4.6(a) may be modeled using the angles θ_1 and θ_2 shown in Fig. 2.4.6(b). Since the geometry of this mechanism is the same as that of Fig. 2.4.5, with $\mathbf{s}_1'^Q = [1, 0]^T$ and

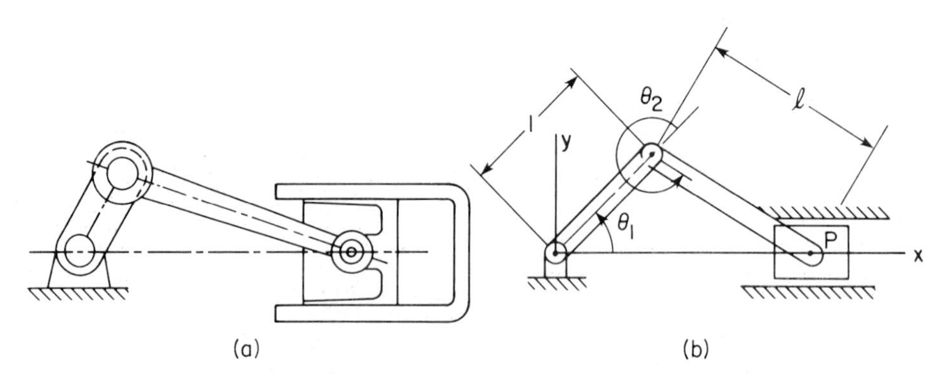

Figure 2.4.6 Slider–crank mechanism. (a) Physical system. (b) Kinematic model.

$\mathbf{s}_2'^{QP} = [\ell, 0]^T$, from Eq. 2.4.13,

$$\mathbf{r}^P = \begin{bmatrix} \cos\theta_1 + \ell\cos(\theta_1 + \theta_2) \\ \sin\theta_1 + \ell\sin(\theta_1 + \theta_2) \end{bmatrix}$$

Since the piston wrist pin at point P is constrained to slide along the x axis,

$$r_y^P \equiv \Phi(\theta_1, \theta_2) = \sin\theta_1 + \ell\sin(\theta_1 + \theta_2) = 0 \qquad \textbf{(2.4.15)}$$

Thus, angles θ_1 and θ_2 are not independent; that is, they must satisfy the constraint equation of Eq. 2.4.15. These variables are typical of *generalized coordinates* that are used to define the geometry of a mechanism.

2.5 VECTOR AND MATRIX DIFFERENTIATION

In kinematics and dynamics of mechanical systems, vectors that represent the positions of points on bodies or equations that describe the geometry of the system are functions of time or some other variables. In analyzing these equations, time derivatives and partial derivatives are needed. In this section these derivatives are defined, and the *matrix calculus notation* that is used throughout the text is introduced.

In analyzing velocities and accelerations, time derivatives of vectors that locate points must be calculated. Consider a vector $\vec{a}(t)$ with components $\mathbf{a} \equiv \mathbf{a}(t) = [a_x(t), a_y(t)]^T$ in a *stationary Cartesian reference frame,* that is, a frame with $\vec{i}$ and $\vec{j}$ constant. The *time derivative of a vector* $\vec{a}$ is

$$\dot{\vec{a}}(t) \equiv \frac{d}{dt}\vec{a}(t) = \frac{d}{dt}[a_x(t)\vec{i} + a_y(t)\vec{j}]$$

$$= \left[\frac{d}{dt}a_x(t)\right]\vec{i} + \left[\frac{d}{dt}a_y(t)\right]\vec{j}$$

Note that this is valid only if $\vec{i}$ and $\vec{j}$ are not time dependent. In matrix notation, this is

$$\dot{\mathbf{a}} \equiv \frac{d}{dt}\mathbf{a}$$
$$= \left[\frac{d}{dt}a_x, \frac{d}{dt}a_y\right]^T = [\dot{a}_x, \dot{a}_y]^T \tag{2.5.1}$$

where an overdot denotes derivative with respect to time. Thus, for vectors that are written in terms of their components in a stationary Cartesian reference frame, the derivative of a vector is obtained by differentiating its components.

The derivative of the sum of two vectors $\mathbf{a} = \mathbf{a}(t)$ and $\mathbf{b} = \mathbf{b}(t)$ is

$$\frac{d}{dt}(\mathbf{a} + \mathbf{b}) = \dot{\mathbf{a}} + \dot{\mathbf{b}} \tag{2.5.2}$$

which is analogous to the ordinary differentiation rule that the derivative of a sum is the sum of the derivatives. The following vector forms of the *product rule of differentiation* are also valid (Prob. 2.5.1):

$$\frac{d}{dt}(\alpha\mathbf{a}) = \dot{\alpha}\mathbf{a} + \alpha\dot{\mathbf{a}} \tag{2.5.3}$$

$$\frac{d}{dt}(\mathbf{a}^T\mathbf{b}) = \dot{\mathbf{a}}^T\mathbf{b} + \mathbf{a}^T\dot{\mathbf{b}} \tag{2.5.4}$$

Many uses may be made of these derivative formulas. For example, if the length of a vector $\mathbf{a}(t)$ is fixed; that is, $\mathbf{a}(t)^T\mathbf{a}(t) = c$, then Eq. 2.5.4 yields

$$\dot{\mathbf{a}}^T\mathbf{a} = 0 \tag{2.5.5}$$

If $\mathbf{a}$ is a position vector that locates a point in a stationary Cartesian reference frame, then $\dot{\mathbf{a}}$ is the *velocity* of that point. Hence, Eq. 2.5.5 indicates that the velocity of a point whose distance from the origin is constant is orthogonal to the position vector of the point. The second time derivative of $\mathbf{a}(t)$ is the *acceleration* of the point, denoted as

$$\ddot{\mathbf{a}} = \frac{d}{dt}\left(\frac{d}{dt}\mathbf{a}(t)\right) = [\ddot{a}_x, \ddot{a}_y]^T \tag{2.5.6}$$

Thus, for vectors that are written in terms of their components in a stationary Cartesian reference frame, the second time derivative of the vector may be calculated in terms of second time derivatives of components of the vector.

Example 2.5.1: If $\mathbf{s}_1'^Q = [1, 0]^T$ and $\mathbf{s}_2'^{QP} = [1, 0]^T$ in the positioning mechanism of Example 2.4.1, from Eq. 2.4.13,

$$\mathbf{r}^P = \begin{bmatrix} \cos\theta_1 + \cos(\theta_1 + \theta_2) \\ \sin\theta_1 + \sin(\theta_1 + \theta_2) \end{bmatrix}$$

Let the actuators be driven with constant angular velocities ω_1 and ω_2, so that $\theta_1 = \omega_1 t$ and $\theta_2 = \omega_2 t$. Then

$$\mathbf{r}^P = \begin{bmatrix} \cos \omega_1 t + \cos(\omega_1 + \omega_2)t \\ \sin \omega_1 t + \sin(\omega_1 + \omega_2)t \end{bmatrix}$$

From Eqs. 2.5.1 and 2.5.6, the velocity and acceleration of point P are obtained by direct differentiation as

$$\dot{\mathbf{r}}^P = \begin{bmatrix} -\omega_1 \sin \omega_1 t - (\omega_1 + \omega_2)\sin(\omega_1 + \omega_2)t \\ \omega_1 \cos \omega_1 t + (\omega_1 + \omega_2)\cos(\omega_1 + \omega_2)t \end{bmatrix}$$

$$\ddot{\mathbf{r}}^P = \begin{bmatrix} -\omega_1^2 \cos \omega_1 t - (\omega_1 + \omega_2)^2 \cos(\omega_1 + \omega_2)t \\ -\omega_1^2 \sin \omega_1 t - (\omega_1 + \omega_2)^2 \sin(\omega_1 + \omega_2)t \end{bmatrix}$$

Just as in differentiation of a vector whose components are functions of t, the *derivative of a matrix* whose components depend on t may be defined. Consider a matrix $\mathbf{B}(t) = [b_{ij}(t)]$. The derivative of $\mathbf{B}(t)$ is defined as

$$\dot{\mathbf{B}} \equiv \frac{d}{dt}\mathbf{B} = \left[\frac{d}{dt} b_{ij}\right] \tag{2.5.7}$$

With this definition and elementary rules of differentiation (Prob. 2.5.3),

$$\frac{d}{dt}(\mathbf{B} + \mathbf{C}) = \dot{\mathbf{B}} + \dot{\mathbf{C}} \tag{2.5.8}$$

$$\frac{d}{dt}(\mathbf{BC}) = \dot{\mathbf{B}}\mathbf{C} + \mathbf{B}\dot{\mathbf{C}} \tag{2.5.9}$$

$$\frac{d}{dt}(\alpha\mathbf{B}) = \dot{\alpha}\mathbf{B} + \alpha\dot{\mathbf{B}} \tag{2.5.10}$$

where $\alpha = \alpha(t)$ is a scalar function of time.

In dealing with systems of nonlinear differential and algebraic equations in several variables, which govern the kinematics and dynamics of mechanical systems, it is essential that a *matrix calculus notation* be employed. To introduce the notation used here, let $\mathbf{q} = [q_1, \ldots, q_k]^T$ be a k vector of real variables, $a(\mathbf{q})$ be a scalar differentiable function of $\mathbf{q}$, and $\mathbf{\Phi}(\mathbf{q}) = [\Phi_1(\mathbf{q}), \ldots, \Phi_n(\mathbf{q})]^T$ be an n vector of differentiable functions of $\mathbf{q}$. Using i as row index and j as column index, the following matrix calculus notations are defined:

$$a_{\mathbf{q}} \equiv \frac{\partial a}{\partial \mathbf{q}} \equiv \left[\frac{\partial a}{\partial q_j}\right]_{1\times k} \tag{2.5.11}$$

$$\mathbf{\Phi}_{\mathbf{q}} \equiv \frac{\partial \mathbf{\Phi}}{\partial \mathbf{q}} \equiv \left[\frac{\partial \Phi_i}{\partial q_j}\right]_{n\times k} \tag{2.5.12}$$

Note that the derivative of a scalar function with respect to a vector variable in

Eq. 2.5.11 is a row matrix. This is one of the few matrix symbols in the text that is a row matrix, rather than the more common column matrix. Note also that, as defined by Eq. 2.5.12, the derivative of a vector function $\mathbf{\Phi}$, whose elements are functions of the vector variable $\mathbf{q}$, is a matrix. The subscript notation used here to denote differentiation is helpful in deriving needed relations without becoming entangled in cumbersome index and partial differentiation notation. To take advantage of this notation, however, it is critically important that the correct matrix definition of derivatives be used.

Example 2.5.2: If $\mathbf{q} = [\theta_1, \theta_2]^T$, then the derivative of the slider–crank constraint function Φ of Eq. 2.4.15 in Example 2.4.3 is

$$\Phi_{\mathbf{q}} = [\cos\theta_1 + \ell\cos(\theta_1 + \theta_2),\ \ell\cos(\theta_1 + \theta_2)]$$

Similarly, the matrix of derivatives of $\mathbf{r}^P$ in the same example is

$$\mathbf{r}^P_{\mathbf{q}} = \begin{bmatrix} -\sin\theta_1 - \ell\sin(\theta_1 + \theta_2) & -\ell\sin(\theta_1 + \theta_2) \\ \cos\theta_1 + \ell\cos(\theta_1 + \theta_2) & \ell\cos(\theta_1 + \theta_2) \end{bmatrix}$$

The partial derivative of the scalar product of two n vector functions $\mathbf{g}(\mathbf{q}) = [g_1(\mathbf{q}), \ldots, g_n(\mathbf{q})]^T$ and $\mathbf{h}(\mathbf{q}) = [h_1(\mathbf{q}), \ldots, h_n(\mathbf{q})]^T$, by careful manipulation, is the *product rule of differentiation*

$$\begin{aligned}
\frac{\partial}{\partial \mathbf{q}}(\mathbf{g}^T\mathbf{h}) &= (\mathbf{g}^T\mathbf{h})_{\mathbf{q}} \\
&= \frac{\partial}{\partial \mathbf{q}}\left(\sum_{k=1}^{n} g_k h_k\right) \\
&= \left[\frac{\partial}{\partial q_j}\left(\sum_{k=1}^{n} g_k h_k\right)\right] \\
&= \left[\sum_{k=1}^{n}\left(\frac{\partial g_k}{\partial q_j} h_k + g_k \frac{\partial h_k}{\partial q_j}\right)\right] \\
&= \left[\sum_{k=1}^{n}\left(h_k \frac{\partial g_k}{\partial q_j}\right) + \sum_{k=1}^{n}\left(g_k \frac{\partial h_k}{\partial q_j}\right)\right] \\
&= \mathbf{h}^T \frac{\partial \mathbf{g}}{\partial \mathbf{q}} + \mathbf{g}^T \frac{\partial \mathbf{h}}{\partial \mathbf{q}} \\
&= \mathbf{h}^T\mathbf{g}_{\mathbf{q}} + \mathbf{g}^T\mathbf{h}_{\mathbf{q}}
\end{aligned} \tag{2.5.13}$$

Note that what might have intuitively appeared to be the appropriate product rule of differentiation is not even defined, much less valid; that is,

$$\frac{\partial}{\partial \mathbf{q}}(\mathbf{g}^T\mathbf{h}) \neq \mathbf{g}^T_{\mathbf{q}}\mathbf{h} + \mathbf{g}^T\mathbf{h}_{\mathbf{q}}$$

If $\mathbf{\Phi}(\mathbf{g}) = [\Phi_1(\mathbf{g}), \ldots, \Phi_m(\mathbf{g})]^T$ and $\mathbf{g} = \mathbf{g}(\mathbf{q}) = [g_1(\mathbf{q}), \ldots, g_n(\mathbf{q})]^T$ are vector functions of vector variables, the *chain rule of differentiation,* in matrix calculus form, is obtained as

$$\begin{aligned}\mathbf{\Phi}_{\mathbf{q}} &\equiv \left[\frac{\partial \Phi_i(\mathbf{g}(\mathbf{q}))}{\partial q_j}\right]_{m\times k} \\ &= \left[\sum_{\ell=1}^{n}\left(\frac{\partial \Phi_i}{\partial g_\ell}\frac{\partial g_\ell}{\partial q_j}\right)\right]_{m\times k} \\ &= \frac{\partial \mathbf{\Phi}}{\partial \mathbf{g}}\frac{\partial \mathbf{g}}{\partial \mathbf{q}} \\ &= \mathbf{\Phi}_{\mathbf{g}}\mathbf{g}_{\mathbf{q}}\end{aligned} \tag{2.5.14}$$

If $\mathbf{B}$ is a constant $m \times n$ matrix and $\mathbf{p}$ and $\mathbf{q}$ are m and n vectors of variables, respectively, the following useful relations may be verified (Prob. 2.5.7):

$$\frac{\partial}{\partial \mathbf{q}}(\mathbf{Bq}) = \mathbf{B} \tag{2.5.15}$$

$$\frac{\partial}{\partial \mathbf{p}}(\mathbf{p}^T\mathbf{Bq}) = \mathbf{q}^T\mathbf{B}^T \tag{2.5.16}$$

$$\frac{d}{dt}(\mathbf{p}^T\mathbf{Bq}) = \mathbf{q}^T\mathbf{B}^T\dot{\mathbf{p}} + \mathbf{p}^T\mathbf{B}\dot{\mathbf{q}} \tag{2.5.17}$$

Example 2.5.3: If $\ell > 1$ for the slider–crank mechanism of Example 2.4.3, then $|\sin(\theta_1 + \theta_2)| = |\sin\theta_1|/\ell < 1$ and Eq. 2.4.15 can be solved for θ_2 as

$$\theta_2 = 2\pi - \operatorname{Arcsin}\left(\frac{\sin\theta_1}{\ell}\right) - \theta_1 \tag{2.5.18}$$

where $-\pi/2 \leq \operatorname{Arcsin}\alpha \leq \pi/2$. Thus, θ_2 may be treated as a function of θ_1, and the motion of the slider–crank mechanism may be controlled by specifying the time history $\theta_1(t)$ of the crank angle. By direct differentiation,

$$\dot{\theta}_2 = -\left[\frac{\cos\theta_1}{\sqrt{\ell^2 - \sin^2\theta_1}} + 1\right]\dot{\theta}_1 \tag{2.5.19}$$

An alternative approach to calculating $\dot{\theta}_2$ in terms of $\dot{\theta}_1$ that is useful in kinematic analysis, even when complicated nonlinear constraint equations cannot be solved in closed form, is to differentiate both sides of Eq. 2.4.15, to obtain

$$\dot{\theta}_1\cos\theta_1 + \ell(\dot{\theta}_1 + \dot{\theta}_2)\cos(\theta_1 + \theta_2) = 0 \tag{2.5.20}$$

This can be solved for $\dot{\theta}_2$ as

$$\dot{\theta}_2 = -\left[\frac{\cos\theta_1}{\ell\cos(\theta_1 + \theta_2)} + 1\right]\dot{\theta}_1 \tag{2.5.21}$$

Similarly, Eq. 2.5.20 can be differentiated to obtain

$$\ddot{\theta}_1 \cos\theta_1 - \dot{\theta}_1^2 \sin\theta_1 + \ell(\ddot{\theta}_1 + \ddot{\theta}_2)\cos(\theta_1+\theta_2) - \ell(\dot{\theta}_1+\dot{\theta}_2)^2 \sin(\theta_1+\theta_2) = 0 \quad \textbf{(2.5.22)}$$

which can be solved for $\ddot{\theta}_2$ as a function of $\dot{\theta}_1$ and $\ddot{\theta}_1$, using Eq. 2.5.21 to eliminate $\dot{\theta}_2$ (Prob. 2.5.9).

2.6 VELOCITY AND ACCELERATION OF A POINT FIXED IN A MOVING FRAME

Quite often in applications, an x'-y' Cartesian reference frame is fixed in a moving body to define its position and orientation, relative to a stationary global x-y reference frame. Consider a point P that is fixed in an x'-y' frame, as shown in Fig. 2.6.1. The vector that locates P in the x-y frame is given by Eq. 2.4.8 as

$$\mathbf{r}^P = \mathbf{r} + \mathbf{A}\mathbf{s}'^P \quad \textbf{(2.6.1)}$$

where $\mathbf{s}'^P$ is the constant vector of coordinates of P in the x'-y' frame and $\mathbf{A}$ is the transformation matrix from the x'-y' frame to the x-y frame.

Since the x'-y' frame is moving and changing its orientation with time, the vector $\mathbf{r}$ and the transformation matrix $\mathbf{A}$ are functions of time. Differentiation results of Section 2.5 can be used to obtain the time derivative of $\mathbf{r}^P$ as

$$\dot{\mathbf{r}}^P = \dot{\mathbf{r}} + \dot{\mathbf{A}}\mathbf{s}'^P \quad \textbf{(2.6.2)}$$

From Eq. 2.4.4,

$$\dot{\mathbf{A}} = \dot{\phi}\frac{d}{d\phi}\mathbf{A} = \dot{\phi}\begin{bmatrix} -\sin\phi & -\cos\phi \\ \cos\phi & -\sin\phi \end{bmatrix} \equiv \dot{\phi}\mathbf{B} \quad \textbf{(2.6.3)}$$

Thus, Eq. 2.6.2 becomes

$$\dot{\mathbf{r}}^P = \dot{\mathbf{r}} + \dot{\phi}\mathbf{B}\mathbf{s}'^P \quad \textbf{(2.6.4)}$$

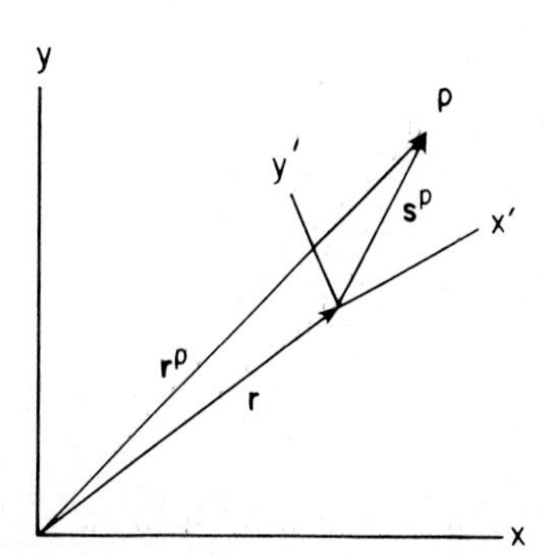

Figure 2.6.1 Point P fixed in an x'-y' frame.

Note that, with the orthogonal rotation matrix **R** of Eq. 2.3.6,

$$\mathbf{AR} = \begin{bmatrix} \cos\phi & -\sin\phi \\ \sin\phi & \cos\phi \end{bmatrix} \begin{bmatrix} 0 & -1 \\ 1 & 0 \end{bmatrix} = \begin{bmatrix} -\sin\phi & -\cos\phi \\ \cos\phi & -\sin\phi \end{bmatrix} = \mathbf{B} \tag{2.6.5}$$

Also,

$$\mathbf{AR} = \mathbf{RA} \tag{2.6.6}$$

Thus, Eq. 2.6.4 may be written as

$$\begin{aligned} \dot{\mathbf{r}}^P &= \dot{\mathbf{r}} + \dot{\phi}\mathbf{ARs}'^P \\ &= \dot{\mathbf{r}} + \dot{\phi}\mathbf{As}'^{P\perp} \\ &= \dot{\mathbf{r}} + \dot{\phi}\mathbf{s}^{P\perp} \end{aligned} \tag{2.6.7}$$

Finally, note that

$$\dot{\mathbf{B}} = \dot{\phi}\frac{d}{d\phi}\mathbf{B} = \dot{\phi}\begin{bmatrix} -\cos\phi & \sin\phi \\ -\sin\phi & -\cos\phi \end{bmatrix} = -\dot{\phi}\mathbf{A} \tag{2.6.8}$$

Taking the time derivative of Eq. 2.6.4 yields the acceleration of point P as

$$\begin{aligned} \ddot{\mathbf{r}}^P &= \ddot{\mathbf{r}} + \ddot{\phi}\mathbf{Bs}'^P + \dot{\phi}\dot{\mathbf{B}}\mathbf{s}'^P \\ &= \ddot{\mathbf{r}} + \ddot{\phi}\mathbf{Bs}'^P - \dot{\phi}^2\mathbf{As}'^P \end{aligned} \tag{2.6.9}$$

Alternative forms of this relation are obtained, using Eqs. 2.6.5 and 2.6.6, as

$$\begin{aligned} \ddot{\mathbf{r}}^P &= \ddot{\mathbf{r}} + \ddot{\phi}\mathbf{As}'^{P\perp} - \dot{\phi}^2\mathbf{As}'^P \\ &= \ddot{\mathbf{r}} + \ddot{\phi}\mathbf{s}^{P\perp} - \dot{\phi}^2\mathbf{s}^P \end{aligned} \tag{2.6.10}$$

Example 2.6.1 The position vector of point P on body 2 of the positioning mechanism of Example 2.4.1 was derived in Example 2.4.2 in the form (Eq. 2.4.14)

$$\mathbf{r}^P = \mathbf{A}_1\mathbf{s}_1'^Q + \mathbf{A}_1\mathbf{A}_{12}\mathbf{s}_2'^{QP}$$

where $\mathbf{A}_1 = \mathbf{A}(\theta_1)$, $\mathbf{A}_{12} = \mathbf{A}(\theta_2)$, and $\mathbf{s}_1'^Q$ and $\mathbf{s}_2'^{QP}$ are constant vectors. Using the chain and product rules of differentiation and Eq. 2.6.3,

$$\dot{\mathbf{r}}^P = \dot{\theta}_1(\mathbf{B}_1\mathbf{s}_1'^Q + \mathbf{B}_1\mathbf{A}_{12}\mathbf{s}_2'^{QP}) + \dot{\theta}_2\mathbf{A}_1\mathbf{B}_{12}\mathbf{s}_2'^{QP} \tag{2.6.11}$$

and, using Eq. 2.6.8 (Prob. 2.6.1),

$$\begin{aligned} \ddot{\mathbf{r}}^P = {} & \ddot{\theta}_1(\mathbf{B}_1\mathbf{s}_1'^Q + \mathbf{B}_1\mathbf{A}_{12}\mathbf{s}_2'^{QP}) + \ddot{\theta}_2\mathbf{A}_1\mathbf{B}_{12}\mathbf{s}_2'^{QP} \\ & - \dot{\theta}_1^2\mathbf{r}^P - \dot{\theta}_2^2\mathbf{A}_1\mathbf{A}_{12}\mathbf{s}_2'^{QP} + 2\dot{\theta}_1\dot{\theta}_2\mathbf{B}_1\mathbf{B}_{12}\mathbf{s}_2'^{QP} \end{aligned} \tag{2.6.12}$$

where $\mathbf{B}_1 = \mathbf{B}(\theta_1)$ and $\mathbf{B}_{12} = \mathbf{B}(\theta_2)$.

PROBLEMS

Section 2.1

2.1.1. Using the component forms of vectors $\vec{a}$, $\vec{b}$, and $\vec{c}$, show that Eq. 2.1.5 holds.

2.1.2. Show that the component form of Eq. 2.1.15 is

$$\theta(\vec{a}, \vec{b}) = \text{sgn}(-a_y b_x + a_x b_y) \text{ Arccos}\left(\frac{a_x b_x + a_y b_y}{\sqrt{a_x^2 + a_y^2}\sqrt{b_x^2 + b_y^2}}\right)$$

2.1.3. Find $\theta(\vec{a}, \vec{b}_k)$, $k = 1, 2$, for $\vec{a} = \vec{i} + \vec{j}$, $\vec{b}_1 = \vec{j}$, and $\vec{b}_2 = \vec{i}$. Draw a figure to display and interpret the result.

Section 2.2

2.2.1. Use the definitions of equality and addition of matrices of equal dimension to show that Eq. 2.2.5 is valid.

2.2.2. Show that Eq. 2.2.6 is valid.

2.2.3. Calculate the sum $\mathbf{C} = \mathbf{A} + \mathbf{B}$, difference $\mathbf{D} = \mathbf{A} - \mathbf{B}$, and product $\mathbf{E} = \mathbf{AB}$ of the 3×3 matrices

$$\mathbf{A} = \begin{bmatrix} 1 & 1 & 0 \\ -1 & 0 & 1 \\ 0 & 1 & -1 \end{bmatrix}, \quad \mathbf{B} = \begin{bmatrix} 1 & 2 & 1 \\ 1 & 3 & 0 \\ 2 & 1 & 0 \end{bmatrix}$$

2.2.4. Carry out the operations on both sides of Eqs. 2.2.12 and 2.2.13 with the matrices

$$\mathbf{A} = \begin{bmatrix} 1 & 0 \\ 0 & 1 \end{bmatrix}, \quad \mathbf{B} = \begin{bmatrix} 1 & 2 \\ 2 & 1 \end{bmatrix}, \quad \mathbf{C} = \begin{bmatrix} 1 & 1 \\ 1 & 1 \end{bmatrix}$$

2.2.5. Show that the expansion of $\mathbf{A}\boldsymbol{\alpha}$ as a linear combination of columns of $\mathbf{A}$ in Eq. 2.2.17 is valid. Similarly verify that the expansion in Eq. 2.2.18 is valid.

2.2.6. Show that the rank of matrix $\mathbf{B}$ of Example 2.2.5 is 2 by finding a 2×2 submatrix with nonzero determinant and showing that $|\mathbf{B}| = 0$.

2.2.7. Use Eqs. 2.2.16 and 2.2.19 to show that Eq. 2.2.20 is valid.

2.2.8. Show that $(\mathbf{AB})(\mathbf{B}^{-1}\mathbf{A}^{-1}) = \mathbf{I}$, to verify that Eq. 2.2.21 is valid.

2.2.9. Show that the following 3×3 matrix is orthogonal for any value of ϕ:

$$\mathbf{A} = \begin{bmatrix} 1 & 0 & 0 \\ 0 & \cos\phi & -\sin\phi \\ 0 & \sin\phi & \cos\phi \end{bmatrix}$$

2.2.10. Show that if $n \times n$ matrices $\mathbf{A}$ and $\mathbf{B}$ are orthogonal then $\mathbf{A}^T\mathbf{B}$ is orthogonal (*Hint*: Use Eq. 2.2.16).

Section 2.3

2.3.1. Show that $\mathbf{c}$ of Eq. 2.3.3 is the algebraic representation of $\vec{c} = \vec{a} + \vec{b}$.

2.3.2. Show that $\mathbf{a}^T\mathbf{a}^\perp = \mathbf{a}^T\mathbf{Ra} = 0$.

2.3.3. Show that $\mathbf{RRa} \equiv \mathbf{R}^2\mathbf{a} = -\mathbf{a}$, $\mathbf{RRRa} \equiv \mathbf{R}^3\mathbf{a} = -\mathbf{a}^\perp$, and $\mathbf{RRRRa} \equiv \mathbf{R}^4\mathbf{a} = \mathbf{a}$.

2.3.4. Show that $(\mathbf{a}+\mathbf{b})^{\perp} = \mathbf{a}^{\perp} + \mathbf{b}^{\perp}$ (*Hint*: Use the matrix property of Eq. 2.2.12 and $\mathbf{a}^{\perp} = \mathbf{Ra}$).

Section 2.4

2.4.1. Show that $\mathbf{R} = \mathbf{A}(\pi/2)$ and, more generally, $\mathbf{R}^n = \mathbf{A}(n\pi/2)$, $n = 2, 3, 4$.

2.4.2. Show that the alternative representations of $\mathbf{r}^P$ in Examples 2.4.1 and 2.4.2 yield identical expressions in θ_1 and θ_2.

2.4.3. Write the vector $\mathbf{r}^P$ from the origin of the x-y frame to point P, in terms of θ, for the parallelogram mechanism shown in Fig. P2.4.3, where $\mathbf{s}'^P$ is a known constant vector in the x'-y' frame that is fixed in the coupler.

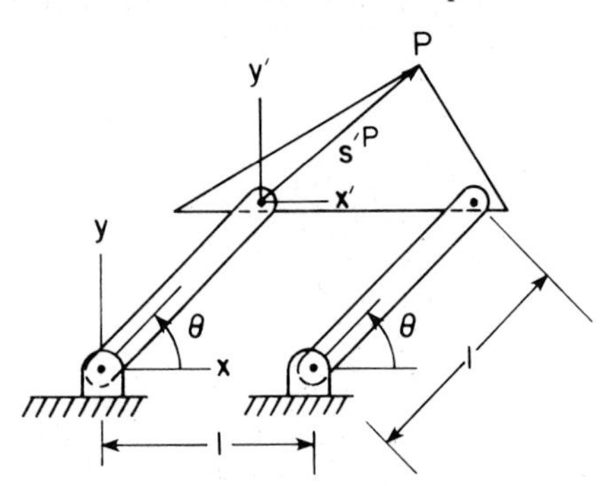

Figure P2.4.3

2.4.4. Pins in a body of unit length are constrained to slide in horizontal and vertical slots as shown in Fig. P2.4.4. Write the vector $\mathbf{r}^P$ from the origin of the x-y frame to point P in terms of generalized coordinates d and θ. Write the constraint equation in d and θ require that point P to slide in the vertical slot.

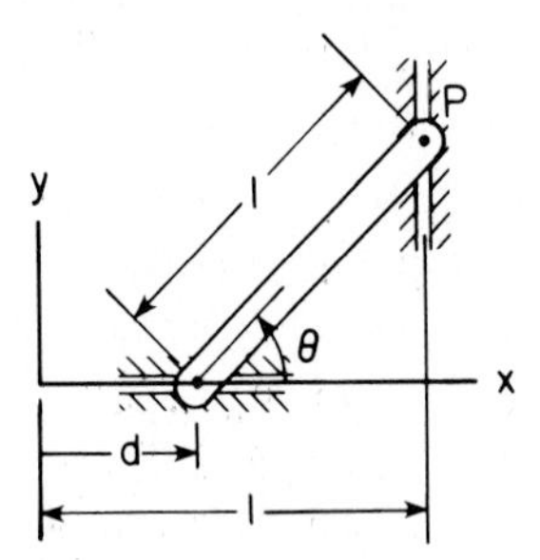

Figure P2.4.4

Section 2.5

2.5.1. Show that Eqs. 2.5.3 and 2.5.4 are valid.

2.5.2. Show that if $\omega_2 = 0$ in Example 2.5.1, then

$$\dot{\mathbf{r}}^P = \omega_1 \mathbf{r}^{P\perp}$$

$$\ddot{\mathbf{r}}^P = -\omega_1^2 \mathbf{r}^P$$

Interpret these results (*Hint*: If $\omega_2 = 0$, then there is no rotation of body 2 relative to body 1).

2.5.3. Show that Eqs. 2.5.8, 2.5.9, and 2.5.10 are valid.

2.5.4. Use the definition of Eq. 2.5.7 and Eq. 2.5.9 to evaluate $\dot{\mathbf{r}}^P$ in Eq. 2.4.13 in terms of $\dot{\theta}_1$ and $\dot{\theta}_2$.

2.5.5. Use Eq. 2.5.9 and the definition of $\mathbf{A}_1$ and $\mathbf{A}_{12}$, as functions of θ_1 and θ_2, to evaluate $\dot{\mathbf{r}}^P$ from Eq. 2.4.14 in terms of $\dot{\theta}_1$ and $\dot{\theta}_2$. Verify that the result is the same as obtained in Prob. 2.5.4.

2.5.6. Use the chain rule of differentiation and the results of Example 2.5.2 to evaluate $\dot{\mathbf{\Phi}}$ and $\dot{\mathbf{r}}^P$ (*Hint*: $\dot{\mathbf{\Phi}} = \mathbf{\Phi}_{\mathbf{q}}\dot{\mathbf{q}}$ and $\dot{\mathbf{r}}^P = \mathbf{r}^P_{\mathbf{q}}\dot{\mathbf{q}}$).

2.5.7. Verify that Eqs. 2.5.15, 2.5.16, and 2.5.17 are valid.

2.5.8. Verify that Eqs. 2.5.19 and 2.5.21 give equivalent results.

2.5.9. Solve Eq. 2.5.22, using Eq. 2.5.21, to obtain $\ddot{\theta}_2$ as a function of $\dot{\theta}_1$ and $\ddot{\theta}_1$.

2.5.10. Let $\mathbf{q}$ be an n vector of real variables. Show that

$$\frac{\partial \mathbf{q}}{\partial \mathbf{q}} = \mathbf{I}$$

where $\mathbf{I}$ is the $n \times n$ identity matrix.

2.5.11. Let $\mathbf{q}$ be an n vector of real variables and $\mathbf{A}$ be a real $n \times n$ constant matrix. Show that

$$\frac{\partial}{\partial \mathbf{q}}(\mathbf{q}^T\mathbf{A}\mathbf{q}) = \mathbf{q}^T(\mathbf{A}^T + \mathbf{A})$$

2.5.12. If the matrix $\mathbf{A}$ in Prob. 2.5.11 is symmetric, show that

$$\frac{\partial}{\partial \mathbf{q}}(\mathbf{q}^T\mathbf{A}\mathbf{q}) = 2\mathbf{q}^T\mathbf{A}$$

Section 2.6

2.6.1. Verify that Eqs. 2.6.11 and 2.6.12 are valid.

2.6.2 With $\mathbf{s}_1'^Q$ and $\mathbf{s}_2'^{QP}$ of Example 2.5.1 and with $\theta_1 = \omega_1 t$ and $\theta_2 = \omega_2 t$, expand Eqs. 2.6.11 and 2.6.12 and verify that $\dot{\mathbf{r}}^P$ and $\ddot{\mathbf{r}}^P$ derived by direct differentiation in Example 2.5.1 are correct.

SUMMARY OF KEY FORMULAS

Matrix Algebra

$$\mathbf{A} + \mathbf{B} = [a_{ij} + b_{ij}] \qquad \textbf{(2.2.4)}$$

$$(\mathbf{A} + \mathbf{B}) + \mathbf{C} = \mathbf{A} + (\mathbf{B} + \mathbf{C}), \qquad \mathbf{A} + \mathbf{B} = \mathbf{B} + \mathbf{A} \qquad \textbf{(2.2.5, 6)}$$

$$\mathbf{AB} = \left[\sum_{k=1}^{p} a_{ik}b_{kj}\right], \qquad \mathbf{AB} \neq \mathbf{BA} \qquad \textbf{(2.2.8, 11)}$$

$$(\mathbf{A}+\mathbf{B})\mathbf{C}=\mathbf{AC}+\mathbf{BC},\qquad (\mathbf{AB})\mathbf{C}=\mathbf{A}(\mathbf{BC}),\quad \alpha\mathbf{A}=[\alpha a_{ij}] \tag{2.2.12, 13, 14}$$

$$(\mathbf{A}+\mathbf{B})^T=\mathbf{A}^T+\mathbf{B}^T,\qquad (\mathbf{AB})^T=\mathbf{B}^T\mathbf{A}^T \tag{2.2.15, 16}$$

$$\mathbf{AA}^{-1}=\mathbf{A}^{-1}\mathbf{A}=\mathbf{I},\quad (\mathbf{A}^{-1})^T=(\mathbf{A}^T)^{-1},\quad (\mathbf{AB})^{-1}=\mathbf{B}^{-1}\mathbf{A}^{-1} \tag{2.2.19, 20, 21}$$

Algebraic Vectors

$$\vec{a}=a_x\vec{i}+a_y\vec{j},\qquad \mathbf{a}=[a_x,\,a_y]^T \tag{2.3.1, 2}$$

$$\mathbf{a}^T\mathbf{b}=\vec{a}\cdot\vec{b},\qquad \mathbf{a}^\perp=\mathbf{Ra}=[-a_y,\,a_x]^T \tag{2.3.4, 5}$$

$$\mathbf{R}=\begin{bmatrix}0 & -1\\ 1 & 0\end{bmatrix},\qquad \mathbf{RR}=-\mathbf{I} \tag{2.3.6, 7}$$

Transformation of Coordinates

$$\mathbf{s}=[s_x,\,s_y]^T,\qquad \mathbf{s}'=[s_{x'},\,s_{y'}]^T \tag{2.4.1}$$

$$\mathbf{s}=\mathbf{As}',\qquad \mathbf{s}'=\mathbf{A}^T\mathbf{s} \tag{2.4.3, 7}$$

$$\mathbf{A}=\begin{bmatrix}\cos\phi & -\sin\phi\\ \sin\phi & \cos\phi\end{bmatrix},\qquad \mathbf{A}^T\mathbf{A}=\mathbf{I},\qquad \mathbf{A}^{-1}=\mathbf{A}^T \tag{2.4.4, 5, 6}$$

$$\mathbf{r}^P=\mathbf{r}+\mathbf{s}^P=\mathbf{r}+\mathbf{As}'^P \tag{2.4.8}$$

$$\mathbf{s}'_i=\mathbf{A}_{ij}\mathbf{s}'_j,\qquad \mathbf{A}_{ij}=\mathbf{A}_i^T\mathbf{A}_j=\begin{bmatrix}\cos(\phi_j-\phi_i) & -\sin(\phi_j-\phi_i)\\ \sin(\phi_j-\phi_i) & \cos(\phi_j-\phi_i)\end{bmatrix} \tag{2.4.10, 11, 12}$$

Vector and Matrix Differentiation

$$\frac{d}{dt}\mathbf{a}=\dot{\mathbf{a}}=[\dot a_x,\,\dot a_y]^T,\qquad \ddot{\mathbf{a}}=[\ddot a_x,\,\ddot a_y]^T \tag{2.5.1, 6}$$

$$\frac{d}{dt}(\mathbf{a}+\mathbf{b})=\dot{\mathbf{a}}+\dot{\mathbf{b}},\qquad \frac{d}{dt}(\alpha\mathbf{a})=\dot\alpha\mathbf{a}+\alpha\dot{\mathbf{a}},\qquad \frac{d}{dt}(\mathbf{a}^T\mathbf{b})=\dot{\mathbf{a}}^T\mathbf{b}+\mathbf{a}^T\dot{\mathbf{b}} \tag{2.5.2, 3, 4}$$

$$\dot{\mathbf{B}}=[\dot b_{ij}],\qquad \frac{d}{dt}(\mathbf{B}+\mathbf{C})=\dot{\mathbf{B}}+\dot{\mathbf{C}},\qquad \frac{d}{dt}(\mathbf{BC})=\dot{\mathbf{B}}\mathbf{C}+\mathbf{B}\dot{\mathbf{C}} \tag{2.5.7, 8, 9}$$

$$\boldsymbol{\Phi}=\boldsymbol{\Phi}(\mathbf{q})=[\Phi_1(\mathbf{q}),\ldots,\Phi_n(\mathbf{q})],\qquad \boldsymbol{\Phi}_{\mathbf{q}}=\left[\frac{\partial\Phi_i}{\partial q_j}\right] \tag{2.5.12}$$

$$(\mathbf{g}^T\mathbf{h})_{\mathbf{q}}=\mathbf{h}^T\mathbf{g}_{\mathbf{q}}+\mathbf{g}^T\mathbf{h}_{\mathbf{q}},\qquad \boldsymbol{\Phi}_{\mathbf{q}}=\boldsymbol{\Phi}_{\mathbf{g}}\mathbf{g}_{\mathbf{q}} \tag{2.5.13, 14}$$

Velocity and Acceleration of a Point

$$\dot{\mathbf{r}}^P = \dot{\mathbf{r}} + \dot{\phi}\mathbf{B}\mathbf{s}'^P, \qquad \dot{\mathbf{A}} = \dot{\phi}\mathbf{B}, \qquad \frac{d}{d\phi}\mathbf{A} = \mathbf{B} \tag{2.6.4, 3}$$

$$\mathbf{B} = \begin{bmatrix} -\sin\phi & -\cos\phi \\ \cos\phi & -\sin\phi \end{bmatrix} = \mathbf{AR} = \mathbf{RA}, \qquad \dot{\mathbf{B}} = -\dot{\phi}\mathbf{A}, \qquad \frac{d}{d\phi}\mathbf{B} = -\mathbf{A} \tag{2.6.5, 6, 8}$$

$$\ddot{\mathbf{r}}^P = \ddot{\mathbf{r}} + \ddot{\phi}\mathbf{B}\mathbf{s}'^P - \dot{\phi}^2\mathbf{A}\mathbf{s}'^P = \ddot{\mathbf{r}} + \ddot{\phi}\mathbf{s}^{P\perp} - \dot{\phi}^2\mathbf{s}^P \tag{2.6.9, 10}$$

CHAPTER THREE

Planar Cartesian Kinematics

The position and orientation of bodies in a plane are easily defined in terms of two position coordinates and one rotation coordinate. Both absolute constraints on the motion of a single body and relative constraints on the motion of pairs of bodies, due to joints between bodies, are formulated in terms of these Cartesian generalized coordinates. The constraint library developed includes absolute and relative point and orientation constraints, revolute and translational constraints, gear constraints, cam constraints, and composite constraints that are used to define the effect of special-purpose couplers that need not be represented as bodies in models of mechanical systems. In addition, kinematic drivers are introduced to control the motion of a system. Position, velocity, and acceleration equations are derived and their use in kinematic analysis is discussed. Elementary mechanism models are used to illustrate the formulation and solution of equations that determine position, velocity, and acceleration of all bodies in a system. Both well-behaved and singular configurations are studied to illustrate the pitfalls that will arise in kinematic analysis if mechanisms are poorly designed.

3.1. BASIC CONCEPTS IN PLANAR KINEMATICS

Kinematics, as the study of motion, is useful in two important ways. First, it is frequently necessary to generate, transmit, or control motion by the use of cams, gears, and linkages. Second, it is often necessary to determine the dynamic response of a system of interconnected bodies that results from applied forces. To formulate equations of motion, the kinematics of the system must be quantitatively defined.

A *rigid body* that is used to model a component of a mechanism is defined as a system of particles the distances between which remain unchanged. Each particle in a rigid body is located by its constant position vector in a reference frame that is attached to and moves with the body, called the *body-fixed reference frame.* In reality, all solids deform to some extent when forces are applied. Nevertheless, if movement associated with deformation is small, compared with the overall movement of a body, then the concept of a rigid body is acceptable.

For example, displacements due to the elastic vibration of the connecting rod in an engine may be of no consequence in the description of engine dynamics as a whole, so the rigid-body assumption is clearly in order. On the other hand, if stress in the connecting rod is to be evaluated, then deformation of the connecting rod is of primary importance and cannot be neglected. In this text, essentially all analysis is based on the assumption that bodies are rigid.

A *mechanism* is a collection of rigid bodies that are arranged to allow relative motion. This definition includes classical linkages, as well as interconnected bodies that make up a vehicle, vending machine, aircraft landing gear, engine, and so on. *Kinematics* is the study of the position, velocity, and acceleration of a system of interconnected bodies that make up a mechanism, independent of the forces that produce the motion.

Kinematic synthesis is the process of finding the geometry of a mechanism that will yield desired motion characteristics. *Kinematic analysis,* on the other hand, is the process of predicting position, velocity, and acceleration, once a design is specified. The processes of kinematic synthesis and analysis are intertwined. To do synthesis, the engineer needs an analysis capability to evaluate designs that are under consideration. This text is devoted to the kinematic and dynamic analysis of mechanical systems. For a treatment of kinematic synthesis and further literature on this subject, see Reference 23.

Any set of variables that uniquely specifies the position and orientation of all bodies in a mechanism, that is, the *configuration* of the mechanism, is called a set of *generalized coordinates.* Generalized coordinates may be *independent* (i.e., free to vary arbitrarily) or *dependent* (i.e., required to satisfy equations of constraint). For systems in motion, generalized coordinates vary with time. Generalized coordinates are designated in this text by a column vector $\mathbf{q} \equiv [q_1, q_2, \ldots, q_{nc}]^T$, where nc is the total number of generalized coordinates used to describe the configuration of the system.

To specify the configuration of a planar system, a *body-fixed x'-y' reference frame* is embedded in each body of the system, as shown in Fig. 3.1.1. Body i (i is an identifying number assigned to each body) can be located by specifying the global coordinates $\mathbf{r}_i = [x, y]_i^T$ of the origin of the body-fixed x_i'-y_i' frame and the angle ϕ_i of rotation of this frame relative to the global x-y frame. The column vector $\mathbf{q}_i \equiv [x, y, \phi]_i^T$ is the vector of planar *Cartesian generalized coordinates* for body i. Using Cartesian generalized coordinates for each body, a maximal set of coordinates is defined to specify the position and orientation of each body in the system.

If a planar mechanism is made up of nb rigid bodies, the number of planar Cartesian generalized coordinates is $nc = 3 \times nb$. The vector of generalized coordinates for the system is denoted by $\mathbf{q} = [\mathbf{q}_1^T, \mathbf{q}_2^T, \ldots, \mathbf{q}_{nb}^T]^T$. Since rigid bodies that make up a mechanism are interconnected by joints, there are equations of constraint that relate the generalized coordinates. Therefore, Cartesian generalized coordinates are generally dependent.

A kinematic constraint between two bodies imposes conditions on the

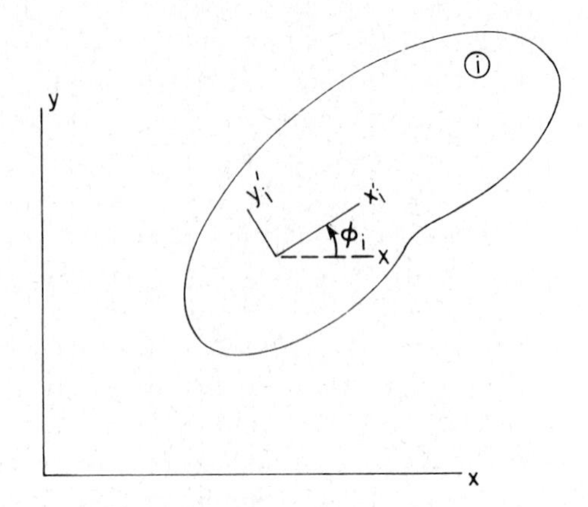

Figure 3.1.1 Planar Cartesian generalized coordinates.

relative motion between the pair of bodies. When these conditions are expressed as algebraic equations in terms of generalized coordinates, they are called *holonomic kinematic constraint equations.* A system of nh holonomic kinematic constraint equations that does not depend explicitly on time can be expressed as

$$\mathbf{\Phi}^K(\mathbf{q}) = [\Phi_1^K(\mathbf{q}), \ldots, \Phi_{nh}^K(\mathbf{q})]^T = \mathbf{0} \tag{3.1.1}$$

If time appears explicitly, as in the case of time-dependent kinematic couplings,

$$\mathbf{\Phi}^K(\mathbf{q}, t) = \mathbf{0} \tag{3.1.2}$$

where t is time. If t does not enter explicitly into the equation of constraint, as in Eq. 3.1.1, the constraint is called a *stationary constraint.* Constraints specified by equations in the form of Eq. 3.1.2 are called *time-dependent constraints.* More general constraint equations that contain inequalities or relations between velocity components are called *nonholonomic constraints.* In this text, the term *constraint* will mean holonomic constraint unless otherwise specified.

The analytical foundation of computer-aided kinematic analysis is a library of joints that restrict the motion of a body or relative motion of a pair of bodies. Algebraic equations of constraint must be derived that are equivalent to the physical joints. This equivalence is a critically important concept that is not generally given adequate attention. Constraint equations are often derived from the geometry of a joint; that is, they are implied by the geometry of the joint. This is not enough. *The equations of constraint must imply the geometry of the joint.* This is required in computer simulation, since it is the equations of constraint that are being satisfied. If they can be satisfied without yielding the geometrical relations due to the joints, then the mathematical model fails to define the motion of the system. This subtle but important point will be highlighted in the developments that follow.

If $nc > nh$, then the constraint equations of Eq. 3.1.1 or 3.1.2 are not sufficient in number to determine $\mathbf{q}$. This is the usual case, since a mechanical system is usually designed to permit motion, distinguishing it from a *structure*, whose function is to transmit load and resist motion. If the constraints of Eq. 3.1.1 or 3.1.2 are consistent and independent (concepts to be defined precisely in Section 3.6), then the system is said to have $nc - nh$ *degrees of freedom*, abbreviated DOF $= nc - nh$. To determine the motion of a system, the engineer must define either (1) DOF additional driving conditions that uniquely determine $\mathbf{q}(t)$ algebraically (*kinematic analysis*), or (2) forces that act on the system, in which case $\mathbf{q}(t)$ is the solution of differential equations of motion (*dynamic analysis*). Kinematic analysis of planar systems is treated in Chapters 3 through 5 and dynamic analysis of planar systems is treated in Chapters 6 through 8.

If DOF independent *driving constraints* are specified for kinematic analysis, denoted

$$\mathbf{\Phi}^D(\mathbf{q}, t) = \mathbf{0} \tag{3.1.3}$$

then the configuration of the system as a function of time can be determined. That is, the combined constraints of Eqs. 3.1.2 and 3.1.3,

$$\mathbf{\Phi}(\mathbf{q}, t) = \begin{bmatrix} \mathbf{\Phi}^K(\mathbf{q}, t) \\ \mathbf{\Phi}^D(\mathbf{q}, t) \end{bmatrix} = \mathbf{0} \tag{3.1.4}$$

can be solved for $\mathbf{q}(t)$. Such a system is called *kinematically driven*.

Example 3.1.1: The simple pendulum of Fig. 3.1.2 is pivoted at point O, the origin of the x-y reference frame. This kinematic constraint is prescribed by the pair of equations

$$\mathbf{\Phi}^K(\mathbf{q}) \equiv \begin{bmatrix} x_1 - \sin \phi_1 \\ y_1 + \cos \phi_1 \end{bmatrix} = \mathbf{0} \tag{3.1.5}$$

where $\mathbf{q} = [x_1, y_1, \phi_1]^T$. Clearly, Eq. 3.1.5 can be solved for x_1 and y_1 as a function of ϕ_1, so one independent variable can specify the motion of the system. Thus, the system has one degree of freedom (DOF = 1).

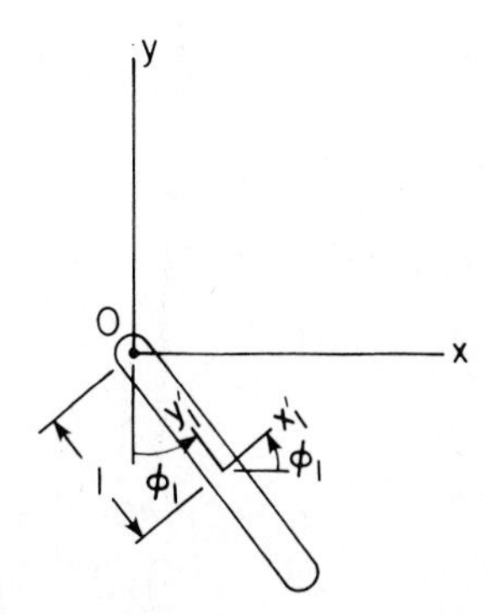

Figure 3.1.2 Simple pendulum.

Even though the elementary form of Eq. 3.1.5 permits easy solution for some variables as functions of others, this is not possible for most mechanisms and machines that are of practical interest. To set the stage for generally applicable methods that are developed in the following sections of this chapter, the conceptually trivial simple pendulum is analyzed here. To specify the motion of the simple pendulum, a driving constraint must be introduced; for example, specifying the time history of ϕ_1,

$$\mathbf{\Phi}^D(\mathbf{q}, t) \equiv \phi_1 - f(t) = 0 \tag{3.1.6}$$

Combining Eqs. 3.1.5 and 3.1.6 yields the system kinematic constraint equation

$$\mathbf{\Phi}(\mathbf{q}, t) = \begin{bmatrix} \mathbf{\Phi}^K(\mathbf{q}) \\ \mathbf{\Phi}^D(\mathbf{q}, t) \end{bmatrix}$$

$$= \begin{bmatrix} x_1 - \sin \phi_1 \\ y_1 + \cos \phi_1 \\ \phi_1 - f(t) \end{bmatrix} = 0 \tag{3.1.7}$$

It is shown in Section 3.6 that if $\mathbf{\Phi}_\mathbf{q}$ is nonsingular, that is, $|\mathbf{\Phi}_\mathbf{q}| \neq 0$ at some value of $\mathbf{q}$ that satisfies Eq. 3.1.7, then Eq. 3.1.7 can be solved for $\mathbf{q}$ as a function of time. To test this condition, note that

$$|\mathbf{\Phi}_\mathbf{q}| = \begin{vmatrix} 1 & 0 & -\cos \phi_1 \\ 0 & 1 & -\sin \phi_1 \\ 0 & 0 & 1 \end{vmatrix} = 1 \tag{3.1.8}$$

Under this condition, the numerical methods presented in Chapter 4 permit effective computation of the solution of Eq. 3.1.7 for $\mathbf{q}$ at discrete instants in time, even for complicated nonlinear equations for which a closed-form solution is extremely messy, if not impossible to obtain.

Assume that numerical methods have been used to solve Eq. 3.1.4 for $\mathbf{q}$ at discrete instants in time. Since $\mathbf{q}$ is not known as an explicit function of time, it cannot be differentiated to obtain $\dot{\mathbf{q}}$ or $\ddot{\mathbf{q}}$. An alternative that is well suited for numerical computation is to use the chain rule of differentiation to evaluate derivatives of both sides of Eq. 3.1.4 with respect to time to obtain the *velocity equation*

$$\dot{\mathbf{\Phi}} = \mathbf{\Phi}_\mathbf{q}\dot{\mathbf{q}} + \mathbf{\Phi}_t = 0$$

or

$$\mathbf{\Phi}_\mathbf{q}\dot{\mathbf{q}} = -\mathbf{\Phi}_t \equiv \mathbf{v} \tag{3.1.9}$$

If $\mathbf{\Phi}_\mathbf{q}$ is nonsingular, Eq. 3.1.9 can be solved for $\dot{\mathbf{q}}$ at discrete instants in time. Similarly, both sides of Eq. 3.1.9 can be differentiated with respect to time, using the chain rule of differentiation, to obtain

$$\mathbf{\Phi_q \ddot{q}} + (\mathbf{\Phi_q \dot{q}})_\mathbf{q} \mathbf{\dot{q}} + \mathbf{\Phi}_{\mathbf{q}t} \mathbf{\dot{q}} = -\mathbf{\Phi}_{t\mathbf{q}} \mathbf{\dot{q}} - \mathbf{\Phi}_{tt}$$

where for purposes of applying the chain rule of differentiation all variables are treated as independent; in particular, $\mathbf{\dot{q}_q} = \mathbf{0}$. Since $\mathbf{\Phi}_{t\mathbf{q}} = \mathbf{\Phi}_{\mathbf{q}t}$, this result can be rearranged to obtain the *acceleration equation*

$$\mathbf{\Phi_q \ddot{q}} = -(\mathbf{\Phi_q \dot{q}})_\mathbf{q} \mathbf{\dot{q}} - 2\mathbf{\Phi}_{\mathbf{q}t} \mathbf{\dot{q}} - \mathbf{\Phi}_{tt} \equiv \gamma \qquad \textbf{(3.1.10)}$$

Since $\mathbf{\Phi_q}$ is nonsingular, Eq. 3.1.10 can be solved for $\mathbf{\ddot{q}}$ at discrete instants in time (Prob. 3.1.1).

In Example 3.1.1, $\mathbf{\Phi}_{\mathbf{q}t} = \mathbf{0}$, $\mathbf{\Phi}_{tt} = [0, 0, -\ddot{f}(t)]^T$, and

$$(\mathbf{\Phi_q \dot{q}})_\mathbf{q} \mathbf{\dot{q}} = \begin{bmatrix} \dot{x} - \dot{\phi}_1 \cos \phi_1 \\ \dot{y}_1 - \dot{\phi}_1 \sin \phi_1 \\ \dot{\phi}_1 \end{bmatrix}_\mathbf{q} \mathbf{\dot{q}} = \begin{bmatrix} \dot{\phi}_1^2 \sin \phi_1 \\ -\dot{\phi}_1^2 \cos \phi_1 \\ 0 \end{bmatrix}$$

so Eq. 3.1.10 is

$$\mathbf{\Phi_q \ddot{q}} = \begin{bmatrix} -\dot{\phi}_1^2 \sin \phi_1 \\ \dot{\phi}_1^2 \cos \phi_1 \\ \ddot{f}(t) \end{bmatrix}$$

The matrix $\mathbf{\Phi_q}$ that arises in the velocity and accelerations, Eqs. 3.1.9 and 3.1.10, plays a central role in the theory and numerical methods of kinematics and dynamics. It is called the *Jacobian matrix,* or simply the *Jacobian.* While it may be assembled in an ad hoc way using the definitions of matrix calculus of Chapter 2, it is assembled in a systematic way for each of the contraints considered later in this chapter. The importance of the Jacobian will become clear as the reader proceeds through the remainder of the text. Without question, it is the most important matrix that is used in the kinematics and dynamics of constrained mechanical systems.

Evaluation of the right side of Eq. 3.1.10 involves calculation of second derivatives, as outlined in Example 3.1.1. To be more specific,

$$\mathbf{\Phi}_{tt} = \left[\frac{\partial^2 \Phi_i}{\partial t^2} \right]_{nh \times 1}$$

$$\mathbf{\Phi}_{\mathbf{q}t} = \left[\frac{\partial^2 \Phi_i}{\partial q_j \, \partial t} \right]_{nh \times nc}$$

and

$$(\mathbf{\Phi_q \dot{q}})_\mathbf{q} = \left[\frac{\partial}{\partial q_j} \left(\sum_{k=1}^{nc} \frac{\partial \Phi_i}{\partial q_k} \dot{q}_k \right) \right]_{nc \times nc}$$

$$= \left[\sum_{k=1}^{nc} \frac{\partial^2 \Phi_i}{\partial q_j \, \partial q_k} \dot{q}_k \right]_{nc \times nc}$$

where nh is the number of holonomic constraints and nc is the number of generalized coordinates.

Example 3.1.2: Consider next an elementary two-body model of the slider–crank mechanism of Example 2.4.3. Rather than using the relative angle generalized coordinates that were used there, consider the Cartesian coordinates shown in Fig. 3.1.3.

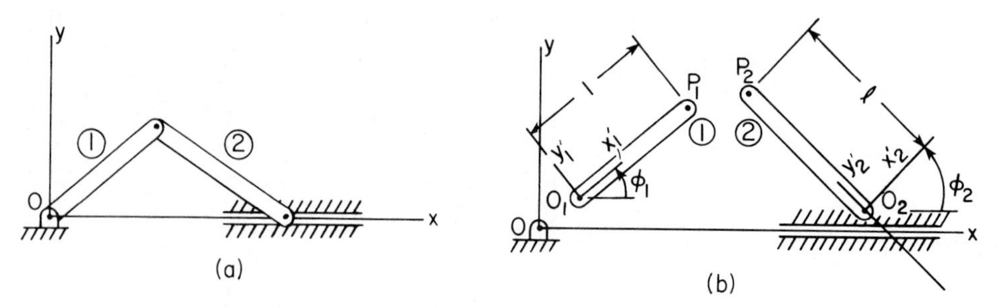

Figure 3.1.3 Cartesian coordinates for two-body slider–crank model. (a) Assembled. (b) Dissected.

To assemble the mechanism, three geometric constraint conditions must be imposed, as follows:

1. Point O_1 on the crank (body 1) must coincide with the origin O of the x-y frame; that is, $x_1 = y_1 = 0$.
2. Point O_2 on the coupler (body 2) must lie on the x axis; that is, $y_2 = 0$.
3. Points P_1 and P_2 on the crank and coupler, respectively, must coincide; that is,

$$\mathbf{r}^{P_1} \equiv \begin{bmatrix} x_1 + \cos\phi_1 \\ y_1 + \sin\phi_1 \end{bmatrix} = \begin{bmatrix} x_2 - \ell\sin\phi_2 \\ y_2 + \ell\cos\phi_2 \end{bmatrix} \equiv \mathbf{r}^{P_2}$$

In terms of $\mathbf{q} = [x_1, y_1, \phi_1, x_2, y_2, \phi_2]^T$, the combined system of kinematic constraint equations is thus

$$\mathbf{\Phi}^K(\mathbf{q}) = \begin{bmatrix} x_1 \\ y_1 \\ y_2 \\ x_1 - x_2 + \cos\phi_1 + \ell\sin\phi_2 \\ y_1 - y_2 + \sin\phi_1 - \ell\cos\phi_2 \end{bmatrix} \tag{3.1.11}$$

Presuming the constraint equations of Eq. 3.1.11 are independent (to be verified in Section 3.6), $\text{DOF} = nc - nh = 6 - 5 = 1$. Thus, one driving constraint is required; for example, the crank is to be rotated with constant angular velocity ω, so

$$\Phi^D(\mathbf{q}, t) = \phi_1 - \omega t \tag{3.1.12}$$

The kinematic equations that determine the motion of this system are thus

$$\mathbf{\Phi}(\mathbf{q}, t) = \begin{bmatrix} \mathbf{\Phi}^K(\mathbf{q}) \\ \mathbf{\Phi}^D(\mathbf{q}) \end{bmatrix} = \mathbf{0} \tag{3.1.13}$$

At any instant t_i in time, Eq. 3.1.13 may be solved by numerical methods (see Chapter 4) for $\mathbf{q}(t_i)$. Differentiating both sides of Eq. 3.1.13 with respect to time yields the velocity equation

$$\mathbf{\Phi}_{\mathbf{q}}(\mathbf{q}(t_i), t_i)\dot{\mathbf{q}}(t_i)$$

$$= \begin{bmatrix} 1 & 0 & 0 & 0 & 0 & 0 \\ 0 & 1 & 0 & 0 & 0 & 0 \\ 0 & 0 & 0 & 0 & 1 & 0 \\ 1 & 0 & -\sin\phi_1(t_1) & -1 & 0 & \ell\cos\phi_2(t_i) \\ 0 & 1 & \cos\phi_1(t_1) & 0 & -1 & \ell\sin\phi_2(t_i) \\ 0 & 0 & 1 & 0 & 0 & 0 \end{bmatrix} \begin{bmatrix} \dot{x}_1(t_i) \\ \dot{y}_1(t_i) \\ \dot{\phi}_1(t_i) \\ \dot{x}_2(t_i) \\ \dot{y}_2(t_i) \\ \dot{\phi}_2(t_i) \end{bmatrix} = \begin{bmatrix} 0 \\ 0 \\ 0 \\ 0 \\ 0 \\ \omega \end{bmatrix} = -\mathbf{\Phi}_t \tag{3.1.14}$$

Since $|\mathbf{\Phi}_{\mathbf{q}}| = -\ell \sin\phi_2$ and, with $\ell > 1$, $|\phi_2 - \pi/2| \leq \text{Arcsin}(1/\ell)$, $|\mathbf{\Phi}_{\mathbf{q}}| \neq 0$ and Eq. 3.1.14 may be solved for $\dot{\mathbf{q}}(t_i)$ (Prob. 3.1.3). Similarly, both sides of Eq. 3.1.14 may be differentiated, to obtain the acceleration equation

$$\mathbf{\Phi}_{\mathbf{q}}(\mathbf{q}(t_i), t_i)\ddot{\mathbf{q}}(t_i) = \begin{bmatrix} 0 \\ 0 \\ 0 \\ \dot{\phi}_1^2(t_i)\cos\phi_1(t_i) + \ell\dot{\phi}_2^2(t_i)\sin\phi_2(t_i) \\ \dot{\phi}_1^2(t_i)\sin\phi_1(t_i) - \ell\dot{\phi}_2^2(t_i)\cos\phi_2(t_i) \\ 0 \end{bmatrix} \tag{3.1.15}$$

where the coefficient matrix on the left is the same as in Eq. 3.1.14. Since $\mathbf{q}(t_i)$ is found from Eq. 3.1.13 and $\dot{\mathbf{q}}(t_i)$ is found from Eq. 3.1.14, Eq. 3.1.15 may be solved for $\ddot{\mathbf{q}}(t_i)$ (Prob. 3.1.4).

Examples 3.1.1 and 3.1.2, while addressing elementary mechanisms, illustrate an approach that is used throughout this text for the formulation and solution of kinematic equations of general mechanical systems. The remainder of Chapter 3 is devoted to deriving a library of kinematic constraints between pairs of bodies that can be used to assemble kinematic equations for broad classes of mechanisms and machines. The resulting kinematic constraint equations are coupled with driving constraints that uniquely determine the motion of the system. Nonlinear equations of constraint are solved for position and orientation and linear equations in velocity and acceleration are solved, using the methods employed in Examples 3.1.1 and 3.1.2. To place the detailed developments of subsequent sections into the context of system kinematic analysis, the reader may wish to refer to Examples 3.1.1 and 3.1.2.

As an additional aid in interpreting the detailed developments that follow, the reader should keep in mind the basic objective of computer-aided kinematic analysis, which is to create a systematic method for both formulating and solving kinematic equations that can be implemented on a digital computer. Only if such a systematic approach is adopted can the burden of extensive analytical derivation be taken from the shoulders of the engineer and delegated to the computer. It is this objective that leads to what might appear to be excessively large matrices, such as those encountered in Example 3.1.2. The reader might be interested to note that the constraint Jacobian $\mathbf{\Phi_q}$ of Eq. 3.1.14 for the slider–crank mechanism contains many zeros and ones. This elementary structure is exploited in Chapter 4 to efficiently solve position, velocity, and acceleration equations of kinematic analysis.

Example 3.1.3: To illustrate a potential problem that must be avoided in computer-aided kinematics, consider the sliding body shown in Fig. 3.1.4,

Constraint equations that are consistent with constraints on body 1 may be written as the condition that P_1 remain on the slide axis $y = x$, that is,

$$\Phi_1 \equiv y_1 - x_1 = 0 \tag{3.1.16}$$

and the condition that the y' axis remain orthogonal to the slide axis, or $\mathbf{r}_1$ since it is along the slide, that is,

$$\begin{aligned}\Phi_2 &\equiv \mathbf{r}_1^T\mathbf{j}_1 = \mathbf{r}_1^T\mathbf{A}(\phi_1)\begin{bmatrix}0\\1\end{bmatrix}\\ &= y_1\cos\phi_1 - x_1\sin\phi_1\end{aligned} \tag{3.1.17}$$

To illustrate the use of these constraint equations, consider the driver

$$\Phi^D \equiv x_1 - vt = 0 \tag{3.1.18}$$

where v is the horizontal component of velocity of P_1. To check for the solvability of Eqs. 3.1.16 through 3.1.18, the determinant of the Jacobian is

$$|\mathbf{\Phi}_{\mathbf{q}_1}| = -(y_1\sin\phi_1 + x_1\cos\phi_1)$$

Thus, all is well if $\mathbf{r}_1 \neq \mathbf{0}$. However, $\mathbf{\Phi}_{\mathbf{q}_1}$ is singular when the center of the block is at the origin of the x-y frame, and the kinematic and driving constraints fail to

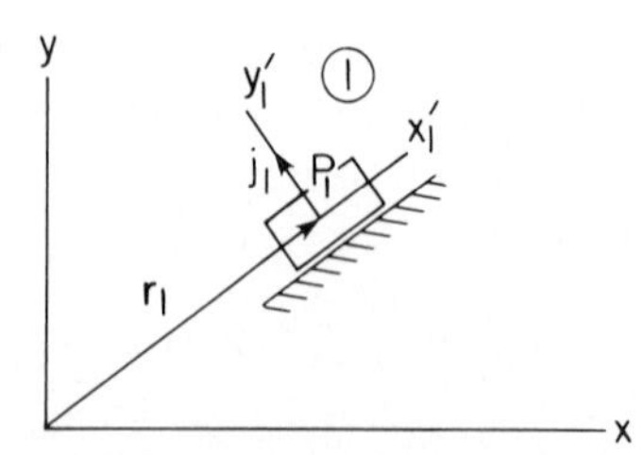

Figure 3.1.4 Inclined slider.

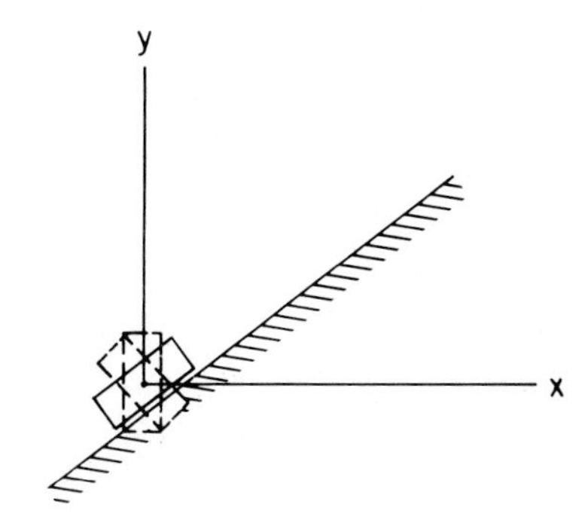

Figure 3.1.5 Inclined slider at singular point.

uniquely define motion at this point. In fact, computer solution methods will fail due to the singular Jacobian.

This mathematical breakdown is puzzling, since there is no physical difficulty when $\mathbf{r}_1 = \mathbf{0}$ (Prob. 3.2.3). The problem has been created by choosing constraint equations that are implied by the geometry of the system, but which fail to imply its geometry when $\mathbf{r}_1 = \mathbf{0}$. To see this more clearly, note that Eq. 3.1.17 is satisfied by arbitrary ϕ_1 when $\mathbf{r}_1 = \mathbf{0}$, as shown in Fig. 3.1.5.

Even in this trivial example, it is possible to make a fatal formulation error by deriving constraint equations that fail to uniquely define the kinematics of a system. The danger of making such errors is greater in more complex kinematic joints, particularly for the spatial systems in Part II of this text. The engineer must, therefore, be careful to assure that the equations used in computer-aided analysis in fact reliably describe the behavior of the system. If they fail at even one point, as in this example, the entire system simulation will break down.

3.2. CONSTRAINTS BETWEEN A BODY AND GROUND (ABSOLUTE CONSTRAINTS)

In many mechanisms, the motion of a body is constrained relative to ground, that is, relative to the stationary x-y reference frame. Constraint equations may be expressed as relationships among the generalized coordinates of a single body, since no other body is involved. Consider for example the case in which point P_i on body i, shown in Fig. 3.2.1, can only move on a circular path. This may be due to a link of fixed length with revolute joints in body i and the x-y plane, as shown in Fig. 3.2.1. The *absolute distance constraint* that defines this physical limitation on motion, using Eq. 2.4.8, is

$$\begin{aligned}\Phi^{ad(i)} &\equiv (\mathbf{r}_i^P - \mathbf{C})^T(\mathbf{r}_i^P - \mathbf{C}) - C_3^2 \\ &= (x_i + x_i'^P \cos\phi_i - y_i'^P \sin\phi_i - C_1)^2 \\ &\quad + (y_i + x_i'^P \sin\phi_i + y_i'^P \cos\phi_i - C_2)^2 - C_3^2 = 0 \qquad \textbf{(3.2.1)}\end{aligned}$$

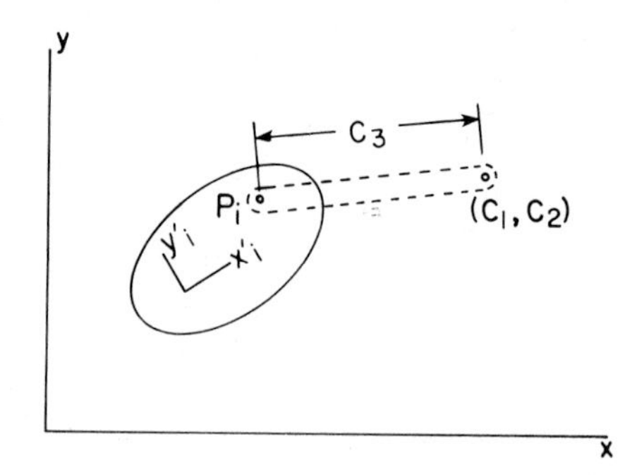

Figure 3.2.1 Constraint that distance from point P to (C_1, C_2) equals C_3.

where $\mathbf{C} = [C_1, C_2]^T$ is the given point at the center of the circle in the x-y plane and $C_3 > 0$ is its given radius. Here, x_i^P and y_i^P are the coordinates of point P_i in the x-y frame and $x_i'^P$ and $y_i'^P$ are the constant coordinates of P_i in the x_i'-y_i' frame, since P_i is fixed to body i. Using Eq. 2.4.8, x_i^P and y_i^P are written as functions of $\mathbf{q}_i = [x_i, y_i, \phi_i]^T$, as in the right side of Eq. 3.2.1.

The Jacobian of $\Phi^{ad(i)}$ with respect to $\mathbf{q}_i$ is

$$\begin{aligned}\mathbf{\Phi}_{\mathbf{q}_i}^{ad(i)} &= 2[(\mathbf{r}_i^P - \mathbf{C})^T, \mathbf{s}_i'^{P^T}\mathbf{B}_i^T(\mathbf{r}_i^P - \mathbf{C})] \\ &= 2[(x_i^P - C_1), (y_i^P - C_2), -(x_i^P - C_1)(x_i'^P \sin\phi_i - y_i'^P \cos\phi_i) \\ &\quad + (y_i^P - C_2)(x_i'^P \cos\phi_i - y_i'^P \sin\phi_i)]\end{aligned} \tag{3.2.2}$$

where the chain rule of differentiation and $d\mathbf{A}/d\phi = \mathbf{B}$ have been used.

Since $\Phi^{ad(i)}$ does not depend explicitly on time, the right side of the velocity equation of Eq. 3.1.9 is

$$\nu^{ad(i)} = 0$$

With a somewhat more intricate calculation, the right side of the acceleration equation of Eq. 3.1.10 is obtained, using the Jacobian of Eq. 3.2.2, as

$$\begin{aligned}\gamma^{ad(i,j)} &= -\frac{\partial}{\partial \mathbf{q}_i}\{2(\mathbf{r}_i^P - \mathbf{C})^T\dot{\mathbf{r}}_i + 2\mathbf{s}_i'^{P^T}\mathbf{B}_i^T(\mathbf{r}_i^P - \mathbf{C})\dot{\phi}_i\}\dot{\mathbf{q}}_i \\ &= -2[\dot{\mathbf{r}}_i^T\dot{\mathbf{r}}_i + 2\mathbf{s}_i'^{P^T}\mathbf{B}_i^T\dot{\mathbf{r}}_i\dot{\phi}_i + \mathbf{s}_i'^{P^T}\mathbf{B}_i^T\mathbf{B}_i\mathbf{s}_i'^P\dot{\phi}_i^2 - \mathbf{s}_i'^{P^T}\mathbf{A}_i^T(\mathbf{r}_i^P - \mathbf{C})\dot{\phi}_i^2] \\ &= -2[\dot{\mathbf{r}}_i^{P^T}\dot{\mathbf{r}}_i^P - \mathbf{s}_i'^{P^T}\mathbf{A}_i^T(\mathbf{r}_i^P - \mathbf{C})\dot{\phi}_i^2]\end{aligned}$$

where Eqs. 2.6.3 and 2.6.8 have been used.

The restriction $C_3 > 0$ is required so that Eq. 3.2.1 represents a single constraint. If $C_3 = 0$, Eq. 3.2.1 requires that both $x_i^P = C_1$ and $y_i^P = C_2$, which is a pair of constraint equations. Furthermore, if $C_3 = 0$, the Jacobian $\Phi_{\mathbf{q}}^{ad(i)}$ of Eq. 3.2.2 is zero, leading to later computational trouble.

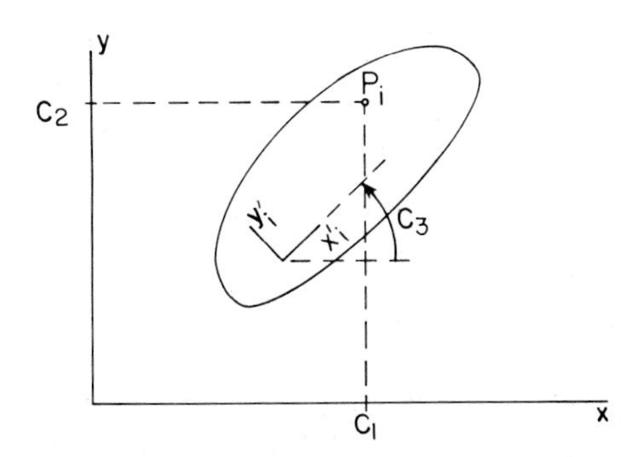

Figure 3.2.2 Constraints on absolute coordinates of point P_i and on angular orientation.

An *absolute position constraint* on point P_i of body i in the x or y direction might be imagined as the condition that a pin on body i at point P_i slide in a slot in the x-y plane that is parallel to the x or y axis, respctively. As shown in Fig. 3.2.2, the following constraint equations define this condition, using Eq. 2.4.8:

$$\begin{aligned}\Phi^{ax(i)} &\equiv x_i^P - C_1 \\ &= x_i + x_i'^P \cos \phi_i - y_i'^P \sin \phi_i - C_1 = 0 \qquad \textbf{(3.2.3)} \\ \Phi^{ay(i)} &\equiv y_i^P - C_2 \\ &= y_i + x_i'^P \sin \phi_i + y_i'^P \cos \phi_i - C_2 = 0 \qquad \textbf{(3.2.4)}\end{aligned}$$

where C_1 and C_2 are given constants and $x_i'^P$ and $y_i'^P$ are the constant body-fixed coordinates of point P_i on body i. The Jacobians of these constraint functions with respect to $\mathbf{q}_i$ are

$$\begin{aligned}\Phi_{\mathbf{q}_i}^{ax(i)} &= [1,\ 0,\ -x_i'^P \sin \phi_i - y_i'^P \cos \phi_i] \\ \Phi_{\mathbf{q}_i}^{ay(i)} &= [0,\ 1,\ x_i'^P \cos \phi_i - y_i'^P \sin \phi_i]\end{aligned} \qquad \textbf{(3.2.5)}$$

An *absolute angular constraint* on a body is defined by the following equation:

$$\Phi^{a\phi(i)} \equiv \phi_i - C_3 = 0 \qquad \textbf{(3.2.6)}$$

where C_3 is a given constant. Figure 3.2.2 illustrates the constraint of Eq. 3.2.6. The Jacobian of this constraint function with respect to $\mathbf{q}_i$ is simply

$$\Phi_{\mathbf{q}_i}^{a\phi(i)} = [0,\ 0,\ 1] \qquad \textbf{(3.2.7)}$$

Since these constraints are not explicitly time dependent, from Eq. 3.1.9,

$$\nu^{ax(i)} = \nu^{ay(i)} = \nu^{a\phi(i)} = 0$$

Expanding terms in Eq. 3.1.10, using the Jacobians of Eqs. 3.2.5 and 3.2.6,

$$\gamma^{ax(i)} = (x_i'^P \cos \phi_i - y_i'^P \sin \phi_i)\dot{\phi}_i^2$$
$$\gamma^{ay(i)} = (x_i'^P \sin \phi_i + y_i'^P \cos \phi_i)\dot{\phi}_i^2$$
$$\gamma^{a\phi(i)} = 0$$

Example 3.2.1: Figure 3.2.3 shows a simple pendulum with its pivot pin P_1 located at $(C_1, C_2) = (1, 1)$ in the x-y plane. With the origin of the x_1'-y_1' frame at the center of the pendulum, whose length is two units, $x_1'^P = 1$ and $y_1'^P = 0$. Equations 3.2.3 and 3.2.4 thus describe the kinematics of the pendulum. Note that the constraint equations for the simple pendulum of Example 3.1.1 could have been written in this way.

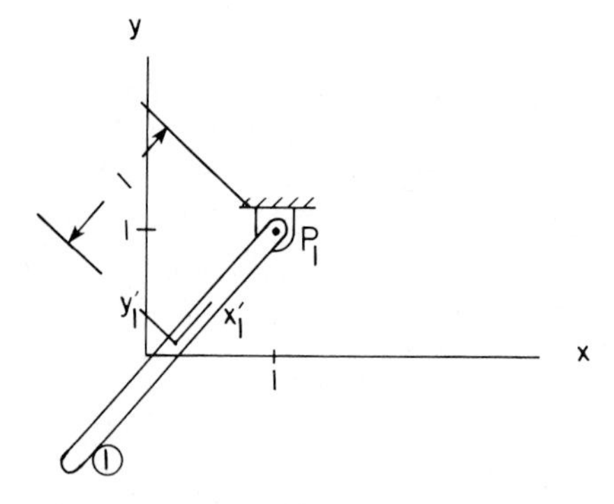

Figure 3.2.3 Simple pendulum with absolute constraints.

Example 3.2.2: As an example of an absolute angular constraint, Fig. 3.2.4 shows a joint that allows body 1 to slide along the x axis, with the body-fixed x_1' axis and the x axis coinciding. The kinematic constraint between this slider and ground is defined by Eq. 3.2.4, with $x_1'^P = y_1'^P = 0$ and $C_2 = 0$, and by Eq. 3.2.6 with $C_3 = 0$.

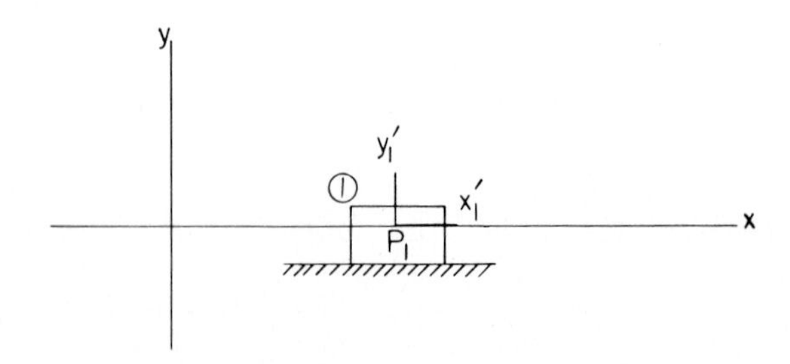

Figure 3.2.4 Slider along x axis.

3.3 CONSTRAINTS BETWEEN PAIRS OF BODIES (RELATIVE CONSTRAINTS)

The constraints described in Section 3.2 impose restrictions on the motion of a single body relative to the global reference frame. In most kinematic systems, constraints are imposed on the relative position and orientation of pairs of bodies that are connected by joints. In this section, commonly used kinematic pairs are formulated. The technique employed to formulate kinematic constraint equations for these joints may be applied to most other commonly used or special-purpose joints.

The objective for each joint treated is to define a set of algebraic constraint equations that are "equivalent to the physical joint." Since the physical joint is to be represented by constraint equations, it is important that the equations employed imply the relative position and orientation restrictions imposed by the physical joint. It is easy to write down equations that are implied by the geometry of the joint, as in Example 3.1.3, but that do not imply the geometry of the joint. If this is done, the computer model will fail to represent the real physical kinematics and numerical difficulties will arise.

3.3.1 Relative Coordinate Constraints

Simple equations can be derived to define the following constraints:

(1) A *relative x constraint* requires that the difference between the x coordinates of point P_j on body j and point P_i on body i equal a given constant C_1 (see Fig. 3.3.1); that is, using Eq. 2.4.8,

$$\begin{aligned}\Phi^{rx(i,j)} &\equiv x_j^P - x_i^P - C_1 \\ &= x_j + x_j'^P \cos\phi_j - y_j'^P \sin\phi_j - x_i - x_i'^P \cos\phi_i + y_i'^P \sin\phi_i - C_1 = 0 \qquad \textbf{(3.3.1)}\end{aligned}$$

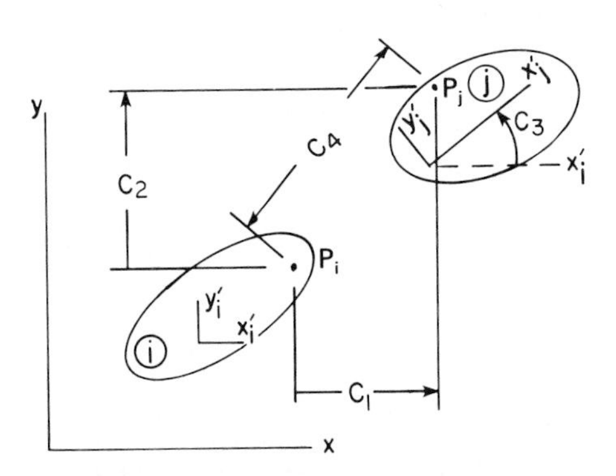

Figure 3.3.1 Simple constraints.

The Jacobians of this constraint with respect to $\mathbf{q}_i$ and $\mathbf{q}_j$ are

$$\begin{aligned}\Phi_{\mathbf{q}_i}^{rx(i,j)} &= [-1,\ 0,\ x_i'^P \sin\phi_i + y_i'^P \cos\phi_i] \\ \Phi_{\mathbf{q}_j}^{rx(i,j)} &= [1,\ 0,\ -x_j'^P \sin\phi_j - y_j'^P \cos\phi_j]\end{aligned} \tag{3.3.2}$$

From Eqs. 3.1.9 and 3.1.10, using the Jacobians of Eq. 3.3.2,

$$\nu^{rx(i,j)} = 0$$

$$\gamma^{rx(i,j)} = -(x_i'^P \cos\phi_i - y_i'^P \sin\phi_i)\dot{\phi}_i^2 + (x_j'^P \cos\phi_j - y_j'^P \sin\phi_j)\dot{\phi}_j^2$$

(2) A *relative y constraint* requires that the difference between the y coordinates of point P_j on body j and point P_i on body i equal a given constant C_2 (see Fig. 3.3.1); that is, using Eq. 2.4.8,

$$\begin{aligned}\Phi^{ry(i,j)} &= y_j^P - y_i^P - C_2 \\ &= y_j + x_j'^P \sin\phi_j + y_j'^P \cos\phi_j - y_i - x_i'^P \sin\phi_i - y_i'^P \cos\phi_i - C_2 = 0\end{aligned} \tag{3.3.3}$$

The Jacobians of this constraint with respect to $\mathbf{q}_i$ and $\mathbf{q}_j$ are

$$\begin{aligned}\Phi_{\mathbf{q}_i}^{ry(i,j)} &= [0,\ -1,\ -x_i'^P \cos\phi_i + y_i'^P \sin\phi_i] \\ \Phi_{\mathbf{q}_j}^{ry(i,j)} &= [0,\ 1,\ x_j'^P \cos\phi_j - y_j'^P \sin\phi_j]\end{aligned} \tag{3.3.4}$$

Again, from Eqs. 3.1.9 and 3.1.10,

$$\nu^{ry(i,j)} = 0$$

$$\gamma^{ry(i,j)} = -(x_i'^P \sin\phi_i + y_i'^P \cos\phi_i)\dot{\phi}_i^2 + (x_j'^P \sin\phi_j + y_j'^P \cos\phi_j)\dot{\phi}_j^2$$

(3) A *relative ϕ constraint* requires that the difference between the rotation angles of bodies i and j equal a given constant C_3 (see Fig. 3.3.1); that is,

$$\Phi^{r\phi(i,j)} \equiv \phi_j - \phi_i - C_3 = 0 \tag{3.3.5}$$

The Jacobians of this constraint with respect to $\mathbf{q}_i$ and $\mathbf{q}_j$ are simply

$$\begin{aligned}\Phi_{\mathbf{q}_i}^{r\phi(i,j)} &= [0,\ 0,\ -1] \\ \Phi_{\mathbf{q}_j}^{r\phi(i,j)} &= [0,\ 0,\ 1]\end{aligned} \tag{3.3.6}$$

Finally, from Eqs. 3.1.9 and 3.1.10,

$$\nu^{r\phi(i,j)} = 0$$

$$\gamma^{r\phi(i,j)} = 0$$

(4) A *relative distance constraint* between point P_i on body i and point P_j on body j requires that the distance between these points equal a given constant

$C_4 > 0$ (see Fig. 3.3.1); that is, using Eq. 2.4.8,

$$\begin{aligned}\Phi^{rd(i,j)} &\equiv (\mathbf{r}_i^P - \mathbf{r}_j^P)^T(\mathbf{r}_i^P - \mathbf{r}_j^P) - C_4^2 \\ &= (x_i + x_i'^P \cos\phi_i - y_i'^P \sin\phi_i - x_j - x_j'^P \cos\phi_j + y_j'^P \sin\phi_j)^2 \\ &\quad + (y_i + x_i'^P \sin\phi_i + y_i'^P \cos\phi_i - y_j - x_j'^P \sin\phi_j - y_j'^P \cos\phi_j)^2 \\ &\quad - C_4^2 = 0\end{aligned} \tag{3.3.7}$$

The Jacobians of this constraint with respect to $\mathbf{q}_i$ and $\mathbf{q}_j$ are

$$\begin{aligned}\Phi_{\mathbf{q}_i}^{rd(i,j)} &= 2[(\mathbf{r}_i^P - \mathbf{r}_j^P)^T,\ \mathbf{s}_i'^{PT}\mathbf{B}_i^T(\mathbf{r}_i^P - \mathbf{r}_j^P)] \\ \Phi_{\mathbf{q}_j}^{rd(i,j)} &= 2[-(\mathbf{r}_i^P - \mathbf{r}_j^P)^T,\ -\mathbf{s}_j'^{PT}\mathbf{B}_j^T(\mathbf{r}_i^P - \mathbf{r}_j^P)]\end{aligned} \tag{3.3.8}$$

Since there is no explicit time dependence in Eq. 3.3.7, from Eq. 3.1.9,

$$\nu^{rd(i,j)} = 0$$

Expanding terms in Eq. 3.1.10, using the Jacobians of Eq. 3.3.8, leads to calculations similar to those encountered in expanding $\gamma^{ad(i)}$; that is,

$$\begin{aligned}\gamma^{rd(i,j)} &= -2\{(\dot{\mathbf{r}}_i - \dot{\mathbf{r}}_j)^T(\dot{\mathbf{r}}_i - \dot{\mathbf{r}}_j) \\ &\quad + (\dot{\mathbf{r}}_i - \dot{\mathbf{r}}_j)^T(\mathbf{B}_i\mathbf{s}_i'^P\dot{\phi}_i - \mathbf{B}_j\mathbf{s}_j'^P\dot{\phi}_j) \\ &\quad - (\mathbf{s}_i'^{PT}\mathbf{A}_i^T\dot{\phi}_i^2 - \mathbf{s}_j'^P\mathbf{A}_j^T\dot{\phi}_j^2)(\mathbf{r}_i^P - \mathbf{r}_j^P) \\ &\quad + (\mathbf{s}_i'^{PT}\mathbf{B}_i^T\dot{\phi}_i - \mathbf{s}_j'^{PT}\mathbf{B}_j^T\dot{\phi}_j)(\dot{\mathbf{r}}_i - \dot{\mathbf{r}}_j + \mathbf{B}_i\mathbf{s}_i'^P\dot{\phi}_i - \mathbf{B}_j\mathbf{s}_j'^P\dot{\phi}_j)\} \\ &= -2\{(\dot{\mathbf{r}}_i^P - \dot{\mathbf{r}}_j^P)^T(\dot{\mathbf{r}}_i^P - \dot{\mathbf{r}}_j^P) \\ &\quad - (\mathbf{s}_i'^{PT}\mathbf{A}_i^T\dot{\phi}_i^2 - \mathbf{s}_j'^{PT}\mathbf{A}_j^T\dot{\phi}_j^2)(\mathbf{r}_i^P - \mathbf{r}_j^P)\}\end{aligned}$$

As noted following Eq. 3.2.1, $C_4 > 0$ is required to rule out both physical and mathematical singularities.

Example 3.3.1: The two-body slider-crank mechanism of Examples 2.4.3 and 3.1.2 (Fig. 3.1.3) can be modeled using an absolute x position constraint on point O_1 (Eq. 3.2.3) with $C_1 = 0$, absolute y position constraints on points O_1 and O_2 (Eq. 3.2.4) with $C_2 = 0$, and relative x and y constraints between P_1 and P_2 (Eqs. 3.3.1 and 3.3.3) with $C_1 = C_2 = 0$ (Prob. 3.3.1). Many other combinations of bodies and relative and absolute constraints can be used to model the slider-crank mechanism (Prob. 3.3.2).

Example 3.3.2: The four-bar mechanism (three moving bodies and ground) of Fig. 3.3.2(a) may be modeled in many ways. Using the points defined in Fig. 3.3.2(b), the following are kinematically equivalent models (Prob. 3.3.3):

1. Bodies 1, 2, and 3
   Constraints: (i) Absolute x and y constraints on points P_1 and Q_3

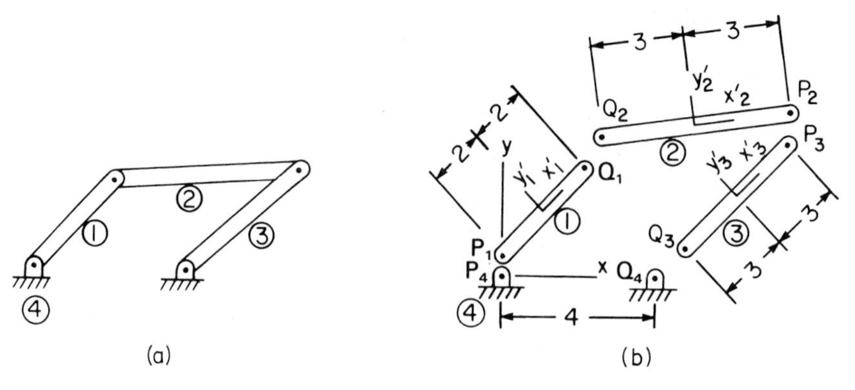

Figure 3.3.2 Four-bar mechanism. (a) Assembled. (b) Dissected.

(ii) Relative x and y constraints between points Q_1 and Q_2 and between points P_2 and P_3

$$\text{DOF} = 3nb - nh = 9 - 8 = 1$$

2. Bodies 1 and 3
Constraints: (i) Absolute x and y constraints on points P_1 and Q_3
(ii) Relative distance constraint between points Q_1 and P_3

$$\text{DOF} = 3nb - nh = 6 - 5 = 1$$

3. Bodies 1 and 2
Constraints: (i) Absolute x and y constraints on point P_1
(ii) Absolute distance constraint on points P_2 from point Q_4
(iii) Relative x and y constraints between points Q_1 and Q_2

$$\text{DOF} = 3nb - nc = 6 - 5 = 1$$

4. Bodies 2
Constraints: (i) Absolute distance constraint on point Q_2 from point P_4 and on point P_2 from point Q_4

$$\text{DOF} = 3nb - nh = 3 - 2 = 1$$

While these models are kinematically equivalent, if the angular orientation of body 1 is to be driven, model 4 cannot be used, and if the angular orientation of body 3 is to be driven, models 3 and 4 cannot be used.

3.3.2 Revolute and Translational Joints

In more complex constraints between bodies, it is helpful to draw figures to aid in developing vector equations that define the constraints. The vector equations may then be manipulated into more useful algebraic forms for numerical

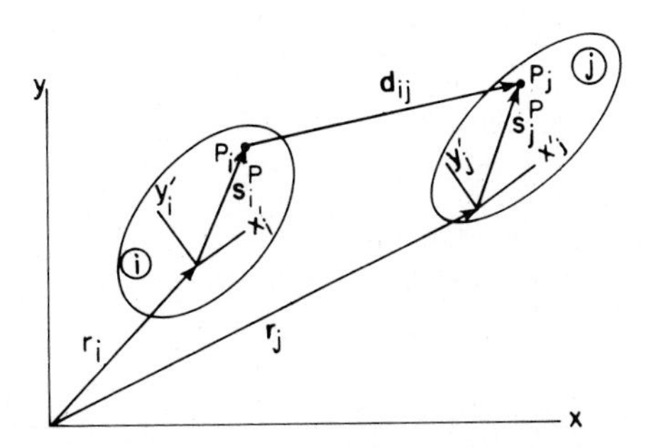

Figure 3.3.3 Relative position of two bodies.

computation and analysis. For example, Fig. 3.3.3 illustrates adjacent bodies i and j. The origins of their corresponding body-fixed reference frames are located by vectors $\mathbf{r}_i$ and $\mathbf{r}_j$ with respect to the global x-y frame. Points P_i and P_j on bodies i and j are located by vectors $\mathbf{s}_i^P$ and $\mathbf{s}_j^P$ (constant vectors $\mathbf{s}_i'^P$ and $\mathbf{s}_j'^P$ in the x_i'-y_i' and x_j'-y_j' frames), respectively. Points P_i and P_j are connected by a vector $\mathbf{d}_{ij}$:

$$\mathbf{d}_{ij} = \mathbf{r}_j + \mathbf{s}_j^P - \mathbf{r}_i - \mathbf{s}_i^P \tag{3.3.9}$$

A *revolute joint* allows relative rotation about a point P that is common to bodies i and j, as shown in Fig. 3.3.4. Physically, such a joint is a rotational bearing between the bodies. If one body is held fixed, the other body has only a single rotational degree of freedom. Thus, a revolute joint eliminates two degrees of freedom from the pair. This joint is defined by locating point P_i on body i by $\mathbf{s}_i'^P$ in the x_i'-y_i' frame and P_j on body j by $\mathbf{s}_j'^P$ in the $x_j' - y_j'$ frame.

Constraint equations that define a revolute joint are obtained by requiring that points P_i and P_j coincide. Employing Eq. 3.3.9 and setting $\mathbf{d}_{ij} = \mathbf{0}$, the constraint equations can be written as

$$\begin{aligned}\mathbf{\Phi}^{r(i,j)} &= \mathbf{r}_i + \mathbf{s}_i^P - \mathbf{r}_j - \mathbf{s}_j^P \\ &= \mathbf{r}_i + \mathbf{A}_i\mathbf{s}_i'^P - \mathbf{r}_j - \mathbf{A}_j\mathbf{s}_j'^P = \mathbf{0}\end{aligned} \tag{3.3.10}$$

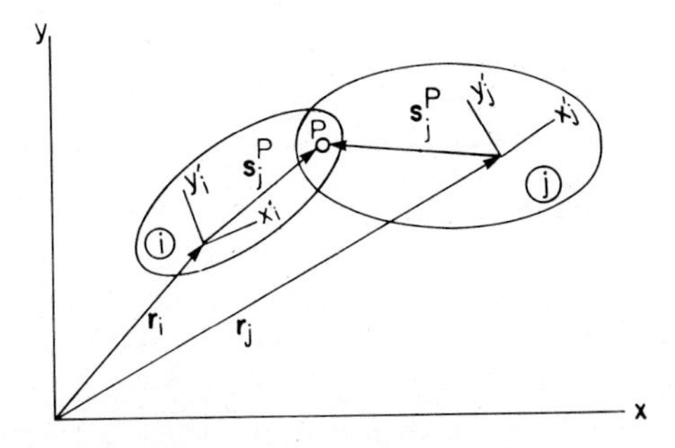

Figure 3.3.4 Revolute joint.

More explicitly,

$$\mathbf{\Phi}^{r(i,j)} \equiv \begin{bmatrix} x_i + x_i'^P \cos\phi_i - y_i'^P \sin\phi_i - x_j - x_j'^P \cos\phi_j + y_j'^P \sin\phi_j \\ y_i + x_i'^P \sin\phi_i + y_i'^P \cos\phi_i - y_j - x_j'^P \sin\phi_j - y_j'^P \cos\phi_j \end{bmatrix} = \mathbf{0} \quad \textbf{(3.3.11)}$$

To confirm that Eq. 3.3.11 is equivalent to the physical revolute joint, it must be shown that it implies the geometry of the joint. Since Eq. 3.3.11 implies $\mathbf{d}_{ij} = \mathbf{0}$, points P_i and P_j coincide (see Fig. 3.3.4), which is the definition of the physical joint.

The Jacobians of this constraint with respect to $\mathbf{q}_i$ and $\mathbf{q}_j$ are, from Eqs. 3.3.10 and 2.6.3,

$$\begin{aligned} \mathbf{\Phi}_{\mathbf{q}_i}^{r(i,j)} &= [\mathbf{I}, \mathbf{B}_i \mathbf{s}_i'^P] \\ \mathbf{\Phi}_{\mathbf{q}_j}^{r(i,j)} &= [-\mathbf{I}, -\mathbf{B}_j \mathbf{s}_j'^P] \end{aligned} \quad \textbf{(3.3.12)}$$

Since Eq. 3.3.11 does not depend explicitly on time, from Eq. 3.1.9,

$$\mathbf{v}^{r(i,j)} = \mathbf{0}$$

Using matrix algebra and the Jacobians of Eq. 3.3.12, Eq. 3.1.10 yields

$$\boldsymbol{\gamma}^{r(i,j)} = \mathbf{A}_i \mathbf{s}_i'^P \dot{\phi}_i^2 - \mathbf{A}_j \mathbf{s}_j'^P \dot{\phi}_j^2$$

where Eq. 2.6.8 has been used.

Example 3.3.3: The four-bar mechanism of Fig. 3.3.2 can be modeled by designating body 4 as ground and using the absolute x, y, and ϕ constraints of Eqs. 3.2.3, 3.2.4, and 3.2.6, with $\mathbf{s}'^P = \mathbf{0}$ and $C_k = 0$, $k = 1, 2, 3$, to cause the $x_4' - y_4'$ frame to coincide with the x-y frame. With ground modeled as a body, the four rotational couplings in the mechanism can be modeled as four revolute joints defined by Eq. 3.3.11, as follows:

(i) Between bodies 1 and 4 (ground), with $\mathbf{s}_1'^P = [-2, 0]^T$ and $\mathbf{s}_4'^P = \mathbf{0}$.
(ii) Between bodies 1 and 2, with $\mathbf{s}_1'^Q = [2, 0]^T$ and $\mathbf{s}_2'^Q = [-3, 0]^T$.
(iii) Between bodies 2 and 3, with $\mathbf{s}_2'^P = [3, 0]^T$ and $\mathbf{s}_3'^P = [3, 0]^T$.
(iv) Between bodies 3 and 4, with $\mathbf{s}_3'^Q = [-3, 0]^T$ and $\mathbf{s}_4'^Q = [4, 0]^T$.

Including the three constraints on body 4 and four revolute joints, providing all constraints are independent,

$$\text{DOF} = 3nb - nh = 3 \times 4 - 3 - 4 \times 2 = 12 - 11 = 1$$

A *translational joint* allows relative translation of a pair of bodies along a common axis, but no relative rotation of the bodies. Physically, such a joint may be defined as a straight block (or key) on one body that fits precisely in a straight slot (or keyway) on the second body and allows relative translation along their common center lines. If one body were held fixed, the other body has only a single translational degree of freedom. Thus, a translational joint eliminates two degrees of freedom from the pair.

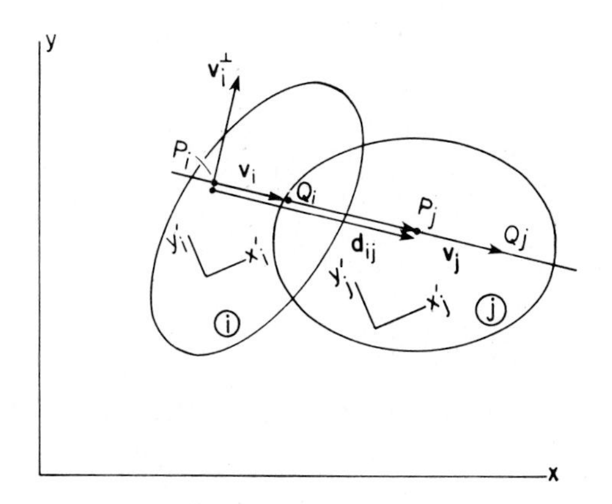

Figure 3.3.5 Translational joint.

For the translational joint shown in Fig. 3.3.5, let points P_i, Q_i, P_j, and Q_j be specified on a line that is parallel to or on the path of relative translation between bodies i and j. Noncoincident points P_i and Q_i are located on body i, and P_j and Q_j are located on body j. Vector $\mathbf{v}_i$ in body i connects points P_i and Q_i, and vector $\mathbf{v}_j$ in body j connects points P_j and Q_j; that is, $\mathbf{v}_i = \mathbf{A}_i\mathbf{v}_i'$, where $\mathbf{v}_i' = [x_i^P - x_i^Q, y_i^P - y_i^Q]^T$ and, similarly, $\mathbf{v}_j$ is defined by data that locate P_j and Q_j on body j. The vector $\mathbf{d}_{ij}$ that connects points P_i and P_j has variable length. Vectors $\mathbf{v}_i$ and $\mathbf{v}_j$ must remain collinear at all times. To have these three vectors collinear, it is necessary that $\mathbf{d}_{ij}$ and $\mathbf{v}_j$ be perpendicular to $\mathbf{v}_i^\perp$, which is perpendicular to $\mathbf{v}_i$. Using Eqs. 2.3.5 and 2.6.6, this is (Prob. 3.3.4)

$$\begin{aligned}
\mathbf{\Phi}^{t(i,j)} &= \begin{bmatrix} (\mathbf{v}_i^\perp)^T\mathbf{d}_{ij} \\ (\mathbf{v}_i^\perp)^T\mathbf{v}_j \end{bmatrix} \\
&= \begin{bmatrix} \mathbf{v}_i'^T\mathbf{R}^T\mathbf{A}_i^T(\mathbf{r}_j + \mathbf{A}_j\mathbf{s}_j'^P - \mathbf{r}_i - \mathbf{A}_i\mathbf{s}_i'^P) \\ \mathbf{v}_i'^T\mathbf{R}^T\mathbf{A}_i^T\mathbf{A}_j\mathbf{v}_j' \end{bmatrix} \\
&= \begin{bmatrix} \mathbf{v}_i'^T\mathbf{B}_i^T(\mathbf{r}_j - \mathbf{r}_i) - \mathbf{v}_i'^T\mathbf{B}_{ij}\mathbf{s}_j'^P - \mathbf{v}_i'^T\mathbf{R}^T\mathbf{s}_i'^P \\ -\mathbf{v}_i'^T\mathbf{B}_{ij}\mathbf{v}_j' \end{bmatrix} = \mathbf{0}
\end{aligned} \tag{3.3.13}$$

As long as $\mathbf{d}_{ij} \neq \mathbf{0}$, Eq. 3.3.13 implies that the three vectors are parallel. Since they have points in common, they are collinear. If $\mathbf{d}_{ij} = \mathbf{0}$, which is possible, then $\mathbf{v}_i$ and $\mathbf{v}_j$ emanate from the same point and, by the second of Eqs. 3.3.13, they are parallel. Thus they are collinear and Eq. 3.3.13 is equivalent to the geometry of the translational joint.

Using Eq. 2.6.8, the Jacobians of $\mathbf{\Phi}^{t(i,j)}$ with respect to $\mathbf{q}_i$ and $\mathbf{q}_j$ are (Prob. 3.3.5)

$$\begin{aligned}
\mathbf{\Phi}_{\mathbf{q}_i}^{t(i,j)} &= \begin{bmatrix} -\mathbf{v}_i'^T\mathbf{B}_i^T, & -\mathbf{v}_i'^T\mathbf{A}_i^T(\mathbf{r}_j - \mathbf{r}_i) - \mathbf{v}_i'^T\mathbf{A}_{ij}\mathbf{s}_j'^P \\ \mathbf{0}, & -\mathbf{v}_i'^T\mathbf{A}_{ij}\mathbf{v}_j' \end{bmatrix} \\
\mathbf{\Phi}_{\mathbf{q}_j}^{t(i,j)} &= \begin{bmatrix} \mathbf{v}_i'^T\mathbf{B}_i^T, & \mathbf{v}_i'^T\mathbf{A}_{ij}\mathbf{s}_j'^P \\ \mathbf{0}, & \mathbf{v}_i'^T\mathbf{A}_{ij}\mathbf{v}_j' \end{bmatrix}
\end{aligned} \tag{3.3.14}$$

Since Eq. 3.3.13 does not depend explicitly on time,

$$\mathbf{v}^{t(i,j)} = \mathbf{0}$$

Using matrix algebra and the Jacobians of Eq. 3.3.14, Eq. 3.1.10 may be expanded as

$$\boldsymbol{\gamma}^{t(i,j)} = -[\mathbf{G}^t_{\mathbf{r}_i}\dot{\mathbf{r}}_i + \mathbf{G}^t_{\phi_i}\dot{\phi}_i + \mathbf{G}^t_{\mathbf{r}_j}\dot{\mathbf{r}}_j + \mathbf{G}^t_{\phi_j}\dot{\phi}_j]$$

where

$$\mathbf{G}^t \equiv \mathbf{\Phi}^{t(i,j)}_{\mathbf{q}}\dot{\mathbf{q}}$$
$$= \begin{bmatrix} \mathbf{v}_i'^T\mathbf{B}_i^T(\dot{\mathbf{r}}_j - \dot{\mathbf{r}}_i) + \mathbf{v}_i'^T\mathbf{A}_{ij}\mathbf{s}_j'^P(\dot{\phi}_j - \dot{\phi}_i) - \mathbf{v}_i'^T\mathbf{A}_i^T(\mathbf{r}_j - \mathbf{r}_i)\dot{\phi}_i \\ \mathbf{v}_i'^T\mathbf{A}_{ij}\mathbf{v}_j'(\dot{\phi}_j - \dot{\phi}_i) \end{bmatrix}$$

Using Eqs. 2.4.12 and 2.6.8,

$$\boldsymbol{\gamma}^{t(i,j)} = -\begin{bmatrix} \mathbf{v}_i'^T[\mathbf{B}_{ij}\mathbf{s}_j'^P(\dot{\phi}_j - \dot{\phi}_i)^2 - \mathbf{B}_i^T(\mathbf{r}_j - \mathbf{r}_i)\dot{\phi}_i^2 - 2\mathbf{A}_i^T(\dot{\mathbf{r}}_j - \dot{\mathbf{r}}_i)\dot{\phi}_i] \\ 0 \end{bmatrix}$$

where the second term on the right is zero, because of Eq. 3.3.13.

Example 3.3.4: Rather than the simplified two-body model of the slider–crank mechanism in Examples 2.4.3 and 3.1.2, the full four-body model of Fig. 3.3.6 can be employed.

To cause the x_4'-y_4' frame to coincide with the x-y frame, Eqs. 3.2.3, 3.2.4, and 3.2.6 can be used with $\mathbf{s}'^P = \mathbf{0}$ and $C_k = 0$, $k = 1, 2, 3$. Three revolute joints are defined by Eq. 3.3.11, as follows:

(i) Between bodies 1 and 4, with $\mathbf{s}_1'^Q = [-2, 0]^T$ and $\mathbf{s}_4'^Q = \mathbf{0}$.
(ii) Between bodies 1 and 2, with $\mathbf{s}_1'^P = [2, 0]^T$ and $\mathbf{s}_2'^Q = [-3, 0]^T$.
(iii) Between bodies 2 and 3, with $\mathbf{s}_2'^P = [3, 0]^T$ and $\mathbf{s}_3'^P = \mathbf{0}$.

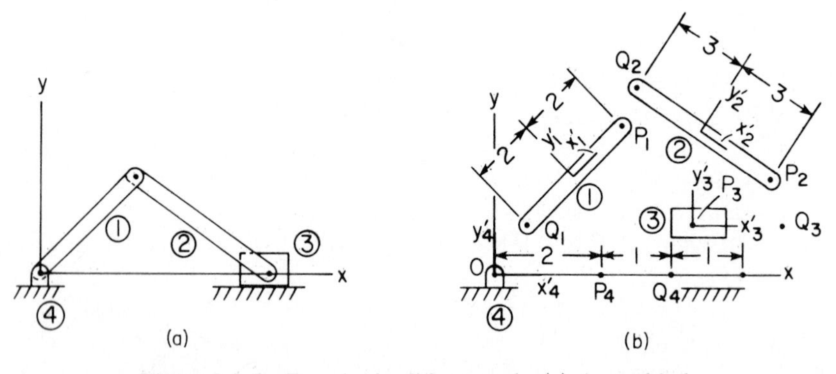

Figure 3.3.6 Four-body slider–crank. (a) Assembled. (b) Dissected.

The final constraint is a translational joint between bodies 3 and 4, which is defined by Eq. 3.3.13 with

$$\mathbf{s}_3'^P = \mathbf{0}, \qquad \mathbf{v}_3' = [1, 0]^T, \qquad \mathbf{s}_4'^P = [2, 0]^T, \qquad \mathbf{v}_4' = [1, 0]^T$$

Accounting for four bodies and all constraints, providing all constraints are independent,

$$\text{DOF} = 3nb - nh = 3 \times 4 - 3 - 3 \times 2 - 2 = 1$$

3.3.3 Composite Joints

In many kinematics applications, the only function of a body is to connect two other bodies, using a combination of revolute and translational joints. Such a connection is called a *coupler,* which need not be treated as a body. The resulting combination may be viewed as a *composite joint,* yielding elementary constraint equations and avoiding the need to introduce generalized coordinates for the coupler. In dynamics, such a body is called a *massless link,* since it is used to represent the kinematic effects of parts that may have insignificant mass. In kinematics, of course, the concept of mass need not arise.

Figure 3.3.7 illustrates a pair of rigid bodies that are connected by a coupler with two revolute joints, called a *revolute–revolute composite joint.* If this special joint were modeled using an additional rigid body and two revolute joints, three additional generalized coordinates and four constraint equations would be required. As shown next, only one constraint equation is required for this composite joint, which simply requires that the length of vector $\mathbf{d}_{ij}$ remain constant and equal to $C > 0$. The constraint equation is thus

$$\Phi^{rr(i,j)} \equiv (x_i^P - x_j^P)^2 + (y_i^P - y_j^P)^2 - C^2 = 0 \tag{3.3.15}$$

The revolute–revolute composite constraint is just a special case of Eq. 3.3.7, with $C_4 = C$. Its Jacobians are given by Eq. 3.3.8: $\nu^{rr(i,j)} = 0$ and $\gamma^{rr(i,j)} = \gamma^{rd(i,j)}$.

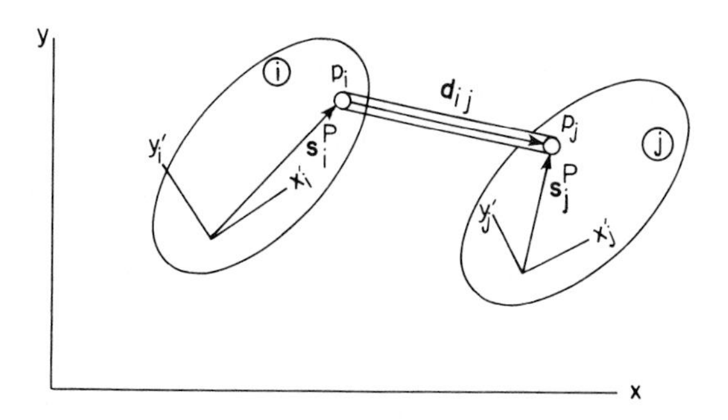

Figure 3.3.7 Revolute–revolute composite joint.

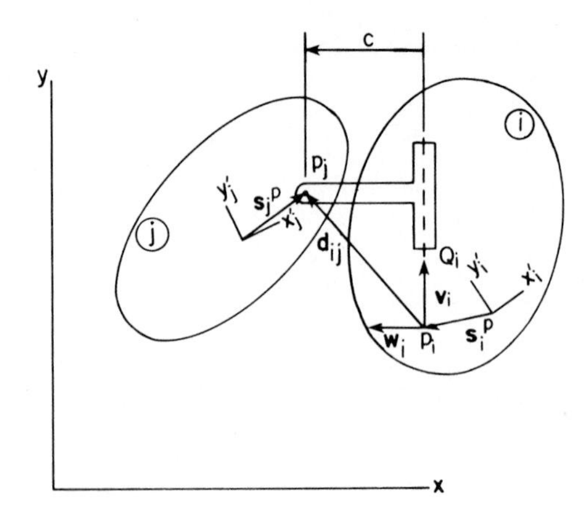

Figure 3.3.8 Revolute–translational composite joint.

Figure 3.3.8 illustrates a *revolute–translational composite joint.* Bodies i and j are connected by a coupler, with a revolute joint on body j and a translational joint on body i. Two noncoincident points P_i and Q_i are chosen on the line of translation in body i. The revolute joint has a given constant directed distance C from the line of translation, with C positive if the revolute joint is to the left of the line of translation, viewed from P_i to Q_i, and negative otherwise.

Note that if body j is fixed, then the coupler, and hence body i, can rotate. Furthermore, body i can translate in the direction $\mathbf{v}_i$. Thus, it has two remaining degrees of freedom. Since the revolute–translational joint eliminates only one degree of freedom, it should be defined by a single scalar constraint equation; that is, the directed distance from the line of translation to the revolute joint should be C.

To formulate this condition, a unit vector $\mathbf{w}_i$, directed to the left of the line of translation, is formed:

$$\mathbf{w}_i = \frac{1}{v_i}\mathbf{v}_i^{\perp} = \frac{1}{v_i}\mathbf{A}_i\mathbf{R}\mathbf{v}'_i \tag{3.3.16}$$

where v_i is the constant distance between P_i and Q_i. The directed distance C must thus be equal to the scalar product of $\mathbf{w}_i$ and $\mathbf{d}_{ij}$, yielding the desired constraint equation:

$$\begin{aligned}\Phi^{rt(i,j)} &= \frac{1}{v_i}\mathbf{v}_i'^T\mathbf{R}^T\mathbf{A}_i^T(\mathbf{r}_j + \mathbf{A}_j\mathbf{s}_j'^P - \mathbf{r}_i - \mathbf{A}_i\mathbf{s}_i'^P) - C \\ &= \frac{1}{v_i}[\mathbf{v}_i'^T\mathbf{B}_i^T(\mathbf{r}_j - \mathbf{r}_i) - \mathbf{v}_i'^T\mathbf{B}_{ij}\mathbf{s}_j'^P - \mathbf{v}_i'^T\mathbf{R}^T\mathbf{s}_i'^P] - C = 0\end{aligned} \tag{3.3.17}$$

The Jacobian matrices for this constraint with respect to $\mathbf{q}_i$ and $\mathbf{q}_j$ are

$$\Phi^{rt(i,j)}_{\mathbf{q}_i} = \frac{1}{v_i}[-\mathbf{v}_i'^T\mathbf{B}_i^T, -\mathbf{v}_i'^T\mathbf{A}_i^T(\mathbf{r}_j - \mathbf{r}_i) - \mathbf{v}_i'^T\mathbf{A}_{ij}\mathbf{s}_j'^P]$$
$$\Phi^{rt(i,j)}_{\mathbf{q}_j} = \frac{1}{v_i}[\mathbf{v}_i'^T\mathbf{B}_i^T, \mathbf{v}_i'^T\mathbf{A}_{ij}\mathbf{s}_j'^P] \tag{3.3.18}$$

Since there is no explicit time dependence in Eq. 3.3.17,

$$\nu^{rt(i,j)} = 0$$

Using matrix algebra and the Jacobians of Eq. 3.3.18,

$$\gamma^{rt(i,j)} = -[G^{rt}_{\mathbf{r}_i}\dot{\mathbf{r}}_i + G^{rt}_{\phi_i}\dot{\phi}_i + G^{rt}_{\mathbf{r}_j}\dot{\mathbf{r}}_j + G^{rt}_{\phi_i}\dot{\phi}_i]$$

where

$$\begin{aligned}G^{rt} &\equiv \Phi^{rt(i,j)}_{\mathbf{q}}\dot{\mathbf{q}}\\ &= \frac{1}{v_i}[\mathbf{v}_i'^T\mathbf{B}_i^T(\dot{\mathbf{r}}_j - \dot{\mathbf{r}}_i) - \mathbf{v}_i'^T\mathbf{A}_i^T(\mathbf{r}_j - \mathbf{r}_i)\dot{\phi}_i\\ &\quad + \mathbf{v}_i'^T\mathbf{A}_{ij}\mathbf{s}_j'^P(\dot{\phi}_j - \dot{\phi}_i)]\end{aligned}$$

Using Eqs. 2.4.12 and 2.6.8,

$$\begin{aligned}\gamma^{rt(i,j)} &= \frac{1}{v_i}\mathbf{v}_i'^T[2\mathbf{A}_i^T(\dot{\mathbf{r}}_j - \dot{\mathbf{r}}_i)\dot{\phi}_i\\ &\quad + \mathbf{B}_i^T(\mathbf{r}_j - \mathbf{r}_i)\dot{\phi}_i^2 - \mathbf{B}_{ij}\mathbf{s}_j'^P(\dot{\phi}_j - \dot{\phi}_i)^2]\end{aligned}$$

Example 3.3.5: The slider–crank mechanism of Example 3.3.4, shown in Fig. 3.3.6, can be modeled by replacing body 3 and its associated revolute joint with body 2 and translational joint with body 4 by a revolute–translational composite joint between bodies 4 and 2. The constraint is defined by Eq. 3.3.17, with

$$\mathbf{s}_4'^P = [2, 0]^T, \qquad \mathbf{v}_4' = [1, 0]^T, \qquad \mathbf{s}_2'^P = [3, 0]^T, \qquad C = 0$$

Since only bodies 1, 2, and 4 remain, with three absolute constraints on body 4, revolute joints (i) and (ii) from Example 3.3.4, and the single constraint of Eq. 3.3.17,

$$\text{DOF} = 3nb - nh = 3 \times 3 - 3 - 2 \times 2 - 1 = 1$$

3.4 GEARS AND CAM–FOLLOWERS

3.4.1 Gears

A *convex–convex gear set* or simply *gear set* consists of a pair of gears on two bodies that are constrained so that the distance $R_i + R_j$ between their centers is

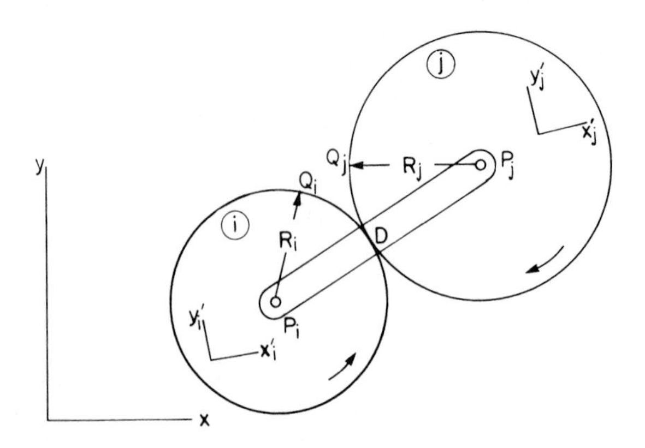

Figure 3.4.1 Gear set.

fixed, as shown in Fig. 3.4.1. Gear teeth on the periphery of the gears cause the pitch circles shown to roll relative to each other, without slip. Thus, the arc lengths DQ_i and DQ_j of contact on the gears must be equal, where points Q_i and Q_j are defined so that they coincide at the initial assembly of the gear set.

Based on the geometry of the gear set shown in Fig. 3.4.2,

$$\begin{aligned} \alpha_i &= \phi_i + \theta_i - \theta \\ \alpha_j &= -(\phi_j + \theta_j - \theta - \pi) \\ \alpha_i R_i &= \alpha_j R_j \end{aligned} \tag{3.4.1}$$

where θ_i and θ_j are data that define the assembly of the gear set, and the third equation is obtained by equating lengths of arcs DQ_i and DQ_j. These are the defining equations for the gear set.

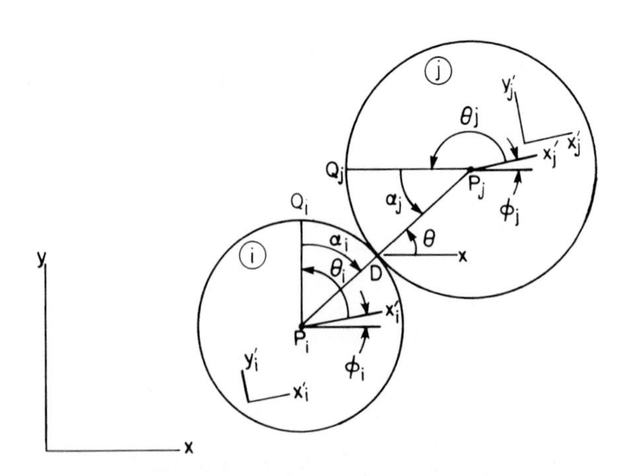

Figure 3.4.2 Geometry of gear set.

Solving Eqs. 3.4.1 for θ,

$$\theta = \frac{R_i(\phi_i + \theta_i) + R_j(\phi_j + \theta_j) - R_j\pi}{R_i + R_j} \tag{3.4.2}$$

Since the distance between gear centers is established by other constraints, the *gear constraint* is just

$$\begin{aligned} \Phi^{g(i,j)} &= (\mathbf{r}_j^P - \mathbf{r}_i^P)\mathbf{u}^\perp \\ &= (x_j^P - x_i^P)\sin\theta - (y_j^P - y_i^P)\cos\theta = 0 \end{aligned} \tag{3.4.3}$$

where θ is given by Eq. 3.4.2 and $\mathbf{u}^\perp \equiv [-\sin\theta, \cos\theta]^T$; that is, $\mathbf{u} = [\cos\theta, \sin\theta]^T$ is a unit vector along the line from P_i to P_j in Fig. 3.4.2.

It is important to confirm that Eq. 3.4.3, with additional constraints that assure the distance between points P_i and P_j is $R_i + R_j$, imply that the gear set geometric relations of Eq. 3.4.1 hold. Since the distance between P_i and P_j is $R_i + R_j$ and at initial assembly Q_i and Q_j coincide, then, for $|t - t_0|$ small, Eq. 3.4.3 implies that θ of Eq. 3.4.2 is the angle shown in Fig. 3.4.2. Thus α_i and α_j are related to θ as defined in the first two of Eqs. 3.4.1. Substituting θ of Eq. 3.4.2 into these relations and expanding,

$$\begin{aligned} R_i\alpha_i - R_j\alpha_j &= R_i(\phi_i + \theta_i - \theta) + R_j(\phi_j + \theta_j - \theta - \pi) \\ &= R_i(\phi_i + \theta_i) + R_j(\phi_j + \theta_j - \pi) - (R_i + R_j)\theta \\ &= 0 \end{aligned}$$

Thus, providing $\mathbf{q}_i$ and $\mathbf{q}_j$ are continuous functions of time, Eqs. 3.4.1 hold for all time, and Eq. 3.4.3 plus the distance constraint between P_i and P_j implies the geometry of the convex–convex gear set.

The Jacobians of the constraint of Eq. 3.4.3, with respect to $\mathbf{q}_i$ and $\mathbf{q}_j$, are

$$\begin{aligned} \Phi^{g(i,j)}_{\mathbf{q}_i} &= \left[-\mathbf{u}^T,\ -\mathbf{s}_i'^{PT}\mathbf{B}_i^T\mathbf{u} + (\mathbf{r}_j^P - \mathbf{r}_i^P)^T\mathbf{u}^\perp\left(\frac{R_i}{R_i + R_j}\right)\right] \\ \Phi^{g(i,j)}_{\mathbf{q}_j} &= \left[\mathbf{u}^T,\ \mathbf{s}_j'^{PT}\mathbf{B}_j^T\mathbf{u} + (\mathbf{r}_j^P - \mathbf{r}_i^P)^T\mathbf{u}^\perp\left(\frac{R_j}{R_i + R_j}\right)\right] \end{aligned} \tag{3.4.4}$$

where $d\mathbf{u}/d\theta = [-\sin\theta, \cos\theta]^T = \mathbf{u}^\perp$ (see Eq. 2.3.5) has been used.

As in all previous constraints, there is no explicit time dependence in Eq. 3.4.3, so

$$\nu^g = 0$$

Using the Jacobians of Eq. 3.4.4 and Eqs. 3.4.2 and 3.4.3, Eq. 3.1.10 yields

$$\gamma^g = \left[2(\dot{\mathbf{r}}_j^P - \dot{\mathbf{r}}_i^P)^T\mathbf{u}\left(\frac{R_i\dot{\phi}_i + R_j\dot{\phi}_j}{R_i + R_j}\right) + (\mathbf{A}_j\mathbf{s}_j'^P\dot{\phi}_j^2 - \mathbf{A}_i\mathbf{s}_i'^P\dot{\phi}_i^2)^T\mathbf{u}^\perp\right]$$

Example 3.4.1: Consider the pair of gears shown in Fig. 3.4.3, with gear 1 fixed to ground. Let $\phi_1 = 0$, $\theta_1 = \pi/6$, $\theta_2 = 7\pi/6$, $R_1 = 1$, and $R_2 = 2$. A revolute–revolute composite joint of length 3 connects the centers of the gears. Find angle ϕ_2 as gear 2 falls to the position shown, in which P_2 is on the vertical line through P_1.

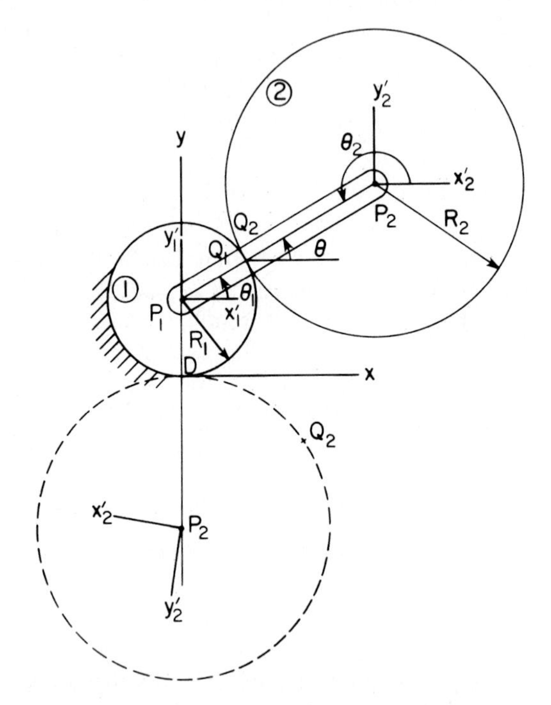

Figure 3.4.3 Gear joint with distance constraint.

From Eq. 3.4.3,

$$-(-2-1)\cos\theta = 0$$

Thus,

$$\theta = -\frac{\pi}{2}$$

Substituting into Eq. 3.4.2,

$$-\frac{\pi}{2} = \frac{1(0+\pi/6) + 2(\phi_2 + 7\pi/6) - 2\pi}{1+2}$$

Thus, $\phi_2 = -\pi$.

Consider next the *concave–convex gear set* shown in Fig. 3.4.4, in which the smaller gear on body j with center at P_j makes rolling contact inside the larger

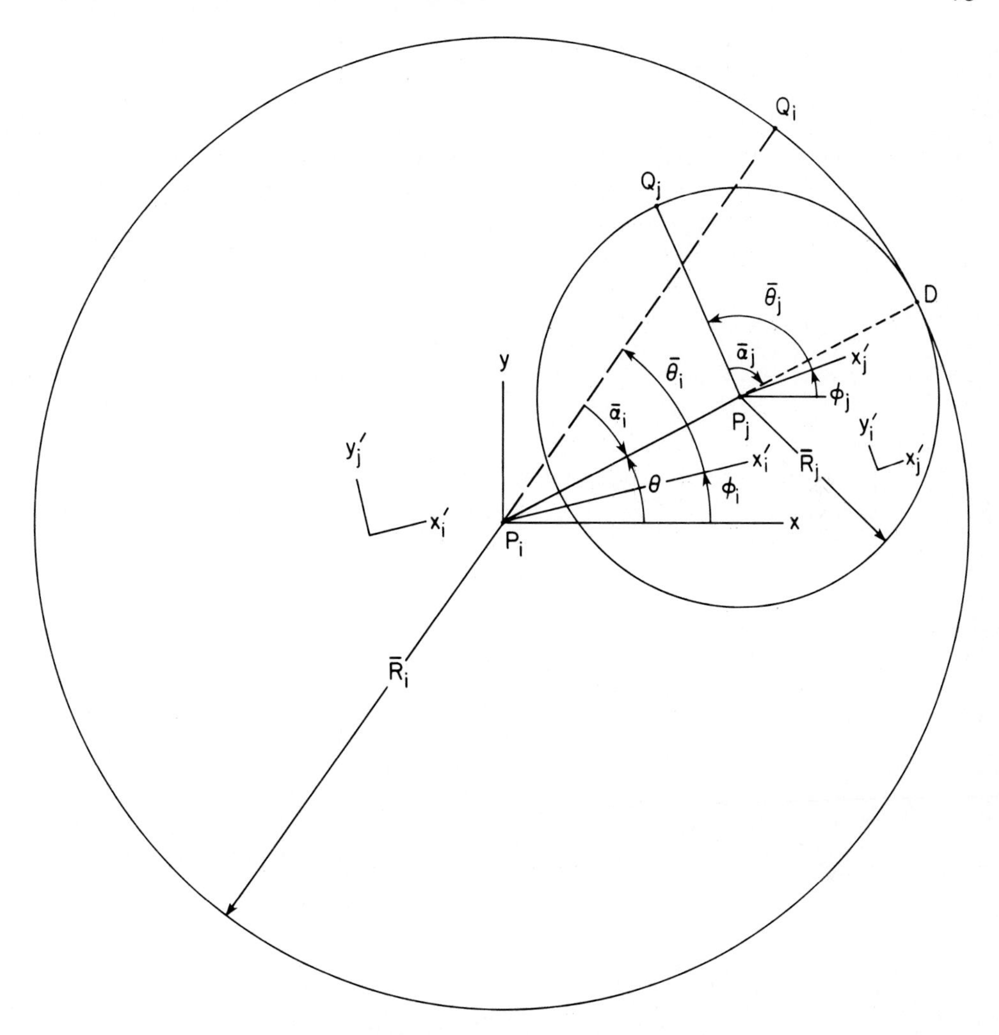

Figure 3.4.4 Concave–convex gear set.

interior gear with center at point P_i on body i. A bar (or distance constraint) of length $\bar{R}_i - \bar{R}_j > 0$ separates the centers of the gears.

Based on the geometry of Fig. 3.4.4,

$$\begin{aligned} \bar{\alpha}_i &= \phi_i + \bar{\theta}_i - \theta \\ \bar{\alpha}_j &= \phi_j + \bar{\theta}_j - \theta \\ \bar{\alpha}_i \bar{R}_i &= \bar{\alpha}_j \bar{R}_j \end{aligned} \tag{3.4.5}$$

where $\bar{\theta}_i$ and $\bar{\theta}_j$ define the assembly of the gear set. Solving Eq. 3.4.5 for θ,

$$\theta = \frac{-\bar{R}_i(\phi_i + \bar{\theta}_i) + \bar{R}_j(\phi_j + \bar{\theta}_j)}{-\bar{R}_i + \bar{R}_j} \tag{3.4.6}$$

From the geometry of the gear set, with the distance between P_i and P_j fixed by other constraints, Eq. 3.4.3 remains valid as an equation of constraint for the concave–convex gear set, but with θ defined by Eq. 3.4.6. A repetition of the argument that follows Eq. 3.4.3 shows that Eqs. 3.4.3 and 3.4.6, taken with additional constraints that define the distance between P_i and P_j, imply the geometry of the gear set.

Note that, if the following conventions are adopted,

$$\begin{aligned} R_i &= -\bar{R}_i < 0 \\ R_j &= \bar{R}_j > 0 \\ \theta_i &= \bar{\theta}_i \\ \theta_j &= \bar{\theta}_j - \pi \end{aligned} \tag{3.4.7}$$

then Eq. 3.4.6 takes exactly the form of Eq. 3.4.2. With this notation, Eq. 3.4.3 is unchanged, and Eq. 3.4.4 gives the constraint Jacobian for the concave–convex gear set.

If the radius of the gear on body i becomes infinite, a straight gear profile called a *rack* results. The gear on body j is then called a *pinion,* and the gear pair is called a *rack and pinion,* as shown in Fig. 3.4.5. If the pinion is on the left of the rack, as viewed from point P_i to point Q_i, then the radius of the pinion is taken as positive. If the pinion is on the right, its radius is negative.

This pair may be interpreted as a special case of the revolute–translational composite joint in which the advance s of the pinion, that is, the distance between points Q_i and D in Fig. 3.4.5, is dictated by the condition that lengths DQ_i and DQ_j in Fig. 3.4.5 be equal. Thus, the revolute–translational equation of Eq. 3.3.17, with i and j interchanged and $C = R_j$, must hold. The equal arc length

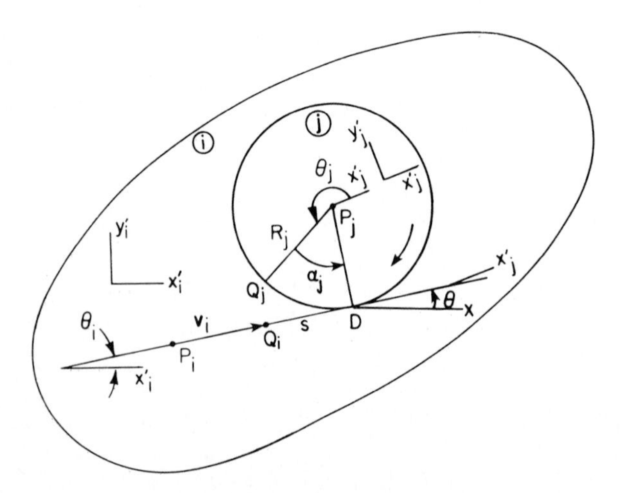

Figure 3.4.5 Rack and pinion.

condition is

$$s - R_j\alpha_j = 0 \tag{3.4.8}$$

To write Eq. 3.4.8 in terms of generalized coordinates, note from Fig. 3.4.5 that

$$s = \frac{1}{v_i}\mathbf{v}_i^T(\mathbf{r}_j^P - \mathbf{r}_i^Q) \tag{3.4.9}$$

$$\theta = \phi_i + \theta_i \tag{3.4.10}$$

and

$$\phi_j + \theta_j + \alpha_j + \frac{\pi}{2} = \theta + 2\pi \tag{3.4.11}$$

From Eqs. 3.4.10 and 3.4.11,

$$\alpha_j = \phi_i + \theta_i - \phi_j - \theta_j + \frac{3\pi}{2} \tag{3.4.12}$$

Substituting Eqs. 3.4.9 and 3.4.12 into Eq. 3.4.8 and writing the revolute–translational constraint of Eq. 3.3.17 in vector form, the constraint equations for this pair are

$$\mathbf{\Phi}^{rp(i,j)} = \begin{bmatrix} (\mathbf{r}_j^P - \mathbf{r}_i^Q)^T\mathbf{v}_i - v_iR_j\alpha_j \\ (\mathbf{r}_j^P - \mathbf{r}_i^P)^T\mathbf{v}_i^\perp - v_iR_j \end{bmatrix} = \mathbf{0} \tag{3.4.13}$$

Derivation of the Jacobians, $\mathbf{v}^{rp(i,j)}$, and $\boldsymbol{\gamma}^{rp(i,j)}$ for this constraint are left as an exercise (Prob. 3.4.4).

Example 3.4.2: Rotation of the gear in Fig. 3.4.6 is controlled by the horizontal motion of point P_1 of the rack. For analytical simplicity, let $\mathbf{q} = [x_1, \phi_1, \phi_2]^T$ be a reduced set of generalized coordinates. Let the distance from

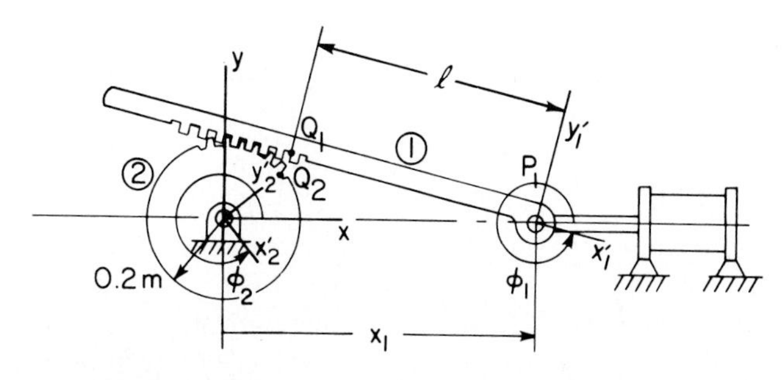

Figure 3.4.6 Slider–crank with a rack and pinion.

point P_1 to point Q_1 be ℓ. Then the vector $\mathbf{v}_1$ in Fig. 3.4.5 is

$$\mathbf{v}_1 = \begin{bmatrix} \cos\phi_1 & -\sin\phi_1 \\ \sin\phi_1 & \cos\phi_1 \end{bmatrix} \begin{bmatrix} -\ell \\ 0 \end{bmatrix} = \begin{bmatrix} -\ell\cos\phi_1 \\ -\ell\sin\phi_1 \end{bmatrix}$$

From Eq. 3.4.12,

$$\alpha_2 = \phi_1 + 0 - \phi_2 - \frac{3\pi}{2} + \frac{3\pi}{2} = \phi_1 - \phi_2$$

With this result and from the first of Eqs. 3.4.13,

$$-\left\{ \begin{bmatrix} x_1 \\ 0 \end{bmatrix} + \begin{bmatrix} \cos\phi_1 & -\sin\phi_1 \\ \sin\phi_1 & \cos\phi_1 \end{bmatrix} \begin{bmatrix} -\ell \\ 0 \end{bmatrix} \right\}^T \begin{bmatrix} -\ell\cos\phi_1 \\ -\ell\sin\phi_1 \end{bmatrix} - \ell(0.2)(\phi_1 - \phi_2) = 0$$

By simplifying,

$$x_1\cos\phi_1 - \ell = 0.2(\phi_1 - \phi_2) = 0$$

From the second of Eqs. 3.4.13,

$$-[x_1, 0] \begin{bmatrix} 0 & -1 \\ 1 & 0 \end{bmatrix} \begin{bmatrix} -\ell\cos\phi \\ -\ell\sin\phi_1 \end{bmatrix} - 0.2\ell = -x_1\ell\sin\phi_1 - 0.2\ell = 0$$

Thus the constraint equation of the rack and pinion joint, Eq. 3.4.13, is

$$\mathbf{\Phi}(\mathbf{q}) = \begin{bmatrix} x_1\cos\phi_1 - \ell - 0.2(\phi_1 - \phi_2) \\ x_1\sin\phi_1 + 0.2 \end{bmatrix} = \mathbf{0}$$

Differentiating both sides,

$$\begin{bmatrix} \cos\phi_1 & -(x_1\sin\phi_1 + 0.2) & 0.2 \\ \sin\phi_1 & x_1\cos\phi_1 & 0 \end{bmatrix} \begin{bmatrix} \dot{x}_1 \\ \dot{\phi}_1 \\ \dot{\phi}_2 \end{bmatrix} = \mathbf{0} \tag{3.4.14}$$

If the piston has velocity $\dot{x}_1 = 0.3\ \text{m/s}$ when $x_1 = 0.8\ \text{m}$, then $\dot{\phi}_1$ and $\dot{\phi}_2$ can be found by partitioning Eq. 3.4.14:

$$\begin{bmatrix} -x_1\sin\phi_1 - 0.2 & 0.2 \\ x_1\cos\phi_1 & 0 \end{bmatrix} \begin{bmatrix} \dot{\phi}_1 \\ \dot{\phi}_2 \end{bmatrix} = \begin{bmatrix} -\cos\phi_1 \\ -\sin\phi_1 \end{bmatrix} \dot{x}_1$$

The angular velocities are thus

$$[\dot{\phi}_1, \dot{\phi}_2]^T = [0.0968, -1.4524]^T$$

3.4.2 Cam–Followers

A *cam–follower* pair is shown in its most general form in Fig. 3.4.7, where body i is the cam and body j is the follower. The two bodies are in contact at point P, but sliding is permitted, unlike the case of a gear set. It is assumed that the cam and its follower always remain in contact with each other; that is, no chattering is allowed. In addition, it is assumed that the contact surfaces (outlines) are either

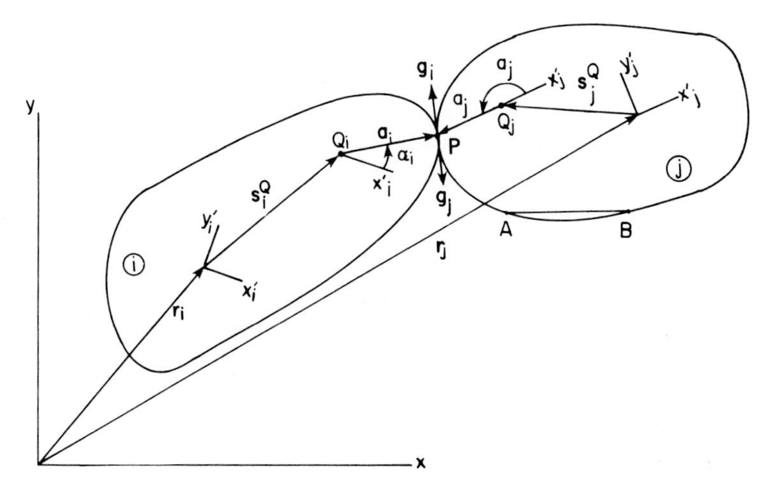

Figure 3.4.7 Cam–follower pair.

convex shapes or flat. An outline is *convex* if the straight line between any pair of points on the outline lies inside the body (e.g., points A and B on body j in Fig. 3.4.7).

A curve that defines the boundary of a body may have a general convex outline, as shown in Fig. 3.4.8. A point P on the boundary can be located in a polar coordinate system that is fixed in the body, with origin at point Q. The angle α is measured relative to the body-fixed x' axis, as shown in Fig. 3.4.8.

The location of point P in the x'-y' frame is given by

$$\mathbf{r}'^{P} = \mathbf{s}'^{Q} + \mathbf{a}' \tag{3.4.15}$$

In terms of the angle α,

$$\mathbf{a}' = \rho(\alpha)\mathbf{u}'(\alpha) \tag{3.4.16}$$

where $\rho(\alpha)$ is the length of $\mathbf{a}(\alpha)$ and $\mathbf{u}'(\alpha) = [\cos\alpha, \sin\alpha]^T$.

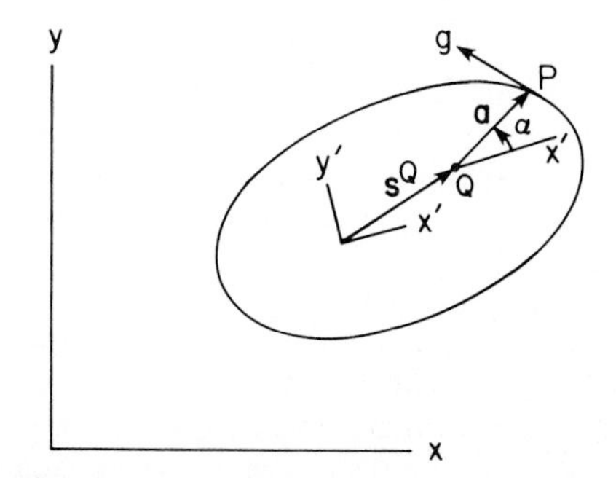

Figure 3.4.8 Body outline representation.

The tangent to the outline curve at point P in Fig. 3.4.8 is expressed in the x'-y' frame as

$$\begin{aligned}\mathbf{g}' &= \frac{d\mathbf{a}'}{d\alpha} = \rho_\alpha \mathbf{u}' + \rho \mathbf{u}'_\alpha \\ &= \rho_\alpha \mathbf{u}' + \rho \mathbf{u}'^{\perp}\end{aligned} \tag{3.4.17}$$

A cam outline $\rho(\alpha)$ is often described by a table of data points. A numerical method of curve fitting and interpolation must then be applied. Cubic spline interpolation is suggested for this purpose, since it provides continuous first and second derivatives [24].

Since the two bodies must remain in contact, from Fig. 3.4.7, a loop equation yields

$$\mathbf{r}_i + \mathbf{s}_i^Q + \mathbf{a}_i - \mathbf{a}_j - \mathbf{s}_j^Q - \mathbf{r}_j = \mathbf{0} \tag{3.4.18}$$

Equation 3.4.18 provides two algebraic constraint equations that involve the outline curve parameters α_i and α_j and the Cartesian coordinates of the bodies. Since the cam and follower are in contact at only one point, the tangents to both outlines at point P, that is, $\mathbf{g}_i$ and $\mathbf{g}_j$, must be collinear; so

$$\mathbf{g}_i^{\perp T} \mathbf{g}_j = (\mathbf{R}\mathbf{g}_i)^T \mathbf{g}_j = \mathbf{g}_i^T \mathbf{R}^T \mathbf{g}_j = -\mathbf{g}_i'^T \mathbf{B}_{ij} \mathbf{g}_j' = 0 \tag{3.4.19}$$

where Eqs. 2.6.5 and 2.6.6 have been used. The expanded form of Eqs. 3.4.18 and 3.4.19 yield three constraint equations:

$$\mathbf{\Phi}^{cf(i,j)} \equiv \begin{bmatrix} \mathbf{r}_i + \mathbf{A}_i(\mathbf{s}_i'^Q + \rho_i \mathbf{u}_i') - \mathbf{A}_j(\rho_j \mathbf{u}_j' + \mathbf{s}_j'^Q) - \mathbf{r}_j \\ -\mathbf{g}_i'^T \mathbf{B}_{ij} \mathbf{g}_j' \end{bmatrix} = \mathbf{0} \tag{3.4.20}$$

The outline parameters α_i and α_j appear in Eq. 3.4.20. They are treated as additional unknown generalized coordinates that locate the point P of contact. Hence, the vector of generalized coordinates for a mechanism with nb rigid bodies and one cam–follower joint has $nc = 3nb + 2$ components; that is,

$$\mathbf{q} \equiv [x_1, y_1, \phi_1, \ldots, x_{nb}, y_{nb}, \phi_{nb}, \alpha_i, \alpha_j]^T$$

Since it is defined by three constraint equations and two excess variables, each cam–follower pair eliminates only one relative degree of freedom between the bodies it connects.

The Jacobians of the constraint of Eq. 3.4.20, with respect to $\mathbf{q}_i$, α_i, $\mathbf{q}_j$, and α_j, using Eqs. 2.4.11, 2.4.12, and 2.6.8, are

$$\begin{aligned}
\mathbf{\Phi}_{\mathbf{q}_i}^{cf(i,j)} &= \begin{bmatrix} \mathbf{I} & \mathbf{B}_i(\mathbf{s}_i'^Q + \rho_i \mathbf{u}_i') \\ \mathbf{0} & -\mathbf{g}_i'^T \mathbf{A}_{ij} \mathbf{g}_j' \end{bmatrix} &
\mathbf{\Phi}_{\alpha_i}^{cf(i,j)} &= \begin{bmatrix} \mathbf{A}_i \mathbf{g}_i' \\ -(\mathbf{g}_i')_{\alpha_i}^T \mathbf{B}_{ij} \mathbf{g}_j' \end{bmatrix} \\
\mathbf{\Phi}_{\mathbf{q}_j}^{cf(i,j)} &= \begin{bmatrix} -\mathbf{I} & -\mathbf{B}_j(\mathbf{s}_j'^Q + \rho_j \mathbf{u}_j') \\ \mathbf{0} & \mathbf{g}_i'^T \mathbf{A}_{ij} \mathbf{g}_i' \end{bmatrix} &
\mathbf{\Phi}_{\alpha_j}^{cf(i,j)} &= \begin{bmatrix} -\mathbf{A}_j \mathbf{g}_j' \\ -\mathbf{g}_i'^T \mathbf{B}_{ij} (\mathbf{g}_j')_{\alpha_j} \end{bmatrix}
\end{aligned} \tag{3.4.21}$$

Since there is no explicit dependence on time in Eq. 3.4.20,

$$\mathbf{v}^{cf(i,j)} = \mathbf{0}$$

Carrying out the expansion of Eq. 3.1.10, using the Jacobians of Eqs. 3.4.21 and 3.4.19 and recalling that α_i and α_j are generalized coordinates,

$$\boldsymbol{\gamma}^{cf(i,j)} = \begin{bmatrix} \mathbf{A}_i(\mathbf{s}_i'^Q + \rho_i\mathbf{u}_i')\dot{\phi}_i^2 - \mathbf{A}_j(\mathbf{s}_j'^Q + \rho_j\mathbf{u}_j')\dot{\phi}_j^2 - \mathbf{A}_i(\mathbf{g}_i')_{\alpha_i}\dot{\alpha}_i^2 \\ + \mathbf{A}_j(\mathbf{g}_j')_{\alpha_j}\dot{\alpha}_j^2 - 2\mathbf{B}_i\mathbf{g}_i'\dot{\phi}_i\dot{\alpha}_i + 2\mathbf{B}_j\mathbf{g}_j'\dot{\phi}_j\dot{\alpha}_j \\ (\mathbf{g}_i')^T_{\alpha_i\alpha_i}\mathbf{B}_{ij}\mathbf{g}_j'\dot{\alpha}_i^2 + \mathbf{g}_i'^T\mathbf{B}_{ij}(\mathbf{g}_j')_{\alpha_j\alpha_j}\dot{\alpha}_j^2 - 2[(\mathbf{g}_i')^T_{\alpha_i}\mathbf{A}_{ij}\mathbf{g}_j'\dot{\alpha}_i \\ + \mathbf{g}_i'^T\mathbf{A}_{ij}(\mathbf{g}_j')_{\alpha_j}\dot{\alpha}_j](\dot{\phi}_j - \dot{\phi}_i) + 2(\mathbf{g}_i')^T_{\alpha_i}\mathbf{B}_{ij}(\mathbf{g}_j')_{\alpha_j}\dot{\alpha}_i\dot{\alpha}_j \end{bmatrix} \tag{3.4.22}$$

Example 3.4.3: A cam–follower joint that is used in a valve operating system of an internal combustion engine is shown in Fig. 3.4.9. Let the outline of the cam consist of a cosine curve and part of a circle:

$$\rho_1(\alpha_1) = \begin{cases} -\frac{1}{4}\cos 3\alpha_1 + \frac{5}{4}, & \text{if } 0 \leq \alpha_1 \leq \dfrac{2\pi}{3} \\ 1, & \text{if } \dfrac{2\pi}{3} \leq \alpha_1 < 2\pi \end{cases}$$

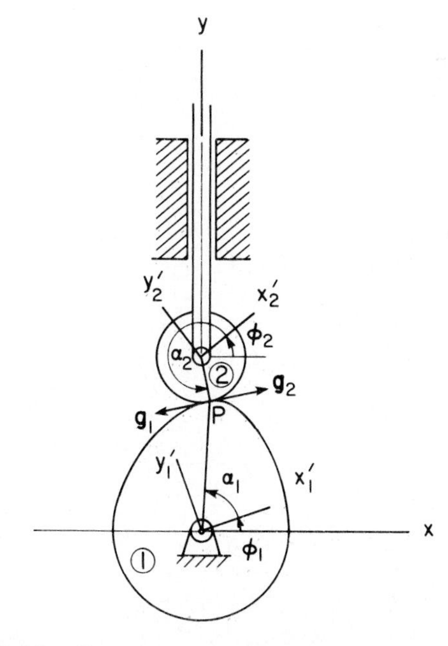

Figure 3.4.9 Cam–follower joint in an internal combustion engine.

Let the radius of the roller be $\frac{1}{4}$, so the outline of the follower is simply

$$\rho_2(\alpha_2) = \tfrac{1}{4}$$

In the configuration shown in Fig. 3.4.9, $0 \leq \alpha_1 \leq (2\pi/3)$. Hence, from Eqs. 3.4.16 and 3.4.17, the tangent vectors to the outlines are

$$\mathbf{g}_1' = \begin{bmatrix} (\frac{3}{4}\sin 3\alpha_1)\cos\alpha_1 - (-\frac{1}{4}\cos 3\alpha_1 + \frac{5}{4})\sin\alpha_1 \\ (\frac{3}{4}\sin 3\alpha_1)\sin\alpha_1 + (-\frac{1}{4}\cos 3\alpha_1 + \frac{5}{4})\cos\alpha_1 \end{bmatrix}$$
$$\mathbf{g}_2' = [-\tfrac{1}{4}\sin\alpha_2, \tfrac{1}{4}\cos\alpha_2]^T \tag{3.4.23}$$

Thus, the explicit form of the third constraint equation of Eq. 3.4.20 is

$$\begin{aligned} &\{[\tfrac{3}{4}\sin 3\alpha_1\cos\alpha_1 + (\tfrac{1}{4}\cos 3\alpha_1 - \tfrac{5}{4})\sin\alpha_1]\cos\phi_1 \\ &- [\tfrac{3}{4}\sin 3\alpha_1\sin\alpha_1 - (\tfrac{1}{4}\cos 3\alpha_1 - \tfrac{5}{4})\cos\alpha_1]\sin\phi_1\} \\ &\times\{-\tfrac{1}{4}\sin\alpha_2\sin\phi_2 + \tfrac{1}{4}\cos\alpha_2\cos\phi_2\} \\ &- \{[\tfrac{3}{4}\sin 3\alpha_1\cos\alpha_1 + (\tfrac{1}{4}\cos 3\alpha_1 - \tfrac{5}{4})\sin\alpha_1]\sin\phi_1 \\ &+ [\tfrac{3}{4}\sin 3\alpha_1\sin\alpha_1 - (\tfrac{1}{4}\cos 3\alpha_1 - \tfrac{5}{4})\cos\alpha_1]\cos\phi_1\} \\ &\times\{-\tfrac{1}{4}\sin\alpha_2\cos\phi_2 - \tfrac{1}{4}\cos\alpha_2\sin\phi_2\} = 0 \end{aligned}$$

A *cam–flat-faced follower pair* is shown in Fig. 3.4.10. The cam, which is shown as body i, may have a convex outline that is parameterized by α_i, as in Eq. 3.4.16. The contacting surface of the follower on body j is flat. The unit vector $\mathbf{u}_j$ defines the line of contact. Since the point P is the point of contact,

$$\mathbf{r}_i^P - \mathbf{a}_j - \mathbf{r}_j^Q = \mathbf{0} \tag{3.4.24}$$

The x_j'-y_j' component form of vector $\mathbf{u}_j$ is

$$\mathbf{u}_j' \equiv \frac{1}{|\mathbf{r}_j'^Q - \mathbf{r}_j'^C|}(\mathbf{r}_j'^Q - \mathbf{r}_j'^C) \tag{3.4.25}$$

Then $\mathbf{u}_j = \mathbf{A}_j\mathbf{u}_j'$.

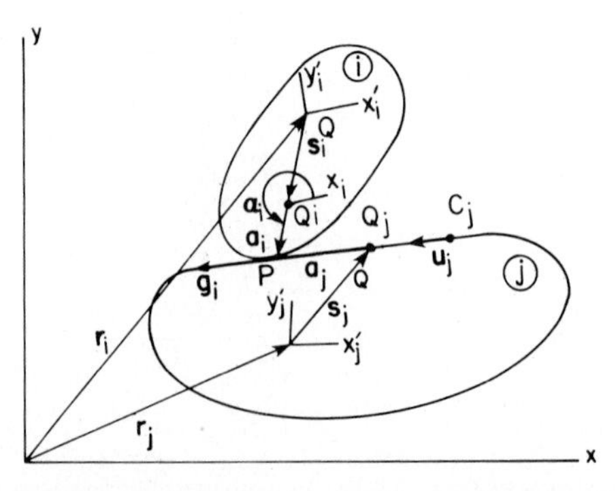

Figure 3.4.10 Cam–flat-faced follower pair.

Since $\mathbf{a}_j$ must be parallel to $\mathbf{u}_j$,

$$\mathbf{a}_j = \rho_j \mathbf{u}_j = \rho_j \mathbf{A}_j \mathbf{u}'_j \tag{3.4.26}$$

where ρ_j is the length of $\mathbf{a}_j$. Taking the scalar product of both sides of Eq. 3.4.24 with

$$\mathbf{u}_j^{\perp} = \mathbf{R}\mathbf{u}_j = \mathbf{R}\mathbf{A}_j\mathbf{u}'_j = \mathbf{B}_j\mathbf{u}'_j$$

yields

$$(\mathbf{r}_i^P - \mathbf{r}_j^Q)^T \mathbf{u}_j^{\perp} = (\mathbf{r}_i^P - \mathbf{r}_j^Q)^T \mathbf{B}_j \mathbf{u}'_j \tag{3.4.27}$$

where $\mathbf{u}'_j$ is constant.

A second constraint equation follows from the condition that the tangent $\mathbf{g}_i$ at point P to body i must be parallel to $\mathbf{u}_j$. Hence

$$\mathbf{g}_i^{\perp T}\mathbf{u}_j = (\mathbf{R}\mathbf{g}_i)^T\mathbf{u}_j = \mathbf{g}_i^T\mathbf{R}^T\mathbf{u}_j = \mathbf{g}_i'^T\mathbf{B}_{ij}\mathbf{u}'_j = 0 \tag{3.4.28}$$

where $\mathbf{g}'_i$ is given by Eq. 3.4.17.

To show that Eqs. 3.4.27 and 3.4.28 imply the geometry of the joint, note first that Eq. 3.4.27 requires that $\mathbf{r}_i^P - \mathbf{r}_j^Q$ be parallel to $\mathbf{u}_j$. Since they have point Q in common, they are collinear. Thus, point P on the cam surface of body i lies on the flat face on body j. Since Eq. 3.4.28 requires that the cam surface of body i be tangent to the flat face on body j at point P, these equations imply the geometry of the joint. For convenience, they are summarized as

$$\mathbf{\Phi}^{cff(i,j)} \equiv \begin{bmatrix} (\mathbf{r}_i^P - \mathbf{r}_j^Q)^T\mathbf{B}_j\mathbf{u}'_j \\ \mathbf{g}_i'^T\mathbf{B}_{ij}\mathbf{u}'_j \end{bmatrix} = \mathbf{0} \tag{3.4.29}$$

This formulation eliminates one relative degree of freedom between the two bodies, since α_i appears in $\mathbf{r}_i^P$, through Eqs. 3.4.15, and is retained as an additonal generalized coordinate. For a system with nb rigid bodies and one cam–flat-faced follower pair, the vector of generalized coordinates will have $nc = 3nb + 1$ components:

$$\mathbf{q} = [x_1, y_1, \phi_1, \ldots, x_{nb}, y_{nb}, \phi_{nb}, \alpha_i]^T$$

The Jacobians of the constraint of Eq. 3.4.29, with respect to $\mathbf{q}_i$, α_i, and $\mathbf{q}_j$, are calculated using the vector and matrix relations of Chapter 2 as

$$\begin{aligned} \mathbf{\Phi}_{\mathbf{q}_i}^{cff(i,j)} &= \begin{bmatrix} \mathbf{u}_j'^T\mathbf{B}_j^T & (\mathbf{s}_i'^P + \rho_i\mathbf{u}'_i)^T\mathbf{A}_{ij}\mathbf{u}'_j \\ \mathbf{0} & \mathbf{g}_i'^T\mathbf{A}_{ij}\mathbf{u}'_j \end{bmatrix} \\ \mathbf{\Phi}_{\alpha_i}^{cff(i,j)} &= \begin{bmatrix} \mathbf{g}_i'^T\mathbf{B}_{ij}\mathbf{u}'_j \\ (\mathbf{g}'_i)_{\alpha_i}^T\mathbf{B}_{ij}\mathbf{u}'_j \end{bmatrix} \\ \mathbf{\Phi}_{\mathbf{q}_j}^{cff(i,j)} &= \begin{bmatrix} -\mathbf{u}_j'^T\mathbf{B}_j^T & -(\mathbf{r}_i^P - \mathbf{r}_j)^T\mathbf{A}_j\mathbf{u}'_j \\ \mathbf{0} & -\mathbf{g}_i'^T\mathbf{A}_{ij}\mathbf{u}'_j \end{bmatrix} \end{aligned} \tag{3.4.30}$$

Since there is no explicit time dependence in Eq. 3.4.29, from Eq. 3.1.9,

$$\mathbf{v}^{cff(i,j)} = \mathbf{0}$$

Carrying out the expansion for $\boldsymbol{\gamma}$ in Eq. 3.1.10, using the Jacobians of Eq. 3.4.30,

$$\boldsymbol{\gamma}^{cff(i,j)} = \begin{bmatrix} 2(\dot{\mathbf{r}}_i^P - \dot{\mathbf{r}}_j)^T \mathbf{A}_j \mathbf{u}_j' \dot{\phi}_j + (\mathbf{s}_i'^P + \rho_i \mathbf{u}_i')^T \mathbf{B}_{ij} \mathbf{u}_j' \dot{\phi}_i^2 + (\mathbf{r}_i^P - \mathbf{r}_j)^T \mathbf{B}_j \mathbf{u}_j' \dot{\phi}_j^2 \\ - (\mathbf{g}_i')_{\alpha_i}^T \mathbf{B}_{ij} \mathbf{u}_j' \dot{\alpha}_i^2 - 2\mathbf{g}_i'^T \mathbf{A}_{ij} \mathbf{u}_j' \dot{\alpha}_i \dot{\phi}_i \\ \mathbf{g}_i'^T \mathbf{B}_{ij} \mathbf{u}_j' (\dot{\phi}_j - \dot{\phi}_i)^2 - (\mathbf{g}_i')_{\alpha_i \alpha_i}^T \mathbf{B}_{ij} \mathbf{u}_j' \dot{\alpha}_i^2 \\ + 2(\mathbf{g}_i')_{\alpha_i}^T \mathbf{A}_{ij} \mathbf{u}_j' (\dot{\phi}_j - \dot{\phi}_i) \dot{\alpha}_i \end{bmatrix}$$

Example 3.4.4: The roller follower of the cam–follower joint in Example 3.4.3 is replaced by the cam–flat-faced follower joint shown in Fig. 3.4.11. The outline of the cam is the same as in Example 3.4.2, that is, Eq. 3.4.21.

The vector $\mathbf{u}_2$ that is parallel to the flat face is simply

$$\mathbf{u}_2 = [-1, 0]^T$$

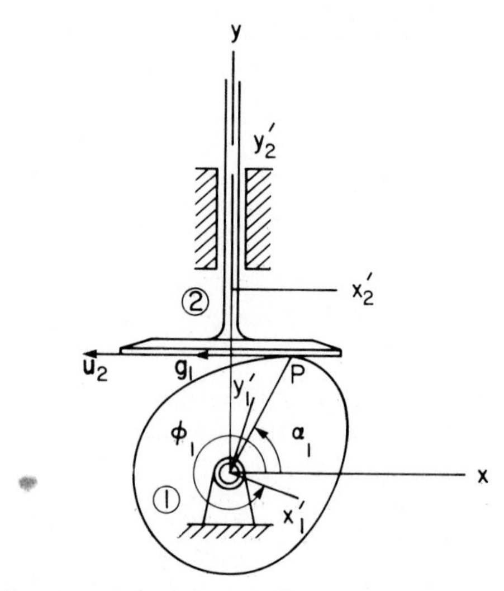

Figure 3.4.11 Cam–flat-faced follower in an internal combustion engine.

When $0 \leq \alpha_1 \leq (2\pi/3)$, as shown in Fig. 3.4.11, the second constraint equation of Eq. 3.4.29 is, from Eq. 3.4.23,

$$-\{\tfrac{3}{4}\sin 3\alpha_1 \cos\alpha_1 - (\tfrac{1}{4}\cos 3\alpha_1 - \tfrac{5}{4})\sin\alpha_1\}\cos\phi_1$$
$$+ \{\tfrac{3}{4}\sin 3\alpha_1 \sin\alpha_1 - (\tfrac{1}{4}\cos 3\alpha_1 - \tfrac{5}{4})\cos\alpha_1\}\sin\phi_1 = 0$$

3.4.3 Point–Follower

A *point–follower* joint between bodies i and j is shown schematically in Fig. 3.4.12. Pin P is attached to body i and can slide and rotate in a slot on body j.

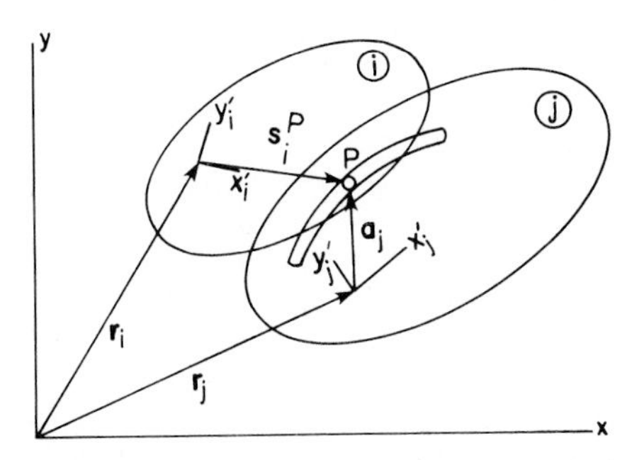

Figure 3.4.12 Point–follower pair.

The coordinates of any point on the slot, relative to the body-fixed reference frame on body j, can be described by $\mathbf{a}_j = \mathbf{A}_j \rho_j(\alpha_j)\mathbf{u}_j'(\alpha_j)$, as in Eq. 3.4.16.

Constraint equations for this joint are similar to the constraint equations of a revolute joint. Note that the position of point P on body j is not constant, but varies as a function of the variable α_j. Hence the constraint equations are

$$\mathbf{\Phi}^{pf(i,j)} \equiv \mathbf{r}_i + \mathbf{A}_i \mathbf{s}_i'^P - \mathbf{A}_j \rho_j \mathbf{u}_j' - \mathbf{r}_j = \mathbf{0} \tag{3.4.31}$$

In this case, α_j is added to the vector of generalized coordinates, and the constraint equations of Eq. 3.4.31 are employed to describe the point–follower joint. Thus, a point–follower joint eliminates only one degree of freedom.

The Jacobians of this constraint with respect to $\mathbf{q}_i$, $\mathbf{q}_j$, and α_j are

$$\begin{aligned} \mathbf{\Phi}_{\mathbf{q}_i}^{pf(i,j)} &= [\mathbf{I}, \mathbf{B}_i \mathbf{s}_i'^P] \\ \mathbf{\Phi}_{\mathbf{q}_j}^{pf(i,j)} &= [-\mathbf{I}, -\mathbf{B}_j \rho_j \mathbf{u}_j'] \\ \mathbf{\Phi}_{\alpha_j}^{pf(i,j)} &= -\mathbf{A}_j \mathbf{g}_j' \end{aligned} \tag{3.4.32}$$

Since the constraint of Eq. 3.4.31 does not depend explicitly on time, from Eq. 3.1.9,

$$\mathbf{v}^{pf(i,j)} = \mathbf{0}$$

Using the Jacobians of Eq. 3.4.32 in Eq. 3.1.10,

$$\boldsymbol{\gamma}^{pf(i,j)} = \mathbf{A}_i \mathbf{s}_i'^P \dot{\phi}_i^2 - \mathbf{A}_j \rho_j \mathbf{u}_j' \dot{\phi}_j^2 + \mathbf{A}_j (\mathbf{g}_j')_{\alpha_j} \dot{\alpha}_j^2 + 2\mathbf{B}_j \mathbf{g}_j' \dot{\phi}_j \dot{\alpha}_j$$

Example 3.4.5: The cam-follower joint in Example 3.4.3 may be modeled as a point–follower joint, as shown in Fig. 3.4.13. The equation of the slot is the same as given in Example 3.4.3.

For a minimal set of generalized coordinates in this example, $\mathbf{q} = [\phi_1, y_2, \alpha_1]^T$, and the constraint equation of Eq. 3.4.31 is simply

$$y_2 - \ell - \rho_1(\alpha_1) = y_2 - \ell + \tfrac{1}{4}\cos 3\alpha_1 - \tfrac{5}{4} = 0$$

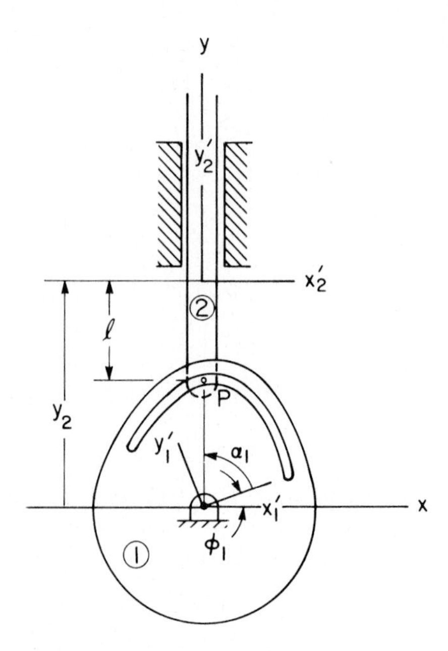

Figure 3.4.13 Point–follower model of valve lifter.

Substituting $\phi_1 + \alpha_1 = \pi/2$ into this equation,

$$y_2 - \ell + \tfrac{1}{4}\cos\left(\frac{3\pi}{2} - 3\phi_1\right) - \tfrac{5}{4} = y_2 - \ell - \tfrac{1}{4}\sin 3\phi_1 - \tfrac{5}{4} = 0$$

Differentiating with respect to time,

$$\dot{y}_2 = \tfrac{3}{4}\cos 3\phi_1$$

3.5 DRIVING CONSTRAINTS

The kinematic constraints presented in Sections 3.2 through 3.4 represent physical connections between bodies; hence they impose limitations on the relative motion of connected bodies. These kinematic constraints are functions of the system generalized coordinates, but they do not depend explicitly on time. Such constraints describe the physical structure of a machine and normally provide one or more degrees of freedom that allow movement of the system.

In addition to kinematic structure, the motion of many mechanical systems is described by actuator input that specifies the time history of some position coordinates or the relative position of pairs of bodies. Robot and numerically controlled machine tool applications are typical of this mode of operation. To uniquely determine the time history of motion of a mechanism, a number of inputs must be specified, equal in number to the number of degrees of freedom of the system. As will be shown in Section 3.6, it is important that these additional input conditions be independent of the kinematic constraint equations. While it is not sufficient to simply count the number of variables and constraint equations to determine the number of degrees of freedom, this is a useful initial computation to guide the engineer in specifying an appropriate number of time-dependent driving conditions.

A family of standard drivers, called *driving constraints,* is presented in this section for use in kinematic analysis.

3.5.1 Absolute Drivers

Analogous to absolute constraints that may be imposed on the coordinates of a point on body i or on rotation of body i, time-dependent absolute coordinate constraints, called *drivers,* may be imposed. Allowing the parameters C_1, C_2, and C_3 in the constraints of Eqs. 3.2.3 through 3.2.6 to be time dependent, the following *absolute coordinate drivers* are obtained:

$$\Phi^{axd(i)} = x_i^P - C_1(t) = 0 \qquad \textbf{(3.5.1)}$$

$$\Phi^{ayd(i)} = y_i^P - C_2(t) = 0 \qquad \textbf{(3.5.2)}$$

$$\Phi^{a\phi d(i)} = \phi_i - C_3(t) = 0 \qquad \textbf{(3.5.3)}$$

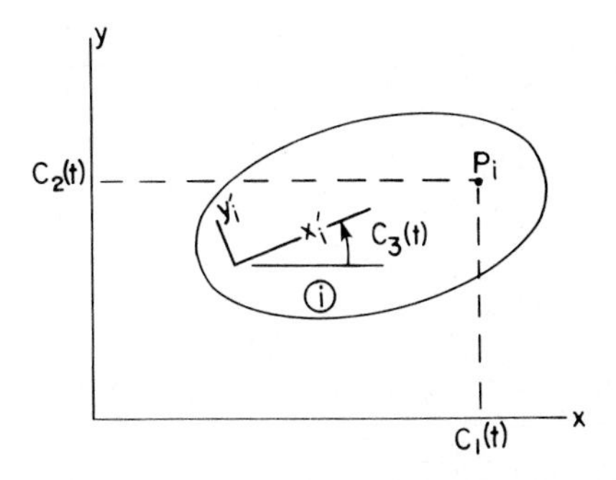

Figure 3.5.1 Absolute coordinate drivers.

The geometry of these drivers is shown in Fig. 3.5.1. The engineer must describe the time-dependent functions $C_k(t)$ and the location of point P on body i in order to complete definition of these driving constraints. Note that the Jacobians of these driving constraints are the same as in Eqs. 3.2.5 and 3.2.7.

Example 3.5.1: The two-body slider–crank shown in Fig. 3.5.2 models a rocker mechanism in which motion of the piston is controlled by hydraulic flow,

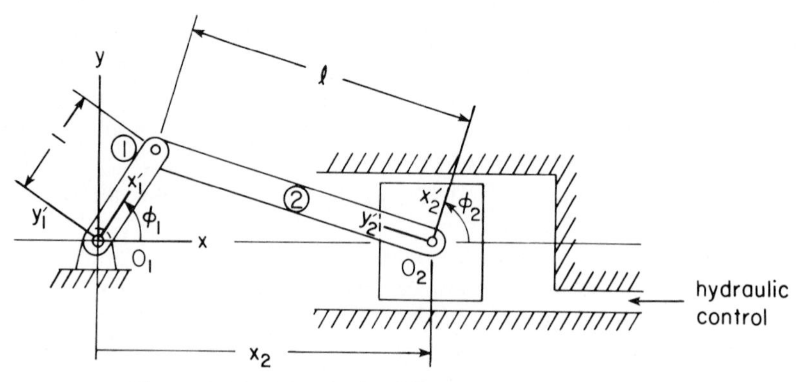

Figure 3.5.2 Two-body slider–crank model with an absolute driver.

specified as

$$x_2 = \sqrt{\ell^2 - 1} + \tfrac{1}{2}\sin\omega t = C(t) \tag{3.5.4}$$

where $\ell > 1$ and ω are constants.

For analytical simplicity, let a reduced set of generalized coordinates be $\mathbf{q} = [x_2, \phi_1, \phi_2]^T$. Equation 3.5.4 represents an absolute driver of the form of Eq. 3.5.1. The composite constraint equation is, from Example 3.1.2 and Eq. 3.5.4,

$$\mathbf{\Phi}(\mathbf{q}) \equiv \begin{bmatrix} -x_2 + \cos\phi_1 + \ell\sin\phi_2 \\ \sin\phi_1 - \ell\cos\phi_2 \\ x_2 - \sqrt{\ell^2 - 1} - \tfrac{1}{2}\sin\omega t \end{bmatrix} = \mathbf{0}$$

Taking the time derivative of both sides yields the velocity equation

$$\begin{bmatrix} -1 & -\sin\phi_1 & \ell\cos\phi_2 \\ 0 & \cos\phi_1 & \ell\sin\phi_2 \\ 1 & 0 & 0 \end{bmatrix} \begin{bmatrix} \dot{x}_2 \\ \dot{\phi}_1 \\ \dot{\phi}_2 \end{bmatrix} = \begin{bmatrix} 0 \\ 0 \\ \tfrac{1}{2}\omega\cos\omega t \end{bmatrix}$$

The solution for $\dot{\phi}_1$ is

$$\dot{\phi}_1 = -\frac{\omega\sin\phi_2\cos\omega t}{2\cos(\phi_1 - \phi_2)}$$

Note that if the numerator is not zero, $\dot{\phi}_1$ approaches infinity as $\cos(\phi_1 - \phi_2)$ approaches zero. This singular configuration occurs when $\phi_1 = \phi_2 + (\tfrac{1}{2} + k)\pi$, where k is an integer. Singular behavior of mechanisms is discussed in Section 3.7.

Example 3.5.2: The four-bar mechanism of Fig. 3.3.2 has one degree of freedom, as shown in Examples 3.3.2 and 3.3.3. Hence a rotational driver $\phi_1 = \omega t$ determines the complete motion. Note that when ground is not modeled as a body [Example 3.3.2(1)], the driver is an absolute rotational driver. A reduced set of generalized coordinates is chosen as $\mathbf{q} = [\phi_1, x_2, y_2, \phi_2, \phi_3]^T$. Dimensions of the mechanism are shown in Fig. 3.5.3.

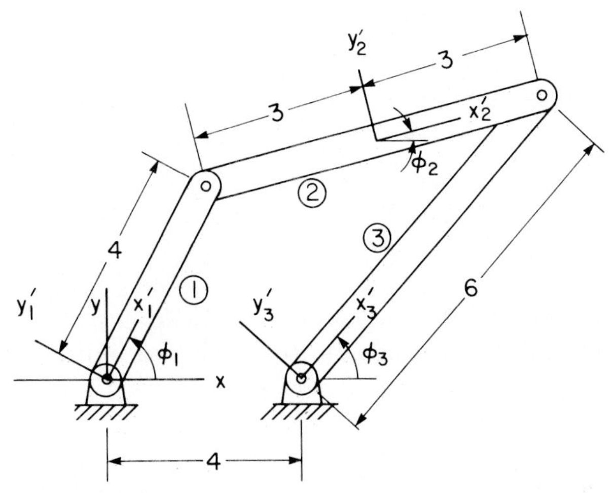

Figure 3.5.3 Four-bar mechanism with an absolute rotational driver.

Composite constraint equations, derived directly in terms of the reduced generalized coordinates, are

$$\mathbf{\Phi}(\mathbf{q}) \equiv \begin{bmatrix} x_2 - 4\cos\phi_1 - 3\cos\phi_2 \\ y_2 - 4\sin\phi_1 - 3\sin\phi_2 \\ 4 - 4\cos\phi_1 - 6\cos\phi_2 + 6\cos\phi_3 \\ 4\sin\phi_1 + 6\sin\phi_2 - 6\sin\phi_3 \\ \phi_1 - \omega t \end{bmatrix} = \mathbf{0}$$

Differentiating with respect to time yields the velocity equation

$$\begin{bmatrix} 4\sin\phi_1 & 1 & 0 & 3\sin\phi_2 & 0 \\ -4\cos\phi_1 & 0 & 1 & -3\cos\phi_2 & 0 \\ 4\sin\phi_1 & 0 & 0 & 6\sin\phi_2 & -6\sin\phi_3 \\ 4\cos\phi_1 & 0 & 0 & 6\cos\phi_2 & -6\cos\phi_3 \\ 1 & 0 & 0 & 0 & 0 \end{bmatrix} \begin{bmatrix} \dot{\phi}_1 \\ \dot{x}_2 \\ \dot{y}_2 \\ \dot{\phi}_2 \\ \dot{\phi}_3 \end{bmatrix} = \begin{bmatrix} 0 \\ 0 \\ 0 \\ 0 \\ \omega \end{bmatrix}$$

whose solution is

$$\dot{\phi}_1 = \omega$$

$$\dot{x}_2 = -4\omega\sin\phi_1 - 2\omega\sin\phi_2 \frac{\sin(\phi_3 - \phi_1)}{\sin(\phi_2 - \phi_3)}$$

$$\dot{y}_2 = 4\omega\cos\phi_1 + 2\omega\cos\phi_2 \frac{\sin(\phi_3 - \phi_1)}{\sin(\phi_2 - \phi_3)}$$

$$\dot{\phi}_2 = \frac{2\omega\sin(\phi_3 - \phi_1)}{3\sin(\phi_2 - \phi_3)}$$

$$\dot{\phi}_3 = -\frac{2\omega\sin(\phi_1 - \phi_2)}{\sin(\phi_2 - \phi_3)}$$

3.5.2 Relative Drivers

Just as in Section 3.3.1, constraints between coordinates of points on pairs of bodies, or between the orientation of bodies in Eqs. 3.3.1, 3.3.3, and 3.3.5 can be specified as functions of time, in the form of *relative coordinate drivers,*

$$\Phi^{rxd(i,j)} = x_j^P - x_i^P - C_1(t) = 0 \tag{3.5.5}$$

$$\Phi^{ryd(i,j)} = y_j^P - y_i^P - C_2(t) = 0 \tag{3.5.6}$$

$$\Phi^{r\phi d(i,j)} = \phi_j - \phi_i - C_3(t) = 0 \tag{3.5.7}$$

The geometry of these driving constraints is shown in Fig. 3.5.4. The Jacobians of these driving constraints are the same as in Eqs. 3.3.2, 3.3.4, and 3.3.6.

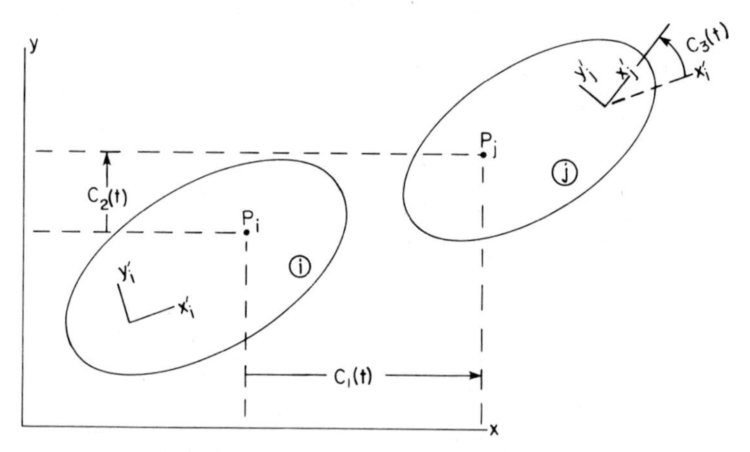

Figure 3.5.4 Relative coordinate drivers.

Analogous to the distance constraint of Eq. 3.3.7 in Section 3.3.1, the length $C_4(t) > 0$ of the actuator can be specified as a function of time, to obtain the *relative distance driver*

$$\Phi^{rdd(i,j)} = (x_j^P - x_i^P)^2 + (y_j^P - y_i^P)^2 - (C_4(t))^2 = 0 \tag{3.5.8}$$

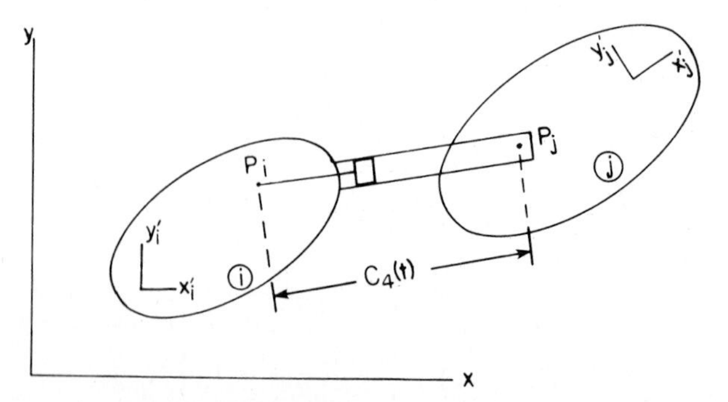

Figure 3.5.5 Relative distance driver.

between a pair of points on two bodies, as shown in Fig. 3.5.5. As in the case of the distance constraint, it is important that $C_4(t) > 0$, since, if $C_4(t) = 0$, Eq. 3.5.8 is equivalent to a pair of scalar constraint equations that describes a revolute joint. The Jacobian of this driver is given in Eq. 3.3.8.

As suggested by Fig. 3.5.5, the distance driver may be defined physically by the action of a hydraulic actuator, in which the volume of fluid inserted in a chamber is controlled to specify the time history of length of the strut.

Example 3.5.3: An excavator boom assembly with two hydraulic actuators (for simplicity the shovel actuator is not included) is shown in Fig. 3.5.6. Let a reduced set of generalized coordinates be $\mathbf{q} = [\phi_1, x_2, y_2, \phi_2]^T$ and let the distance driver inputs be

$$C_{41}(t) = \tfrac{1}{5}t + 1.8$$
$$C_{12}(t) = \tfrac{1}{10}t + 1.9$$

The relative distance driver constraint equations of Eq. 3.5.8, with the revolute constraint equations, yield the composite constraint equations

$$\mathbf{\Phi}(\mathbf{q}) \equiv \begin{bmatrix} x_2 - 2\sqrt{3}\cos\phi_1 \\ y_2 - 2\sqrt{3}\sin\phi_1 \\ (\sqrt{3}\cos\phi_1 - 0.5)^2 + (\sqrt{3}\sin\phi_1 + 0.5)^2 - (C_{41}(t))^2 \\ u^2 + v^2 - (C_{12}(t))^2 \end{bmatrix} = \mathbf{0} \qquad (3.5.9)$$

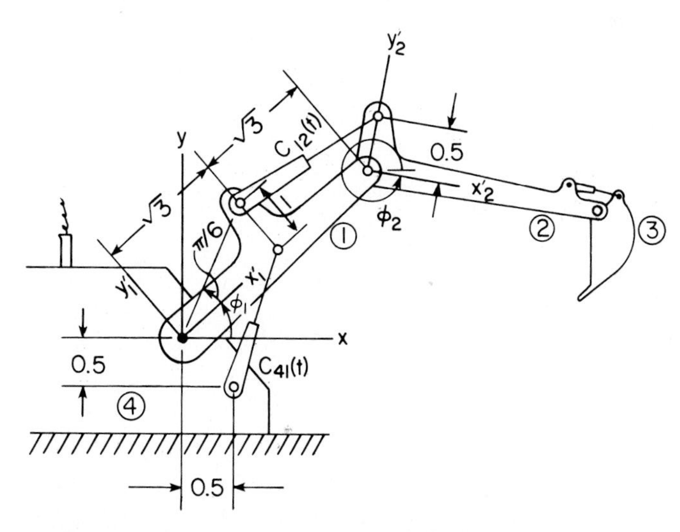

Figure 3.5.6 Excavator boom assembly with two distance drivers.

where

$$u = \left\{2\cos\left(\phi_1 + \frac{\pi}{6}\right) - (x_2 - 0.5\sin\phi_2)\right\}$$

$$v = \{2\sin\left(\phi_1 + \frac{\pi}{6}\right) - (y_2 + 0.5\cos\phi_2)\right\}$$

Differentiating with respect to time, the velocity equation is

$$\begin{bmatrix} 2\sqrt{3}\sin\phi_1 & 1 & 0 & 0 \\ -2\sqrt{3}\cos\phi_1 & 0 & 1 & 0 \\ \begin{matrix} -2\sqrt{3}(\sqrt{3}\cos\phi_1 - 0.5)\sin\phi_1 \\ +2\sqrt{3}(\sqrt{3}\sin\phi_1 + 0.5)\cos\phi_1 \end{matrix} & 0 & 0 & 0 \\ \begin{matrix} -4u\sin\left(\phi_1 + \frac{\pi}{6}\right) \\ +4v\cos\left(\phi_1 + \frac{\pi}{6}\right) \end{matrix} & -2u & -2v & \begin{matrix} u\cos\phi_2 \\ +v\sin\phi_2 \end{matrix} \end{bmatrix} \begin{bmatrix} \dot{\phi}_1 \\ \dot{x}_2 \\ \dot{y}_2 \\ \dot{\phi}_2 \end{bmatrix}$$

$$= \begin{bmatrix} 0 \\ 0 \\ \dfrac{2(\frac{1}{5}t + 1.8)}{5} \\ \dfrac{(\frac{1}{10}t + 1.9)}{5} \end{bmatrix} \qquad \textbf{(3.5.10)}$$

The solution for $\dot{\phi}_1$ and $\dot{\phi}_2$ is

$$\dot{\phi}_1 = \frac{t + 9}{-25\sqrt{3}\{(\sqrt{3}\cos\phi_1 - 0.5)\sin\phi_1 - (\sqrt{3}\sin\phi_1 + 0.5)\cos\phi_1\}}$$

$$\dot{\phi}_2 = \frac{1}{B}\{C - A + 4\sqrt{3}\,u\sin\phi_1 - 4\sqrt{3}\,v\cos\phi_1)\,\dot{\phi}_1\}$$

where

$$A = -4\left\{2\cos\left(\phi_1 + \frac{\pi}{6}\right) - x_2 + 0.5\sin\phi_2\right\}\sin\left(\phi_1 + \frac{\pi}{6}\right)$$

$$+ 4\left\{2\sin\left(\phi_1 + \frac{\pi}{6}\right) - y_2 - 0.5\cos\phi_2\right\}\cos\left(\phi_1 + \frac{\pi}{6}\right)$$

$$B = \left\{2\cos\left(\phi_1 + \frac{\pi}{6}\right) - x_2 + 0.5\sin\phi_2\right\}\cos\phi_2$$

$$+ \left\{2\sin\left(\phi_1 + \frac{\pi}{6}\right) - y_2 - 0.5\cos\phi_2\right\}\sin\phi_2$$

$$C = \frac{t + 19}{50}$$

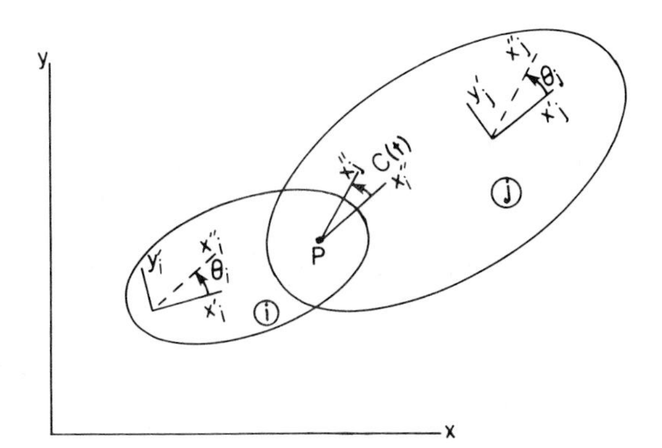

Figure 3.5.7 Revolute–rotational driver.

A common form of driver that is encountered in mechanical systems is an electrical or hydraulic actuator that controls the relative angular orientation of a pair of bodies at a revolute joint. The geometry of a *revolute–rotational driver* is shown in Fig. 3.5.7, where the attachment angles θ_i and θ_j of the actuator on bodies i and j, respectively, are defined by the physical mounting of the actuator on the two bodies. The analytical form of this driving constraint is simply

$$\Phi^{rrd(i,j)} = (\phi_j + \theta_j) - (\phi_i + \theta_i) - C(t) = 0 \tag{3.5.11}$$

where $C(t)$ is the driving angle of the body-fixed x_j'' axis on body j, relative to the body-fixed x_i'' axis on body i, with counterclockwise positive. The Jacobian of this driver is given in Eq. 3.3.6.

Example 3.5.4: One of the distance drivers in Example 3.5.3 is replaced by a revolute–rotational driver

$$C(t) = \tfrac{1}{10}t$$

as shown in Fig. 3.5.8.

Let the generalized coordinate vector again be $\mathbf{q} = [\phi_1, x_2, y_2, \phi_2]^T$. Then the composite constraint equation is Eq. 3.5.9, with the fourth equation replaced by

$$\phi_2 - \phi_1 - \tfrac{1}{10}t = 0$$

Similarly, the fourth equation of Eq. 3.5.10 is replaced by

$$[-1, 0, 0, 1][\dot{\phi}_1, \dot{x}_2, \dot{y}_2, \dot{\phi}_2]^T = \tfrac{1}{10}t$$

The solution for $\dot{\phi}_2$ is

$$\dot{\phi}_2 = \frac{t+9}{-25\sqrt{3}\,\{(\sqrt{3}\cos\phi_1 - 0.5)\sin\phi_1 - (\sqrt{3}\sin\phi_1 + 0.5)\cos\phi_1\}} + \frac{1}{10}$$

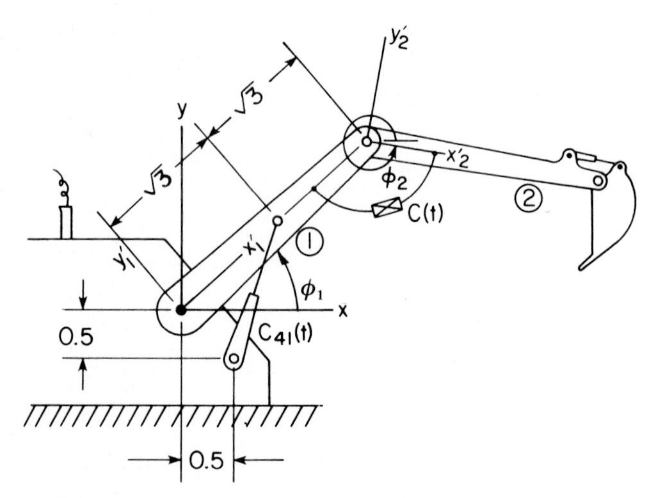

Figure 3.5.8 Excavator boom assembly with a distance driver and a revolute–rotational driver.

A common form of relative position driver that is encountered in robots and numerically controlled machine tools, as well as in other machine applications, is the relative translation of a pair of bodies along a translational joint, as shown in Fig. 3.5.9. Electrical and hydraulic actuators may act directly or through a gearing system that controls translation of body j relative to body i.

The mathematical form of the *translational–distance driver* is, using the notation of Fig. 3.5.9,

$$\frac{\mathbf{v}_i^T \mathbf{d}_{ij}}{v_i} - C(t) = 0$$

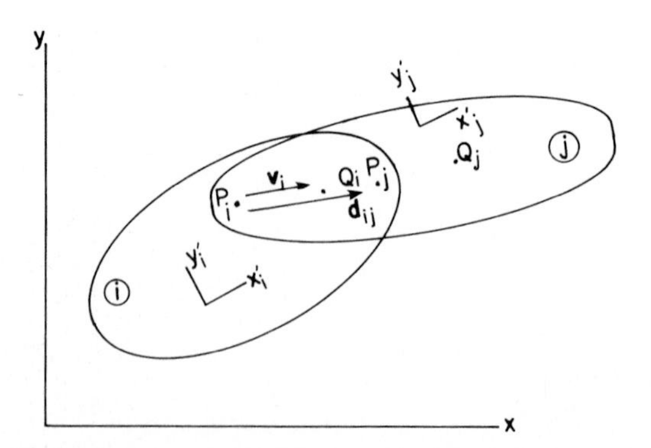

Figure 3.5.9 Translational–distance driver.

where $C(t)$, which can be zero, is the time history of the directed distance from P_i and P_j, and v_i is the constant distance on body i between points P_i and Q_i. This equation can be expanded as

$$\Phi^{tdd(i,j)} = \mathbf{v}_i'^T\mathbf{A}_i^T(\mathbf{r}_j - \mathbf{r}_i) + \mathbf{v}_i'^T\mathbf{A}_{ij}\mathbf{s}_j'^P - \mathbf{v}_i'^T\mathbf{s}_i'^P - v_iC(t) = 0 \qquad \textbf{(3.5.12)}$$

The Jacobians of this driver are

$$\begin{aligned} \Phi_{\mathbf{q}_i}^{tdd(i,j)} &= [-\mathbf{v}_i'^T\mathbf{A}_i^T,\ \mathbf{v}_i'^T\mathbf{B}_i^T(\mathbf{r}_j - \mathbf{r}_i)] \\ \Phi_{\mathbf{q}_j}^{tdd(i,j)} &= [\mathbf{v}_i'^T\mathbf{A}_i^T,\ \mathbf{0}] \end{aligned} \qquad \textbf{(3.5.13)}$$

where the fact that $\mathbf{A}_{ij}$ is constant, due to the translational joint between bodies i and j, is used.

Example 3.5.5: The boom of a crane is shown in Fig. 3.5.10. Body 1 is hinged to ground at point O and is driven by the rotational driver

$$\phi_1 = 0.025t \qquad \textbf{(3.5.14)}$$

The relative distance between bodies 1 and 2 is driven by the translational–distance driver

$$\frac{\mathbf{v}_1^T\mathbf{d}_{12}}{v_1} = C(t) = 0.1t \qquad \textbf{(3.5.15)}$$

where $\mathbf{v}_1$ and $\mathbf{d}_{12}$ are shown in Fig. 3.5.10. Let $\mathbf{q} = [\phi_1, x_2, y_2, \phi_2]^T$ be a reduced set of generalized coordinates. With Eq. 3.5.15, Eq. 3.5.12 is

$$[v_1 \cos\phi_1,\ v_1 \sin\phi_1]\begin{bmatrix} x_2 \\ y_2 \end{bmatrix} - v_1(0.1t) = 0 \qquad \textbf{(3.5.16)}$$

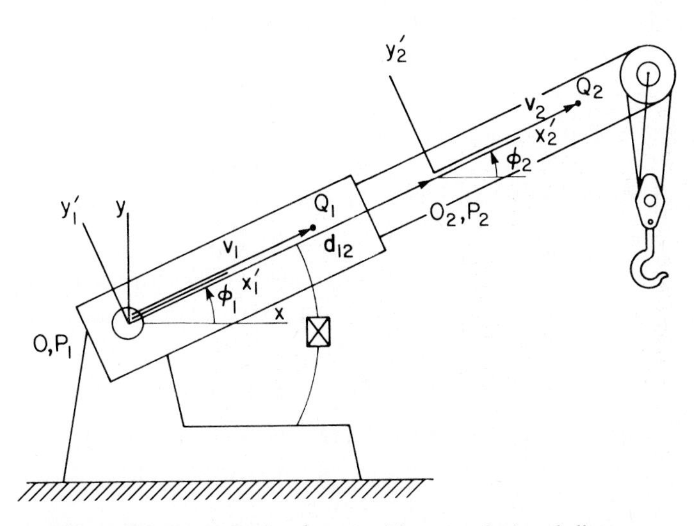

Figure 3.5.10 Wrecker boom with a translational-distance driver.

The translational joint constraint of Eq. 3.3.13 is

$$\begin{bmatrix} \left(\begin{bmatrix} 0 & -1 \\ 1 & 0 \end{bmatrix}\begin{bmatrix} v_1 \cos\phi_1 \\ v_1 \sin\phi_1 \end{bmatrix}\right)^T \begin{bmatrix} x_2 \\ y_2 \end{bmatrix} \\ \left(\begin{bmatrix} 0 & -1 \\ 1 & 0 \end{bmatrix}\begin{bmatrix} v_1 \cos\phi_1 \\ v_1 \sin\phi_1 \end{bmatrix}\right)^T \begin{bmatrix} v_2 \cos\phi_2 \\ v_2 \sin\phi_2 \end{bmatrix} \end{bmatrix} = \mathbf{0}$$

or, in simplified form,

$$\begin{bmatrix} -v_1(\sin\phi_1)x_2 + v_1(\cos\phi_1)y_2 \\ -v_1 v_2 \sin\phi_1 \cos\phi_2 + v_1 v_2 \cos\phi_1 \sin\phi_2 \end{bmatrix} = \mathbf{0} \tag{3.5.17}$$

From Eqs. 3.5.14, 3.5.16, and 3.5.17, the composite constraint equation is

$$\mathbf{\Phi}(\mathbf{q}) \equiv \begin{bmatrix} -x_2 \sin\phi_1 + y_2 \cos\phi_1 \\ \sin(\phi_1 - \phi_2) \\ \phi_1 - 0.025t \\ x_2 \cos\phi_1 + y_2 \sin\phi_1 - 0.1t \end{bmatrix} = \mathbf{0}$$

Differentiating with respect to time, the velocity equation is

$$\begin{bmatrix} -x_2 \cos\phi_1 - y_2 \sin\phi_1 & -\sin\phi_1 & \cos\phi_1 & 0 \\ \cos(\phi_1 - \phi_2) & 0 & 0 & -\cos(\phi_1 - \phi_2) \\ 1 & 0 & 0 & 0 \\ -x_2 \sin\phi_1 + y_2 \cos\phi_1 & \cos\phi_1 & \sin\phi_1 & 0 \end{bmatrix} \begin{bmatrix} \dot{\phi}_1 \\ \dot{x}_2 \\ \dot{y}_2 \\ \dot{\phi}_2 \end{bmatrix} = \begin{bmatrix} 0 \\ 0 \\ 0.025 \\ 0.1 \end{bmatrix}$$

The solution is

$$\begin{aligned} \dot{\phi}_1 &= 0.025 \\ \dot{x}_2 &= -y_2\dot{\phi}_1 = -0.025y_2 + 0.1\cos\phi_1 \\ \dot{y}_2 &= x_2\dot{\phi}_1 = 0.025x_2 + 0.1\sin\phi_1 \\ \dot{\phi}_2 &= \dot{\phi}_1 = 0.025 \end{aligned}$$

3.5.3 Right Sides of Velocity and Acceleration Equations

With the exception of the translational–distance driver, the equations for driving constraints derived in this section are of the form of constraint equations of Sections 3.2 and 3.3, but with isolated time-dependent terms; that is, for a kinematic constraint

$$\mathbf{\Phi}(\mathbf{q}) = \mathbf{0} \tag{3.5.18}$$

the associated driver is of the form

$$\mathbf{\Phi}^d(\mathbf{q}, t) = \mathbf{\Phi}(\mathbf{q}) - \mathbf{C}(t) = \mathbf{0} \tag{3.5.19}$$

Thus, the right sides of velocity and acceleration equations for Eqs. 3.1.9 and

TABLE 3.5.1 Right Sides of Velocity and Acceleration Equations

Φ	Eq.	$\nu = -\Phi_t$	$\gamma = -(\Phi_{\mathbf{q}}\dot{\mathbf{q}})_{\mathbf{q}}\dot{\mathbf{q}} - \Phi_{tt}$
$\Phi^{axd(i)}$	3.5.1	$\dot{C}_1(t)$	$\gamma^{ax(i)} + \ddot{C}_1(t)$
$\Phi^{ayd(i)}$	3.5.2	$\dot{C}_2(t)$	$\gamma^{ay(i)} + \ddot{C}_2(t)$
$\Phi^{a\phi d(i)}$	3.5.3	$\dot{C}_3(t)$	$\ddot{C}_3(t)$
$\Phi^{rxd(i,j)}$	3.5.5	$\dot{C}_1(t)$	$\gamma^{rx(i,j)} + \ddot{C}_1(t)$
$\Phi^{ryd(i,j)}$	3.5.6	$\dot{C}_2(t)$	$\gamma^{ry(i,j)} + \ddot{C}_2(t)$
$\Phi^{r\phi d(i,j)}$	3.5.7	$\dot{C}_3(t)$	$\ddot{C}_3(t)$
$\Phi^{rdd(i,j)}$	3.5.8	$2C_4(t)\dot{C}_4(t)$	$\gamma^{rd(i,j)} + 2C_4(t)\ddot{C}_4(t)$ $+ 2(\dot{C}_4(t))^2$
$\Phi^{rrd(i,j)}$	3.5.11	$\dot{C}(t)$	$\ddot{C}(t)$
$\Phi^{tdd(i,j)}$	3.5.12	$v_i\dot{C}(t)$	$\mathbf{v}_i'^T[\mathbf{A}_i^T(\mathbf{r}_j - \mathbf{r}_i)\dot{\phi}_i^2$ $-2\mathbf{B}_i(\dot{\mathbf{r}}_j - \dot{\mathbf{r}}_i)\dot{\phi}_i] + v_i\ddot{C}(t)$

3.1.10, respectively, are

$$\boldsymbol{\nu}^d = \dot{\mathbf{C}}(t) \tag{3.5.20}$$

$$\boldsymbol{\gamma}^d = \boldsymbol{\gamma} + \ddot{\mathbf{C}}(t) \tag{3.5.21}$$

where $\boldsymbol{\gamma}$ is the right side of the acceleration equation for the kinematic constraint of Eq. 3.5.18.

Using the form of $\boldsymbol{\nu}^d$ and $\boldsymbol{\gamma}^d$ of Eqs. 3.5.20 and 3.5.21, with the right sides of velocity and acceleration equations derived for the kinematic constraints of Sections 3.2 and 3.3, the right sides of driving constraint velocity and acceleration equations can be obtained. The results of these calculations are summarized in Table 3.5.1.

3.6 POSITION, VELOCITY, AND ACCELERATION ANALYSIS

The kinematic constraints presented in Sections 3.2 through 3.4 and the driving constraints presented in Section 3.5 provide the analytical foundation for the analysis of position, velocity, and acceleration of planar mechanical systems. The forms of position, velocity, and acceleration equations are presented in this section, and the computational approach that is developed in detail in Chapter 4 is outlined.

3.6.1 Position Analysis

The kinematic constraints formulated in Sections 3.2 through 3.4 provide the geometric definition of the permissible motion of a system. For a particular system, they may be written in matrix form as

$$\boldsymbol{\Phi}^K(\mathbf{q}) \equiv [\Phi_1^K(\mathbf{q}), \ldots, \Phi_{nh}^K(\mathbf{q})]^T = \mathbf{0} \tag{3.6.1}$$

where $\mathbf{q}$ is the vector of nc generalized coordinates for the complete system and nh is the number of holonomic constraint equations.

In addition to the kinematic constraints, the driving constraints of Section 3.5 may be assembled in matrix form as

$$\mathbf{\Phi}^D(\mathbf{q}, t) \equiv [\Phi_1^D(\mathbf{q}, t), \ldots, \Phi_{\mathrm{DOF}}^D(\mathbf{q}, t)]^T = \mathbf{0} \tag{3.6.2}$$

These equations are time dependent and provide impetus for motion of the system. It is presumed that the number of driving constraints in Eq. 3.6.2 is equal to the number of degrees of freedom of the physical system and that $nh + \mathrm{DOF} = nc$. Therefore, Eqs. 3.6.1 and 3.6.2 comprise nc equations in nc generalized coordinates.

Assembling the kinematic and driving constraints of Eqs. 3.6.1 and 3.6.2 in matrix form, the system of constraint equations is

$$\mathbf{\Phi}(\mathbf{q}, t) = \begin{bmatrix} \mathbf{\Phi}^K(\mathbf{q}) \\ \mathbf{\Phi}^D(\mathbf{q}, t) \end{bmatrix}_{nc \times 1} = \mathbf{0} \tag{3.6.3}$$

The objective of *position analysis* is to solve this system of equations for the vector $\mathbf{q}$ of system generalized coordinates as a function of time. Since these equations are highly nonlinear, finding an analytical solution is generally impossible. Furthermore, it is easy to formulate a set of system kinematic constraints and drivers that cannot be physically satisfied, and hence for which no solution exists. Position analysis must therefore seriously consider questions of the existence of solutions of Eq. 3.6.3, as well as the numerical methods of solving the equations when a solution exists.

Example 3.6.1: Body 1, shown in Fig. 3.6.1, is to slide along the x axis of the x-y frame without rotation. A pair of constraint equations that accomplishes this objective is

$$\mathbf{\Phi}^1 = \begin{bmatrix} y_1 + \sin \phi_1 \\ y_1 - \sin \phi_1 \end{bmatrix} = \mathbf{0}$$

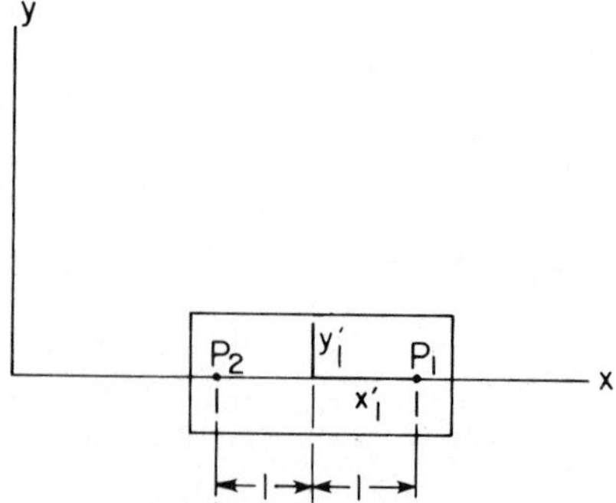

Figure 3.6.1 Block sliding along x axis.

Since this requires $\sin\phi_1 = 0$, $\phi_1 = 0$. The Jacobian of this constraint is

$$\mathbf{\Phi}_{\mathbf{q}}^1 = \begin{bmatrix} 0 & 1 & \cos\phi_1 \\ 0 & 1 & -\cos\phi_1 \end{bmatrix}$$

Since $\cos\phi_1 = 1$, this matrix has full row rank.

If the additional constraint $\phi_1 = 0$ is inadvertently imposed, the constraint equations become

$$\mathbf{\Phi}^2 \equiv \begin{bmatrix} y_1 + \sin\phi_1 \\ y_1 - \sin\phi_1 \\ \phi_1 \end{bmatrix} = \mathbf{0}$$

with Jacobian

$$\mathbf{\Phi}_{\mathbf{q}}^2 = \begin{bmatrix} 0 & 1 & \cos\phi_1 \\ 0 & 1 & -\cos\phi_1 \\ 0 & 0 & 1 \end{bmatrix}$$

Since $|\mathbf{\Phi}_{\mathbf{q}}^2| = 0$, $\mathbf{\Phi}_{\mathbf{q}}^2$ is now rank deficient and constraints $\mathbf{\Phi}^2 = \mathbf{0}$ are not independent. In this simple case, it is clear that the third constraint equation is a *redundant constraint* and is automatically satisfied if the first two constraint equations are satisfied. It thus represents a *consistent redundancy.*

Finally, let the constraint $\phi_1 = 0.1$ be added to the original pair of constraint equations, in which case the constraint equations are

$$\mathbf{\Phi}^3 = \begin{bmatrix} y_1 + \sin\phi_1 \\ y_1 - \sin\phi_1 \\ \phi_1 - 0.1 \end{bmatrix} = \mathbf{0}$$

Since $\mathbf{\Phi}_{\mathbf{q}}^3 = \mathbf{\Phi}_{\mathbf{q}}^2$, $\mathbf{\Phi}_{\mathbf{q}}^3$ is row rank deficient and it appears that the last constraint equation is again redundant. In this case, however, the last constraint equation represents an *inconsistent redundancy*; that is, $\mathbf{\Phi}^3 = \mathbf{0}$ cannot be satisfied for any value of $\mathbf{q}$.

While Example 3.6.1 is trivial, it illustrates difficulties that can arise in complex mechanism models. It is surprisingly easy to inadvertently include redundant constraints in a model, especially for the spatial systems treated in Part II of the text. Computational checks are required to assure that the mechanism defined by constraints that are imposed can in fact be assembled, that is, to assume the constraints are consistent. If the constraints cannot be satisfied, they are said to be inconsistent. If constraints are consistent, the row rank of the constraint Jacobian $\mathbf{\Phi}_{\mathbf{q}}$ must be confirmed to be equal to the number of constraint equations. If not, one or more constraint equations are redundant.

A crucial ingredient in solving Eq. 3.6.3 and in determining configurations in which singular behavior occurs is the Jacobian

$$\mathbf{\Phi}_{\mathbf{q}}(\mathbf{q}, t) = \left[\frac{\partial \Phi_i(\mathbf{q}, t)}{\partial q_j}\right]_{nc \times nc} \tag{3.6.4}$$

where matrix calculus notation of Section 2.5 is employed. An important result from advanced calculus [25, 26] is the following:

Implicit Function Theorem: Let $\mathbf{q}^0$ be a solution of Eq. 3.6.3 at $t = t_0$, and let the function $\mathbf{\Phi}(\mathbf{q}, t)$ be twice continuously differentiable with respect to its arguments. Then, if the Jacobian of Eq. 3.6.4 is nonsingular at $(\mathbf{q}^0, t_0)$, there exists a unique solution

$$\mathbf{q} = \mathbf{f}(t) \tag{3.6.5}$$

of Eq. 3.6.3, in some interval of time about t_0, such that $\mathbf{f}(t_0) = \mathbf{q}^0$. That is, for some $\delta > 0$,

$$\mathbf{\Phi}(\mathbf{f}(t), t) = \mathbf{0}, \qquad \text{for } |t - t_0| \leq \delta \tag{3.6.6}$$

Furthermore, the solution of Eq. 3.6.5 is twice continuously differentiable with respect to time; that is, velocity $\dot{\mathbf{q}} = \dot{\mathbf{f}}(t)$ and acceleration $\ddot{\mathbf{q}} = \ddot{\mathbf{f}}(t)$ are continuous.

While the implicit function theorem provides concrete conditions under which a solution of the kinematic equations exists, it does not provide a method for analytically constructing such a solution. The condition that the Jacobian of Eq. 3.6.4 be nonsingular as a sufficient condition for existence of a solution of the kinematic equations can be used to test for the onset of singular configurations that are associated with lock-up of a mechanism, which may be mathematically interpreted as loss of the physical existence of a solution.

In addition to being a valuable theoretical tool, the constraint Jacobian of Eq. 3.6.4 plays a key role in the numerical solution of the kinematic equations. The most common method used in solving nonlinear equations of the form of Eq. 3.6.3 is the *Newton–Raphson method,* which is discussed in detail in Chapter 4. The Newton–Raphson method is an iterative technique that begins with an estimate $\mathbf{q}^{(0)}$ of a configuration that satisfies Eq. 3.6.3 at time t. At a typical iteration k, the following equation is solved for a correction $\Delta\mathbf{q}^{(k)}$:

$$\mathbf{\Phi}_{\mathbf{q}}(\mathbf{q}^{(k)}, t)\Delta\mathbf{q}^{(k)} = -\mathbf{\Phi}(\mathbf{q}^{(k)}, t) \tag{3.6.7}$$

which is then added to the estimate $\mathbf{q}^{(k)}$ to obtain an improved estimate; that is,

$$\mathbf{q}^{(k+1)} = \mathbf{q}^{(k)} + \Delta\mathbf{q}^{(k)}, \qquad k = 0, 1, \ldots \tag{3.6.8}$$

Iteration is continued until an error tolerance in satisfying Eq. 3.6.3 is met.

The Newton–Raphson method has the attractive property that it is *quadratically convergent*; that is, the solution error in a given iteration is proportional to the square of the error in the preceding iteration. The method, however, may diverge if poor initial estimates of the position of all bodies in the system are given or if no solution of the kinematic equations exists. There is a close relationship between the implicit function theorem, which indicates the need for the Jacobian to be nonsingular, and the requirement that the Jacobian be nonsingular in solving Newton–Raphson equations in Eq. 3.6.7.

Failure of the Newton–Raphson method to converge may indicate that a mechanism has been specified that cannot be physically assembled. However, since the method can also diverge due to poor initial estimates, some other more robust method of solving nonlinear equations is desired to provide a reliable indication that an infeasible mechanism has been specified. One such technique is to seek a vector $\mathbf{q}$ of generalized coordinates at t_0 that minimizes error in satisfying the constraint equations and is as close as possible to the initial estimate $\mathbf{q}(0)$ given. One such technique is to successively minimize

$$\Psi_0(\mathbf{q}, t_0, r) \equiv (\mathbf{q} - \mathbf{q}^{(0)})^T(\mathbf{q} - \mathbf{q}^{(0)}) + r\boldsymbol{\Phi}^T(\mathbf{q}, t_0)\boldsymbol{\Phi}(q, t_0) \tag{3.6.9}$$

with increasing values of the parameter $r > 0$ to more heavily weight satisfying the constraint equations. Numerical methods of minimizing such functions are discussed in Chapter 4. Such a minimization technique provides a valuable tool to *assemble* a mechanism or to discover that a design has been specified that cannot be assembled. If this optimization method converges as r becomes large to a value of $\mathbf{q}$ for which Eq. 3.6.3 is not satisfied, then the engineer should be concerned that the mechanism cannot be assembled.

3.6.2 Velocity Analysis

Presuming that the Jacobian matrix of Eq. 3.6.4 is nonsingular and that a solution of the kinematic equations has been obtained numerically, the implicit function theorem guarantees that velocity and acceleration of the system exist. The challenge is to calculate these quantities numerically.

Since Eq. 3.6.3 must hold for all time, both sides may be differentiated with respect to time and rearranged to obtain the *velocity equation*

$$\boldsymbol{\Phi}_{\mathbf{q}}\dot{\mathbf{q}} = -\boldsymbol{\Phi}_t \equiv \mathbf{v} \tag{3.6.10}$$

Since the constraint Jacobian is nonsingular, this equation uniquely determines the velocity $\dot{\mathbf{q}}$. This computation is efficient, as well as direct, since the Jacobian must already have been formed and factored (see Chapter 4) to solve the position equations using the Newton–Raphson method in Eq. 3.6.7.

The right side $\mathbf{v}$ of the velocity equation of Eq. 3.6.10 may be assembled using the right sides of velocity equations for individual kinematic and driving constraints in Sections 3.2 through 3.5. Methods for computer generation of $\mathbf{v}$ and the Jacobian $\boldsymbol{\Phi}_{\mathbf{q}}$ are discussed in Chapter 4.

3.6.3 Acceleration Analysis

Just as the velocity equations of Eq. 3.6.10 were obtained by differentiating the kinematic equations of Eq. 3.6.3, differentiating both sides of Eq. 3.6.10 yields the *acceleration equation*

$$\mathbf{\Phi_q \ddot{q}} = -(\mathbf{\Phi_q \dot{q}})_\mathbf{q}\mathbf{\dot{q}} - 2\mathbf{\Phi}_{\mathbf{q}t}\mathbf{\dot{q}} - \mathbf{\Phi}_{tt} \equiv \boldsymbol{\gamma} \tag{3.6.11}$$

that determines the acceleration $\ddot{\mathbf{q}}$. Just as in the case of the velocity equation, the right side of the acceleration equation of Eq. 3.6.11 can be assembled using the right sides of acceleration equations for the constraints of Sections 3.2 through 3.5. For details, see Chapter 4.

Note that, as in the velocity equation, the constraint Jacobian plays a dominant role in acceleration analysis. The right side of Eq. 3.6.11 can be evaluated after the solution for velocities is obtained, leading to direct and efficient computation of acceleration.

Example 3.6.2: Consider the one degree of freedom double pendulum shown in Fig. 3.6.2. Body 1 is driven by an absolute angle driver

$$\phi_1 = \frac{5\pi}{3} + \frac{\pi}{6}t, \qquad 0 \leq t \leq \tfrac{3}{2}$$

With a maximal set of Cartesian generalized coordinates, $\mathbf{q} = [x_1, y_1, \phi_1, x_2, y_2, \phi_2]^T$, the system of kinematic and driving constraint equations, using Eq. 3.6.3, is

$$\mathbf{\Phi}(\mathbf{q}, t) \equiv \begin{bmatrix} x_1 - \cos\phi_1 \\ y_1 - \sin\phi_1 \\ x_2 - \cos\phi_1 - \cos\phi_2 \\ y_2 - \sin\phi_1 - \sin\phi_2 \\ y_2 + 1 \\ \phi_1 - \dfrac{\pi}{6}t - \dfrac{5\pi}{3} \end{bmatrix} = \mathbf{0}$$

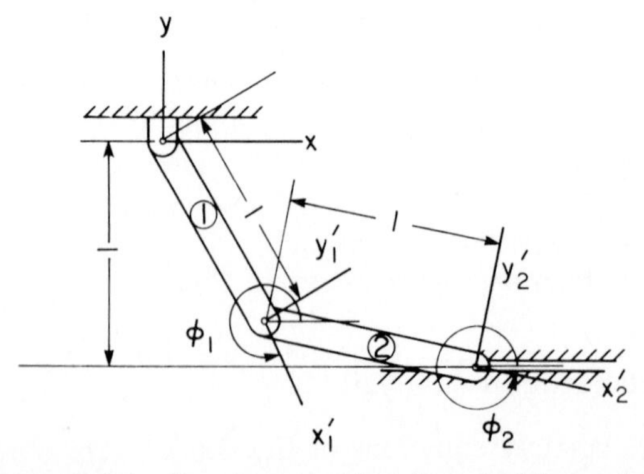

Figure 3.6.2 One degree of freedom double pendulum.

The constraint Jacobian is

$$\mathbf{\Phi_q}(\mathbf{q}, t) \equiv \begin{bmatrix} 1 & 0 & \sin\phi_1 & 0 & 0 & 0 \\ 0 & 1 & -\cos\phi_1 & 0 & 0 & 0 \\ 0 & 0 & \sin\phi_1 & 1 & 0 & \sin\phi_2 \\ 0 & 0 & -\cos\phi_1 & 0 & 1 & -\cos\phi_2 \\ 0 & 0 & 0 & 0 & 1 & 0 \\ 0 & 0 & 1 & 0 & 0 & 0 \end{bmatrix}$$

The determinant of this Jacobian is $|\mathbf{\Phi_q}| = -\cos\phi_2$. Unless body 2 is in an upright position, that is, $\phi_2 = \pi/2 + k\pi$, where k is an integer, the determinant is never zero. When $\cos\phi_2 = 0$, the system is in a singular configuration, which is discussed in Section 3.7.

When $t = 0$, $\phi_1 = 5\pi/3$. Let the error tolerance for the Newton–Raphson method be $\varepsilon = 1.0 \times 10^{-4}$ and estimate $\mathbf{q}^{(0)} = [0.5, -0.8660, 5\pi/3, 1.5, -1.0, 6.14]^T$. Equation 3.6.7 becomes

$$\begin{bmatrix} 1 & 0 & -0.8660 & 0 & 0 & 0 \\ 0 & 1 & -0.5 & 0 & 0 & 0 \\ 0 & 0 & -0.8660 & 1 & 0 & -0.1427 \\ 0 & 0 & -0.5 & 0 & 1 & -0.9898 \\ 0 & 0 & 0 & 0 & 1 & 0 \\ 0 & 0 & 1 & 0 & 0 & 0 \end{bmatrix} \begin{bmatrix} \Delta x_1 \\ \Delta y_1 \\ \Delta\phi_1 \\ \Delta x_2 \\ \Delta y_2 \\ \Delta\phi_2 \end{bmatrix} = \begin{bmatrix} 0 \\ 0 \\ -0.0102 \\ -0.0087 \\ 0 \\ 0 \end{bmatrix}$$

The solution of this matrix equation is

$$\Delta\mathbf{q}^{(1)} = [0, 0, 0, -0.0089, 0, 0.0088]^T$$

From Eq. 3.6.8,

$$\mathbf{q}^{(2)} = [0.5, -0.8660, 5\pi/3, 1.4911, -1.0, 6.1488]^T$$

Substituting $\mathbf{q}^{(2)}$ into Eq. 3.6.7 and solving for $\Delta\mathbf{q}^{(2)}$,

$$\Delta\mathbf{q}^{(2)} = [0, 0, 0, -0.0001, 0, 0]^T$$

Hence

$$\mathbf{q}^{(3)} = [0.5, -0.8660, 5\pi/3, 1.4910, -1.0, 6.1488]^T$$

Substituting $\mathbf{q}^{(2)}$ into Eq. 3.6.7 and solving for $\Delta\mathbf{q}^{(2)}$,

$$\Delta\mathbf{q}^{(2)} = [0, 0, 0, -0.0001, 0, 0]^T$$

After only two iterations, the constraint violation is smaller than the prescribed error tolerance.

Equation 3.6.10 for velocity is

$$\begin{bmatrix} 1 & 0 & -0.8660 & 0 & 0 & 0 \\ 0 & 1 & -0.5 & 0 & 0 & 0 \\ 0 & 0 & -0.8660 & 1 & 0 & -0.1340 \\ 0 & 0 & -0.5 & 0 & 1 & -0.9910 \\ 0 & 0 & 0 & 0 & 1 & 0 \\ 0 & 0 & 1 & 0 & 0 & 0 \end{bmatrix} \begin{bmatrix} \dot{x}_1 \\ \dot{y}_1 \\ \dot{\phi}_1 \\ \dot{x}_2 \\ \dot{y}_2 \\ \dot{\phi}_2 \end{bmatrix} = \begin{bmatrix} 0 \\ 0 \\ 0 \\ 0 \\ 0 \\ 0.5236 \end{bmatrix}$$

and the solution is

$$\dot{\mathbf{q}} = [0.4534, 0.2618, 0.5236, 0.4180, 0, -0.2642]^T$$

Since $\mathbf{\Phi}_{\mathbf{q}t} = \mathbf{0}$ and $\mathbf{\Phi}_{tt} = \mathbf{0}$, the acceleration equation of Eq. 3.6.11 is

$$\begin{bmatrix} 1 & 0 & -0.8660 & 0 & 0 & 0 \\ 0 & 1 & -0.5 & 0 & 0 & 0 \\ 0 & 0 & -0.8660 & 1 & 0 & -0.1340 \\ 0 & 0 & -0.5 & 0 & 1 & -0.9910 \\ 0 & 0 & 0 & 0 & 1 & 0 \\ 0 & 0 & 1 & 0 & 0 & 0 \end{bmatrix} \begin{bmatrix} \ddot{x}_1 \\ \ddot{y}_1 \\ \ddot{\phi}_1 \\ \ddot{x}_2 \\ \ddot{y}_2 \\ \ddot{\phi}_2 \end{bmatrix} = \begin{bmatrix} -0.1371 \\ 0.2374 \\ -0.2063 \\ -0.2468 \\ 0 \\ 0 \end{bmatrix}$$

and the solution is

$$\ddot{\mathbf{q}} = [-0.1371, 0.2374, 0, -0.2397, 0, -0.2490]^T$$

3.6.4 Kinematic Analysis on a Time Grid

Rather than calculating position, velocity, and acceleration at isolated times, it is often desired to solve for these variables at a grid of time points t_i, $i = 1, \ldots, m$, that is, to calculate

$$\left.\begin{aligned} \mathbf{q}_i &= \mathbf{q}(t_i) \\ \dot{\mathbf{q}}_i &= \dot{\mathbf{q}}(t_i) \\ \ddot{\mathbf{q}}_i &= \ddot{\mathbf{q}}(t_i) \end{aligned}\right\}, \qquad i = 1, \ldots, m \tag{3.6.12}$$

Use of the Newton–Raphson method of Eqs. 3.6.7 and 3.6.8 to solve for position is enhanced by starting the iterative computation at a good estimate of $\mathbf{q}$. Presuming that position, velocity, and acceleration are known at time t_i, $\mathbf{q}$ at time t_{i+1} may be approximated using the second-order Taylor expansion [25, 26]

$$\mathbf{q}_{i+1} \approx \mathbf{q}_i + (t_{i+1} - t_i)\dot{\mathbf{q}}_i + \tfrac{1}{2}(t_{i+1} - t_i)^2\ddot{\mathbf{q}}_i \tag{3.6.13}$$

This estimate can be used to begin Newton–Raphson iteration and, if the increment in time points is not large, rapid convergence may be expected.

3.7 SINGULAR CONFIGURATIONS

The preceding sections are devoted to formulating and solving constraint equations for the kinematic analysis of mechanical systems in which motion proceeds smoothly with time. For trial designs, however, singular configurations may be encountered, beyond which motion cannot continue or more than one possible motion can occur. To see that such pathological behavior is not restricted to peculiar cases invented by a mathematician, consider the following elementary slider–crank example.

Example 3.7.1: Consider first the elementary model of a slider–crank mechanism in Example 3.1.2, with $\ell < 1$, which is driven in the positive ϕ_1 direction. As motion continues from the position shown in Fig. 3.1.3, the system reaches a configuration in which $\sin \phi_1 = \ell$, as shown in Fig. 3.7.1. It is physically clear that in this configuration the crank can no longer be driven in the positive ϕ_1 direction. This is called a *lock-up configuration.*

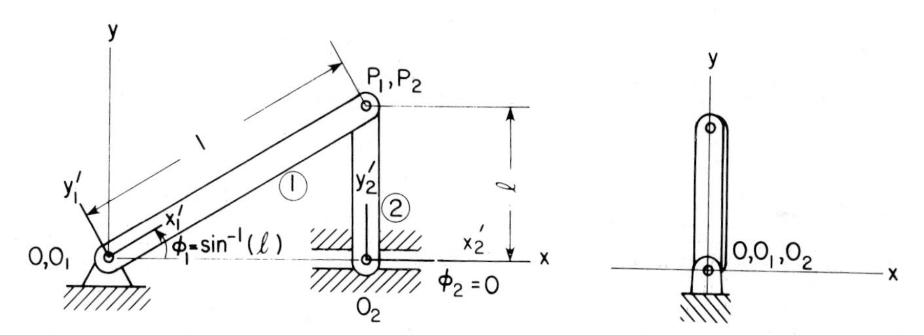

Figure 3.7.1 Slider–crank at lock-up configuration.

Figure 3.7.2 Slider–crank at bifurcation configuration.

An interesting observation can be made when the system is driven in the negative ϕ_1 direction from the configuration shown in Fig. 3.7.1. Two possible motions can occur: point O_2 could move either to the left or to the right. This branching of motion to two possible paths is called *bifurcation.*

Consider next the slider–crank of Example 3.1.2, but with $\ell = 1$. When $\phi_1 = \pi/2$, $\phi_2 = 0$, as illustrated in Fig. 3.7.2. As ϕ_1 is driven in the positive sense, O_2 may either stay at O, and hence bodies 1 and 2 are locked together and move as a pendulum pivoted at O [Fig. 3.7.3(a)], or O_2 passes point O and moves along the negative x axis [Fig. 3.7.3(b)]. This is another example of bifurcation.

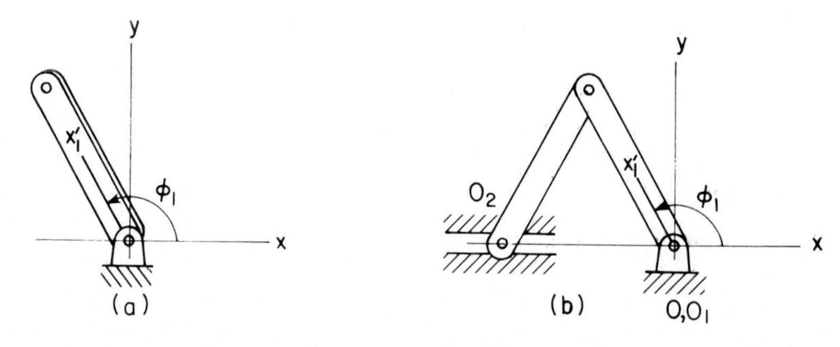

Figure 3.7.3 Bifurcation of slider–crank.

Example 3.7.2: A second example of bifurcation can be illustrated with a parallelogram four-bar linkage with two pairs of equal-length members, shown in Fig. 3.7.4. If body 1 is driven in the positive ϕ_1 direction from the flat configuration, that is, $\phi_1 = \phi_3 = 0$, two types of motion are possible, as illustrated in Fig. 3.7.5.

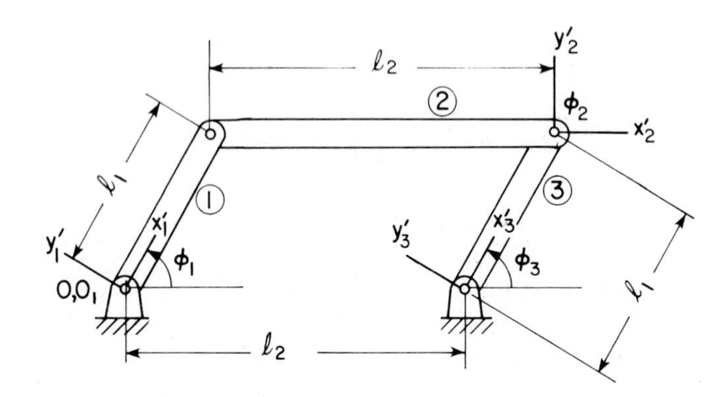

Figure 3.7.4 Parallelogram four-bar linkage.

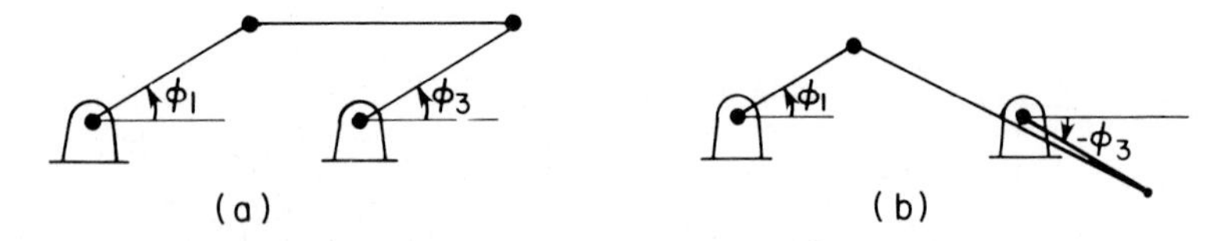

Figure 3.7.5 Bifurcation motion of four-bar linkage.

Example 3.7.3: As a first step in developing analytical methods for finding and analyzing singular configurations, consider the alternative two-body model of the slider–crank mechanism shown in Fig. 3.7.6, which was studied with the aid

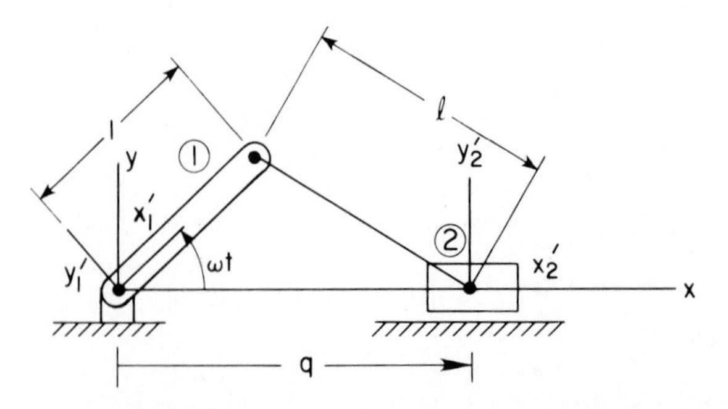

Figure 3.7.6 Two-body slider–crank model.

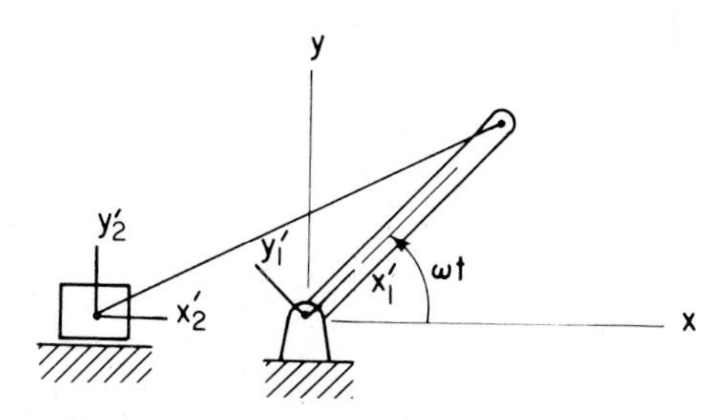

Figure 3.7.7 Alternative assembly of two-body slider–crank.

of physical arguments in Example 3.7.1. The crank is driven at constant angular velocity ω; that is, $\phi_1 = \omega t$. If absolute constraints are used to define coupling between bodies 1 and 2 and ground, $x_1 = y_1 = y_2 = \phi_2 = 0$. Thus, only $x_2 \equiv q$ remains as a generalized coordinate. The distance constraint imposed by the coupler of length ℓ may be written in terms of q as

$$\Phi(q, t) = (q - \cos \omega t)^2 + \sin^2 \omega t - \ell^2 = 0 \tag{3.7.1}$$

Expanding Eq. 3.7.1 as a quadratic equation in q and solving it,

$$q = \cos \omega t \mp \sqrt{\cos^2 \omega t + (\ell^2 - 1)} \tag{3.7.2}$$

where the plus sign corresponds to the assembled configuration shown in Fig. 3.7.6 and the minus sign corresponds to the configuration shown in Fig. 3.7.7. Note, from Example 3.7.1, that if $\ell > 1$ then the mechanism will stay in only one of the distinct configurations for all values of ϕ_1. If $\ell < 1$, on the other hand, then at singular points $\sin \phi_1 = \ell$ the configuration can change.

From Eq. 3.7.2, it is clear that if $\ell > 1$ there is a valid solution for all time. If $\ell < 1$, however, then when $\cos^2 \omega t = 1 - \ell^2$ (equivalently when $\sin \omega t = \ell$) loss of existence (lock-up) is encountered. Finally, if $\ell = 1$, a solution exists, but when $\cos \omega t = 0$ (equivalently when $\sin \omega t = \ell$), two solution branches (bifurcation) are possible.

Direct differentiation of Eq. 3.7.2 shows that if $\ell < 1$ then $\dot{\mathbf{q}}$ and $\ddot{\mathbf{q}}$ approach infinity as $\sin \omega t$ approaches ℓ. This behavior is typical at lock-up points and can serve as a warning that a lock-up configuration is being approached.

It is comforting to see that the properties of the analytical solution of kinematic equations agree with reality at lock-up and bifurcation points in Example 3.7.3. However, since analytical (closed-form) solutions are seldom possible, it is important to develop criteria for singular points that can be applied to large-scale systems and evaluated numerically. Prior to presenting a general method, the elementary model of Example 3.7.3 is investigated from a different point of view.

Example 3.7.4: Taking the time derivative of the constraint equation of Example 3.7.3 (Eq. 3.7.1),

$$\begin{aligned}\Phi_q \dot{q} &\equiv 2(q - \cos \omega t)\dot{q} \\ &= -2\omega q \sin \omega t \equiv -\Phi_t\end{aligned} \quad \textbf{(3.7.3)}$$

If $\ell > 1$, $q > \cos \omega t$ for all t and no difficulties arise. If $\ell < 1$, then $q = \cos \omega t \neq 0$ when $\sin \omega t = \ell$. Thus, Eq. 3.7.3 shows that $\dot{q}$ approaches infinity as the lock-up point is approached, since there $\Phi_q = 0$ and $\Phi_t \neq 0$. This is, in fact, a general criteria.

The approach illustrated in Example 3.7.4 is typical of analysis that can be carried out using the constraint Jacobian and the implicit function theorem. In general, if $\mathbf{q}$ is an n vector of generalized coordinates and if the combination of kinematic and driving constraints is n equations, then $\mathbf{\Phi}(\mathbf{q}, t) = \mathbf{0}$ is to determine $\mathbf{q}(t)$. Presume an assembled configuration $\mathbf{q}^*$ is known at time t^*. Then, if the Jacobian $\mathbf{\Phi}_{\mathbf{q}}(\mathbf{q}^*, t^*)$ were nonsingular, the implicit function theorem would guarantee a smooth unique solution $\mathbf{q}(t)$ near $\mathbf{q}^*$ in the time interval $|t - t^*| < \delta$, for some $\delta > 0$. If lock-up or bifurcation occurs at t^*, however, then arbitrarily close to t^* either existence or uniqueness is lost. Thus, $\mathbf{\Phi}_{\mathbf{q}}(\mathbf{q}^*, t^*)$ must be singular at lock-up and bifurcation configurations. This property and the associated behavior of velocities and accelerations permit systematic detection and interpretation of singular points.

Example 3.7.5: Suppressing absolute constraints and the associated variables in Eq. 3.1.11 and adding the driving constraint of Eq. 3.1.12, constraint equations for the slider–crank in Example 3.1.2 reduce to

$$\mathbf{\Phi}(\mathbf{q}, t) = \begin{bmatrix} -x_2 + \cos \phi_1 + \ell \sin \phi_2 \\ \sin \phi_1 - \ell \cos \phi_2 \\ \phi_1 - \omega t \end{bmatrix} = \mathbf{0} \quad \textbf{(3.7.4)}$$

where $\mathbf{q} = [\phi_1, x_2, \phi_2]^T$. The determinant of the Jacobian is

$$|\mathbf{\Phi}_{\mathbf{q}}| = \begin{vmatrix} -\sin \phi_1 & -1 & \ell \cos \phi_2 \\ \cos \phi_1 & 0 & \ell \sin \phi_2 \\ 1 & 0 & 0 \end{vmatrix} = -\ell \sin \phi_2 \quad \textbf{(3.7.5)}$$

which is zero when $\phi_2 = 0$.

Two possible situations can be considered for the singular point $\phi_2 = 0$, as in Examples 3.1.2 and 3.7.1.

1. With $\ell < 1$, the singular configuration $\phi_1 = \sin^{-1}(\ell)$ and $x_2 = \sqrt{1 - \ell^2}$ satisfies the constraint equation of Eq. 3.7.4 and represents the lock-up configuration shown in Fig. 3.7.1.
2. With $\ell = 1$, the singular configuration $\phi_1 = \pi/2$ and $x_2 = 0$ satisfies Eq. 3.7.1 and represents the bifurcation configuration shown in Fig. 3.7.3.

Some physical insights into singular behavior can be gained by studying the velocity equation of Eq. 3.6.10:

$$\mathbf{\Phi_q \dot{q}} = -\mathbf{\Phi}_t \tag{3.7.6}$$

If $\mathbf{\Phi_q}$ is singular at a time t^*, Eq. 3.7.6 has a solution if and only if, by the *theorem of the alternative* [22],

$$\boldsymbol{\beta}^T \mathbf{\Phi}_t = 0 \tag{3.7.7}$$

for all solutions $\boldsymbol{\beta}$ of

$$\mathbf{\Phi_q}^T \boldsymbol{\beta} = \mathbf{0} \tag{3.7.8}$$

This result clearly shows that Eq. 3.7.6 has a finite solution for velocity only if $\mathbf{\Phi_q}$ and $\mathbf{\Phi}_t$ are properly related.

Consider next a small *virtual displacement* $\delta\mathbf{q}$ that satisfies constraints to "first order" (a concept that is developed more thoroughly in Chapter 6), with time held fixed. Expanding the constraint equations to first order, $\delta\mathbf{q}$ satisfies

$$\mathbf{\Phi_q}\delta\mathbf{q} = \mathbf{0} \tag{3.7.9}$$

Since $\mathbf{\Phi_q}$ is singular at t^*, there exists a solution $\delta\mathbf{q} \neq \mathbf{0}$ that may be viewed as tangent to solution trajectories that emanate from $\mathbf{q}(t^*)$. Note that, if a finite $\dot{\mathbf{q}}$ satisfies Eq. 3.7.6 and $\delta\mathbf{q}$ satisfies Eq. 3.7.9, then

$$\mathbf{\Phi_q}(\dot{\mathbf{q}} \pm \delta\mathbf{q}) = \mathbf{\Phi_q}\dot{\mathbf{q}} \pm \mathbf{\Phi_q}\delta\mathbf{q} = -\mathbf{\Phi}_t \tag{3.7.10}$$

Thus, two distinct velocities may occur after t^*, which characterizes a bifurcation point. Since the determinant $|\mathbf{\Phi_q}|$ is zero at t^*, except in cases for which the rank deficiency of $\mathbf{\Phi_q}$ is greater than 1, its sign must change as a bifurcation is passed. This can serve as a computational test for a bifurcation configuration. Only if no solution of Eq. 3.7.6 exists beyond t^*, that is, only if $\dot{\mathbf{q}}$ approaches infinity as t approaches t^*, is there no possibility of continuing a solution. This characterizes a lock-up configuration. Thus, $\dot{\mathbf{q}}$ and/or $\ddot{\mathbf{q}}$ approaching infinity serves as a criteria for lock-up.

A final note on the effect of a design change on performance of a mechanism near a singular point may help in evaluating the engineering feasibility of a design. Let $\mathbf{b} = [b_1, \ldots, b_k]^T$ be a vector of design parameters, such as dimensions. Since these parameters arise in kinematic constraint equations, consider the constraints

$$\mathbf{\Phi}(\mathbf{q}, t; \mathbf{b}) = \mathbf{0} \tag{3.7.11}$$

There is a relationship between admissible virtual displacements $\delta\mathbf{q}$ and small variations $\delta\mathbf{b}$ in design. To first order, Eq. 3.7.11 may be expanded (linearized) as

$$\mathbf{\Phi_q}\delta\mathbf{q} = -\mathbf{\Phi_b}\delta\mathbf{b} \tag{3.7.12}$$

At a singular point t^*, $\mathbf{\Phi_q}$ is singular. Thus, for some $\delta\mathbf{b}$ Eq. 3.7.12 may have multiple solutions for $\delta\mathbf{q}$, and for other $\delta\mathbf{b}$ it may have no solutions. Physically, this may be interpreted as indicating that, for some design variations, multiple changes in configuration are possible, somewhat as in bifurcation, and for other design variations, the mechanism cannot be assembled. This observation will be of value in evaluating the engineering realism of models in Chapter 5.

Example 3.7.6: Consider the slider–crank model of Example 3.7.1 in the lock-up configuration shown in Fig. 3.7.1. If ℓ is a design parameter, a positive $\delta\ell$ leads to two values of δx_2, as shown in Fig. 3.7.8(a). A negative $\delta\ell$, however, leads to a design that cannot be assembled with $\phi_1 = \sin^{-1}\ell$, as shown in Fig. 3.7.8(b).

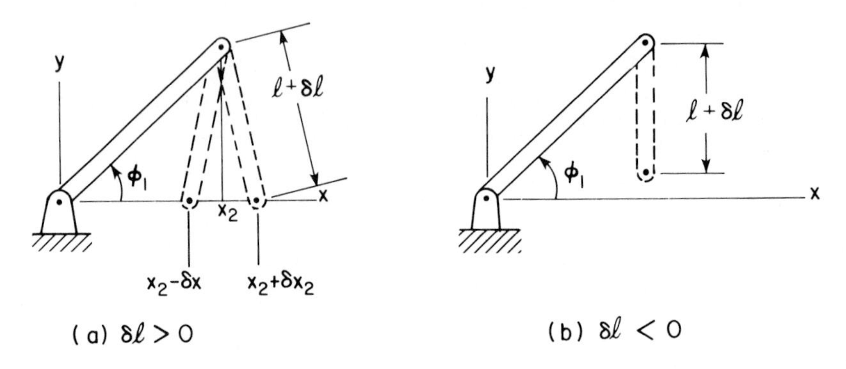

Figure 3.7.8 Effect of design change on slider–crank.

Singular points that define lock-up or bifurcation are *isolated singular points,* that is, intervals of time in which a unique solution of the equations of kinematics occurs on one or both sides of an isolated singular point t^*. Thus, except at t^*, if the mechanism can be assembled, then it is well behaved. Another form of singularity that is commonly encountered is a *redundant constraint.* If a mechanism is modeled using the standard joints of Sections 3.2 through 3.4, it is possible that one or more of the kinematic constraint equations are automatically satisfied if the remaining constraints are satisfied, in which case the automatically satisfied constraints are redundant and consistent.

Example 3.7.7: The five-body parallelogram mechanism of Fig. 3.7.9 is modeled as a single bar, with three absolute distance constraints that represent three bars of unit length that are pivoted in ground. With $\mathbf{q} = [x_1, y_1, \phi_1]^T \equiv [x, y, \phi]^T$, the kinematic constraint equations are

$$\mathbf{\Phi}^K(\mathbf{q}) = \begin{bmatrix} x^2 + y^2 - 1 \\ (x + \cos\phi - 1)^2 + (y + \sin\phi)^2 - 1 \\ (x + 2\cos\phi - 2)^2 + (y + 2\sin\phi)^2 - 1 \end{bmatrix} = \mathbf{0} \tag{3.7.13}$$

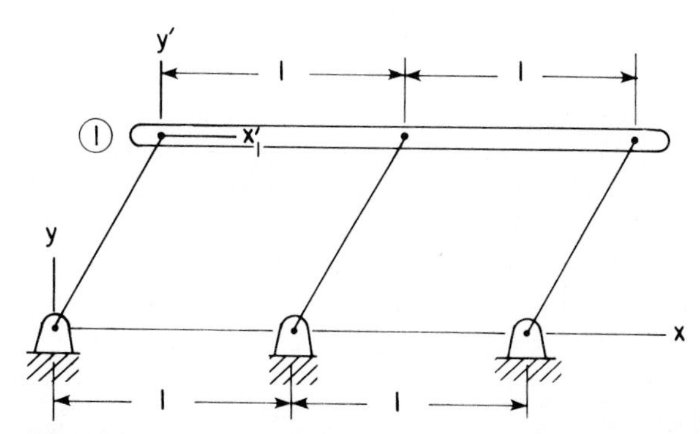

Figure 3.7.9 Five-bar parallelogram mechanism model.

The constraint Jacobian of Eq. 3.7.13 is

$$\Phi_{\mathbf{q}}^{K} = \begin{bmatrix} 2x & 2y & 0 \\ 2(x + \cos\phi - 1) & 2(y + \sin\phi) & -2(x-1)\sin\phi + 2y\cos\phi \\ 2(x + 2\cos\phi - 2) & 2(y + 2\sin\phi) & -4(x-2)\sin\phi + 4y\cos\phi \end{bmatrix} \quad \textbf{(3.7.14)}$$

Subtracting the first row from two times the second row and subtracting the result from the third row leads to a matrix with a zero third row for all values of $\mathbf{q}$. Thus, the rank of the constraint Jacobian is only two, and the third of the constraint equations of Eq. 3.7.13 is dependent. This is clearly true from Fig. 3.7.9, since the third distance constraint is automatically satisfied by the parallelogram that is defined by the first two distance constraints.

A numerical method is presented in Section 4.6 for the identification and elimination of redundant constraints.

PROBLEMS

Section 3.1

3.1.1. Let $f(t) = \sin 10t$ in Example 3.1.1. At instants $t_i = i/10$, $i = 0, 1, \ldots, 10$, in time.

(a) Solve Eq. 3.1.7 for $\mathbf{q}(t_i)$.

(b) Use the result of part a in Eq. 3.1.9 to determine $\dot{\mathbf{q}}(t_i)$.

(c) Use the results of parts a and b in Eq. 3.1.10 to determine $\ddot{\mathbf{q}}(t_i)$.

Report the results in the form of the following table:

i	0	1	2	3	4	5	6	7	8	9	10
$x_1(t_i)$											
$y_1(t_i)$											
$\phi_1(t_i)$											
$\dot{x}_1(t_i)$											
$\dot{y}_1(t_i)$											
$\dot{\phi}_1(t_i)$											
$\ddot{x}_1(t_i)$											
$\ddot{y}_1(t_i)$											
$\ddot{\phi}_1(t_i)$											

3.1.2. Flowchart a digital computer program that could be used to calculate and report the results for Prob. 3.1.1.

3.1.3. Verify that $|\mathbf{\Phi_q}| = -\ell \sin \phi_2$ for the matrix of Eq. 3.1.14. If $\ell < 1$, then for some angle $\phi_1 < \pi/2$, $\phi_2 = 0$ and $\mathbf{\Phi_q}$ is singular, leading to infinite velocities and accelerations from Eqs. 3.1.14 and 3.1.15, respectively. Explain this pathological behavior from a physical point of view.

3.1.4. Verify that Eqs. 3.1.14 and 3.1.15 are correct. Can the same flowchart of Prob. 3.1.2 be used to construct a digital computer program to solve the equations of Example 3.1.2?

Section 3.2

3.2.1. Model the system of Prob. 2.4.4 using absolute position constraints. Write velocity and acceleration equations using the approach of Examples 3.1.1 and 3.1.2.

3.2.2 Model the slider–crank mechanism of Example 3.1.2, with body 2 as the only body, using absolute position and distance constraints. Write velocity and acceleration equations using the approach of Examples 3.1.1 and 3.1.2.

3.2.3. Replace the constraint equation of Eq. 3.1.17 in Example 3.1.3 with $\Phi_2 \equiv \phi_1 - \pi/4 = 0$ and show that this equation and Eqs. 3.1.16 and 3.1.18 uniquely define the motion of the system.

Section 3.3

3.3.1. Write out the constraint equations of Example 3.3.1 and show that they are identical to those derived in Example 3.1.2.

3.3.2. Use absolute and relative constraints to create three different but equivalent models of the three-body slider–crank mechanism shown in Fig. P3.3.2.

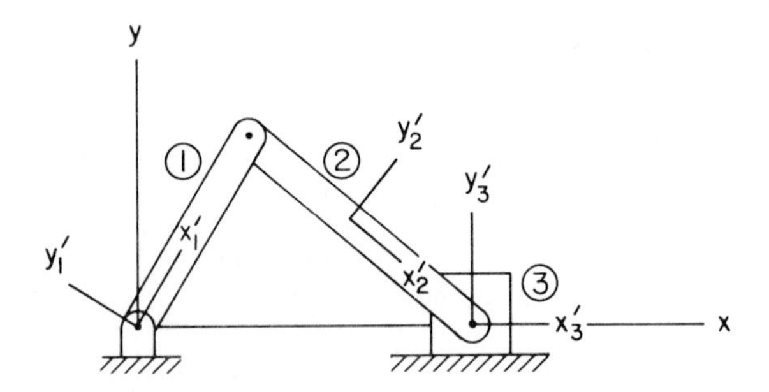

Figure P3.3.2

3.3.3. Identify the specific constraint equations that implement the four alternative models of Example 3.3.2 and define values of $x_i'^P$, $y_i'^P$, and C_k that are used for each.

3.3.4. Verify that the manipulations carried out in deriving the final form of Eq. 3.3.13 are correct.

3.3.5. Derive Eq. 3.3.14.

Section 3.4

3.4.1. Consider the gear set in Fig. 3.4.2. Let the centers P_i and P_j be fixed to ground; hence θ is constant. Show that the rotation on the rates of the gears are inversely proportional to their radii (*Hint:* Use Eq. 3.4.1).

3.4.2. The small rail trolley shown in Fig. P3.4.2 is driven by a 0.3-m diameter gear that has a constant angular acceleration of 5 revs/s^2. Find the vehicle's translational acceleration when the angular velocity of the driving gear is 600 rpm (*Hint:* Use Eqs. 3.4.1 and 3.4.8).

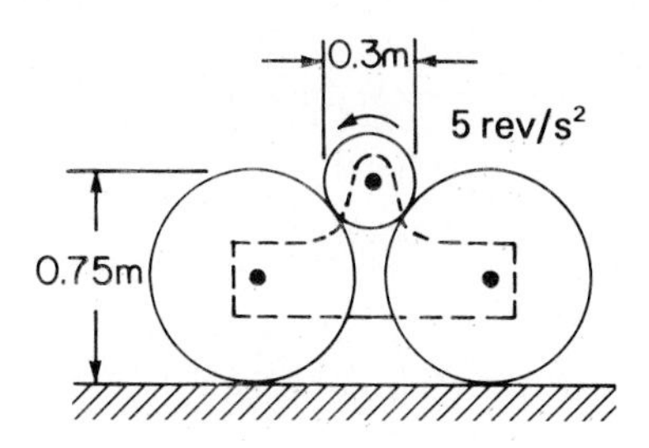

Figure P3.4.2

3.4.3. Consider the oil pumping rig shown in Fig. P3.4.3. The flexible pump rod, body 2, is fastened to a curved sector on body 1 at point Q_1 and is always vertical as it enters the well. This connection between the sector and flexible pump rod can be modeled as a rack and pinion constraint. Choose proper coordinate systems and write out the constraint equation of Eq. 3.4.13.

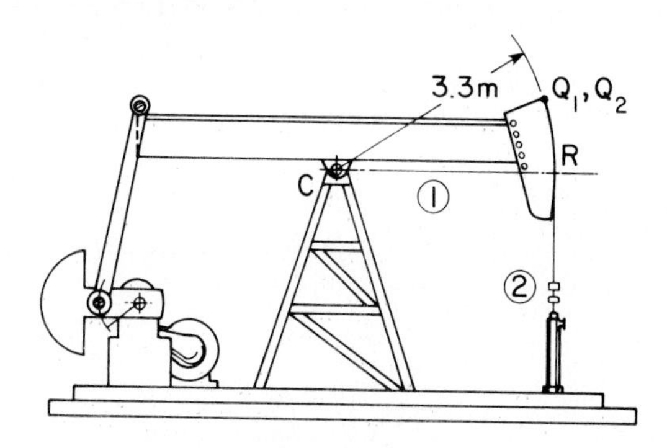

Figure P3.4.3

3.4.4. Derive the Jacobians **v** and **γ** for the rack and pinion constraint equation of Eq. 3.4.13.

3.4.5. Show that the gear and follower in Example 3.4.3 are always in contact at one point (*Hint:* Calculate the radius of curvature of the follower).

3.4.6. Are the first two constraint equations of Eq. 3.4.20 equivalent to the revolute joint constraint equation of Eq. 3.3.11? If not, give an explanation.

3.4.7. Repeat the analysis of Example 3.4.3 with the follower curve

$$\rho_1(\alpha_1) = \begin{cases} 1 + 2\alpha_1^2 - 2\alpha_1^3 + \frac{1}{2}\alpha_1^4, & 0 \le \alpha_1 \le 2 \\ 1, & 2 < \alpha_1 < 2\pi \end{cases}$$

3.4.8. Derive the Jacobian for the cam–follower constraint equation of Eq. 3.4.20.

3.4.9. Derive the Jacobian for the cam–flat-faced follower constraint equation of Eq. 3.4.29.

3.4.10. The slider–crank in Example 3.1.2 can be modeled as the three-body mechanism shown in Fig. P3.4.10. Body 2 and ground (body 3) are connected by a point–follower joint. Write the point–follower constraint equations of Eq. 3.4.31 using explicit expressions for $x_3'^P(\alpha_3)$ and $y_3'^P(\alpha_3)$.

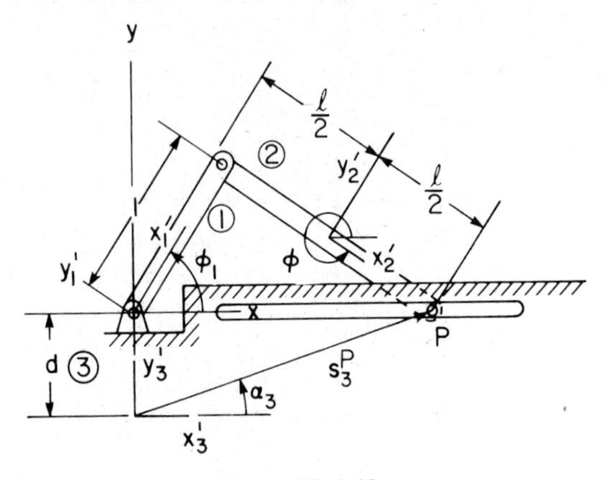

Figure P3.4.10

3.4.11. Derive the Jacobian for the point–follower constraint equation of Eq. 3.4.31.

Section 3.5

3.5.1. By taking the time derivative of the velocity equation in Example 3.5.1, find the angular acceleration $\ddot{\phi}_1$.

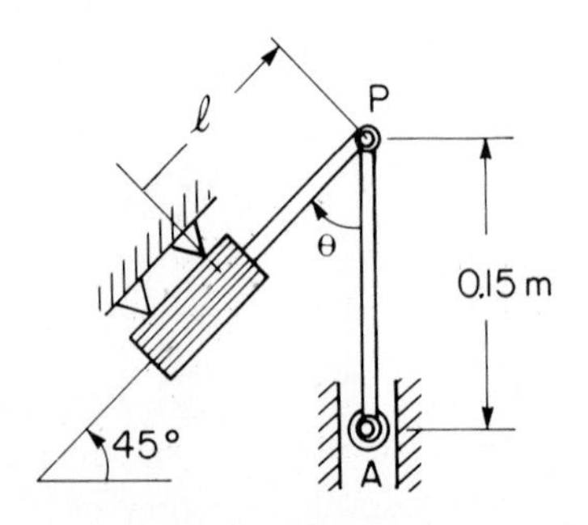

Figure P3.5.3

3.5.2. By taking the time derivative of the velocity equation in Example 3.5.2, find the acceleration vector $\ddot{\mathbf{q}} = [\ddot{\phi}_1, \ddot{x}_2, \ddot{y}_2, \ddot{\phi}_2, \ddot{\phi}_3]^T$.

3.5.3. The piston of the hydraulic cylinder shown in Fig. P3.5.3 moves point P with $\ell = \ell_0 + 0.08t$. Find the acceleration of point A when $\theta = 45°$.

3.5.4. Rod AB shown in Fig. P3.5.4 slides through a collar that is pivoted at point C as A moves along the slot. Rotation of the collar is driven so that $\phi_1 = \pi/3 + t + t^2$. Find the velocity and the acceleration of point A when $t = 0.25$ s.

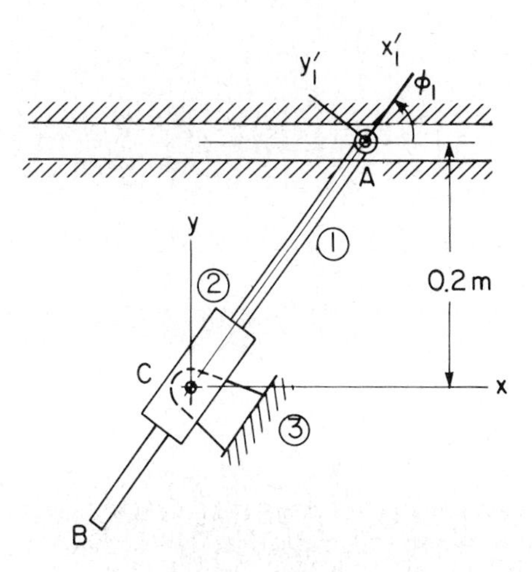

Figure P3.5.4

3.5.5. Consider the two-body mechanism shown in Fig. P3.5.5. The relative angle between bodies 1 and 2 is driven as $\phi_1 - \phi_2 = (\pi/18)t$, $0 \leq t \leq 8$. Find the velocity and acceleration of point O_2 when $\phi_1 - \phi_2 = \pi/3$.

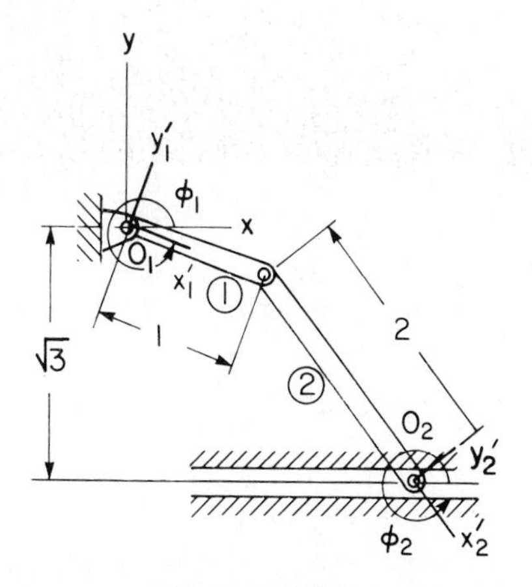

Figure P3.5.5

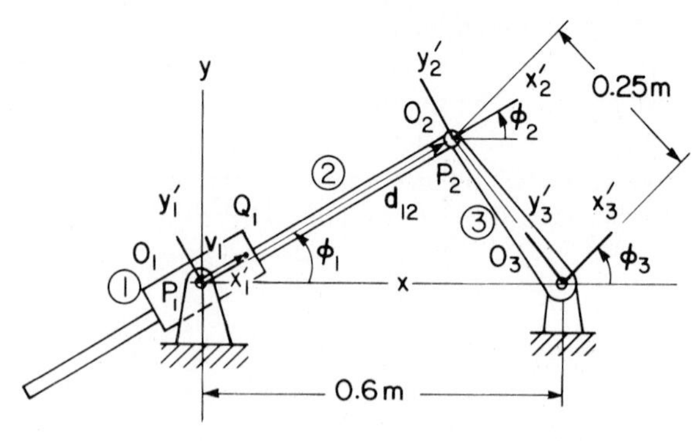

Figure P3.5.6

3.5.6. Consider the mechanism shown in Fig. P3.5.6, which is driven by

$$\frac{\mathbf{v}_1^T \mathbf{d}_{12}}{v_1} + 0.4 - \tfrac{1}{10}t = 0, \qquad 0 \leq t \leq 10$$

Find $\dot{\phi}_3$ and $\ddot{\phi}_3$ when $\phi_3 = 0$.

Section 3.6

3.6.1. Carry out position, velocity, and acceleration analysis for the simple mechanism in Fig. P3.6.1, with rotation of body 1 driven as

$$\phi_1 - \frac{5\pi}{3} - \frac{\pi}{15}t = 0, \qquad 0 \leq t \leq 4$$

Let $\mathbf{q} = [x_1, y_1, \phi_1]^T$ and estimate $\mathbf{q}^{(0)} = [1.05, 1.73, 5\pi/3]^T$.

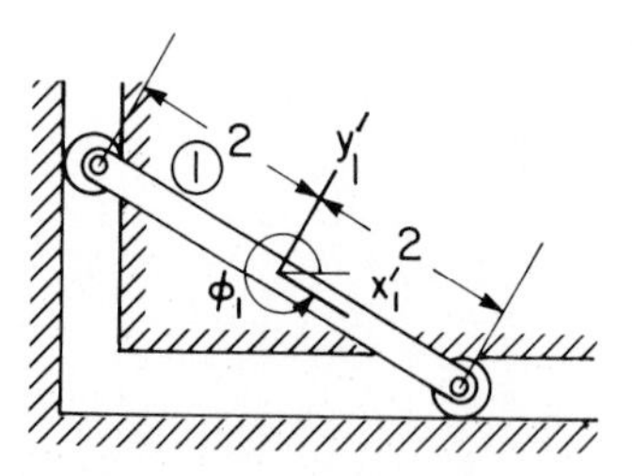

Figure P3.6.1

3.6.2. The pendulum of length 2 is suspended from a slider as shown in Fig. P3.6.2. The slider and pendulum are driven by the following absolute and revolute–rotational drivers:

$$x_1 - 0.2t = 0$$

$$\phi_1 - \phi_2 + \frac{\pi}{6} + \tfrac{1}{10}t = 0$$

Carry out position, velocity, and acceleration analysis for $\mathbf{q} = [x_1, \phi_1, x_2, y_2, \phi_2]^T$, with $\mathbf{q}^{(0)} = [0, 0.001, 0.45, -1.73, \pi/6]^T$, in the interval $0 \leq t \leq 5$.

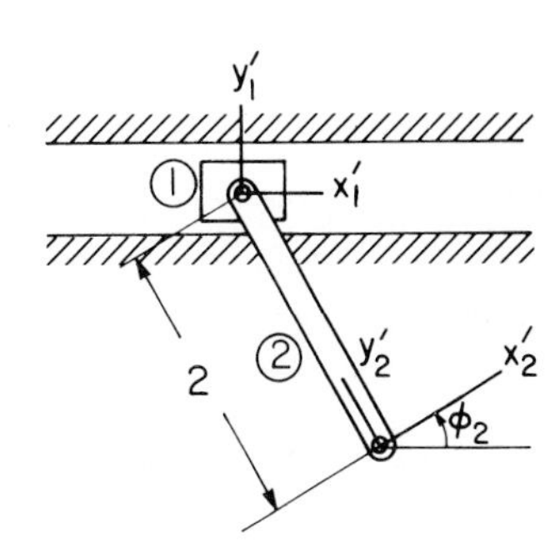

Figure P3.6.2

Section 3.7

3.7.1. Consider the slider–crank mechanism of Fig. 3.3.6. Since the connecting rod (body 2) is longer than the crank (body 1), no singularity occurs when the crank is driven by ϕ_1. A singularity may occur, however, when the slider (body 3) is driven. Use physical arguments to analyze singular configurations of the system when the slider is driven.

3.7.2. Write out the constraint equations, with a driver for the slider, for the slider–crank of Prob. 3.7.1. Calculate the Jacobian and use the implicit function theorem to confirm the singular configurations identified in Prob. 3.7.1.

3.7.3. Let $\ell = \frac{1}{2}$ and $\omega = 1$ for the slider–crank of Example 3.7.3. Write the velocity equation of Eq. 3.6.10 and the acceleration equation of Eq. 3.6.11, and verify that $\dot{q}$ and $\ddot{q}$ approach ∞ as the system approaches the lock-up configuration by filling in the following table:

t	q	Φ_q	$-\Phi_t$	$\dot{q}$	γ	$\ddot{q}$
0.4						
0.45						
0.5						
0.52						
0.523						
0.5235						

3.7.4. Write the constraint equation for the slider–crank of Example 3.7.1, with $\ell = 1$. Construct the composite constraint equation, including a driving constraint $\phi_1 - \omega t = 0$. Evaluate the Jacobian at $\phi_1 = 1.4835$ (85°) and at $\phi_1 = 1.6581$ (95°), and verify that the sign of the determinant of the Jacobian changes when the system passes the bifurcation point.

CHAPTER FOUR

Numerical Methods in Kinematics

The governing equations of kinematics developed in Chapter 3 may be assembled and solved to determine the position, velocity, and acceleration of a system, provided adequate driving conditions are specified. While the kinematic equations that determine the position of the system are highly nonlinear, the velocity and acceleration equations are linear and may be solved using matrix factorization and solution techniques that are well suited for computer implementation. In addition to the inherently nonlinear nature of kinematic equations, the large number of variables involved makes it impractical to even write the explicit governing equations for large-scale systems, much less solve them analytically. Numerical methods that permit both computer assembly and solution of kinematic equations are presented in this chapter.

Four distinct modes of kinematic computation are considered. The first is the determination of an assembled configuration of a system, given only estimates for the position and orientation of each component and the governing constraint equations. While the basic theory of system assembly is similar to position analysis, that is, requiring solution of kinematic constraint and driving conditions, it differs from position analysis that is begun from an assembled configuration in that a good estimate of the assembled configuration may not be available. Furthermore, since kinematic position equations are highly nonlinear, a mechanical system may be specified that cannot be physically assembled or for which constraints are not independent, hence leading to severe mathematical and numerical difficulties.

Once an assembled configuration is obtained with independent kinematic and driving constraints, rapidly convergent and efficient numerical methods may be used to solve the kinematic position equations at each instant on a time grid. This is the second mode of kinematic analysis. Once position is known, the velocity equations may be solved and the results used, in conjunction with position information, to evaluate and solve the acceleration equations. These are the third and fourth modes of kinematic analysis.

Prior to launching into a technical treatment of each of the four modes of kinematic analysis, Sections 4.1 and 4.2 are devoted to the important task of organizing computations and assembling the required equations of kinematics.

This aspect of computer-aided kinematic analysis is the key to harnessing the power of the high-speed computer and relieving the engineer of the burden of equation formulation.

4.1 ORGANIZATION OF COMPUTATIONS

A large-scale kinematics and dynamics computer code, called the Dynamic Analysis and Design System (DADS) [27], has been developed to implement the theory presented in this text for both the kinematic and dynamic analysis of planar and spatial systems. The purpose of this section is to outline the organization of computations within the DADS code that carry out the four phases of kinematic analysis. The DADS code is used to formulate and solve planar kinematics and dynamics examples in Chapters 5 and 8. Spatial kinematics and dynamics applications are studied in Chapters 10 and 12.

Illustrations of the Cartesian coordinate planar kinematics formulation presented in Chapter 3 were limited to small-scale mechanisms. For simple systems, it is practical to manually define body numbering and kinematic constraint equation ordering, evaluate Jacobian matrices and right sides for velocity and acceleration equations, and solve the equations. For large-scale applications, however, even the bookkeeping required for organizing kinematic equations can be quite time consuming and prone to error. The computational flow within the DADS computer code that implements these organizational calculations and forms and solves the resulting equations is briefly summarized in Fig. 4.1.1. For support of kinematic analysis, the DADS code consists of three basic elements: (1) a preprocessor that collects problem definition information and organizes data for computation, (2) a kinematic analysis program that constructs and solves the equations of kinematics, and (3) a postprocessor that prepares output information and displays the results of a kinematic simulation. The DADS code uses many subroutines and modules to support both kinematic and dynamic analysis. Only those functions carried out for kinematic analysis are outlined in this section.

Logic is programmed into the preprocessor to permit the user to enter body names or numbers and define types of joints that connect pairs of bodies. This defines the structure of the kinematic system. The user then enters detailed data for each of the joints, which are used in the formulation of the equations of kinematics. If the user desires a graphical portrayal of motion of the system as output, then the outline geometry of individual bodies must also be entered. Finally, the user enters the desired simulation time interval, defines data that control the numerical solution process, and defines the form of output information desired for subsequent control of the postprocessor. The output of preprocessor computation is a kinematic analysis data set that is read and implemented by the kinematic analysis program.

During kinematic analysis, equations are automatically assembled and

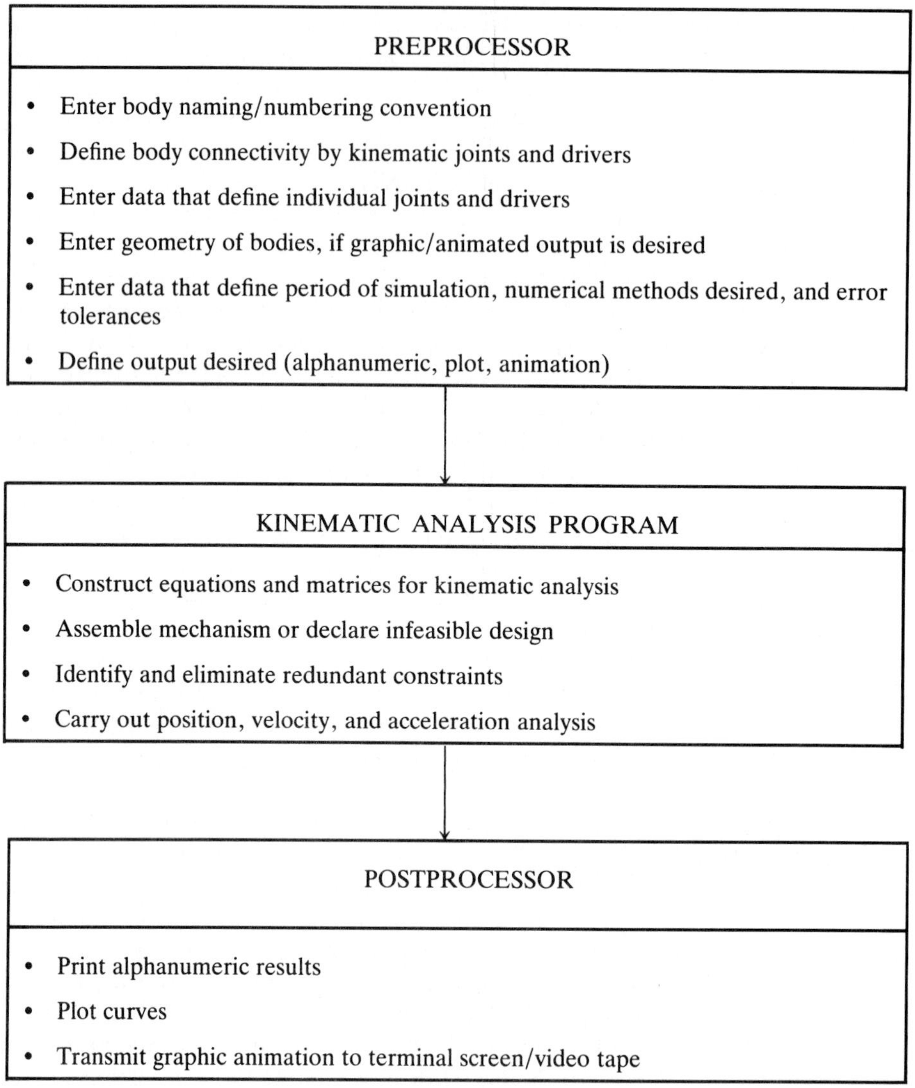

PREPROCESSOR
• Enter body naming/numbering convention • Define body connectivity by kinematic joints and drivers • Enter data that define individual joints and drivers • Enter geometry of bodies, if graphic/animated output is desired • Enter data that define period of simulation, numerical methods desired, and error tolerances • Define output desired (alphanumeric, plot, animation)

↓

KINEMATIC ANALYSIS PROGRAM
• Construct equations and matrices for kinematic analysis • Assemble mechanism or declare infeasible design • Identify and eliminate redundant constraints • Carry out position, velocity, and acceleration analysis

↓

POSTPROCESSOR
• Print alphanumeric results • Plot curves • Transmit graphic animation to terminal screen/video tape

Figure 4.1.1 DADS computational flow.

solved. The assembly analysis outlined in Section 3.6 is carried out, beginning with estimates provided during preprocessing. If the system assembles, kinematic analysis proceeds. If the system fails to assemble, the user is informed that poor initial estimates have been provided or that an infeasible design has been specified. If the system assembles, computational checks are carried out with the constraint Jacobian to determine if redundant constraints have been defined in the model by evaluating the row rank of the Jacobian $\mathbf{\Phi_q}$. If the Jacobian does not have full row rank, the user is informed that redundant constraints are present and that a reformulation of the model is required. Redundant constraints are automatically removed and, if the proper number of drivers has been specified, analysis is carried out. Once a feasible model is established, position, velocity, and acceleration analysis are executed to predict the kinematic performance of the system over the time interval specified by the user. An output data set that defines the results of analysis is transmitted to the postprocessor for display.

Finally, the postprocessor controls the printing of tabular output and construction of plots of variables of interest, or transmits graphic instructions to a high-speed graphics device that displays an animation of system motion on a terminal screen and/or video tape for subsequent playback and evaluation.

Implementation of the extensive logical and numerical computations indicated by the computational flow in Fig. 4.1.1 requires a large-scale computer code, the details of which are beyond the scope of this text. It is important, however, that the user of such software understand the basic ideas and computational methods that are employed, in order to be able to interpret results and create meaningful and appropriate models that represent reality. Prior to delving into the numerical methods that are used to carry out these computations in later sections of this chapter, it is instructive to review the method that is used to control the flow of computation.

The structure of the DADS kinematic analysis program that carries out the numerical computation is shown in Fig. 4.1.2. The kinematic analysis program consists of three basic elements: (1) an analysis program that controls the process of kinematic analysis, (2) a junction program that assigns tasks to a set of computation modules (subroutines), and (3) a library of modules associated with bodies and joint types that are supported by the formulation.

The analysis program use flags (integers) to define the phase of analysis that is being carried out and passes instructions to the junction subroutine to assemble terms in the required equations during each computational step. The junction subroutine subsequently assigns evaluation tasks to computation modules that evaluate needed terms in the kinematic equations. When these terms are returned to the junction subroutine, it assembles them into arrays that are subsequently passed to the analysis program. The analysis program uses these arrays to carry out computations that solve assigned equations. The process is repeated as kinematic analysis proceeds. The computational aspects of these steps are presented in subsequent sections of this chapter.

A more detailed view of the computational flow and information that is generated during analysis is presented in Fig. 4.1.3. Input data are initially read from the preprocessor and the problem setup phase is initiated under control of the analysis program. Data on individual bodies and joints are passed to modules through the junction subroutine. The modules then accumulate the number of generalized coordinates, the number of constraint equations, and the number and addresses (row and column numbers) of nonzero entries in Jacobian matrices.

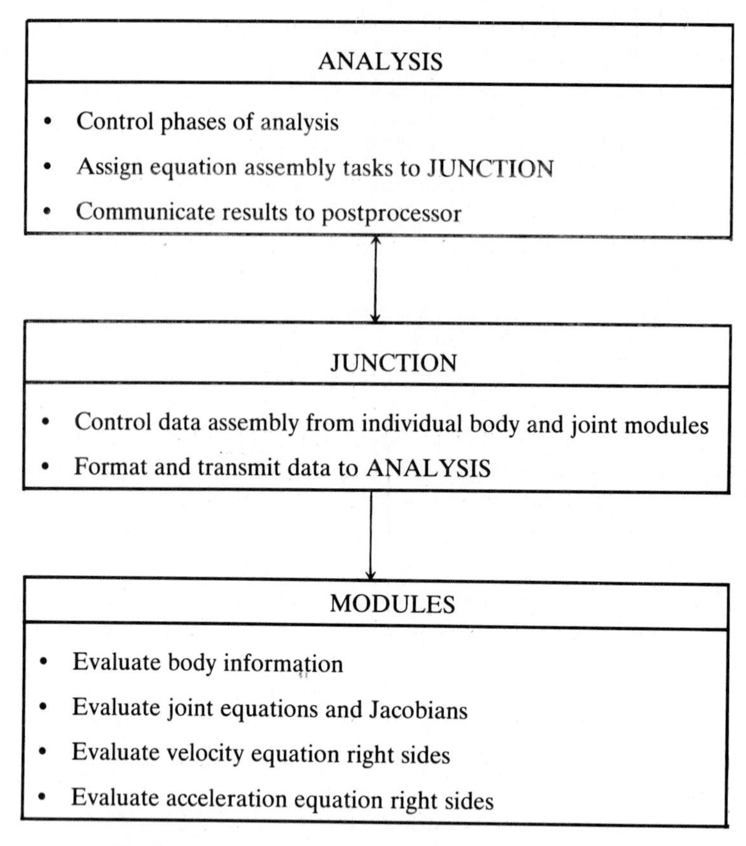

Figure 4.1.2 Structure of DADS kinematic analysis program.

This information is passed back to analysis through the junction subroutine. The problem setup phase is complete when the analysis program establishes array dimensions and addresses that are required for the execution of kinematic analysis.

The model assembly and feasibility analysis phase is carried out under the control of analysis using the assembly minimization method outlined in Section 3.6. At each iteration of the minimization process, modules generate the constraint equations and Jacobian information that are required for the minimization computation that is presented in Section 4.3. Following successful assembly, an analysis subroutine carries out a computational check on the rank of the constraint Jacobian, using the Gaussian elimination method that is presented in Section 4.4, to identify any redundant constraints that may exist. If a feasible model has been specified, analysis proceeds. If not, the user is informed that refinement of the model or estimated data is required.

Position analysis is carried out at each time step in the analysis program

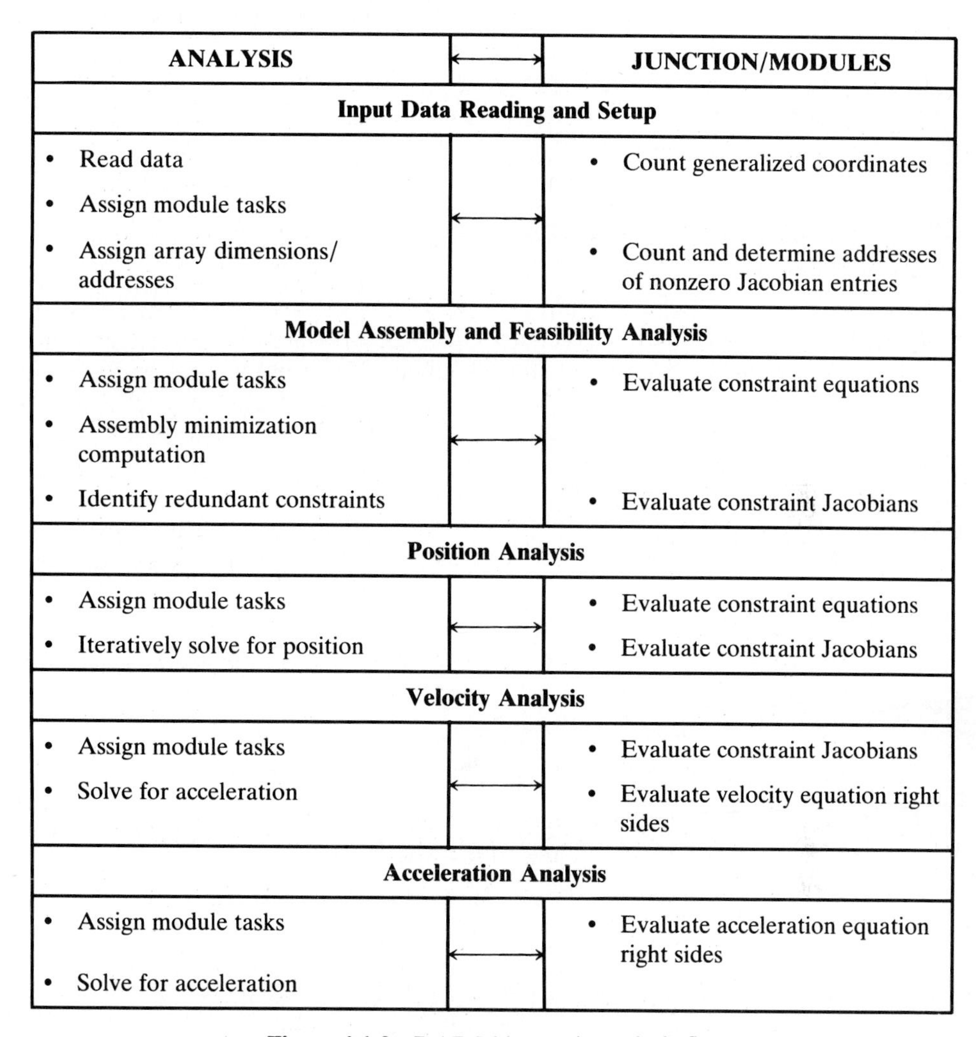

Figure 4.1.3 DADS kinematic analysis flow.

using the iterative Newton–Raphson method that is presented in Section 4.5. During each iteration, modules provide information on constraint equation violations and Jacobians. Upon completion of position iteration at a given time step, velocity analysis is initiated, carrying out the computations defined in Section 3.6. During velocity analysis, modules evaluate constraint Jacobian entries and the right side of the velocity equation. Following completion of velocity analysis at the given time step, acceleration analysis is carried out. Modules need provide only the right side of the acceleration equation, since the constraint Jacobian is identical to that constructed during velocity analysis. Upon completion of acceleration analysis, if the final time has been reached, analysis terminates. If not, time is indexed to the next time step and control passes back to position analysis. This process continues until the kinematic simulation is complete.

4.2 EVALUATION OF CONSTRAINT EQUATIONS AND JACOBIAN

When the constraint Jacobian needs to be evaluated, a loop is entered that calls one module for each joint that is defined in the input data, as indicated in Fig. 4.1.2. A vector of nonzero Jacobian entries is evaluated and sent, along with the row and column indexes (pointers), to a subroutine for matrix factoring.

Following assembly, three distinct phases are executed during kinematic analysis over an interval of time. Position analysis involves solving the set of nonlinear constraint equations to find a set of generalized coordinates that satisfies all constraints, as shown in Fig. 4.1.3. A Newton–Raphson algorithm is used to iterate for a solution, taking advantage of the structure of the Jacobian. The following system of linear equations is solved for updated coordinates, the Jacobian is recalculated, and the equations are solved again, until each of the coordinate errors is less than a specified tolerance:

$$\begin{aligned} \mathbf{\Phi}_{\mathbf{q}}^{(i+1)}\Delta\mathbf{q}^{(i)} &= -\mathbf{\Phi}^{(i)} \\ \mathbf{q}^{(i)} &= \mathbf{q}^{(i)} + \Delta\mathbf{q}^{(i)} \end{aligned} \tag{4.2.1}$$

After finding a solution that satisfies constraints within tolerance, the Jacobian must be recalculated prior to solving for velocities from

$$\mathbf{\Phi}_{\mathbf{q}}\dot{\mathbf{q}} = -\mathbf{\Phi}_t \equiv \mathbf{v} \tag{4.2.2}$$

The last phase solves the acceleration equations

$$\mathbf{\Phi}_{\mathbf{q}}\ddot{\mathbf{q}} = -[(\mathbf{\Phi}_{\mathbf{q}}\dot{\mathbf{q}})_{\mathbf{q}}\dot{\mathbf{q}} + 2\mathbf{\Phi}_{\mathbf{q}t}\dot{\mathbf{q}} + \mathbf{\Phi}_{tt}] \equiv \boldsymbol{\gamma} \tag{4.2.3}$$

for the acceleration vector. Note that each of these equations uses the same Jacobian matrix. Throughout all calculations, the equation and variable ordering that were established during preprocessing are retained.

A specific example of how the nonzero entry scheme works shows each step involved. For the revolute joint, there are two constraint equations that arise from the vector relation

$$\mathbf{\Phi}^{r(i,j)} \equiv \mathbf{r}_j + \mathbf{A}_j\mathbf{s}_j'^P - (\mathbf{r}_i + \mathbf{A}_i\mathbf{s}_i'^P) = \mathbf{0}$$

or, in scalar form,

$$\begin{aligned} \Phi_1^{r(i,j)} &\equiv x_j + x_j'^P \cos\phi_j - y_j'^P \sin\phi_j - x_i - x_i'^P \cos\phi_i + y_i'^P \sin\phi_i = 0 \\ \Phi_2^{r(i,j)} &\equiv y_j + x_j'^P \sin\phi_j + y_j'^P \cos\phi_j - y_i - x_i'^P \sin\phi_i - y_i'^P \cos\phi_i = 0 \end{aligned} \tag{4.2.4}$$

The Jacobian for this joint has 8 nonzero entries out of 12 possible Jacobian matrix positions associated with variables of bodies i and j. The nonzero Jacobian entries are

$$\begin{aligned}
&\text{ENTRY}\,(1) = -1.0\\
&\text{ENTRY}\,(2) = x_i'^P \sin\phi_i + y_i'^P \cos\phi_i\\
&\text{ENTRY}\,(3) = 1.0\\
&\text{ENTRY}\,(4) = -x_j'^P \sin\phi_j - y_j'^P \cos\phi_j\\
&\text{ENTRY}\,(5) = -1.0\\
&\text{ENTRY}\,(6) = -x_i'^P \cos\phi_i + y_i'^P \sin\phi_i\\
&\text{ENTRY}\,(7) = 1.0\\
&\text{ENTRY}\,(8) = x_j'^P \cos\phi_j - y_j'^P \sin\phi_j
\end{aligned} \tag{4.2.5}$$

organized as shown in Fig. 4.2.1.

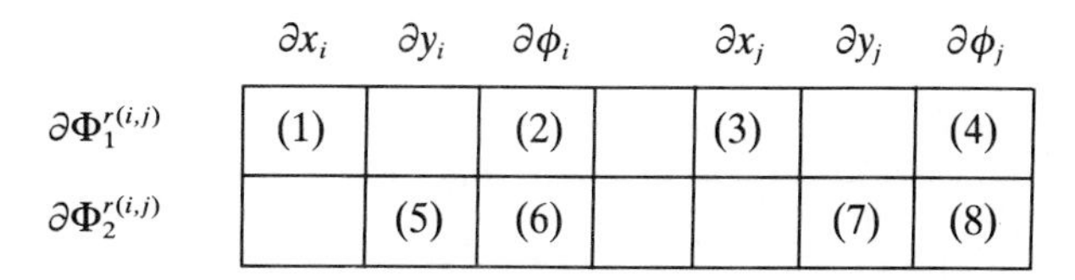

Figure 4.2.1 Jacobian entries for revolute joint.

Regardless of how many bodies there are in a model, all entries in columns of the Jacobian associated with variables for bodies other than i and j are zero, since these variables do not appear in this constraint equation. If, for example, there are 15 bodies in a system, only 4 of 45 entries in each revolute Jacobian row are nonzero. Thus, only about 10% of the entries are other than zero. Use of this fact saves computer memory and computer time by suppressing operations with zeros.

Example 4.2.1: The four-bar mechanism of Fig. 4.2.2 is made up of four bodies, one of which is ground (body 4). There are four revolute joints, numbered R_1 through R_4, with constraint equations $\mathbf{\Phi}_i^r = 0$, $i = 1, 2, 3$, and 4, of Eq. 4.2.4. Three absolute constraints of Eqs. 3.2.3, 3.2.4, and 3.2.6 are imposed on body 4 to fix it to ground. Finally, an absolute angle driving constraint of Eq. 3.5.3 is imposed on body 1, with $C_3(t) = \omega t$, to cause it to rotate with angular velocity ω (Prob. 4.2.1).

Denoting by $\partial\Phi_1^{ri}$ and $\partial\Phi_2^{ri}$ the rows of the Jacobian corresponding to joint ri, $i = 1, 2, 3$, and 4, placing the Jacobian entries of the ground constraints in rows 9, 10, and 11, and placing the Jacobian entries for the driving constraint

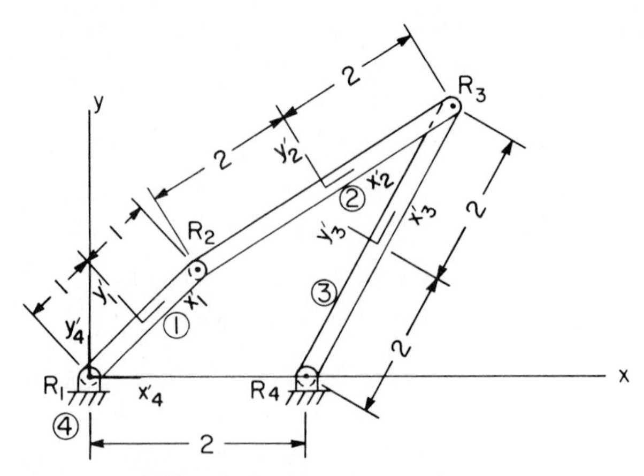

Figure 4.2.2 Four-bar mechanism

in row 12, using Eq. 4.2.5, the Jacobian matrix is (Prob. 4.2.2)

$\mathbf{\Phi_q} =$

	∂x_1	∂y_1	$\partial \phi_1$	∂x_2	∂y_2	$\partial \phi_2$	∂x_3	∂y_3	$\partial \phi_3$	∂x_4	∂y_4	$\partial \phi_4$
$\partial \Phi_1^{r1}$	-1	0	$-\sin \phi_1$	0	0	0	0	0	0	1	0	0
$\partial \Phi_2^{r1}$	0	-1	$\cos \phi_1$	0	0	0	0	0	0	0	1	0
$\partial \Phi_1^{r2}$	-1	0	$\sin \phi_1$	1	0	$2 \sin \phi_2$	0	0	0	0	0	0
$\partial \Phi_2^{r2}$	0	-1	$-\cos \phi_1$	0	1	$-2 \cos \phi_2$	0	0	0	0	0	0
$\partial \Phi_1^{r3}$	0	0	0	-1	0	$2 \sin \phi_2$	1	0	$-2 \sin \phi_3$	0	0	0
$\partial \Phi_2^{r3}$	0	0	0	0	-1	$-2 \cos \phi_2$	0	1	$2 \cos \phi_3$	0	0	0
$\partial \Phi_1^{r4}$	0	0	0	0	0	0	-1	0	$-2 \sin \phi_3$	1	0	$-2 \sin \phi_4$
$\partial \Phi_2^{r4}$	0	0	0	0	0	0	0	-1	$2 \cos \phi_3$	0	1	$2 \cos \phi_4$
$\partial \Phi^{ax}$	0	0	0	0	0	0	0	0	0	1	0	0
$\partial \Phi^{ay}$	0	0	0	0	0	0	0	0	0	0	1	0
$\partial \Phi^{a\phi}$	0	0	0	0	0	0	0	0	0	0	0	1
$\partial \Phi^{a\phi d}$	0	0	1	0	0	0	0	0	0	0	0	0

(4.2.6)

Note that only 34 of the 144 entries in this 12×12 matrix are nonzero.

The right side $\mathbf{v}$ of the velocity equation of Eq. 4.2.2 has $v_{12} = \omega$ and $v_i = 0$, $i = 1, \ldots, 11$. Since the last four constraints are linear in $\mathbf{q}$ and t, $\gamma_i = 0$, $i = 9$, 10, 11, 12, in Eq. 4.2.3. The first eight components of $\boldsymbol{\gamma}$ in Eq. 4.2.3 are made up of contributions from the four revolute joint constraints, which may be evaluated using Eq. 4.2.4 (Prob. 4.2.3).

4.3 ASSEMBLY OF A SYSTEM

Recall the system constraint equations of Eq. 3.6.3 at the initial time t_0:

$$\mathbf{\Phi}(\mathbf{q}, t_0) = \begin{bmatrix} \mathbf{\Phi}^K(\mathbf{q}) \\ \mathbf{\Phi}^D(\mathbf{q}, t_0) \end{bmatrix} = \mathbf{0} \quad \textbf{(4.3.1)}$$

The number of kinematic and driving constraints is normally selected to be equal to the number of generalized coordinates in the system. If the constraints are independent, the Jacobian will be nonsingular. However, it is possible to attempt assembly of the system without specifying an adequate number of driving constraints and perhaps to have specified kinematic or driving constraints that are dependent on other constraints that act on the system, in which case the constraint Jacobian will be rank deficient and Newton–Raphson iteration may fail. Furthermore, the system may have a singular configuration at t_0 or it may be impossible to assemble, in which case Newton–Raphson iteration will fail.

Example 4.3.1: The elementary two-body model of a slider–crank mechanism studied in Example 3.7.3, shown in Fig. 4.3.1, is driven by the condition $\phi_1 = \omega t$. Since absolute constraints require that $x_1 = y_1 = y_2 = \phi_2 = 0$, $x_2 = q$ may be taken as the only generalized coordinate, which must satisfy the distance constraint

$$\Phi(q, t) = (q - \cos \omega t)^2 + \sin^2 \omega t - \ell^2 = 0$$

If $\ell < 1$, then no solution exists beyond the time t^* for which $\sin \omega t^* = \ell$ (i.e., when the coupler is vertical). To observe the behavior of Newton–Raphson iteration in attempting to assemble this simple system, let $\ell = \sqrt{2}/2$ and $\omega = 1$. The critical or lock-up time is then $t^* = \pi/4$. For Newton–Raphson iteration,

$$\Phi_q(q, t) = 2(q - \cos t)$$

Computations are carried out for the following three cases:

Case 1: $t_0 = 0$, $q^{(0)} = 2$. Newton–Raphson iteration yields the results of Table

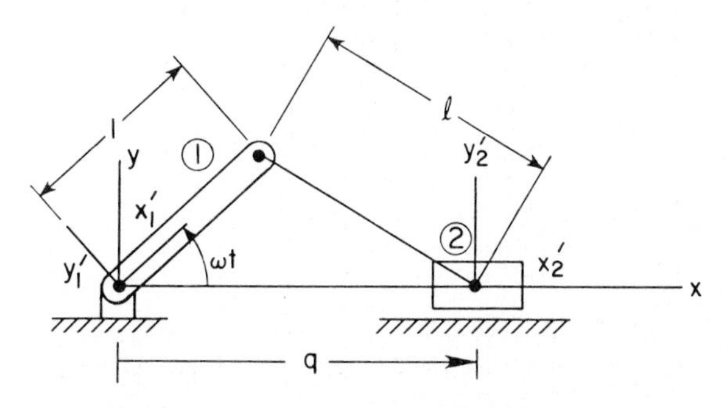

Figure 4.3.1 Two-body slider–crank model.

TABLE 4.3.1 Slider–Crank Results: $t_0 = 0$, $q^{(0)} = 2$

i	$q^{(i)}$	$\Phi(q^{(i)}, 0)$	$\Phi_q(q^{(i)}, 0)$	$\Delta q^{(i)}$	$\left\|q^{(i)} - \frac{2+\sqrt{2}}{2}\right\|$
0	2.000000	0.500000	2.000000	−0.250000	0.292893
1	1.750000	0.062500	1.500000	−0.041667	0.042893
2	1.708333	0.001736	1.416607	−0.001225	0.001226
3	1.707108	0.000000	1.414216	−0.000000	0.000001

TABLE 4.3.2 Slider–Crank Results: $t_0 = \pi/4$, $q^0 = 1$

i	$q^{(i)}$	$\Phi(q^{(i)}, \pi/4)$	$\Phi_q(q^{(i)}, \pi/4)$	$\Delta q^{(i)}$	$\left\|q^{(i)} - \frac{\sqrt{2}}{2}\right\|$
0	1.000000	0.085786	0.585786	−0.146447	0.292893
1	0.853553	0.021447	0.292893	−0.073223	0.146447
2	0.780330	0.005362	0.146447	−0.036612	0.073223
3	0.743718	0.001340	0.073223	−0.018306	0.036612
4	0.725413	0.000335	0.036612	−0.009153	0.018306
5	0.716260	0.000084	0.018306	−0.004576	0.009153
6	0.711683	0.000021	0.009153	−0.002288	0.004576
7	0.709395	0.000005	0.004576	−0.001144	0.002288
8	0.708251	0.000001	0.002288	−0.000572	0.001144
9	0.707679	0.000000	0.001144	−0.000286	0.000572
10	0.707393	0.000000	0.000572	−0.000143	0.000286
11	0.707250	0.000000	0.000286	−0.000072	0.000143
12	0.707178	0.000000	0.000143	−0.000036	0.000072
13	0.707143	0.000000	0.000072	−0.000018	0.000036
14	0.707125	0.000000	0.000036	−0.000009	0.000018
15	0.707116	0.000000	0.000018	−0.000004	0.000009
16	0.707111	0.000000	0.000009	−0.000002	0.000004
17	0.707109	0.000000	0.000004	−0.000001	0.000002
18	0.707108	0.000000	0.000003	−0.000001	0.000002
19	0.707108	0.000000	0.000002	−0.000000	0.000001
20	0.707107	0.000000	0.000001	−0.000000	0.000000

4.3.1. Note that the solution error decreases quadratically; that is,

$$\left|q^{(i+1)} - \frac{2+\sqrt{2}}{2}\right| \leq c \left|q^{(i)} - \frac{2+\sqrt{2}}{2}\right|^2$$

where $c \approx 0.06$ and $q = (2+\sqrt{2})/2$ is the solution. This property is typical of Newton–Raphson iteration (see Section 4.5) if $\mathbf{\Phi_q}$ is nonsingular, which is true in this case.

Case 2: $t_0 = \pi/4$, $q^{(0)} = 1$. Newton–Raphson iteration yields the results of Table 4.3.2. Note that even though the Newton–Raphson algorithm converges to the singular solution $q = \sqrt{2}/2$ the rate of convergence is much slower than in case 1. In fact, for singular equations that have a solution, the Newton–Raphson algorithm converges only linearly. For systems of many equations in many unknowns, numerical error may dominate and create very slow convergence or even failure to converge.

Case 3: $t_0 = 3\pi/8$, $q^{(0)} = 1$. Newton-Raphson iteration yields the results of Table 4.3.3. As may have been expected, since there is no solution in this case, the

TABLE 4.3.3 Slider–Crank Results: $t_0 = 3\pi/8$, $q^0 = 1$

i	$q^{(i)}$	$\Phi(q^{(i)}, 3\pi/8)$	$\Phi_q(q^{(i)}, 3\pi/8)$	$\Delta q^{(i)}$
0	1.000000	0.734633	1.234633	−0.595021
1	0.404979	0.354050	0.044590	−7.940071
2	−7.535093	63.044735	−15.835553	3.981215
3	−3.553878	15.850071	−7.873125	2.013187
4	−1.540691	4.052922	−3.846749	1.053597
5	−0.487094	1.110066	−1.739555	0.638132
6	0.151038	0.407213	−0.463291	0.878958
7	1.029996	0.772566	1.294024	−0.596750
8	0.433246	0.356110	0.101125	−3.521482
9	−3.088236	12.400835	−6.941839	1.780391
10	−1.301846	3.191191	−3.369058	0.947206
11	−0.354640	0.897199	−1.474646	0.608416
12	0.253777	0.370170	−0.257814	1.435806
13	1.689582	2.061537	2.613797	−0.788714
14	0.900868	0.622069	1.030370	−0.600238
15	0.300630	0.360286	−0.164107	2.195438

Newton–Raphson algorithm cannot possibly converge. In this case, it simply meanders. In some cases, it may diverge and give computer overflow.

Since the nonlinear constraint equations of Eq. 4.3.1 may contain redundant equations, they may not uniquely determine the position of the system. In fact, they may have no solution. A stable computational method is required to obtain a good estimate of the assembled position of the system or to determine that no solution exists. The basic idea employed is to minimize error in satisfying Eq. 4.3.1, to find a solution of Eq. 4.3.1 that is as near as possible to the initial estimate $\mathbf{q}^0$. The estimate $\mathbf{q}^0$ of the assembled configuration that is provided by the user should be as reasonably accurate. The user may employ experimental data or graphical estimates of the position and orientation of each body in the system to create an estimate $\mathbf{q}^0$ of the assembled configuration.

As outlined in Section 3.6, the objective is to minimize

$$\psi(\mathbf{q}, t_0, r) = (\mathbf{q} - \mathbf{q}^0)^T(\mathbf{q} - \mathbf{q}^0) + r\mathbf{\Phi}^T(\mathbf{q}, t_0)\mathbf{\Phi}(\mathbf{q}, t_0) \tag{4.3.2}$$

to obtain an assembled configuration $\mathbf{q}^a$. The parameter $r > 0$ is a weighting constant that places relative emphasis on deviation from the initial estimate $\mathbf{q}^0$ and solution of Eq. 4.3.1. Since a solution of Eq. 4.3.1 is desired, if one exists, the parameter r in Eq. 4.3.2 is allowed to grow to a very large number in order to precisely satisfy the constraint equations. The method employed is to find $\mathbf{q}(r)$ to minimize the function ψ in Eq. 4.3.2 for a given r, using an optimization algorithm. The parameter r is then increased and the minimization problem is solved again using $\mathbf{q}(r)$ from the previous solution as a starting point. Incrementally increasing r is important, since if the optimization computation is initiated with a very large value of r, convergence might be difficult to achieve, particularly in cases for which no solution of Eq. 4.3.1 exists. At particular values of r, the function ψ of Eq. 4.3.2 is minimized to obtain $\mathbf{q}(r)$, and the limit as r approaches infinity is approximated to obtain

$$\mathbf{q}^a = \lim_{r \to \infty} \mathbf{q}(r) \tag{4.3.3}$$

In actual computation, r is increased in discrete steps, $r_i > r_{i-1}$, and convergence of the limit of Eq. 4.3.3 is tested by evaluating the difference between $\mathbf{q}(r_i)$ and $\mathbf{q}(r_{i-1})$.

If the sequence $\mathbf{q}(r_i)$ converges, then a check is required to confirm that $\mathbf{q}^a$ satisfies Eq. 4.3.1. If it does, an assembled configuration has been achieved. If it does not, it may be concluded either that the system cannot be assembled or that the iterative optimization algorithm is incapable of converging from the initial estimate of the assembled configuration that was given. In the latter case, the user is encouraged to initiate the minimization process from other estimates. While this is not a precise mathematical process, it can be greatly assisted by engineering intuition and graphically obtained estimates.

If there are fewer than nc (the dimension of $\mathbf{q}$) constraint equations in Eq. 4.3.1, or if there are nc equations and some are dependent, then there may be many solutions of Eq. 4.3.1. In either of these cases, the minimization solution $\mathbf{q}^a$, if it satisfies Eq. 4.3.1, is the solution that minimizes the first term in Eq. 4.3.2. In the least-square sense, it is the solution of Eq. 4.3.1 that is nearest the initial estimate $\mathbf{q}^0$.

A method that is well suited for minimizing the function of Eq. 4.3.2 for large-scale systems is called the *conjugate gradient minimization algorithm* of nonlinear programming [28]. The implementation employed here is presented in reference 28. This and virtually all other iterative optimization methods require the gradient of the function that is to be minimized. For the function ψ in Eq. 4.3.2, this gradient is

$$\psi_{\mathbf{q}} = 2(\mathbf{q} - \mathbf{q}^0)^T + 2r\boldsymbol{\Phi}^T(\mathbf{q}, t_0)\boldsymbol{\Phi}_{\mathbf{q}}(\mathbf{q}, t_0) \tag{4.3.4}$$

It is important to note that evaluation of this gradient requires only the constraint Jacobian and the constraint function. Thus, the computational machinery that is needed to implement the remaining modes of kinematic analysis is directly applicable to carry out numerical minimization in the assembly mode.

The iterative conjugate gradient algorithm employed here was developed by Fletcher and Powell [29] and involves the following sequence of computations [28, 29]:

1. Begin with the estimate $\mathbf{q}^{(1)} = \mathbf{q}^0$ of $\mathbf{q}$ and $\mathbf{H}^{(1)} = \mathbf{H}^0 = \mathbf{I}$ as an estimate of the matrix of second derivatives of ψ.
2. At iteration i, compute

$$\mathbf{s}^i = -\mathbf{H}^{(i+1)}\psi_{\mathbf{q}}^T(\mathbf{q}^{(i)})$$

3. Use a one-dimensional search algorithm [28] to find $\alpha = \alpha^i$ that minimizes $\psi(\mathbf{q}^{(i)} + \alpha\mathbf{s}^i)$.
4. Compute

$$\mathbf{q}^{(i+1)} = \mathbf{q}^{(i)} + \alpha^i\mathbf{s}^i$$
$$\mathbf{H}^{(i+1)} = \mathbf{H}^{(i)} + \mathbf{A}^i + \mathbf{C}^i$$

where

$$\mathbf{A}^i = \left(\frac{\alpha^i}{\mathbf{s}^{iT}\mathbf{y}^i}\right)\mathbf{s}^i\mathbf{s}^{iT}$$
$$\mathbf{y}^i = \psi_{\mathbf{q}}^T(\mathbf{q}^{(i+1)}) - \psi_{\mathbf{q}}^T(\mathbf{q}^{(i)})$$
$$\mathbf{C}^i = -\left(\frac{1}{\mathbf{y}^{iT}\mathbf{H}^{(i)}\mathbf{y}^i}\right)\mathbf{H}^{(i)}\mathbf{y}^i\mathbf{y}^{iT}\mathbf{H}^{(i)}$$

5. If $\psi_{\mathbf{q}}(\mathbf{q}^{(i+1)}) = \mathbf{0}$ or if $\mathbf{q}^{(i+1)} - \mathbf{q}^{(i)}$ is sufficiently small, terminate. Otherwise, return to step 2 with $i \rightarrow i + 1$.

Examples in which this algorithm is used for the assembly of large-scale systems are given in Chapters 5 and 10.

4.4 LINEAR EQUATION SOLUTION AND MATRIX FACTORIZATION

Matrix equations arise in both the theory and numerical solution of kinematics problems. The rank of the constraint Jacobian provides information on the number of degrees of freedom of the system and must usually be determined numerically. Velocity and acceleration equations derived in Chapter 3 are matrix equations that must be solved numerically. Finally, the position equations are generally nonlinear, but are solved iteratively through solution of a sequence of linear equations (see Section 4.5).

The purpose of this section is to provide an introduction to the analysis and solution of matrix equations. The reader who is interested in more detail is referred to the literature on matrix theory and numerical methods (e.g., [22], [30]).

4.4.1 Gaussian Methods

Consider a system of n linear algebraic equations in n unknowns, with real constant coefficients,

$$\begin{aligned} a_{11}x_1 + a_{12}x_2 + \cdots + a_{1n}x_n &= b_1 \\ a_{21}x_1 + a_{22}x_2 + \cdots + a_{2n}x_n &= b_2 \\ &\vdots \\ a_{n1}x_1 + a_{n2}x_2 + \cdots + a_{nn}x_n &= b_n \end{aligned} \tag{4.4.1}$$

which can be written in matrix form as

$$\mathbf{Ax} = \mathbf{b} \tag{4.4.2}$$

where

$$\mathbf{x} = [x_1, x_2, \ldots, x_n]^T$$

$$\mathbf{b} = [b_1, b_2, \ldots, b_n]^T$$

$$\mathbf{A} = \begin{bmatrix} a_{11} & a_{12} & \cdots & a_{2n} \\ a_{21} & a_{22} & \cdots & a_{2n} \\ & & \vdots & \\ a_{n1} & a_{n2} & \cdots & a_{nn} \end{bmatrix}$$

There are many methods of solving Eq. 4.4.1. Cramer's rule [22] is one of the best known but most inefficient methods. Among the more efficient methods is *Gaussian elimination,* which is based on the elementary idea of eliminating variables one at a time. The Gaussian elimination method consists of two major steps, *forward elimination* and *back substitution.*

Forward Elimination First, make the coefficient of x_1 in the first equation unity by dividing that equation by a_{11} (presume for now that $a_{11} \neq 0$). Then eliminate the variable x_1 from the jth equation by multiplying the first modified equation by $-a_{j1}$ and adding the resulting equation to the jth equation, $j = 2, \ldots, n$, to obtain

$$\begin{bmatrix} 1 & a_{12}^{(1)} & a_{13}^{(1)} & \cdots & a_{1n}^{(1)} \\ 0 & a_{22}^{(1)} & a_{23}^{(1)} & \cdots & a_{2n}^{(1)} \\ \vdots & & & \vdots & \\ 0 & a_{n2}^{(1)} & a_{n3}^{(1)} & \cdots & a_{nn}^{(1)} \end{bmatrix} \begin{bmatrix} x_1 \\ x_2 \\ \vdots \\ x_n \end{bmatrix} = \begin{bmatrix} b_1^{(1)} \\ b_2^{(1)} \\ \vdots \\ b_n^{(1)} \end{bmatrix} \tag{4.4.3}$$

Next, make the coefficient of x_2 in the second equation of Eq. 4.4.3 equal to unity by dividing that equation by $a_{22}^{(1)}$, presuming for now that it is not zero. Then eliminate x_2 from the jth equation by adding $-a_{j2}$ times the second modified equation to the jth equation, $j = 3, \ldots, n$, to obtain

$$\begin{bmatrix} 1 & a_{12}^{(1)} & a_{13}^{(1)} & \cdots & a_{1n}^{(1)} \\ 0 & 1 & a_{23}^{(2)} & \cdots & a_{2n}^{(2)} \\ 0 & 0 & a_{(33)}^{(2)} & \cdots & a_{3n}^{(2)} \\ \vdots & & & & \\ 0 & 0 & a_{n3}^{(3)} & \cdots & a_{nn}^{(2)} \end{bmatrix} \begin{bmatrix} x_1 \\ x_2 \\ \vdots \\ \\ x_n \end{bmatrix} = \begin{bmatrix} b_1^{(1)} \\ b_2^{(2)} \\ \vdots \\ \\ b_n^{(2)} \end{bmatrix} \tag{4.4.4}$$

After $n-1$ steps of this algorithm, divide the nth row by $a_{nn}^{(n-1)}$, presuming for now that it is not zero, to obtain the final result of the forward-elimination step:

$$\begin{bmatrix} 1 & a_{12}^{(1)} & a_{13}^{(1)} & \cdots & a_{1n}^{(1)} \\ 0 & 1 & a_{23}^{(2)} & \cdots & a_{2n}^{(2)} \\ 0 & 0 & 1 & \cdots & a_{3n}^{(3)} \\ \vdots & & & & \\ 0 & 0 & 0 & \cdots & 1 \end{bmatrix} \begin{bmatrix} x_1 \\ x_2 \\ \vdots \\ \\ x_n \end{bmatrix} = \begin{bmatrix} b_1^{(1)} \\ b_2^{(2)} \\ \vdots \\ \\ b_n^{(n)} \end{bmatrix} \tag{4.4.5}$$

Observe that all elements below the diagonal of the matrix are zero, so the determinant of the matrix is 1; hence it is nonsingular.

Recall that to complete calculations leading to Eq. 4.4.5, it was presumed that the modified diagonal elements $a_{jj}^{(j-1)}$, obtained after $j-1$ steps, are not zero. Refinements in this algorithm are presented later in this section to overcome this limitation.

Back Substitution Back substitution consists of $n-1$ elementary solution steps, using Eq. 4.4.5. From the nth equation of Eq. 4.4.5, $x_n = b_n^{(n)}$. This value of x_n may be substituted into equation $n-1$. Equation $n-1$ then gives the value of x_{n-1}. This process is repeated, using the preceding equations, to solve for the remaining variables.

Example 4.4.1: Solve the matrix equation

$$\begin{bmatrix} 3 & 1 & -1 \\ -1 & 2 & 1 \\ 2 & -3 & 1 \end{bmatrix} \begin{bmatrix} x_1 \\ x_2 \\ x_3 \end{bmatrix} = \begin{bmatrix} 2 \\ 6 \\ -1 \end{bmatrix}$$

Forward elimination yields

1.
$$\begin{bmatrix} 1 & \frac{1}{3} & -\frac{1}{3} \\ -1 & 2 & 1 \\ 2 & -3 & 1 \end{bmatrix} \begin{bmatrix} x_1 \\ x_2 \\ x_3 \end{bmatrix} = \begin{bmatrix} \frac{2}{3} \\ 6 \\ -1 \end{bmatrix}$$

$$\begin{bmatrix} 1 & \frac{1}{3} & -\frac{1}{3} \\ 0 & \frac{7}{3} & \frac{2}{3} \\ 0 & -\frac{11}{3} & \frac{5}{3} \end{bmatrix} \begin{bmatrix} x_1 \\ x_2 \\ x_3 \end{bmatrix} = \begin{bmatrix} \frac{2}{3} \\ \frac{20}{3} \\ -\frac{7}{3} \end{bmatrix}$$

2.
$$\begin{bmatrix} 1 & \frac{1}{3} & -\frac{1}{3} \\ 0 & 1 & \frac{2}{7} \\ 0 & -\frac{11}{3} & \frac{5}{3} \end{bmatrix} \begin{bmatrix} x_1 \\ x_2 \\ x_3 \end{bmatrix} = \begin{bmatrix} \frac{2}{3} \\ \frac{20}{7} \\ -\frac{7}{3} \end{bmatrix}$$

$$\begin{bmatrix} 1 & \frac{1}{3} & -\frac{1}{3} \\ 0 & 1 & \frac{2}{7} \\ 0 & 0 & \frac{19}{7} \end{bmatrix} \begin{bmatrix} x_1 \\ x_2 \\ x_3 \end{bmatrix} = \begin{bmatrix} \frac{2}{3} \\ \frac{20}{7} \\ \frac{57}{7} \end{bmatrix}$$

3.
$$\begin{bmatrix} 1 & \frac{1}{3} & -\frac{1}{3} \\ 0 & 1 & \frac{2}{7} \\ 0 & 0 & 1 \end{bmatrix} \begin{bmatrix} x_1 \\ x_2 \\ x_3 \end{bmatrix} = \begin{bmatrix} \frac{2}{3} \\ \frac{20}{7} \\ 3 \end{bmatrix}$$

Back substitution yields

1. $x_3 = 3$
2. $x_2 + \frac{2}{7}(3) = \frac{20}{7} \rightarrow x_2 = 2$
3. $x_1 + \frac{1}{3}(2) - \frac{1}{3}(3) = \frac{2}{3} \rightarrow x_1 = 1$

In forward elimination, it is clear that the algorithm fails at the jth step if the *pivot element* $a_{jj}^{(j-1)}$ is zero. Also, when the pivot element $a_{jj}^{(j-1)}$ becomes very small, round-off error in computation may lead to erroneous results. Therefore, the order in which the equations are treated during forward elimination may significantly affect the accuracy of the solution obtained. To circumvent this difficulty, the ordering of equations is determined by the algorithm.

Row Pivoting In the jth forward-elimination step of Gaussian elimination, the equation with the largest coefficient (in absolute value) of x_j on or below the diagonal of the coefficient matrix is chosen. Before the elimination step, this row of the matrix and the right-side vector are interchanged with the jth row of the matrix and right side. This procedure is called *row pivoting* and amounts simply to interchanging the order of equations, which does not change the solution. The following example illustrates this procedure.

Example 4.4.2: Perform Gaussian elimination with row pivoting to solve the following set of equations:

$$\begin{bmatrix} 4 & -3 & 5 & 2 \\ -3 & 1 & 1 & -6 \\ 5 & -5 & 10 & 0 \\ 2 & -3 & 9 & -7 \end{bmatrix} \begin{bmatrix} x_1 \\ x_2 \\ x_3 \\ x_4 \end{bmatrix} = \begin{bmatrix} -1.5 \\ 9 \\ 2.5 \\ 13.5 \end{bmatrix}$$

(1) The largest coefficient in column 1 is 5. Interchanging the first and third equations,

$$\begin{bmatrix} 5 & -5 & 10 & 0 \\ -3 & 1 & 1 & -6 \\ 4 & -3 & 5 & 2 \\ 2 & -3 & 9 & -7 \end{bmatrix} \begin{bmatrix} x_1 \\ x_2 \\ x_3 \\ x_4 \end{bmatrix} = \begin{bmatrix} 2.5 \\ 9 \\ -1.5 \\ 13.5 \end{bmatrix}$$

Forward elimination now yields

$$\begin{bmatrix} 1 & -1 & 2 & 0 \\ 0 & -2 & 7 & -6 \\ 0 & 1 & -3 & 2 \\ 0 & -1 & 5 & -7 \end{bmatrix} \begin{bmatrix} x_1 \\ x_2 \\ x_3 \\ x_4 \end{bmatrix} = \begin{bmatrix} 0.5 \\ 10.5 \\ -3.5 \\ 12.5 \end{bmatrix}$$

(2) The largest coefficient in column 2 on or below the diagonal is -2, which is on the diagonal, so no interchange is necessary. Forward elimination yields

$$\begin{bmatrix} 1 & -1 & 2 & 0 \\ 0 & 1 & -3.5 & 3 \\ 0 & 0 & 0.5 & -1 \\ 0 & 0 & 1.5 & -4 \end{bmatrix} \begin{bmatrix} x_1 \\ x_2 \\ x_3 \\ x_4 \end{bmatrix} = \begin{bmatrix} 0.5 \\ -5.25 \\ 7.25 \\ 1.75 \end{bmatrix}$$

(3) The largest coefficient in column 3 on or below the diagonal is in the fourth row. Interchanging the third and fourth equations,

$$\begin{bmatrix} 1 & -1 & 2 & 0 \\ 0 & 1 & -3.5 & 3 \\ 0 & 0 & 1.5 & -4 \\ 0 & 0 & 0.5 & -1 \end{bmatrix} \begin{bmatrix} x_1 \\ x_2 \\ x_3 \\ x_4 \end{bmatrix} = \begin{bmatrix} 0.5 \\ -5.25 \\ 7.25 \\ 1.75 \end{bmatrix}$$

Forward elimination yields

$$\begin{bmatrix} 1 & -1 & 2 & 0 \\ 0 & 1 & -3.5 & 3 \\ 0 & 0 & 1 & -2.66 \\ 0 & 0 & 0 & 0.33 \end{bmatrix} \begin{bmatrix} x_1 \\ x_2 \\ x_3 \\ x_4 \end{bmatrix} = \begin{bmatrix} 0.5 \\ -5.25 \\ 4.83 \\ -0.66 \end{bmatrix}$$

(4) Finally,

$$\begin{bmatrix} 1 & -1 & 2 & 0 \\ 0 & 1 & 3.5 & 3 \\ 0 & 0 & 1 & -2.66 \\ 0 & 0 & 0 & 1 \end{bmatrix} \begin{bmatrix} x_1 \\ x_2 \\ x_3 \\ x_4 \end{bmatrix} = \begin{bmatrix} 0.5 \\ -5.25 \\ 4.83 \\ -2 \end{bmatrix}$$

Back substitution yields $\mathbf{x} = [0.5, -1, -0.5, -2]^T$.

Full Pivoting *Full pivoting* is the selection of the largest element (in absolute value) from among the diagonal element and all elements below and to the right of the diagonal element as the pivot for the next stage of forward elimination. In full pivoting, both row and column interchanges are required to bring the largest element to the diagonal. When columns of the coefficient matrix are interchanged, the corresponding variables in vector **x** are interchanged without changing the solution.

Example 4.4.3: Apply Gaussian elimination with full pivoting to solve the following set of equations:

$$\begin{bmatrix} 2 & -1 & 1 \\ -1 & 0 & 2 \\ 1 & 4 & -2 \end{bmatrix} \begin{bmatrix} x_1 \\ x_2 \\ x_3 \end{bmatrix} = \begin{bmatrix} 0 \\ 5 \\ -5 \end{bmatrix}$$

(1) The largest element in the matrix is 4. Interchanging columns 2 and 1 and the associated variables to bring the largest element into the first column,

$$\begin{bmatrix} -1 & 2 & 1 \\ 0 & -1 & 2 \\ 4 & 1 & -2 \end{bmatrix} \begin{bmatrix} x_2 \\ x_1 \\ x_3 \end{bmatrix} = \begin{bmatrix} 0 \\ 5 \\ -5 \end{bmatrix}$$

Interchanging equations 3 and 1 to bring the largest element to the pivot position,

$$\begin{bmatrix} 4 & 1 & -2 \\ 0 & -1 & 2 \\ -1 & 2 & 1 \end{bmatrix} \begin{bmatrix} x_2 \\ x_1 \\ x_3 \end{bmatrix} = \begin{bmatrix} -5 \\ 5 \\ 0 \end{bmatrix}$$

Forward elimination yields

$$\begin{bmatrix} 1 & 0.25 & -0.5 \\ 0 & -1 & 2 \\ 0 & 2.25 & 0.5 \end{bmatrix} \begin{bmatrix} x_2 \\ x_1 \\ x_3 \end{bmatrix} = \begin{bmatrix} -1.25 \\ 5 \\ -1.25 \end{bmatrix}$$

(2) The largest eligible coefficient for the second pivot element is 2.25. Since it is in the second column, interchanging equations 2 and 3 yields

$$\begin{bmatrix} 1 & 0.25 & -0.5 \\ 0 & 2.25 & 0.5 \\ 0 & -1 & 2 \end{bmatrix} \begin{bmatrix} x_2 \\ x_1 \\ x_3 \end{bmatrix} = \begin{bmatrix} -1.25 \\ -1.25 \\ 5 \end{bmatrix}$$

Forward elimination yields

$$\begin{bmatrix} 1 & 0.25 & -0.5 \\ 0 & 1 & 0.22 \\ 0 & 0 & 2.22 \end{bmatrix} \begin{bmatrix} x_2 \\ x_1 \\ x_3 \end{bmatrix} = \begin{bmatrix} -1.25 \\ -0.55 \\ 4.44 \end{bmatrix}$$

(3) Finally,

$$\begin{bmatrix} 1 & 0.25 & -0.5 \\ 0 & 1 & 0.22 \\ 0 & 0 & 1 \end{bmatrix} \begin{bmatrix} x_2 \\ x_1 \\ x_3 \end{bmatrix} = \begin{bmatrix} -1.25 \\ -0.55 \\ 2 \end{bmatrix}$$

Back substitution yields $\mathbf{x} = [-1, 0, 2]^T$.

Forward Elimination with Nonsquare Matrices The forward-elimination step of the Gaussian method, with or without pivoting, can be applied to nonsquare matrices. Consider the case of m equations in n unknowns, with $m < n$; that is,

$$\begin{aligned} a_{11}x_1 + \cdots + a_{m}x_n &= b_1 \\ &\vdots \\ a_{m1}x_1 + \cdots + a_{mn}x_n &= b_m \end{aligned} \tag{4.4.6}$$

If no zero pivots are encountered in forward elimination, after m steps the system may be reduced to the form

$$\begin{bmatrix} 1 & a_{12}^{(m)} & \cdots & a_{1m}^{(m)} & a_{1m+1}^{(m)} & \cdots & a_{1n}^{(m)} \\ 0 & 1 & \cdots & a_{2m}^{(m)} & \vdots & & \vdots \\ \vdots & \vdots & & \vdots & & & \\ 0 & 0 & \cdots & 1 & a_{mm+1}^{(m)} & \cdots & a_{mn}^{(m)} \end{bmatrix} \begin{bmatrix} u_1 \\ \vdots \\ u_m \\ v_1 \\ \vdots \\ v_{n-m} \end{bmatrix} = \begin{bmatrix} b_1^{(m)} \\ \vdots \\ \\ b_m^{(m)} \end{bmatrix} \tag{4.4.7}$$

where $\mathbf{u} = [u_1, \ldots, u_m]^T$ and $\mathbf{v} = [v_1, \ldots, v_{n-m}]^T$ contain elements of $\mathbf{x}$ that are reordered due to column pivoting. Equation 4.4.7 may be written in partitioned form as

$$\mathbf{Uu} + \mathbf{Rv} = \hat{\mathbf{b}} \tag{4.4.8}$$

where

$$\mathbf{U} = \begin{bmatrix} 1 & a_{12}^{(m)} & \cdots & a_{1m}^{(m)} \\ 0 & 1 & \cdots & a_{2m}^{(m)} \\ \vdots & \vdots & & \vdots \\ 0 & 0 & \cdots & 1 \end{bmatrix}$$

$$\mathbf{R} = \begin{bmatrix} a_{1m+1}^{(m)} & \cdots & a_{1n}^{(m)} \\ \vdots & & \vdots \\ a_{mm+1}^{(m)} & \cdots & a_{mn}^{(m)} \end{bmatrix} \tag{4.4.9}$$

$$\hat{\mathbf{b}} = \begin{bmatrix} b_1^{(m)} \\ \vdots \\ b_m^{(m)} \end{bmatrix}$$

Note that for any value of $\mathbf{v}$ in Eq. 4.4.8 $\mathbf{u}$ may be determined through back substitution, that is, from

$$\mathbf{Uu} = \hat{\mathbf{b}} - \mathbf{Rv}$$

Thus, $\mathbf{v}$ may be treated as a vector of *independent variables* and $\mathbf{u}$ as a vector of *dependent variables*. The variable $\mathbf{x}$ is said to be *partitioned* into independent and dependent coordinates.

The factorization of Eq. 4.4.7 is possible, using full pivoting, if and only if the coefficient matrix $\mathbf{A}$ of Eq. 4.4.6 has full row rank [22]. Thus, a foolproof method of determining the rank of $\mathbf{A}$ is to carry out Gaussian forward elimination, using full pivoting, to obtain

$$\begin{bmatrix} 1 & a_{12}^{(r)} & \cdots & a_{1r}^{(r)} & a_{1r+1}^{(r)} & \cdots & a_{1n}^{(r)} \\ 0 & 1 & \cdots & a_{2r}^{(r)} & a_{2r+1}^{(r)} & \cdots & a_{2n}^{(r)} \\ \vdots & \vdots & & \vdots & \vdots & & \vdots \\ 0 & 0 & \cdots & 1 & a_{rr+1}^{(r)} & \cdots & a_{rn}^{(r)} \\ 0 & 0 & \cdots & 0 & 0 & \cdots & 0 \\ \vdots & \vdots & & \vdots & \vdots & & \vdots \\ 0 & 0 & \cdots & 0 & 0 & \cdots & 0 \end{bmatrix} \begin{bmatrix} u_1 \\ \vdots \\ u_r \\ v_1 \\ \vdots \\ v_{n-r} \end{bmatrix} = \begin{bmatrix} b_1^{(r)} \\ \vdots \\ b_r^{(r)} \\ b_{r+1}^{(r)} \\ \vdots \\ b_m^{(r)} \end{bmatrix} \tag{4.4.10}$$

In partitioned matrix form, this is

$$\begin{aligned} \mathbf{U}_r\mathbf{u} + \mathbf{R}_r\mathbf{v} &= \hat{\mathbf{b}}_r \\ \mathbf{0} &= \hat{\mathbf{b}}_{m-r} \end{aligned} \tag{4.4.11}$$

where $\mathbf{U}_r$, $\mathbf{R}_r$, and $\hat{\mathbf{b}}_r$ are defined as in Eq. 4.4.9, with m replaced by r, and

$$\hat{\mathbf{b}}_{m-r} = \begin{bmatrix} b_{r+1}^{(r)} \\ \vdots \\ b_m^{(r)} \end{bmatrix} \tag{4.4.12}$$

Thus, the row rank of matrix $\mathbf{A}$ is r and, if $r < m$, the second of Eqs. 4.4.11 shows that Eq. 4.4.10, and hence Eq. 4.4.6, has a solution if and only if $\hat{\mathbf{b}}_{m-r} = \mathbf{0}$. If $\hat{\mathbf{b}}_{m-r} = \mathbf{0}$, then the equations of Eq. 4.4.6 that were interchanged into the last $m - r$ equations (rows) of Eq. 4.4.7 are dependent and may be eliminated.

Example 4.4.4: Use Gaussian forward elimination to determine the rank and a dependent and independent variable partitioning for the following set of equations:

$$\mathbf{Ax} \equiv \begin{bmatrix} 1 & 0 & 2 & 1 & 4 \\ 1 & 1 & 2 & 0 & 3 \\ 3 & 1 & 6 & 2 & 11 \end{bmatrix} \begin{bmatrix} x_1 \\ x_2 \\ x_3 \\ x_4 \\ x_5 \end{bmatrix} = \begin{bmatrix} 3 \\ 2 \\ 8 \end{bmatrix} \equiv \mathbf{b}$$

The solution will be carried out without pivoting until a zero diagonal entry appears that requires pivoting.

1. Forward elimination in the first column yields

$$\begin{bmatrix} 1 & 0 & 2 & 1 & 4 \\ 0 & 1 & 0 & -1 & -1 \\ 0 & 1 & 0 & -1 & -1 \end{bmatrix} \begin{bmatrix} x_1 \\ x_2 \\ x_3 \\ x_4 \\ x_5 \end{bmatrix} = \begin{bmatrix} 3 \\ -1 \\ -1 \end{bmatrix}$$

2. Forward elimination in the second column yields

$$\begin{bmatrix} 1 & 0 & 2 & 1 & 4 \\ 0 & 1 & 0 & -1 & -1 \\ 0 & 0 & 0 & 0 & 0 \end{bmatrix} \begin{bmatrix} x_1 \\ x_2 \\ x_3 \\ x_4 \\ x_5 \end{bmatrix} = \begin{bmatrix} 3 \\ -1 \\ 0 \end{bmatrix}$$

Thus, the rank of the coefficient matrix is 2 and, since $b_3 = 0$, the equations are consistent. The partitioning defined by this process is $\mathbf{u} = [x_1, x_2]^T$ and $\mathbf{v} = [x_3, x_4, x_5]^T$. The solution for x_1 and x_2, for any values of x_3, x_4, and x_5, is thus

$$x_1 = 3 - 2x_3 - x_4 - 4x_5$$

$$x_2 = -1 + x_4 + x_5$$

If a different right-side vector had been chosen, the rank would still be 2, but the equations could be *inconsistent*; that it, there may exist no solution (Prob. 4.4.4). If full pivoting had been employed in the forward-elimination process, the rank of **A** would still be determined to be 2, but a different partitioning of variables would result (Prob. 4.4.5).

On a digital computer, the bottom right block of zeros in the coefficient matrix of Eq. 4.4.10 will not contain precisely zero entries. The entries will be zero only to within the round-off error that has accumulated during forward elimination. Similarly, $\hat{\mathbf{b}}_{m-r}$ can only be zero to within the same numerical precision.

A perplexing and numerically challenging situation arises when the bottom right factors in the coefficient matrix in Eq. 4.4.10 are nearly zero, but not to within round-off error. If the entries in $\hat{\mathbf{b}}_{m-r}$ are a few orders of magnitude larger than elements in the corresponding rows of the matrix, numerical difficulties arise. In this case, the coefficient matrix is said to be *ill-conditioned,* and severe numerical difficulties may be encountered.

4.4.2 L–U Factorization

While the Gaussian methods of Section 4.4.1 are easy to understand and are reliable, better methods exist as regards efficiency and computer memory utilization. One such method is introduced here. Given any nonsingular matrix **A**, there exists [22] an upper triangular matrix **U** with nonzero diagonal elements and a lower triangular matrix **L** with unit diagonal elements, such that

$$\mathbf{A} = \mathbf{LU} \tag{4.4.13}$$

Factorization of **A** into the product **LU** is called *L–U factorization.* Once the **L** and **U** factors are obtained, by whatever method, the equation

$$\mathbf{Ax} = \mathbf{LUx} = \mathbf{b} \tag{4.4.14}$$

is solved by transforming it into

$$\mathbf{Ly} = \mathbf{b} \tag{4.4.15}$$

and

$$\mathbf{Ux} = \mathbf{y} \tag{4.4.16}$$

Equation 4.4.15 is solved first for **y** and then Eq. 4.4.16 is solved for **x**. Since both Eqs. 4.4.15 and 4.4.16 have triangular coefficient matrices, their solutions are easily obtained by back substitution.

Crout's method calculates elements of **L** and **U** recursively. To illustrate how Crout's method generates elements of **L** and **U**, consider a matrix **A** of rank $n = 3$ that requires no row or column interchanges (i.e., no pivoting). Matrix **A** can be written as

$$\begin{bmatrix} 1 & 0 & 0 \\ \ell_{21} & 1 & 0 \\ \ell_{31} & \ell_{32} & 1 \end{bmatrix} \begin{bmatrix} u_{11} & u_{12} & u_{13} \\ 0 & u_{22} & u_{23} \\ 0 & 0 & u_{33} \end{bmatrix} = \begin{bmatrix} a_{11} & a_{12} & a_{13} \\ a_{21} & a_{22} & a_{23} \\ a_{31} & a_{32} & a_{33} \end{bmatrix} \quad \textbf{(4.4.17)}$$

An *auxiliary matrix* **B** can be defined, consisting of elements of **L** and **U**, such that

$$\mathbf{B} \equiv \begin{bmatrix} u_{11} & u_{12} & u_{13} \\ \ell_{21} & u_{22} & u_{23} \\ \ell_{31} & \ell_{32} & u_{33} \end{bmatrix} \quad \textbf{(4.4.18)}$$

Elements of **B** are to be calculated in the order indicated by the diagram

$$\begin{Bmatrix} ① & ② & ③ \\ ④ & ⑥ & ⑦ \\ ⑤ & ⑧ & ⑨ \end{Bmatrix} \quad \textbf{(4.4.19)}$$

where ⓚ indicates the kth element to be calculated. The elements of **L** and **U** are calculated simply by equating elements in Eq. 4.4.17 to a_{jk} successively, according to the order shown in Eq. 4.4.19, using the product of the jth row of **L** and the kth column of **U**. For a 3×3 matrix, Eq. 4.4.17 yields

① $a_{11} = u_{11}$, ② $a_{12} = u_{12}$, ③ $a_{13} = u_{13}$

④ $a_{21} = \ell_{21} u_{11}$, so $\ell_{21} = a_{21}/u_{11}$

⑤ $a_{31} = \ell_{31} u_{11}$, so $\ell_{31} = a_{31}/u_{11}$

⑥ $a_{22} = \ell_{21} u_{12} + u_{22}$, so $u_{22} = a_{22} - \ell_{21} u_{12}$

⑦ $a_{23} = \ell_{23} = \ell_{21} u_{13} + u_{23}$, so $u_{23} = a_{23} - \ell_{21} u_{13}$

⑧ $a_{32} = \ell_{31} u_{12} + \ell_{32} u_{22}$, so $\ell_{32} = (a_{32} - \ell_{31} u_{12})/u_{22}$

⑨ $a_{33} = \ell_{31} u_{13} + \ell_{32} u_{23} + u_{33}$, so $u_{33} = a_{33} - \ell_{31} u_{13} - \ell_{32} u_{23}$

Note that calculation of element ⓚ of the auxiliary matrix **B**, which is an element of either **L** or **U**, involves only the element of **A** in the same position and elements of **B** that have already been calculated. As element ⓚ is obtained, it is recorded in matrix **B**. In fact, it may be recorded in the corresponding position of the original **A** matrix, if there is no need to keep **A**.

Crout's method for a general $n \times n$ matrix **A** can be stated as follows [30, 31]:

1. Initially set the iteration counter to $i = 1$.
2. Use the matrix conversion diagram

$$\begin{bmatrix} a_{ii} & \mathbf{a}_{ik}^T \\ \mathbf{a}_{ki} & \mathbf{A}_{kk} \end{bmatrix} \rightarrow \begin{bmatrix} u_{ii} & \mathbf{u}_{ik}^T \\ \mathbf{l}_{ki} & \mathbf{D}_{kk} \end{bmatrix} \quad \textbf{(4.4.20)}$$

to obtain

$$u_{ii} = a_{ii}$$
$$\mathbf{u}_{ik}^T = \mathbf{a}_{ik}^T$$
$$\mathbf{l}_{ki} = \frac{\mathbf{a}_{ki}}{u_{ii}}$$
$$\mathbf{D}_{kk} = \mathbf{A}_{kk} - \mathbf{l}_{ki}\mathbf{u}_{ik}^T$$

where a_{ii} is the ith diagonal element of $\mathbf{A}$, $\mathbf{a}_{ik}^T$ is the part of the ith row of $\mathbf{A}$ to the right of a_{ii}, $\mathbf{a}_{ki}$ is the part of the ith column of $\mathbf{A}$ below a_{ii}, and $\mathbf{A}_{kk}$ is the $(n-i) \times (n-i)$ submatrix to the bottom right of $\mathbf{A}$.

3. Increment i; that is, $i \rightarrow i+1$. If $i = n$, L–U factorization is complete. Otherwise, go to step 2.

Example 4.4.5: Apply L–U factorization to matrix $\mathbf{A}$ and solve the set of algebraic equations $\mathbf{Ax} = \mathbf{b}$ for the unknown $\mathbf{x}$, where

$$\mathbf{A} = \begin{bmatrix} 2 & 1 & -2 \\ -1 & 2.5 & -5 \\ 3 & -4.5 & 0 \end{bmatrix}, \qquad \mathbf{b} = \begin{bmatrix} 1 \\ -9.5 \\ 10.5 \end{bmatrix}$$

Following the algorithm, $i = 1$:

$$a_{11} = 2, \qquad \mathbf{a}_{ik}^T = [1, -2], \qquad \mathbf{a}_{ki} = [-1, 3]^T, \qquad \mathbf{A}_{kk} = -\begin{bmatrix} 2.5 & -5 \\ 4.5 & 0 \end{bmatrix}$$

This step can be performed by overwriting on matrix $\mathbf{A}$ to obtain

$$\begin{bmatrix} 2 & 1 & -2 \\ -1 & 2.5 & -5 \\ 3 & -4.5 & 0 \end{bmatrix} \rightarrow \begin{bmatrix} 2 & 1 & -2 \\ -0.5 & 3 & -6 \\ 1.5 & -6 & 3 \end{bmatrix}$$

$i = 2$: At this step, row 1 and column 1 remain intact. Step 2 applies only to the 2×2 matrix $\begin{bmatrix} 3 & -6 \\ -6 & 3 \end{bmatrix}$; that is, $a_{22} = 3$, $\mathbf{a}_{2k}^T = [-6]$, $\mathbf{a}_{k2} = [-6]$, and $\mathbf{A}_{kk} = [3]$, which yields $u_{22} = 3$, $\mathbf{u}_{2k}^T = [-6]$, $\mathbf{l}_{k2} = [-2]$, and $\mathbf{D}_{kk} = [-9]$. In matrix form,

$$\begin{bmatrix} 2 & 1 & -2 \\ -0.5 & 3 & -6 \\ 1.5 & -6 & 3 \end{bmatrix} \rightarrow \begin{bmatrix} 2 & 1 & -2 \\ -0.5 & 3 & -6 \\ 1.5 & -2 & -9 \end{bmatrix}$$

$i = 3$: At this point, since a 1×1 matrix remains (i.e., $[-9]$), the process is complete. The $\mathbf{L}$ and $\mathbf{U}$ matrices are, respectively,

$$\mathbf{L} = \begin{bmatrix} 1 & 0 & 0 \\ -0.5 & 1 & 0 \\ 1.5 & -2 & 1 \end{bmatrix}, \qquad \mathbf{U} = \begin{bmatrix} 2 & 1 & -2 \\ 0 & 3 & -6 \\ 0 & 0 & -9 \end{bmatrix}$$

The solution of $\mathbf{Ax} = \mathbf{b}$ can be obtained first by solving $\mathbf{Ly} = \mathbf{b}$ for y_1, then y_2, and then y_3; that is,

$$\begin{bmatrix} 1 & 0 & 0 \\ -0.5 & 1 & 0 \\ 1.5 & -2 & 1 \end{bmatrix} \begin{bmatrix} y_1 \\ y_2 \\ y_3 \end{bmatrix} = \begin{bmatrix} 1 \\ -9.5 \\ 10.5 \end{bmatrix} \rightarrow \mathbf{y} = \begin{bmatrix} 1 \\ -9 \\ -9 \end{bmatrix}$$

Solving $\mathbf{Ux} = \mathbf{y}$ for x_3, then x_2, and then x_1,

$$\begin{bmatrix} 2 & 1 & -2 \\ 0 & 3 & -6 \\ 0 & 0 & -9 \end{bmatrix} \begin{bmatrix} x_1 \\ x_2 \\ x_3 \end{bmatrix} = \begin{bmatrix} 1 \\ -9 \\ -9 \end{bmatrix} \rightarrow \mathbf{x} = \begin{bmatrix} 2 \\ -1 \\ 1 \end{bmatrix}$$

4.5 NEWTON–RAPHSON METHOD FOR NONLINEAR EQUATIONS

One of the most frequently occurring problems in kinematics is to find solutions of nonlinear algebraic equations of the form

$$\mathbf{\Phi}(\mathbf{q}) = \mathbf{0} \tag{4.5.1}$$

When $\mathbf{\Phi}(\mathbf{q})$ and its Jacobian $\mathbf{\Phi_q}$ can be evaluated, the iterative *Newton–Raphson method* may be used to find a solution of Eq. 4.5.1.

Newton–Raphson Method for One Equation in One Unknown Consider the single equation

$$\Phi(q) = 0 \tag{4.5.2}$$

that is nonlinear in the scalar variable q. Let $q = q^*$ be a solution of Eq. 4.5.1 and let $q^{(i)}$ be an approximation of q^*. The Taylor series expansion of $\Phi(q)$ about the point $q = q^{(i)}$ is [25, 26]

$$\Phi(q) = \Phi(q^{(i)}) + \Phi_q(q^{(i)})(q - q^{(i)}) + \text{higher-order terms} \tag{4.5.3}$$

Let $q = q^{(i+1)}$ be an improved approximation that is to be determined from the condition

$$\Phi(q^{(i+1)}) \approx \Phi(q^{(i)}) + \Phi_q(q^{(i)})(q^{(i+1)} - q^{(i)}) = 0 \tag{4.5.4}$$

where, if $q^{(i+1)} - q^{(i)}$ is small, higher-order terms are neglected. If $\Phi_q(q^{(i)}) \neq 0$, Eq. 4.5.4 can be solved to obtain

$$q^{(i+1)} = \frac{q^{(i)} - \Phi(q^{(i)})}{\Phi_q(q^{(i)})} \tag{4.5.5}$$

Equation 4.5.5 defines the *Newton–Raphson algorithm* for scalar equations that can be employed iteratively to produce a sequence of approximate solutions, beginning with an initial estimate of the solution of Eq. 4.5.2.

1. Make an estimate $q^{(0)}$ of the solution of Eq. 4.5.2.
2. In iterations $i = 0, 1, \ldots,$ evaluate $\Phi(q^{(i)})$ and $\Phi_q(q^{(i)})$. If $|\Phi(q^{(i)})| < \varepsilon_e$, where ε_e is the equation error tolerance, and $|q^{(i)} - q^{(i-1)}| < \varepsilon_s$, where ε_s is the solution error tolerance, terminate. If $\Phi_q(q^{(i)}) = 0$ and $\Phi(q^{(i)}) \neq 0$, return to step 1 with a new estimate. Otherwise, go to step 3.
3. Calculate $q^{(i+1)}$ from Eq. 4.5.5 and return to step 2, with $i+1$ in place of i.

The sequence of approximate solutions generated by the Newton–Raphson algorithm, in many problems, will approach a solution of $\Phi(q) = 0$. However, if $q^{(0)}$ is not adequately close to q^*, the sequence may diverge. If the algorithm converges and the Jacobian is nonsingular, it is *quadratically convergent* [31]; that is, there is a constant c such that

$$|q^{(k+1)} - q^*| < c\,|q^{(k)} - q^*|^2 \tag{4.5.6}$$

As an illustration of this behavior, see Example 4.3.1. As might be expected, if $|q^{(0)} - q^*|$ is large, the error may grow quadratically.

The geometry of Newton–Raphson iteration is illustrated in Fig. 4.5.1 for a case in which it converges. The illustration in Fig. 4.5.2 shows that the method may diverge when the solution is an inflection point. If there are multiple solutions, the solution obtained is a function of the initial estimate, as shown in Fig. 4.5.3. If the initial estimate is near a local minimum or maximum, as shown in Fig. 4.5.4, the algorithm may fail to converge. It is essential that a reasonable estimate of the solution be used to initiate the iterative process.

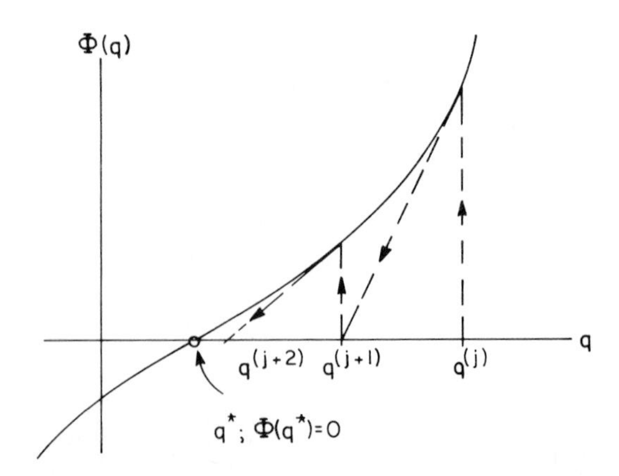

Figure 4.5.1 Graphic representation of Newton–Raphson method.

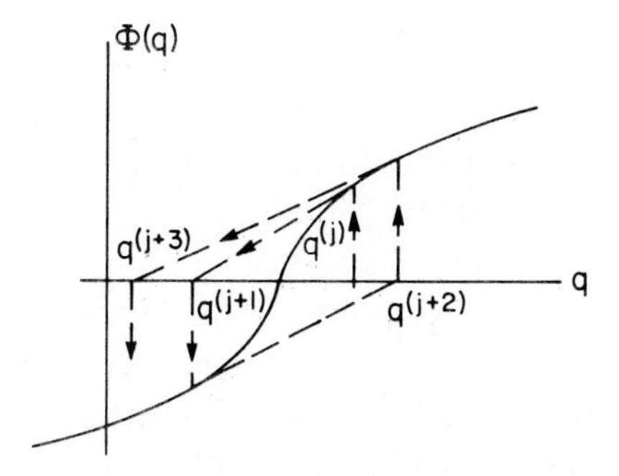

Figure 4.5.2 Root at inflection point.

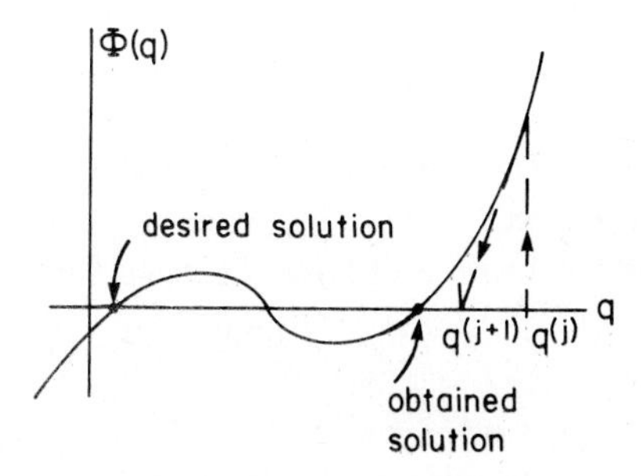

Figure 4.5.3 Multiple roots.

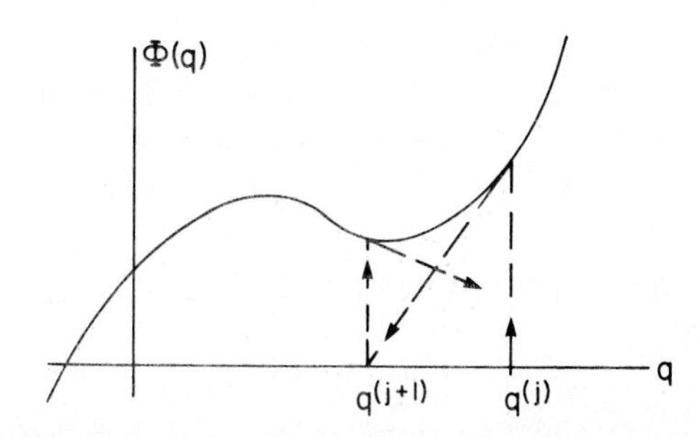

Figure 4.5.4 Divergence near a local minimum or maximum.

Newton–Raphson Method for n Equations in n Unknowns. Consider n nonlinear algebraic equations in n unknowns,

$$\begin{aligned} \Phi_1(\mathbf{q}) &\equiv \Phi_1(q_1, q_2, \ldots, q_n) = 0 \\ \Phi_2(\mathbf{q}) &\equiv \Phi_2(q_1, q_2, \ldots, q_n) = 0 \\ &\vdots \\ \Phi_n(\mathbf{q}) &\equiv \Phi_n(q_1, q_2, \ldots, q_n) = 0 \end{aligned} \tag{4.5.7}$$

where a solution vector is denoted as $\mathbf{q}^* \equiv [q_1^*, q_2^*, \ldots, q_n^*]^T$. A first-order Taylor expansion about an estimate $\mathbf{q}^{(i)}$ of the solution can be applied to Eqs. 4.5.7. Neglecting higher-order terms, the first-order Taylor approximation [25, 26] for an improved approximation $\mathbf{q}^{(i+1)}$ is

$$\mathbf{\Phi}(\mathbf{q}^{(i)}) + \mathbf{\Phi}_{\mathbf{q}}(\mathbf{q}^{(i)})[\mathbf{q}^{(i+1)} - \mathbf{q}^{(i)}] = \mathbf{0}$$

If the Jacobian matrix $\mathbf{\Phi}_{\mathbf{q}}(\mathbf{q}^{(i)})$ is nonsingular, this equation may be solved for $\mathbf{q}^{(i+1)}$ by solving

$$\mathbf{\Phi}_{\mathbf{q}}(\mathbf{q}^{(i)})\Delta\mathbf{q}^{(i)} = -\mathbf{\Phi}(\mathbf{q}^{(i)}) \tag{4.5.8}$$

and setting $\mathbf{q}^{(i+1)} = \mathbf{q}^{(i)} + \Delta\mathbf{q}^{(i)}$.

The *Newton–Raphson algorithm* for systems of equations is as follows:

1. Make an initial estimate $\mathbf{q}^{(0)}$ of the solution of Eq. 4.5.7.
2. In iteration $i = 0, 1, 2, \ldots$, evaluate $\mathbf{\Phi}(\mathbf{q}^{(i)})$ and $\mathbf{\Phi}_{\mathbf{q}}(\mathbf{q}^{(i)})$. If the magnitudes of all errors and changes in approximate solutions satisfy

$$\left.\begin{aligned} |\Phi_k(\mathbf{q}^{(i)})| &\le \varepsilon_e \\ |q_k^{(i)} - q_k^{(i-1)}| &\le \varepsilon_s \end{aligned}\right\} \quad k = 1, \ldots, n$$

   terminate. If $\mathbf{\Phi}_{\mathbf{q}}(\mathbf{q}^{(i)})$ is singular and $\mathbf{\Phi}(\mathbf{q}^{(i)}) \neq \mathbf{0}$, return to step 1 and restart the process with a new estimate. Otherwise, go to step 3. Here, ε_e is the equation error tolerance and ε_s is the solution error tolerance.
3. Solve $\mathbf{\Phi}_{\mathbf{q}}(\mathbf{q}^{(i)})\Delta\mathbf{q}^{(i)} = -\mathbf{\Phi}(\mathbf{q}^{(i)})$, set $\mathbf{q}^{(i+1)} = \mathbf{q}^{(i)} + \Delta\mathbf{q}^{(i)}$, and return to step 2 with $i + 1$ in place of i.

Convergence properties of this algorithm for systems of equations are essentially the same as for scalar equations [31].

Numerical examples of the application of the Newton–Raphson method for kinematic position analysis have been presented, for two different mechanisms, in Examples 3.6.2 and 4.3.1.

4.6 DETECTION AND ELIMINATION OF REDUNDANT CONSTRAINTS

If redundant constraint equations exist in the nh kinematic constraint equations

$$\mathbf{\Phi}^K(\mathbf{q}) = \mathbf{0} \tag{4.6.1}$$

the number of system degrees of freedom is not $3nb - nh$. For a simple mechanical system, careful examination of the system may determine the number of degrees of freedom. However, this is not easy for even a moderately complicated system, so the correct number of independent constraint equations is often difficult to define.

Assume that a system is in an assembled configuration obtained by the minimization procedure of Section 4.3. If the constraint Jacobian of Eq. 4.6.1 has full row rank, then an $nh \times nh$ submatrix associated with dependent variables is nonsingular and, by the implicit function theorem, these dependent variables can be expressed in terms of the remaining $3nb - nh$ independent variables. To determine the row rank of a matrix, Gaussian elimination of Section 4.4 can be used with either row or full pivoting. Row and full pivoting theoretically give the same rank of the matrix, but full pivoting is numerically more reliable.

As physical motivation for the use of Gaussian elimination, consider a *virtual displacement* $\delta\mathbf{q}$ (a small variation in $\mathbf{q}$ with time held fixed) that is consistent with the kinematic constraint equations of Eq. 4.6.1; that is, it satisfies the linearized constraint equations,

$$\mathbf{\Phi}_{\mathbf{q}}^{K}\delta\mathbf{q} = \mathbf{0} \tag{4.6.2}$$

Gaussian elimination can be applied to the Jacobian matrix on the left of Eq. 4.6.2. The resulting form of the modified Jacobian matrix and equations is

$$\begin{bmatrix} \mathbf{\Phi}_{\mathbf{u}}^{KI} & \mathbf{\Phi}_{\mathbf{v}}^{KI} \\ \mathbf{0} & \mathbf{\Phi}_{\mathbf{v}}^{KR} \end{bmatrix} \begin{bmatrix} \delta\mathbf{u} \\ \delta\mathbf{v} \end{bmatrix} = \mathbf{0} \tag{4.6.3}$$

where $\mathbf{\Phi}_{\mathbf{u}}^{KI}$ is a triangular $nh' \times nh'$ matrix with ones in its diagonal, $nh' \leq nh$, and $\mathbf{\Phi}_{\mathbf{v}}^{KR}$ is a zero matrix, to within computation tolerance. If Gaussian elimination of Eq. 4.6.2 is done with finite digit arithmetic, errors are involved during the elimination procedure, and all elements of $\mathbf{\Phi}_{\mathbf{v}}^{KR}$ will not be exactly zero. It is thus necessary to set elements that are close to machine precision zero to exactly zero during the elimination procedure. Equation 4.6.2 is thus reduced to the independent equations

$$\mathbf{\Phi}_{\mathbf{u}}^{KI}\delta\mathbf{u} + \mathbf{\Phi}_{\mathbf{v}}^{KI}\delta\mathbf{v} = \mathbf{0} \tag{4.6.4}$$

Since $|\mathbf{\Phi}_{\mathbf{u}}^{KI}| = 1$, $\delta\mathbf{u}$ can be expressed as

$$\delta\mathbf{u} = -(\mathbf{\Phi}_{\mathbf{u}}^{KI})^{-1}\mathbf{\Phi}_{\mathbf{v}}^{KI}\delta\mathbf{v}$$

By the implicit function theorem, nh' constraint equations corresponding to $\mathbf{\Phi}_{\mathbf{u}}^{KI}$ are independent, that is,

$$\mathbf{\Phi}^{KI}(\mathbf{q}) = \mathbf{0} \tag{4.6.5}$$

and the remaining constraint equations $\mathbf{\Phi}^{KR}$ are redundant. If there exists an isolated singular point at the initial configurations, that is, a bifurcation point that has multiple solutions, an apparently redundant constraint can be detected. In this case, by perturbing a few generalized coordinates and reassembling the system with the fixed perturbed variables, if the Jacobian has increased row rank, it can be concluded that the constraint is redundant only at an isolated singular point. By checking redundant constraints at the reassembled configuration, the correct number of redundant constraints for the system can be determined. These dependent constraints are eliminated, yielding nh' independent kinematic constraints.

The system has $3nb - nh'$ degrees of freedom; so, for kinematic analysis, $3nb - nh' = d$ driving constraint equations

$$\mathbf{\Phi}^D(\mathbf{q}, t) = \mathbf{0} \tag{4.6.6}$$

must be independent of each other and must also be independent of the retained kinematic constraints of Eq. 4.6.1. If there exists a redundant driving constraint, it can be dependent on other driving constraint equations or it could be dependent on the kinematic constraint equations. Since the reduced nh' kinematic constraint equations of Eq. 4.6.5 are independent, it is desired to determine and eliminate only redundant driving constraint equations from Eq. 4.6.6.

To detect redundant driving constraint equations, Gaussian elimination of the Jacobian matrix of the combined constraints

$$\mathbf{\Phi}(\mathbf{q}, t) = \begin{bmatrix} \mathbf{\Phi}^K(\mathbf{q}) \\ \mathbf{\Phi}^D(\mathbf{q}, t) \end{bmatrix} = \mathbf{0} \tag{4.6.7}$$

can be used. If full pivoting Gaussian elimination is used, then kinematic constraint equations might be selected as dependent. To prevent this, only column exchange is allowed during Gaussian elimination through the diagonal elements of $\mathbf{\Phi}^I_{\mathbf{u}'}$. Beyond this point in Gaussian elimination, full pivoting is allowed. The resulting matrix can be expressed as

$$\begin{bmatrix} \mathbf{\Phi}^I_{\mathbf{u}'} & \mathbf{\Phi}^I_{\mathbf{v}'} \\ \mathbf{0} & \mathbf{\Phi}^D_{\mathbf{v}'} \end{bmatrix} \tag{4.6.8}$$

where the upper triangular matrix $\mathbf{\Phi}^I_{\mathbf{u}'}$ has ones on its diagonal and the upper triangular matrix $\mathbf{\Phi}^D_{\mathbf{v}'}$ can have zero rows. Driving constraint equations corresponding to zero rows of $\mathbf{\Phi}^D_{\mathbf{v}'}$ are identified as redundant driving constraints. Redundant driving constraints must be removed and replaced by independent driving constraints prior to kinematic analysis.

The numerical procedure used to automate the *redundant constraint elimination algorithm* may be summarized as follows:

1. From initial estimates, assemble the system to minimize kinematic constraint equation violation, as in Section 4.3. If assembly fails, advise the user that the design may be infeasible or a better estimate of the assembled configuration is needed.
2. After successful assembly, evaluate the Jacobian matrix of the kinematic constraint equations.

3. Using full pivoting Gaussian elimination, determine the rank of the kinematic constraint Jacobian. During the elimination procedure, record row and column interchanges. The resulting matrix has the form of Eq. 4.6.3. Eliminate redundant kinematic constraint equations $\mathbf{\Phi}_{\mathbf{v}}^{KR}$ from the system.
4. From step 3, a good choice of independent generalized coordinates can be reported from the column indexes of submatrix $\mathbf{\Phi}_{\mathbf{v}}^{KI}$ in Eq. 4.6.3. Report this suggested choice of independent coordinates and the number of degrees of freedom to the user, to help in the selection of driving constraints.
5. Add driving constraint equations at the bottom of the independent kinematic constraint equations, as in Eq. 4.6.7. Perform Gaussian elimination, with column exchanges only through the rows of $\mathbf{\Phi}_{\mathbf{q}}^{K}$ and full pivoting beyond. Identify redundant driving constraints that correspond to zero rows in $\mathbf{\Phi}_{\mathbf{v}'}^{D}$ in Eq. 4.6.8. Eliminate redundant driving constraint equations from the system.
6. If the number of independent kinematic and driving constraint equations remaining $(nh' + d')$ is less than the number nc of generalized coordinates, then write a message that $nc - nh' - d'$ additional driving constraints are required to proceed with kinematic analysis.

Example 4.6.1: Consider the four-bar parallelogram mechanism shown in Fig. 4.6.1, with a unit distance constraint between points A and B. Equation 4.6.1, with $\mathbf{q} = [\phi_1, x_2, y_2, \phi_2, \phi_3]^T$, is

$$\mathbf{\Phi}^K(\mathbf{q}) = \begin{bmatrix} x_2 - \cos\phi_1 - \cos\phi_2 \\ y_2 - \sin\phi_1 - \sin\phi_2 \\ x_2 - \cos\phi_3 - 1 \\ y_2 - \sin\phi_3 \\ (x_2 + \cos\phi_2 - 2)^2 + (y_2 + \sin\phi_2)^2 - 1 \end{bmatrix} = \mathbf{0} \tag{4.6.9}$$

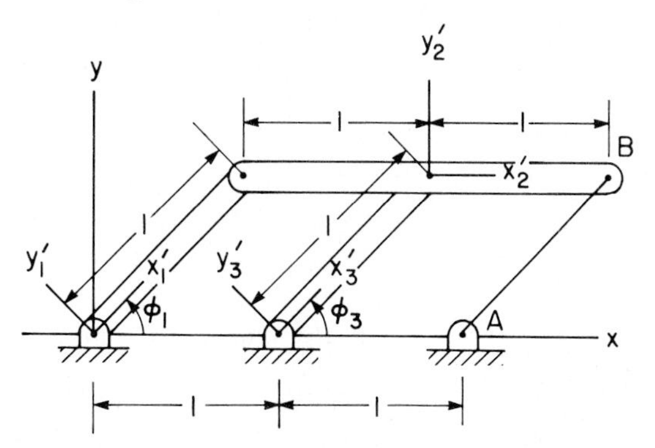

Figure 4.6.1 Four-bar crank mechanism with a redundant constraint.

As is geometrically clear, $\phi_1 = \phi_3$ and $\phi_2 = 0$. Thus, from the first and second of Eqs. 4.6.9, $x_2 = 1 + \cos\phi_1$ and $y_2 = \sin\phi_1$. Substituting these results into the last of Eqs. 4.6.9 shows that it is identically satisfied; that is, it is redundant.

To see that the constraint Jacobian row rank criterion identifies this redundancy, the Jacobian of Eq. 4.6.1 is evaluated at the assembled configuration as

$$\boldsymbol{\Phi}_{\mathbf{q}}^{K}(\mathbf{q}) = \begin{bmatrix} \sin\phi_1 & 1 & 0 & \sin\phi_2 & 0 \\ -\cos\phi_1 & 0 & 1 & -\cos\phi_2 & 0 \\ 0 & 1 & 0 & 0 & \sin\phi_3 \\ 0 & 0 & 1 & 0 & -\cos\phi_3 \\ 0 & 2(x_2 + \cos\phi_2 - 2) & 2(y_2 + \sin\phi_2) & \begin{matrix} -2\sin\phi_2(x_2 + \cos\phi_2 - 2) \\ +2\cos\phi_2(y_2 + \sin\phi_2) \end{matrix} & 0 \end{bmatrix} \tag{4.6.10}$$

Let $\phi_1^0 = \phi_3^0 = 1.0472$ (60°) and $\phi_2^0 = 0$, so $x_2 = 1.5$ and $y_2 = 0.866$, and evaluate the Jacobian of Eq. 4.6.10. After Gaussian elimination, without pivoting on the last row, the Jacobian is reduced to

$$\begin{bmatrix} 1 & 0 & 0 & 0.866 & 0 \\ 0 & 1 & -1 & -0.5 & 0 \\ 0 & 0 & 1 & 0.5 & -0.5 \\ 0 & 0 & 0 & 1 & -1 \\ 0 & 0 & 0 & 0 & 0 \end{bmatrix}$$

During the elimination, the last row did not change its position; that is, the reduced Jacobian matrix is rank deficient and the last constraint is identified as a redundant constraint. Even if a small change is made in ϕ_1^0, and hence ϕ_3^0, the same conclusion follows, so the redundancy is not an isolated singular point.

As a second check, consider the Jacobian of Eq. 4.6.10 at $\phi_1^0 = \phi_3^0 = 0$ and $\phi_2^0 = 0$. After Gaussian elimination, using full pivoting, the Jacobian is reduced to

$$\begin{bmatrix} 1 & 0 & -1 & 1 & 0 \\ 0 & 1 & 0 & 0 & 0 \\ 0 & 0 & 1 & 0 & -1 \\ 0 & 0 & 0 & 0 & 0 \\ 0 & 0 & 0 & 0 & 0 \end{bmatrix}$$

which has a row rank deficiency of 2. This suggests that two constraints may be redundant. However, an arbitrarily small perturbation of ϕ_1^0, and hence ϕ_3^0, yields a matrix with row rank deficiency of only 1. Thus, in addition to a redundant constraint, at $\phi_1^0 = \phi_3^0 = 0$ the mechanism has an isolated singular point.

PROBLEMS

Section 4.2

4.2.1. Write the explicit form of the kinematic and driving constraints of Example 4.2.1.

4.2.2. Verify that the Jacobian matrix of Eq. 4.2.6 is correct by direct differentiation of the constraints of Prob. 4.2.1.

4.2.3. Evaluate γ for Example 4.2.1.

4.2.4. Write the position, velocity, and acceleration equations for the simple pendulum shown in Fig. P4.2.4, with angle driver ωt and

(a) modeling ground as body 2, with absolute constraints, and imposing a revolute constraint between ground and the pendulum (body 1).

(b) modeling the pivot between the global frame and pendulum using absolute x and y constraints (i.e., a one-body model).

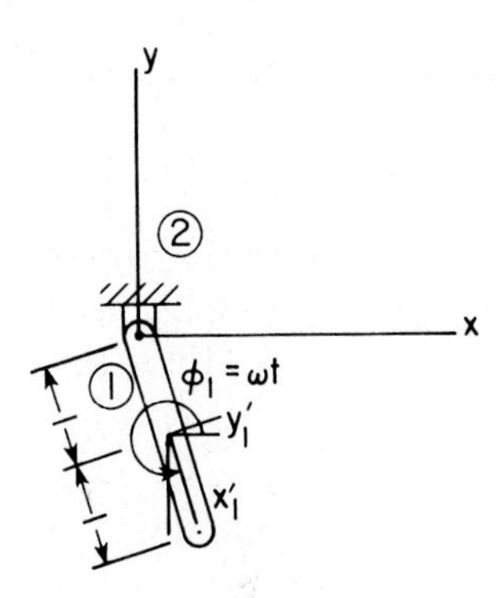

Figure P4.2.4

Section 4.4

4.4.1. Use Gaussian elimination without pivoting to solve the equations

$$\begin{bmatrix} 1 & 2 & 1 \\ 2 & 1 & 1 \\ 0 & 1 & 4 \end{bmatrix} \begin{bmatrix} x_1 \\ x_2 \\ x_3 \end{bmatrix} = \begin{bmatrix} 1 \\ 0 \\ 2 \end{bmatrix}$$

4.4.2. Solve Prob. 4.4.1 using Gaussian elimination with row pivoting.

4.4.3. Solve Prob. 4.4.1 using Gaussian elimination with full pivoting.

4.4.4. Repeat the forward elimination of Example 4.4.4 with a right side $\mathbf{b} = [3, 2, 5]^T$ and determine whether a solution exists.

4.4.5. Repeat the forward elimination of Example 4.4.4 with full pivoting. What dependent–independent variable partitioning is obtained?

4.4.6. Apply **L–U** factorization to matrix **A** and solve the set of algebraic equations

$\mathbf{Ax} = \mathbf{b}$ for the unknown $\mathbf{x}$, where

$$\mathbf{A} = \begin{bmatrix} 1 & 2 & 0 & 1 \\ 0 & 1 & 0 & 0 \\ 1 & 0 & 3 & 1 \\ 2 & 1 & 0 & 2 \end{bmatrix}, \qquad \mathbf{b} = \begin{bmatrix} 1 \\ 1 \\ 2 \\ 0 \end{bmatrix}$$

Section 4.5

4.5.1. Use the Newton–Raphson method to solve $\Phi(q) = (q-1)^{1/3} = 0$, with $q^{(0)} = 2$. Confirm the solution behavior of Fig. 4.5.2.

4.5.2. Use the Newton–Raphson method to solve $\Phi(q) = q^3 - 4q + 4 = 0$, with $q^{(0)} = 1.5$. Confirm the solution behavior of Fig. 4.5.4.

4.5.3. Consider the four-bar mechanism of Example 3.5.2, with $\omega = \pi$ rad/s. Solve the constraint equation of Eq. 4.5.1 at $t = \frac{1}{3}$ s using the Newton–Raphson method with $\mathbf{q}^{(0)} = [\pi/3,\ 5.0,\ 4.0,\ 0.2,\ 0.9]^T$ and $\varepsilon_e = \varepsilon_s = 10^{-5}$. Monitor $\|\mathbf{q}^{(i)} - \mathbf{q}^*\|$, where $\mathbf{q}^* = [\pi/3,\ 4.94949,\ 4.01222,\ 0.18376,\ 0.86344]^T$ is the exact solution, and confirm that the Newton–Raphson method converges quadratically.

Section 4.6

4.6.1. Consider the one degree of freedom double pendulum of Example 3.6.2. If, instead of the constraint equations of Example 3.6.2, the kinematic constraint equations

$$\mathbf{\Phi}^K(\mathbf{q}) = \begin{bmatrix} x_1 - \cos\phi_1 \\ y_1 - \sin\phi_1 \\ x_2 - \cos\phi_1 - \cos\phi_2 \\ x_2 - x_1 - \cos\phi_2 \\ y_2 + 1 \end{bmatrix} = \mathbf{0}$$

are used, determine if there is a redundant constraint by evaluating the Jacobian and performing Gaussian elimination at $\phi_1 = 5\pi/3$ and $\phi_2 = 6.14881$ rad. By monitoring row index, decide which constraint is redundant.

CHAPTER FIVE

Planar Kinematic Modeling and Analysis

The mathematical and numerical aspects of kinematic analysis were stressed in Chapters 3 and 4. Examples involving even very simple systems have shown that the nonlinear character of the equations of kinematics leads to potential analytical, computational, and practical performance difficulties. The objective of this chapter is to present and illustrate the use of modeling and analysis concepts that exploit and reinforce physical intuition. A judicious combination of complementary physical and mathematical reasoning can be invaluable in the kinematic analysis of realistic mechanical systems. This is especially the case when a computer code such as DADS is employed for automated formulation and solution of kinematic equations. The availability of such computer codes encourages engineers to analyze actual systems, with large-scale realistic models. Effective utilization of such a resource, however, requires that the engineer create models that represent reality, recognize and correct modeling errors, recognize infeasible or marginal designs, and use analysis tools to create and test improved designs. Happily, many of the intuitive aids that are used in conventional design analysis are also valuable in computer-aided analysis and design.

A general approach to the art of kinematic modeling and analysis is suggested in Section 5.1. Analyses of specific systems, with the aid of the DADS code [27], are presented in the remaining sections of the chapter. As in most arts, learning to use kinematic analysis tools is enhanced by practice. Project-oriented problems are given at the end of the chapter. Readers are encouraged to try the methods presented and illustrated in this chapter on these problems or, even better, on applications of their own choosing.

5.1 MODELING AND ANALYSIS TECHNIQUES

Kinematic modeling of a mechanism involves the selection of bodies that make up the mechanism, kinematic constraints that act between pairs of bodies, and time-dependent kinematic drivers. A key requirement of kinematic modeling is that the combination of bodies, kinematic constraints, and drivers must have no free degrees of freedom. That is, the number of generalized coordinates in the model must equal the number of independent constraint equations. Since composite joints define constraints that are imposed by coupler components, without introducing generalized coordinates for the coupler, they reduce the number of computations required to solve position, velocity, and acceleration equations.

Once the bodies that make up a system model have been selected, a body-fixed reference frame must be attached to each body. Since kinematic analysis is not concerned with forces and inertias, the locations of the origins of body-fixed reference frames are arbitrary. They can be located at points that make the model easy to develop. If subsequent dynamic analysis is anticipated, however, body-fixed reference frames with their origins at centers of mass should be selected (see Chapter 6).

A careful sketch of each body should be made and used to define data for analysis. The old adage "a picture is worth a thousand words" is as true in kinematic modeling as it is in the art of communication.

In addition to specifying joints, an initial estimate of the position and orientation of each body must be provided in order to assemble the system. Estimates of x, y, and ϕ for each body can be obtained from a reasonably scaled diagram of the system. While estimates should be reasonably accurate, extreme precision is not required, since the assembly mode of analysis uses an optimization algorithm to assemble the system.

After specification of bodies and kinematic constraints, one or more degrees of freedom will remain. To complete the model, a number of drivers equal to the number of degrees of freedom must be specified. Drivers usually define relative or absolute motion that is imposed by actuators or by specifying some characteristics of the motion that is desired, regardless of the prime mover that is used to generate the motion.

Upon completion of kinematic analysis, the positions, velocities, and accelerations of each body are available at each time step in the time interval under consideration. This can be a "curse of riches," since the quantity of output data can be overwhelming, making analysis of the data difficult. The use of a postprocessor to tabulate, graph, or animate data is especially valuable.

While automated computational checks are carried out in the DADS code to warn the user that an infeasible design or redundant constraints may have been specified, it is good engineering practice to formulate a reasonable model of the system that is intended. The process of kinematic modeling is partially an engineering art that requires physical insight and intuition. To assist the engineer in modeling, the following *kinematic modeling recipe* is suggested, making use of theoretical information and intuition to arrive at a reasonable model:

1. Draw a clear diagram of the system to be modeled and select and name or number bodies and kinematic joints between pairs of bodies. As preliminary checks on reasonableness of the model:

a. Count the number nh of constraint equations that have been defined and calculate $d = 3nb - nh$, where nb is the number of bodies in the model. If the constraints specified are independent, d is the number of degrees of freedom of the system. Since the engineer generally knows how many degrees of freedom are intended, this is a good check on the reasonableness of the model. If $d \leq 0$, or if d is less than the number of degrees of freedom anticipated, very likely redundant constraints have been defined. While agreement of d with the number of degrees of freedom intended does not guarantee a good model, it is certainly necessary.

b. Inspect the model for feasibility by asking the question, "If there is any manufacturing imperfection in the model; can the system be assembled?" The foundation for this check is explained in Section 3.7. If the system can be assembled with the data defined, but not with imperfectly fabricated joints, then redundant constraints are present. If it cannot be assembled with even the nominal data, then an infeasible design has been specified.

2. Presuming redundant constraints have been eliminated, define $d = 3nb - nh$ drivers to specify the motion of the system. Repeat the analysis of step 1b to evaluate the feasibility and redundancy of the driving constraints that have been specified.

3. Make drawings of each body in the system, defining the body-fixed x'-y' frames and data required for each of the kinematic joints and drivers selected, according to the formulations presented in Chapter 3. Tabulate the data required to specify bodies and joints.

4. Use a reasonable sketch or diagram of the system to estimate the position and orientation variables to initiate the assembly process of analysis.

While this suggested procedure is not guaranteed to find all flaws in models, it can provide a foundation on which to gain engineering confidence in the model selected. To illustrate the use of the procedure, consider the following elementary example.

Example 5.1.1: The five-bar parallelogram mechanism shown in Fig. 5.1.1, with cross bars of unit length, might be modeled with six revolute joints and three absolute constraints to fix body 5 as ground. To check the reasonableness of this model, first count to see if the single degree of freedom is $1 = 3nb - nh$. Here $nb = 5$ and $nh = 3 + 6 \times 2 = 15$. Thus, $3nb - nh = 15 - 15 = 0 \neq 1$, so something is wrong.

To see what is wrong, note that if ϕ_1 is fixed then ϕ_2 is known and the revolute joint definition point on body 4, for the joint between bodies 3 and 4, is

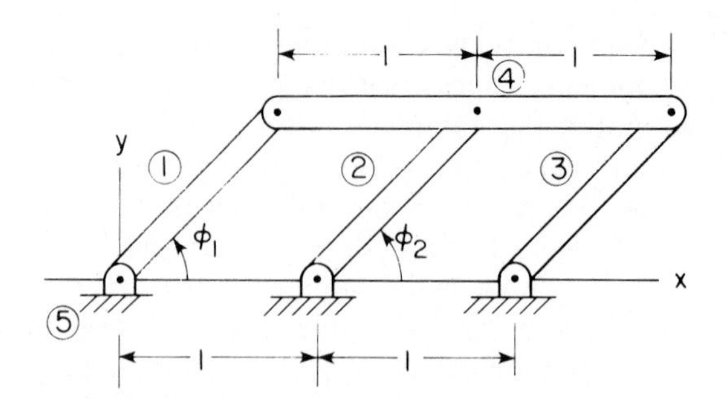

Figure 5.1.1 Five-bar parallelogram mechanism.

fixed in the x-y plane. If the length of bar 3 is slightly greater than 1, it cannot be inserted into the mechanism. Thus, the revolute joint between bars 3 and 4 contains a redundancy. If it is replaced by a revolute–translational joint between bars 3 and 4, then the mechanism can be assembled, as indicated in Fig. 5.1.2. In this model, $nh = 3 + 5 \times 2 + 1 = 14$, so $3nb - nh = 15 - 14 = 1$ and the counting criteria is satisfied.

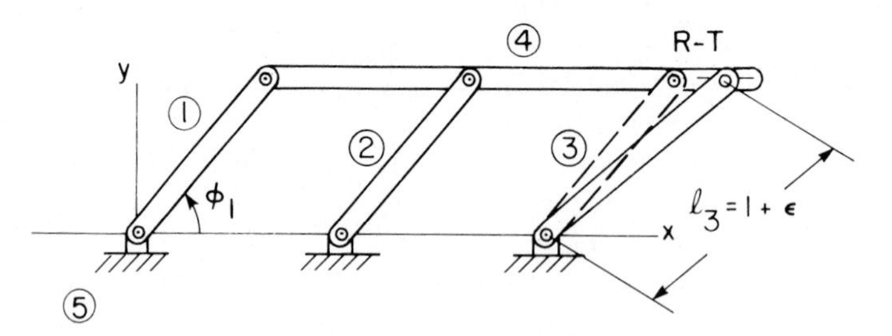

Figure 5.1.2 Five-bar imperfect parallelogram mechanism.

With $\phi_1 = \omega t$ as a driver, as long as $\phi_1 \neq k\pi$, $k = 0, 1, \ldots,$ if $\ell_3 > 1$ the mechanism of Fig. 5.1.2 performs well. If $\ell_3 = 1$, however, then a problem is encountered at $\phi_1 = \pi/2$, as seen physically in Fig. 5.1.3. As ϕ_1 passes just beyond $\pi/2$, the slider point on body 3 can move either to the left or to the

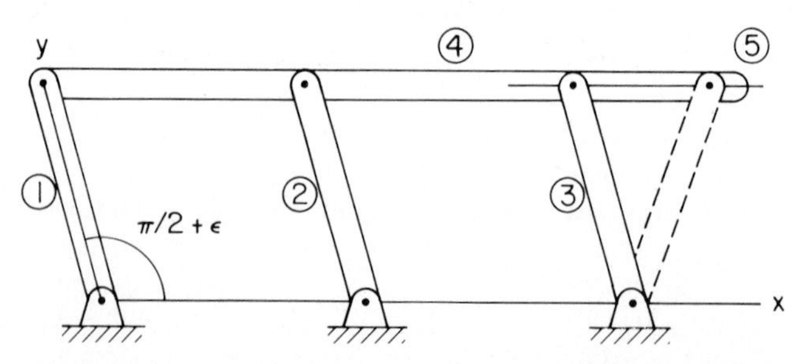

Figure 5.1.3 Bifurcation point for five-bar parallelogram mechanism.

right. Thus, $\phi_1 = \omega t^* = \pi/2$ is a bifurcation point and the constraint Jacobian that contains the driver must be singular. It is interesting that this physical reasoning shows that the 15×15 Jacobian matrix is singular, without evaluating its determinant or calculating its rank using Gaussian elimination. This illustrates the value of complementary physical and mathematical reasoning in kinematic modeling and analysis. Finally, if $\ell_3 < 1$, a lock-up singularity will be encountered prior to t^*.

5.2 KINEMATIC ANALYSIS OF A SLIDER–CRANK MECHANISM

The slider–crank mechanism of Fig. 2.4.6 is one of the most commonly used machine subsystems in mechanical system design. It is employed as the principal element of internal combustion engines (Fig. 1.1.1), compressors, fly-ball governors (Fig. 1.1.3), stamping machines (Fig. 1.1.2), and many other machines. Analytical kinematic analyses of simplified models of slider–cranks have been carried out in Chapters 2 and 3. The purpose of this section is to systematically analyze the kinematic performance of a slider–crank mechanism, illustrate use of models, investigate the effect of design variations on kinematic performance, and identify lock-up or singular configurations that may occur.

Data used in this section and throughout the text are given in SI units (meter, kilogram, seconds) unless noted otherwise. The specific values used are not intended to represent any particular machine element.

5.2.1 Alternative Models

The slider–crank mechanisms can be modeled in many different ways. Two models are developed here to demonstrate the preparation and analysis of the data required for kinematic analysis. In model 1 (see Fig. 5.2.1), each link and ground is modeled as a body. Joints can then be modeled as revolute and translational joints, as follows:

Model 1		
Bodies		
Four bodies		$nc = 12$
Constraints		
Revolute joints:	A	2
	B	2
	C	2
Translational joint:	D	2
Ground constraints		3
		$nh = 11$
Thus, DOF = 12 − 11 = 1.		

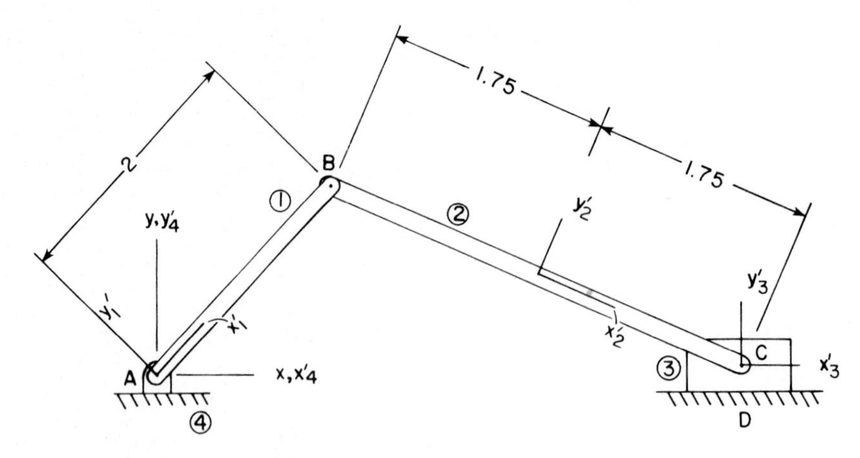

Figure 5.2.1 Slider–crank, model 1.

To eliminate the one remaining DOF, a driver must be introduced. Typically, this would be an angle driver on ϕ_1.

To define each revolute joint, the common point P between bodies connected must be defined in the body-fixed frame. The three revolute joints in model 1 are defined in Table 5.2.1.

TABLE 5.2.1 Revolute Joint Data, Model 1

Joint	1	2	3
Common point P	A	B	C
Body i	4	1	2
$x_i'^P$	0.0	2.0	1.75
$y_i'^P$	0.0	0.0	0.0
Body j	1	2	3
$x_j'^P$	0.0	−1.75	0.0
$y_j'^P$	0.0	0.0	0.0

To define a translational joint, two parallel vectors, one on each body connected, must be defined. Vectors $\mathbf{v}_4$ and $\mathbf{v}_3$ are fixed in bodies 4 and 3, respectively (see Fig. 5.2.2). Coordinates of the points that define these two vectors, in their respective body-fixed frames, are given in Table 5.2.2.

Figure 5.2.2 Translational joint, model 1.

TABLE 5.2.2 Translational Joint Data, Model 1

Vector	$\mathbf{v}_4$	$\mathbf{v}_3$
$x_i'^P$	$x_4'^P = 0.0$	$x_3'^P = 0.0$
$y_i'^P$	$y_4'^P = 0.0$	$y_3'^P = 0.0$
$x_i'^Q$	$x_4'^Q = 1.0$	$x_3'^Q = 1.0$
$y_i'^Q$	$y_4'^Q = 0.0$	$y_3'^Q = 0.0$

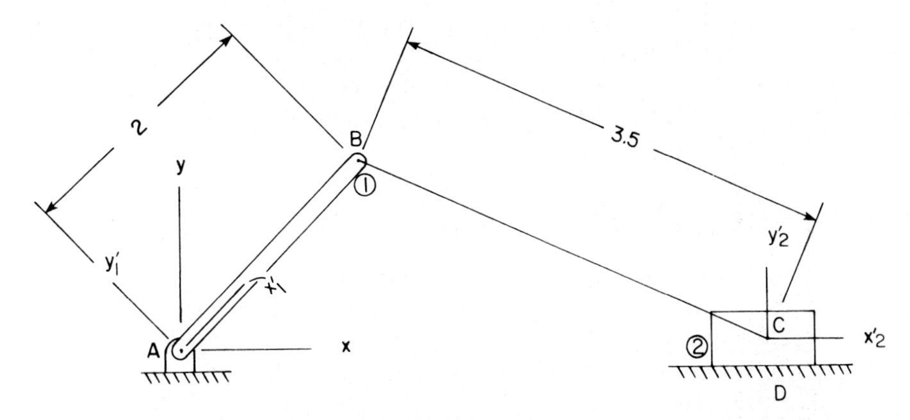

Figure 5.2.3 Slider–crank, model 2.

In model 2 (see Fig. 5.2.3), the connecting rod *BC* is modeled as a composite revolute–revolute joint and no ground body is introduced. The advantage of the composite joint is that it reduces the dimension of the matrices that must be handled by eliminating three generalized coordinates that would normally be required for the connecting rod. This mechanism is modeled using absolute constraints to represent the rotational joint at point *A* and the translational joint at point *C*, as follows:

Model 2		
Bodies		
Two bodies		$nc = 6$
Constraints		
Body 1 absolute constraints:	$x_1 = 0$	1
	$y_1 = 0$	1
Body 2 absolute constraints:	$y_2 = 0$	1
	$\phi_2 = 0$	1
Revolute–revolute joint		1
		$nh = 5$
Thus, DOF $= 6 - 5 = 1$.		

TABLE 5.2.3 Revolute–Revolute Joint *B–C*, Model 2

Length of $BC = 3.5$	
Point B (body 1)	Point C (body 2)
$x_1'^B = 2.0$	$x_2'^C = 0.0$
$y_1'^B = 0.0$	$y_2'^C = 0.0$

The absolute constraints on body 2 are equivalent to a translational joint, prohibiting rotation and y translation of the body. Similarly, the absolute constraints on body 1 are equivalent to a revolute joint. To specify a revolute–revolute joint, the end points of the coupler on the bodies that are connected must be specified in their respective body-fixed frames, and the length of the link must be provided. Table 5.2.3 contains data for this joint.

5.2.2 Assembly Analysis

For model 1, Fig. 5.2.1 can be used as the source for position and orientation estimates. Assuming a scale of 1:1, the position and angular orientation of each body can be measured. To show the power of the assembly technique discussed in Section 4.3, the poor initial estimate given in Table 5.2.4 and shown in Fig. 5.2.4 is used.

TABLE 5.2.4 Initial Estimate, Model 1

Body no.	1	2	3	4
x	−0.5	3.0	5.0	0.0
y	0.0	1.0	0.0	0.0
ϕ	0.5	−0.4	0.0	0.0

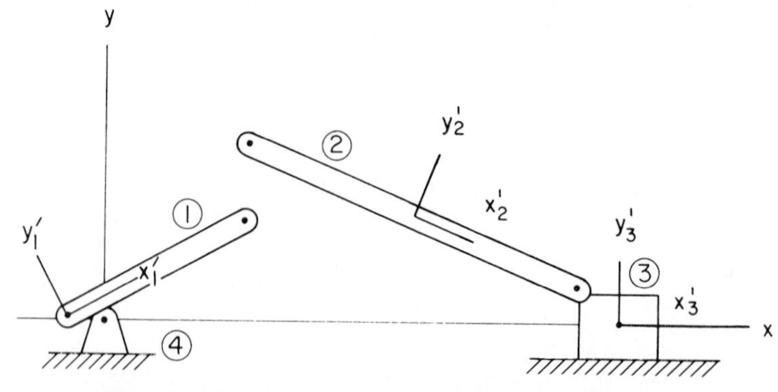

Figure 5.2.4 Initial estimate for slider–crank, model 1.

TABLE 5.2.5 Assembled Configuration without Driver, Model 1

Body no.	1	2	3	4
x	0.0	3.1702	4.992	0.0
y	0.0	0.6387	0.0	0.0
ϕ	0.6916	−0.3728	0.0	0.0

The system is first assembled without consideration of a driver. Since the system has one degree of freedom, the kinematic constraint equations have infinitely many solutions. The optimization algorithm converges to a solution that is near to the initial estimate. The solution found is tabulated in Table 5.2.5 and illustrated as the solid line drawing in Fig. 5.2.5.

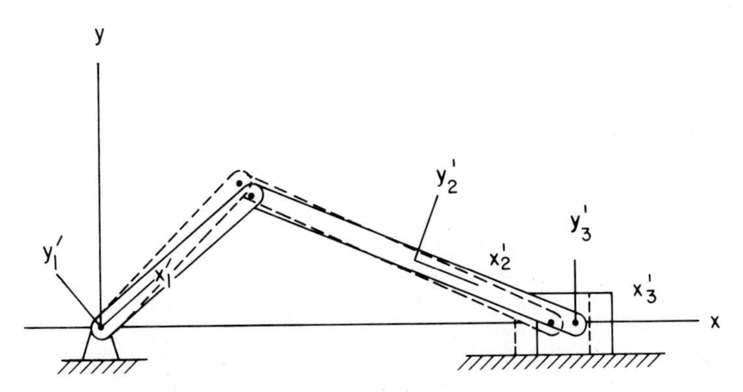

Figure 5.2.5 Assembled slider–crank, model 1.

The next step is to assemble the slider–crank mechanism with the driver $\phi_1 = \pi/4 + \omega t$. The initially ($t = 0$) assembled configuration is tabulated in Table 5.2.6 and illustrated as the dashed line drawing in Fig. 5.2.5.

Consider another initial estimate for which the user made a sign error in estimating the x coordinate of the third body, as shown in Fig. 5.2.6. All other data are the same as in Table 5.2.4. The assembled configuration found by the

TABLE 5.2.6 Assembled Configuration with Driver at $t = 0$, Model 1

Body no.	1	2	3	4
x	0.0	3.0152	4.6162	0.0
y	0.0	0.7066	0.0	0.0
ϕ	0.7852	−0.4158	0.0	0.0

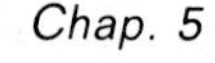

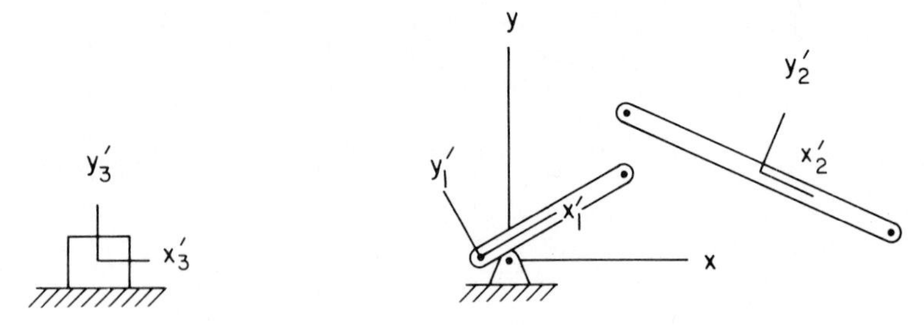

Figure 5.2.6 Erroneous initial estimate, model 1.

optimization algorithm is tabulated in Table 5.2.7 and shown in Fig. 5.2.7 for the same driver. This shows successful assembly of the system, satisfying the driving constraint. However, this is not the configuration that is desired. These variations of initial estimates show that care must be taken to confirm that the assembled configuration is physically meaningful and is the desired configuration.

TABLE 5.2.7 Undesirable Assembled Configuration, Model 1

Body no.	1	2	3	4
x	0.0	−0.1863	−1.7873	0.0
y	0.0	0.7080	0.0	0.0
ϕ	0.7858	−2.7253	0.0	0.0

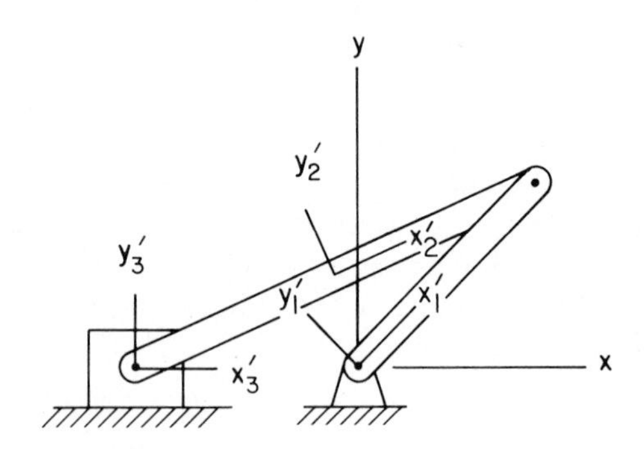

Figure 5.2.7 Undesirable assembled configuration, model 1.

5.2.3 Driver Specification

Since the model has one kinematic degree of freedom, one driver must be specified in order to complete the model. Taking ϕ_1 as the driven coordinate, it will be specified that body 1 rotates at ω rad/s, with an initial orientation angle of $\phi_1(0) = \pi/4$ rad. The driver is thus specified by the condition

$$\phi_1 = \frac{\pi}{4} + \omega t$$

where ω is the constant angular velocity of the crank (body 1).

5.2.4 Analysis

Kinematic analysis of the slider–crank mechanism, using model 1 with the dimensions of Fig. 5.2.4, is carried out using the DADS code [27]. Two constant driving angular velocities, $\omega_1 = 2\pi$ rad/s and $\omega_2 = 4\pi$ rad/s, are considered (i.e., one and two revolutions per second respectively). Since the motion of the slider (body 3) is of greatest concern in applications, the position (x), velocity ($\dot{x}$), and acceleration ($\ddot{x}$) of the slider are plotted in Figs. 5.2.8, 5.2.9, and 5.2.10, respectively. Identical results are obtained with model 2.

As expected, the mechanism goes through two cycles of motion in one second with ω_2, and the natures of the responses with two driving velocities are

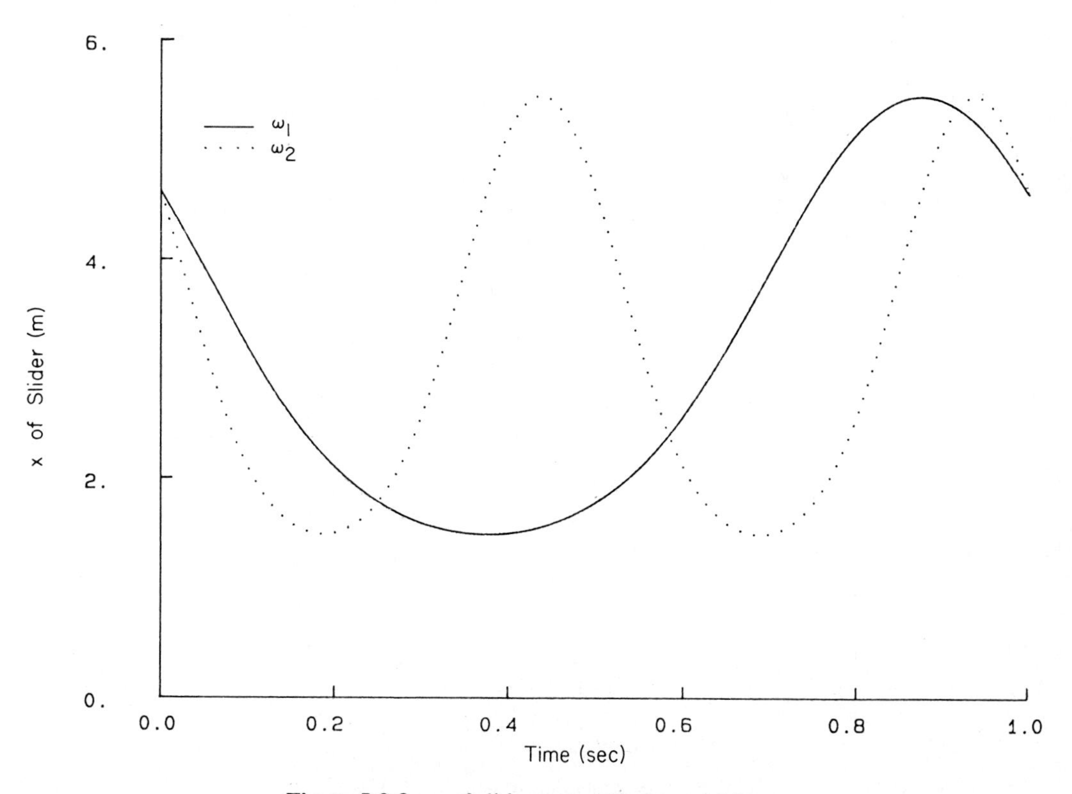

Figure 5.2.8 x of slider versus time, variable ω.

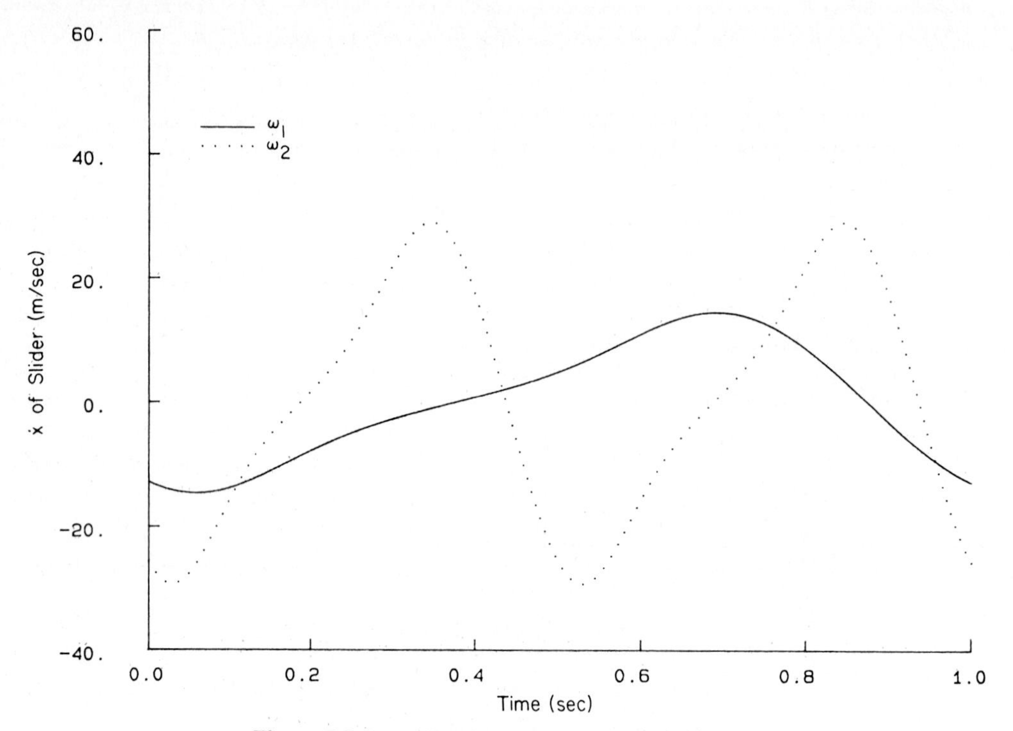

Figure 5.2.9 $\dot{x}$ of slider versus time, variable ω.

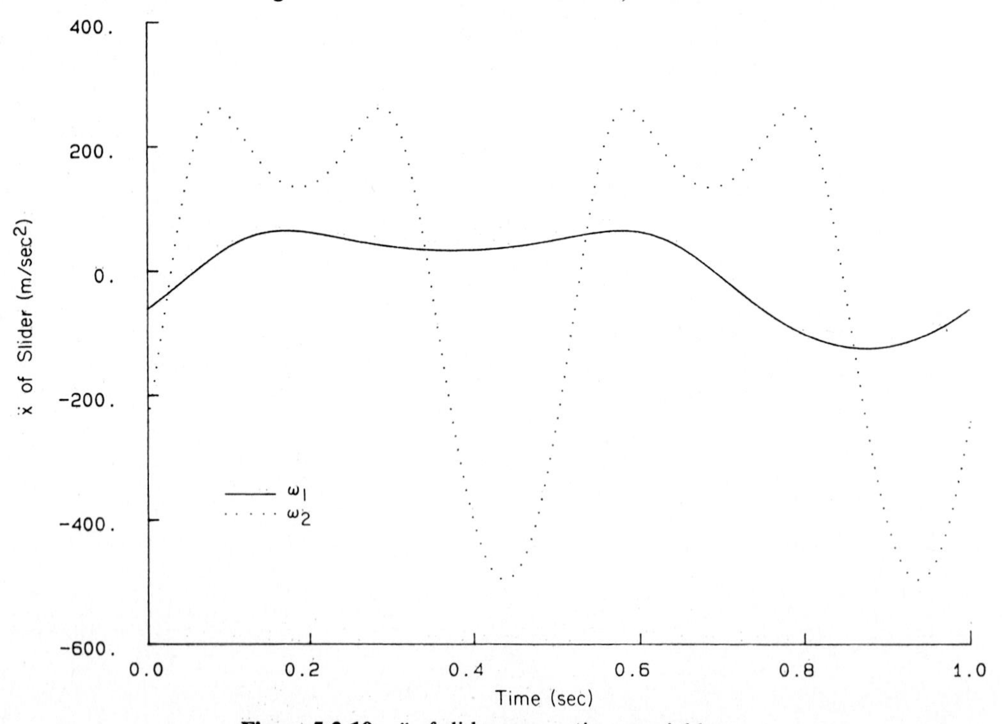

Figure 5.2.10 $\ddot{x}$ of slider versus time, variable ω.

similar. Note, however, that while the stroke lengths are equal for both driving velocities the ratios of maximum velocity and acceleration are approximately 2 and 4, respectively. This suggests that the quadratic terms in velocity on the right side of the acceleration equation are dominating acceleration; that is, these terms grow by a factor of $(2)^2 = 4$ due to the doubled velocity. Since the Jacobian is not changed, acceleration increases by a factor of 4. Thus, increasing operating speed has a much more marked effect on slider acceleration, and hence on bearing loads, than on position and velocity.

To consider the effect of variations in design on the kinematic performance of the slider–crank, let ℓ denote the length of the connecting rod (body 2 in model 1), where $\ell_1 = 3.5$ m is the nominal design considered thus far. Let $\ell_2 = 2.5$, $\ell_3 = 2.2$, and $\ell_4 = 2.1$ be a sequence of decreasing connecting rod lengths. The numerical results of simulations for x, $\dot{x}$, and $\ddot{x}$ of the slider, for each of the four lengths, are shown in Figs. 5.2.11, 5.2.12, and 5.2.13, respectively. Note that position (apart from a translation) and velocity are only moderately sensitive to variation in connecting rod length. Acceleration, on the other hand, shows extreme sensitivity to variation in ℓ, for the smallest values of ℓ, near $t = 0.15$ and 0.6 s. These are times when the slider is near its extreme

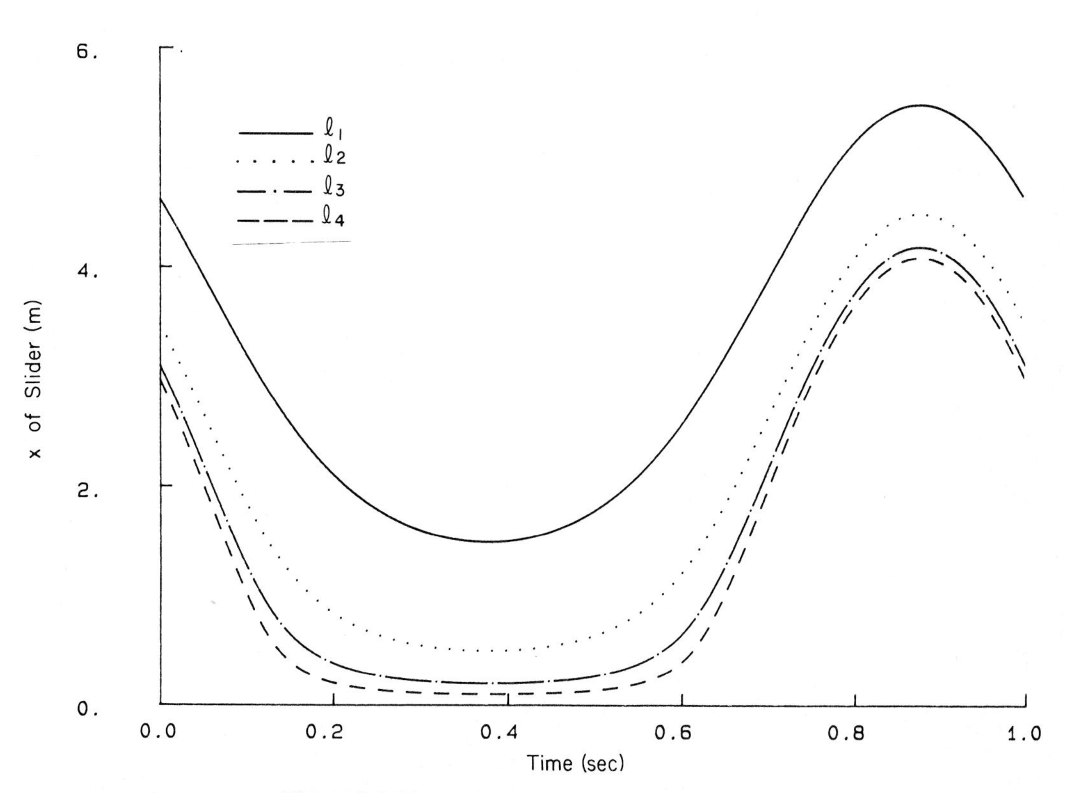

Figure 5.2.11 x of slider versus time, variable ℓ.

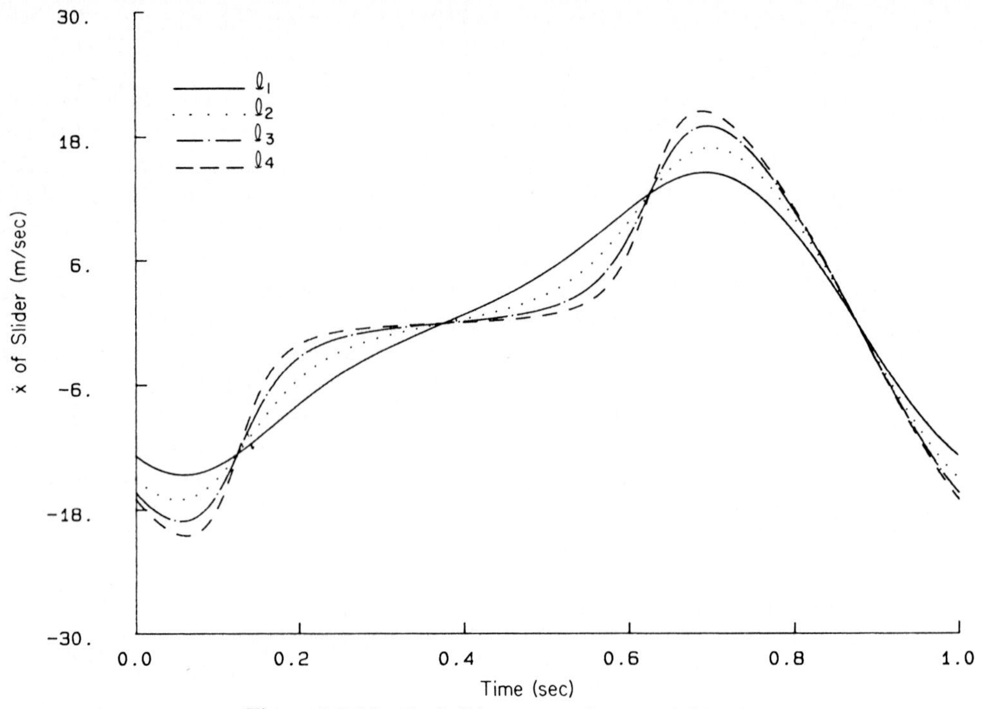

Figure 5.2.12 $\dot{x}$ of slider versus time, variable ℓ.

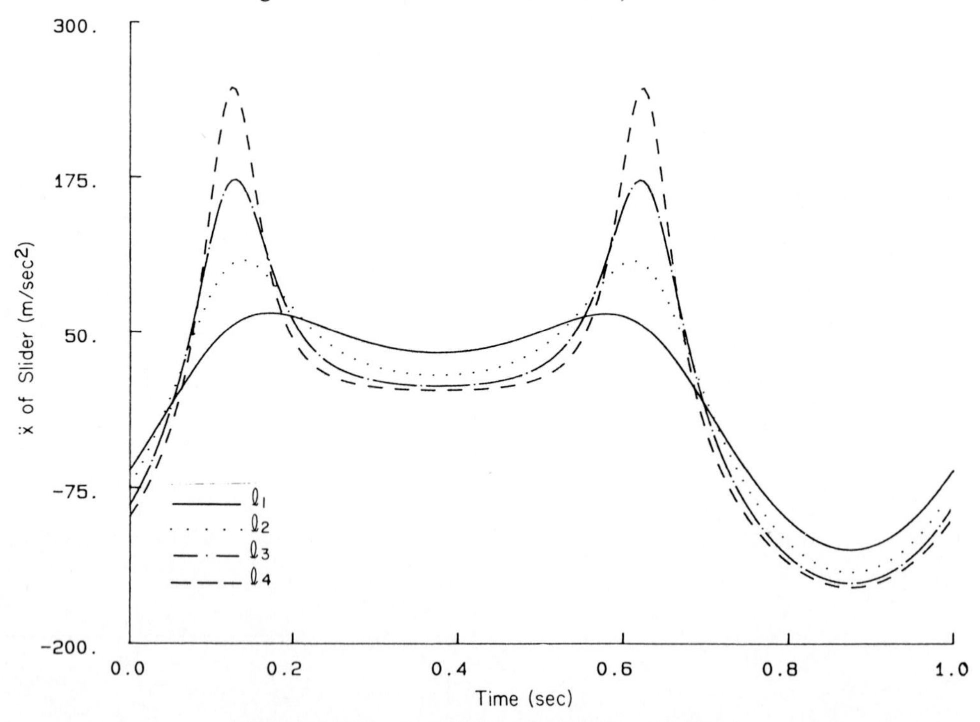

Figure 5.2.13 $\ddot{x}$ of slider versus time, variable ℓ.

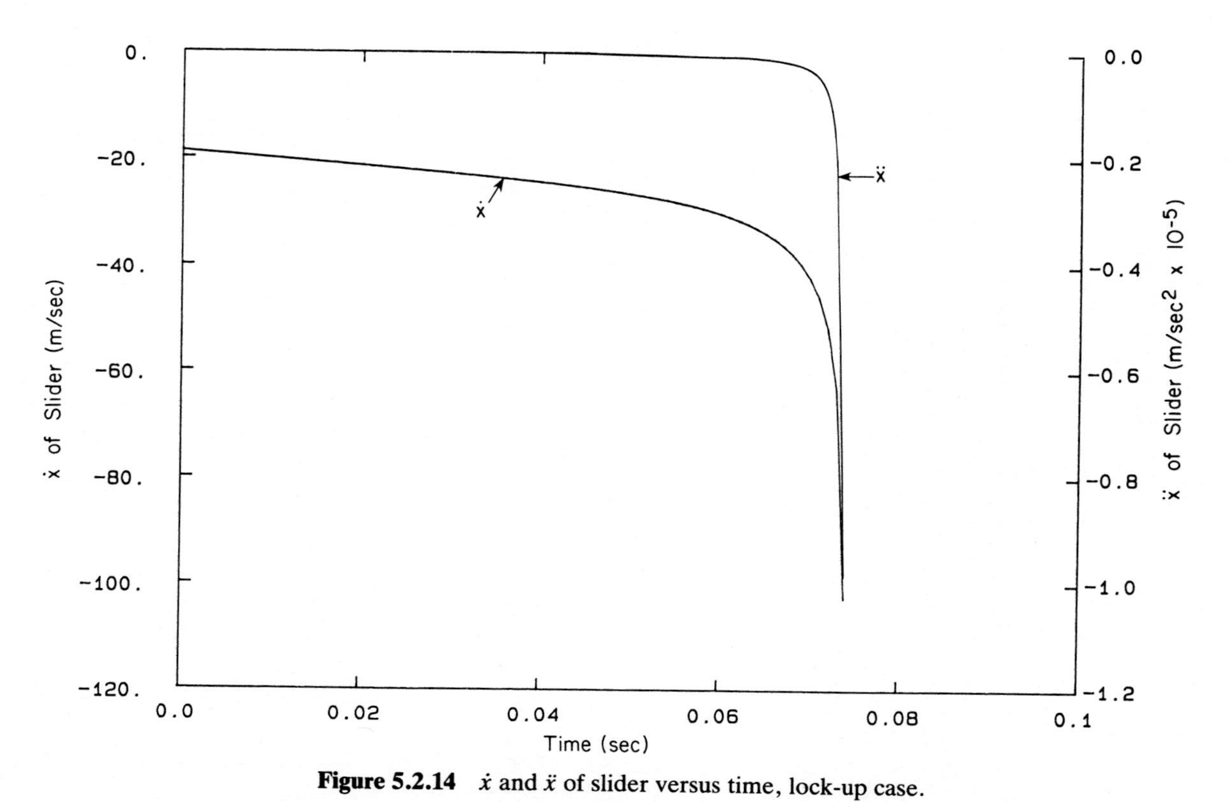

Figure 5.2.14 $\dot{x}$ and $\ddot{x}$ of slider versus time, lock-up case.

right and left positions, respectively. The extreme character of this sensitivity is highlighted by the fact that the peak variation in acceleration between $\ell_3 = 2.2$ and $\ell_4 = 2.1$ is greater than that between $\ell_1 = 3.5$ and $\ell_2 = 2.5$. These results should be expected, based on the singular configuration analysis of Example 3.7.3 and the general guideline that accelerations become very large near singular configurations.

5.2.5 Lock-up Configuration

As shown analytically in Example 3.7.3, the slider–crank mechanism will *lock-up* if the crank arm (body 1) is longer than the connecting rod (body 2). Figure 5.2.14 shows the velocity and acceleration of the slider for a simulation with the length of the slider reduced to $\ell_5 = 1.9$, which is shorter than the length of the crank. Note that as t approaches 0.074 s both the velocity and acceleration of the slider approach infinity. As has been suggested by theoretical considerations in Chapter 3, acceleration shows the more rapid divergence to infinity.

5.3 KINEMATIC ANALYSIS OF A FOUR-BAR MECHANISM

One of the most commonly encountered of the basic mechanisms is the four-bar mechanism shown schematically in Fig. 5.3.1. Examples of the use of this

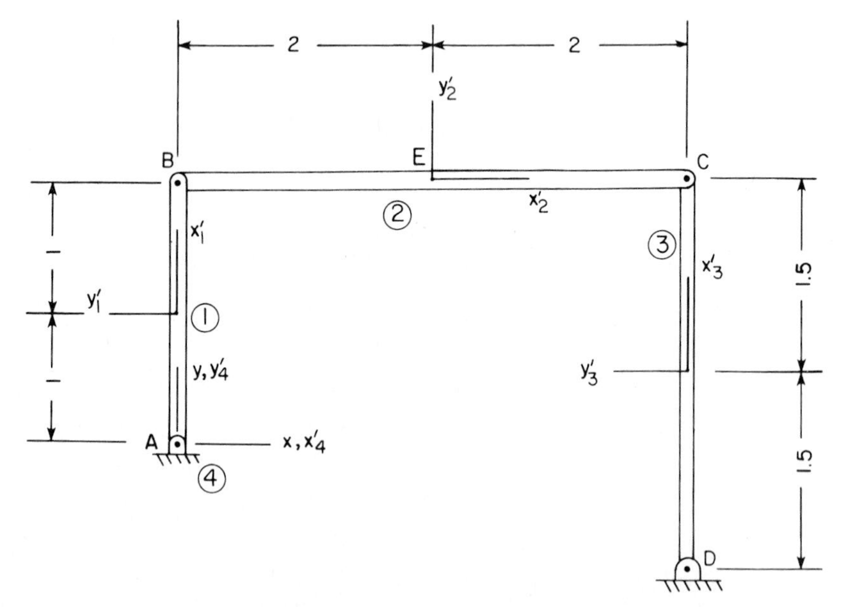

Figure 5.3.1 Four-bar linkage, model 1.

mechanism are the vehicle suspension subsystem (Fig. 1.1.4), windshield wiper (Fig. 1.1.5), and material-handling linkage (Fig. 1.1.6). In some applications, the motion of the coupler (body 2 in Fig. 5.3.1) is controlled by the input angle history of the crank (body 1 in Fig. 5.3.1). In other applications, it is the rotation of the follower (body 3 in Fig. 5.3.1).

5.3.1 Alternative Models

A four-bar mechanism with four revolute joints is modeled here in two different ways. In model 1 (Fig. 5.3.1), each link and ground is modeled as a body.

Four revolute joints complete the model, as follows:

Model 1	
Bodies	
Four bodies	$nc = 12$
Constraints	
Revolute joints: A	2
B	2
C	2
D	2
Ground constraints	3
	$nh = 11$
Thus, DOF = 12 − 11 = 1.	

Revolute joint definition data for this model are given in Table 5.3.1.

TABLE 5.3.1 Revolute Joint Data, Model 1

Joint no.	1	2	3	4
Common point P	A	B	C	D
Body i	4	1	2	3
$x_i'^P$	0.0	1.0	2.0	−1.5
$y_i'^P$	0.0	0.0	0.0	0.0
Body j	1	2	3	4
$x_j'^P$	−1.0	−2.0	1.5	4.0
$y_j'^P$	0.0	0.0	0.0	−1.0

In model 2 (Fig. 5.3.2), link CD is modeled as a revolute–revolute composite joint. Also, the body-fixed frames have been relocated. The mechan-

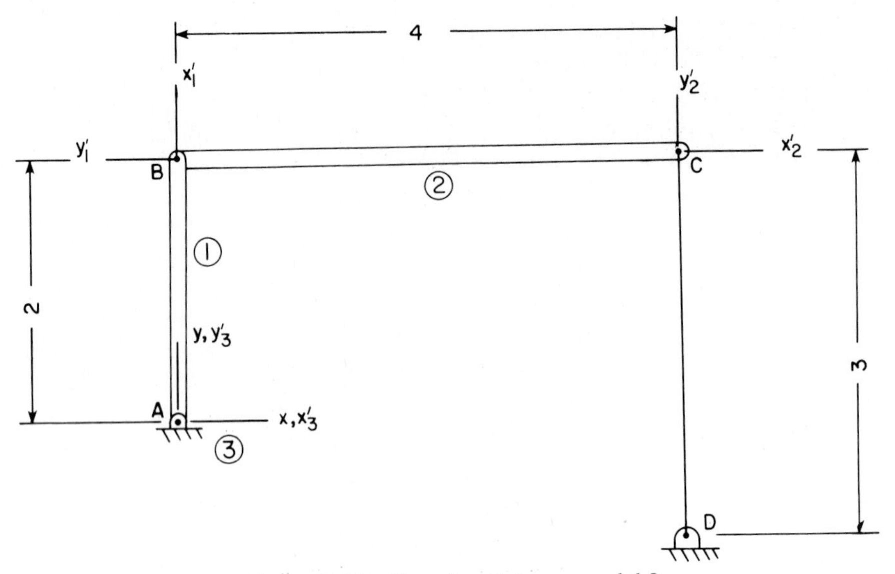

Figure 5.3.2 Four-bar linkage, model 2.

ism can be modeled as follows:

Model 2

Bodies		
Three bodies		$nc = 9$
Constraints		
Revolute joint:	*A*	2
	B	2
Revolute–revolute *CD*		1
Ground constraints		3
		$nh = 8$

Thus, DOF $= 9 - 8 = 1$.

Revolute joint data for model 2 are shown in Table 5.3.2. Table 5.3.3 contains data for the revolute–revolute composite joint.

TABLE 5.3.2 Revolute Joint Data, Model 2

Joint no.	1	2
Common point P	A	B
Body i	3	1
$x_i'^P$	0.0	0.0
$y_i'^P$	0.0	0.0
Body j	1	2
$x_j'^P$	−2.0	−4.0
$y_j'^P$	0.0	0.0

TABLE 5.3.3 Revolute–Revolute Joint Data, Model 2

Length of $CD = 3.0$	
Point C (body 2)	Point D (body 3)
$x_2'^C = 0.0$	$x_3'^D = 4.0$
$y_2'^C = 0.0$	$y_3'^D = -1.0$

5.3.2 Assembly

For model 1, initial position estimates can be measured from Fig. 5.3.1. An intentionally poor initial estimate is tabulated in Table 5.3.4 and shown in Fig. 5.3.3.

TABLE 5.3.4 Initial Estimate, Model 1

Body no.	1	2	3	4
x	0.5	2.1	4.2	0.0
y	0.5	1.9	0.0	0.0
ϕ	1.3	0.1	1.4	0.0

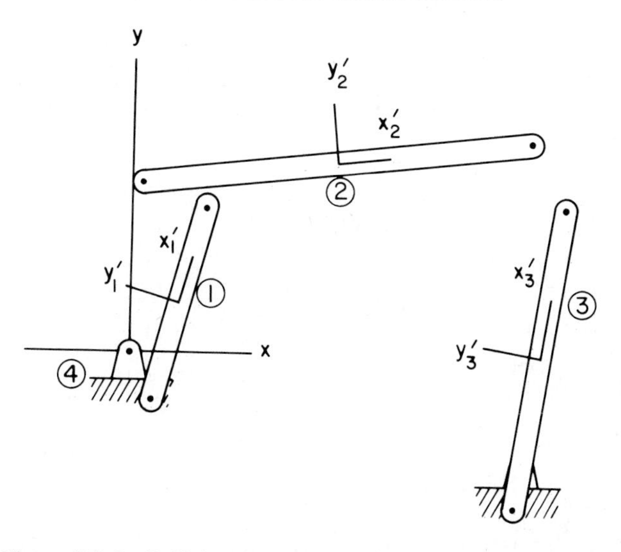

Figure 5.3.3 Initial estimate for four-bar linkage, model 1.

The assembled configuration is tabulated in Table 5.3.5 and shown as a solid line drawing in Fig. 5.3.4. The configuration shown as a dashed line drawing in Fig. 5.3.4 is the assembled configuration at $t = 0$, with the driving constraint $\phi_1 = \pi/2 + 2\pi t$. Numerical values are tabulated in Table 5.3.6.

TABLE 5.3.5 Assembled Configuration without Driver, Model 1

Body no.	1	2	3	4
x	0.2009	2.4002	4.2001	0.0
y	0.9798	1.9664	0.4866	0.0
ϕ	1.3693	0.0034	1.4370	0.0

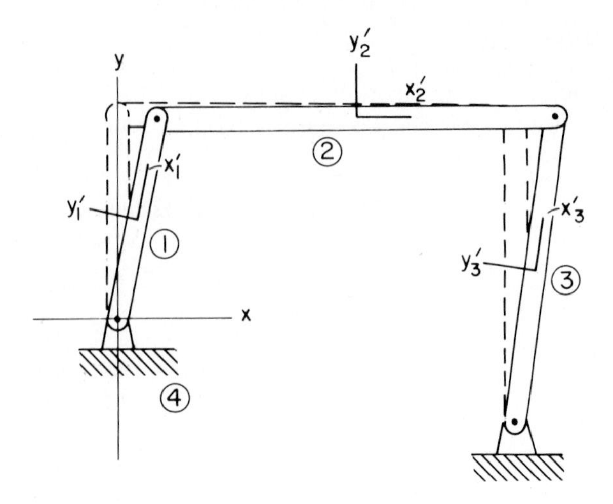

Figure 5.3.4 Assembled four-bar linkage, model 1.

TABLE 5.3.6 Assembled Configuration with Driver, Model 1

Body no.	1	2	3	4
x	0.0	2.0	4.0	0.0
y	1.0	2.0	0.5	0.0
ϕ	1.5708	0.0	1.5708	0.0

5.3.3 Driver Specification

Since the model has only one degree of freedom, one driver must be specified. Taking ϕ_1 as the driven coordinate, it may be driven at 2π rad/s. The driver is specified by

$$\phi_1 = \frac{\pi}{2} + 2\pi t$$

5.3.4 Analysis

Three aspects of the kinematic performance of the mechanism are considered; (1) the path in the x-y plane that is swept out by point E on body 2 (see Fig. 5.3.1), (2) the relation between the follower angle (ϕ_3) and the input angle (ϕ_1), and (3) the follower angular velocity and acceleration as functions of time. To gain insights into these relations as the geometry of the mechanism is changed, consider four lengths of the follower (body 3 in model 1); $\ell_1 = 3$, $\ell_2 = 2.5$, $\ell_3 = 2.2$, and $\ell_4 = 2.1$. As the length of the follower is changed, the y coordinate

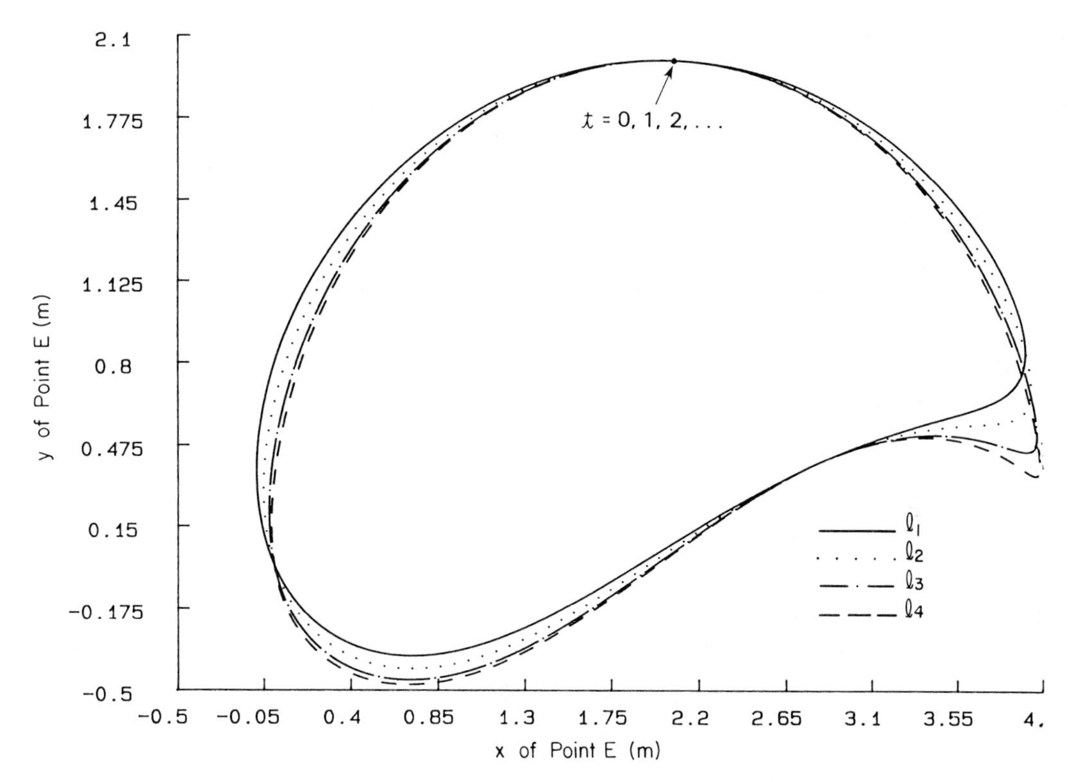

Figure 5.3.5 y versus x coordinates of point E, model 1.

of point D in Fig. 5.3.1 is changed so that when $\phi_1 = \phi_3 = \pi/2$ (as shown in Fig. 5.3.1), $\phi_2 = 0$; that is, the y coordinate of point D is moved upward by the amount the length of body 3 is reduced.

Plots of the y coordinate of point E versus its x coordinate, for each length of body 3, are presented in Fig. 5.3.5. Note that all four of the trajectories have points in common at $t = 0, 1, 2, \ldots$. Note also that, as the length of the follower approaches $\ell = 2$, extreme curvature of the path of point E occurs near the point of the trajectory when E is to the far right. This suggests that some form of singular behavior is impending as the mechanism approaches a parallelogram. This is consistent with the singular behavior noted in Example 3.7.2.

Next, the follower angle (ϕ_3) is plotted versus the crank angle (ϕ_1) for each length of body 3 in Fig. 5.3.6. The extreme variation of ϕ_3 versus ϕ_1 near $\phi_1 = 6.5$ rad is associated with the high curvature noted in Fig. 5.3.5 as ℓ approaches 2.

Finally, plots of the follower angular velocity and angular acceleration as functions of time are presented in Figs. 5.3.7 and 5.3.8 for each length of the follower. The follower angle versus time is not plotted, since it is proportional to

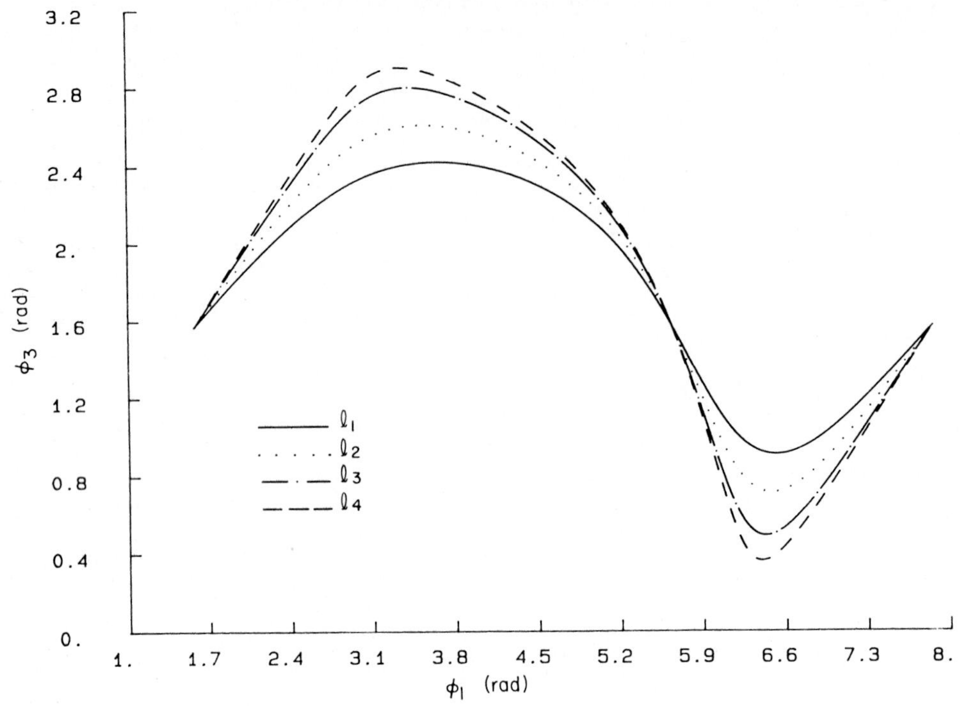

Figure 5.3.6 ϕ of body 3 versus ϕ of body 1, model 1.

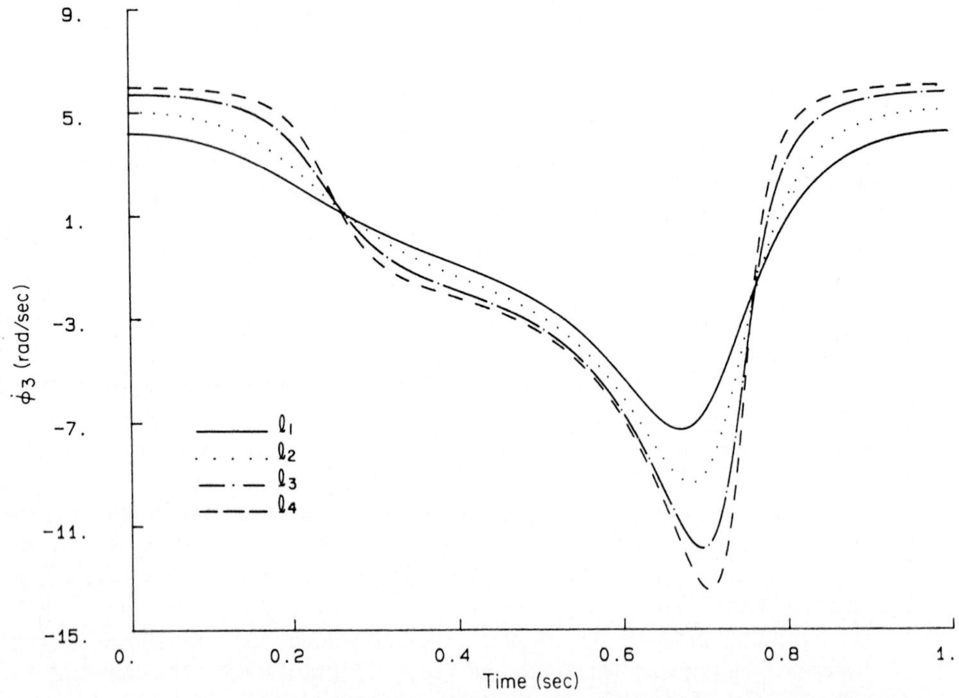

Figure 5.3.7 x of body 1 versus time, model 1.

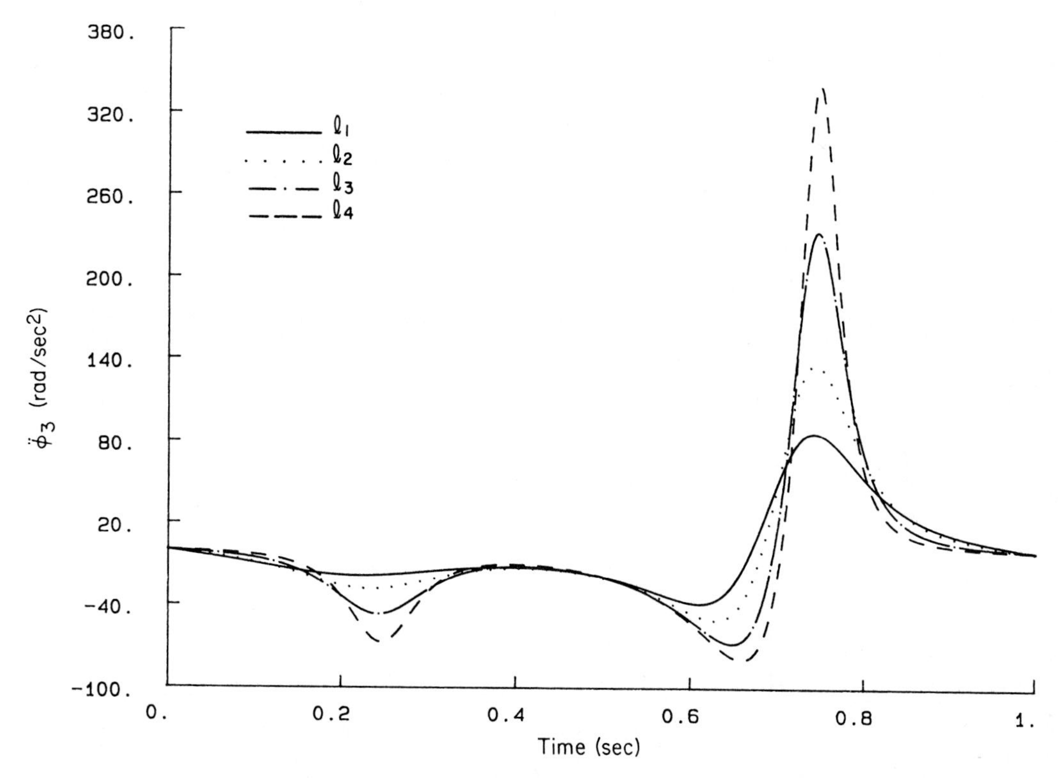

Figure 5.3.8 $\ddot{\phi}$ of follower versus time, model 1.

the curve of Fig. 5.3.6. The extreme behavior of velocity and especially acceleration near $t = 0.75$ s is a clear indication of impending singularity as the length of the follower approaches 2. The results obtained are the same with model 2, as expected.

5.3.5 Lock-up Configuration

If the length of the follower is reduced to $\ell_3 = 1.9$, moving point D upward as in the previous cases, it is shorter than the crank. It is geometrically clear that trouble is in the offing. Plots of $\dot{\phi}_3$ and $\ddot{\phi}_3$ versus time in Fig. 5.3.9 show that lock-up occurs near $t = 0.18$ s, where both approach infinity.

5.4 KINEMATIC ANALYSIS OF A QUICK-RETURN MECHANISM

As an example of the many compound mechanisms that arise in practice (e.g., the shaper of Fig. 1.1.8, the serial manipulator of Fig. 1.1.9, or the vehicle of

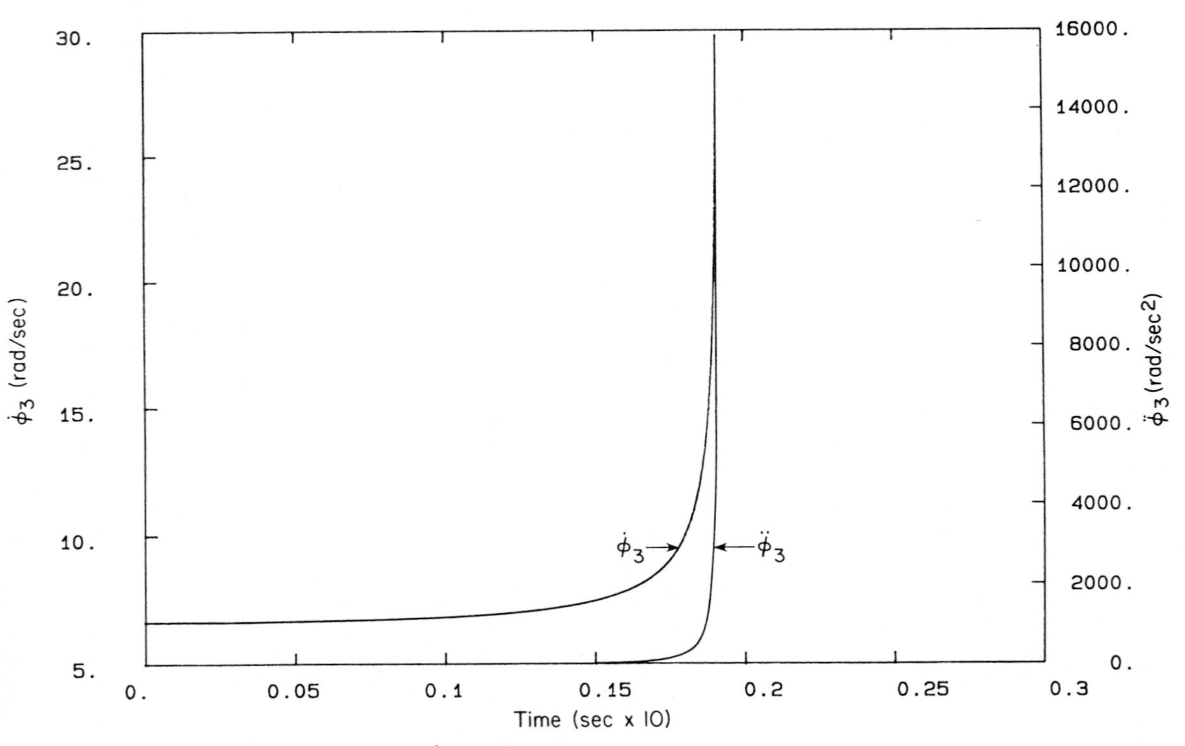

Figure 5.3.9 $\dot{\phi}$ and $\ddot{\phi}$ of follower versus time, lock-up case, model 1.

Figs. 1.1.10, 1.1.11, and 1.1.12), the quick-return mechanism of Fig. 5.4.1 that represents a shaper is considered. With counterclockwise rotation of the crank (body 3), cutting occurs as the tool (body 6) moves to the left through the workpiece (see Fig. 1.1.8). The quick-return stroke of the tool occurs as it moves to the right.

5.4.1 Alternative Models

A quick-return mechanism is modeled here in two different ways. In the model of Fig. 5.4.1, each link is modeled as a body. Elements of the model are as follows:

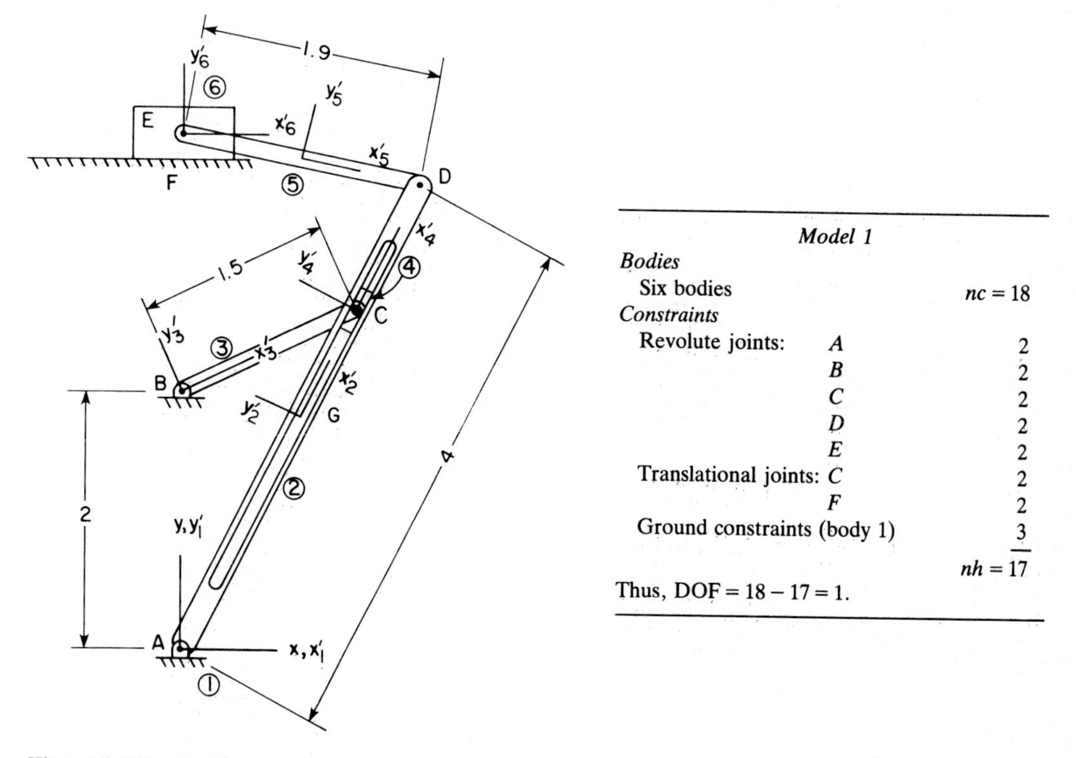

Model 1		
Bodies		
Six bodies		$nc = 18$
Constraints		
Revolute joints:	A	2
	B	2
	C	2
	D	2
	E	2
Translational joints:	C	2
	F	2
Ground constraints (body 1)		3
		$nh = 17$
Thus, DOF $= 18 - 17 = 1$.		

Figure 5.4.1 Quick-return mechanism, model 1.

Body 1 is ground, and the body-fixed frames of bodies 1 to 6 are shown in Fig. 5.4.1. Revolute joint definition data are given in Table 5.4.1.

Vectors that define the two translational joints are shown in Fig. 5.4.2. Point data that define these vectors, and hence the translational joints, are tabulated in Table 5.4.2.

TABLE 5.4.1 Revolute Joint Data, Model 1

Joint no.	1	2	3	4	5
Common point P	A	B	C	D	E
Body i	1	1	3	2	5
$x_i'^P$	0.0	0.0	1.5	2.0	−0.95
$y_i'^P$	0.0	2.0	0.0	0.0	0.0
Body j	2	3	4	5	6
$x_j'^P$	−2.0	0.0	0.0	0.95	0.0
$y_j'^P$	0.0	0.0	0.0	0.0	0.0

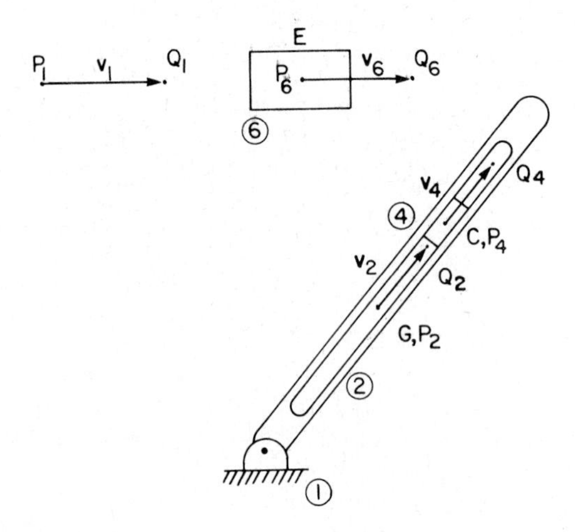

Figure 5.4.2 Translational joint definition, model 1.

TABLE 5.4.2 Translational Joint Data

Vector	$\mathbf{v}_1$	$\mathbf{v}_6$	$\mathbf{v}_2$	$\mathbf{v}_4$
$x_i'^P$	$x_1'^P = -3.0$	$x_6'^P = 0.0$	$x_2'^P = 0.0$	$x_4'^P = 0.0$
$y_i'^P$	$y_1'^P = 4.0$	$y_6'^P = 0.0$	$y_2'^P = 0.0$	$y_4'^P = 0.0$
$x_i'^Q$	$x_1'^Q = -2.0$	$x_6'^Q = 1.0$	$x_2'^Q = 1.0$	$x_4'^Q = 1.0$
$y_i'^Q$	$y_1'^Q = 4.0$	$y_6'^Q = 0.0$	$y_2'^Q = 0.0$	$y_4'^Q = 0.0$

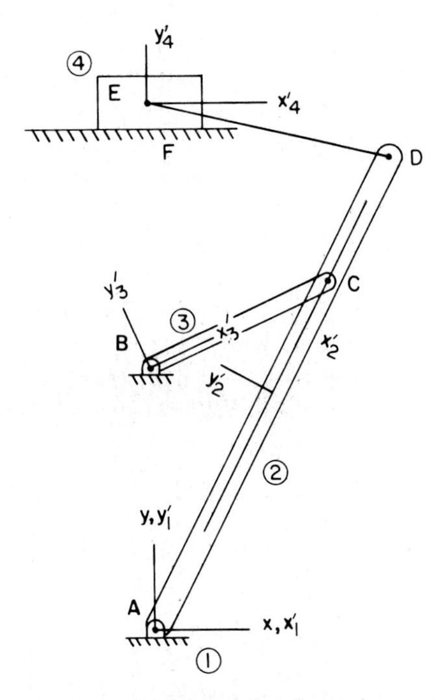

Figure 5.4.3 Quick-return mechanism, model 2.

In model 2 of Fig. 5.4.3, link DE is replaced by a revolute–revolute composite joint, and the slider block at point C is replaced by a revolute–translational joint. The mechanism is thus modeled as follows:

Model 2		
Bodies		
Four bodies		$nc = 12$
Constraints		
Revolute joints:	A	2
	B	2
Translational joint:	F	2
Revolute–revolute joint:	DE	1
Revolute–translational joint:	C	1
Ground constraints (body 1)		3
		$nh = 11$

Thus, DOF = 12 − 11 = 1.

The same lengths and data used to define the revolute joints at A and B in model 1 are used to define these joints in model 2 (see Table 5.4.1). Similarly, data used to define the translational joint at F in model 1 are used to define the corresponding joint in model 2 (see Table 5.4.2). The revolute–revolute joint DE is defined in Table 5.4.3, and the revolute–translational joint at C is defined in Table 5.4.4. (see Fig. 5.4.2).

TABLE 5.4.3 Revolute–Revolute Joint *DE*, Model 2

Length of $DE = 1.90$	
Point E (body 4)	Point D (body 2)
$x_4'^E = 0.0$	$x_2'^D = 2.0$
$y_4'^E = 0.0$	$y_2'^D = 0.0$

TABLE 5.4.4 Revolute–Translational Joint at *C*, Model 2

Translational component	Revolute component
(Vector $\mathbf{v}_2$)	
$x_2'^P = 0.0$	$x_3'^P = 1.5$
$y_2'^P = 0.0$	$y_3'^P = 0.0$
$x_2'^Q = 1.0$	
$y_2'^Q = 0.0$	

5.4.2 Assembly

Initial assembly is to be carried out so that the x coordinate of the slider (body 6 in model 1) is zero, that is, with the initial condition $x_6 = 0$. For model 1, initial position and orientation estimates are measured from Fig. 5.4.1, assuming a scale of 1:1. The results of assembly are tabulated in Table 5.4.5.

TABLE 5.4.5 Assembled Configuration, Model 1

Body no.	x	y	ϕ
1	0.0	0.0	0.0
2	0.9190	1.7768	1.0934
3	0.0	2.0	0.4356
4	1.3606	2.6320	1.0938
5	0.9144	3.7766	−0.2368
6	−0.0088	3.9997	0.0

For model 2, position data are obtained from Fig. 5.4.3. The results of assembly are tabulated in Table 5.4.6.

TABLE 5.4.6 Assembled Configuration, Model 2

Body no.	x	y	ϕ
1	0.0	0.0	0.0
2	0.9190	1.7768	1.0934
3	0.0	2.0	0.4356
4	−0.0088	3.9997	0.0

5.4.3 Driver Specification

In both models 1 and 2, there is one kinematic degree of freedom. By placing a driver on ϕ_3, the kinematic model is completed. If body 3 rotates counterclockwise at ω rad/s, the driver takes the form

$$\phi_3 = 0.44 + \omega t$$

5.4.4 Analysis

The quick-return mechanism is first analyzed with two driving angular velocities; $\omega_1 = 2\pi$ rad/s and $\omega_2 = 4\pi$ rad/s. Plots of x, $\dot{x}$, and $\ddot{x}$ for the slider (body 6 in model 1) versus time are given in Figs. 5.4.4, 5.4.5, and 5.4.6, respectively.

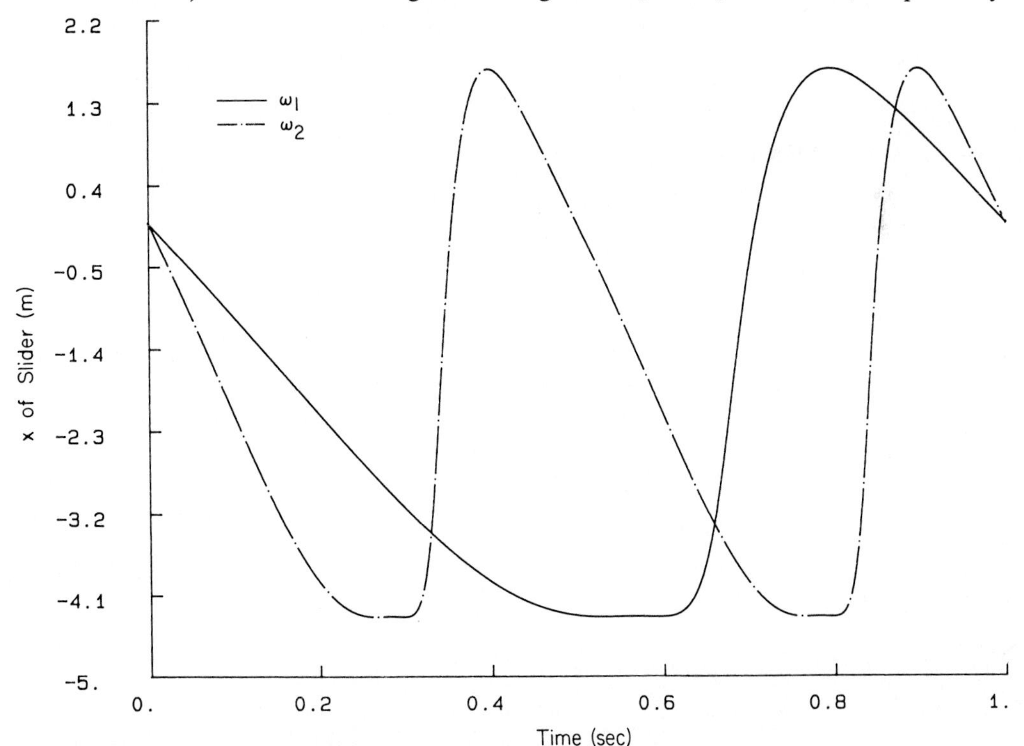

Figure 5.4.4 x of slider versus time.

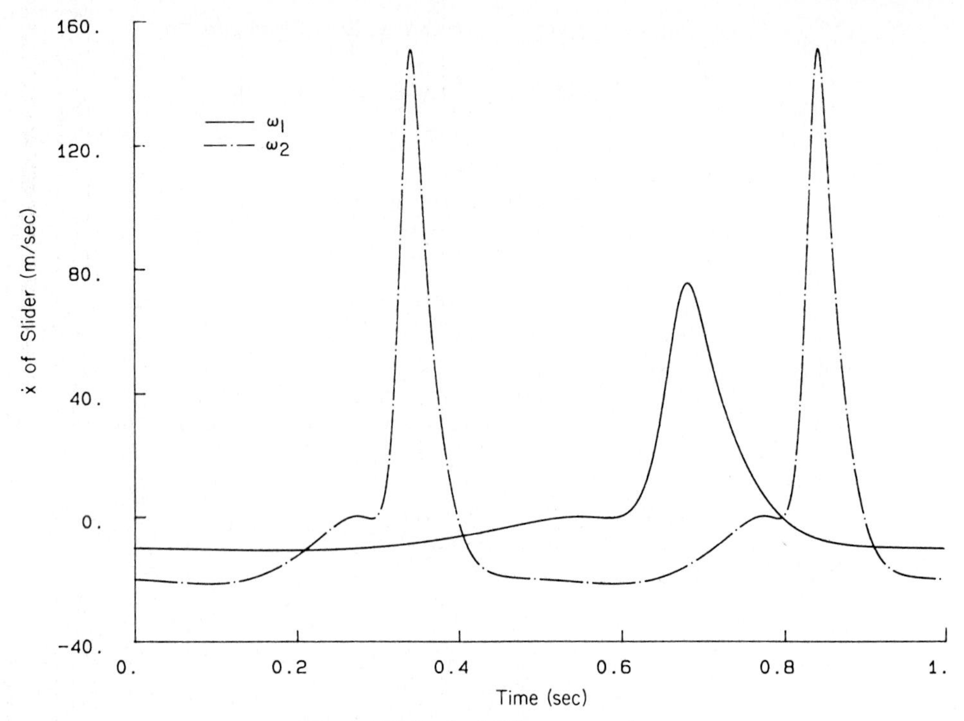

Figure 5.4.5 $\dot{x}$ of slider versus time.

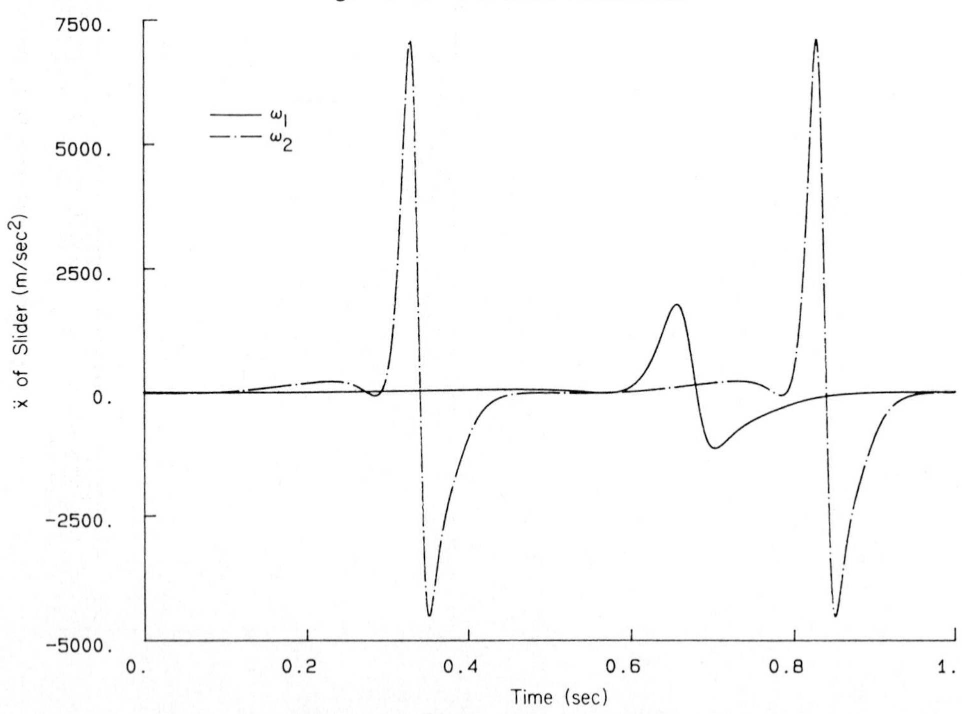

Figure 5.4.6 $\ddot{x}$ of slider versus time.

Identical results are obtained with model 2. Since motion is periodic, the amplitude of motion for both driving speeds is the same. Figures 5.4.5 and 5.4.6, however, show that the ratios of peak velocities and peak accelerations, due to doubling the driving velocity, are approximately 2 and 4, respectively. This behaviour is due to the quadratic dependence of the right side of the acceleration equation on velocity, as noted for the slider–crank analysis of Section 5.2.4.

Note that for ω_1 the leftward stroke of the slider (the cutting tool in a shaper) begins at 0.8 s and continues to about 1.5 s (a repetition of the curve from 0.0 to 0.5 s), with nearly constant velocity (confirmed by the velocity plot of Fig. 5.4.5). This means that 70% of the cycle time of the mechanism is devoted to the cutting stroke and only 30% to the return stroke; hence the name quick-return mechanism. Of further interest, note the nearly constant position of the slider (called *dwell*) during the period 0.5 to 0.6 s. This period would permit an auxiliary mechanism to lift the tool from the workpiece, prior to the return

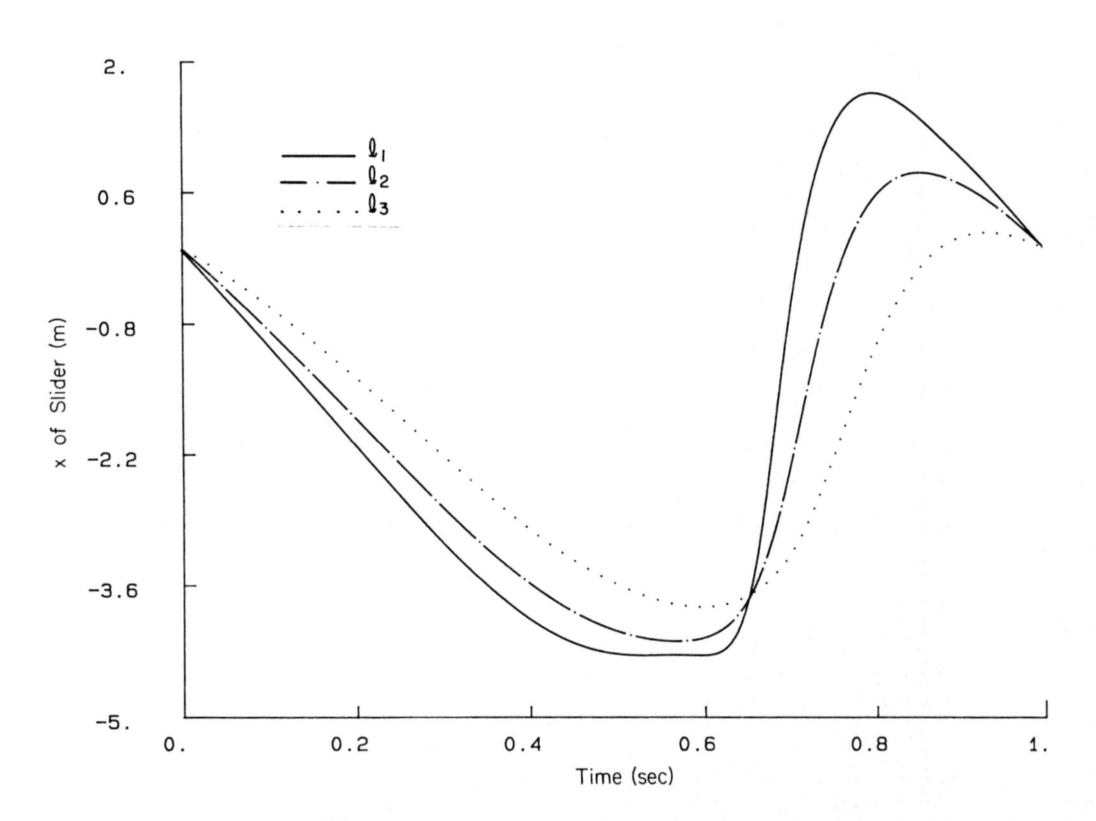

Figure 5.4.7 x of slider versus time with design variations.

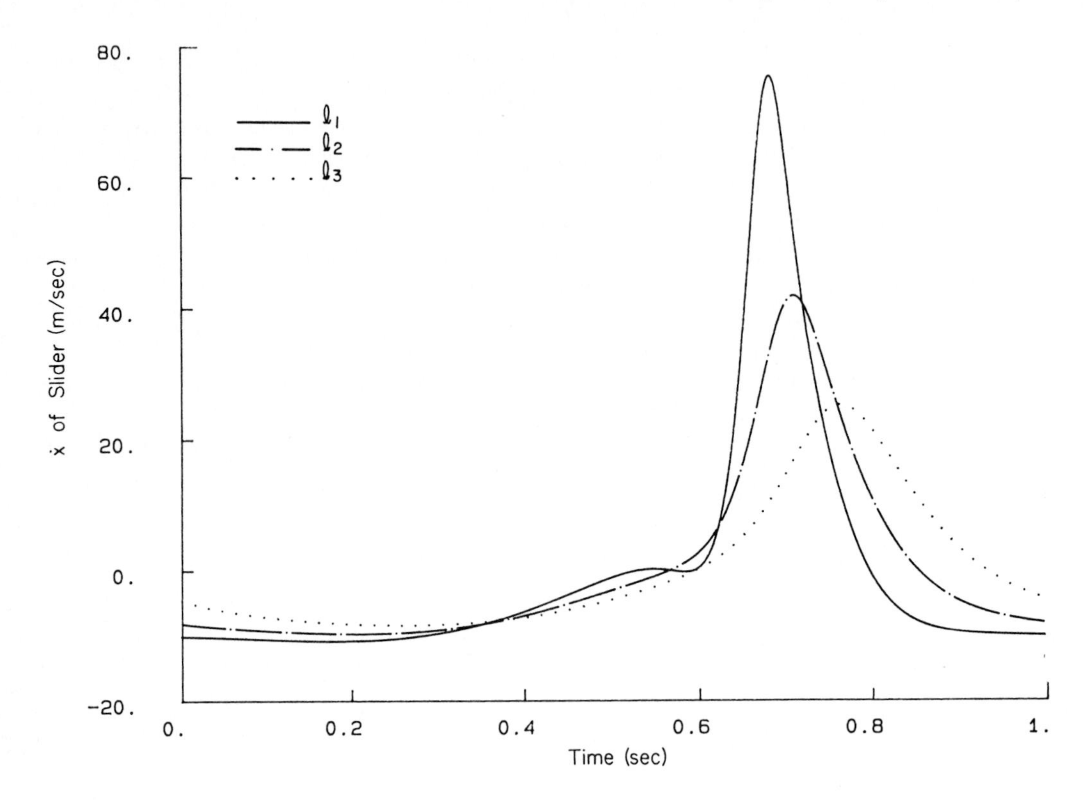

Figure 5.4.8 $\dot{x}$ of slider versus time with design variations.

stroke. It is the essentially nonlinear nature of the kinematic equations that yields these peculiar and often valuable nonlinear characteristics of machines.

As a final consideration, the effect of reducing the length of the crank (body 3 in model 1) is analyzed by carrying out simulations with $\ell_1 = 1.5$ (the previous design), $\ell_2 = 1.25$, and $\ell_3 = 1.0$ For each design, initial assembly is carried out with the slider in the same position to facilitate comparison of performance. The driver is adjusted to rotate the crank at $\omega_1 = 2\pi$ rad/s from the initially assembled configuration. Plots of x, $\dot{x}$, and $\ddot{x}$ for the slider, with each of the designs, are presented in Figs. 5.4.7, 5.4.8, and 5.4.9, respectively.

Note that the reduced crank lengths lead to somewhat reduced stroke lengths, but a comparable fraction of the total stroke is devoted to cutting. They do, however, lead to marked reduction in peak accelerations of the slider, and hence significant reductions in reaction forces in joints. Finally, note that only the nominal design (ℓ_1) exhibits the dwell in position at the left extreme position of the slider.

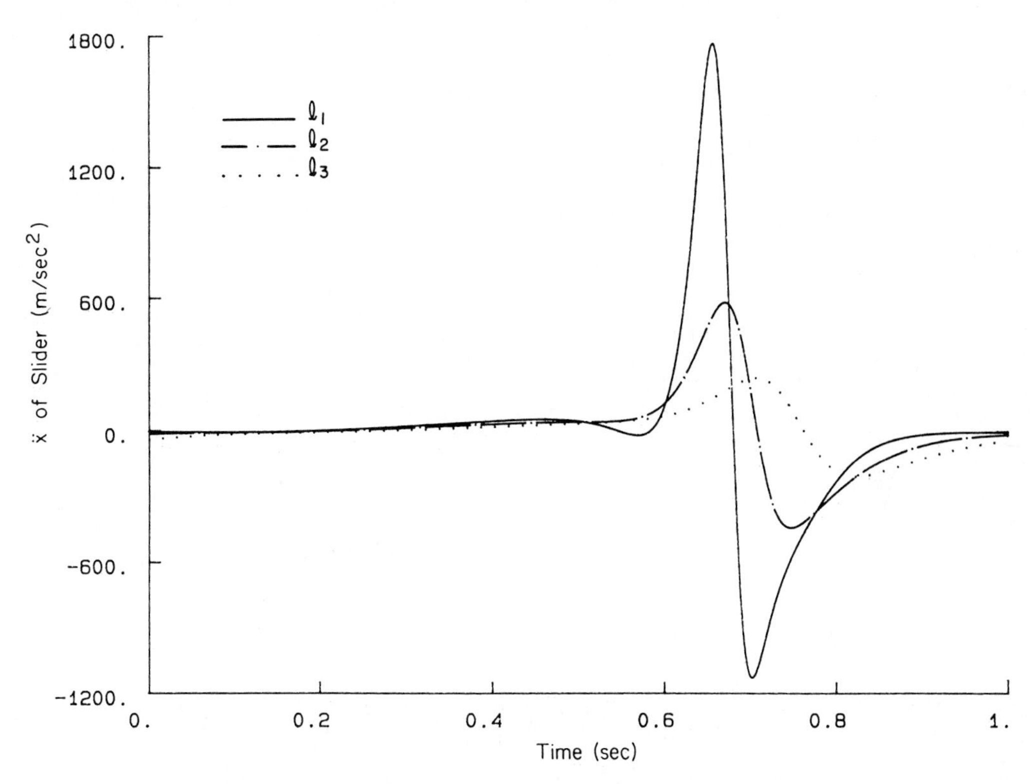

Figure 5.4.9 $\ddot{x}$ of slider versus time with design variations.

5.5 KINEMATIC ANALYSIS OF A GEAR–SLIDER MECHANISM

A compound mechanism that contains a gear set that drives a variable stroke slider mechanism is shown in Fig. 5.5.1. The three-gear arrangement permits input rotation of the first gear (body 2) to be transmitted to the third gear (body 4), even when body 5 is rotated to control the stroke of the slider (body 6).

5.5.1 Kinematic Model

The gear–slider mechanism is modeled as shown in Fig. 5.5.1. This mechanism is composed of six bodies. Bodies 1 to 6 are ground (body 1), three gears (bodies 2, 3, and 4), a stroke control link (body 5), and a slider (body 6).

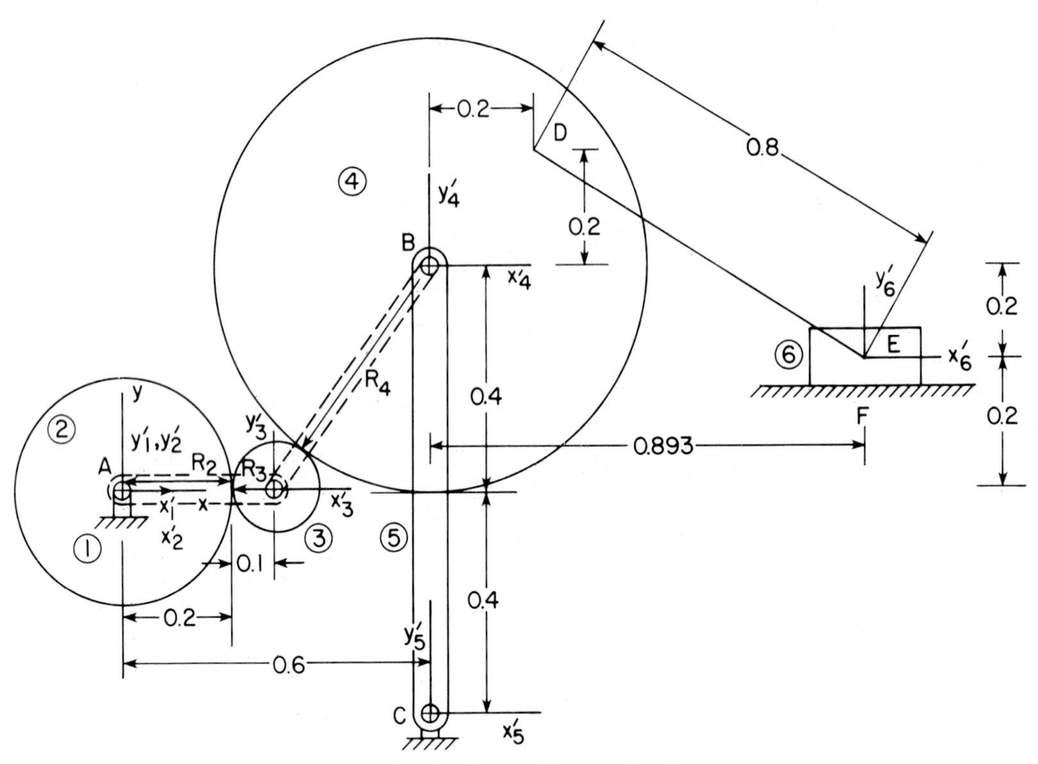

Figure 5.5.1 Gear–slider mechanism.

Kinematic joints between bodies are modeled as follows:

Bodies	
Six bodies	$nc = 18$
Constraints	
Three revolute joints	$3 \times 2 = 6$
Ground–gear 2 (Point A)	
Gear 4–link 5 (Point B)	
Link 5–ground (Point C)	
Two gear set joints	$2 \times 1 = 2$
Gear 2–Gear 3	
Gear 3–Gear 4	
Three distance constraints	$3 \times 1 = 3$
Gear 2–gear 3	
Gear 3–gear 4	
Gear 4–slider	
One translational joint	$2 \times 1 = 2$
Slider–ground (point F)	
One ground constraint (point A)	$3 \times 1 = 3$
	$nh = 16$
Thus, DOF $= 18 - 16 = 2$	

TABLE 5.5.1 Revolute Joint Data

Joint no.	1	2	3
Common point P	A	B	C
Body i	1	4	5
$x_i'^P$	0.0	0.0	0.0
$y_i'^P$	0.0	0.0	0.0
Body j	2	5	1
$x_j'^P$	0.0	0.0	0.6
$y_j'^P$	0.0	0.8	−0.4

TABLE 5.5.2 Gear Set Data

Joint no.	1	2
Reference points	R_1	R_2
Body i	2	3
Reference angle	0.0	0.9273
Body j	3	4
Reference angle	3.1416	4.0689

TABLE 5.5.3 Distance Constraint Data

Constraint no.	1	2	3
Body i	2	3	4
x_i'	0.0	0.0	0.2
y_i'	0.0	0.0	0.2
Body j	3	4	6
x_j'	0.0	0.0	0.0
y_j'	0.0	0.0	0.0
Distance	0.3	0.5	0.8

TABLE 5.5.4 Translational Joint Data

Body no.	6	1
x'^P	0.0	0.0
y'^P	0.0	0.2
x'^Q	1.0	1.0
y'^Q	0.0	0.2

The radii of gears 2, 3, and 4 are 0.2, 0.1, and 0.4, respectively. The length of link 5 is 0.8. Kinematic joint definition data are summarized in Tables 5.5.1 to 5.5.4.

5.5.2 Assembly

Initial position and orientation estimates are measured from Fig. 5.5.1. The results of assembly calculations are tabulated in Table 5.5.5.

TABLE 5.5.5 Assembled Configuration

Body no.	x	y	ϕ
1	0.0	0.0	0.0
2	0.0	0.0	0.0
3	0.3	0.0	0.0
4	0.6	0.4	0.0
5	0.6	−0.4	0.0
6	1.4928	0.2	0.0

5.5.3 Driver Specification

Two driving constraints are imposed on the gear of body 2 and the stroke control link (body 5). The first driving constraint is a constant angular velocity of the gear of body 2,

$$\phi_2 = 4\pi t$$

The second is a fixed angular position of link 5,

$$\phi_5 = \theta$$

By changing the angle θ of the second driving constraint, the range of slider motion can be varied. Three values of θ are considered: $\theta_1 = -0.1047$, $\theta_2 = 0.0$, and $\theta_3 = 0.5236$ rad. To avoid a lock-up configuration, θ must be in the range $0.7330 < \theta < -0.1204$.

5.5.4 Analysis

Figures 5.5.2 to 5.5.4 are plots of x positions, velocities, and accelerations of the slider for three different values of θ. Note that, even though the stroke of the slider varies significantly with θ, the time histories of velocity and acceleration of the slider do not significantly vary.

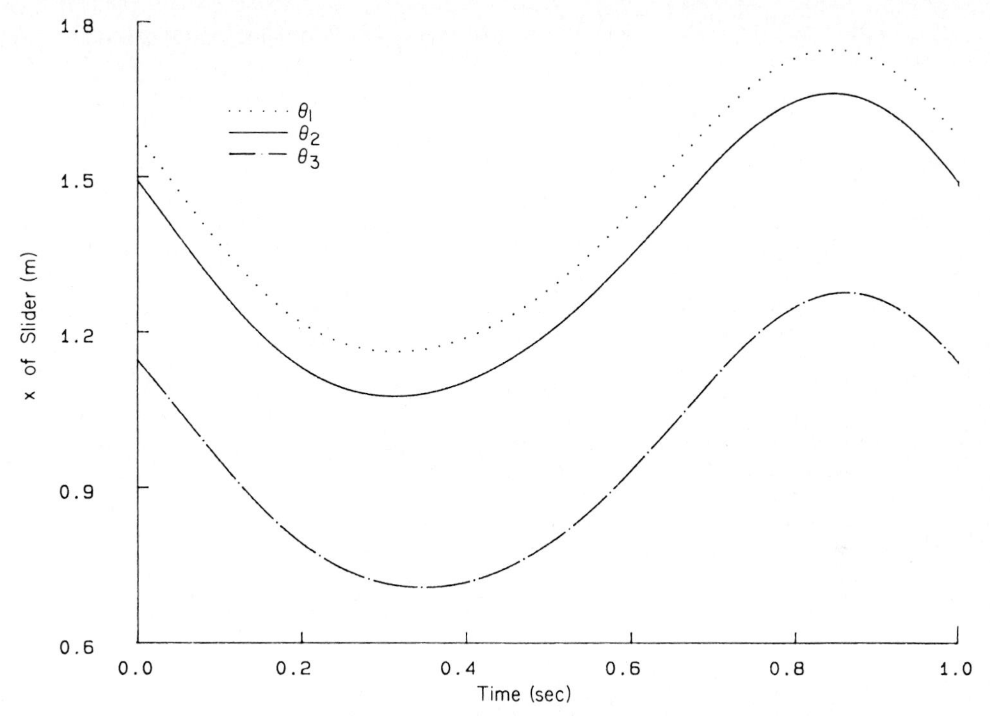

Figure 5.5.2 x of slider versus time.

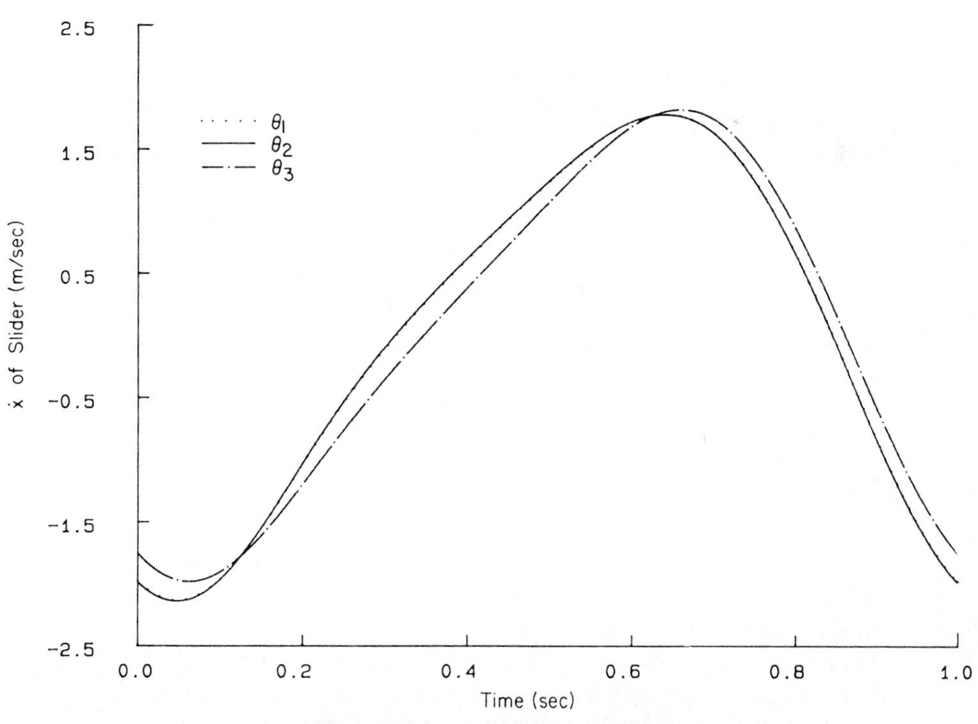

Figure 5.5.3 $\dot{x}$ of slider versus time.

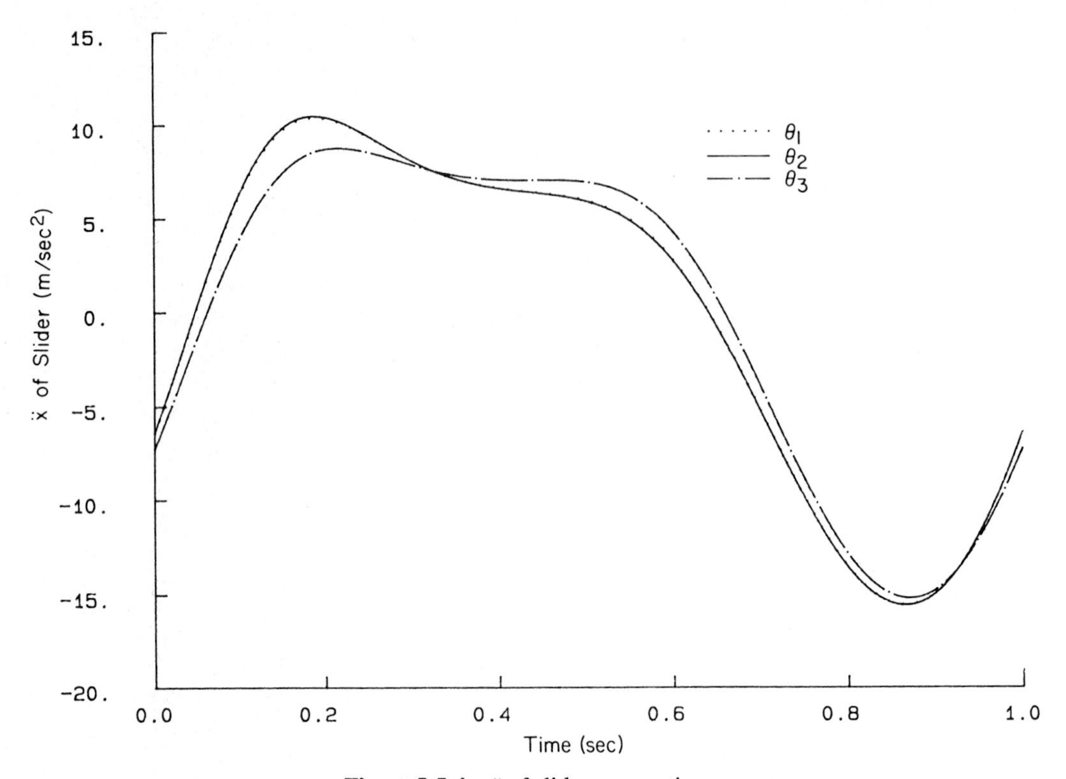

Figure 5.5.4 $\ddot{x}$ of slider versus time.

5.6 KINEMATIC ANALYSIS OF A VALVE-LIFTER MECHANISM

Cam-driven mechanisms are commonly used to create desired time histories of motion, such as in the valve-lifter mechanism of an engine (as in Fig. 1.1.1), shown schematically in Fig. 5.6.1. A cam–flat-faced follower is studied in this section, driving a rocker arm mechanism that controls valve position in an overhead cam engine.

5.6.1 Valve-Lifter Mechanism Model

As an example of a valve-lifter mechanism model, a simple circular cam–flat-faced follower is shown in Fig. 5.6.1, with the cam center offset from the pivot at point B in ground to induce follower motion due to imposed angular motion of the cam. A rocker arm is pivoted with a revolute joint in ground at point C. Revolute–translational composite joints are used to couple the rocker to the follower at point F and the valve stem at point G. The follower and valve stem are constrained to translate relative to ground, as shown in Fig. 5.6.1.

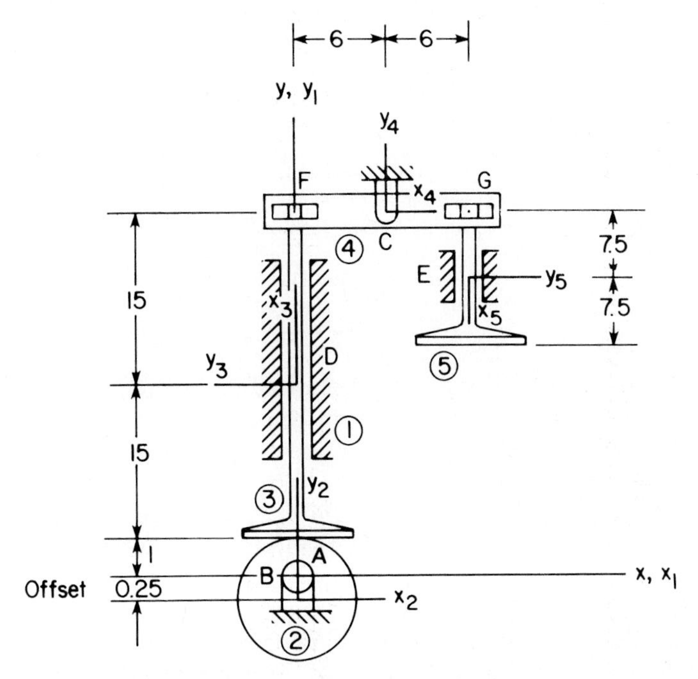

Figure 5.6.1 Valve-lifter mechanism.

Five bodies, including ground, are employed in this model, with constraints defined as follows:

Model		
Bodies		
Five bodies		$nc = 15$
Constraints		
Cam–flat-faced follower joint:	*A*	1
Revolute joints:	*B*	2
	C	2
Translational joints:	*D*	2
	E	2
Revolute–translational joints:	*F*	1
	G	1
Ground constraints		3
		$nh = 14$
Thus, DOF = 15 − 14 = 1.		

To prescribe motion, the one remaining degree of freedom is eliminated by specifying the angular velocity of the cam shaft; that is, by placing an angle driver on ϕ_2.

TABLE 5.6.1 Cam–Flat-faced Follower Joint Data

Shape center Q of body 2 (cam)	
$x_2'^Q$	0
$y_2'^Q$	0
Shape perfect circle with radius	1.25
Revolute joint offset	0.25
P on body 3 (follower): $x_3'^P$	−15
$y_3'^P$	0
Q on body 3 (follower): $x_3'^Q$	−15
$y_3'^Q$	1

As an elementary numerical example, a circular cam of radius 1.25 cm (all data in this example are given in centimeters) is employed. Data defining the cam–flat-faced follower between bodies 2 and 3 are given in Table 5.6.1.

Revolute joints between the cam and ground at point B and between the rocker and ground at point C are imposed, with data specified in Table 5.6.2. Translational joints between the follower and valve stem and ground are specified by data given in Table 5.6.3. Vector $\mathbf{v}_{11}$ is in ground, located between point D and a point one unit above D. Vector $\mathbf{v}_3$ is fixed in the follower between the point at the origin of the x_3-y_3 frame and a point one unit above. Similarly, vectors $\mathbf{v}_{12}$ in ground and $\mathbf{v}_5$ in the valve stem define the translational joint at point E.

Finally, revolute–translational joints between the follower and valve stem and the rocker at points F and G are specified by the vectors shown in Fig. 5.6.2

TABLE 5.6.2 Revolute Joint Data

Joint no.	1	2
Common point P	B	C
Body i	1	4
$x_i'^P$	0.0	0.0
$y_i'^P$	0.0	0.0
Body j	2	1
$x_j'^P$	0.0	6.0
$y_j'^P$	0.25	31.0

TABLE 5.6.3 Translational Joint Data

Vector	$\mathbf{v}_{11}$	$\mathbf{v}_3$	$\mathbf{v}_{12}$	$\mathbf{v}_5$
$x_i'^P$	$x_{11}'^P = 0.0$	$x_3'^P = 0.0$	$x_{12}'^P = 12.0$	$x_5'^P = 0.0$
$y_i'^P$	$y_{11}'^P = 16.0$	$y_3'^P = 0.0$	$y_{12}'^P = 23.5$	$y_5'^P = 0.0$
$x_i'^Q$	$x_{11}'^Q = 0.0$	$x_3'^Q = 1.0$	$x_{12}'^Q = 12.0$	$x_5'^Q = 1.0$
$y_i'^Q$	$y_{11}'^Q = 17.0$	$y_3'^Q = 0.0$	$y_{12}'^Q = 22.5$	$y_5'^Q = 0.0$

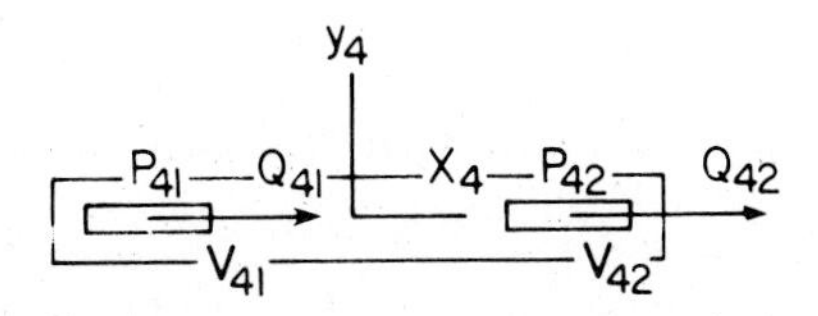

Figure 5.6.2 Revolute–translational joints.

on the rocker. Data that define the locations of points P and Q, and hence the vectors $\mathbf{v}_{41}$ and $\mathbf{v}_{42}$, are given in Table 5.6.4. The locations of revolute joints of the couplers in the follower and valve stem, bodies 3 and 5, respectively, are also prescribed in Table 5.6.4.

TABLE 5.6.4 Revolute–Translational Joint Data

Translational components		
Vector	$\mathbf{v}_{41}$	$\mathbf{v}_{42}$
$x_i'^P$	$x_{41}'^P = -6.0$	$x_{42}'^P = 6.0$
$y_i'^P$	$y_{41}'^P = 0.0$	$y_{42}'^P = 0.0$
$x_i'^Q$	$x_{41}'^Q = -5.0$	$x_{42}'^Q = 7.0$
$y_i'^Q$	$y_{41}'^Q = 0.0$	$y_{42}'^Q = 0.0$
Revolute components		
Body 3	Body 5	
$x_3'^P = 15.0$	$x_5'^P = -7.5$	
$y_3'^P = 0.0$	$y_3'^P = 0.0$	

5.6.2 Assembly

The configuration of the valve-lifter mechanism shown in Fig. 5.6.1 permits easy generation of accurate initial estimates for the location and orientation of bodies. The results of numerical assembly analysis, holding $\phi_2 = 0$, are presented in Table 5.6.5.

TABLE 5.6.5 Assembled Configuration

Body no.	x	y	ϕ
1	0.0	0.0	0.0
2	0.0	−0.2500	0.0
3	0.0	16.0000	1.5708
4	6.0000	31.0000	0.0
5	12.0000	23.5000	4.7124

5.6.3 Driver Specification

The mechanism is driven by specifying a cam angular velocity of 20 revolutions per second; that is, with the driver

$$\phi_2 = 40\pi t$$

5.6.4 Analysis

To investigate the effect of variations in cam geometry on motion of the valve lifter, the eccentricity of the circular cam used in this study is varied from the nominal value $e_1 = 0.25$ cm shown in Fig. 5.6.1 to $e_2 = 0.35$ cm and $e_3 = 0.45$ cm. To maintain the geometry shown in Fig. 5.6.1, the height of the revolute joint at point B between the cam (body 2) and ground is set equal to the eccentricity in each case, leaving the initial height of the follower at point A unchanged for each design.

Valve stem position, velocity, and acceleration versus time, for each of the three cam eccentricities, are plotted in Figs. 5.6.3 to 5.6.5, respectively. As expected on physical grounds, the cyclic behavior of each of the three designs is

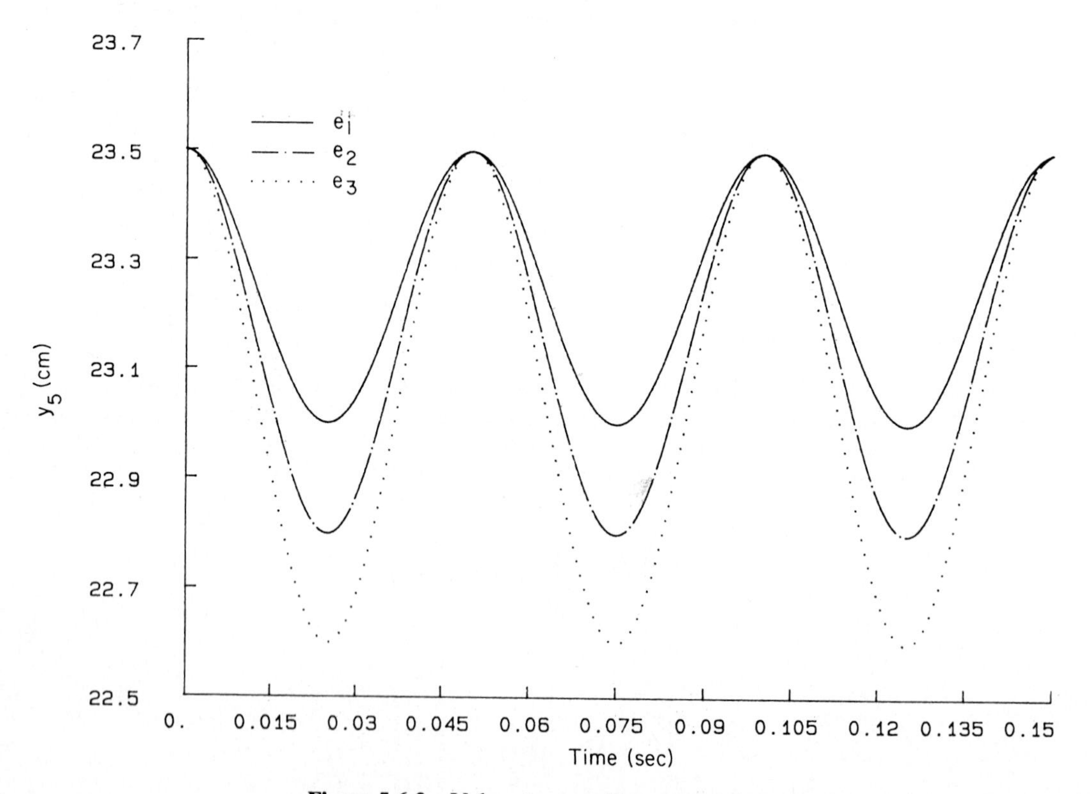

Figure 5.6.3 Valve stem position versus time.

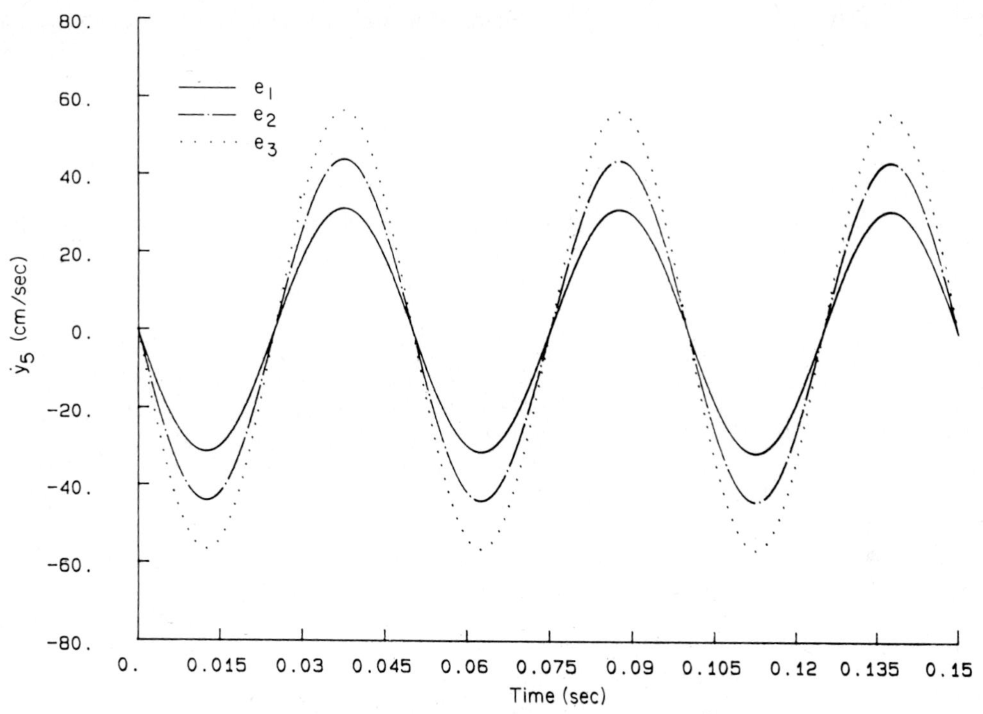

Figure 5.6.4 Valve stem velocity versus time.

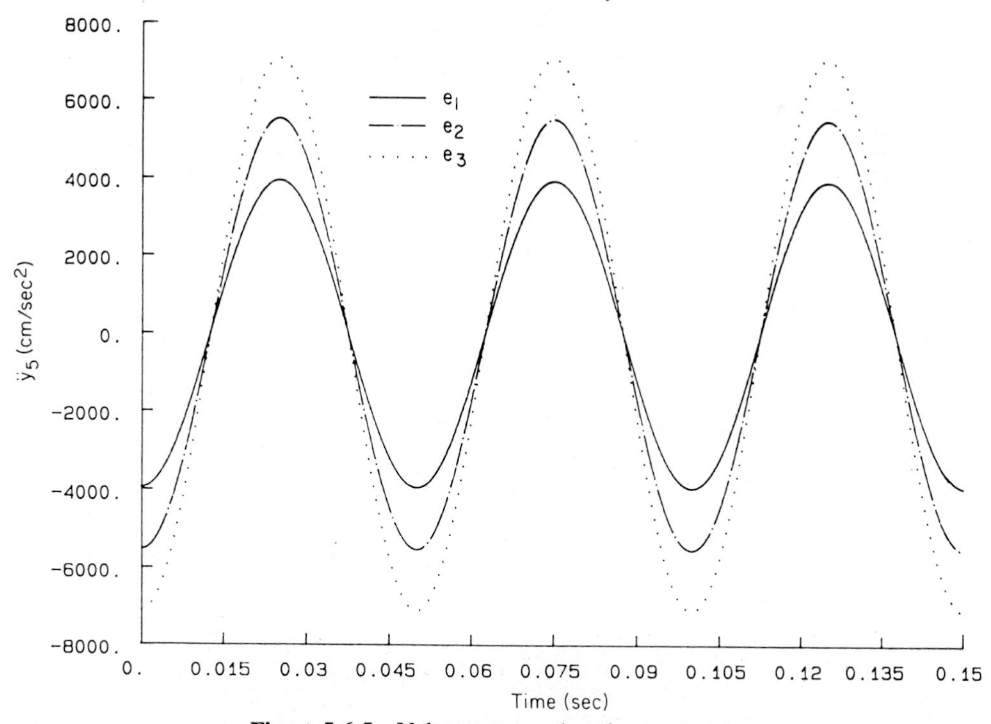

Figure 5.6.5 Valve stem acceleration versus time.

very similar, with only amplitude varying. The essentially sinusoidal character of the response of the valve lifter in this elementary example is associated with the circular cam. In automotive applications, an essentially circular cam surface, with center at the pivot point B, is employed, with a distinct lobe that causes valve motion. As a result, a shorter and more distinct response of the valve stem occurs, with a dwell in the closed position for most of the rotation of the cam.

DADS PROJECTS

5.1. Consider the film follower mechanism shown in Fig. P5.1. The mechanism can be modeled using four bodies: ground, a crank, a rocker arm, and a follower. A point of interest P is defined on the end of the follower. It is this point that engages and disengages the film as the mechanism moves. The crank rotates counterclockwise at 1000 rpm. Construct a DADS model for kinematic analysis, using the following procedures:

(a) Count the number of generalized coordinates and constraints to find the number of DOF and check physically.

(b) Draw each body separately and define body-fixed frames, joint definition points, and point of interest.

(c) Create DADS input data and run.

(d) Plot

(i) y^P of the point of interest versus x^P of the point of interest.

(ii) $\dot{x}^P$ of the point of interest versus time.

(iii) $\ddot{x}^P$ of the point of interest versus time.

(iv) $\dot{y}^P$ of the point of interest versus time.

(v) $\ddot{y}^P$ of the point of interest versus time.

(e) Give a physical interpretation of the plots obtained and discuss kinematic singularities, if any exist.

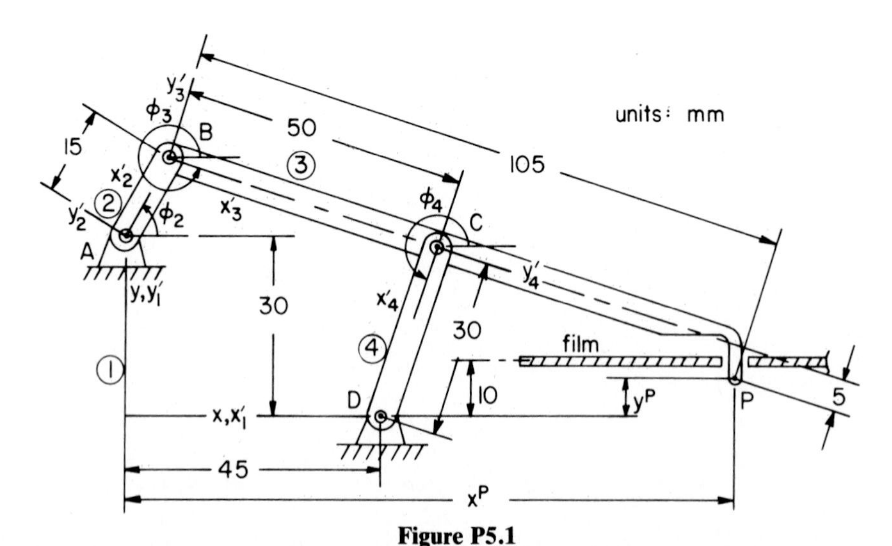

Figure P5.1

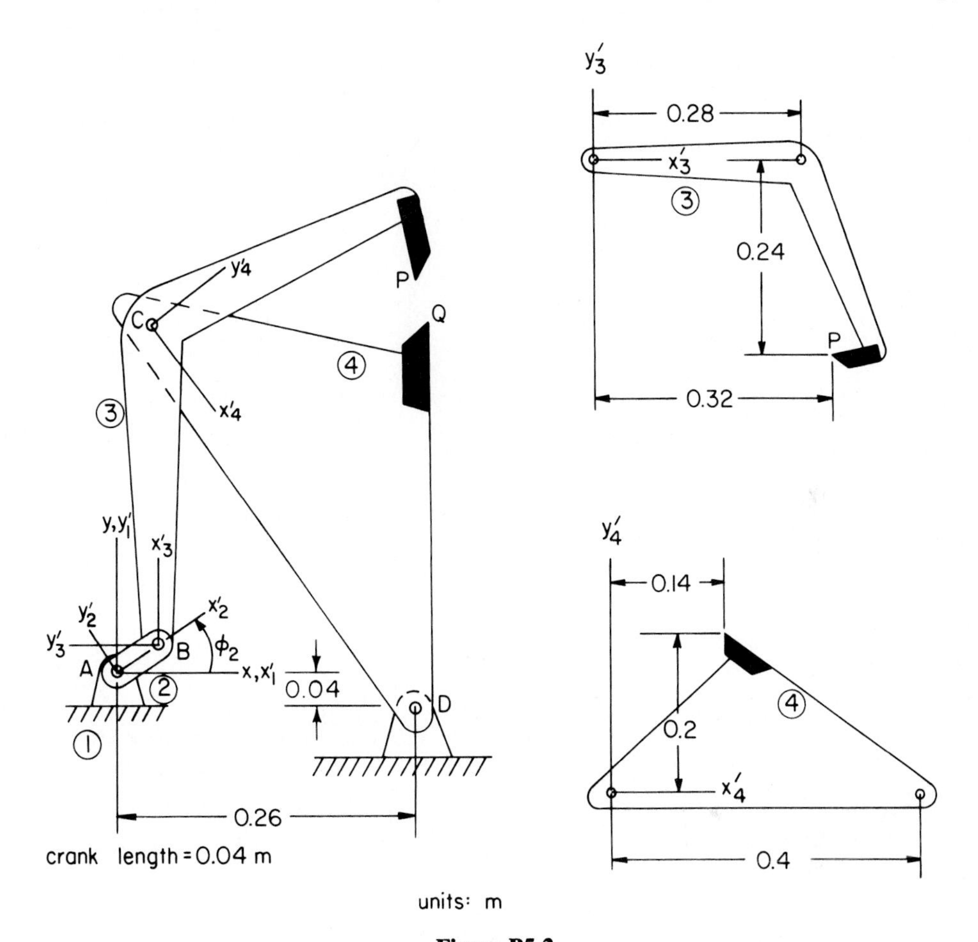

Figure P5.2

5.2. Consider the web cutting mechanism shown in Fig. P5.2. The model has four bodies: ground, a crank, a knife-arm, and a rocker arm. Points of interest P and Q are defined on the tips of the cutting blades. The x velocities of these points are of particular interest. In an actual mechanism, a sheet of fabric or metal is fed between the blades and the blade tips come together periodically to slice off a section of the web. The machine is designed such that the x velocity of the blade tips matches very closely the velocity of the web. This allows for a clean cut, with little or no tearing. The crank is driven with an angular velocity of 60 rpm. Construct a DADS model for kinematic analysis, following steps (a) through (c) of DADS Project 5.1, and

(d) Draw two curves on one plot for:

(i) x^P and x^Q versus time.

(ii) $\dot{x}^P$ and $\dot{x}^Q$ versus time.

(iii) $\ddot{x}^P$ and $\ddot{x}^Q$ versus time.

(iv) y^P and y^Q versus time.

(v) $\dot{y}^P$ and $\dot{y}^Q$ versus time.

(vi) $\ddot{y}^P$ and $\ddot{y}^Q$ versus time.

(e) Monitor how close the x positions of points P and Q agree throughout the simulation. If there is a time interval over which x^P and x^Q differ, discuss this position difference concerning y^P and y^Q. Suggest any design changes that may be necessary.

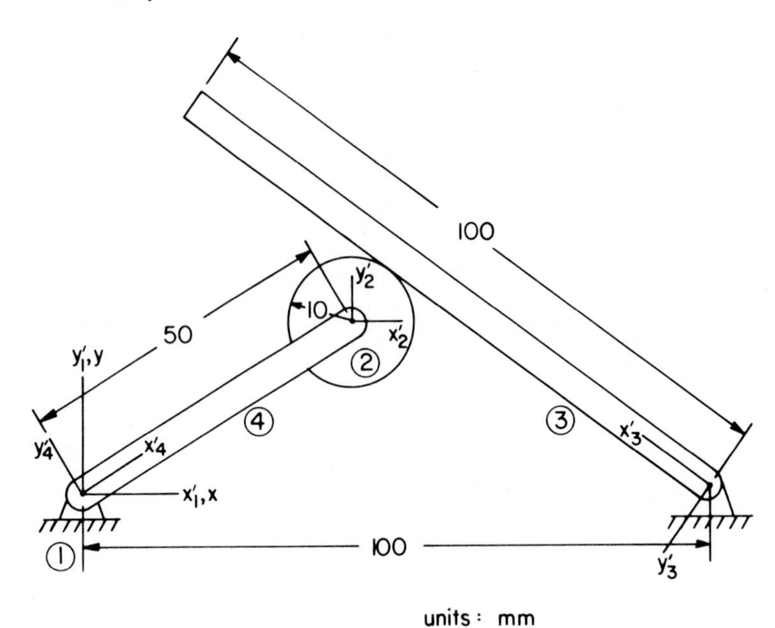

Figure P5.3

5.3. Consider the mechanism shown in Fig. P5.3, which uses a rack and pinion joint. The four bodies used are ground, the rack, the pinion, and the coupler. The pinion rotates clockwise at 60 rpm. Construct a DADS model for kinematic analysis, following steps (a) to (c) of DADS Project 5.1, and

(d) Plot:

(i) x, $\dot{x}$, and $\ddot{x}$ of the pinion versus time.

(ii) y, $\dot{y}$, and $\ddot{y}$ of the pinion versus time.

(iii) ϕ, $\dot{\phi}$, and $\ddot{\phi}$ of the rack versus time.

(e) Create an alternative three-body model by replacing the coupler with a revolute–revolute joint. Compare results with those of the four-body model.

CHAPTER SIX

Dynamics of Planar Systems

The equations of motion of a rigid body and constrained systems of rigid bodies are derived beginning with Newton's laws for a particle and a model of rigid bodies. The *virtual work* approach is introduced to develop a self-contained derivation of both variational and differential equation formulations. The variational, or virtual work, formulation is best suited for multibody dynamics. The Lagrange multiplier form of constrained equations of motion is derived, and properties of the resulting differential–algebraic equations of motion that are important for numerical solution are developed. Special forms of the equations of motion that are suitable for inverse dynamic and equilibrium analysis are developed. Finally, reaction forces that act on bodies due to joints are derived from Lagrange multipliers.

6.1 EQUATIONS OF MOTION OF A PLANAR RIGID BODY

Consider the rigid body shown in Fig. 6.1.1, located in the x-y plane by the vector $\mathbf{r}$ and angle of rotation ϕ. A differential mass $dm(P)$ at point P is located on the body by the vector $\mathbf{s}^P$. Forces acting on this differential element of mass include external forces $\mathbf{F}(P)$ per unit of mass at point P and internal forces $\mathbf{f}(P, R)$ per units of mass located at points P and R, as shown in Fig. 6.1.1. As a *model of a rigid body,* let a distance constraint (a massless bar with revolute joints at both ends) act between each pair of differential elements (thought of as particles) in the body. With this model, physical internal forces $\mathbf{f}(P, R)\, dm(R)\, dm(P)$ and $\mathbf{f}(R, P)\, dm(P)\, dm(R)$ of interaction on $dm(P)$ and $dm(R)$ act along the massless links between points P and R, due to the distance constraint between points P and R, so they are equal in magnitude, opposite in direction, and colinear.*

* Colinearity of $\mathbf{f}(P, R)$ and $\mathbf{f}(R, P)$ in the model used here for a rigid body is a function of the model. While $\mathbf{f}(P, R)\, dm(R)\, dm(P) = -\mathbf{f}(R, P)\, dm(P)\, dm(R)$ follows from Newton's law of action–reaction, in general Newton said nothing about colinearity. This matter has been studied by many scholars and is elegantly summarized by Truesdell's essay "Whence the Law of Moment of Momentum?" in Reference 32. The rigid-body model used here is adequate to represent internal

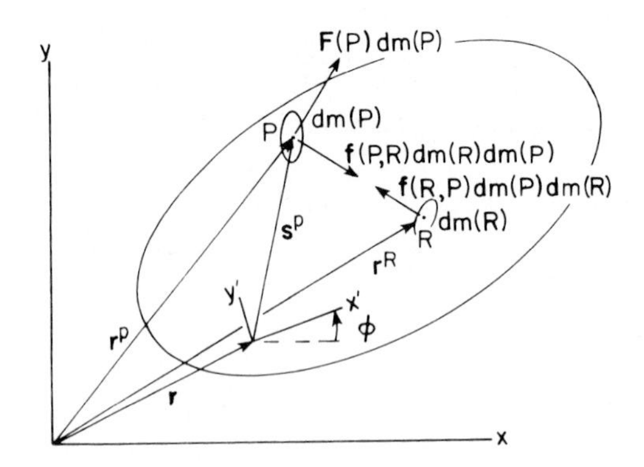

Figure 6.1.1 Forces acting on a rigid body.

6.1.1 Variational Equations of Motion from Newton's Equations

Newton's equations of motion [7, 9] for differential mass $dm(P)$ are

$$\ddot{\mathbf{r}}^P\, dm(P) = \mathbf{F}(P)\, dm(P) + \left[\int_m \mathbf{f}(P, R)\, dm(R)\right] dm(P) \tag{6.1.1}$$

where integration of the internal forces acting on $dm(P)$ is taken over the total mass of the body. Equation 6.1.1 is difficult to use, since it explicitly involves the internal forces that act within the body. Furthermore, it is written for every differential element of mass, yielding a ridiculously large number of equations of motion. The reason for this difficulty is that Eq. 6.1.1 fails to take advantage of the kinematic characteristics of the rigid body, which relate motion of all differential elements in the body. A unifying concept that can be used to resolve this dilemma and yield broadly applicable tools for mechanical system dynamics [35] is the variational or *virtual work* approach.

Let $\delta\mathbf{r}^P$ denote an arbitrary *virtual displacement* of point P; that is, a small variation in the location of point P that is permitted to occur with time held fixed. The variation of a vector may be thought of as the partial differential of the vector with only physical coordinates varied and time held fixed. The δ operator is just the differential operator of calculus with time held fixed. Thus, it is

---

gravitational attraction and the effects of stress distributions (as a limit of an elastic body) with symmetric stress tensors. The model used here requires that, if forces of interaction occur in a body due to electric or magnetic field effects, they must be accounted for as external forces. This model of a rigid body is adequate for commonly used engineering materials and machine dynamics applications. For more philosophical considerations, the reader may wish to consult Reference 32 and references cited therein.

possible to relate the virtual displacement $\delta\mathbf{r}^P$ of point P to variations $\delta\mathbf{q}$ in generalized coordinates $\mathbf{q}$ of a body, once the relation $\mathbf{r}^P = \mathbf{r}^P(\mathbf{q})$ is established. This calculation is temporarily postponed to permit development of equations of dynamics in which it is of value.

Premultiplying both sides of Eq. 6.1.1 by $\delta\mathbf{r}^{PT}$ and integrating over the total mass m of the body yields

$$\int_m \delta\mathbf{r}^{PT}\ddot{\mathbf{r}}^P\,dm(P) = \int_m \delta\mathbf{r}^{PT}\mathbf{F}(P)\,dm(P) + \iint_{mm} \delta\mathbf{r}^{PT}\mathbf{f}(P,R)\,dm(R)\,dm(P) \qquad \textbf{(6.1.2)}$$

which must hold for arbitrary $\delta\mathbf{r}^P$.

Manipulation of the double integral appearing on the right of Eq. 6.1.2 yields

$$\begin{aligned}\iint_{mm} \delta\mathbf{r}^{PT}\mathbf{f}(P,R)\,dm(R)\,dm(P) &= \frac{1}{2}\iint_{mm} [\delta\mathbf{r}^{PT}\mathbf{f}(P,R) + \delta\mathbf{r}^{PT}\mathbf{f}(P,R)]\,dm(R)\,dm(P)\\ &= \frac{1}{2}\iint_{mm} \delta\mathbf{r}^{PT}\mathbf{f}(P,R)\,dm(R)\,dm(P) + \frac{1}{2}\iint_{mm} \delta\mathbf{r}^{PT}\mathbf{f}(P,R)\,dm(R)\,dm(P)\end{aligned}$$

Since P and R are dummy variables of integration, they may be renamed without affecting the value of the integral; that is,

$$\iint_{mm} \delta\mathbf{r}^{PT}\mathbf{f}(P,R)\,dm(R)\,dm(P) = \iint_{mm} \delta\mathbf{r}^{RT}\mathbf{f}(R,P)\,dm(P)\,dm(R)$$

Furthermore, the order of carrying out the integration can be reversed without affecting the value of the integral [26], and $\mathbf{f}(P,R)\,dm(R)\,dm(P) = -\mathbf{f}(R,P)\,dm(P)\,dm(R)$ can be used to obtain

$$\begin{aligned}&\iint_{mm} \delta\mathbf{r}^{PT}\mathbf{f}(P,R)\,dm(R)\,dm(P)\\ &\qquad = \frac{1}{2}\iint_{mm} \delta\mathbf{r}^{PT}\mathbf{f}(P,R)\,dm(R)\,dm(P) - \frac{1}{2}\iint_{mm} \delta\mathbf{r}^{RT}\mathbf{f}(P,R)\,dm(R)\,dm(P)\end{aligned}$$

Combining the integrals on the right and using the fact from differential calculus [25, 26, 35] that $\delta\mathbf{r}^P - \delta\mathbf{r}^R = \delta(\mathbf{r}^P - \mathbf{r}^R)$,

$$\iint_{mm} \delta\mathbf{r}^{PT}\,\mathbf{f}(P,R)\,dm(R)\,dm(P) = \frac{1}{2}\iint_{mm} \delta(\mathbf{r}^P - \mathbf{r}^R)^T\mathbf{f}(P,R)\,dm(R)\,dm(P)$$

Recall that the definition of a rigid body and the model used here require that the distance between points P and R be constant; that is,

$$(\mathbf{r}^P - \mathbf{r}^R)^T(\mathbf{r}^P - \mathbf{r}^R) = c$$

Taking the differential of both sides, using the rules of differential calculus (Prob. 6.1.1), yields

$$\delta(\mathbf{r}^P - \mathbf{r}^R)^T(\mathbf{r}^P - \mathbf{r}^R) = 0$$

Thus, $\delta(\mathbf{r}^P - \mathbf{r}^R)$ is orthogonal to $\mathbf{r}^P - \mathbf{r}^R$. Since $\mathbf{f}(P, R)$ acts along the line between P and R in the model used here, it is also orthogonal to $\delta(\mathbf{r}^P - \mathbf{r}^R)$, so $\delta(\mathbf{r}^P - \mathbf{r}^R)\mathbf{f}(P, R) = 0$. Thus,

$$\iint_{mm} \delta\mathbf{r}^{PT}\mathbf{f}(P, R)\, dm(R)\, dm(P) = 0 \tag{6.1.3}$$

Using the result of Eq. 6.1.3, Eq. 6.1.2 simplifies to

$$\int_m \delta\mathbf{r}^{PT}\ddot{\mathbf{r}}^P\, dm(P) = \int_m \delta\mathbf{r}^{PT}\mathbf{F}(P)\, dm(P) \tag{6.1.4}$$

which must hold for all $\delta\mathbf{r}^P$ that are consistent with constraints on $\mathbf{r}^P$ that define rigid-body motion. It is important to note that while Eq. 6.1.2 holds for arbitrary $\delta\mathbf{r}^P$, Eq. 6.1.3 does not. The price for eliminating the second term on the right of Eq. 6.1.2 is that $\delta\mathbf{r}^P$ must be consistent with the definition of rigid-body motion. This result is sometimes called *D'Alembert's principle* [4, 35] or the *principle of virtual work*. To take full advantage of Eq. 6.1.4, the virtual displacement $\delta\mathbf{r}^P$ of point P must be written in terms of variations in generalized coordinates of the body.

The vector $\mathbf{r}^P$ that locates point P may be written, using Eq. 2.4.8, as

$$\mathbf{r}^P = \mathbf{r} + \mathbf{A}\mathbf{s}'^P \tag{6.1.5}$$

where the matrix $\mathbf{A}$ is

$$\mathbf{A} = \begin{bmatrix} \cos\phi & -\sin\phi \\ \sin\phi & \cos\phi \end{bmatrix} \tag{6.1.6}$$

and $\mathbf{r}$ and ϕ are generalized coordinates that locate the body in the plane. An expression for the virtual displacement of point P, in terms of variations in the generalized coordinates, may be obtained by taking the differential of Eq. 6.1.5 (Prob. 6.1.2):

$$\delta\mathbf{r}^P = \delta\mathbf{r} + \delta\phi\mathbf{B}\mathbf{s}'^P \tag{6.1.7}$$

where, as in Eq. 2.6.3,

$$\mathbf{B} = \begin{bmatrix} -\sin\phi & -\cos\phi \\ \cos\phi & -\sin\phi \end{bmatrix} \tag{6.1.8}$$

To write the acceleration term on the left of Eq. 6.1.4 in terms of time derivatives of the generalized coordinates, Eq. 6.1.5 may be differentiated with respect to time to obtain, as in Eq. 2.6.4,

$$\dot{\mathbf{r}}^P = \dot{\mathbf{r}} + \dot{\phi}\mathbf{B}\mathbf{s}'^P \tag{6.1.9}$$

Differentiating this result with respect to time yields, as in Eq. 2.6.9,

$$\ddot{\mathbf{r}}^P = \ddot{\mathbf{r}} + \ddot{\phi}\mathbf{B}\mathbf{s}'^P - \dot{\phi}^2\mathbf{A}\mathbf{s}'^P \tag{6.1.10}$$

Equations 6.1.7 and 6.1.10 may be used to expand Eq. 6.1.4, obtaining the general *variational form of the equations of planar motion* (Prob. 6.1.3):

$$\begin{aligned}\delta\mathbf{r}^T\ddot{\mathbf{r}}\int_m dm(P) &+ [\delta\mathbf{r}^T(\ddot{\phi}\mathbf{B} - \dot{\phi}^2\mathbf{A}) + \delta\phi\ddot{\mathbf{r}}^T\mathbf{B}]\int_m \mathbf{s}'^P\, dm(P) \\ &+ \delta\phi\int_m \mathbf{s}'^{PT}\mathbf{B}^T[\ddot{\phi}\mathbf{B} - \dot{\phi}^2\mathbf{A}]\mathbf{s}'^P\, dm(P) \\ &= \delta\mathbf{r}^T\int_m \mathbf{F}(P)\, dm(P) + \delta\phi\int_m \mathbf{s}'^{PT}\mathbf{B}^T\mathbf{F}(P)\, dm(P)\end{aligned} \tag{6.1.11}$$

Even though Eq. 6.1.4 must hold for all $\delta\mathbf{r}^P$ that are consistent with constraints that define rigid-body motion, Eq. 6.1.11 holds for arbitrary $\delta\mathbf{r}$ and $\delta\phi$. This is true since $\delta\mathbf{r}^P$ given by Eq. 6.1.7 is consistent with Eq. 6.1.5, which defines rigid-body motion, for arbitrary $\delta\mathbf{r}$ and $\delta\phi$.

6.1.2 Variational Equations of Motion with Centroidal Coordinates

The general form of the equations of motion of a planar body of Eq. 6.1.11 can be substantially simplified if special features of the body-fixed x'-y' frame are exploited. The first integral on the left of Eq. 6.1.11 is simply the total mass of the body; that is,

$$m = \int_m dm(P) \tag{6.1.12}$$

regardless of the location of the x'-y' frame. If the origin of the x'-y' frame is located at the *center of mass* or *centroid* of the body, then, by definition of the centroid,

$$\int_m \mathbf{s}'^P\, dm(P) = \mathbf{0} \tag{6.1.13}$$

and the second integral on the left of Eq. 6.1.11 vanishes.

Direct matrix multiplications verify that

$$\begin{aligned}\mathbf{B}^T\mathbf{B} &= \mathbf{I} \\ \mathbf{B}^T\mathbf{A} &= \begin{bmatrix} 0 & 1 \\ -1 & 0 \end{bmatrix}\end{aligned} \tag{6.1.14}$$

Thus, the first term in the third integral on the left of Eq. 6.1.11, after $\ddot{\phi}$ is moved outside the integral, becomes simply the *polar moment of inertia* of the body with respect to the origin of the body-fixed x'-y' frame; that is,

$$J' \equiv \int_m \mathbf{s}'^{PT}\mathbf{s}'^P \, dm(P) \tag{6.1.15}$$

A direct expansion of the second term in the integrand of the third integral on the left of Eq. 6.1.11 shows that (Prob. 6.1.4)

$$\mathbf{s}'^{PT}\mathbf{B}^T\mathbf{A}\mathbf{s}'^P = 0$$

To interpret terms on the right of Eq. 6.1.11, note first that the resultant force acting on the body, including applied forces and forces of contact with other bodies, is

$$\mathbf{F} = \int_m \mathbf{F}(P) \, dm(P) \tag{6.1.16}$$

The second integral on the right of Eq. 6.1.11 may be expanded to see that it defines the *torque,* or *moment* of the forces acting on the body, about an axis perpendicular to the x-y plane, counterclockwise taken as positive; that is, writing $\mathbf{F}$ in terms of body-fixed components and using Eq. 6.1.14, the torque acting on the body is

$$\begin{aligned} n &= \int_m \mathbf{s}'^{PT}\mathbf{B}^T\mathbf{F}(P) \, dm(P) \\ &= \int_m \mathbf{s}'^{PT}\mathbf{B}^T\mathbf{A}\mathbf{F}' \, dm(P) \\ &= \int_m [-s_{y'}^P F_{x'}(P) + s_{x'}^P F_{y'}(P)] \, dm(P) \end{aligned} \tag{6.1.17}$$

Substituting Eqs. 6.1.12 through 6.1.17 in Eq. 6.1.11,

$$\delta\mathbf{r}^T[m\ddot{\mathbf{r}} - \mathbf{F}] + \delta\phi[J'\ddot{\phi} - n] = 0 \tag{6.1.18}$$

for arbitrary virtual displacements $\delta\mathbf{r}$ and $\delta\phi$. Equation 6.1.18 is called the *variational equation of motion* of a rigid body with a *centroidal body-fixed reference frame* in the plane. It may be thought of as an extension of the *principle of virtual work,* in that the first term is the virtual work of the unbalanced force $m\ddot{\mathbf{r}} - \mathbf{F}$ and the second term is the virtual work of the unbalanced torque $J'\ddot{\phi} - n$. Recall that Eq. 6.1.18 is valid only if the origin of the body-fixed x'-y' frame is at the centroid of the body.

6.1.3 Differential Equations of Motion

If all forces acting on a body have been accounted for, since $\delta\mathbf{r}$ and $\delta\phi$ are arbitrary in Eq. 6.1.18, their coefficients must be zero and the conventional differential equations for motion of a rigid body in the plane are obtained as

$$\begin{aligned} m\ddot{\mathbf{r}} &= \mathbf{F} \\ J'\ddot{\phi} &= n \end{aligned} \tag{6.1.19}$$

To see that this is true, first set $\delta\mathbf{r} = 0$ and $\delta\phi = 0.001$ in Eq. 6.1.18. This shows that the last of Eqs. 6.1.19 holds. A similar argument can be given with $\delta\phi = 0$ and each component of $\delta\mathbf{r}$, in turn, equal to zero to obtain the first of Eqs. 6.1.19.

It is important to remember that Eqs. 6.1.19 are valid only if the vector $\mathbf{F}$ of forces and the torque n acting on the body represent all force effects, that is, both applied and constraint forces.

Example 6.1.1: The tractor shown in Fig. 6.1.2 has no suspension and can be modeled as one body. It has mass m and polar moment of inertia J'. A driving force T_r is generated at the rear wheels, and ground reaction forces F_f and F_r are applied at the front and rear wheels, respectively.

A body fixed x_1'-y_1' frame is located at the center of mass 0_1 of the tractor. Tire force due to penetration into the soil is modeled as

$$F = f(d) = \begin{cases} kd, & d \geq 0 \\ 0, & d < 0 \end{cases}$$

where k is a spring constant and d is tire deflection.

For a small pitch angle ϕ_1, tire deflections d_f and d_r for the front and rear wheels are, respectively,

$$d_f = h_0 - (y_1 + \ell_f \phi_1)$$

$$d_r = h_0 - (y_1 - \ell_r \phi_1)$$

where h_0 is the height of the center of mass when the tractor is level and there is

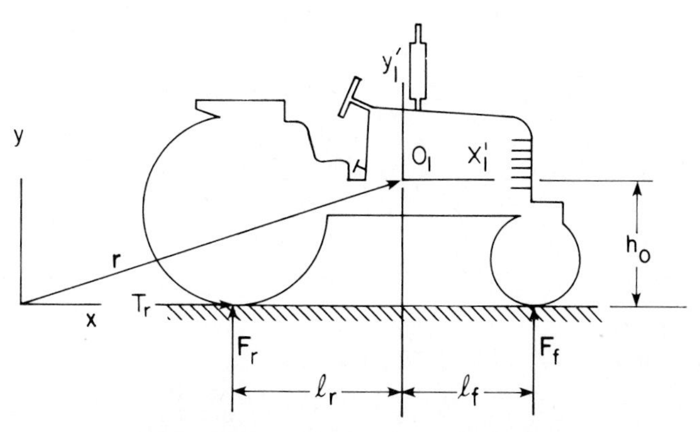

Figure 6.1.2 Plane motion of a tractor.

no tire deflection. Hence, the supporting forces are

$$F_f = k(h_0 - (y_1 + \ell_f \phi_1))$$
$$F_r = k(h_0 - (y_1 - \ell_r \phi_1))$$

Since there are no kinematic constraints acting on the tractor and all applied forces have been accounted for, the equations of motion of Eq. 6.1.19 are

$$m\begin{bmatrix} \ddot{x}_1 \\ \ddot{y}_1 \end{bmatrix} = \begin{bmatrix} T_r \\ k(h_0 - y_1 - \ell_f \phi_1) + k(h_0 - y_1 + \ell_r \phi_1) - mg \end{bmatrix}$$
$$J\ddot{\phi}_1 = \ell_f k(h_0 - y_1 - \ell_f \phi_1) - \ell_r k(h_0 - y_1 + \ell_r \phi_1) + T_r h_0$$

Note that these are linear differential equations. The small pitch angle approximation used in deriving tire deflection leads to the linear form. Most often, the equations of machine dynamics are nonlinear.

6.1.4 Properties of the Centroid and Polar Moment of Inertia

Consider the planar body shown in Fig. 6.1.3, with a noncentroidal x''-y'' body-fixed reference frame. Dimensions and mass distribution in the body are known, and the location of the centroid in the x''-y'' frame is to be found.

Let vector $\boldsymbol{\rho}''$ in the x''-y'' frame locate the centroid of the body and define an x'-y' frame with its origin at the centroid (a body-fixed *centroidal reference frame*) and with axes parallel to the x''-y'' frame. To determine $\boldsymbol{\rho}''$, the equation

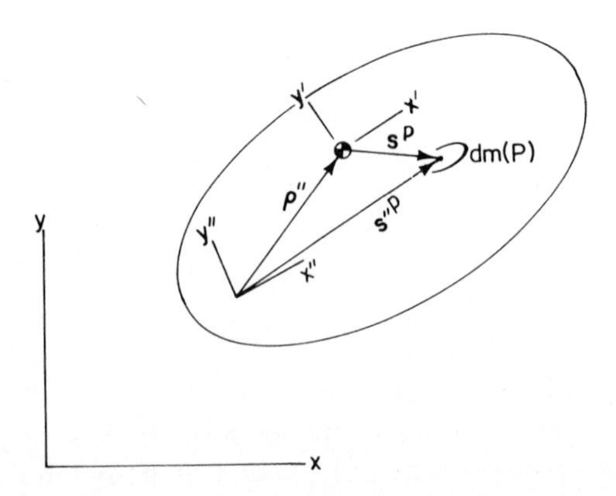

Figure 6.1.3 Location of a centroid.

that defines the centroid, Eq. 6.1.13, is

$$\begin{aligned}\mathbf{0} &= \int_m \mathbf{s}'^P \, dm(P) \\ &= \int_m (\mathbf{s}''^P - \boldsymbol{\rho}'') \, dm(P) \\ &= \int_m \mathbf{s}''^P \, dm(P) - \boldsymbol{\rho}'' \int_m dm(P) \\ &= \int_m \mathbf{s}''^P dm(P) - m\boldsymbol{\rho}''\end{aligned}$$

since $\boldsymbol{\rho}''$ does not depend on P. Thus, the vector $\boldsymbol{\rho}''$ that locates the centroid in the x''-y'' frame is

$$\boldsymbol{\rho}'' = \frac{1}{m} \int_m \mathbf{s}''^P \, dm(P) \qquad \textbf{(6.1.20)}$$

Let the polar moment of inertia J'' with respect to the x''-y'' frame be calculated as

$$J'' = \int_m \mathbf{s}''^{PT} \mathbf{s}''^P \, dm(P) \qquad \textbf{(6.1.21)}$$

The goal now is to use this information and the location $\boldsymbol{\rho}''$ of the centroid in the x''-y'' frame to calculate the polar moment of inertia with respect to the x'-y' frame.

Substituting $\mathbf{s}''^P = \boldsymbol{\rho}'' + \mathbf{s}'^P$ from Fig. 6.1.3 into Eq. 6.1.21 and manipulating,

$$\begin{aligned}J'' &= \int_m (\boldsymbol{\rho}'' + \mathbf{s}'^P)^T (\boldsymbol{\rho}'' + \mathbf{s}'^P) \, dm(P) \\ &= \boldsymbol{\rho}''^T \boldsymbol{\rho}'' \int_m dm(P) + 2\boldsymbol{\rho}''^T \int_m \mathbf{s}'^P \, dm(P) + \int_m \mathbf{s}'^{PT} \mathbf{s}'^P \, dm(P)\end{aligned}$$

Since the x'-y' frame has its origin at the centroid, the next to last term on the right is zero. Furthermore, the last term is J'. Thus,

$$J'' = J' + m \, |\boldsymbol{\rho}''|^2 \qquad \textbf{(6.1.22)}$$

This relation is called the *parallel axis theorem* [4, 7, 9] for polar moment of inertia.

Masses, polar moments of inertia, and centroid locations for a few commonly encountered homogeneous plane bodies are given in Table 6.1.1.

Consider a body that has an axis of both physical and mass distribution symmetry that passes through the origin of the x''-y'' frame, with unit normal

TABLE 6.1.1 Mass and Polar Moment of Inertia of Planar Bodies

Body	Mass and polar moment of inertia (γ = mass per unit area)
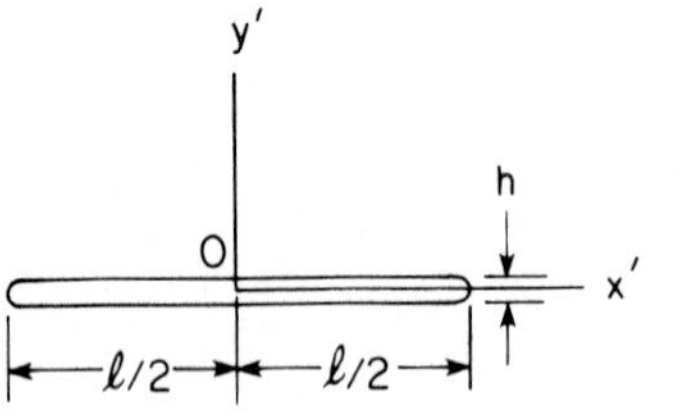 Thin Rod	$m = \gamma \ell h$, h = height $J' = \dfrac{m}{12}\ell^2$
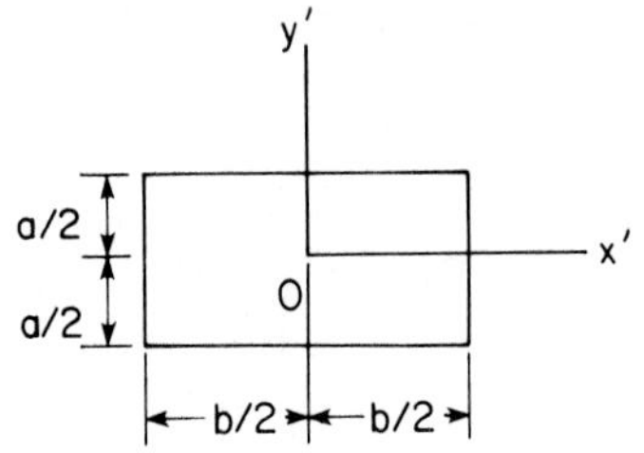 Rectangle	$m = \gamma ab$ $J' = \dfrac{1}{12}m(a^2 + b^2)$
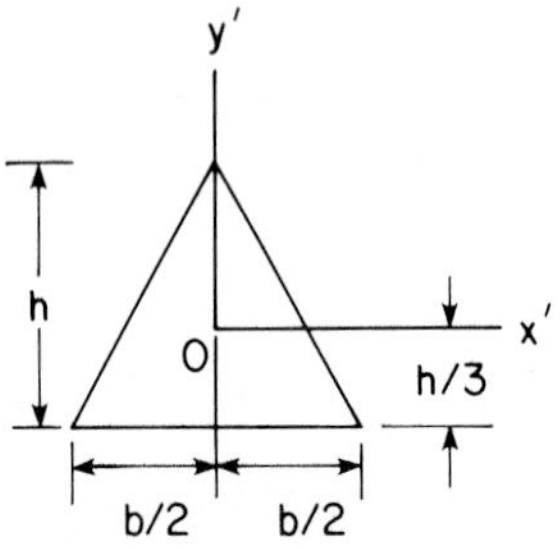 Isosceles Triangle	$m = \dfrac{1}{2}\gamma bh$ $J' = m\left(\dfrac{b^2}{24} + \dfrac{h^2}{18}\right)$
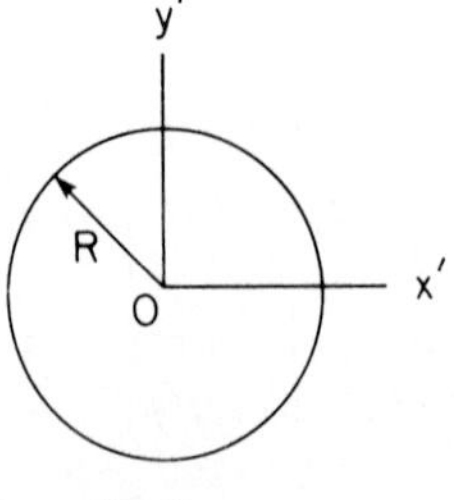 Circle	$m = \gamma \pi R^2$ $J' = \dfrac{1}{2}mR^2$

TABLE 6.1.1 (continued)

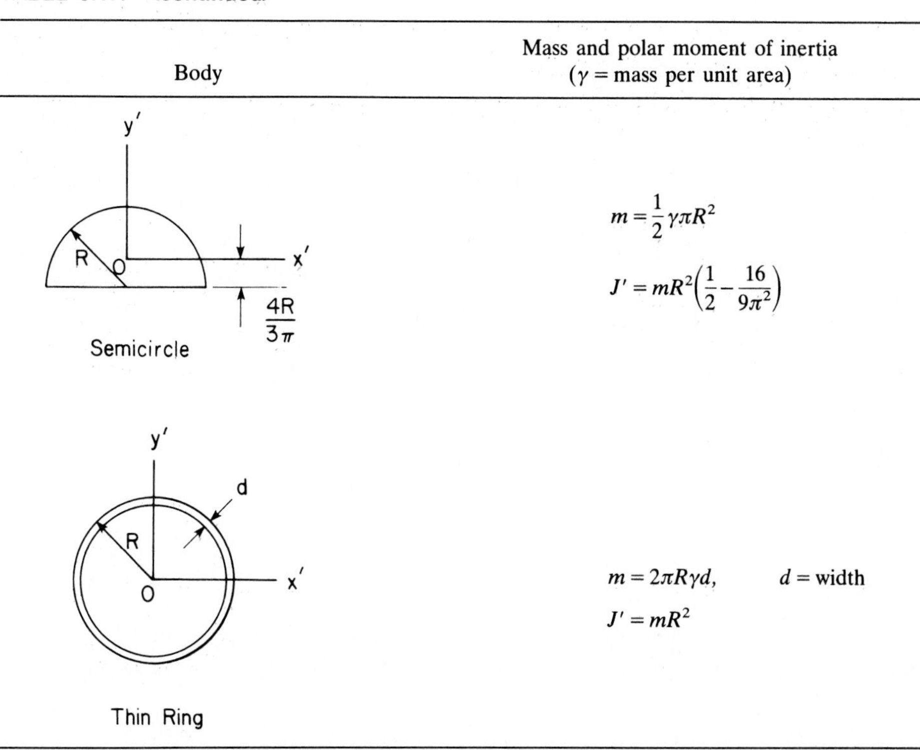

Body	Mass and polar moment of inertia (γ = mass per unit area)
Semicircle	$m = \frac{1}{2}\gamma\pi R^2$ $J' = mR^2\left(\frac{1}{2} - \frac{16}{9\pi^2}\right)$
Thin Ring	$m = 2\pi R\gamma d, \qquad d = \text{width}$ $J' = mR^2$

vector $\mathbf{k}$, as shown by the dashed line in Fig. 6.1.4. Then, for any point P there is a point P^R that is symmetrically placed across the line of symmetry. Thus,

$$\mathbf{k}^T\mathbf{s}''^P\, dm(P) = -\mathbf{k}^T\mathbf{s}''^{P^R}\, dm(P^R)$$

Premultiplying both sides of Eq. 6.1.20 by $\mathbf{k}^T$,

$$\begin{aligned}\mathbf{k}^T\boldsymbol{\rho}'' &= \frac{1}{m}\mathbf{k}^T\int_m \mathbf{s}''^P\, dm(P)\\ &= \frac{1}{m}\int_m \mathbf{k}^T\mathbf{s}''^P\, dm(P)\end{aligned}$$

Denoting by m^+ the mass on the side of the line of symmetry toward which $\mathbf{k}$ points and by m^- the mass on the other side, $m^+ = m^- = m/2$ and

$$\begin{aligned}\mathbf{k}^T\boldsymbol{\rho}'' &= \frac{1}{m}\left(\int_{m^+} \mathbf{k}^T\mathbf{s}''^P\, dm(P) + \int_{m^-} \mathbf{k}^T\mathbf{s}''^{P^R}\, dm(P^R)\right)\\ &= \frac{1}{m}\left(\int_{m^+} \mathbf{k}^T\mathbf{s}''^P\, dm(P) - \int_{m^+} \mathbf{k}^T\mathbf{s}''^P\, dm(P)\right)\\ &= 0\end{aligned}$$

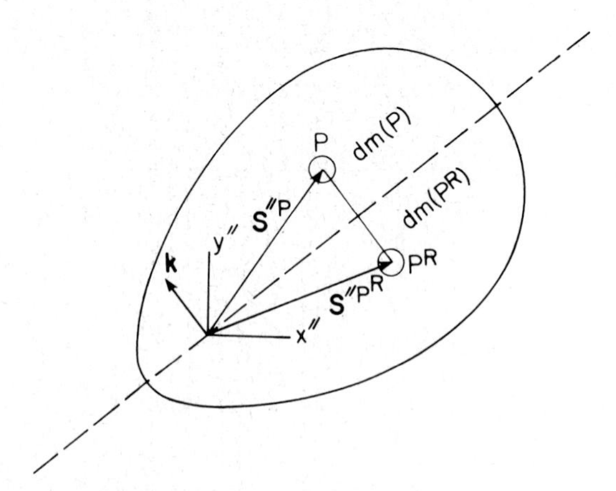

Figure 6.1.4 Body with axis of symmetry.

Thus, the centroid of the body lies on each axis of symmetry of the body. In particular, if a body has two distinct axes of symmetry, the centroid must lie at their intersection. Note that the standard shapes of Table 6.1.1 have one or more axes of symmetry, on which their centroids lie.

Note that even if a body has a geometric axis of symmetry, if its mass is not symmetrically distributed, the centroid will not be on the geometric axis of symmetry.

6.1.5 Inertial Properties of Composite Bodies

Components of machines are often made up of combinations of subcomponents that have standard shapes, such as rods, circles, rings, rectangles, and triangles. A typical example of such a *component* (or *composite body*) is shown in Fig. 6.1.5, in which all subcomponents and voids have some standard shape typical of those resulting from common manufacturing processes. The objective of this subsection

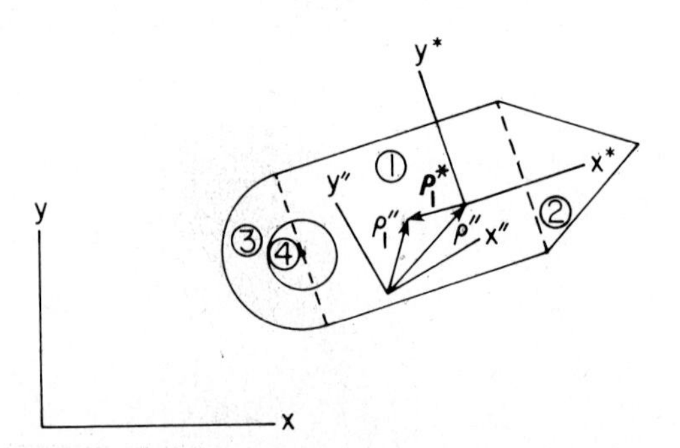

Figure 6.1.5 Body made up of subcomponents.

is to develop expressions for inertia properties of composite bodies using easily calculated properties of individual subcomponents.

Let an x''-y'' frame be fixed to the composite body in a convenient location (e.g., as shown in Fig. 6.1.5). Using this frame, the centroid of the complex body may be obtained by applying the definition of Eq. 6.1.20, employing the property that an integral over the entire mass may be written as the sum of integrals over masses m_i of subcomponents to obtain

$$\begin{aligned}\boldsymbol{\rho}'' &= \frac{1}{m}\sum_{i=1}^{k}\int_{m_i}\mathbf{s}''^{P}\,dm_i(P)\\ &= \frac{1}{m}\sum_{i=1}^{k}m_i\boldsymbol{\rho}_i''\end{aligned} \tag{6.1.23}$$

where $m=\sum_{i=1}^{k}m_i$. To use this result, the centroids $\boldsymbol{\rho}_i''$ are first located in the x''-y'' frame using mass and centroid location information such as that found in Table 6.1.1 or from direct numerical calculation. Equation 6.1.23 is then used to locate the centroid of the composite body in the x''-y'' frame.

Denote by x^*-y^* the centroidal frame of the composite body, as shown in Fig. 6.1.5, to avoid confusion with centroidal x_i'-y_i' frames on each subcomponent. To calculate the polar moment of inertia J^* with respect to the x^*-y^* composite body centroidal frame, the defining equation of Eq. 6.1.15 may be written and its integral evaluated as a sum of integrals over subcomponents m_i that make up the composite body to obtain

$$\begin{aligned}J^* &= \int_m \mathbf{s}^{*PT}\mathbf{s}^{*P}\,dm(P)\\ &= \sum_{i=1}^{k}\left[\int_{m_i}\mathbf{s}_i^{*PT}\mathbf{s}_i^{*P}\,dm_i(P)\right]\\ &= \sum_{i=1}^{k}J_i^*\end{aligned} \tag{6.1.24}$$

For each subcomponent i, the polar moment of inertia J_i' with respect to its centroidal x_i'-y_i' frame is often known. Using Eq. 6.1.22, the polar moment of inertia J_i^* with respect to the centroidal x^*-y^* frame of the composite body, which is noncentroidal for each of the individual subcomponents, is

$$J_i^* = J_i' + m_i\,|\boldsymbol{\rho}_i^*|^2 \tag{6.1.25}$$

where $\boldsymbol{\rho}_i^*$ is the vector that locates the centroid of the subcomponent in the x^*-y^* frame, as shown in Fig. 6.1.5.

If there are voids in a composite body, Eq. 6.1.24 may be used by including material that would have been in the void in a subcomponent with the void removed. The void is then treated as a subcomponent, and $-J_i^*$ is assigned as the polar moment of inertia of the void, where J_i^* is the polar moment of inertia of a body that would occupy the void with the same material density as the subcomponent from which it is removed.

Example 6.1.2: The pendulum shown in Fig. 6.1.6 consists of a slender rod and a disk with a hole in it. The rod has a mass of 2.8 kg and the disk has a density of 8000 kg/m³ and a thickness of 0.01 m. Find the centroid and polar moment of inertia of the pendulum.

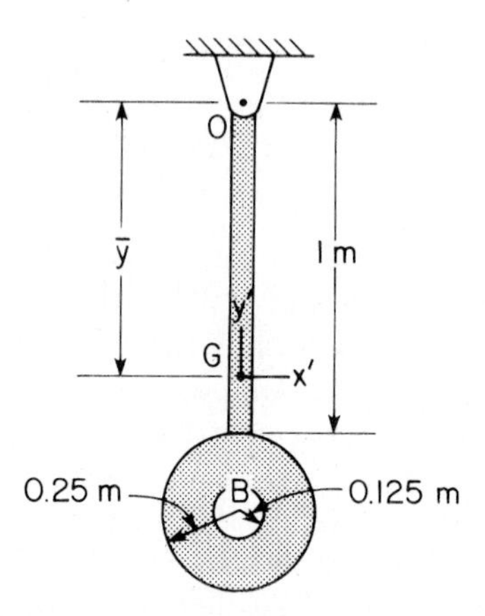

Figure 6.1.6 Composite pendulum.

Let m_d denote the mass of a disk that does not have the hole, and let m_h be the mass of the disk that would be removed to create the hole. Then

$$m_d = 8000[\pi(0.25)^2(0.01)] = 15.71 \text{ kg}$$

$$m_h = 8000[\pi(0.125)^2(0.01)] = 3.93 \text{ kg}$$

Since the y axis is an axis of symmetry of the composite body, the centroid lies on it. Let the distance between the centroid of the composite pendulum and point O be $\bar{y}$. From Eq. 6.1.23,

$$\bar{y} = \frac{\sum_i m_i y_i}{\sum_i m_i} = \frac{m_r y_r + (m_d - m_h) y_d}{m_r + (m_d - m_h)}$$

$$= \frac{2.80(0.5) + 15.71(1.25) - (3.93)(1.25)}{2.80 + 15.71 - 3.93} = 1.11 \text{ m}$$

where m_r, m_d, and m_h are masses of the rod, disk, and material that would have been in the hole, and y_r, y_d, and y_h are the distances from point O to the centroids of the rod, disk, and hole, respectively. From Table 6.1.1, the polar moment of inertia of the rod, with respect to its centroid, is

$$J_r = \frac{m_r \ell^2}{12} = \frac{2.8(1)^2}{12} = 0.233 \text{ kg} \cdot \text{m}^2$$

From Eq. 6.1.25, the polar moment of inertia of the rod, with respect to the centroid, point G of the assembly, is

$$\begin{aligned} J'_r &= 0.233 + m_r\rho^2 \\ &= 0.233 + 2.8(1.11 - 0.5)^2 = 1.275 \text{ kg} \cdot \text{m}^2 \end{aligned}$$

where $\rho = |\boldsymbol{\rho}|$.

Similarly, from Table 6.1.1 and Eq. 6.1.25, the polar moment of inertia of the disk, with respect to point G, is

$$\begin{aligned} J'_d &= \tfrac{1}{2}m_d R^2 + m_d\rho^2 \\ &= 0.5(15.71)(0.25)^2 + (15.71)(1.25 - 1.11)^2 \\ &= 0.799 \text{ kg} \cdot \text{m}^2 \end{aligned}$$

Similarly, for the hole,

$$\begin{aligned} J'_h &= \tfrac{1}{2}m_h R^2 + m_h\rho^2 \\ &= 0.5(3.93)(0.125)^2 + (3.93)(1.25 - 1.11)^2 \\ &= 0.108 \text{ kg} \cdot \text{m}^2 \end{aligned}$$

The polar moment of inertia of the pendulum about point G is, therefore,

$$\begin{aligned} J' &= J'_r + J'_d - J'_h \\ &= 1.275 + 0.799 - 0.108 \\ &= 1.966 \text{ kg} \cdot \text{m}^2 \end{aligned}$$

6.2 VIRTUAL WORK AND GENERALIZED FORCE

The variational equation of motion of a rigid body in the plane derived in Section 6.1 is used to assemble the variational equation of motion of a constrained multibody system in Section 6.3. Concepts of virtual work and generalized force are introduced in this section as a foundation for the multibody variational method.

The *virtual work* of a force $\mathbf{F}$ that acts at the origin of the x'-y' frame and a torque n that acts on the body in the plane may be written in terms of its virtual displacement $\delta\mathbf{r}$ and virtual rotation $\delta\phi$ as

$$\begin{aligned} \delta W &= \delta\mathbf{r}^T\mathbf{F} + \delta\phi n \\ &\equiv \delta\mathbf{q}^T\mathbf{Q} \end{aligned} \qquad \textbf{(6.2.1)}$$

where

$$\begin{aligned} \mathbf{q} &= [\mathbf{r}^T, \phi]^T \\ \mathbf{Q} &= [\mathbf{F}^T, n]^T \end{aligned} \qquad \textbf{(6.2.2)}$$

The basic idea in defining *generalized force* $\mathbf{Q}$ associated with generalized

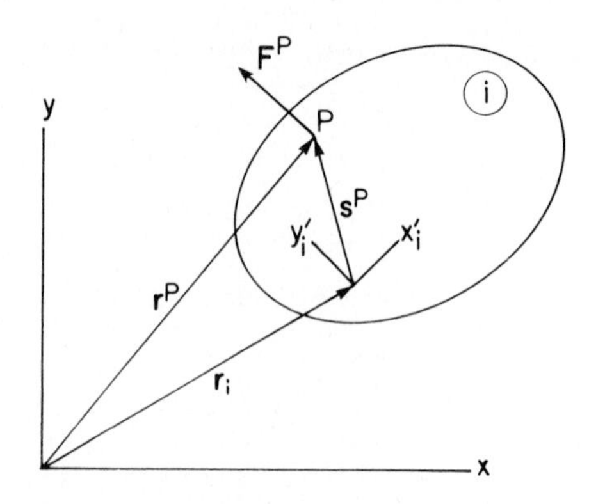

Figure 6.2.1 Force at point P on body i.

coordinate $\mathbf{q}$ is to first write the virtual work of a set of forces and moments that act on a body in terms of products of physical virtual displacements and rotations and physical forces and torques. Next, virtual displacements and rotations are written in terms of variations $\delta\mathbf{q}$ in generalized coordinates. Finally, coefficients of all variations of generalized coordinates that are used in the formulation are collected. These coefficients are defined to be generalized forces associated with the corresponding generalized coordinates. Note that virtual work is the key quantity that is preserved in defining generalized force.

Consider a force $\mathbf{F}^P$ that acts at point P on body i, as shown in Fig. 6.2.1. Since

$$\mathbf{r}_i^P = \mathbf{r}_i + \mathbf{A}_i\mathbf{s}'^P$$

and

$$\delta\mathbf{r}_i^P = \delta\mathbf{r}_i + \delta\phi_i\mathbf{B}_i\mathbf{s}'^P$$

the virtual work of force $\mathbf{F}^P$ is

$$\begin{aligned}\delta W &= \delta\mathbf{r}_i^{PT}\mathbf{F}^P \\ &= \delta\mathbf{r}_i^T\mathbf{F}^P + \delta\phi_i\mathbf{s}'^{PT}\mathbf{B}_i^T\mathbf{F}^P\end{aligned}$$

Thus, for a general force $\mathbf{F}^P$ acting at point P on body i, the corresponding generalized force is

$$\mathbf{Q} = \begin{bmatrix}\mathbf{F}^P \\ \mathbf{s}'^{PT}\mathbf{B}_i^T\mathbf{F}^P\end{bmatrix} \tag{6.2.3}$$

Note that the third component of generalized force $\mathbf{Q}$ in Eq. 6.2.3 is just the moment of $\mathbf{F}^P$ about the origin of the x_i'-y_i' frame, as in Eq. 6.1.17.

If force $\mathbf{F}^P$ were fixed in body i (e.g., in the case of a rocket thrustor attached to a spacecraft), then $\mathbf{F}^P = \mathbf{A}\cdot\mathbf{F}'^P$, and Eq. 6.2.3 becomes

$$\mathbf{Q} = \begin{bmatrix}\mathbf{A}_i\mathbf{F}'^P \\ \mathbf{s}'^{PT}\mathbf{B}_i^T\mathbf{A}_i\mathbf{F}'^P\end{bmatrix}$$

Since $\mathbf{B}_i = \mathbf{A}_i\mathbf{R}$, $\mathbf{s}'^{PT}\mathbf{B}_i^T\mathbf{A}_i = \mathbf{s}'^{PT}\mathbf{R}^T = (\mathbf{R}\mathbf{s}'^P)^T$. Thus, for a body-fixed force $\mathbf{F}'^P$,

$$\mathbf{Q} = \begin{bmatrix} \mathbf{A}_i \\ (\mathbf{R}\mathbf{s}'^P)^T \end{bmatrix} \mathbf{F}'^P \tag{6.2.4}$$

Using the notation of Eqs. 6.2.2, 6.2.3, or 6.2.4, $\delta\mathbf{q} = [\delta\mathbf{r}^T, \delta\phi]^T$, and

$$\mathbf{M} = \text{diag}(m, m, J')$$

the variational equation of motion of Eq. 6.1.18 may be written in the form

$$\delta\mathbf{q}^T[\mathbf{M}\ddot{\mathbf{q}} - \mathbf{Q}] = 0 \tag{6.2.5}$$

A common force element that is encountered in mechanical system dynamics is a compliant connection between points P_i and P_j on bodies i and j, respectively, as shown in Fig. 6.2.2. A compliant element in applications may consist of a spring, a damper, or a force actuator such as a hydraulic cylinder that exerts a force along the vector between points P_i and P_j. Such an element is called a *translational spring–damper–actuator.*

Since this force element exerts equal and opposite forces on bodies i and j, rather than calculating generalized forces on each body separately, as in the case of external applied forces, it is instructive to make direct use of the virtual work definition of generalized force. Consider first the vector $\mathbf{d}_{ij}$ from point P_i to point P_j:

$$\mathbf{d}_{ij} = \mathbf{r}_j + \mathbf{A}_j\mathbf{s}_j' - \mathbf{r}_i - \mathbf{A}_i\mathbf{s}_i' \tag{6.2.6}$$

The length of this element is calculated from

$$\ell^2 = \mathbf{d}_{ij}^T\mathbf{d}_{ij} \tag{6.2.7}$$

Note that ℓ may be positive or negative, depending on the relative orientation at assembly of the system. Since mechanical force elements are usually designed with care so that their length can never be zero, they should be modeled to

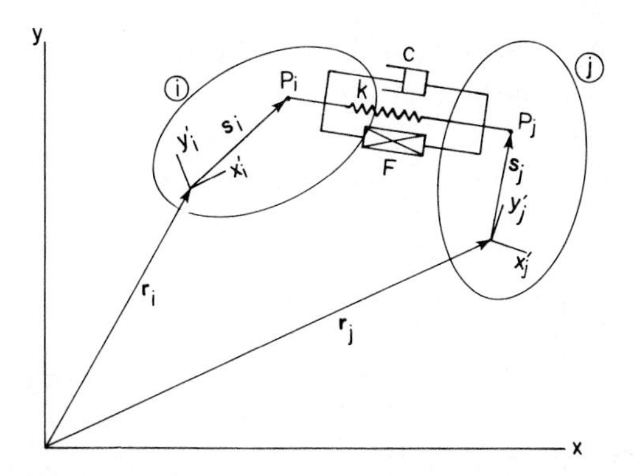

Figure 6.2.2 Translational spring–damper–actuator.

preclude the pathological case $\ell = 0$. The reader is encouraged to adopt the convention $\ell > 0$.

Differentiating Eq. 6.2.7,

$$2\ell\dot{\ell} = 2\mathbf{d}_{ij}^T\dot{\mathbf{d}}_{ij}$$

As long as ℓ is not equal to zero, dividing both sides by ℓ and differentiating Eq. 6.2.6 yields

$$\dot{\ell} = \left(\frac{\mathbf{d}_{ij}}{\ell}\right)^T (\dot{\mathbf{r}}_j + \mathbf{B}_j\mathbf{s}_j'\dot{\phi}_j - \dot{\mathbf{r}}_i - \mathbf{B}_i\mathbf{s}_i'\dot{\phi}_i) \qquad \textbf{(6.2.8)}$$

In case ℓ approaches zero, the limit of the right side of Eq. 6.2.8 may be taken as ℓ approaches zero, using L'Hospital's rule [26], to obtain

$$\lim_{\ell \to 0} \frac{\mathbf{d}_{ij}}{\ell} = \lim_{\ell \to 0} \frac{\dot{\mathbf{d}}_{ij}}{\dot{\ell}} = \left.\frac{\dot{\mathbf{d}}_{ij}}{\dot{\ell}}\right|_{\ell = 0}$$

providing $\dot{\ell} \neq 0$ when $\ell = 0$. Substituting from this result into Eq. 6.2.8 and solving for $(\dot{\ell})^2$

$$(\dot{\ell})^2 = \dot{\mathbf{d}}_{ij}^{\prime T}\mathbf{d}_{ij} \qquad \textbf{(6.2.9)}$$

when $\ell = 0$. The sign of $\dot{\ell}$ is selected as the value it had just prior to the occurrence of $\ell = 0$.

Defining tension in the compliant element as positive; that is, a force that tends to draw the bodies together is positive, the force acting in the element is written as

$$f = k(\ell - \ell_0) + c\dot{\ell} + F(\ell, \dot{\ell}, t) \qquad \textbf{(6.2.10)}$$

where k is the spring coefficient, ℓ_0 is the free length of the spring, c is the damping coefficient, and F is a general actuator force that may depend on ℓ, $\dot{\ell}$, or time t.

Making direct use of the definition of virtual work, the virtual work of the force f is simply the product of the force and the variation $\delta\ell$ in the length of the element. Since a positive $\delta\ell$ tends to separate the bodies and the sign convention for f is that tension is positive, tending to pull the bodies together, for positive f and $\delta\ell$, the work done on the bodies is negative. Thus, the virtual work is

$$\delta W = -f\delta\ell \qquad \textbf{(6.2.11)}$$

Taking the differential of both sides of Eq. 6.2.7 and manipulating, as in the derivation of the velocity relation of Eq. 6.2.8,

$$\delta\ell = \left(\frac{\mathbf{d}_{ij}}{\ell}\right)^T (\delta\mathbf{r}_j + \mathbf{B}_j\mathbf{s}_j'\,\delta\phi_j - \delta\mathbf{r}_i - \mathbf{B}_i\mathbf{s}_i'\,\delta\phi_i) \qquad \textbf{(6.2.12)}$$

where $\mathbf{B}_i$ is defined in Eq. 6.1.8. Substituting this result into Eq. 6.2.11, the virtual work of the spring–damper–actuator force is

$$\delta W = -\frac{f}{\ell}\mathbf{d}_{ij}^T(\delta\mathbf{r}_j + \mathbf{B}_j\mathbf{s}_j'\,\delta\phi_j - \delta\mathbf{r}_i - \mathbf{B}_i\mathbf{s}_i'\,\delta\phi_i) \tag{6.2.13}$$

By definition, the generalized forces acting on bodies i and j are simply the coefficients of variations in generalized coordinates that define the position and orientation of each body; that is,

$$\begin{aligned}\mathbf{Q}_i &= \frac{f}{\ell}\begin{bmatrix}\mathbf{d}_{ij}\\ \mathbf{d}_{ij}^T\mathbf{B}_i\mathbf{s}_i'\end{bmatrix}\\ \mathbf{Q}_j &= -\frac{f}{\ell}\begin{bmatrix}\mathbf{d}_{ij}\\ \mathbf{d}_{ij}^T\mathbf{B}_j\mathbf{s}_j'\end{bmatrix}\end{aligned} \tag{6.2.14}$$

Note that as ℓ approaches zero in Eq. 6.2.14 an indeterminate fraction occurs, which may be resolved by applying L'Hospital's rule as ℓ approaches zero to obtain

$$\left.\begin{aligned}\mathbf{Q}_i &= \frac{f}{\ell}\left[\frac{\dot{\mathbf{d}}_{ij}}{\dot{\mathbf{d}}_{ij}^T\mathbf{B}_i\mathbf{s}_i'}\right]\\ \mathbf{Q}_j &= -\frac{f}{\ell}\left[\frac{\dot{\mathbf{d}}_{ij}}{\dot{\mathbf{d}}_{ij}^T\mathbf{B}_j\mathbf{s}_j'}\right]\end{aligned}\right., \qquad \ell = 0 \tag{6.2.15}$$

if $\dot{\ell} \neq 0$, where the fact that $\mathbf{d}_{ij} = \mathbf{0}$ when $\ell = 0$ has been used.

Consider next the pair of bodies shown in Fig. 6.2.3, with a revolute joint defined at point P and a torsional compliant element acting between the

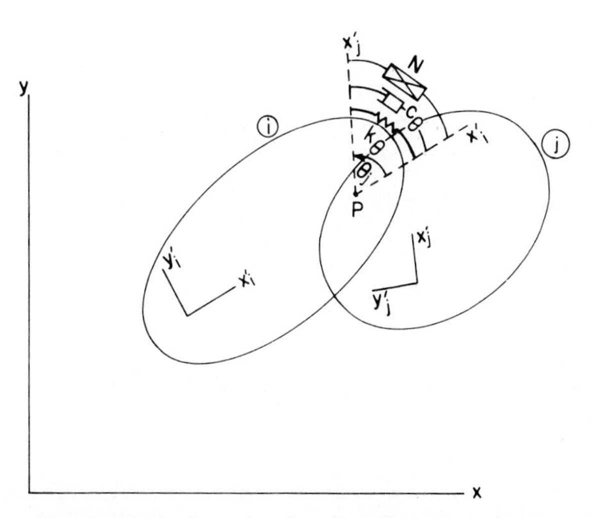

Figure 6.2.3 Rotational spring–damper–actuator.

body-fixed x_i' and x_j' axes. This compliant element exerts torques of equal magnitude but opposite orientation on bodies i and j; hence it is called a *rotational spring–damper–actuator*. A positive torque is one that acts counterclockwise on body i and clockwise on body j.

The angle θ_{ij} from the x_i' axis to the x_j' axis, counterclockwise taken as positive, is

$$\theta_{ij} = \phi_j - \phi_i \tag{6.2.16}$$

Taking the time derivative of this relation yields

$$\dot{\theta}_{ij} = \dot{\phi}_j - \dot{\phi}_i \tag{6.2.17}$$

The torque that acts between the bodies is

$$n = k_\theta(\theta_{ij} - \theta_0) + c_\theta \dot{\theta}_{ij} + N(\theta_{ij}, \dot{\theta}_{ij}, t) \tag{6.2.18}$$

where k_θ is a torsional spring coefficient, θ_0 is the free angle of the spring, c_θ is a torsional damping coefficient, and N is a general actuator torque that may depend on θ_{ij}, $\dot{\theta}_{ij}$, and time t. The virtual work of the torque n is simply the negative of its product with the relative virtual rotation $\delta\theta_{ij}$; that is,

$$\delta W = -n\,\delta\theta_{ij} \tag{6.2.19}$$

since when $\theta_{ij} > 0$, a positive torque draws the axes together, but a positive $\delta\theta_{ij}$ separates them. From Eq. 6.2.16, the relative virtual rotation is just

$$\delta\theta_{ij} = \delta\phi_j - \delta\phi_i \tag{6.2.20}$$

Substituting this result into Eq. 6.2.19, the virtual work is

$$\delta W = -n(\delta\phi_j - \delta\phi_i) \tag{6.2.21}$$

By definition, the generalized forces acting on bodies i and j are simply the coefficients of their respective generalized coordinate variations; that is,

$$\begin{aligned} \mathbf{Q}_i &= \begin{bmatrix} 0 \\ 0 \\ n \end{bmatrix} \\ \mathbf{Q}_j &= -\begin{bmatrix} 0 \\ 0 \\ n \end{bmatrix} \end{aligned} \tag{6.2.22}$$

6.3 EQUATIONS OF MOTION OF CONSTRAINED PLANAR SYSTEMS

Consider now mechanical systems that are made up of a collection of rigid bodies in the plane, with kinematic constraints between bodies. The variational and differential equations of motion of Eqs. 6.1.18 and 6.1.19 are valid for each body in the system, provided that all forces that act on each body are accounted for, including constraint reaction forces. In this section, the variational formulation of the equations of dynamics is extended to include the effect of kinematic constraints between bodies. Lagrange multipliers are introduced to account for the effect of kinematic constraints.

6.3.1 Variational Equations of Motion for Planar Systems

The variational equations of motion for each body i in a planar multibody system, given by Eq. 6.2.5, may be summed to obtain the *system variational equation of motion,*

$$\sum_{i=1}^{nb} \delta \mathbf{q}_i^T [\mathbf{M}_i \ddot{\mathbf{q}}_i - \mathbf{Q}_i] = 0 \tag{6.3.1}$$

for arbitrary $\delta \mathbf{q}_i$, $i = 1, \ldots, nb$, provided all forces that act are included in $\mathbf{Q}$. To treat a multibody system, it is convenient to define a composite state variable vector, a composite mass matrix, and a composite vector of generalized forces as

$$\begin{aligned} \mathbf{q} &= [\mathbf{q}_1^T, \mathbf{q}_2^T, \ldots, \mathbf{q}_{nb}^T]^T \\ \mathbf{M} &= \text{diag}(\mathbf{M}_1, \mathbf{M}_2, \ldots, \mathbf{M}_{nb}) \\ \mathbf{Q} &= [\mathbf{Q}_1^T, \mathbf{Q}_2^T, \ldots, \mathbf{Q}_{nb}^T]^T \end{aligned} \tag{6.3.2}$$

Using this notation, the variational equation of Eq. 6.3.1 may be written in the more compact form:

$$\delta \mathbf{q}^T [\mathbf{M}\ddot{\mathbf{q}} - \mathbf{Q}] = 0$$

Equations 6.2.5 for each body or Eq. 6.3.1 for the entire system are difficult to apply, since the generalized forces must include both *constraint forces* and *applied forces* (defined here to be all forces acting on or between bodies in the system except forces of constraint). Thus, forces due to gravity and spring–damper–actuators are treated as applied forces. To make systematic distinction between applied and constraint forces, let $\mathbf{F}_i^A$ and n_i^A denote applied force and torque on body i and $\mathbf{F}_i^C$ and n_i^C denote constraint force and torque on body i. Similarly, $\mathbf{Q}_i^A$ and $\mathbf{Q}_i^C$ denote generalized applied and constraint forces on body i. Finally, consistent with the composite vector notation of Eq. 6.3.2, $\mathbf{Q}^A$ and $\mathbf{Q}^C$ denote composite vectors of generalized applied and constraint forces acting on the system. In each case,

$$\begin{aligned} \mathbf{F}_i &= \mathbf{F}_i^A + \mathbf{F}_i^C, \qquad n_i = n_i^A + n_i^C \\ \mathbf{Q}_i &= \mathbf{Q}_i^A + \mathbf{Q}_i^C \\ \mathbf{Q} &= \mathbf{Q}^A + \mathbf{Q}^C \end{aligned}$$

Using the preceding notation, Eq. 6.3.1 may be written as

$$\sum_{i=1}^{nb} \delta \mathbf{q}_i^T[\mathbf{M}_i \ddot{\mathbf{q}}_i - \mathbf{Q}_i^A] - \sum_{i=1}^{nb} \delta \mathbf{q}_i^T \mathbf{Q}_i^C = 0$$

or, using the notation of Eq. 6.3.2, Eq. 6.3.1 is

$$\delta \mathbf{q}^T[\mathbf{M}\ddot{\mathbf{q}} - \mathbf{Q}^A] - \delta \mathbf{q}^T \mathbf{Q}^C = 0$$

These equations must hold for arbitrary $\delta \mathbf{q}_i$ and $\delta \mathbf{q}$, respectively.

Note that, whereas $\delta \mathbf{q}_i^T \mathbf{Q}_i^C$ is the virtual work of constraint forces that act on only body *i*,

$$\sum_{i=1}^{nb} \delta \mathbf{q}_i^T \mathbf{q}_i^C = \delta \mathbf{q}^T \mathbf{Q}^C$$

is the total virtual work of constraint forces that act on all bodies of the system. By Newton's law of action and reaction, if there is no friction in kinematic joints, constraint forces act perpendicular to contacting surfaces and are equal in magnitude. Thus, if attention is restricted to *virtual displacements that are consistent with the constraints* that act on the system, then the virtual work of all constraint forces is zero; that is,

$$\sum_{i=1}^{nb} \delta \mathbf{q}_i^T \mathbf{Q}_i^C = \delta \mathbf{q}^T \mathbf{Q}^C = 0$$

It is important to note that this is true even though $\delta \mathbf{q}_i^T \mathbf{Q}_i^C \neq 0$ for an individual body, even for virtual displacements that are consistent with constraints. It is only after the variational equations for individual bodies of Eq. 6.2.5 are summed to obtain the system variational equation of Eq. 6.3.1 that the effect of constraint forces can be eliminated.

To summarize, the system variational equation of motion of Eq. 6.3.1 may be written in either of the following forms of *constrained variational equations of motion*:

$$\sum_{i=1}^{nb} \delta \mathbf{q}_i^T[\mathbf{M}_i \ddot{\mathbf{q}}_i - \mathbf{Q}_i^A] = 0$$
$$\delta \mathbf{q}^T[\mathbf{M}\ddot{\mathbf{q}} - \mathbf{Q}^A] = 0 \qquad \textbf{(6.3.3)}$$

for all virtual displacements $\delta \mathbf{q}$ that are *consistent with constraints* that act on the system.

The combined set of kinematic and driving constraints is written in the form

$$\mathbf{\Phi}(\mathbf{q}, t) = \mathbf{0} \qquad \textbf{(6.3.4)}$$

where the specific form of constraint equations is presented in Chapter 3. Since generalized coordinate variations (or virtual displacements) $\delta \mathbf{q}$ are considered to occur with time held fixed, the condition for a *kinematically admissible virtual displacement* $\delta \mathbf{q}$ is obtained by taking the differential of Eq. 6.3.4 with time held fixed; that is,

$$\mathbf{\Phi_q}\, \delta\mathbf{q} = \mathbf{0} \tag{6.3.5}$$

where the Jacobian is evaluated at a state $\mathbf{q}$ that satisfies Eq. 6.3.4. Thus, the *constrained variational equations of motion* are that Eq. 6.3.3 hold for all virtual displacements $\delta\mathbf{q}$ that satisfy Eq. 6.3.5.

Example 6.3.1: Consider the simple pendulum shown in Fig. 6.3.1. The generalized coordinate vector is $\mathbf{q} = [x_1, y_1, \phi_1]^T$. The virtual work of the gravitational force (applied force), from Eq. 6.2.1, is

$$\delta W = [\delta x_1, \delta y_1]^T \begin{bmatrix} 0 \\ -mg \end{bmatrix}$$

$$= [\delta x_1, \delta y_1, \delta\phi_1]^T \begin{bmatrix} 0 \\ -mg \\ 0 \end{bmatrix}$$

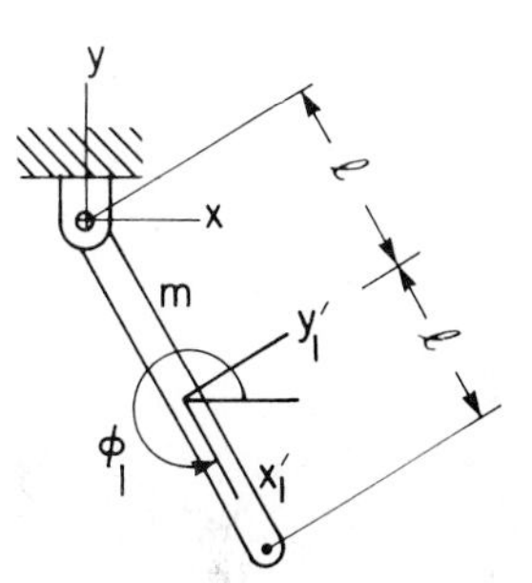

Figure 6.3.1 Simple pendulum.

where m is mass of the pendulum. Therefore, the generalized applied force $\mathbf{Q}^A$ is

$$\mathbf{Q}^A = [0, -mg, 0]^T$$

From Eq. 6.3.3,

$$\delta\mathbf{q}^T[\mathbf{M}\ddot{\mathbf{q}} - \mathbf{Q}^A] = [\delta x_1, \delta y_1, \delta\phi_1] \begin{bmatrix} m\ddot{x}_1 \\ m\ddot{y}_1 + mg \\ (m\ell^2/3)\ddot{\phi}_1 \end{bmatrix} = 0 \tag{6.3.6}$$

where the polar moment of inertia of the pendulum is $m\ell^2/3$. The constraint equation is

$$\mathbf{\Phi}(\mathbf{q}) = \begin{bmatrix} x_1 - \ell\cos\phi_1 \\ y_1 - \ell\sin\phi_1 \end{bmatrix} = \mathbf{0} \tag{6.3.7}$$

The condition for kinematically admissible virtual displacements of Eq. 6.3.5 is

$$\begin{bmatrix} 1 & 0 & \ell\sin\phi_1 \\ 0 & 1 & -\ell\cos\phi_1 \end{bmatrix} \begin{bmatrix} \delta x_1 \\ \delta y_1 \\ \delta\phi_1 \end{bmatrix} = \mathbf{0} \tag{6.3.8}$$

Hence, the variational equation of motion of the pendulum of Eq. 6.3.6 must hold for all virtual displacements $\delta\mathbf{q}$ that satisfy Eq. 6.3.8.

Example 6.3.2: Consider the two-body slider–crank model shown in Fig. 6.3.2. The virtual work of gravitational forces, from Eq. 6.2.1, is

$$\delta W = \delta\mathbf{r}_1^T\mathbf{F}_1 + \delta\mathbf{r}_2^T\mathbf{F}_2$$

$$= \begin{bmatrix} \delta x_1 \\ \delta y_1 \\ \delta\phi_1 \\ \delta x_2 \\ \delta y_2 \\ \delta\phi_2 \end{bmatrix}^T \begin{bmatrix} 0 \\ -m_1 g \\ 0 \\ 0 \\ -m_2 g \\ 0 \end{bmatrix}$$

Hence the generalized applied force vector is

$$\mathbf{Q}^A = [0, -m_1 g, 0, 0, -m_2 g, 0]^T$$

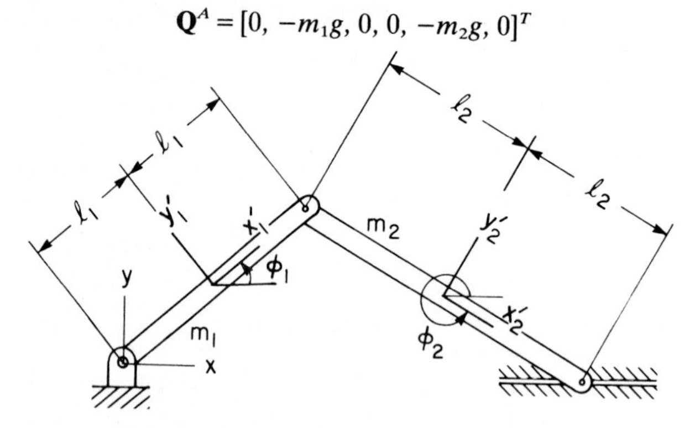

Figure 6.3.2 Two-body slider–crank.

From Eq. 6.3.3,

$$\delta\mathbf{q}^T[\mathbf{M}\ddot{\mathbf{q}} - \mathbf{Q}^A] = \begin{bmatrix} \delta x_1 \\ \delta y_1 \\ \delta\phi_1 \\ \delta x_2 \\ \delta y_2 \\ \delta\phi_2 \end{bmatrix}^T \begin{bmatrix} m_1\ddot{x}_1 \\ m_1\ddot{y}_1 + m_1 g \\ \left(\dfrac{m_1\ell_1^2}{3}\right)\ddot{\phi}_1 \\ m_2\ddot{x}_2 \\ m_2\ddot{y}_2 + m_2 g \\ \left(\dfrac{m_2\ell_2^2}{3}\right)\ddot{\phi}_2 \end{bmatrix} = 0 \qquad \textbf{(6.3.9)}$$

The constraint equation is

$$\mathbf{\Phi}(\mathbf{q}) = \begin{bmatrix} x_1 - \ell_1\cos\phi_1 \\ y_1 - \ell_1\sin\phi_1 \\ x_2 - 2\ell_1\cos\phi_1 - \ell_2\cos\phi_2 \\ y_2 - 2\ell_1\sin\phi_1 - \ell_2\sin\phi_2 \\ 2\ell_1\sin\phi_1 + 2\ell_2\sin\phi_2 \end{bmatrix} = \mathbf{0} \qquad \textbf{(6.3.10)}$$

The condition for kinematically admissible virtual displacements of Eq. 6.3.5 is

$$\begin{bmatrix} 1 & 0 & \ell_1 \sin\phi_1 & 0 & 0 & 0 \\ 0 & 1 & -\ell_1 \cos\phi_1 & 0 & 0 & 0 \\ 0 & 0 & 2\ell_1 \sin\phi_1 & 1 & 0 & \ell_2 \sin\phi_2 \\ 0 & 0 & -2\ell_1 \cos\phi_1 & 0 & 1 & -\ell_2 \cos\phi_2 \\ 0 & 0 & 2\ell_1 \cos\phi_1 & 0 & 0 & 2\ell_2 \cos\phi_2 \end{bmatrix} \begin{bmatrix} \delta x_1 \\ \delta y_1 \\ \delta\phi_1 \\ \delta x_2 \\ \delta y_2 \\ \delta\phi_2 \end{bmatrix} = \mathbf{0} \qquad \textbf{(6.3.11)}$$

The equation of motion of Eq. 6.3.9 thus holds for all virtual displacements $\delta\mathbf{q}$ that satisfy Eq. 6.3.11.

6.3.2 Lagrange Multipliers

A classical method in mechanics is to introduce Lagrange multipliers to reduce the variational equation of Eq. 6.3.3 to a mixed system of differential–algebraic equations. The conventional method of introducing Lagrange multipliers in this context may be found in References 7 and 9. A more mathematically precise introduction of Lagrange multipliers is employed here, using a theorem of optimization theory [28, 33].

Lagrange Multiplier Theorem: Let $\mathbf{b}$ be an n vector of constants, $\mathbf{x}$ be an n vector of variables, and $\mathbf{A}$ be an $m \times n$ constant matrix. If

$$\mathbf{b}^T\mathbf{x} = 0 \qquad \textbf{(6.3.12)}$$

holds for all $\mathbf{x}$ that satisfy

$$\mathbf{A}\mathbf{x} = \mathbf{0} \qquad \textbf{(6.3.13)}$$

then there exists an m vector $\boldsymbol{\lambda}$ of *Lagrange multipliers* such that

$$\mathbf{b}^T\mathbf{x} + \boldsymbol{\lambda}^T\mathbf{A}\mathbf{x} = 0 \qquad \textbf{(6.3.14)}$$

for arbitrary $\mathbf{x}$.

Example 6.3.3: Consider $\mathbf{b} = [1, 3, 2]^T$, $\mathbf{x} = [x_1, x_2, x_3]^T$, and $\mathbf{A} = \begin{bmatrix} -1 & 0 & 1 \\ 2 & 3 & 1 \end{bmatrix}$. Note that

$$\begin{aligned} \mathbf{b}^T\mathbf{x} &= [1, 3, 2][x_1, x_2, x_3]^T \\ &= [-1, 0, 1][x_1, x_2, x_3]^T + [2, 3, 1][x_1, x_2, x_3]^T \end{aligned}$$

So $\mathbf{b}^T\mathbf{x} = 0$ for all $\mathbf{x}$ such that

$$\mathbf{A}\mathbf{x} = \begin{bmatrix} -1 & 0 & 1 \\ 2 & 3 & 1 \end{bmatrix} \begin{bmatrix} x_1 \\ x_2 \\ x_3 \end{bmatrix} = \mathbf{0}$$

Then, by the Lagrange multiplier theorem, there exists $\boldsymbol{\lambda} = [\lambda_1, \lambda_2]^T$ that satisfies

$$[1, 3, 2]\begin{bmatrix} x_1 \\ x_2 \\ x_3 \end{bmatrix} + [\lambda_1, \lambda_2]\begin{bmatrix} -1 & 0 & 1 \\ 2 & 3 & 1 \end{bmatrix}\begin{bmatrix} x_1 \\ x_2 \\ x_3 \end{bmatrix}$$

$$= x_1 + 3x_2 + 2x_3 + \lambda_1(-x_1 + x_3) + \lambda_2(2x_1 + 3x_2 + x_3)$$

$$= (1 - \lambda_1 + 2\lambda_2)x_1 + (3 + 3\lambda_2)x_2 + (2 + \lambda_1 + \lambda_2)x_3 = 0$$

Since this equation holds for all $\mathbf{x}$, it is equivalent to a system of three equations; that is,

$$1 - \lambda_1 + 2\lambda_2 = 0$$

$$3 + 3\lambda_2 = 0$$

$$2 + \lambda_1 + \lambda_2 = 0$$

The solution is

$$\boldsymbol{\lambda} = [-1, -1]^T$$

It is important to note that the Lagrange multiplier theorem does *not* say that $\mathbf{b}^T\mathbf{x} = 0$ for all $\mathbf{x}$ such that $\mathbf{Ax} = \mathbf{0}$ for *any* $\mathbf{b}$ and $\mathbf{A}$. In fact, there must be a close relation between $\mathbf{A}$ and $\mathbf{b}$ in order for this to be true (Prob. 6.3.2).

Equation 6.3.3 corresponds to Eq. 6.3.12, and the kinematic admissibility condition of Eq. 6.3.5 corresponds to Eq. 6.3.13. Since Eq. 6.3.3 must hold for all $\delta\mathbf{q}$ that satisfy Eq. 6.3.5, the Lagrange multiplier theorem guarantees the existence of a Lagrange multiplier vector $\boldsymbol{\lambda}$ such that

$$[\mathbf{M}\ddot{\mathbf{q}} - \mathbf{Q}^A]^T\,\delta\mathbf{q} + \boldsymbol{\lambda}^T\boldsymbol{\Phi}_{\mathbf{q}}\,\delta\mathbf{q} = [\mathbf{M}\ddot{\mathbf{q}} + \boldsymbol{\Phi}_{\mathbf{q}}^T\boldsymbol{\lambda} - \mathbf{Q}^A]^T\,\delta\mathbf{q} = 0 \qquad \textbf{(6.3.15)}$$

for arbitrary $\delta\mathbf{q}$. Therefore, the coefficient of $\delta\mathbf{q}$ in Eq. 6.3.15 must be $\mathbf{0}$, yielding the *Lagrange multiplier form of the equations of motion*:

$$\mathbf{M}\ddot{\mathbf{q}} + \boldsymbol{\Phi}_{\mathbf{q}}^T\boldsymbol{\lambda} = \mathbf{Q}^A \qquad \textbf{(6.3.16)}$$

In addition to these equations of motion, recall the velocity and acceleration equations of Eqs. 3.6.10 and 3.6.11:

$$\begin{aligned} \boldsymbol{\Phi}_{\mathbf{q}}\dot{\mathbf{q}} &= -\boldsymbol{\Phi}_t \equiv \boldsymbol{\nu} \\ \boldsymbol{\Phi}_{\mathbf{q}}\ddot{\mathbf{q}} &= -(\boldsymbol{\Phi}_{\mathbf{q}}\dot{\mathbf{q}})_{\mathbf{q}}\dot{\mathbf{q}} - 2\boldsymbol{\Phi}_{\mathbf{q}t}\dot{\mathbf{q}} - \boldsymbol{\Phi}_{tt} \equiv \boldsymbol{\gamma} \end{aligned} \qquad \textbf{(6.3.17)}$$

These equations, together with Eq. 6.3.16, comprise the complete set of *constrained equations of motion* for the system.

6.3.3 Mixed Differential–Algebraic Equations of Motion

Equation 6.3.16 and the second of Eqs. 6.3.17 may be written in matrix form as

$$\begin{bmatrix} \mathbf{M} & \boldsymbol{\Phi}_{\mathbf{q}}^T \\ \boldsymbol{\Phi}_{\mathbf{q}} & \mathbf{0} \end{bmatrix}\begin{bmatrix} \ddot{\mathbf{q}} \\ \boldsymbol{\lambda} \end{bmatrix} = \begin{bmatrix} \mathbf{Q}^A \\ \boldsymbol{\gamma} \end{bmatrix} \qquad \textbf{(6.3.18)}$$

This is a *mixed system of differential–algebraic equations* (DAE), since no derivatives of the Lagrange multiplier $\boldsymbol{\lambda}$ appear. Furthermore, Eq. 6.3.4 and the first of Eqs. 6.3.17 must be satisfied.

Example 6.3.4: Consider the simple pendulum of Example 6.3.1. By successive differentiation of Eq. 6.3.7 with respect to time, the velocity and acceleration equations are

$$\boldsymbol{\Phi}_{\mathbf{q}}\dot{\mathbf{q}} = \begin{bmatrix} 1 & 0 & \ell \sin\phi_1 \\ 0 & 1 & -\ell\cos\phi_1 \end{bmatrix} \begin{bmatrix} \dot{x}_1 \\ \dot{y}_1 \\ \dot{\phi}_1 \end{bmatrix} = \mathbf{0} \equiv \mathbf{v}$$

$$\boldsymbol{\Phi}_{\mathbf{q}}\ddot{\mathbf{q}} = -\begin{bmatrix} \ell\cos\phi_1\dot{\phi}_1^2 \\ \ell\sin\phi_1\dot{\phi}_1^2 \end{bmatrix} \equiv \boldsymbol{\gamma} \tag{6.3.19}$$

From Eqs. 6.3.6, 6.3.8, and 6.3.19, the equations of motion are

$$\begin{bmatrix} m & 0 & 0 & 1 & 0 \\ 0 & m & 0 & 0 & 1 \\ 0 & 0 & m\ell^2/3 & \ell\sin\phi_1 & -\ell\cos\phi_1 \\ 1 & 0 & \ell\sin\phi_1 & 0 & 0 \\ 0 & 1 & -\ell\cos\phi & 0 & 0 \end{bmatrix} \begin{bmatrix} \ddot{x}_1 \\ \ddot{y}_1 \\ \ddot{\phi}_1 \\ \lambda_1 \\ \lambda_2 \end{bmatrix} = \begin{bmatrix} 0 \\ -mg \\ 0 \\ -\ell\cos\phi_1\dot{\phi}_1^2 \\ -\ell\sin\phi_1\dot{\phi}_1^2 \end{bmatrix} \tag{6.3.20}$$

Example 6.3.5: In Example 6.3.2, the variational equation of motion and kinematic constraint equations of the two-body slider–crank are written as Eqs. 6.3.9 and 6.3.10. The acceleration equation is

$$\boldsymbol{\Phi}_{\mathbf{q}}\ddot{\mathbf{q}} = -\begin{bmatrix} \ell_1\cos\phi_1\dot{\phi}_1^2 \\ \ell_1\sin\phi_1\dot{\phi}_1^2 \\ 2\ell_1\cos\phi_1\dot{\phi}_1^2 + \ell_2\cos\phi_2\dot{\phi}_2^2 \\ 2\ell_1\sin\phi_1\dot{\phi}_1^2 + \ell_2\sin\phi_2\dot{\phi}_2^2 \\ -2\ell_1\sin\phi_1\dot{\phi}_1^2 - 2\ell_2\sin\phi_2\dot{\phi}_2^2 \end{bmatrix} \equiv \boldsymbol{\gamma}$$

Applying the Lagrange multiplier theorem, the equations of motion are as given on page 226.

It is important to know when Eq. 6.3.18 has a unique solution for accelerations and Lagrange multipliers. A practical condition for the coefficient matrix in Eq. 6.3.18 to be nonsingular is the following theorem:

Constrained Dynamic Existence Theorem: Let $\boldsymbol{\Phi}_{\mathbf{q}}$ have full row rank and let the kinetic energy of the system be positive for any nonzero virtual velocity that is consistent with the constraints; that is,

$$\delta\dot{\mathbf{q}}^T\mathbf{M}\,\delta\dot{\mathbf{q}} > 0 \tag{6.3.21}$$

$$\begin{bmatrix} m_1 & 0 & 0 & 0 & 0 & 0 & 1 & 0 & 0 & 0 & 0 \\ 0 & m_1 & 0 & 0 & 0 & 0 & 0 & 1 & 0 & 0 & 0 \\ 0 & 0 & m_1\ell_1^2/3 & 0 & 0 & 0 & \ell_1 \sin\phi_1 & -\ell_1\cos\phi_1 & 2\ell_1\sin\phi_1 & -2\ell_1\cos\phi_1 & 2\ell_1\cos\phi_1 \\ 0 & 0 & 0 & m_2 & 0 & 0 & 0 & 0 & 1 & 0 & 0 \\ 0 & 0 & 0 & 0 & m_2 & 0 & 0 & 0 & 0 & 1 & 0 \\ 0 & 0 & 0 & 0 & 0 & m_2\ell_2^2/3 & 0 & 0 & \ell_2\sin\phi_2 & -2\ell_2\cos\phi_2 & 2\ell_2\cos\phi_2 \\ 1 & 0 & \ell_1\sin\phi_1 & 0 & 0 & 0 & 0 & 0 & 0 & 0 & 0 \\ 0 & 1 & -\ell_1\cos\phi_1 & 0 & 0 & 0 & 0 & 0 & 0 & 0 & 0 \\ 0 & 0 & 2\ell_1\sin\phi_1 & 1 & 0 & \ell_2\sin\phi_2 & 0 & 0 & 0 & 0 & 0 \\ 0 & 0 & -2\ell_1\cos\phi_1 & 0 & 1 & -\ell_2\cos\phi_2 & 0 & 0 & 0 & 0 & 0 \\ 0 & 0 & 2\ell_1\cos\phi_1 & 0 & 0 & 2\ell_2\cos\phi_2 & 0 & 0 & 0 & 0 & 0 \end{bmatrix} \begin{bmatrix} \ddot{x}_1 \\ \ddot{y}_1 \\ \ddot{\phi}_1 \\ \ddot{x}_2 \\ \ddot{y}_2 \\ \ddot{\phi}_2 \\ \lambda_1 \\ \lambda_2 \\ \lambda_3 \\ \lambda_4 \\ \lambda_5 \end{bmatrix} = \begin{bmatrix} 0 \\ -m_1 g \\ 0 \\ 0 \\ -m_2 g \\ 0 \\ -\ell_1\cos\phi_1\dot{\phi}_1^2 \\ -\ell_1\sin\phi_1\dot{\phi}_1^2 \\ -2\ell_1\cos\phi_1\dot{\phi}_1^2 - \ell_2\cos\phi_2\dot{\phi}_2^2 \\ -2\ell_1\sin\phi_1\dot{\phi}_1^2 - \ell_2\sin\phi_2\dot{\phi}_2^2 \\ 2\ell_1\sin\phi_1\dot{\phi}_1^2 + 2\ell_2\sin\phi_2\dot{\phi}_2^2 \end{bmatrix}$$

for all $\delta\dot{\mathbf{q}} \neq \mathbf{0}$ that satisfy

$$\mathbf{\Phi_q}\,\delta\dot{\mathbf{q}} = \mathbf{0} \tag{6.3.22}$$

Then the coefficient matrix in Eq. 6.3.18 is nonsingular and $\ddot{\mathbf{q}}$ and $\boldsymbol{\lambda}$ are uniquely determined.

If the mass matrix is positive definite [22], which is normally the case, and if the constraints are independent, then the preceding conditions are satisfied and Eq. 6.3.18 has a unique solution. To prove this result, use the fact that the coefficient matrix in Eq. 6.3.18 is nonsingular if [22] the only solution of the homogeneous equation

$$\begin{bmatrix} \mathbf{M} & \mathbf{\Phi_q^T} \\ \mathbf{\Phi_q} & \mathbf{0} \end{bmatrix} \begin{bmatrix} \mathbf{y} \\ \mathbf{z} \end{bmatrix} = \mathbf{0}$$

is $\mathbf{y} = \mathbf{0}$ and $\mathbf{z} = \mathbf{0}$. To show this, the preceding matrix equation can be written as two separate equations:

$$\mathbf{My} + \mathbf{\Phi_q^T z} = \mathbf{0} \tag{6.3.23}$$

$$\mathbf{\Phi_q y} = \mathbf{0} \tag{6.3.24}$$

Multiplying Eq. 6.3.23 on the left by $\mathbf{y}^T$,

$$\mathbf{y}^T\mathbf{My} + \mathbf{y}^T\mathbf{\Phi_q^T z} = \mathbf{y}^T\mathbf{My} = 0$$

where the fact that $\mathbf{y}$ satisfies Eq. 6.3.24 has been used. Therefore, $\mathbf{y} = \mathbf{0}$. Thus, Eq. 6.3.23 reduces to

$$\mathbf{\Phi_q^T z} = \mathbf{0}$$

But the left side of this equation is just a linear combination of columns of the matrix $\mathbf{\Phi_q^T}$, with the components of $\mathbf{z}$ as coefficients in the expansion. Since $\mathbf{\Phi_q}$ has full row rank, its transpose is made up of linearly independent columns. Hence, the preceding equation requires that $\mathbf{z} = \mathbf{0}$. This completes the proof.

6.3.4 Initial Conditions

A set of additional conditions that, together with the kinematic constraints, determines initial position and velocity must be defined to initiate motion of the system. Recall from Chapter 3 that the system constraint equations involve both kinematic and driving constraints of the form

$$\mathbf{\Phi}(\mathbf{q}, t) = \begin{bmatrix} \mathbf{\Phi}^K(\mathbf{q}) \\ \mathbf{\Phi}^D(\mathbf{q}, t) \end{bmatrix} = \mathbf{0} \tag{6.3.25}$$

which must hold for all time.

A system of initial conditions on position and orientation may be used to define the position of the system at the initial time t_0. *Initial position conditions* are defined in the form

$$\mathbf{\Phi}^I(\mathbf{q}(t_0), t_0) = \mathbf{0} \tag{6.3.26}$$

where it is required that the number of equations in Eq. 6.3.26 be equal to the number of system degrees of freedom. Furthermore, Eq. 6.3.25 at t_0 and Eq. 6.3.26 must be independent to uniquely determine the *initial position* $\mathbf{q}(t_0)$.

Initial velocity conditions may be given in the form

$$\mathbf{B}^I\dot{\mathbf{q}}(t_0) = \mathbf{v}^I \tag{6.3.27}$$

which must contain the same number of equations as the number of system degrees of freedom. Furthermore, it is required that the first of Eqs. 6.3.17 at t_0 and Eq. 6.3.27 uniquely determine the *initial velocity* $\dot{\mathbf{q}}(t_0)$. Thus, a complete set of initial conditions that are consistent with constraints is obtained.

Example 6.3.6: Consider again the simple pendulum of Example 6.3.1. Let the initial orientation of the pendulum be $\phi_1 = 3\pi/2$. Hence, Eq. 6.3.26 is written, with $t_0 = 0$, as

$$\Phi^I(\mathbf{q}(0), 0) = \phi_1(0) - \frac{3\pi}{2} = 0 \tag{6.3.28}$$

Equation 6.3.7 at $t = 0$ and Eq. 6.3.28 determine the initial position $\mathbf{q}(0)$, since they are independent. Let the initial angular velocity of the simple pendulum be

$$\mathbf{B}^I\dot{\mathbf{q}}(0) = \dot{\phi}_1(0) - 2\pi = 0 \tag{6.3.29}$$

The velocity equation can be obtained by differentiating Eq. 6.3.7 with respect to time, yielding

$$\mathbf{\Phi}_{\mathbf{q}}\dot{\mathbf{q}}(0) = \begin{bmatrix} 1 & 0 & \ell \sin \phi_1(0) \\ 0 & 1 & -\ell \cos \phi_1(0) \end{bmatrix} \begin{bmatrix} \dot{x}_1(0) \\ \dot{y}_1(0) \\ \dot{\phi}_1(0) \end{bmatrix} = \mathbf{0} \tag{6.3.30}$$

Equations 6.3.29 and 6.3.30 uniquely determine the initial velocity $\dot{\mathbf{q}}(0)$.

While it is not necessary that initial position and velocity conditions be given on the same variables, as is done in Example 6.3.6, this is often the most natural situation in applications. The essential considerations are (1) that the number of initial conditions on both position and velocity be equal to the number of degrees of freedom of the system after the constraints of Eq. 6.3.25 are specified and (2) that Eqs. 6.3.25 and 6.3.26 uniquely determine $\mathbf{q}(t_0)$ and the velocity equation associated with Eqs. 6.3.25 and 6.3.27 uniquely determine $\dot{\mathbf{q}}(t_0)$. These conditions will be met if the following matrices are nonsingular at t_0:

$$\begin{bmatrix} \mathbf{\Phi}_{\mathbf{q}} \\ \mathbf{\Phi}^I_{\mathbf{q}} \end{bmatrix}, \quad \begin{bmatrix} \mathbf{\Phi}_{\mathbf{q}} \\ \mathbf{B}^I \end{bmatrix}$$

These considerations are identical to the selection of a complete set of drivers in kinematic analysis (see Section 3.5).

6.4 INVERSE DYNAMICS OF KINEMATICALLY DRIVEN SYSTEMS

The equations of motion of a constrained dynamic system derived in Section 6.3, specifically Eqs. 6.3.18, are valid regardless of the number of degrees of freedom of the system. If driving constraints equal in number to the number of kinematic degrees of freedom of the system are appended to the kinematic constraints, then the system is kinematically determined and the position, velocity, and acceleration equations of Chapter 3 completely determine the motion of the system. In this special case, the constraint Jacobian of Eq. 6.3.4 is square and nonsingular; that is,

$$|\mathbf{\Phi_q}(\mathbf{q}, t)| \neq 0 \tag{6.4.1}$$

This very special situation yields simplified results that are applicable for the analysis of kinematically driven systems.

Expanding the equations of motion of Eq. 6.3.18,

$$\mathbf{M\ddot{q}} + \mathbf{\Phi_q^T}\boldsymbol{\lambda} = \mathbf{Q}^A \tag{6.4.2}$$

$$\mathbf{\Phi_q}\mathbf{\ddot{q}} = \boldsymbol{\gamma} \tag{6.4.3}$$

Since the Jacobian is nonsingular, Eq. 6.4.3 may be solved to obtain

$$\mathbf{\ddot{q}} = \mathbf{\Phi_q^{-1}}\boldsymbol{\gamma} \tag{6.4.4}$$

In reality, the algebraic equations of Eq. 6.4.3 are solved numerically for the acceleration $\mathbf{\ddot{q}}$, but the notation of Eq. 6.4.4 emphasizes this direct relationship.

Once the acceleration vector is known, Eq. 6.4.2 may be solved for $\boldsymbol{\lambda}$ as

$$\boldsymbol{\lambda} = \mathbf{\Phi_q^{-1T}}[\mathbf{Q}^A - \mathbf{M\ddot{q}}] \tag{6.4.5}$$

since the transpose of a nonsingular matrix is nonsingular. In reality, Eq. 6.4.2 is solved numerically for $\boldsymbol{\lambda}$ after $\mathbf{\ddot{q}}$ is found from Eq. 6.4.3.

As is shown in Section 6.6, the Lagrange multiplier $\boldsymbol{\lambda}$ uniquely determines the constraint forces and torques that act in the system. This fact motivates the use of the terminology *inverse dynamics* for systems that are kinematically determined. For such systems, the engineer may specify desired driving constraint relations and solve Eq. 6.4.3 for $\mathbf{\ddot{q}}$. Equation 6.4.2 is then solved for the Lagrange multipliers. The reaction force relations of Section 6.6 are then used to determine forces that would be required to impose the driving constraints that have been specified. Such analysis is valuable in assessing the availability of actuators that can provide forces and torques that are required to achieve the desired motion.

Example 6.4.1: Consider the simple pendulum of Examples 6.3.1, 6.3.4, and 6.3.6 with the kinematic driver $\Phi^D = \phi_1 - 3\pi/2 - 2\pi t_0$. From Eq. 6.3.20, Eqs. 6.4.2 and 6.4.3 are

$$\begin{bmatrix} m & 0 & 0 \\ 0 & m & 0 \\ 0 & 0 & \frac{m\ell^2}{3} \end{bmatrix} \begin{bmatrix} \ddot{x}_1 \\ \ddot{y}_1 \\ \ddot{\phi}_1 \end{bmatrix} + \begin{bmatrix} 1 & 0 & 0 \\ 0 & 1 & 0 \\ \ell \sin \phi_1 & -\ell \cos \phi_1 & 1 \end{bmatrix} \begin{bmatrix} \lambda_1 \\ \lambda_2 \\ \lambda_3 \end{bmatrix} = \begin{bmatrix} 0 \\ -mg \\ 0 \end{bmatrix} \tag{6.4.6}$$

$$\begin{bmatrix} 1 & 0 & \ell \sin \phi_1 \\ 0 & 1 & -\ell \cos \phi_1 \\ 0 & 0 & 1 \end{bmatrix} \begin{bmatrix} \ddot{x}_1 \\ \ddot{y}_1 \\ \ddot{\phi}_1 \end{bmatrix} = -\begin{bmatrix} \ell \cos \phi_1 \dot{\phi}_1^2 \\ \ell \sin \phi_1 \dot{\phi}_1^2 \\ 0 \end{bmatrix} \tag{6.4.7}$$

The solution of Eq. 6.4.7 is

$$\begin{bmatrix} \ddot{x}_1 \\ \ddot{y}_1 \\ \ddot{\phi}_1 \end{bmatrix} = \begin{bmatrix} -\ell \cos \phi_1 \dot{\phi}_1^2 \\ -\ell \sin \phi_1 \dot{\phi}_1^2 \\ 0 \end{bmatrix} \tag{6.4.8}$$

Substituting Eq. 6.4.8 into Eq. 6.4.6, the Lagrange multipliers are determined as

$$\begin{bmatrix} \lambda_1 \\ \lambda_2 \\ \lambda_3 \end{bmatrix} = \begin{bmatrix} m\ell \cos \phi_1 \dot{\phi}_1^2 \\ -mg + m\ell \sin \phi_1 \dot{\phi}_1^2 \\ -mg\ell \cos \phi_1 \end{bmatrix}$$

Even though the reaction force can be calculated using the general method presented in Section 6.6, notice that λ_1 is the x component of the centripetal force on the pendulum, λ_2 is the y component of the centripetal force minus the weight of the pendulum, and λ_3 is the moment due to the weight.

6.5 EQUILIBRIUM CONDITIONS

A system is said to be in *equilibrium* if it remains stationary under the action of applied forces; that is, if

$$\ddot{\mathbf{q}} = \dot{\mathbf{q}} = \mathbf{0} \tag{6.5.1}$$

If there is adequate damping in a system, the system equations of motion may be formulated and integrated until all motion damps out and an equilibrium state is obtained.

An alternative approach is to substitute the equilibrium definition of Eq. 6.5.1 into the equations of motion, in this case Eq. 6.4.2, to obtain

$$\mathbf{\Phi}_{\mathbf{q}}^T \boldsymbol{\lambda} = \mathbf{Q}^A \tag{6.5.2}$$

These equations constitute the system *equilibrium equations.* The state $\mathbf{q}$ and Lagrange multiplier $\boldsymbol{\lambda}$ are determined by Eq. 6.5.2 and the constraint equations of

Eq. 6.3.4. While this approach is computationally feasible, it suffers from the requirement that a good estimate of the equilibrium position is needed as a starting point for iterative computation. Furthermore, the equilibrium equations of Eq. 6.5.2 are valid for both stable and unstable states of equilibrium. Thus, an algorithm that is based on solving the equilibrium equations alone can converge to either a stable or an unstable state of equilibrium, depending on which is nearest to the initial estimate.

To avoid the difficulty associated with unstable equilibrium configurations, for *conservative mechanical systems* [7, 9], the *principle of minimum total potential energy* [34] may be employed. It states that a system is in a state of stable equilibrium if the total potential energy takes on a strict relative minimum at that position.

The total potential energy of a system is defined as

$$TPE = SE - W(F) \tag{6.5.3}$$

where SE is the strain energy of compliant components and $-W(F)$ is the potential energy of all forces that act on the system. In the case of linear translational and rotational springs, the strain energy is

$$SE = \tfrac{1}{2}k(\ell - \ell_0)^2 \quad \text{and} \quad \tfrac{1}{2}k_\theta(\theta - \theta_0)^2 \tag{6.5.4}$$

respectively. For a constant force $\mathbf{F}^P$ that acts at a point P and a torque n on the body, the potential energy is

$$-W(F) = -(x^P F_x^P + y^P F_y^P + n\phi) \tag{6.5.5}$$

For conservative systems, the condition that defines a state $\mathbf{q}_0$ of stable equilibrium is that the total potential energy be minimized; that is,

$$TPE(\mathbf{q}_0) \leqslant TPE(\mathbf{q}) \tag{6.5.6}$$

for all states $\mathbf{q}$ near the state of stable equilibrium; that is, for some small positive ε, for all $\mathbf{q}$ such that

$$(\mathbf{q} - \mathbf{q}_0)^T(\mathbf{q} - \mathbf{q}_0) \leqslant \varepsilon \tag{6.5.7}$$

A valuable fact that is exploited in numerical minimization techniques that determine $\mathbf{q}_0$ to satisfy Eq. 6.5.6, consistent with the constraints, is that the generalized applied force $\mathbf{Q}^A$ is the negative of the gradient of the total potential energy [7, 9]; that is

$$\frac{\partial TPE}{\partial \mathbf{q}} = -\mathbf{Q}^{A^T} \tag{6.5.8}$$

Thus, if it is known that forces acting on a system are conservative, the generalized forces provide the gradient of the total potential energy, which is needed by numerical minimization algorithms. A numerical method of implementing this idea is presented in Chapter 7.

Example 6.5.1: Consider a simple pendulum with a torsional spring attached, as shown in Fig. 6.5.1. Let m and 2ℓ be the mass and length of the pendulum, respectively. The total potential energy is

$$TPE = \tfrac{1}{2}k_\theta(\phi_1 - \phi_0)^2 + mgy_1 \tag{6.5.9}$$

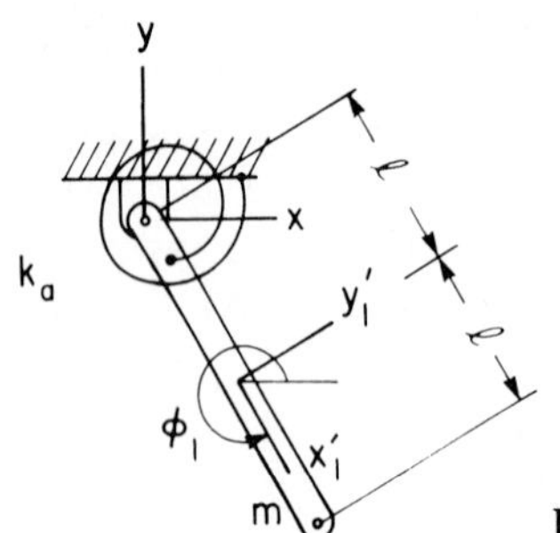

Figure 6.5.1 Simple pendulum with a torsional spring.

where k_θ is the torsional spring constant and ϕ_0 is the angle at which the spring is not deflected. The constraint equation is

$$\mathbf{\Phi} = \begin{bmatrix} x_1 - \ell \cos \phi_1 \\ y_1 - \ell \sin \phi_1 \end{bmatrix} = \mathbf{0} \tag{6.5.10}$$

Substituting the second equation of Eq. 6.5.10 into Eq. 6.5.9,

$$TPE = \tfrac{1}{2}k_\theta(\phi_1 - \phi_0)^2 + mg\ell \sin \phi_1$$

By the minimum potential energy theorem, the pendulum is in equilibrium when

$$\frac{\partial TPE}{\partial \phi_1} = k_\theta(\phi_1 - \phi_0) + mg\ell \cos \phi_1 = 0 \tag{6.5.11}$$

This is a nonlinear equation, which cannot be easily solved analytically. Equation 6.5.11 can be written as

$$-k_\theta(\phi_1 - \phi_0) = mg\ell \cos \phi_1$$

which shows that the pendulum is in equilibrium when the spring torque is equal to the moment due to gravity. With the generalized coordinate $\mathbf{q} = [x_1, y_1, \phi_1]^T$, the gradient of total potential energy with respect to $\mathbf{q}$ of Eq. 6.5.8 gives the generalized force:

$$\frac{\partial TPE}{\partial \mathbf{q}} = [0, mg, k_\theta(\phi_1 - \phi_0)] = -\mathbf{Q}^T$$

Example 6.5.2: Consider the double pendulum shown in Fig. 6.5.2. At the tip of the second body, a constant force F is applied horizontally. Let m_1 and m_2 be masses of the first and second bars, respectively, and the lengths be as shown. The total potential energy of Eq. 6.5.3 may be written in terms of ϕ_1 and ϕ_2,

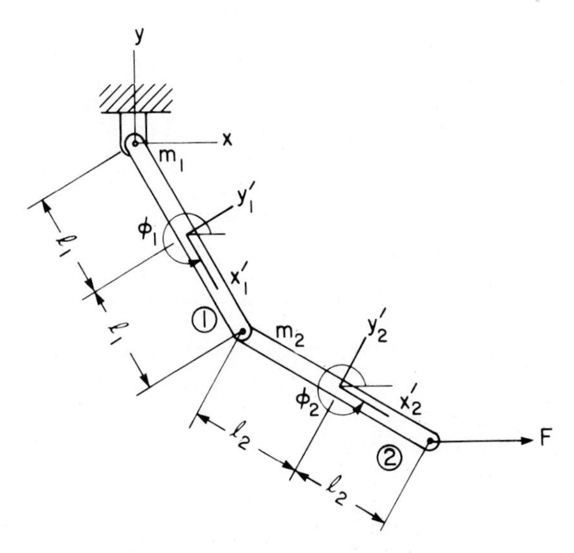

Figure 6.5.2 Double pendulum.

which can be regarded as independent, as

$$TPE = m_1 g \ell_1 \sin \phi_1 + m_2 g(2\ell_1 \sin \phi_1 + \ell_2 \sin \phi_2) - F(2\ell_1 \cos \phi_1 + 2\ell_2 \cos \phi_2)$$

By the principle of minimum total potential energy, the system is in equilibrium when

$$\frac{\partial TPE}{\partial \phi_1} = \frac{\partial TPE}{\partial \phi_2} = 0$$

$$\frac{\partial TPE}{\partial \phi_1} = m_1 g \ell_1 \cos \phi_1 + 2m_2 g \ell_1 \cos \phi_1 + 2F\ell_1 \sin \phi_1 = 0$$

$$\frac{\partial TPE}{\partial \phi_2} = m_2 g \ell_2 \cos \phi_2 + 2F\ell_2 \sin \phi_2 = 0$$

Solving these equations,

$$\phi_1 = \text{Arctan}\left[-\frac{(m_1 + 2m_2)g}{2F} \right]$$

$$\phi_2 = \text{Arctan}\left[-\frac{m_2 g}{2F} \right]$$

6.6 CONSTRAINT REACTION FORCES

Consider a kinematic or driving constraint between bodies i and j of a multibody system, numbered as constraint k. Denote by $\mathbf{\Phi}^k = \mathbf{0}$ its kinematic constraint

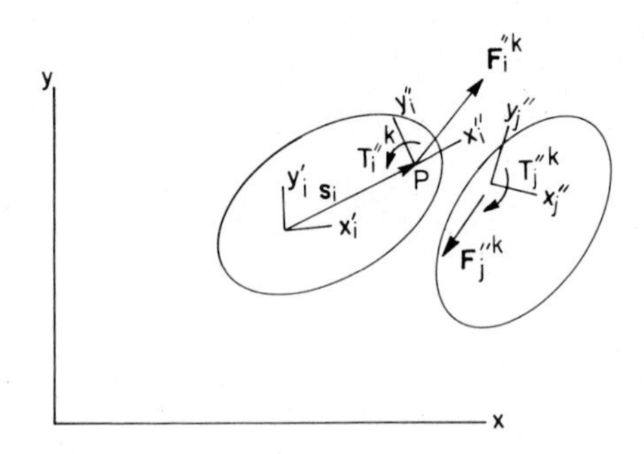

Figure 6.6.1 Constraint reaction force.

equation. The variational equation of motion associated with body i may be written, from Eq. 6.3.15, by setting $\delta\mathbf{q}_j = \mathbf{0}$ for $j \neq i$ to obtain

$$\delta\mathbf{q}_i^T\mathbf{M}_i\ddot{\mathbf{q}}_i + \sum_{\ell\neq k} \delta\mathbf{q}_i^T\mathbf{\Phi}_{\mathbf{q}_i}^{\ell T}\boldsymbol{\lambda}^\ell + \delta\mathbf{q}_i^T\mathbf{\Phi}_{\mathbf{q}_i}^{kT}\boldsymbol{\lambda}^k = \delta\mathbf{q}_i^T\mathbf{Q}_i^A \quad \textbf{(6.6.1)}$$

where $\boldsymbol{\lambda}^\ell$ is the subvector of $\boldsymbol{\lambda}$ corresponding to constraint ℓ and constraint k that acts between bodies i and j has been singled out for special attention. As a consequence of Eq. 6.3.15, Eq. 6.6.1 holds for arbitrary $\delta\mathbf{q}_i$.

If constraint k were broken and replaced by reaction forces in the joint that occur during dynamics in the system, as shown schematically in Fig. 6.6.1, then the resulting motion will be identical to that predicted by Eq. 6.6.1. If the equations of motion with reaction forces due to constraint k included were formulated and constraint k were deleted, the last term on the left of Eq. 6.6.1 would disappear and be replaced by the virtual work of the constraint reaction force and torque; that is,

$$-(\delta\mathbf{r}_i^T\mathbf{\Phi}_{\mathbf{r}_i}^{kT}\boldsymbol{\lambda}^k + \delta\phi_i\mathbf{\Phi}_{\phi_i}^{kT}\boldsymbol{\lambda}^k) = \delta\mathbf{r}_i^{\prime\prime PT}\mathbf{F}_i^{\prime\prime k} + \delta\phi_i T_i^{\prime\prime k} \quad \textbf{(6.6.2)}$$

where the expression on the left of Eq. 6.6.1 has been expanded in terms of physical generalized coordinates. Equation 6.6.2 must hold for arbitrary $\delta\mathbf{r}_i$ and $\delta\phi_i$. For a more detailed derivation of Eq. 6.6.2, which shows that the multipliers associated with each joint are unaffected by replacing a joint by its reaction forces, see Reference 35.

The objective is to use Eq. 6.6.2 to develop relations for joint reaction forces $\mathbf{F}_i^{\prime\prime k}$ and torques $T_i^{\prime\prime k}$ in terms of the Lagrange multipliers. First, recall that

$$\mathbf{r}_i^P = \mathbf{r}_i + \mathbf{A}_i\mathbf{s}_i^{\prime P} \quad \textbf{(6.6.3)}$$

whose differential yields

$$\delta\mathbf{r}_i^P = \delta\mathbf{r}_i + \mathbf{B}_i\mathbf{s}_i^{\prime P}\,\delta\phi_i \quad \textbf{(6.6.4)}$$

Since virtual displacement is a vector quantity and the constant transformation matrix $\mathbf{C}_i$ transforms vectors in the x_i''-y_i'' frame to the x_i'-y_i' frame,

$$\delta \mathbf{r}_i^P = \mathbf{A}_i \mathbf{C}_i \, \delta \mathbf{r}_i''^P \tag{6.6.5}$$

Substituting from Eq. 6.6.5 into Eq. 6.6.4 yields

$$\delta \mathbf{r}_i = \mathbf{A}_i \mathbf{C}_i \, \delta \mathbf{r}_i''^P - \mathbf{B}_i \mathbf{s}_i'^P \, \delta \phi_i \tag{6.6.6}$$

Substituting this result into Eq. 6.6.2,

$$-\delta \mathbf{r}_i''^{PT} \mathbf{C}_i^T \mathbf{A}_i^T \mathbf{\Phi}_{\mathbf{r}_i}^{kT} \boldsymbol{\lambda}^k - \delta \phi_i (\mathbf{\Phi}_{\phi_i}^{kT} - \mathbf{s}_i'^{PT} \mathbf{B}_i^T \mathbf{\Phi}_{\mathbf{r}_i}^{kT}) \boldsymbol{\lambda}^k = \delta \mathbf{r}_i''^{PT} \mathbf{F}_i''^k + \delta \phi_i T_i''^k \tag{6.6.7}$$

Since virtual displacements $\delta \mathbf{r}_i''^P$ and rotations $\delta \phi_i$ are arbitrary, their coefficients on both sides of Eq. 6.6.7 must be equal, yielding the desired results:

$$\mathbf{F}_i''^k = -\mathbf{C}_i^T \mathbf{A}_i^T \mathbf{\Phi}_{\mathbf{r}_i}^{kT} \boldsymbol{\lambda}^k \tag{6.6.8}$$

$$T_i''^k = (\mathbf{s}_i'^{PT} \mathbf{B}_i^T \mathbf{\Phi}_{\mathbf{r}_i}^{kT} - \mathbf{\Phi}_{\phi_i}^{kT}) \boldsymbol{\lambda}^k \tag{6.6.9}$$

Note that the product of the last two terms on the right of Eq. 6.6.8 must be the negative of the joint reaction force in the global x-y frame. As in Eq. 6.6.8, the first term on the right of Eq. 6.6.9 is the negative of the moment of the joint reaction force about the origin of the x'-y' frame; that is, it is associated with transfer of the force from the origin of the body-fixed x'-y' frame to the origin P of the x''-y'' frame, as shown in Fig. 6.6.1.

Equations 6.6.8 and 6.6.9 may be readily programmed to provide joint reaction forces and torques at a joint in the body-fixed joint x''-y'' reference frame.

Example 6.6.1: Consider the kinematically driven simple pendulum of Example 6.4.1, shown in Fig. 6.6.2. The Lagrange multiplier vector is obtained in Example 6.4.1 as

$$\boldsymbol{\lambda} = \begin{bmatrix} \boldsymbol{\lambda}^r \\ \boldsymbol{\lambda}^d \end{bmatrix} = \begin{bmatrix} m\ell \cos \phi_1 \dot{\phi}_1^2 \\ -mg + m\ell \sin \phi_1 \dot{\phi}_1^2 \\ -mg\ell \cos \phi_1 \end{bmatrix}$$

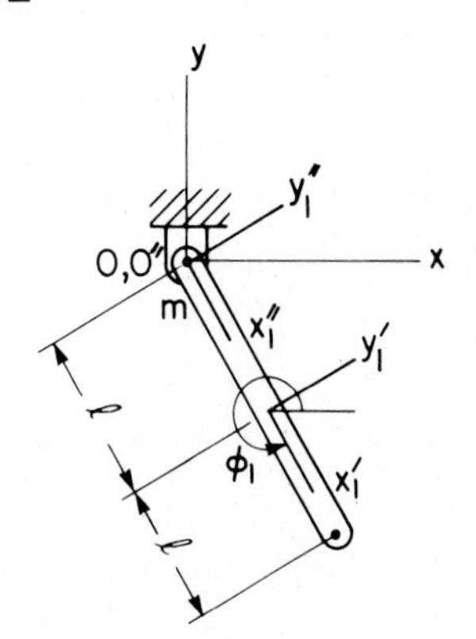

Figure 6.6.2 Reaction force on a simple pendulum.

Since the x_1'-y_1' and x_1''-y_1'' frames are parallel, $\mathbf{C}$ is an identity matrix. From Eq. 6.6.8, the joint reaction force at point O, in the x_1''-y_1'' frame, is

$$\mathbf{F}'' = -\begin{bmatrix} \cos\phi_1 & \sin\phi_1 \\ -\sin\phi_1 & \cos\phi_1 \end{bmatrix}\begin{bmatrix} 1 & 0 & 0 \\ 0 & 1 & 0 \end{bmatrix}\begin{bmatrix} m\ell\cos\phi_1\dot{\phi}_1^2 \\ -mg + m\ell\sin\phi_1\dot{\phi}_1^2 \\ -mg\ell\cos\phi_1 \end{bmatrix}$$

$$= \begin{bmatrix} -m\ell\dot{\phi}_1^2 + mg\sin\phi_1 \\ mg\cos\phi_1 \end{bmatrix} \tag{6.6.10}$$

The reaction torque from Eq. 6.6.9, in this case the driving torque required to create the specified motion of the pendulum, is

$$T = \left\{[-\ell, 0]\begin{bmatrix} -\sin\phi_1 & \cos\phi_1 \\ -\cos\phi_1 & -\sin\phi_1 \end{bmatrix}\begin{bmatrix} 1 & 0 & 0 \\ 0 & 1 & 0 \end{bmatrix} - [\ell\sin\phi_1, -\ell\cos\phi_1, 1]\right\}\boldsymbol{\lambda}$$

$$= [0, 0, -1]\begin{bmatrix} m\cos\phi_1\dot{\phi}_1^2 \\ -mg + m\sin\phi_1\dot{\phi}_1^2 \\ -mg\ell\cos\phi_1 \end{bmatrix}$$

$$= mg\ell\cos\phi_1 \tag{6.6.11}$$

The force of Eq. 6.6.10 is the reaction force in the revolute joint at point O in the x_1''-y_1'' frame. The torque of Eq. 6.6.11 is necessary to generate the motion specified by the driving constraint $\phi_1 = 3\pi/2 + 2\pi t$.

Since the driver implies that $\ddot{\phi}_1 = 0$, the summation of moments about point O must be zero; that is,

$$T - mg\ell\cos\phi_1 = 0$$

This direct application of the equations of motion for the body confirms that the result of Eq. 6.6.11 is correct. Similarly, using the fact that the acceleration of the center of mass is $\ell\dot{\phi}_1^2$ in the negative x_1' direction, equating mass times this acceleration to the sum of the gravitational and reaction forces, Eq. 6.6.10 is verified.

It is instructive to expand Eqs. 6.6.8 and 6.6.9 for a few of the standard joints of Chapter 3. Consider first an absolute x constraint of Eq. 3.2.3. Using the Jacobian of Eq. 3.2.5, the reaction force and torque on body i, from Eqs. 6.6.8 and 6.6.9, are

$$\mathbf{F}_i''^{ax(i)} = -\mathbf{C}_i^T\mathbf{A}_i^T\begin{bmatrix} 1 \\ 0 \end{bmatrix}\lambda^{ax(i)}$$

$$T_i''^{ax(i)} = [-x_i'^P\sin\phi_i - y_i'^P\cos\phi_i - (-x_i'^P\sin\phi_i - y_i'^P\cos\phi_i)]\lambda^{ax(i)} = 0$$

As expected, the constraint supports no reaction torque at point P. Transforming $\mathbf{F}_i''^{ax(i)}$ to the global frame, by multiplying by $\mathbf{C}_i$ and then $\mathbf{A}_i$ yields

$$\mathbf{F}_i^{ax(i)} = -\begin{bmatrix} 1 \\ 0 \end{bmatrix}\lambda^{ax(i)}$$

Thus, there is no reaction force in the global y direction and the global x component of reaction force is just the Lagrange multiplier $\lambda^{ax(i)}$.

For a revolute joint of Eq. 3.3.10, using the Jacobian of Eq. 3.3.12 in Eqs. 6.6.8 and 6.6.9,

$$\mathbf{F}_i''^{r(i,j)} = -\mathbf{C}_i^T\mathbf{A}_i^T(\mathbf{I})\boldsymbol{\lambda}^{r(i,j)} = -\mathbf{C}_i^T\mathbf{A}_i^T\boldsymbol{\lambda}^{r(i,j)}$$

$$T_i''^{k(i,j)} = [\mathbf{s}_i'^{PT}\mathbf{B}_i^T(\mathbf{I}) - (\mathbf{s}_i'^{PT}\mathbf{B}_i^T)]\boldsymbol{\lambda}^{r(i,j)} = 0$$

As expected, the revolute joint supports no reaction torque. Premultiplying the reaction force by $\mathbf{C}_i$ and then $\mathbf{A}_i$ yields the reaction force in the global x-y frame as

$$\mathbf{F}_i^{r(i,j)} = -\boldsymbol{\lambda}^{r(i,j)}$$

Thus, for the revolute joint, the negative of the Lagrange multiplier is the reaction force on body i, in the global x-y frame.

Finally, for the translational joint of Eq. 3.3.13, using the Jacobian of Eq. 3.3.14 in Eqs. 6.6.8 and 6.6.9,

$$\begin{aligned}
\mathbf{F}_i''^{t(i,j)} &= -\mathbf{C}_i^T\mathbf{A}_i^T[-\mathbf{B}_i\mathbf{v}_i', \mathbf{0}]\boldsymbol{\lambda}^{t(i,j)} \\
&= \mathbf{C}_i^T\mathbf{A}_i^T\mathbf{R}\mathbf{A}_i\mathbf{v}_i'\lambda_1^{t(i,j)} \\
&= \lambda_1^{t(i,j)}\mathbf{C}_i^T\mathbf{A}_i^T\mathbf{R}\mathbf{v}_i \\
T_i''^{t(i,j)} &= \{\mathbf{s}_i'^{PT}\mathbf{B}_i^T[-\mathbf{B}_i\mathbf{v}_i', \mathbf{0}] \\
&\quad + [(\mathbf{r}_j - \mathbf{r}_i)^T\mathbf{A}_i\mathbf{v}_i' + \mathbf{s}_j'^{PT}\mathbf{A}_{ij}^T\mathbf{v}_i', \mathbf{v}_j'^T\mathbf{A}_{ij}^T\mathbf{v}_i']\}\boldsymbol{\lambda}^{t(i,j)} \\
&= [(\mathbf{r}_j - \mathbf{r}_i)^T\mathbf{A}_i + \mathbf{s}_j'^{PT}\mathbf{A}_{ij}^T - \mathbf{s}_i'^{PT}, \mathbf{v}_j'^T\mathbf{A}_{ij}^T]\mathbf{v}_i'\boldsymbol{\lambda}^{t(i,j)}
\end{aligned}$$

Note that the reaction force $\mathbf{F}_i^{t(i,j)}$ is perpendicular to the vector $\mathbf{v}_i$ along the translational joint, as should be expected, and that the reaction torque is not generally zero.

PROBLEMS

Section 6.1

6.1.1. A virtual displacement $\delta\mathbf{r}^P$ is just the differential of the vector $\mathbf{r}^P$. Since the differential of calculus obeys rules of differentiation,

$$\begin{aligned}
\delta(\mathbf{r}^{P^T}\mathbf{r}^R) &= \delta(x^Px^R + y^Py^R) \\
&= \delta x^Px^R + x^P\,\delta x^R + \delta y^Py^R + y^P\,\delta y^R \\
&= [\delta x^P, \delta y^P]\begin{bmatrix} x^R \\ y^R \end{bmatrix} + [x^P, y^P]\begin{bmatrix} \delta x^R \\ \delta y^R \end{bmatrix} \\
&\equiv \delta\mathbf{r}^{P^T}\mathbf{r}^R + \mathbf{r}^{P^T}\,\delta\mathbf{r}^R
\end{aligned}$$

Use this result to show that if $(\mathbf{r}^P - \mathbf{r}^R)^T(\mathbf{r}^P - \mathbf{r}^R) = \text{constant}$ then

$$[\delta(\mathbf{r}^P - \mathbf{r}^R)]^T(\mathbf{r}^P - \mathbf{r}^P) = 0$$

6.1.2. With x, y, and ϕ as generalized coordinates, Eq. 6.1.5 may be written as

$$\mathbf{r}^P = \begin{bmatrix} x \\ y \end{bmatrix} + \begin{bmatrix} \cos\phi & -\sin\phi \\ \sin\phi & \cos\phi \end{bmatrix} \begin{bmatrix} s_{x'}^P \\ s_{y'}^P \end{bmatrix}$$

where $s_{x'}^P$ and $s_{y'}^P$ are constants. Use the chain rule of differential calculus to expand

$$\delta\mathbf{r}^P = \frac{\partial \mathbf{r}^P}{\partial x}\delta x + \frac{\partial \mathbf{r}^P}{\partial y}\delta y + \frac{\partial \mathbf{r}^P}{\partial \phi}\delta\phi$$

and verify that Eq. 6.1.7 is correct.

6.1.3. Verify that Eq. 6.1.11 is correct.

6.1.4. Use the second of Eqs. 6.1.14 to verify that for any $\mathbf{s}'^P$, $\mathbf{s}'^{PT}\mathbf{B}^T\mathbf{A}\mathbf{s}'^P = 0$.

6.1.5. Find the centroid and the polar moment of inertia with respect to the centroid of the machine element shown in Fig. P6.1.5, whose density is 8000 kg/m^3 and thickness is 0.01 m.

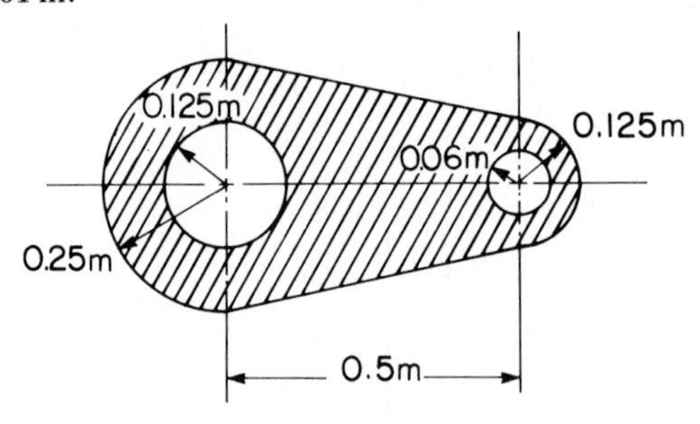

Figure P6.1.5

6.1.6. Find the centroid and the polar moment of inertia with respect to the centroid of the thin metal object shown in Fig. P6.1.6, whose mass per unit area is 80 kg/m^2.

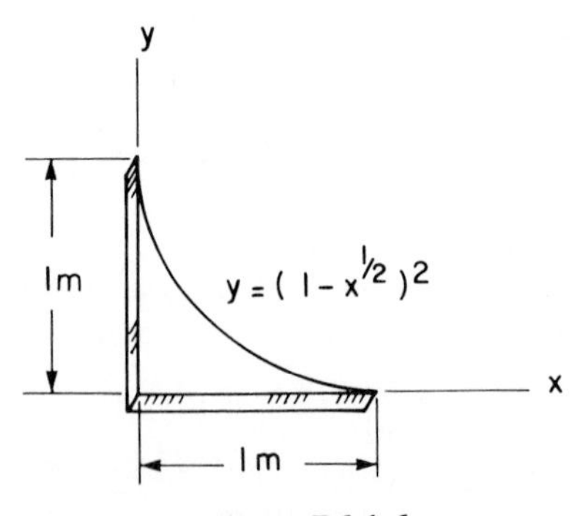

Figure P6.1.6

Section 6.2

6.2.1. In place of the rigid coupler in the revolute–translational composite joint of Fig. 3.3.8, the translational actuator shown in Fig. P6.2.1 is inserted, where the actuator

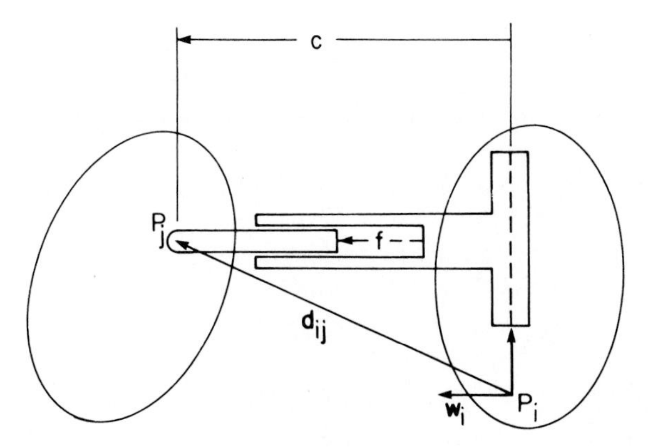

Figure P6.2.1

force is given as a function of the variable distance c as $f = g(c)$. Derive generalized forces $\mathbf{Q}_i$ and $\mathbf{Q}_j$ on bodies i and j, respectively, due to this force element (*Hint*: Use the fact that $c = \mathbf{w}_i^T\mathbf{d}_{ij}$ and use Eqs. 3.3.16, 3.3.17, and 3.3.18 to form $\delta W = f\,\delta c$.)

6.2.2. Derive generalized forces on bodies i and j that are connected by the translational joint of Fig. 3.3.5 or 3.5.9, with a spring–damper–actuator attached between points P_i and P_j. [*Hint*: Use the force expression of Eq. 6.2.10 and the fact that $\ell = C(t)$ in Eq. 3.5.12.]

Section 6.3

6.3.1. Write the variational equations of motion of Eqs. 6.3.3 and 6.3.5 for the double pendulum shown in Fig. P6.3.1, where the masses of the uniform bars are m_1 and m_2.

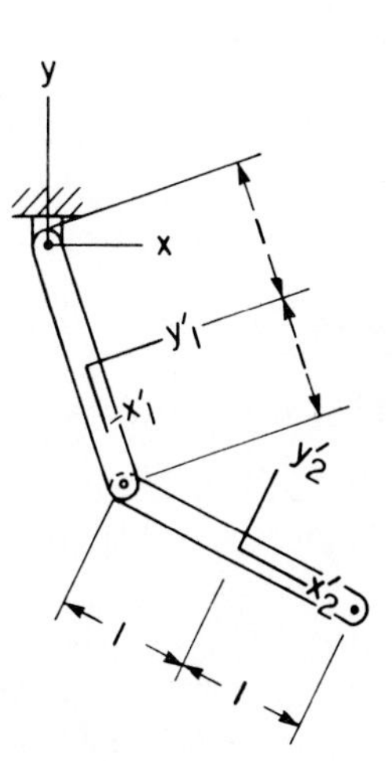

Figure P6.3.1

6.3.2. Show that, if $\mathbf{b}_1 = [1, 3, 1]^T$ and $\mathbf{A}$ is as defined in Example 6.3.3, then there is an $\mathbf{x}$ such that $\mathbf{Ax} = \mathbf{0}$, but $\mathbf{b}_1^T\mathbf{x} \neq 0$. Show further that the conclusion of the Lagrange multiplier theorem fails; that is, there is no $\boldsymbol{\lambda} = [\lambda_1, \lambda_2]^T$ such that $\mathbf{b}_1^T\mathbf{x} + \boldsymbol{\lambda}^T\mathbf{Ax} = 0$ for all $\mathbf{x}$.

6.3.3. Write the mixed differential–algebraic equations of motion of Eq. 6.3.18 for the double pendulum of Prob. 6.3.1.

6.3.4. Define initial conditions for the double pendulum of Prob. 6.3.1, with both bars vertical and with $\dot{\phi}_1 = \dot{\phi}_2 = 1$ rad/s at $t = 0$.

Section 6.4

6.4.1. Consider the mechanism of Prob. 6.3.1. Let the dimensions be in m and the mass of the bar be 20 kg. The system is driven as in Prob. 3.6.1. Write the equations of motion of Eqs. 6.4.2 and 6.4.3 and find the Lagrange multipliers. Give a physical interpretation of the Lagrange multipliers.

6.4.2. Consider the mechanism of Prob. 3.6.2. For the purpose of dynamic analysis, the x_2'-y_2' frame is moved to the centroid of the pendulum, as shown in Fig. P6.4.2. Use the masses of the block and the pendulum and the lengths given in the figure. The drivers are defined as

$$x_1 - at - bt^2 = 0$$

$$\phi_2 - \frac{3\pi}{2} - \omega t = 0$$

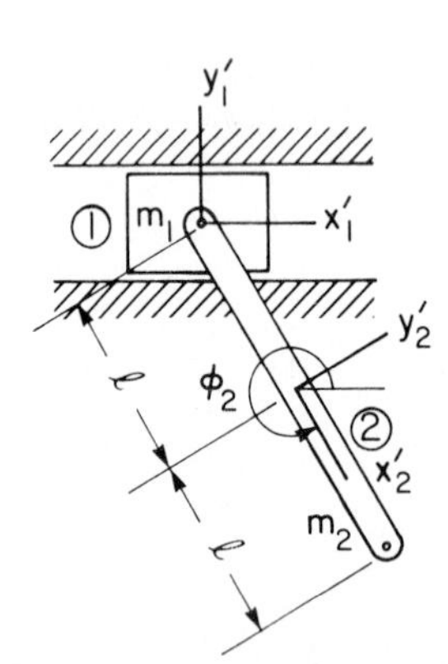

Figure P6.4.2

Write the equations of motion of Eqs. 6.4.2 and 6.4.3 and find the Lagrange multipliers.

6.4.3. Consider the mechanism of Example 3.6.2. Figure 3.6.2 is redrawn here for the purpose of dynamic analysis (Fig. P6.4.3). Use the same driver as in Example 3.6.2 and write the equations of motion of Eqs. 6.4.2 and 6.4.3 and find the Lagrange multipliers.

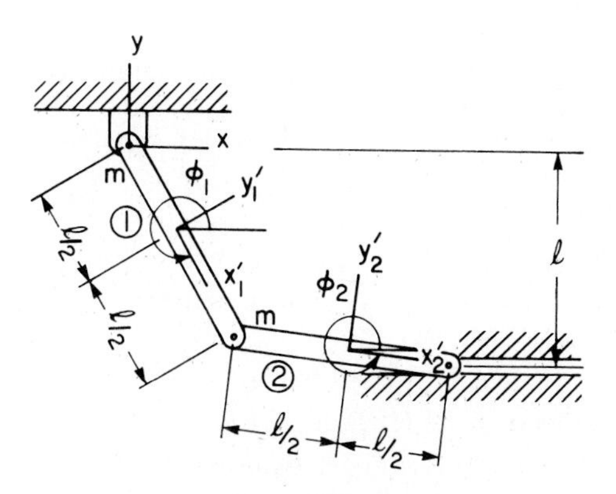

Figure P6.4.3

Section 6.5

6.5.1. Find the equilibrium position of the block shown in Fig. P6.5.1 supported by a linear spring. Use the minimum potential energy theorem and confirm the result by using the equilibrium condition $\sum F = 0$. Let m and k be the mass of the block and spring constant, respectively, and let $\ell_0 = 0$ be the free length of the spring.

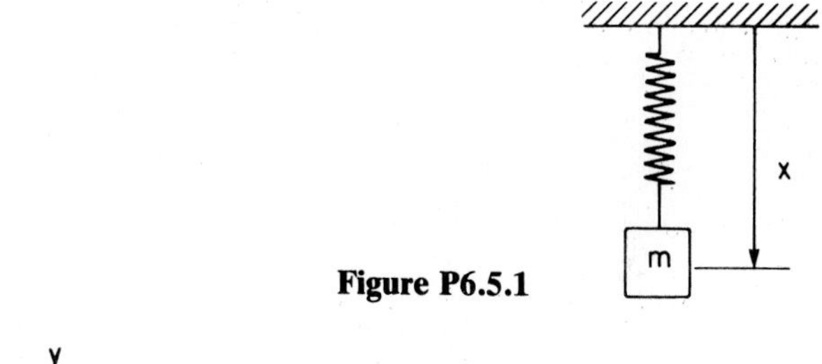

Figure P6.5.1

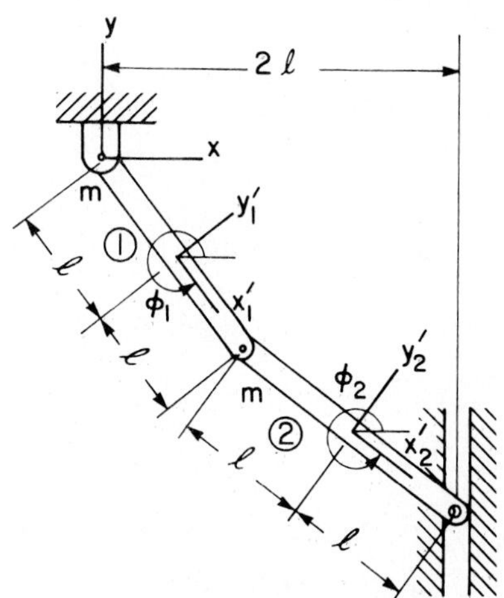

Figure P6.5.2

6.5.2. Consider the double pendulum of Example 6.5.2. The tip of the second bar is constrained by the vertical absolute constraint shown in Fig. P6.5.2. Let $\ell_1 = \ell_2 = \ell$ and $m_1 = m_2 = m$. Write the total potential energy and constraint equation in terms of ϕ_1 and ϕ_2. Confirm that $\phi_1 = 5.0233$ rad and $\phi_2 = 5.4795$ rad is the equilibrium configuration.

Section 6.6

6.6.1. Choose reasonable x''-y'' frames and continue Prob. 6.4.1 to find the constraint reaction forces.

6.6.2. Repeat Prob. 6.6.1 for Prob. 6.4.2.

6.6.3. Repeat Prob. 6.6.1 for Prob. 6.4.3.

6.6.4. Derive expressions for constraint reaction forces and torques, in terms of Lagrange multipliers, for the following joints:

(a) Relative distance constraint of Eq. 3.3.7.

(b) Revolute–translational constraint of Eq. 3.3.17.

(c) Relative angle driver of Eq. 3.5.7.

Interpret the physical significance of the Lagrange multiplier.

CHAPTER SEVEN

Numerical Methods in Dynamics

The equations of motion for constrained multibody systems derived in Chapter 6 must be assembled and solved numerically. Three distinct modes of dynamic analysis may be carried out: (1) equilibrium, (2) inverse dynamic, and (3) dynamic. Section 7.1 presents an overview of the computations required to carry out each of the three modes of analysis. The properties of mixed differential–algebraic equations of motion are briefly studied in Section 7.2 to provide the theoretical foundation needed for numerical solution. Three distinct methods for the solution of mixed differential–algebraic equations are presented in Section 7.3, and a hybrid method that takes advantage of favorable properties of each is presented. The result is a numerical reduction of the mixed differential–algebraic equations to a system of first-order differential equations that can be integrated using the standard numerical integration algorithms outlined in Section 7.4. Finally, a method for the minimization of total potential energy is presented in Section 7.4 for equilibrium analysis.

The theory and numerical methods for integration of initial-value problems of ordinary differential equations is well developed and documented in textbooks [31, 36, 37] and computer codes. A review of computer codes that are applicable for integrating differential equations of dynamics may be found in Reference 38. The presentation of methods in this chapter is only a brief introduction to the subject.

It is fascinating that, even though the Lagrange multiplier form of the equations of motion for constrained mechanical systems has been known since the late 1700s, only in 1981 was it fully recognized [39] that these equations cannot be treated as differential equations. A flurry of research activity on methods of integrating mixed differential–algebraic equations of dynamics has occurred in the 1980s [19, 40–43]. Presentation of methods for integrating differential–algebraic equations of dynamics in this text is limited to algorithms that are specialized for this field. It is anticipated that developments in the late 1980s will yield more generally applicable algorithms and computer codes.

7.1 ORGANIZATION OF COMPUTATIONS

The DADS computer code introduced in Section 4.1 for kinematic analysis also carries out dynamic analysis using the theory presented in Chapter 6 and numerical methods presented in the following sections of this chapter. Many of the organizational aspects of computation discussed in Chapter 4 for kinematic analysis are also used in DADS for dynamic analysis, so they will not be discussed here in detail. This section focuses on the organization of computations for dynamic analysis.

Especially for large-scale dynamics applications, bookkeeping and equation formulation tasks are extensive. As shown by even the simple examples analyzed in Chapter 6, the dimension of constrained equations of motion for multibody systems is about twice that encountered in kinematic analysis and the equations are nonlinear. Furthermore, solution of the equations of dynamics involves the integration of mixed differential–algebraic equations, rather than solution of algebraic equations in kinematic analysis. Automation of computation for multibody system dynamic analysis is required to obtain a general-purpose applications software capability.

DADS computational flow to implement dynamic analysis is summarized in Fig. 7.1.1. For purposes of dynamic analysis, the DADS code employs three basic program segments: (1) a preprocessor that assembles problem definition information and organizes data for computation, (2) a dynamic analysis program that constructs and solves the equations of dynamics, and (3) a postprocessor that prepares output information and displays the results of a dynamic simulation. Many of the subroutines employed for dynamic analysis are also used in kinematic analysis.

As indicated in Fig. 7.1.1, in addition to kinematic data, which are identical to those required for the kinematic analysis in Fig. 4.1.1, inertia, force, and initial condition data are required for dynamic analysis. The analysis mode (equilibrium, dynamic, or inverse dynamic) and output data desired must also be specified. The output of preprocessor computation is a dynamic analysis data set that is read and implemented by the dynamic analysis program.

Prior to dynamic analysis, the system is assembled and checked for feasibility, just as in kinematic analysis. During dynamic analysis, mass-, constraint-, and force-related matrices are assembled and used for either equilibrium analysis, transient dynamic analysis, or inverse dynamic analysis. The nature of the computations carried out during each of these three modes of analysis is quite different. Equilibrium analysis may be accomplished through either dynamic settling or minimization of total potential energy. Dynamic analysis is carried out by numerically integrating mixed differential–algebraic equations of motion. Finally, inverse dynamic analysis involves algebraic solution of the equations of kinematics and subsequent algebraic solution for Lagrange multipliers and forces that act in the kinematically driven system.

Following completion of dynamic analysis, the DADS postprocessor or-

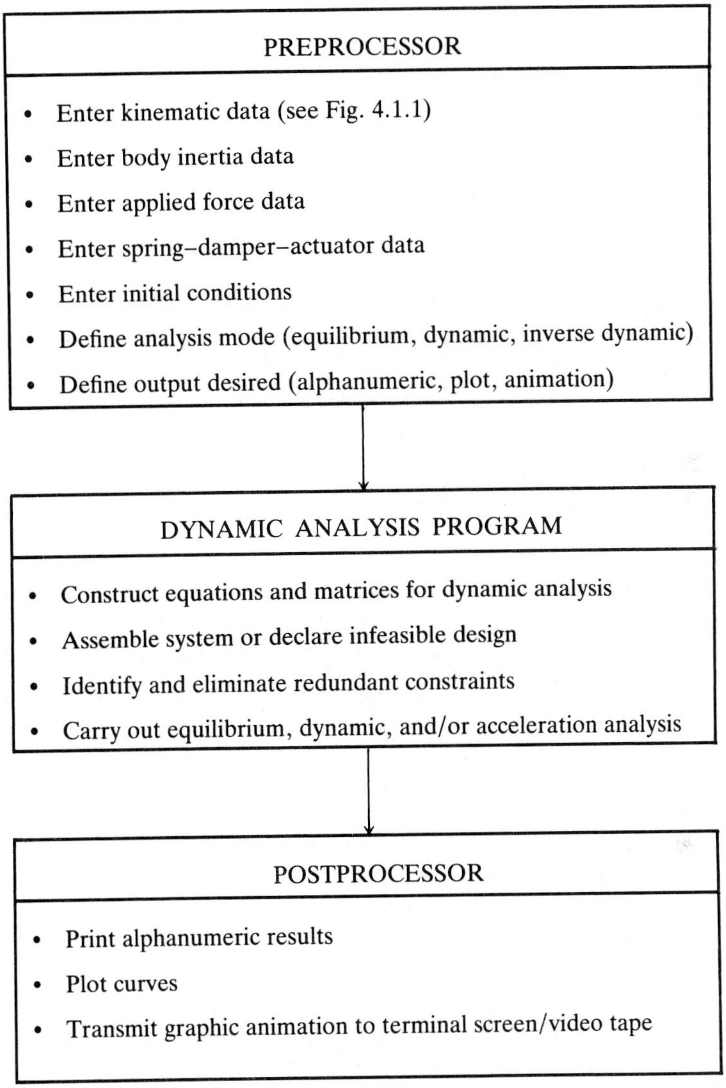

PREPROCESSOR
• Enter kinematic data (see Fig. 4.1.1) • Enter body inertia data • Enter applied force data • Enter spring–damper–actuator data • Enter initial conditions • Define analysis mode (equilibrium, dynamic, inverse dynamic) • Define output desired (alphanumeric, plot, animation)

↓

DYNAMIC ANALYSIS PROGRAM
• Construct equations and matrices for dynamic analysis • Assemble system or declare infeasible design • Identify and eliminate redundant constraints • Carry out equilibrium, dynamic, and/or acceleration analysis

↓

POSTPROCESSOR
• Print alphanumeric results • Plot curves • Transmit graphic animation to terminal screen/video tape

Figure 7.1.1 DADS dynamics computational flow.

ganizes and transmits the results of the simulation to a printer, plotter, or animation workstation.

As in the case of kinematic analysis, implementation of the extensive logical and numerical computations that are identified by the computational flow in Fig. 7.1.1 requires a large-scale computer code, the details of which are beyond the scope of this text. Prior to delving into the numerical methods that are used to carry out each of the modes of dynamic analysis, it is of value to understand the

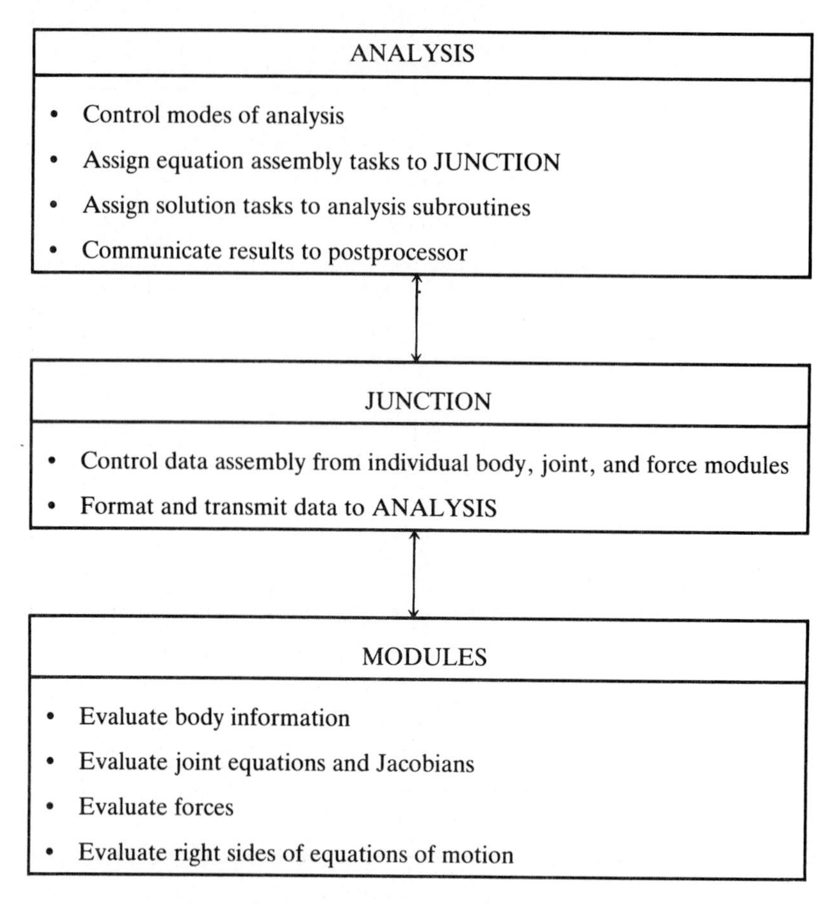

Figure 7.1.2 Structure of DADS dynamic analysis program.

flow of information that must be generated during dynamic analysis. The structure of the DADS dynamic analysis program is illustrated schematically in Fig. 7.1.2. The analysis program defines control over the modes of analysis and assigns equation assembly tasks to the junction program, which in turn calls modules that generate the required information and transmit it to the analysis program. Once equations are generated at each time step, the analysis program assigns solution tasks to analysis subroutines and communicates results to the postprocessor. The same modules used in kinematic analysis are also used in dynamic analysis, but they generate additional data that are required for dynamic analysis. In addition, force element modules generate force data that are required in the equations of motion. Rather than having separate programs for kinematic and dynamic analysis, the functions required for both are embedded in common modules, and only the functions required to support the analysis being undertaken are exercised.

A more detailed definition of computational flow and information that is generated during dynamic analysis are presented in Fig. 7.1.3. Input data reading and problem setup functions and model assembly and feasibility analysis functions

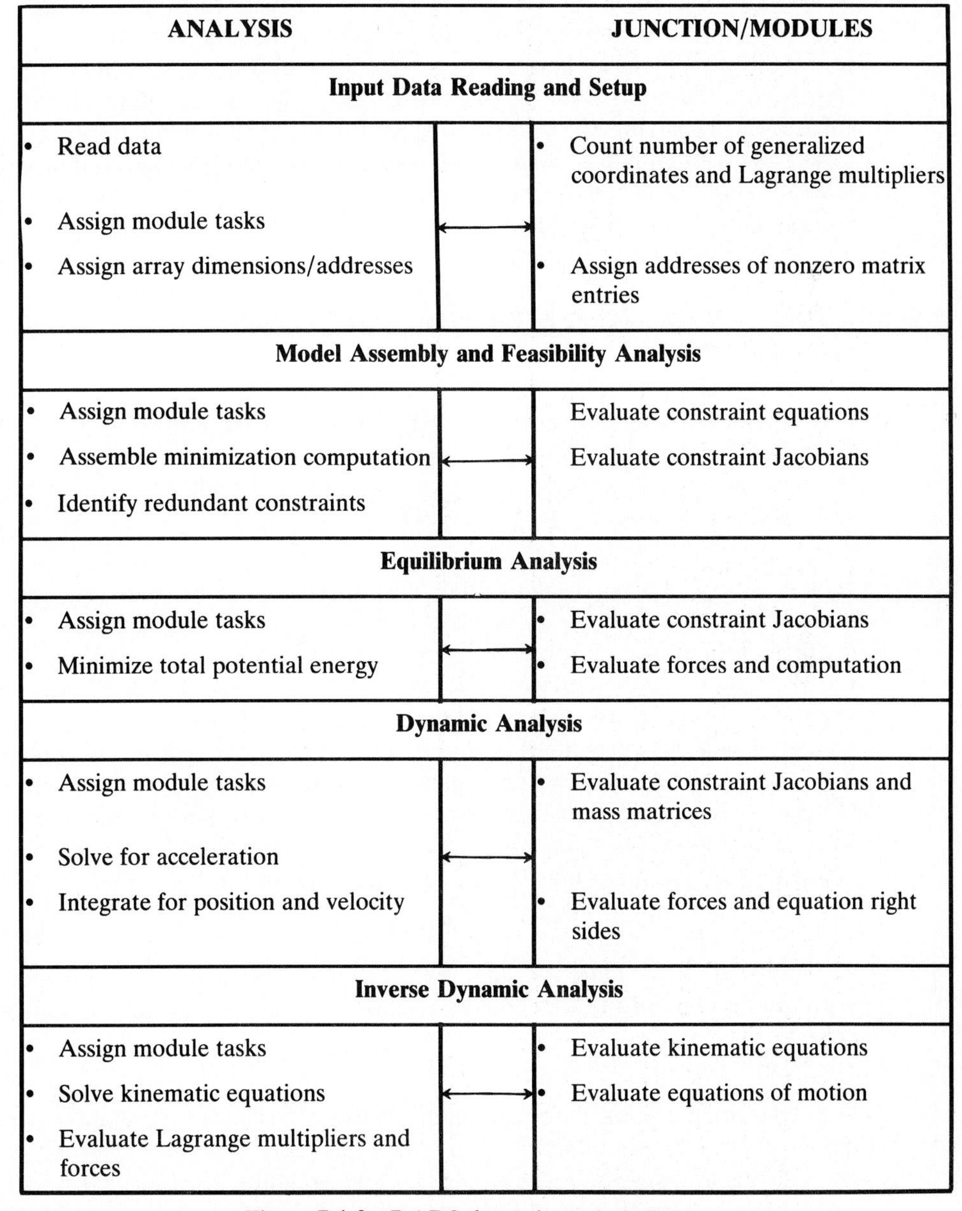

Figure 7.1.3 DADS dynamic analysis flow.

are identical to those encountered in kinematic analysis. Dimensioning and addressing of arrays, of course, requires consideration of new variables, such as Lagrange multipliers, that must be determined during dynamic analysis.

Equilibrium analysis is carried out either by direct application of dynamic analysis for dynamic settling or minimization of total potential energy for conservative mechanical systems. Dynamic analysis is carried out using a hybrid method of reducing mixed differential–algebraic equations to differential equations and subsequent numerical integration for position and velocity of the system throughout the time interval of interest. Finally, inverse dynamic analysis is carried out by solving kinematic equations for a kinematically determined system, assembling the equations of motion, and solving for Lagrange multipliers. The reaction and driving forces and torques that are required to impose the specified motion are subsequently calculated.

7.2 SOLUTION OF MIXED DIFFERENTIAL–ALGEBRAIC EQUATIONS OF MOTION

The equations of motion derived in Chapter 6 are summarized in matrix form as Eq. 6.3.18, which is repeated here as

$$\begin{bmatrix} \mathbf{M} & \mathbf{\Phi}_{\mathbf{q}}^T \\ \mathbf{\Phi}_{\mathbf{q}} & \mathbf{0} \end{bmatrix} \begin{bmatrix} \ddot{\mathbf{q}} \\ \boldsymbol{\lambda} \end{bmatrix} = \begin{bmatrix} \mathbf{Q}^A \\ \boldsymbol{\gamma} \end{bmatrix} \tag{7.2.1}$$

In the case of planar systems, the mass matrix $\mathbf{M}$ is constant and $\mathbf{Q}^A$ is the generalized applied force. Precisely the same form of spatial equations of motion is obtained in Chapter 11, but with a mass matrix that may depend on generalized coordinates and nonlinear terms in velocities and generalized coordinates that represent Coriolis acceleration effects on the right side. For the numerical solution techniques presented in this chapter to be applicable in the general case, it will be presumed that the mass matrix $\mathbf{M}$ and generalized applied force vector $\mathbf{Q}^A$ are nonlinear functions of generalized coordinates, their velocities, and time.

In addition to the equations of motion in Eq. 7.2.1, the kinematic constraints of Eq. 6.3.4 must hold, repeated here as

$$\mathbf{\Phi}(\mathbf{q}, t) = \mathbf{0} \tag{7.2.2}$$

where $\mathbf{\Phi}(\mathbf{q}, t)$ is presumed to have two continuous derivatives with respect to its arguments. In addition, the velocity equations of Eq. 6.3.17 must hold, repeated here as

$$\mathbf{\Phi}_{\mathbf{q}}\dot{\mathbf{q}} = \boldsymbol{\nu} \tag{7.2.3}$$

The differential and algebraic equations of Eqs. 7.2.1 to 7.2.3 comprise the *mixed differential–algebraic equations of motion.*

In addition to the position and velocity equations of Eqs. 7.2.2 and 7.2.3, which must hold at the initial time t_0, additional initial conditions on position and velocity must be introduced, as in Eqs. 6.3.26 and 6.3.27, repeated here as

$$\mathbf{\Phi}^I(\mathbf{q}(t_0), t_0) = \mathbf{0} \tag{7.2.4}$$

$$\mathbf{B}^I\dot{\mathbf{q}}(t_0) = \mathbf{v}^I \tag{7.2.5}$$

It is essential that the initial position condition of Eqs. 7.2.4 and 7.2.2, evaluated at t_0, uniquely determine the initial position $\mathbf{q}(t_0)$. Similarly, it is essential that the initial velocity conditions of Eqs. 7.2.5 and 7.2.3, evaluated at t_0, uniquely determine the initial velocity $\dot{\mathbf{q}}(t_0)$.

As shown in Section 6.3, the coefficient matrix in Eq. 7.2.1 is nonsingular for meaningful physical systems. Therefore, it may theoretically be inverted to obtain

$$\begin{aligned} \ddot{\mathbf{q}} &= \mathbf{f}(\mathbf{q}, \dot{\mathbf{q}}, t) \\ \boldsymbol{\lambda} &= \mathbf{g}(\mathbf{q}, \dot{\mathbf{q}}, t) \end{aligned} \tag{7.2.6}$$

where, by the implicit function theorem, the vector functions on the right are continuously differentiable with respect to their arguments. This result is theoretical, since explicit inversion of a large coefficient matrix with variable terms is impractical. Nevertheless, this theoretical result shows that $\ddot{\mathbf{q}}$ may be uniquely determined.

Direct numerical integration for $\mathbf{q}$ and $\dot{\mathbf{q}}$, from Eq. 7.2.6 and initial conditions, is both theoretically and computationally nontrivial. Since $\mathbf{q}$ and $\dot{\mathbf{q}}$ must satisfy Eqs. 7.2.2 and 7.2.3, all components of $\mathbf{q}$ and $\dot{\mathbf{q}}$ are not independent. One algorithm for integration that is used in some computer programs is to ignore this dependence and simply integrate accelerations that are obtained by numerically solving Eq. 7.2.1. There is no guarantee, however, that the function $\mathbf{q}(t)$ obtained will accurately satisfy the constraints of Eqs. 7.2.2 and 7.2.3. This difficulty arises because the equations of motion are in fact mixed differential–algebraic equations.

To obtain insight into the differential–algebraic equations of motion, consider a virtual displacement $\delta\mathbf{q}$ that satisfies the constraint equations to first order; that is, $\mathbf{\Phi}_{\mathbf{q}}\,\delta\mathbf{q} = \mathbf{0}$. At a value of $\mathbf{q}$ that satisfies the constraints of Eq. 7.2.2, the Gaussian reduction technique of Section 4.4 may be employed to transform this equation to the form of Eq. 4.4.8, that is,

$$\mathbf{U}\,\delta\mathbf{u} + \mathbf{R}\,\delta\mathbf{v} = \mathbf{0} \tag{7.2.7}$$

Since the coefficient matrix $\mathbf{U}$ is triangular with unit values on the diagonal, it is nonsingular, and Eq. 7.2.7 uniquely determines $\delta\mathbf{u}$ once values are selected for $\delta\mathbf{v}$. Therefore, $\mathbf{v}$ is interpreted as a vector of *independent generalized coordinates* and $\mathbf{u}$ is a vector of *dependent generalized coordinates.*

Since $\mathbf{U}$ in Eq. 7.2.7 is a matrix that is obtained by elementary row operations [22] on the matrix $\mathbf{\Phi}_{\mathbf{u}}$, $\mathbf{\Phi}_{\mathbf{u}}$ is nonsingular. Therefore, by the implicit function theorem, Eq. 7.2.2 can theoretically be solved for $\mathbf{u}$ as a function of $\mathbf{v}$

and t; that is,

$$\mathbf{u} = \mathbf{h}(\mathbf{v}, t) \tag{7.2.8}$$

where the vector function $\mathbf{h}(\mathbf{v}, t)$ is twice continuously differentiable in its arguments. Since Eq. 7.2.2 is highly nonlinear, explicit construction of the functional relationship in Eq. 7.2.8 is impractical. The result, however, is of great theoretical importance in analyzing the nature of solutions of mixed differential–algebraic equations of motion.

To analyze the equations of motion of Eq. 7.2.1, they may be reordered and partitioned, according to the decomposition of $\mathbf{q}$ into $\mathbf{u}$ and $\mathbf{v}$, as

$$\begin{aligned} \mathbf{M}^{\mathbf{uu}}\ddot{\mathbf{u}} + \mathbf{M}^{\mathbf{uv}}\ddot{\mathbf{v}} + \mathbf{\Phi}_{\mathbf{u}}^{T}\boldsymbol{\lambda} &= \mathbf{Q}^{A\mathbf{u}} \\ \mathbf{M}^{\mathbf{vu}}\ddot{\mathbf{u}} + \mathbf{M}^{\mathbf{vv}}\ddot{\mathbf{v}} + \mathbf{\Phi}_{\mathbf{v}}^{T}\boldsymbol{\lambda} &= \mathbf{Q}^{A\mathbf{v}} \\ \mathbf{\Phi}_{\mathbf{u}}\ddot{\mathbf{u}} + \mathbf{\Phi}_{\mathbf{v}}\ddot{\mathbf{v}} &= \boldsymbol{\gamma} \end{aligned} \tag{7.2.9}$$

where the mass matrices in the first two equations are submatrices of $\mathbf{M}$, and the vector functions on the right are a partitioning of the generalized applied force vector $\mathbf{Q}^{A}$. Similarly, the velocity equation of Eq. 7.2.3 may be written in the form

$$\mathbf{\Phi}_{\mathbf{u}}\dot{\mathbf{u}} + \mathbf{\Phi}_{\mathbf{v}}\dot{\mathbf{v}} = \boldsymbol{\nu} \tag{7.2.10}$$

For purposes of theoretical analysis, Eq. 7.2.8 may be used to write all expressions that involve the dependent generalized coordinates $\mathbf{u}$ as functions of the independent coordinates $\mathbf{v}$ and time. Therefore, all terms in Eqs. 7.2.9 and 7.2.10 that depend on $\mathbf{u}$ may be interpreted as functions of $\mathbf{v}$. Since the coefficient matrix of $\dot{\mathbf{u}}$ in Eq. 7.2.10 is nonsingular, the dependent velocity $\dot{\mathbf{u}}$ may theoretically be written as

$$\dot{\mathbf{u}} = \mathbf{\Phi}_{\mathbf{u}}^{-1}[\boldsymbol{\nu} - \mathbf{\Phi}_{\mathbf{v}}\dot{\mathbf{v}}] \tag{7.2.11}$$

which, after using Eq. 7.2.8, may be written as a function of only $\mathbf{v}$ and $\dot{\mathbf{v}}$. Similarly, the third of Eqs. 7.2.9 may theoretically be solved for $\ddot{\mathbf{u}}$, to obtain

$$\ddot{\mathbf{u}} = \mathbf{\Phi}_{\mathbf{u}}^{-1}[\boldsymbol{\gamma} - \mathbf{\Phi}_{\mathbf{v}}\ddot{\mathbf{v}}] \tag{7.2.12}$$

which, after using Eqs. 7.2.8 and 7.2.11, may be written in terms of $\mathbf{v}$, $\dot{\mathbf{v}}$, and $\ddot{\mathbf{v}}$.

Since the coefficient matrix of $\boldsymbol{\lambda}$ in the first of Eqs. 7.2.9 is nonsingular, this equation may theoretically be solved for $\boldsymbol{\lambda}$ to obtain

$$\boldsymbol{\lambda} = (\mathbf{\Phi}_{\mathbf{u}}^{-1})^{T}[\mathbf{Q}^{A\mathbf{u}}(\mathbf{v}, \dot{\mathbf{v}}, t) - \mathbf{M}^{\mathbf{uv}}\ddot{\mathbf{v}} - \mathbf{M}^{\mathbf{uu}}\ddot{\mathbf{u}}] \tag{7.2.13}$$

Substituting this result and Eq. 7.2.12 into the second of Eqs. 7.2.9 yields (Prob. 7.2.1)

$$\hat{\mathbf{M}}^{\mathbf{v}}(\mathbf{v}, \dot{\mathbf{v}}, t)\ddot{\mathbf{v}} = \hat{\mathbf{Q}}^{\mathbf{v}}(\mathbf{v}, \dot{\mathbf{v}}, t) \tag{7.2.14}$$

where

$$\hat{\mathbf{M}}^{\mathbf{v}} = \mathbf{M}^{\mathbf{vv}} - \mathbf{M}^{\mathbf{vu}}\mathbf{\Phi}_{\mathbf{u}}^{-1}\mathbf{\Phi}_{\mathbf{v}} - \mathbf{\Phi}_{\mathbf{v}}^{T}(\mathbf{\Phi}_{\mathbf{u}}^{-1})^{T}[\mathbf{M}^{\mathbf{uv}} - \mathbf{M}^{\mathbf{uu}}\mathbf{\Phi}_{\mathbf{u}}^{-1}\mathbf{\Phi}_{\mathbf{v}}] \tag{7.2.15}$$

$$\hat{\mathbf{Q}}^{\mathbf{v}} = \mathbf{Q}^{A\mathbf{v}} - \mathbf{M}^{\mathbf{vu}}\mathbf{\Phi}_{\mathbf{u}}^{-1}\boldsymbol{\gamma} - \mathbf{\Phi}_{\mathbf{v}}^{T}(\mathbf{\Phi}_{\mathbf{u}}^{-1})^{T}[\mathbf{Q}^{A\mathbf{u}} - \mathbf{M}^{\mathbf{uu}}\mathbf{\Phi}_{\mathbf{u}}^{-1}\boldsymbol{\gamma}] \tag{7.2.16}$$

Equation 7.2.14 is a set of differential equations in only the independent generalized coordinates $\mathbf{v}$ that are consistent with the position and velocity constraints that act on the system.

Example 7.2.1: To illustrate the foregoing theoretical reduction process, consider the simple pendulum of Example 6.3.1, with $\ell = 1$. For this elementary system, the constraint equations are

$$\mathbf{\Phi}(\mathbf{q}) = \begin{bmatrix} x_1 - \cos\phi_1 \\ y_1 - \sin\phi_1 \end{bmatrix} = \mathbf{0}$$

The constraint Jacobian is

$$\mathbf{\Phi_q} = \begin{bmatrix} 1 & 0 & \sin\phi_1 \\ 0 & 1 & -\cos\phi_1 \end{bmatrix}$$

and virtual displacements $\delta\mathbf{q}$ satisfy

$$\mathbf{I}\begin{bmatrix} \delta x_1 \\ \delta y_1 \end{bmatrix} + \begin{bmatrix} \sin\phi_1 \\ -\cos\phi_1 \end{bmatrix}\delta\phi_1 = \mathbf{0}$$

which is of the form of Eq. 7.2.7, with $\mathbf{u} = [x_1, y_1]^T$ and $v = \phi_1$. Thus,

$$\mathbf{\Phi_u} = \mathbf{I}$$

$$\mathbf{\Phi}_v = \begin{bmatrix} \sin\phi_1 \\ -\cos\phi_1 \end{bmatrix}$$

In this elementary example, the constraint equations may be trivially solved for $\mathbf{u}$ as

$$\mathbf{u} = \begin{bmatrix} \cos\phi_1 \\ \sin\phi_1 \end{bmatrix} \equiv \mathbf{h}(v)$$

The right sides of the velocity and acceleration equations are $\mathbf{v} = \mathbf{0}$ and

$$\boldsymbol{\gamma} = \begin{bmatrix} -\cos\phi_1 \\ -\sin\phi_1 \end{bmatrix}\dot{\phi}_1^2$$

Thus, Eqs. 7.2.11 and 7.2.12 are

$$\dot{\mathbf{u}} = \begin{bmatrix} -\sin\phi_1 \\ \cos\phi_1 \end{bmatrix}\dot{v}$$

$$\ddot{\mathbf{u}} = \begin{bmatrix} -\cos\phi_1 \\ -\sin\phi_1 \end{bmatrix}\dot{v}^2 - \begin{bmatrix} \sin\phi_1 \\ -\cos\phi_1 \end{bmatrix}\ddot{v}$$

The partitioned form of the equations of motion of Eq. 7.2.9 is (Prob. 7.2.2)

$$\begin{bmatrix} m & 0 \\ 0 & m \end{bmatrix}\ddot{\mathbf{u}} + \boldsymbol{\lambda} = \begin{bmatrix} 0 \\ -mg \end{bmatrix}$$

$$J'\ddot{v} + [\sin\phi_1, -\cos\phi_1]\boldsymbol{\lambda} = 0$$

Carrying out the reduction that leads to Eqs. 7.2.14 through 7.2.16 (or simply evaluating terms in Eqs. 7.2.15 and 7.2.16) yields

$$\hat{M}^v = J' + m$$
$$\hat{Q}^v = -mg \cos v$$

Thus, the reduced differential equation of motion of Eq. 7.2.14 is

$$(J' + m)\ddot{v} = -mg \cos v$$

The reader is cautioned that, while the theoretical reduction presented in this section can be carried out explicitly for the simple pendulum, it is not practical for more realistic systems. Even in the case of the two-body model of a slider–crank mechanism of Example 6.3.2, this reduction procedure is impractical.

To assure the existence of solutions of Eq. 7.2.14, it is important to show that the coefficient matrix on the left is nonsingular. To see that this is true, first observe that *kinematically admissible virtual velocities* $\delta\dot{\mathbf{q}} = [\delta\dot{\mathbf{u}}^T, \delta\dot{\mathbf{v}}^T]^T$ satisfy the velocity equation, with time held fixed, that is;

$$\mathbf{\Phi_u}\, \delta\dot{\mathbf{u}} + \mathbf{\Phi_v}\, \delta\dot{\mathbf{v}} = \mathbf{0} \tag{7.2.17}$$

which is essentially Eq. 7.2.10 with the right side equal to zero. Using the definition of the reduced mass matrix of Eqs. 7.2.15 and 7.2.17, the quadratic form associated with kinetic energy and independent virtual velocities may be expanded to obtain (Prob. 7.2.3)

$$\delta\dot{\mathbf{v}}^T\hat{\mathbf{M}}^{\mathbf{v}}\, \delta\dot{\mathbf{v}} = \delta\dot{\mathbf{q}}^T\mathbf{M}\, \delta\dot{\mathbf{q}} > 0 \tag{7.2.18}$$

which must hold for all nonzero independent virtual velocities $\delta\dot{\mathbf{v}}$. Since independent virtual velocities are arbitrary, the matrix of Eq. 7.2.15 is positive definite, and hence nonsingular. From well-known theory of ordinary differential equations [44], the differential equation of Eq. 7.2.14 and initial conditions on $\mathbf{v}$ defined by Eqs. 7.2.2 through 7.2.5 have a unique solution $\mathbf{v}(t)$. This $\mathbf{v}(t)$ and the associated $\mathbf{u}(t)$ defined by Eq. 7.2.8 satisfy the mixed differential–algebraic equations of motion (Eqs. 7.2.1 through 7.2.3).

The preceding argument guarantees the existence of a unique solution of the mixed differential–algebraic equations of motion under modest hypotheses on the kinematic and kinetic structure of the system. It shows that loss of uniqueness, which might be associated with bifurcation behavior, can only occur at states for which the constraint Jacobian is rank deficient or the system mass matrix fails to be positive definite for all kinematically admissible virtual velocities. Explicit implementation of the foregoing reduction to an independent set of differential equations, however, is not practical. Rather, computational algorithms that carry out this reduction numerically are presented in Section 7.3.

7.3 ALGORITHMS FOR SOLVING DIFFERENTIAL–ALGEBRAIC EQUATIONS

Efficient numerical integration algorithms are available for computing solutions of first-order systems of ordinary differential equations, with initial conditions given. Such methods calculate approximate solutions at grid points t_i in time that are specified by the user, or under the control of a computer program, based on various forms of integration error control. One such class of methods is introduced in Section 7.4. Prior to delving into the details of numerical integration algorithms for first-order systems, however, it is instructive to first see that (1) second-order differential equations can be reduced to systems of first-order differential equations, and (2) mixed differential–algebraic equations can be treated by differential equation methods.

7.3.1 First-Order Initial-Value Problems

Consider first a second-order differential equation and the associated initial conditions, that is, a second-order initial-value problem,

$$\begin{aligned} \ddot{\mathbf{x}} &= \mathbf{f}(\mathbf{x}, \dot{\mathbf{x}}, t) \\ \mathbf{x}(t_0) &= \mathbf{x}^0 \\ \dot{\mathbf{x}}(t_0) &= \dot{\mathbf{x}}^0 \end{aligned} \tag{7.3.1}$$

By defining new variables $\mathbf{s} = \mathbf{x}$ and $\mathbf{r} = \dot{\mathbf{s}} = \dot{\mathbf{x}}$, a first-order initial-value problem that is equivalent to Eq. 7.3.1 may be written as

$$\begin{aligned} \dot{\mathbf{s}} &= \mathbf{r} \\ \dot{\mathbf{r}} &= \mathbf{f}(\mathbf{s}, \mathbf{r}, t) \\ \mathbf{s}(t_0) &= \mathbf{x}^0 \\ \mathbf{r}(t_0) &= \dot{\mathbf{x}}^0 \end{aligned} \tag{7.3.2}$$

This first-order initial-value problem may now be integrated for $\mathbf{s}$ and $\mathbf{r}$, and equivalently for $\mathbf{x}$ and $\dot{\mathbf{x}}$.

Example 7.3.1: The differential equation of motion for the simple pendulum derived in Example 7.2.1, with $\delta\phi_1 = v$, is

$$(J' + m)\ddot{\phi}_1 = -mg\cos\phi_1$$

Let initial conditions on ϕ_1 be (see Fig. 6.3.1)

$$\phi_1(0) = \frac{3\pi}{2}\text{ rad}$$

$$\dot{\phi}_1(0) = 1\text{ rad/s}$$

Defining $s \equiv \phi$, and $r \equiv \dot{s} = \dot{\phi}_1$, the second-order differential equation becomes

$$\dot{s} = r$$

$$\dot{r} = -\frac{mg}{(J' + m)} \cos s$$

and the initial conditions are

$$r(0) = \frac{3\pi}{2}$$

$$s(0) = 1$$

A vast literature is available that presents a relatively complete theory of the existence, uniqueness, and stability of solutions of first-order initial-value problems [44]. A single theorem is given here that provides valuable results, even with modest assumptions on the smoothness of the functions involved. For a more extensive treatment, the reader may consult references such as Reference 44.

Consider a vector variable $\mathbf{x}$ that depends on time and a continuously differentiable vector function $\mathbf{f}$ that depends on $\mathbf{x}$ and time; that is,

$$\mathbf{x}(t) \equiv [x_1(t), \ldots, x_m(t)]^T$$

$$\mathbf{f}(\mathbf{x}, t) \equiv [f_1(\mathbf{x}, t), \ldots, f_m(\mathbf{x}, \mathrm{t})]^T$$

A system of first-order, nonlinear, ordinary differential equations may be written in the general form

$$\dot{\mathbf{x}} = \mathbf{f}(\mathbf{x}, t) \tag{7.3.3}$$

This formulation includes the first-order differential equations of Eq. 7.3.2.

As is known [44], many solutions of such differential equations may exist. To obtain a unique solution, initial conditions may be specified at some time t_0, in the form

$$\mathbf{x}(t_0) = \mathbf{x}^0 \tag{7.3.4}$$

where $\mathbf{x}^0$ is a specified vector of initial values. The combination of the differential equation of Eq. 7.3.3 and initial conditions of Eq. 7.3.4 is an *initial-value problem* of ordinary differential equations. The existence and uniqueness theory for such initial-value problems, under rather weak hypotheses, is presented in the literature. An adequate result, for the purposes of the present study of dynamics, is the following [44]:

> ***Initial-Value Problem Existence Theorem:*** Let $\mathbf{f}$ be continuously differentiable in its arguments, for $|x_i| \leq k$, $i = 1, \ldots, m$ and $t_0 \leq t \leq t_1$, and let $|x_i^0| < k$, $i = 1, \ldots, m$, where $k > 0$ is a constant. Then there exists a unique solution $\mathbf{x}(t)$ of Eqs. 7.3.3 and 7.3.4 in an interval $t_0 \leq t < t^*$, such that $|x_i(t)| \leq k$, $i = 1, \ldots, m$.

It is important to note that this theorem guarantees the existence of a *local solution* of the initial-value problem of Eqs. 7.3.3 and 7.3.4. The local nature of the solution is dictated by the fact that instabilities in solutions of differential equations may arise, leading to solutions that diverge to infinity. Globally valid existence theorems for nonlinear ordinary differential equations are not generally available. Nevertheless, this theorem yields valuable information and permits the use of a numerical integration method based on the methods outlined in Section 7.4. Providing the solution process continues to the desired terminal time, confidence in the validity of the solution is assured. Only if the integration process diverges must deeper theoretical analysis of the properties of the solution be investigated.

The reduction of second-order differential equations to first-order form is the easy part of solving differential–algebraic equations. Four basic approaches that reduce differential–algebraic equations to differential equations, for purposes of numerical integration, are presented in the following subsections. The last algorithm combines the better aspects of the first three to achieve both reliability and efficiency.

7.3.2 Generalized Coordinate Partitioning

It was demonstrated in Section 7.2 that, if the constraint Jacobian has full row rank, it is theoretically possible to reduce the system of differential–algebraic equations of Eqs. 7.2.1 through 7.2.3 to a set of second-order differential equations in independent generalized coordinates. The generalized coordinate partitioning algorithm presented here implements this method using an implicit numerical reduction in place of the explicit reduction derived in Section 7.2. Based on the knowledge that a theoretical reduction exists and has attractive mathematical properties, Eq. 7.2.1 is solved numerically for $\ddot{\mathbf{q}}$ and $\boldsymbol{\lambda}$. The $\ddot{\mathbf{v}}$ components in $\ddot{\mathbf{q}}$ are simply extracted. They are identical to the values that would have been obtained if the reduction that leads to Eq. 7.2.14 had been carried out and the equation solved. An apparent disadvantage in this approach is that the dimension of the coefficient matrix in Eq. 7.2.1 is much greater than that of the coefficient matrix in Eq. 7.2.14. This disadvantage is, however, offset by the fact that (1) the coefficient matrix in Eq. 7.2.1 is sparse [13, 14, 46] and can be efficiently solved by well-developed sparse matrix computer software [47], and (2) the matrix inversions required in forming Eq. 7.2.14 can be avoided. The independent accelerations can then be integrated using the method of Section 7.4, and dependent variables can be obtained by solving the kinematic position and velocity equations.

A computational implementation of this approach, called the *generalized coordinate partitioning algorithm,* may be carried out as follows:

1. Begin with an assembled configuration and initial conditions that satisfy Eqs. 7.2.2 through 7.2.5 at the initial time t_0, that is, with $\mathbf{q}(t_0)$ and $\dot{\mathbf{q}}(t_0)$

that satisfy constraints. Such an initial configuration may be obtained using the methods presented in Section 4.3.

2. Evaluate and factor $\mathbf{\Phi_q}(t_0)$, as in Eq. 4.4.8, to determine dependent and independent variables $\mathbf{u}$ and $\mathbf{v}$, respectively. This partitioning of the generalized coordinate vector $\mathbf{q}$ will be retained until computational tests indicate that a new partitioning is required.
3. At a typical time step t_i, solve Eq. 7.2.1 for $\ddot{\mathbf{u}}$, $\ddot{\mathbf{v}}$, and $\boldsymbol{\lambda}$.
4. Integrate the first-order system

$$\dot{\mathbf{r}} = \ddot{\mathbf{v}}$$

$$\dot{\mathbf{s}} = \mathbf{r}$$

from t_i to t_{i+1}, using the integration algorithm of Section 7.4, to obtain $\mathbf{v}(t_{i+1})$ and $\dot{\mathbf{v}}(t_{i+1})$. Use the same integration algorithm and the first-order system

$$\dot{\mathbf{x}} = \ddot{\mathbf{u}}$$

$$\dot{\mathbf{y}} = \dot{\mathbf{x}}$$

to approximate $\mathbf{u}(t_{i+1})$ without error control (see Section 7.4).

5. Solve Eq. 7.2.2 for $\mathbf{u}(t_{i+1})$ using the Newton–Raphson method, starting with the approximation obtained in step (4). Solve Eq. 7.2.10 for $\dot{\mathbf{u}}(t_{i+1})$.
6. If t_{i+1} exceeds the final time, stop. Otherwise, update i to $i+1$ and continue.
7. If $\mathbf{\Phi_u}$ is ill-conditioned (see Section 4.4.1) or if the numerical integration algorithm requires multiple prediction iterations [31, 36], return to step (2) with t_0 replaced by t_{i+1}. Otherwise, return to step (3).

This implementation of the generalized coordinate partitioning algorithm has been used extensively [19] and has been found to be reliable and accurate. It satisfies constraints to the precision that is specified by the user and maintains good error control. It suffers from somewhat poorer numerical efficiency than alternative algorithms due to the requirement for iterative solution for dependent generalized coordinates $\mathbf{u}$ and dependent velocities $\dot{\mathbf{u}}$ in step (5).

7.3.3 Direct Integration

As observed in Section 7.2, Eq. 7.2.1 can be solved numerically for $\ddot{\mathbf{q}}$. If errors in satisfying constraints are ignored, the methods of Section 7.4 can be applied to integrate for $\mathbf{q}$ and $\dot{\mathbf{q}}$. This approach suffers from an accumulation of constraint error and may lead to substantial violation of the position and velocity constraint equations of Eqs. 7.2.2 and 7.2.3. For moderately regular system dynamics applications and small intervals of time, this approach may be satisfactory. It forms the basis for the *direct integration algorithm*.

1. Begin with an assembled configuration that satisfies Eqs. 7.2.2 through

7.2.5 at t_0, that is, with $\mathbf{q}(t_0)$ and $\dot{\mathbf{q}}(t_0)$, as in step (1) of the generalized coordinate partitioning algorithm.

2. At a typical time step t_i, solve Eq. 7.2.1 for $\ddot{\mathbf{q}}$ and $\boldsymbol{\lambda}$.
3. Integrate the first-order system

$$\begin{aligned}\dot{\mathbf{r}} &= \ddot{\mathbf{q}} \\ \dot{\mathbf{s}} &= \mathbf{r}\end{aligned} \qquad \textbf{(7.3.5)}$$

from t_i to t_{i+1}, using the algorithm of Section 7.4, to obtain $\mathbf{q}(t_{i+1})$ and $\dot{\mathbf{q}}(t_{i+1})$.

4. If t_{i+1} exceeds the final time, terminate. Otherwise, update i to $i+1$ and return to step (2).

The direct integration algorithm is simple, easy to implement, and computationally fast. It suffers, however, from a lack of error control on the constraints and may lead to erroneous results. Since no provision is made to control the accumulation of constraint error, the user must monitor constraint error and terminate the process if unacceptable constraint errors are encountered.

7.3.4 Constraint Stabilization

The second of Eqs. 7.2.1 is obtained by taking two time derivatives of the kinematic constraint equations of Eq. 7.2.2 and may be written in the form

$$\ddot{\boldsymbol{\Phi}} = \boldsymbol{\Phi}_{\mathbf{q}}\ddot{\mathbf{q}} - \boldsymbol{\gamma} = \mathbf{0} \qquad \textbf{(7.3.6)}$$

From the control literature, it is known that numerical solution of the equation $\ddot{\boldsymbol{\Phi}} = \mathbf{0}$ can be unstable; that is, it can lead to values of $\boldsymbol{\Phi}$ and $\dot{\boldsymbol{\Phi}}$ that are far from $\mathbf{0}$. Baumgarte [40] observed, however, that the modified acceleration equation

$$\ddot{\boldsymbol{\Phi}} + 2\alpha\dot{\boldsymbol{\Phi}} + \beta^2\boldsymbol{\Phi} = \mathbf{0} \qquad \textbf{(7.3.7)}$$

with $\alpha > 0$ and $\beta \neq 0$ is stable, hence implying that $\dot{\boldsymbol{\Phi}} \approx \boldsymbol{\Phi} \approx \mathbf{0}$. This observation forms the basis of the *constraint stabilization method* presented by Baumgarte [40], in which Eq. 7.3.7 is used in Eq. 7.2.1 instead of Eq. 7.3.6. To implement this method, Eq. 7.3.7 is written explicitly in the form

$$\boldsymbol{\Phi}_{\mathbf{q}}\ddot{\mathbf{q}} = \boldsymbol{\gamma} - 2\alpha(\boldsymbol{\Phi}_{\mathbf{q}}\dot{\mathbf{q}} + \boldsymbol{\Phi}_t) - \beta^2\boldsymbol{\Phi} \equiv \hat{\boldsymbol{\gamma}} \qquad \textbf{(7.3.8)}$$

where $\hat{\boldsymbol{\gamma}}$ replaces $\boldsymbol{\gamma}$ in Eq. 7.2.1.

With this replacement, the *constraint stabilization integration algorithm* consists simply of carrying out direct integration, as in Section 7.3.3. This approach has been demonstrated to be more stable and accurate than the elementary direct integration algorithm of Section 7.3.3 and is essentially as fast computationally. No general and uniformly valid method of selecting α and β, however, has been found. Furthermore, in the vicinity of kinematically singular configurations, even though Eq. 7.3.6 may have a well-behaved solution, the additional terms on the right side of Eq. 7.3.8 often lead to divergence of the algorithm.

7.3.5 A Hybrid Algorithm

The generalized coordinate partitioning and constraint stabilization algorithms are drastically different and have interesting complementary performance characteristics. A hybrid algorithm has been developed by Park [42] to take advantage of the better features of both methods. This algorithm follows the basic idea of generalized coordinate partitioning in that it defines a set of independent generalized coordinates, based on factorization of the constraint Jacobian $\mathbf{\Phi_q}$. It employs the modified acceleration equation of Eq. 7.3.8 during integration to stabilize small-amplitude oscillations in constraint error. Constraint errors are monitored, and the correction step in the generalized coordinate partitioning algorithm that solves the constraint equations for dependent coordinates $\mathbf{u}$ and $\dot{\mathbf{u}}$ is applied only when position and velocity constraint errors exceed a preset error tolerance. To avoid the instability that may occur with the constraint stabilization algorithm in the neighborhood of a kinematically singular configuration, upon occurrence of ill-conditioning in the coefficient matrix of Eq. 7.2.1, the algorithm reverts to pure generalized coordinate partitioning. Thus, the hybrid algorithm retains the reliability and positive error control characteristics of the generalized coordinate partitioning algorithm and approaches the computational speed that is achievable with the constraint stabilization algorithm.

The *hybrid numerical integration algorithm* is as follows:

1. Begin with an assembled configuration $\mathbf{q}(t_0)$ and $\dot{\mathbf{q}}(t_0)$ that satisfies Eqs. 7.2.2 through 7.2.5 at t_0, as in step (1) of the generalized coordinate partitioning algorithm of Section 7.3.1.
2. Evaluate and factor the matrix $\mathbf{\Phi_q}(t_0)$, as in Eq. 4.4.8, to determine dependent and independent variables $\mathbf{u}$ and $\mathbf{v}$, respectively. This partitioning of the generalized coordinate vector $\mathbf{q}$ will be retained until computational tests indicate the need for a new partitioning.
3. Solve Eq. 7.2.1 for $\ddot{\mathbf{q}}$ and $\boldsymbol{\lambda}$ without stabilization terms.
4. Integrate the first-order system

$$\dot{\mathbf{r}} = \ddot{\mathbf{q}}$$

$$\dot{\mathbf{s}} = \mathbf{r}$$

from t_i to t_{i+1}, as in the algorithm of Section 7.3.2, to obtain $\mathbf{q}(t_{i+1})$ and $\dot{\mathbf{q}}(t_{i+1})$, with error control only on $\mathbf{v}$ and $\dot{\mathbf{v}}$.
5. **a.** If $\mathbf{\Phi}(\mathbf{q}, t)$ at t_{i+1} is within the specified constraint error tolerance, accept the integrated values of $\mathbf{q}(t_{i+1})$ and $\dot{\mathbf{q}}(t_{i+1})$. To compensate for error that may accumulate in continued integration, constraint stabilization terms are used until an unacceptable level of error is encountered. To evaluate the correction terms, multiply β^2 by the

previously evaluated constraint violation. For the velocity violation term, evaluate the Jacobian and $\mathbf{\Phi}_t$ and calculate $2\alpha(\mathbf{\Phi_q}\dot{\mathbf{q}} + \mathbf{\Phi}_t)$. Add these constraint and velocity equation violation terms to $\boldsymbol{\gamma}$ to form $\hat{\boldsymbol{\gamma}}$, and solve Eq. 7.2.1, with $\boldsymbol{\gamma}$ replaced by $\hat{\boldsymbol{\gamma}}$, for $\ddot{\mathbf{q}}$ and $\boldsymbol{\lambda}$.

b. If the integrated values of $\mathbf{q}(t_{i+1})$ do not satisfy the constraint error tolerance, use the Newton–Raphson method to iteratively solve Eq. 7.2.2 for accurate values of $\mathbf{u}(t_{i+1})$. After position analysis, solve Eq. 7.2.3 for $\dot{\mathbf{u}}(t_{i+1})$. Since position and velocity are corrected using the constraint equations, use Eq. 7.2.1 to solve for $\ddot{\mathbf{q}}$ and $\boldsymbol{\lambda}$. Evaluate matrix conditioning and integration performance to check if the independent variable set is still valid [45]. If the independent variable set is to be changed, set a flag so that after the correction step integration will be restarted with the new independent variables from step (2).

6. Provide velocity and acceleration to the integration algorithm and use independent variable error estimates to determine the integration time step and order (see Section 7.4).

7. If t_{i+1} exceeds the final time, terminate. Otherwise, update t_{i+1} and return to step (4).

This method will perform best with an optimized choice of α and β at each time step, but to date there is no general method of selecting α and β. The algorithm uses small fixed values of α and β to damp out a moderate amount of constraint error, to prevent adversely affecting the integration algorithm by the correction term. This algorithm is the most reliable of those discussed in this section. It is employed in the DADS computer code [27].

7.4 NUMERICAL INTEGRATION OF FIRST-ORDER INITIAL-VALUE PROBLEMS

Algorithms for solving differential–algebraic equations of machine dynamics presented in Section 7.3 provide alternative methods of reducing differential–algebraic equations to initial-value problems that can be integrated using standard numerical solution techniques. The purpose of this section is to present basic ideas and algorithms that may be used for this purpose. Regardless of the algorithm selected in Section 7.3, the task remains to integrate an initial-value problem of the form

$$\begin{aligned} \dot{\mathbf{x}} &= \mathbf{f}(\mathbf{x}, t) \\ \mathbf{x}(t_0) &= \mathbf{x}^0 \end{aligned} \tag{7.4.1}$$

where $\mathbf{x} = [x_1, x_2, \ldots, x_m]^T$ is the vector of variables to be integrated and the function $\mathbf{f}(\mathbf{x}, t)$ on the right of Eq. 7.4.1 is defined by the computational sequence of the algorithms employed in Section 7.3. The constrained dynamic existence theorem of Section 6.3.3 and the implicit function theorem guarantee that, if the constraint Jacobian has full row rank and if its entries and the applied generalized force are continuously differentiable, then the hypotheses of the initial-value problem existence theorem are satisfied for each of the algorithms. The vector $\mathbf{x}^0$ defines the initial conditions.

Equation 7.4.1 comprises an initial-value problem that is to be integrated numerically. Methods of polynomial interpolation, upon which most numerical integration algorithms are based, are presented in this section and used to derive and demonstrate properties of numerical integration algorithms. Error estimates are presented to give an indication of the methods used for error control during the integration process. For a more detailed survey of numerical integration methods in dynamics, the reader may consult the survey of Reference 38.

7.4.1 Polynomial Interpolation

The fundamental idea used in numerical integration is the approximation of a function $f(t)$ by a polynomial $P(t)$, that is, a polynomial that agrees with f and perhaps some of its derivatives at one or more points t_i. There are numerous methods by which approximating polynomials can be derived, each with its inherent advantages and disadvantages.

Taylor Expansion The Taylor expansion method is based on approximating a function $f(t)$ near a point t_0 by the *Taylor polynomial* [25, 26]

$$T_k(t) = f(t_0) + f^{(1)}(t_0)(t - t_0) + \frac{f^{(2)}(t_0)(t - t_0)^2}{2!} + \cdots + \frac{f^{(k-1)}(t_0)(t - t_0)^{k-1}}{(k-1)!} \quad \textbf{(7.4.2)}$$

where $f^{(i)}(t_0)$ denotes the ith derivative of $f(t)$ evaluated at t_0. This polynomial agrees with f and its first $k - 1$ derivatives at t_0 (Prob. 7.4.1); that is,

$$T_k^{(j)}(t_0) = f^{(j)}(t_0), \qquad j = 0, 1, \ldots, k - 1 \quad \textbf{(7.4.3)}$$

Example 7.4.1: Consider the use of Eq. 7.4.2 to approximate e^t by expanding $f(t) = e^t$ about $t_0 = 0$. Since $f^{(j)} = e^t$, for all j, Eq. 7.4.2 becomes

$$T_k(t) = e^0 + e^0(t - 0) + \frac{e^0(t-0)^2}{2!} + \cdots + \frac{e^0(t-0)^{k-1}}{(k-1)!}$$

$$= 1 + t + \frac{t^2}{2!} + \frac{t^3}{3!} + \cdots + \frac{t^{(k-1)}}{(k-1)!}$$

If $t = 1$, then $f(1) = e$ is approximated by

$$e \approx T_k(1) = 1 + 1 + \frac{1^2}{2!} + \frac{1^3}{3!} + \cdots + \frac{1^{(k-1)}}{(k-1)!}$$

The value of e to ten-place accuracy is $e = 2.718281828$ so $e^{1/2} = 1.648721271$. Table 7.4.1 lists $T_k(t)$ and the error $E_k(t) = T_k(t) - e^t$ for various values of k and $t = \frac{1}{2}$ and $t = 1$.

TABLE 7.4.1 Taylor Approximate Values of e^t

k	$T_k(0.5)$	$E_k(0.5)$	$T_k(1)$	$E_k(1)$
1	1.000000000	−0.6487212707	1.000000000	−1.7182818285
2	1.500000000	−0.1487212707	2.000000000	−0.7182818285
3	1.625000000	−0.0237212707	2.500000000	−0.2182818285
4	1.645833333	−0.0028879374	2.666666667	−0.0516151618
5	1.648437500	−0.0002837707	2.708333333	−0.0099484951
6	1.648697917	−0.0000233540	2.716666667	−0.0016151618
7	1.648719618	−0.0000016526	2.718055556	−0.0002262729
8	1.648721168	−0.0000001025	2.718253968	−0.0000278602
9	1.648721265	−0.0000000057	2.718278770	−0.0000030586
10	1.648721270	−0.0000000003	2.718281526	−0.0000003029
11	1.648721271	0.0000000000	2.718281801	−0.0000000273
12	1.648721271	0.0000000000	2.718281826	−0.0000000023
13	1.648721271	0.0000000000	2.718281828	−0.0000000002
14	1.648721271	0.0000000000	2.718281828	0.0000000000

From this example it is clear that a Taylor polynomial approximation may be used to evaluate a function at nearby points to any desired accuracy. It also illustrates that accuracy of the Taylor approximation is improved when $|t - t_0|$ is reduced and when more terms are included in the expansion of Eq. 7.4.2, that is, when higher order derivatives are used.

In general, the values of a function $f(t)$ at several values of t are easier to obtain than high-order derivatives of f at a single value of t, so Taylor expansion is not well suited for general-purpose integration algorithms. It is, however, useful in understanding how other interpolating polynomials are developed into integration algorithms and in deriving numerical integration error estimates.

Interpolating Polynomials A polynomial $P_k(t)$ of degree $k-1$ can be constructed that agrees with $f(t)$ at k distinct points $t_{n-k+1}, \ldots, t_{n-1}, t_n$; that is,

$$P_k(t_i) = f(t_i), \qquad i = n-k+1, \ldots, n-1, n$$

as shown in Fig. 7.4.1. There is one and only one polynomial of degree $k-1$ that satisfies these interpolating conditions [31]. With equally spaced points, $t_i - t_{i-1} = h$, this polynomial is constructed in the form of a *Newton backward difference polynomial* [31]

$$P_k(t) = f_n + \sum_{i=1}^{k-1} \frac{\nabla^i f_n}{(i)!\, h^i} \prod_{j=0}^{i-1} (t - t_{n-j}) \tag{7.4.4}$$

where the *product notation* (analogous to $\sum_{j=0}^{\ell}$ summation notation)

$$\prod_{j=0}^{\ell} g_j \equiv g_0 \times g_1 \times \cdots \times g_{\ell-1} \times g_\ell$$

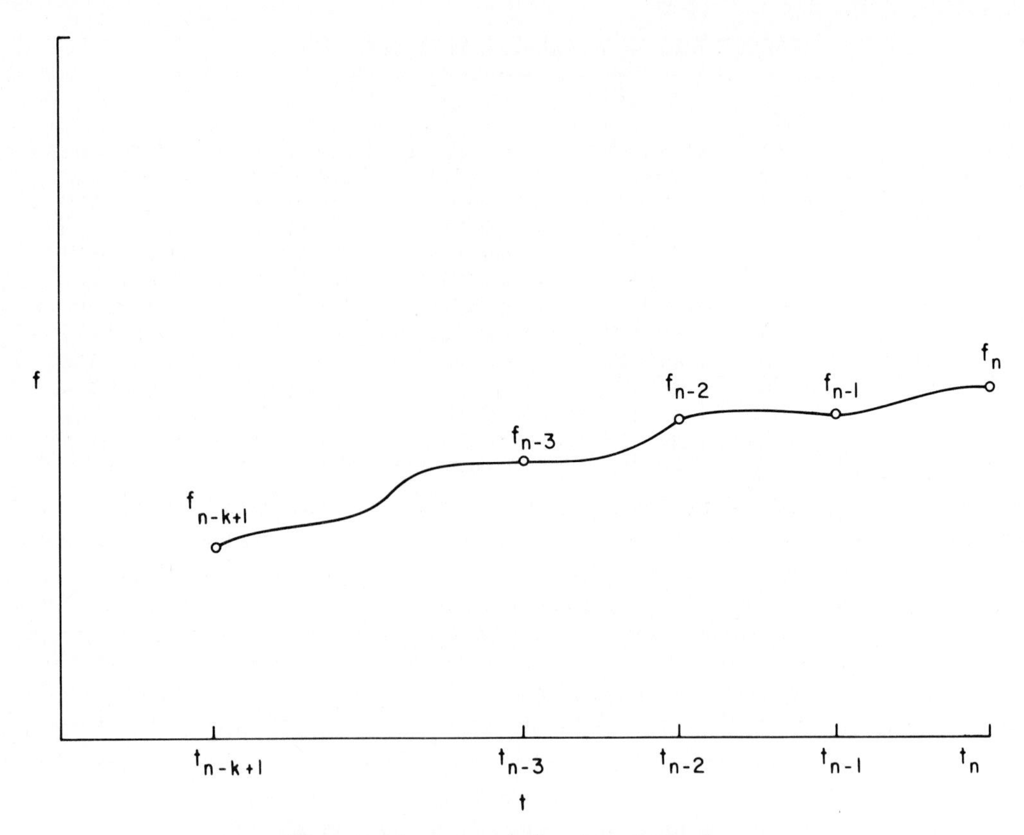

Figure 7.4.1 Interpolation through k points.

is used and the *backward differences* $\nabla^i f_n$ are defined as

$$
\begin{aligned}
f_i &= f(t_i), \qquad i = n-k+1, \ldots, n \\
\nabla^1 f_n &= f_n - f_{n-1} \\
\nabla^2 f_n &= \nabla^1 f_n - \nabla^1 f_{n-1} \\
&\vdots \\
\nabla^{k-1} f_n &= \nabla^{k-2} f_n - \nabla^{k-2} f_{n-1}
\end{aligned}
$$

The reader is encouraged to verify that Eq. 7.4.4 provides the desired values f_i for $k = 1, 2,$ and 3 (Prob. 7.4.2).

Example 7.4.2: Consider the use of Eq. 7.4.4 to approximate e^t near $t = 1$ using Newton backward difference polynomials. Let $t_n = 1$ and $h = 0.1$. $P_k(1.1)$ and $P_k(1.2)$ are evaluated for $k = 1, 2, \ldots, 10$ and tabulated in Table 7.4.2. The error $E_k(t) = P_k(t) - e^t$ is also tabulated. As expected, the higher the order of the polynomial and the smaller $|t-1|$, the better the approximation.

TABLE 7.4.2 Newton Backward Difference Approximation of e^t

k	$P_k(1.1)$	$E_k(1.1)$	$P_k(1.2)$	$E_k(1.2)$
1	2.718281828	−0.285884195	2.718281828	−0.601835094
2	2.976960546	−0.027205478	3.235639263	−0.084477660
3	3.001577080	−0.002588944	3.309488867	−0.010628056
4	3.003919653	−0.000246371	3.318859159	−0.001257764
5	3.004142579	−0.000023445	3.319973785	−0.000143137
6	3.004163793	−0.000002231	3.320101070	−0.000015852
7	3.004165812	−0.000000212	3.320115202	−0.000001721
8	3.004166004	−0.000000020	3.320116739	−0.000000184
9	3.004166022	−0.000000002	3.320116903	−0.000000019
10	3.004166024	0.000000000	3.320116921	−0.000000002
11	3.004166024	0.000000000	3.320116923	0.000000000

Interpolation Error In solving differential equations, an interpolation polynomial on one set of data points needs to be transformed to an interpolation polynomial on another set that is obtained by dropping the left point and adding a new point on the right, to march ahead on a time grid. The form of Taylor and Newton backward difference polynomials suggests that it may be possible to take advantage of the fact that most of the data remain the same on the two grids. To change from approximating f near t_0 using $f_0, \ldots, f_0^{(k-1)}$ to using $f_0, \ldots, f_0^{(k)}$, by Taylor polynomials, Eq. 7.4.2 yields

$$T_{k+1}(t) = T_k(t) + \frac{f_0^{(k)}}{(k)!}(t - t_0)^k$$

To achieve something comparable in the case of increasing the data set from k to $k + 1$ points for Newton backward difference interpolation requires that

$$P_{k+1}(t) = P_k(t) + R_{k+1}(t) \tag{7.4.5}$$

where $R_{k+1}(t)$ must be of degree k. That is, adding a data point amounts to adding a correction term to the current interpolating polynomial that is of higher degree. Substituting Eq. 7.4.4 into Eq. 7.4.5 and solving for $R_{k+1}(t)$,

$$R_{k+1}(t) = \frac{\nabla^k f_n}{(k)!\, h^k} \prod_{j=0}^{k-1} (t - t_{n-j}) \tag{7.4.6}$$

It is shown in References 31, 36, and 37 that

$$\frac{\nabla^k f_k}{(k)!\, h^k} = \frac{f^{(k)}(\xi)}{(k)!} \tag{7.4.7}$$

for some ξ in $[t_{n-k}, t_n]$. The error due to interpolating $f(t)$ by $P_k(t)$ is estimated, assuming $f(t) \approx P_{k+1}(t)$, by combining Eqs. 7.4.6 and 7.4.7,

$$f(t) - P_k(t) \approx R_{k+1}(t) = \frac{f^{(k)}(\xi)}{(k)!} \prod_{j=0}^{k-1} (t - t_{n-j}) \tag{7.4.8}$$

for some ξ in $[t_{n-k}, t_n]$. How good the estimate is depends on how much $f^{(k)}$ varies over the interval $[t_{n-k}, t_n]$.

An important application of Newton backward difference polynomials is the use of past function values to extrapolate ahead in time, as was illustrated in Example 7.4.2. Equation 7.4.8 indicates that, if the kth derivative of the function being approximated grows rapidly, then approximation error increases. Even though the kth derivative of the function is not known, the error estimate on the right of Eq. 7.4.8 may be approximated by the backward difference term on the left of Eq. 7.4.7, which can be monitored during computation. If the error estimate indicates a problem, the time step h may be reduced to decrease the magnitude of error.

Use of Polynomial Interpolation in Numerical Integration Interpolation methods for approximating the solution of Eq. 7.4.1., or

$$\begin{aligned}\dot{\mathbf{x}} \equiv \mathbf{x}^{(1)}(t) &= \mathbf{f}(\mathbf{x}, t)\\ \mathbf{x}(t_0) &= \mathbf{x}^0\end{aligned}$$

on an interval $[a, b]$ employ a mesh of grid points $\{t_0, t_1, \ldots\}$ in $[a, b]$, with $t_0 = a$. It is presumed that these grid points are equally spaced, in which case

$$t_n = t_0 + nh, \qquad n = 0, 1, \ldots$$

where h is the time step. The notation $\mathbf{x}_n$ means an approximation of the solution $\mathbf{x}(t)$ of Eq. 7.4.1 at grid point t_n; that is,

$$\mathbf{x}_n \approx \mathbf{x}(t_n)$$

Because $\mathbf{x}(t)$ satisfies the differential equation, an approximation of $\mathbf{x}(t_n)$ leads to an approximation of $\mathbf{x}_n^{(1)}(t_n)$; that is,

$$\mathbf{f}_n = \mathbf{f}(\mathbf{x}_n, t_n) \approx \mathbf{x}^{(1)}(t_n) = \mathbf{f}(\mathbf{x}(t_n), t_n)$$

The basic computational task in numerical integration is to advance the numerical solution from t_n to t_{n+1}, after having computed $\mathbf{x}_0, \mathbf{x}_1, \ldots, \mathbf{x}_n$. The exact solution can be obtained by integrating both sides of the differential equation from t_n to t_{n+1} to obtain

$$\begin{aligned}\mathbf{x}(t_{n+1}) &= \mathbf{x}(t_n) + \int_{t_n}^{t_{n+1}} \mathbf{x}^{(1)}(t)\, dt\\ &= \mathbf{x}(t_n) + \int_{t_n}^{t_{n+1}} \mathbf{f}(\mathbf{x}(t), t)\, dt \end{aligned} \tag{7.4.9}$$

The *Adams family of numerical integration methods* approximate this solution by replacing $\mathbf{f}(x(\mathbf{t}), t)$ in the interval $[t_n, t_{n+1}]$ by a polynomial that interpolates past computed values $\mathbf{f}_i$, which is then integrated.

Example 7.4.3: The simplest application of polynomial interpolation for numerical integration is to interpolate $\mathbf{f}(\mathbf{x}(t), t)$ by Eq. 7.4.4 with $k = 1$; that is, in $[t_n, t_{n+1}]$,

$$\mathbf{f}(\mathbf{x}(t), t) \approx \mathbf{f}_n \equiv \mathbf{f}(\mathbf{x}_n, t_n)$$

Using this relation in Eq. 7.4.9,

$$\begin{aligned}\mathbf{x}_{n+1} \approx \mathbf{x}(t_{n+1}) &\approx \mathbf{x}(t_n) + \int_{t_n}^{t_{n+1}} \mathbf{f}_n \, dt \\ &= \mathbf{x}(t_n) + \mathbf{f}_n(t_{n+1} - t_n) \\ &\approx \mathbf{x}_n + h\mathbf{f}_n\end{aligned}$$

where $t_{n+1} - t_n \equiv h$. This elementary approximation may be applied to the initial-value problem

$$\dot{x} = x$$

$$x(0) = 1$$

which has the unique solution $x(t) = e^t$. Approximate solutions are obtained with $h_1 = 0.1$ and $h_2 = 0.01$, with results tabulated at $t_i = 0.1i$ in Table 7.4.3.

TABLE 7.4.3 Numerical Approximation of e^t with $k = 1$

i	t_i	e^{t_i}	x_i ($h = 0.1$) (% error)	x_{10i} ($h = 0.01$) (% error)
0	0.0	1.0000000000	1.0000000000 (0.000)	1.0000000000 (0.000)
1	0.1	1.1051709181	1.1000000000 (0.468)	1.1046221254 (0.050)
2	0.2	1.2214027582	1.2100000000 (0.934)	1.2201900399 (0.099)
3	0.3	1.3498588076	1.3310000000 (1.397)	1.3478489153 (0.149)
4	0.4	1.4918246976	1.4641000000 (1.858)	1.4888637336 (0.198)
5	0.5	1.6487212707	1.6105100000 (2.318)	1.6446318218 (0.248)
6	0.6	1.8221188004	1.7715610000 (2.775)	1.8166966986 (0.298)
7	0.7	2.0137527075	1.9487171000 (3.230)	2.0067633684 (0.347)
8	0.8	2.2255409285	2.1435888100 (3.682)	2.2167152172 (0.397)
9	0.9	2.4596031112	2.3579476910 (4.133)	2.4486326746 (0.446)
10	1.0	2.7182818285	2.5937424601 (4.582)	2.7048138294 (0.495)

As expected, the accuracy of the numerical results is much better with the smaller step size. It must be recalled, however, that 100 time steps are required to reach $t = 1.0$, with the step size $h = 0.01$. Roughly speaking, you pay for the accuracy that you get.

7.4.2 Adams–Bashforth Predictor

The Adams–Bashforth method of approximating the solution of the initial-value problem of Eq. 7.4.1 is based on approximating the function $\mathbf{f}(\mathbf{x}(t), t)$ on the right of Eq. 7.4.9 by the Newton backward difference polynomial of Eq. 7.4.4. This presumes that $\mathbf{f}_{n-k+1}, \ldots, \mathbf{f}_n$ are available as approximations of $\mathbf{f}(\mathbf{x}(t), t)$ at $t_{n-k+1}, \ldots, t_n$, with $t_i - t_{i-1} = h$. The algorithm that is used to construct these approximations will become apparent soon. For now, presume they are known. Substituting from Eq. 7.4.4 for $\mathbf{f}(\mathbf{x}(t), t)$ into Eq. 7.4.9 and integrating the polynomial of degree $k-1$ yields a polynomial of degree k in t, which is evaluated at $t = t_{n+1}$. The result is called the *Adams–Bashforth formula of order k.*

A convenient notation for the polynomial of degree $k-1$ that is employed in the Adams–Bashforth formula of order k at t_n, which interpolates the computed values of $\mathbf{f}_i$ at k preceding points, is

$$\mathbf{P}_{k,n}(t_{n+1-j}) = \mathbf{f}_{n+1-j}, \qquad j = 1, 2, \ldots, k$$

An approximation, or prediction, of the solution at t_{n+1} is obtained as

$$\mathbf{x}^p_{n+1} = \mathbf{x}_n + \int_{t_n}^{t_{n+1}} \mathbf{P}_{k,n}(t)\, dt \tag{7.4.10}$$

Newton's backward difference polynomial from Eq. 7.4.4 is

$$\mathbf{P}_{k,n}(t) = \mathbf{f}_n + \sum_{i=1}^{k-1} \frac{\nabla^i \mathbf{f}_n}{(i)!\, h^i} \prod_{j=0}^{i-1} (t - t_{n-j})$$

This may be substituted into Eq. 7.4.10 to obtain the *Adams–Bashforth predictor of order k*:

$$\mathbf{x}^p_{n+1} = \mathbf{x}_n + h \sum_{i=1}^{k} \gamma_{i-1} \nabla^{i-1} \mathbf{f}_n \tag{7.4.11}$$

where $\nabla^0 \mathbf{f}_n = \mathbf{f}_n$ and

$$\begin{aligned} \gamma_0 &= 1 \\ \gamma_i &= \frac{1}{(i)!\, h} \int_{t_n}^{t_{n+1}} \prod_{j=0}^{i-1} \frac{t - t_{n-j}}{h}\, dt, \qquad i = 1, 2, \ldots \end{aligned} \tag{7.4.12}$$

which are called *Adams–Bashforth coefficients.*

To see that the values of γ_i are independent of h, introduce the change of

TABLE 7.4.4 Adams–Bashforth Coefficients

γ_0	γ_1	γ_2	γ_3	γ_4	γ_5
1	$\frac{1}{2}$	$\frac{5}{12}$	$\frac{3}{8}$	$\frac{251}{720}$	$\frac{95}{298}$

variable $s = (t - t_n)/h$ in Eq. 7.4.12 to obtain (Prob. 7.4.3)

$$\gamma_i = \frac{1}{(i)!}\int_0^1 \prod_{j=0}^{i-1} (s+j)\, ds$$

since $(t - t_{n-j})/h = [t - t_n + (t_n - t_{n-j})]/h = s + j$. Numerical values of the first six Adams–Bashforth coefficients are given in Table 7.4.4.

The Adams–Bashforth formula of Eq. 7.4.11 is called an *explicit method,* because the term x_{n+1} appears only on the left side of the equation and is determined explicitly by prior data. Note that at t_0 there are no past data points to interpolate, so the order k must be selected as 1. Thus, from Eq. 7.4.11 with $k = 1$,

$$\mathbf{x}_1 = \mathbf{x}_0 + h\mathbf{f}_0$$

At t_1, there is now one past data point, so the order $k = 2$ can be selected, and, from Eq. 7.4.11 and Table 7.4.4,

$$\begin{aligned}\mathbf{x}_2 &= \mathbf{x}_1 + h(\mathbf{f}_1 + \mathbf{f}_1 - \mathbf{f}_0)\\ &= \mathbf{x}_1 + h(2\mathbf{f}_1 - \mathbf{f}_0)\end{aligned}$$

This process can be repeated to increase the order k of the Adams–Bashforth predictor as more past integration points t_i, where $\mathbf{f}_i$ is known, are available. This process of increasing order k, from 1 up to the desired order, is called *starting the Adams–Bashforth algorithm.*

The approximation $\mathbf{x}^p_{n+1}$ in Eq. 7.4.11 has two sources of error. One source is due to approximating $\mathbf{x}^{(1)}(t)$ by an interpolating polynomial, known as *local truncation error.* The other source of error is errors that are present in the previously calculated values $\mathbf{x}_n, \mathbf{x}_{n-1}, \ldots$. It has been rigorously shown [31, 36, 37] that the latter contribution is insignificant when compared to local truncation error.

It is shown in Reference 31 that the local truncation error is

$$\boldsymbol{\tau}^p_{n+1} = \gamma_k h^{k+1}\mathbf{x}^{(k+1)}(\xi) \approx h\gamma_k \nabla^k \mathbf{f}_n \quad \textbf{(7.4.13)}$$

for constant step size h and some ξ in $[t_{n+1-k}, t_{n+1}]$. Note that if enough differences are kept and if $h < 1$ it is possible to adjust the order and/or to decrease the step size to make the local truncation error less than some desired error tolerance.

Example 7.4.4: Consider again the initial-value problem of Example 7.4.3:

$$\dot{x} = x$$
$$x(0) = 1$$

The solution is $x(t) = e^t$. A numerical approximation of the solution is to be constructed using Eq. 7.4.11 with $h = 0.1$. Backward differences calculated at t_i, using associated f_i, are shown in Table 7.4.5. Notice that at $t = 0$ the only available information is $x(0) = 1$. Hence, k must be 1 at $t = 0$. From Eq. 7.4.11,

$$\begin{aligned} x_1 &= x_0 + h(\gamma_0 \nabla^0 f_0) \\ &= 1 + (0.1)(1)(1) = 1.1 \end{aligned}$$

At $t = 0.1$, x_1 and x_0 are available. Hence, $\nabla^1 f_1$ can be evaluated, as shown in Table 7.4.5. Therefore, from Eq. 7.4.11 with $k = 2$,

$$\begin{aligned} x_2 &= x_1 + h(\gamma_0 \nabla^0 f_1 + \gamma_1 \nabla^1 f_1) \\ &= 1.1 + (0.1)\{(1)(1.1) + (\tfrac{1}{2})(0.1)\} \\ &= 1.215 \end{aligned}$$

Moving to $t_3, t_4, \ldots$, using successively larger values of order k in Eq. 7.4.11,

$$\begin{aligned} x_3 &= x_2 + h(\gamma_0 \nabla^0 f_2 + \gamma_1 \nabla^1 f_2 + \gamma_2 \nabla^2 f_2) \\ &= 1.215 + (0.1)\{(1)(1.215) + (\tfrac{1}{2})(0.115) + (\tfrac{5}{12})(0.015)\} \\ &= 1.342875 \\ x_4 &= x_3 + h(\gamma_0 \nabla^0 f_3 + \gamma_1 \nabla^1 f_3 + \gamma_2 \nabla^2 f_3 + \gamma_3 \nabla^3 f_3) \\ &= 1.342875 + (0.1)\{(1)(1.342875) + (\tfrac{1}{2})(0.127875) \\ &\quad + (\tfrac{5}{12})(0.012875) + (\tfrac{3}{8})(-0.002125)\} \\ &= 1.484013 \end{aligned}$$

TABLE 7.4.5 Newton Backward Differences

k	$\nabla^i f$	t_0	t_1	t_2	t_3	t_4
1	$\nabla^0 f$	1.0				
2	$\nabla^0 f$	1.0	1.1			
	$\nabla^1 f$		0.1			
3	$\nabla^0 f$	1.0	1.1	1.215		
	$\nabla^1 f$		0.1	0.115		
	$\nabla^2 f$			0.015		
4	$\nabla^0 f$	1.0	1.1	1.215	1.342875	
	$\nabla^1 f$		0.1	0.115	0.127875	
	$\nabla^2 f$			0.015	0.012875	
	$\nabla^3 f$				−0.002125	
5	$\nabla^0 f$	1.0	1.1	1.215	1.342875	1.484013
	$\nabla^1 f$		0.1	0.115	0.127875	0.141138
	$\nabla^2 f$			0.015	0.012875	0.132630
	$\nabla^3 f$				−0.002125	0.000385
	$\nabla^4 f$					0.002510

TABLE 7.4.6 Adams–Bashforth Predictor Values of e^{t_i}, $h = 0.1$

i	t_i	e^{t_i}	x_i (% error) $k = 1$	$k = 2$	$k = 3$	$k = 4$	$k = 5$
0	0.0	1.0					
1	0.1	1.105171	1.100000 (0.468)	1.100000 (0.468)	1.100000 (0.468)	1.100000 (0.468)	1.100000 (0.468)
2	0.2	1.221403	1.210000 (0.934)	1.215000 (0.524)	1.215000 (0.524)	1.215000 (0.524)	1.215000 (0.524)
3	0.3	1.349859	1.331000 (1.397)	1.342250 (0.564)	1.342875 (0.517)	1.342875 (0.517)	1.342875 (0.517)
4	0.4	1.491825	1.464100 (1.858)	1.482838 (0.602)	1.484093 (0.518)	1.484013 (0.524)	1.484013 (0.524)
5	0.5	1.648721	1.610510 (2.318)	1.638151 (0.641)	1.640119 (0.522)	1.640038 (0.527)	1.640126 (0.521)
6	0.6	1.822119	1.771561 (2.775)	1.809731 (0.680)	1.812549 (0.525)	1.812525 (0.527)	1.812679 (0.518)
7	0.7	2.013753	1.948717 (3.230)	1.999284 (0.719)	2.003109 (0.529)	2.003146 (0.527)	2.003305 (0.519)
8	0.8	2.225541	2.143589 (3.682)	2.208689 (0.757)	2.213703 (0.532)	2.213811 (0.527)	2.213988 (0.519)
9	0.9	2.459603	2.357948 (4.133)	2.440029 (0.796)	2.446348 (0.535)	2.446631 (0.527)	2.446842 (0.519)
10	1.0	2.718282	2.593742 (4.582)	2.695599 (0.834)	2.703641 (0.539)	2.703938 (0.528)	2.704177 (0.519)

TABLE 7.4.7 Adams–Bashforth Predictor Values of e^{t_i}, $h = 0.01$

i	t_i	e^{t_i}	x_i (% error) $k = 1$	$k = 2$	$k = 3$	$k = 4$	$k = 5$
0	0.0	1.0					
1	0.0	1.010050	1.010000 (0.005)	1.010000 (0.005)	1.010000 (0.005)	1.010000 (0.005)	1.010000 (0.005)
2	0.0	1.020201	1.020100 (0.010)	1.020150 (0.005)	1.020150 (0.005)	1.020150 (0.005)	1.020150 (0.005)
3	0.0	1.030455	1.030301 (0.015)	1.030402 (0.005)	1.030403 (0.005)	1.030403 (0.005)	1.030403 (0.005)
4	0.0	1.040811	1.040604 (0.020)	1.040758 (0.005)	1.040759 (0.005)	1.040748 (0.005)	1.040748 (0.005)
5	0.1	1.051271	1.051010 (0.025)	1.051217 (0.005)	1.051218 (0.005)	1.051218 (0.005)	1.051218 (0.005)
6	0.1	1.061837	1.061520 (0.030)	1.061781 (0.005)	1.061783 (0.005)	1.061783 (0.005)	1.061783 (0.005)
7	0.1	1.072508	1.072135 (0.035)	1.072452 (0.005)	1.072454 (0.005)	1.072454 (0.005)	1.072454 (0.005)
8	0.1	1.083287	1.082857 (0.040)	1.083230 (0.005)	1.083233 (0.005)	1.083233 (0.005)	1.083233 (0.005)
9	0.1	1.094174	1.093685 (0.045)	1.094116 (0.005)	1.094119 (0.005)	1.094119 (0.005)	1.094119 (0.005)
10	0.1	1.105171	1.104622 (0.050)	1.105112 (0.005)	1.105115 (0.005)	1.105115 (0.005)	1.105115 (0.005)

The foregoing process of implementing Eq. 7.4.11, starting with $k = 1$ and increasing the order to some desired value of k that is held fixed as the process proceeds to time grid points $t_{k+1}, t_{k+2}, \ldots$, may be continued until the terminal time is reached. This process is called the *self-starting Adams–Bashforth algorithm of order k*. For purposes of example calculations, it was implemented on a digital computer so that the effect of varying order k of the algorithm and step size h can be studied.

Detailed results for ten steps of Adams–Bashforth approximation of e^t that are generated with a digital computer for $h = 0.1$ and 0.01 are presented in Tables 7.4.6 and 7.4.7, respectively. Note that due to the self-starting procedure the results at t_1 are the same for all k. At t_2, the results are the same for $k = 2$ through 5. At t_3, the results are the same for $k = 3$ through 5. And so on. The substantial improvement in the accuracy of approximate solutions with increased k and reduced step size, as predicted by the error estimate of Eq. 7.4.13, is also evident.

In Table 7.4.8, a summary of approximate solutions at $t = 10$, with $k = 1, 2, \ldots, 5$ and $h = 0.1$, is provided. Note that the error $E_k(10)$ in the approximate solution at $t = 10$ decreases as k increases. These calculations are repeated with $h = 0.01$ and summarized in Table 7.4.9. As expected with a smaller step size, more accurate approximations are obtained. Recall, however, that 100 time steps are used with $h = 0.1$ and 1000 are required with $h = 0.01$, with an associated increase by a factor of 10 in computer cost. The solution is, however, 100 times more accurate, a nice return on investment, if accuracy is needed.

TABLE 7.4.8 Adams–Bashforth Predictor Values of e^{10}, $h = 0.1$

k	$x^p(10)$	$E_k(10)$	$\frac{\lvert E_k(10)\rvert}{e^{10}} \times 100(\%)$
1	13,780.612	−8245.853	37
2	21,090.171	−936.294	4
3	21,841.518	−184.947	0.8
4	21,904.347	−122.118	0.5
5	21,911.635	−114.830	0.5

TABLE 7.4.9 Adams–Bashforth Predictor Values of e^{10}, $h = 0.01$

k	$x^p(10)$	$E_k(10)$	$\frac{\lvert E_k(10)\rvert}{e^{10}} \times 100\ (\%)$
1	20,959.155	−1067.310	5
2	22,016.255	−10.209	0.05
3	22,025.280	−1.185	0.005
4	22,025.356	−1.108	0.005
5	22,025.361	−1.104	0.005

A brief examination of the results of Tables 7.4.8 and 7.4.9 reveals that with $h = 0.1$ the error in the approximation of the solution e^{10} by the first-order algorithm $(k = 1)$ is clearly unacceptable. With $h = 0.1$ and $k = 2$, as well as with $h = 0.01$ and $k = 1$, error is significant, but possibly acceptable for some applications. Acceptable solution accuracy, for many applications, is obtained with $h = 0.1$ and $k = 3$, 4, or 5. Excellent accuracy is obtained with $h = 0.01$ and $k = 3$, 4, or 5.

It is interesting to note that in Example 7.4.4, for $h = 0.01$, accuracy is not significantly improved by increasing the order of the algorithm from $k = 4$ to $k = 5$. Similarly, for $h = 0.01$, accuracy is not significantly improved when the order is increased from $k = 3$ to $k = 4$ or 5. The reason for this apparent contradiction of the error estimate of Eq. 7.4.13 is that error in the starting steps, when $k = 1$ and 2, cannot be eliminated or compensated for by the much more accurate higher-order steps that occur throughout the remainder of the numerical integration process. This suggests that very small time steps should be used during the starting steps until enough time grid points with accurate solution values are available to switch to the intended larger time step and a higher-order algorithm. For example, a starting time step of $h_{\text{start}} = 0.01$ could be used for 40 time steps to reach $t = 0.4$ with order k growing to 5. At that time data at $t = 0.0$, 0.1, 0.2, 0.3, and 0.4 could be used to implement the Adams–Bashforth algorithm with order $k = 5$ and $h = 0.1$ for the remainder of the numerical integration.

7.4.3 Adams–Moulton Corrector

The value $\mathbf{x}_{n+1}^p$ obtained by the Adams–Bashforth predictor may not be the best approximation to $\mathbf{x}(t_{n+1})$, because $\mathbf{f}(\mathbf{x}(t_{n+1}), t_{n+1})$ was not involved in the determination of $\mathbf{x}_{n+1}$. Indeed, if $\mathbf{f}(\mathbf{x}(t_{n+1}), t_{n+1})$ should unexpectedly change in going from t_n to t_{n+1}, the difference between $\mathbf{x}_{n+1}$ and $\mathbf{x}(t_{n+1})$ could be much greater than $\boldsymbol{\tau}_{n+1}^p$. This is because the factor $\mathbf{x}^{(k+1)}(\xi)$ in Eq. 7.4.13, which previously was smooth, has suddenly changed in magnitude in the interval t_n to t_{n+1}.

It seems plausible that a better approximation to $\mathbf{x}(t_{n+1})$ could be obtained if the Adams–Bashforth predicted value $\mathbf{x}_{n+1}^p$ of Eq. 7.4.11 were improved by incorporating it into an improved interpolating polynomial for $\mathbf{f}(\mathbf{x}(t), t)$ in Eq. 7.4.9. The *Adams–Moulton corrector formula of order* $k + 1$ at t_n uses a polynomial $\mathbf{P}_{k+1,n}^*(t)$ that interpolates the previous k values of $\mathbf{f}_i$,

$$\mathbf{P}_{k+1,n}^*(t_{n+1-j}) = \mathbf{f}_{n+1-j}, \qquad j = 1, \ldots, k$$

and the approximation for $\mathbf{f}$ at t_{n+1} that has been obtained from the Adams–Bashforth predictor; that is,

$$\mathbf{P}_{k+1,n}^*(t_{n+1}) = \mathbf{f}(\mathbf{x}_{n+1}^p, t_{n+1}) \equiv \mathbf{f}_{n+1}^p$$

Since $k + 1$ points are being interpolated, $\mathbf{P}_{k+1,n}^*(t)$ is a polynomial of degree k.

Note that this is a polynomial of one higher degree than was used in the associated Adams–Bashforth predictor.

The approximate corrected solution $\mathbf{x}_{n+1}^c$ is obtained from Eq. 7.4.9 as

$$\mathbf{x}_{n+1}^c = \mathbf{x}_n + \int_{t_n}^{t_{n+1}} \mathbf{P}_{k+1,n}^*(t)\, dt \qquad \textbf{(7.4.14)}$$

The backward difference form of Eq. 7.4.14, derived in exactly the same way as Eq. 7.4.11, yields the *Adams–Moulton corrector of order* $k+1$ [31] as

$$\mathbf{x}_{n+1}^c = \mathbf{x}_n + h \sum_{i=1}^{k+1} \gamma_{i-1}^* \nabla^{i-1} \mathbf{f}_{n+1}^p \qquad \textbf{(7.4.15)}$$

where, with $s = (t - t_n)/h$,

$$\gamma_0^* = 1$$

$$\gamma_i^* = \frac{1}{i!} \int_0^1 (s-1)(s) \ldots (s+i-2)\, ds, \qquad i \geq 1$$

and

$$\nabla^0 \mathbf{f}_{n+1}^p = \mathbf{f}_{n+1}^p$$

$$\nabla^1 \mathbf{f}_{n+1}^p = \nabla \mathbf{f}_{n+1}^p = \mathbf{f}_{n+1}^p - \mathbf{f}_n$$

$$\vdots$$

$$\nabla^i \mathbf{f}_{n+1}^p = \nabla^{i-1} \mathbf{f}_{n+1}^p - \nabla^{i-1} \mathbf{f}_n$$

The first five *Adams–Moulton coefficients* are given in Table 7.4.10.

TABLE 7.4.10 Adams–Moulton Coefficients

γ_0^*	γ_1^*	γ_2^*	γ_3^*	γ_4^*	γ_5^*
1	$-\frac{1}{2}$	$-\frac{1}{12}$	$-\frac{1}{24}$	$-\frac{19}{720}$	$-\frac{3}{160}$

Example 7.4.5: Consider the initial-value problem of Example 7.4.4 for application of the Adams–Moulton corrector formula. The predicted value of x_1 is obtained with an Adams–Bashforth predictor of order 1, from Eq. 7.4.11, as

$$x_1^p = x_0 + h(\gamma_0 \nabla^0 f_0)$$

$$= 1 + (0.1)(1)(1) = 1.1$$

This prediction is corrected by the Adams–Moulton corrector of order 2; that is, from Eq. 7.4.15,

$$x_1^c = x_0 + h(\gamma_0^* \nabla^0 f_1 + \gamma_1^* \nabla^1 f_1)$$

$$= 1 + (0.1)\{(1)(1.1) + (-\tfrac{1}{2})(0.1)\} = 1.105$$

The value of x_2 is predicted with order 2 as

$$x_2^p = x_1 + h(\gamma_0 \nabla^0 f_1 + \gamma_1 \nabla^1 f_1)$$
$$= 1.105 + (0.1)\{(1)(1.105) = (0.5)(0.105)\} = 1.22075$$

and is corrected with order 3 as

$$x_2^c = x_1 + h(\gamma_0^* \nabla^0 f_2 + \gamma_1^* \nabla^1 f_2 + \gamma_2^* \nabla^2 f_2)$$
$$= 1.105 + (0.1)\{(1)(1.22075) + (-0.5)(0.11575) + (-\tfrac{1}{12})(0.01075)\}$$
$$= 1.2211979$$

Similarly, the prediction of x_3 with order 3 is

$$x_3^p = x_2 + h(\gamma_0 \nabla^0 f_3 + \gamma_1 \nabla^1 f_3 + \gamma_2 \nabla^2 f_3)$$
$$= 1.2211979 + (0.1)\{(1)(1.2211979) + (0.5)(0.1161979) + (\tfrac{5}{12})(0.0111979)\}$$
$$= 1.3495942$$

and correction with order 4 yields

$$x_3^c = x_2 + h(\gamma_0^* \Delta^0 f_3 + \gamma_1^* \Delta^1 f_3 + \gamma_2^* \Delta^2 f_3 + \gamma_3^* \Delta^3 f_3)$$
$$= 1.2211979 + (0.1)\{(1)(1.3495942) + (-0.5)(0.1283963)$$
$$+ (-\tfrac{1}{12})(0.0121984) + (-\tfrac{1}{24})(0.0010005)\}$$
$$= 1.3496317$$

This sequence starts the *Adams–Bashforth, Adams–Moulton predictor–corrector algorithm,* which is self-starting.

In Table 7.4.11, a summary of the *predictor–corrector approximate solution* at $t = 10$, predicted by Adams–Bashforth formulas and corrected by one-higher-order Adams–Moulton formulas, is provided. The value k given is the order of the Adams–Bashforth prediction that is used after starting is complete. The order of the Adams–Moulton corrector is one higher. Calculations are repeated with $h = 0.01$ and results are summarized in Table 7.4.12. Note that the solutions are substantially more accurate than those obtained in Example 7.4.4 using only Adams–Bashforth predictors.

TABLE 7.4.11 Adams–Moulton Corrector Values of e^{10}, $h = 0.1$

k	$x^c(10)$	$E_k(10)$	$\frac{\lvert E_k(10)\rvert}{e^{10}} \times 100(\%)$
1	21,688.414370387	−338.051452637	2.
2	21,980.499079835	−45.966720581	0.2
3	21,994.743244024	−31.722553253	0.2
4	21,990.708589885	−35.757209778	0.2
5	21,986.702800628	−39.763000488	0.2

TABLE 7.4.12 Adams–Moulton Corrector Values of e^{10}, $h = 0.01$

k	$x^c(10)$	$E_k(10)$	$\frac{\lvert E_k(10)\rvert}{e^{10}} \times 100(\%)$
1	22,022.822441367	−3.643353462	0.02
2	22,026.251549431	−0.214245379	0.001
3	22,026.188149769	−0.277645051	0.001
4	22,026.130130982	−0.335663855	0.002
5	22,026.088790048	−0.377004802	0.002

Similar to the previous development, the local truncation error $\boldsymbol{\tau}_{n+1}^c$ of the corrected value is defined by

$$\mathbf{x}^c(t_{n+1}) = \mathbf{x}(t_n) + \sum_{i=1}^{k+1} \gamma_{i-1}^* \nabla^{i-1} \mathbf{f}_{n+1}^p + \boldsymbol{\tau}_{n+1}^c$$

It is shown in Reference 31 that, for constant step size h, this truncation error is given by

$$\boldsymbol{\tau}_{n+1}^c = \gamma_{k+1}^* h^{k+2} \mathbf{x}^{(k+2)}(\xi) \approx h \gamma_{k+1}^* \nabla^{k+1} \mathbf{f}_{n+1}^p \tag{7.4.16}$$

with ξ in $[t_{n-1}, t_{n+1}]$.

7.4.4 Computer Implementation of Predictor–Corrector Algorithms

The work involved in doing numerical integration on a computer is measured by the number of evaluations of $\mathbf{f}(\mathbf{x}, t)$ that are required. For computing intensive function evaluations, such as those encountered in mechanical system dynamics (see Sections 7.2 and 7.3), the remaining computation is relatively small, so this practice is an effective way of measuring work that is also machine independent. The predictor–corrector procedure costs two function evaluations at each step, so it is natural to ask, "Why not use the Adams–Bashforth method, which costs only one function evaluation at each step?" The predictor–corrector approach is more accurate and is so much better with respect to propagation of error that it can use steps that may be more than twice as large. Also, error estimates are more reliable, which leads to a more effective selection of step size.

The kind of predictor–corrector procedure presented here is called a *PECE method,* an acronym derived from a description of how the computation is done. First, Predict $\mathbf{x}_{n+1}^p$, Evaluate $\mathbf{f}_{n+1}^p$, Correct to obtain $\mathbf{x}_{n+1}^c$, and Evaluate $\mathbf{f}_{n+1}$ to complete the step. Corrector formulas of different orders could be considered, but in a PECE mode, there is no advantage in using a corrector formula that is more than one order higher than the predictor. It is shown in Reference 36 that a corrector of order $k + 1$, together with a predictor of order k, is the best choice.

This choice is also convenient, since the predictor of order k uses the values $\mathbf{x}_n$ and $\mathbf{f}_{n+1-j}$ for $j = 1, \ldots, k$, and the corrector of order $k + 1$ uses the same values, along with $\mathbf{x}_{n+1}^p$. Thus, quantities to be retained are the same for the predictor and the corrector. Reference to PECE formulas of order k will mean the Adams–Bashforth predictor of order k and the Adams–Moulton corrector of order $k + 1$.

Self-starting computer codes have been developed that require the use of only the differential equations and initial conditions. Since a PECE method requires two function evaluations per step, regardless of the order, and since higher-order formulas are expected to be more efficient, starting values for higher-order formulas are developed as in Examples 7.4.4 and 7.4.5. After completion of the first step with a first-order predictor and second-order corrector, $\mathbf{x}_1$, $\mathbf{f}_1$, and $\mathbf{f}_0$ have been stored, which is exactly the information that is needed for a second-order predictor and third-order corrector to take the second step. With $\mathbf{x}_2$, $\mathbf{f}_2$, $\mathbf{f}_1$, and $\mathbf{f}_0$, a third-order predictor and fourth-order corrector can be used, and so on. In this way, the order can be increased as necessary data are computed. The lower-order formulas are generally less accurate, which must be compensated for by taking smaller steps.

Excellent texts are available that expand on the theory outlined here for predictor–corrector solution of initial-value problems [31, 36, 37]. Error estimates are derived and their use in the selection of order k and step size h to maintain specified error tolerances is presented. A number of well-developed and tested computer codes are available that implement these techniques. One of the best of these codes is the DE numerical integration subroutine developed by Shampine and Gordon [36]. The reader who is interested in greater theoretical depth and information on the foundations of numerical integration computer codes is referred to References 36 and 38.

A few comments here may assist the reader in appreciating that the error estimates presented in Sections 7.4.2 and 7.4.3 can be used to adjust order k and step size h to satisfy an error tolerance. The error estimate of Eq. 7.4.16 is evaluated for the order k and step size h that have been used for the past few steps.

If components of $\boldsymbol{\tau}_{n+1}^c$ are greater than the specified error tolerance $\varepsilon > 0$, then the step size is reduced and/or the order k is adjusted. Normally, step size is reduced by a factor of 2 and predictor calculations are repeated with the reduced step size. If the error estimate of Eq. 7.4.16 is still greater than ε, further action is required. The error estimate may now be evaluated for reduced and increased orders $k - 1$ and $k + 1$ to determine if a change in order will reduce error. If significant error reduction can be achieved by a modification of order, the change is made and integration is continued, perhaps with still further reduction in step size.

If all components of $\boldsymbol{\tau}_{n+1}^c$ after a predictor step are much smaller than ε, then the algorithm is too conservative and order and/or step size can be increased to achieve greater efficiency. As before, error estimates are evaluated with

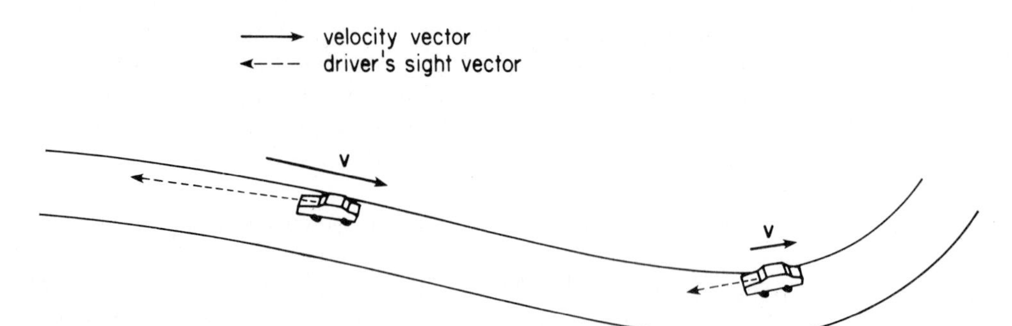

Figure 7.4.2 Backward driving analogy.

reduced and increased order to determine if an order change is beneficial. Only if significant gains are possible are step size and/or order changed.

The process of changing step size and order may be intuitively related to the homely example of driving a vehicle backward, looking only at the road previously traversed. This analogy is meaningful, since numerical integration ahead in time in fact uses only past information. Along a straight stretch of road, as indicated at the left of Fig. 7.4.2, the driver may move with large velocity v, traversing long patches of road in a unit of time. This is analogous to using a large numerical step size h. At the same time, the driver is relatively relaxed and is viewing a long patch of the road over which he has come, as illustrated by the long dashed sight vector in Fig. 7.4.2. This is analogous to using a high integration order k, that is, making use of many previously passed data points.

If the road changes direction abruptly, as shown at the right of Fig. 7.4.2, the prudent driver will take two actions. First, the driver will slow down, which is analogous to reducing step size in the integration process. Second, the driver will concentrate his or her attention on road curvature information near the vehicle, as illustrated by the short dashed sight vector in Fig. 7.4.2, to sense the changing direction of the road. This is analogous to reducing the integration order k, using only the most recently traversed road data, which contains information on changing road direction.

7.5 NUMERICAL METHODS FOR EQUILIBRIUM ANALYSIS

A special mode of dynamic analysis that seeks to find a configuration in which velocity and acceleration are zero is called *equilibrium analysis*. The basic analytical condition that defines equilibrium is that $\dot{\mathbf{q}} = \ddot{\mathbf{q}} = \mathbf{0}$, consistent with the equations of motion of Eq. 6.4.2 and the equations of constraint. Three approaches to determining equilibrium configurations are discussed in this section, two of which are computationally practical. The first involves integrating the equations of motion of the system under the action of applied forces, perhaps adding artificial damping, until the system comes to rest at an equilibrium configuration. The second approach is based on writing the equations of motion with velocity and acceleration equal to zero and attempting to solve the resulting equations of equilibrium for an equilibrium configuration. The third and final method is based on the principle of minimum total potential energy for conservative systems, which states that a conservative system is in a state of stable equilibrium if and only if the total potential energy is at a strict relative minimum.

Dynamic Settling The most universal method for finding a stable equilibrium configuration for systems with nonconservative forces is to integrate the equations of motion until $\dot{\mathbf{q}} = \ddot{\mathbf{q}} = \mathbf{0}$ to within a numerical tolerance. To accelerate convergence to an equilibrium configuration, it is often desirable to add artificial damping to the system. Given a dynamic analysis capability, this method is easy to implement, but requires substantial computing time. In the case of dissipative nonconservative systems, many equilibrium configurations may exist. The one that actually occurs depends on the initial conditions chosen. In such situations, the dynamic settling approach to equilibrium analysis is the only valid method. Even when the system is nonconservative, but has a unique equilibrium configuration, the dynamic settling approach is valid and practical.

Equations of Equilibrium As noted previously, the analytical conditions for equilibrium are

$$\dot{\mathbf{q}} = \ddot{\mathbf{q}} = \mathbf{0} \tag{7.5.1}$$

Substituting these equations in the Lagrange multiplier form of the equations of motion of Eq. 7.2.1 yields the *system equilibrium equations*

$$\begin{aligned} \mathbf{\Phi}_{\mathbf{q}}^{T}(\mathbf{q})\boldsymbol{\lambda} &= \mathbf{Q}^{A}(\mathbf{q}) \\ \mathbf{\Phi}(\mathbf{q}) &= \mathbf{0} \end{aligned} \tag{7.5.2}$$

These equations are nonlinear in the generalized coordinates $\mathbf{q}$ and Lagrange multipliers $\boldsymbol{\lambda}$. Thus, some iterative form of numerical solution is required, such as Newton's method of Section 4.5.

Numerical solution of Eqs. 7.5.2 is not recommended for equilibrium analysis of general mechanical systems for several reasons. First, the Newton–Raphson equations that arise in attempting to solve this system are often ill-conditioned, leading to large perturbations and uncertainty as to convergence or arrival at an unwanted equilibrium configuration. This difficulty is aggravated by poor estimates of unknowns, which included Lagrange multipliers $\boldsymbol{\lambda}$. Intuition is often of little value in making reasonable estimates of $\boldsymbol{\lambda}$. Second, Eqs. 7.5.2 are satisfied at both stable and unstable equilibrium configurations. Therefore, depending on the estimate given to start Newton–Raphson iteration and conditioning of the Newton–Raphson equations, it is entirely possible that the algorithm will converge to an unstable equilibrium configuration that makes little physical sense, but is mathematically valid as a solution of Eqs. 7.5.2. This behavior has been observed to occur often for systems in which applied forces vary by several orders of magnitude, such as in systems with very heavy components and components of modest weight.

Minimum Total Potential Energy A well-known condition that defines a configuration $\mathbf{q}_e$ of stable equilibrium is that the *total potential energy* $V(\mathbf{q})$ of a *conservative system* should be at a strict relative minimum [34]; that is,

$$V(\mathbf{q}_e) < V(\mathbf{q}) \tag{7.5.3}$$

for all $\mathbf{q} \neq \mathbf{q}_e$ in a neighborhood of $\mathbf{q}_e$, such that

$$\mathbf{\Phi}(\mathbf{q}) = \mathbf{0} \tag{7.5.4}$$

This criterion for equilibrium is a *constrained minimization problem*, which can be solved using an algorithm similar to that employed for assembly analysis in Section 4.3. Prior to developing the minimization algorithm, however, an expression must be obtained for the total potential energy $V(\mathbf{q})$ of the system, and its gradient must be evaluated.

Consider first a constant force $\mathbf{F}$ that acts at point P on a rigid body (e.g., a gravitational force). The virtual work of this force is

$$\delta W_F = \mathbf{F}^T \delta \mathbf{r}^P \tag{7.5.5}$$

Since force $\mathbf{F}$ is constant, this is the total differential of the work function

$$W_F = \mathbf{F}^T \mathbf{r}^P \tag{7.5.6}$$

Thus, the potential energy of the constant applied force $\mathbf{F}$ is

$$V_{\mathbf{F}} = -W_{\mathbf{F}} = -\mathbf{F}^T \mathbf{r}^P \tag{7.5.7}$$

where $\mathbf{r}^P$ depends on $\mathbf{q}$.

Consider next a translational spring–damper–actuator that acts between a pair of bodies, which applies a force that is a function of the length ℓ of the element,

$$f = k(\ell - \ell_0) + F(\ell) \tag{7.5.8}$$

where ℓ_0 is the free length of the spring. Since a positive f tends to draw the bodies together and a positive $\delta\ell$ tends to move them apart, the virtual work is

$$\begin{aligned} \delta W_f &= -f\,\delta\ell \\ &= -k(\ell - \ell_0)\,\delta\ell - F(\ell)\,\delta\ell \\ &= -k(\ell - \ell_0)\,\delta(\ell - \ell_0) - F(\ell)\,\delta\ell \end{aligned} \tag{7.5.9}$$

where $\delta\ell_0 = 0$, since ℓ_0 is constant. Integrating Eq. 7.5.9, the work function associated with the translational spring–damper–actuator force is

$$W_f = -\frac{k}{2}(\ell - \ell_0)^2 - \int_{\ell_0}^{\ell} F(\ell)\,d\ell \tag{7.5.10}$$

Thus, the potential energy of the translational spring–damper–actuator force is simply

$$V_f = \frac{k}{2}(\ell - \ell_0)^2 + \int_{\ell_0}^{\ell} F(\ell)\, d\ell \tag{7.5.11}$$

Similarly, for a torsional spring–damper–actuator, the potential energy is

$$V_\tau = \frac{k_\theta}{2}(\theta - \theta_0)^2 + \int_{\theta_0}^{\theta} T(\theta)\, d\theta \tag{7.5.12}$$

The *total potential energy* of the system is the sum of the potential energies of all forces that act on the system; that is,

$$V = V_{\mathbf{F}} + V_f + V_\tau \tag{7.5.13}$$

where the sum is taken over all bodies in the system, so that $V = V(\mathbf{q})$.

For implementation of numerical minimization techniques, it is important to be able to calculate the derivatives of total potential energy with respect to generalized coordinates. It is well known in the theory of conservative force systems [34] that the gradient of total potential energy is the negative of the generalized applied force; that is,

$$V_{\mathbf{q}}^T = -\mathbf{Q}^A \tag{7.5.14}$$

Since the generalized applied force $\mathbf{Q}^A$ has been assembled in the formulation of the equations of motion, the gradient of total potential energy is readily available within the dynamics formulation presented in this text.

Many constrained optimization algorithms are available and can be applied to the minimization problem of Eqs. 7.5.3 and 7.5.4. However, the problem can be transformed to an unconstrained minimization problem using the generalized coordinate partitioning technique of Section 7.2.

For a mechanical system with a vector $\mathbf{q}$ of nc generalized coordinates and nh independent constraint equations $\mathbf{\Phi}(\mathbf{q}) = \mathbf{0}$, the vector $\mathbf{q}$ may be partitioned into $\mathbf{q} = [\mathbf{u}^T, \mathbf{v}^T]^T$, where $\mathbf{u}$ is an nh vector of dependent generalized coordinates and $\mathbf{v}$ is an $nc - nh$ vector of independent generalized coordinates. Accordingly, the Jacobian $\mathbf{\Phi}_{\mathbf{q}}$ is partitioned into $\mathbf{\Phi}_{\mathbf{q}} = [\mathbf{\Phi}_{\mathbf{u}}, \mathbf{\Phi}_{\mathbf{v}}]$. The total differential of the constraint equations is

$$d\mathbf{\Phi} = \mathbf{\Phi}_{\mathbf{u}}\, d\mathbf{u} + \mathbf{\Phi}_{\mathbf{v}}\, d\mathbf{v} = \mathbf{0} \tag{7.5.15}$$

If the nh constraint equations are independent, the submatrix $\mathbf{\Phi}_{\mathbf{u}}$ of $\mathbf{\Phi}_{\mathbf{q}}$ may be selected to be nonsingular. Therefore, Eq. 7.5.15 can be rewritten in the form

$$d\mathbf{u} = -\mathbf{\Phi}_{\mathbf{u}}^{-1}\mathbf{\Phi}_{\mathbf{v}}\, d\mathbf{v} \tag{7.5.16}$$

A matrix $\mathbf{H}$, called the *influence coefficient matrix*, is defined as

$$\mathbf{H} = -\mathbf{\Phi}_{\mathbf{u}}^{-1}\mathbf{\Phi}_{\mathbf{v}} \tag{7.5.17}$$

Then Eq. 7.5.16 becomes

$$d\mathbf{u} = \mathbf{H}\, d\mathbf{v} \tag{7.5.18}$$

The influence coefficient matrix $\mathbf{H}$ is calculated by solving

$$\mathbf{\Phi_u H}^{(i)} = -\mathbf{\Phi}_{\mathbf{v}}^{(i)}, \qquad i = 1, 2, \ldots, nc - nh \tag{7.5.19}$$

where $\mathbf{H}^{(i)}$ is the ith column of $\mathbf{H}$ and $\mathbf{\Phi}_{\mathbf{v}}^{(i)}$ is the ith column of $\mathbf{\Phi_v}$.

The total differential of the potential energy function V may be written as

$$dV = V_{\mathbf{u}}\, d\mathbf{u} + V_{\mathbf{v}}\, d\mathbf{v} \tag{7.5.20}$$

Substitution of Eq. 7.5.18 into Eq. 7.5.20 yields

$$\frac{dV}{d\mathbf{v}} = V_{\mathbf{u}}\mathbf{H} + V_{\mathbf{v}} \tag{7.5.21}$$

Recalling that the implicit function theorem guarantees that $\mathbf{u} = \mathbf{h}(\mathbf{v})$ (see Eq. 7.2.8), the minimization problem can be restated as choosing $\mathbf{v}$ to

$$\text{minimize } V = V(\mathbf{v}) \tag{7.5.22}$$

Equation 7.5.22 is an unconstrained optimization problem in the $nc - nh$ variables $\mathbf{v}$. Newton–Raphson iteration is used to determine dependent coordinates that satisfy the nonlinear kinematic constraint equations, while independent coordinates are fixed at the value determined by an unconstrained minimization algorithm at each step. The gradient $(dV/d\mathbf{v})$ is an $nc - nh$ vector that is needed for most commonly used optimization algorithms (e.g., the Fletcher–Powell algorithm of Section 4.3).

The gradient vector $dV^T/d\mathbf{v}$ can be easily determined at a feasible position $\mathbf{q}$, that is, where $\mathbf{\Phi}(\mathbf{q}) = \mathbf{0}$ is satisfied, as the negative of the generalized applied force $\mathbf{Q}^{AT}(\mathbf{q})$ from Eq. 7.5.14. If $\mathbf{Q}^A$ is partitioned, according to $\mathbf{q} \equiv [\mathbf{u}^T, \mathbf{v}^T]^T$, into $\mathbf{Q}^A = [\mathbf{Q}^{A\mathbf{u}^T}, \mathbf{Q}^{A\mathbf{v}^T}]$, then

$$\begin{aligned} V_{\mathbf{u}}^T &= -\mathbf{Q}^{A\mathbf{u}} \\ V_{\mathbf{v}}^T &= -\mathbf{Q}^{A\mathbf{v}} \end{aligned} \tag{7.5.23}$$

Substituting Eq. 7.5.23 into Eq. 7.5.21 yields the desired result:

$$\frac{dV^T}{d\mathbf{v}} = -\mathbf{H}^T\mathbf{Q}^{A\mathbf{u}} - \mathbf{Q}^{A\mathbf{v}} \tag{7.5.24}$$

The Fletcher–Powell algorithm for unconstrained minimization presented in Section 4.3 may now be applied to minimize $V(\mathbf{v})$.

PROBLEMS

Section 7.2

7.2.1. Verify that Eq. 7.2.14 is valid.

7.2.2. Partition the Lagrange multiplier form of the equations of motion for the simple pendulum derived in Example 6.3.4 to obtain the result given in Example 7.2.1.

7.2.3. Verify that Eq. 7.2.18 is valid.

Section 7.4

7.4.1. Show that $T_k^{(i)}(t_0) = f^{(i)}(t_0)$, where $T_k(t)$ is given by Eq. 7.4.2.

7.4.2. Show that $f(t_i) = P_k(t_i)$, where $P_k(t)$ is given by Eq. 7.4.4, for $k = 1$, 2, and 3, where $t_i = t_n - (n - i)h$, $i = n - k + 1, \ldots, n$.

7.4.3. Make the change of variable $s = (t - t_n)/h$ in Eq. 7.4.12 and verify that the value of γ_i is independent of h.

CHAPTER EIGHT

Planar Dynamic Modeling and Analysis

8.1 MODELING AND ANALYSIS TECHNIQUES

Modeling the dynamics of a planar machine involves selecting a kinematic model; defining the initial conditions, forces, and inertia properties of components of the system; and forming and solving the equations of motion. Three fundamentally different modes of dynamic analysis may be considered; (1) *equilibrium analysis,* in which an equilibrium position is sought under the action of specified forces, (2) *inverse dynamic analysis,* in which drivers define the motion of the system and reaction forces are calculated from Lagrange multipliers, and (3) *dynamic analysis,* in which transient response to applied forces is predicted by integrating the mixed differential–algebraic equations of motion.

The results of a dynamic analysis depend on the inertia properties of a system, as well as on the kinematic model selected. If composite joints, called massless links, are employed in analysis, the engineer must be aware that the mass properties of the couplers are being neglected. In many cases, the mass of a coupler can be distributed as lumped masses at the attachment points on each of the bodies that are connected. This procedure can improve the quality of approximation, but will not be precisely equivalent to results that are obtained by defining the actual mass distribution of the coupler.

It is important that data be generated that are consistent with the conventions used in Chapter 6 to derive the governing equations of motion. In particular, the origin of each body-fixed x'-y' frame must be at the center of mass of the body. Otherwise, fundamental errors will be made in the resulting dynamic predictions. Furthermore, the force and torque system that acts on each body from external sources must be reduced to a force that acts at the center of mass of the body and an associated torque.

The same examples studied kinematically in Chapter 5 are analyzed in this chapter with the aid of the DADS computer code [27]. All data required to formulate the equations of motion derived in Chapter 5 are given in tabular form. These data and kinematic data for the models presented in Chapter 5 are entered into the DADS computer code. The governing equations of motion are automatically formed and solved in the DADS code, which implements the numerical methods presented in Chapter 7. The focus of this chapter is on problem formulation and interpretation of results.

While the analytical techniques presented in the preceding chapters can be effectively learned in a classroom or in self-study, the development of a practical capability in dynamic system modeling and analysis requires considerable application experience. The engineer or engineering analyst must develop intuition that is based on experience in order to evaluate results that are predicted by either a general-purpose computer code such as DADS or by a special-purpose computer program. Even with a well-tested, error-free, general-purpose computer code, if the user makes errors in input data or defines a poor design, undesired behavior of the system will occur. The engineer must thus develop a qualitative understanding of machine dynamics in order to evaluate the reasonableness of the results and to modify a design to improve its dynamic performance.

Examples presented in this chapter are intended to assist the engineer in developing a practical capability in problem formulation and analysis of results. There is no substitute, however, for hands-on experience. Project-oriented problems are given in the problem set at the end of the chapter. It is recommended that the reader obtain access to a general-purpose machine dynamics computer code, such as DADS, and gain experience through the systematic analysis of systems of his or her own design. Critical evaluation of analysis results and study of the effect of alternative models and parameter variations (e.g., dimension changes, mass changes, and force changes) can provide valuable insights and a qualitative understanding of machine dynamics. The value of such experience-based understanding cannot be overestimated.

8.2 DYNAMIC ANALYSIS OF A SLIDER–CRANK MECHANISM

8.2.1 Dynamic Modeling

The dynamic model of a slider–crank mechanism used here for illustration is based on model 1 of Section 5.2.1. Kinematic joint data are the same as provided in Section 5.2.1, with the dimensions in Tables 5.2.1 and 5.2.2 given in units of meters.

Centers of mass of the bodies are at the origins of the body-fixed reference frames in Fig. 5.2.1. Mass and inertia of the ground body are taken as unit quantities, which is unimportant since this body is rigidly connected to the inertial reference frame. Masses and moments of inertia of the four bodies that make up the model are given in Table 8.2.1.

TABLE 8.2.1 Inertia Properties of Slider–Crank

Body no.	1	2	3	4
Mass (kg)	200.0	35.0	25.0	1.0
Moment of inertia (kg · m²)	450.0	35.0	0.02	1.0

If the slider–crank functions as a compressor, the slider moves in the compression chamber, as shown in Fig. 8.2.1. As the slider moves to the right, a resisting force due to compression of gas acts on the slider. This force increases until the exhaust valve opens. Figure 8.2.2 defines gas force F_c on the slider, as a function of position and velocity of the slider. During the compression stroke of the slider, that is, when $\dot{x}_3 > 0$,

$$F_c = \begin{cases} -\dfrac{282{,}857}{6 - x_3} + 62{,}857, & 1.5 \le x_3 \le 5 \\ -110{,}000[1 - \sin 2\pi(x_3 - 5.25)], & 5 < x_3 \le 5.5 \end{cases}$$

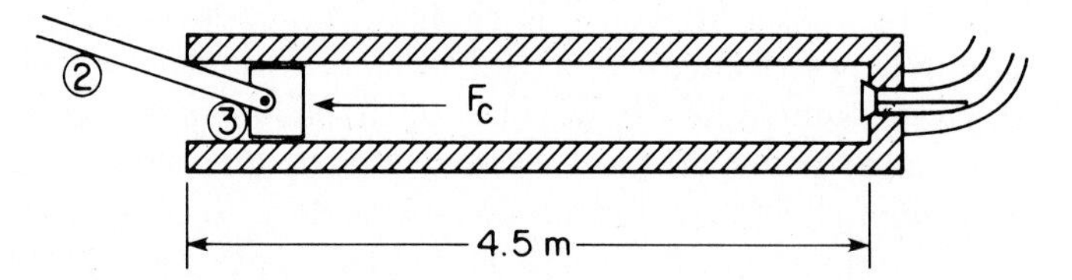

Figure 8.2.1 Slider in a compression gas chamber.

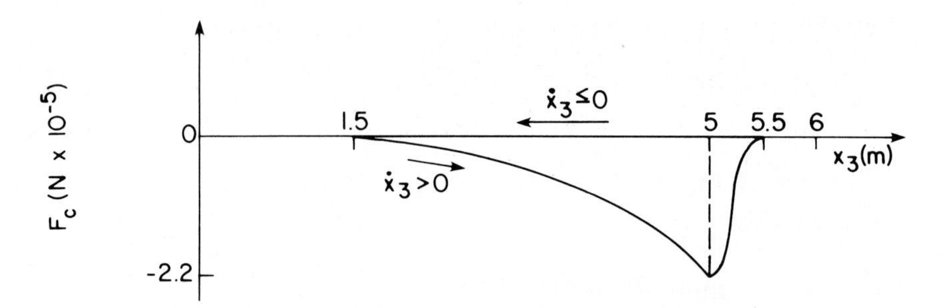

Figure 8.2.2 Gas force versus slider position.

When $x_3 = 5$ m, the valve opens. During the intake stroke of the slider; that is, when $\dot{x}_3 \le 0$, no force acts on the slider.

8.2.2 Equilibrium Analysis

Gravity is taken as acting in the negative y direction. If gravitational force is the only force that acts on the system, then the equilibrium configuration is expected

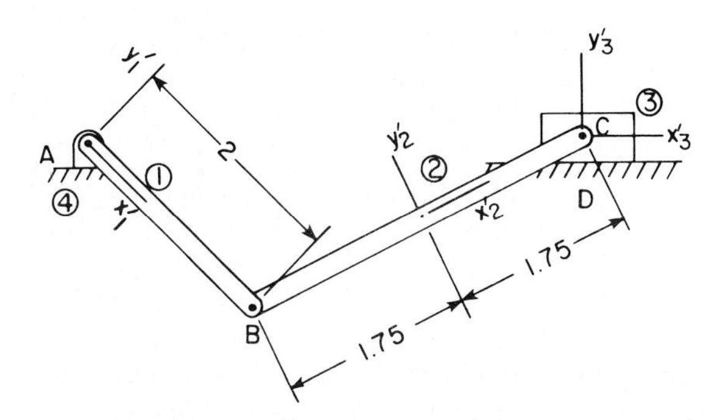

Figure 8.2.3 Estimated equilibrium position.

to be of the form shown in Fig. 8.2.3. An estimate of the equilibrium position used as input is given in Table 8.2.2.

TABLE 8.2.2 Estimated Slider–Crank Equilibrium Position

Body no.	1	2	3	4
x	0.0	1.20	2.67	0.0
y	−0.0	−1.0	0.0	0.0
ϕ	4.6	0.6	0.0	0.0

Equilibrium analysis under only the influence of gravitational force is carried out using the minimum total potential energy method of Section 6.5, with the numerical results presented in Table 8.2.3. Note that the center of mass of the connecting rod is as low as possible, hence minimizing the potential energy of the gravitational force.

TABLE 8.2.3 Slider–Crank Equilibrium Position under Gravitational Force

Body no.	1	2	3	4
x	0.0	1.4635	2.900	0.0
y	0.0	−1.0	0.0	0.0
ϕ	4.726	0.6082	0.0	0.0

If the gas force F_c acts on body 3 in the negative x direction, equilibrium analysis may again be carried out using the minimum total potential energy

TABLE 8.2.4 Slider–Crank Equilibrium Position under Gravitational and Gas Force

Body no.	1	2	3	4
x	0.0	1.2170	2.6575	0.0
y	0.0	−0.9937	0.0	0.0
ϕ	4.6004	0.6039	0.0	0.0

method. Numerical results for this calculation are given in Table 8.2.4. Note that, as expected, body 3 has moved to the left, relative to the position given in Table 8.2.3, and the positions and angles of bodies 1 and 2 are correspondingly changed.

8.2.3 Inverse Dynamic Analysis

The kinematic driving condition prescribed in Section 5.2.3 is used, with the assembled position of Table 5.2.6, to carry out inverse dynamic analysis. Gravitational force is neglected in this analysis. The torque required to achieve the constant angular velocity driving condition and the reaction force at the crank bearing (point A in Fig. 8.2.3) are calculated using Lagrange multipliers, as in Section 6.6, and plotted in Fig. 8.2.4.

8.2.4 Dynamic Analysis

As an illustration of dynamic response analysis, the compressor is analyzed with gravitational force acting in the positive x direction. Motion begins at $t = 0$, with $\phi_1(0) = \pi$ and $\dot{\phi}_1(0) = 30$ rad/s. A constant torque of 41,450 N is applied to the crank, and the force of Fig. 8.2.2 acts on the slider. The applied torque selected does the same amount of work in one revolution as the work done in compressing gas during one cycle, that is, the area under the curve of Fig. 8.2.2. Thus, the compressor should run in steady-state motion.

Three simulations are carried out with different values of polar moment of inertia of the crank, the flywheel in this application: $J_1^1 = 225\ \text{kg} \cdot \text{m}^2$, $J_1^2 = 450\ \text{kg} \cdot \text{m}^2$, and $J_1^3 = 900\ \text{kg} \cdot \text{m}^2$. It is expected that the larger flywheel inertias will lead to less variation in the angular velocity of the flywheel. Plots of angular velocity of the flywheel in Fig. 8.2.5 confirm this behavior. In fact, the smallest inertia J_1^1 leads to failure of the compressor to complete a single cycle. To understand why this happens, note that the work required to compress the gas during the first half-revolution of the flywheel is 260,438 N · m, whereas the sum of the initial kinetic energy of the flywheel and the work done by the applied torque during the same period would be only $225(30^2)/2 + \pi \times 41{,}450 = 231{,}469\ \text{N} \cdot \text{m}$.

Both of the larger moments of inertia of the flywheel provide adequate initial kinetic energy to carry the compressor through a full cycle and to continue

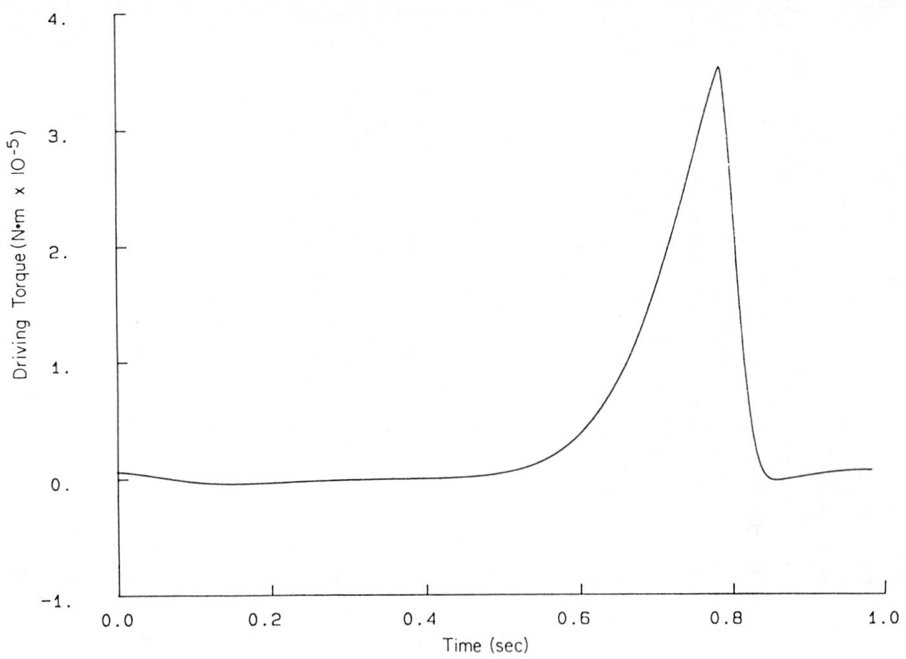

Figure 8.2.4(a) Driving torque: Kinematically driven slider–crank.

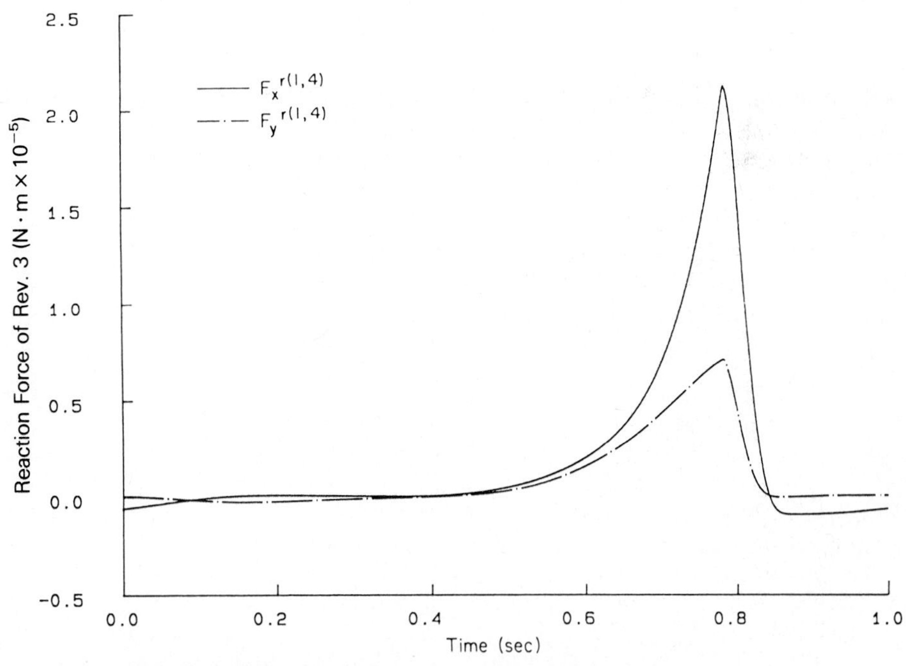

Figure 8.2.4(b) Reaction force: Kinematically driven slider-crank.

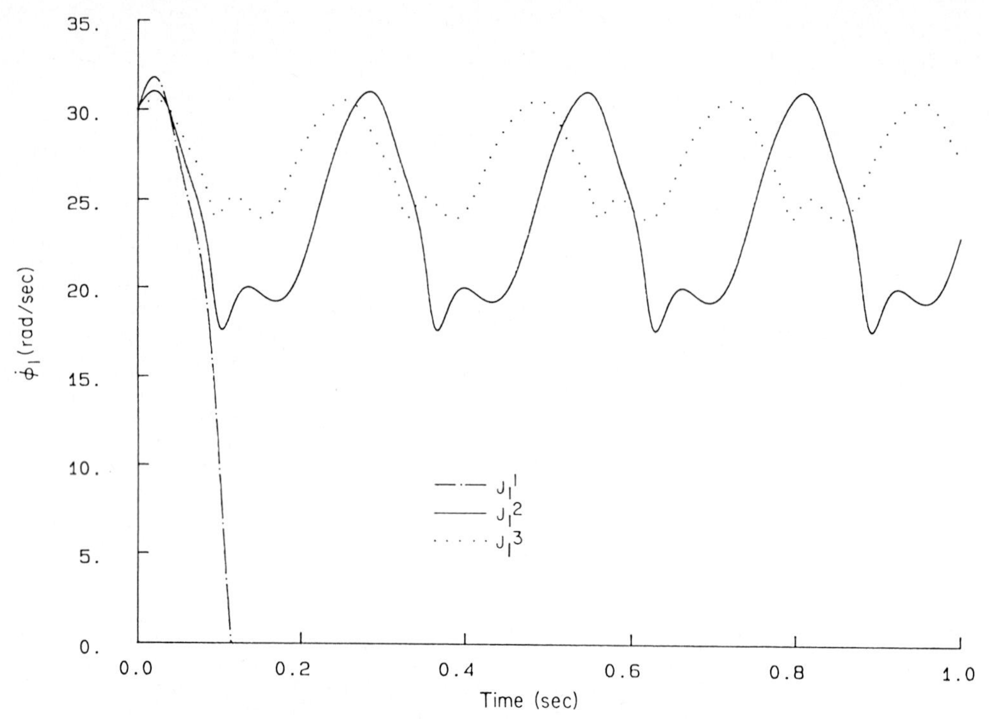

Figure 8.2.5 Angular velocity of flywheel versus time.

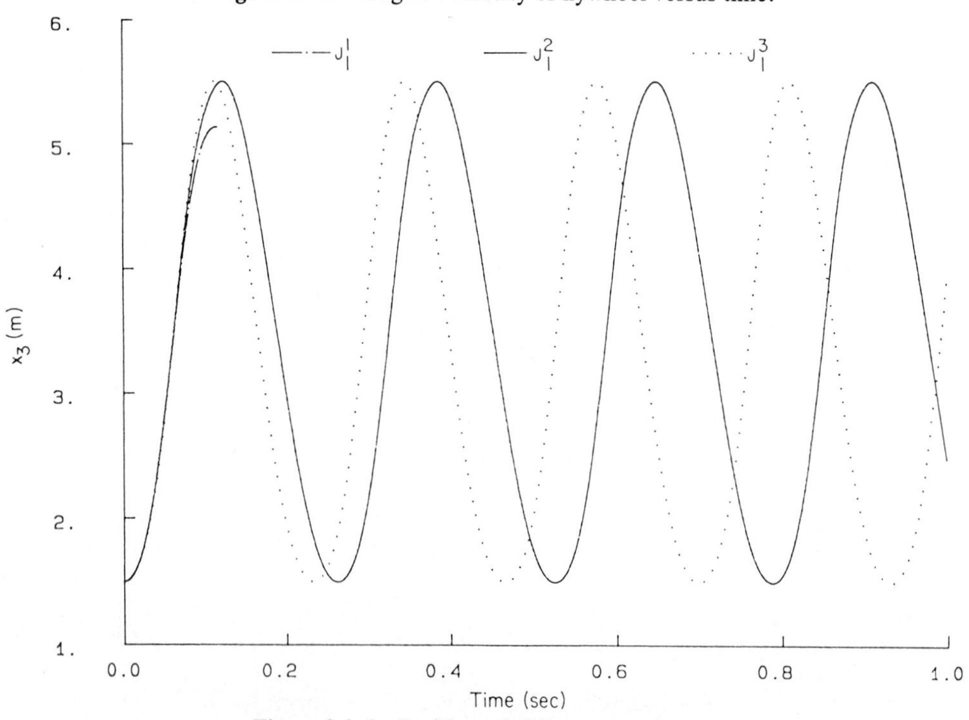

Figure 8.2.6 Position of slider versus time.

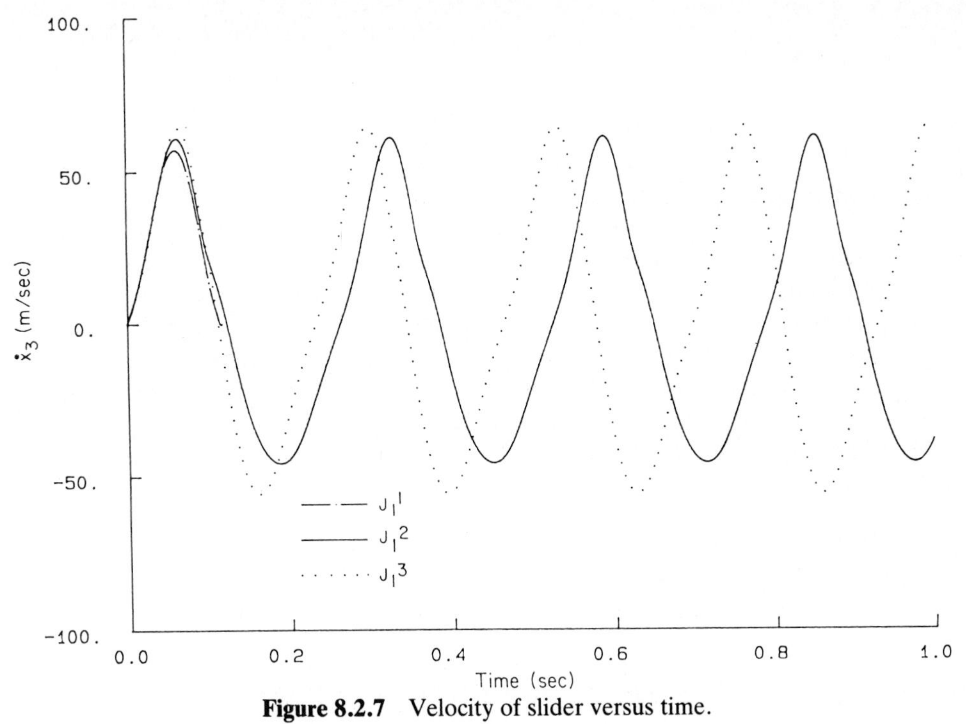

Figure 8.2.7 Velocity of slider versus time.

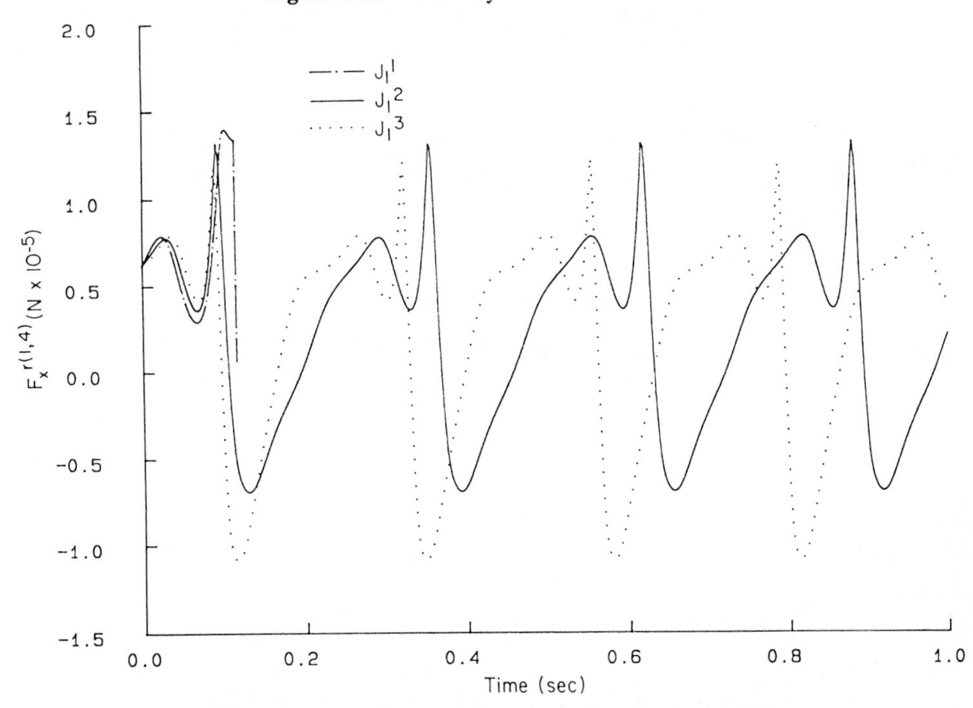

Figure 8.2.8 x-component of reaction force at flywheel bearing versus time.

periodic motion. Note also that the largest flywheel inertia leads to the least variation in flywheel angular velocity, as expected. In fact, from Fig. 8.2.5, the number of compressor cycles per second for J_1^2 and J_1^3 are approximately 3.7 and 4.2 cycles/s, respectively.

Plots of slider position and velocity for each of the flywheel moments of inertia are shown in Figs. 8.2.6 and 8.2.7, respectively. The stroke of the compressor is the same due to identical kinematics. There is a phase shift in velocity, however, due to the difference in flywheel angular velocities. As shown in Fig. 8.2.7, there is also a slight variation in peak slider velocity.

Finally, the x component of reaction force at the flywheel bearing, the revolute joint between bodies 1 and 4, is plotted versus time in Fig. 8.2.8. The more vigorous action of the flywheel with largest inertia leads to a slightly larger variation in reaction force, as might have been expected.

8.3 DYNAMIC ANALYSIS OF A QUICK-RETURN MECHANISM

8.3.1 Dynamic Modeling

The kinematic model 1 of Section 5.4.1 is used here to study the dynamics of the quick-return mechanism of Fig. 5.4.1, which functions as a shaper (see Fig. 1.1.8). A counterclockwise torque is applied to the driving link, body 3. The force between the tool and workpiece required to remove material is given as a function of slider position in Fig. 8.3.1, for the cutting stroke, that is, when $\dot{x}_6 < 0$. When $\dot{x}_6 > 0$, the cutting force is zero. Masses and moments of inertia of the components of the model are presented in Table 8.3.1.

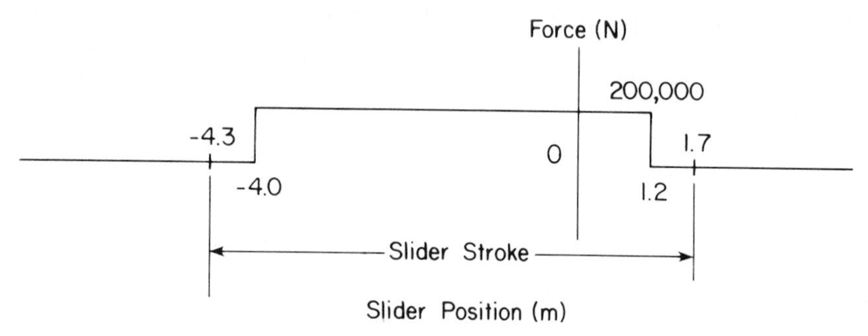

Figure 8.3.1 Cutting force versus slider position.

TABLE 8.3.1 Inertia Properties of Quick-Return Mechanism

Body no.	1	2	3	4	5	6
Mass (kg)	1.0	100.0	1000.0	5.0	30.0	50.0
Moment of inertia (kg · m^2)	1.0	100.0	2000.0	0.05	10.0	1.5

8.3.2 Equilibrium Analysis

With only gravitational force acting in the negative y direction, the equilibrium position of the system might be expected to have body 2 rotated somewhat clockwise from the position shown in Fig. 5.4.1. Equilibrium analysis under gravitational loading only is carried out, using the position estimates provided in Table 5.4.5. The resulting equilibrium configuration is given in Table 8.3.2. As expected, the angle of orientation of body 3 is in the fourth quadrant, and body 2 is rotated clockwise toward the horizontal position.

TABLE 8.3.2 Equilibrium Position of Quick-Return Mechanism under Gravitational Load

Body no.	1	2	3	4	5	6
x	0.0	1.4998	0.0	0.97196	2.3329	1.6663
y	0.0	1.3232	2.0	0.85750	3.3232	4.0
ϕ	0.0	0.72292	−0.86588	0.72292	−0.79302	0.0

8.3.3 Inverse Dynamic Analysis

Inverse dynamic analysis is carried out with a constant angular velocity $\dot{\phi}_3 = 2\pi$ rad/s of the crank, or flywheel. The simulation starts immediately after the cutting stroke; that is, when the slider position is $x_6 = -4.0$ m. Three simulations are carried out, with slider masses $m_6^1 = 25$ kg, $m_6^2 = 50$ kg, and $m_6^3 = 100$ kg. Driving torques that are required to maintain the specified angular velocity of the flywheel are shown in Fig. 8.3.2. Note that the only significant variations in driving torque occur during the period just prior to the cutting stroke, when the slider velocity must be changed from negative to positive (i.e., at the left extreme of the slider stroke).

8.3.4 Dynamic Analysis

Dynamic analysis is carried out with a slider mass $m_6 = 50$ kg, a torque $T_3 = 165{,}521$ N · m applied to the flywheel, and flywheel polar moments of inertia $J_3^1 = 1000$ kg · m^2, $J_3^2 = 2000$ kg · m^2, and $J_3^3 = 4000$ kg · m^2. The applied torque is selected so that the work done in one cycle of operation ($2\pi T_3$) is equal to the work done in cutting the workpiece (the area under the curve of Fig. 8.3.1, 10.4×10^5 N · m).

Plots of slider position and flywheel angular velocity versus time are given in Figs. 8.3.3 and 8.3.4. As expected, the variation in flywheel angular velocity is least for the largest polar moment of inertia. The slider strokes are equal, but phase changes occur due to variations in average angular velocities. The cyclic rates of the shaper, with J_3^1, J_3^2, and J_3^3, are approximately 2.2, 1.8, and 1.5 cycles/s, respectively. It is interesting that in this example the largest flywheel

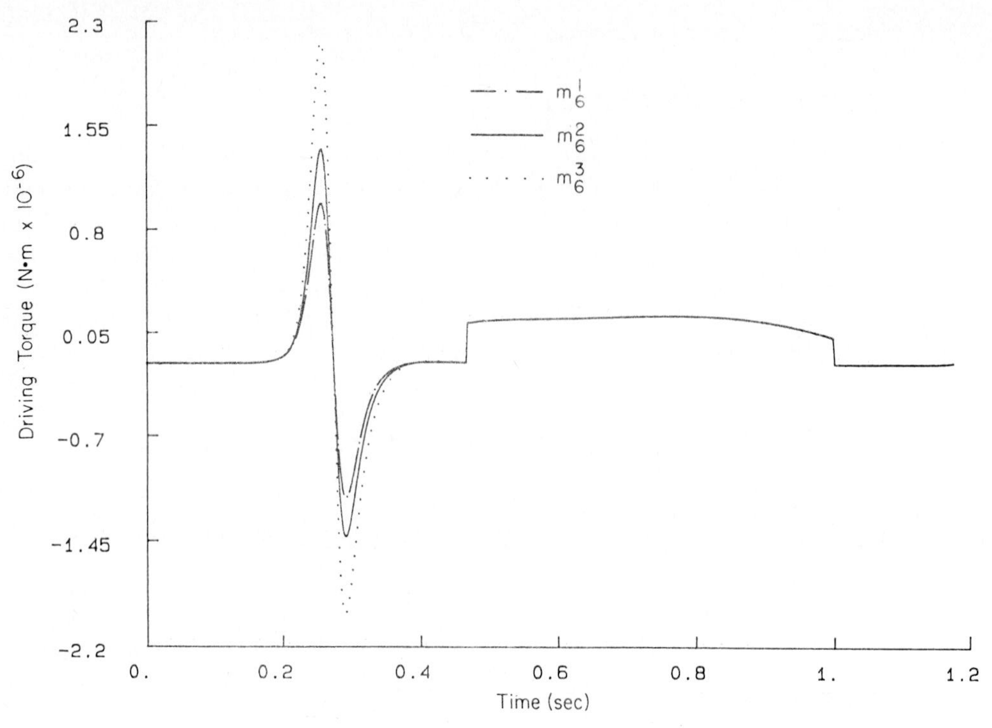

Figure 8.3.2 Driving torque versus time.

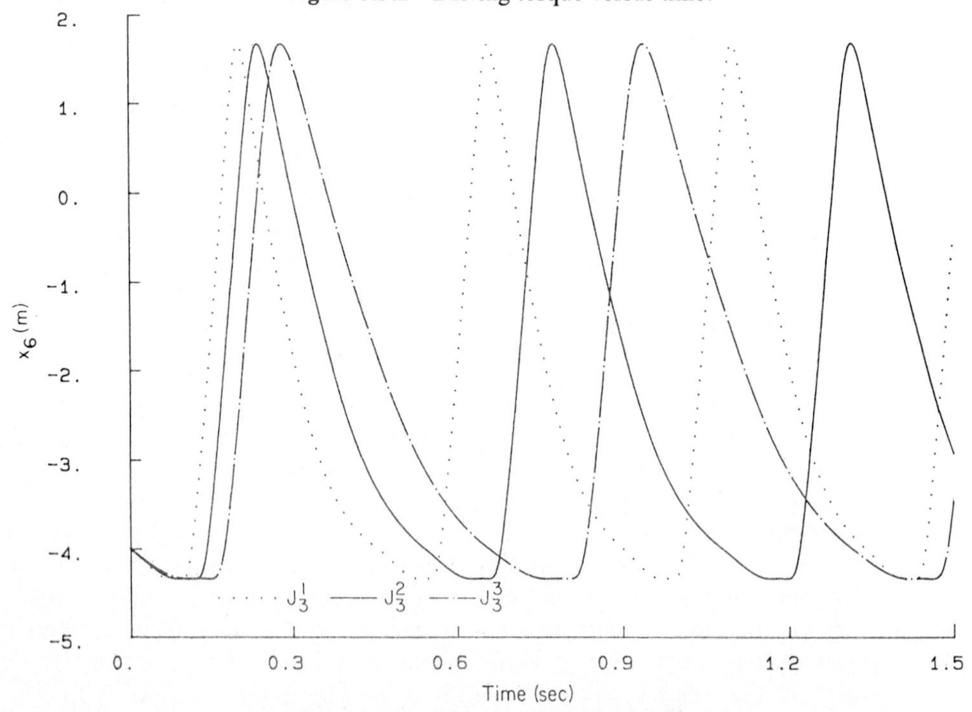

Figure 8.3.3 Slider position versus time.

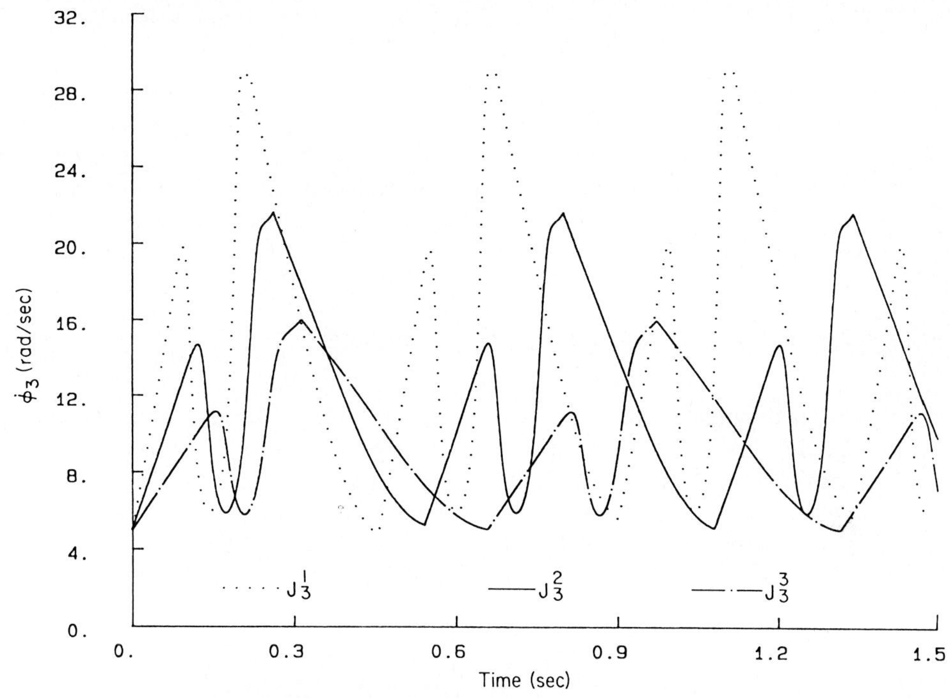

Figure 8.3.4 Flywheel angular velocity versus time.

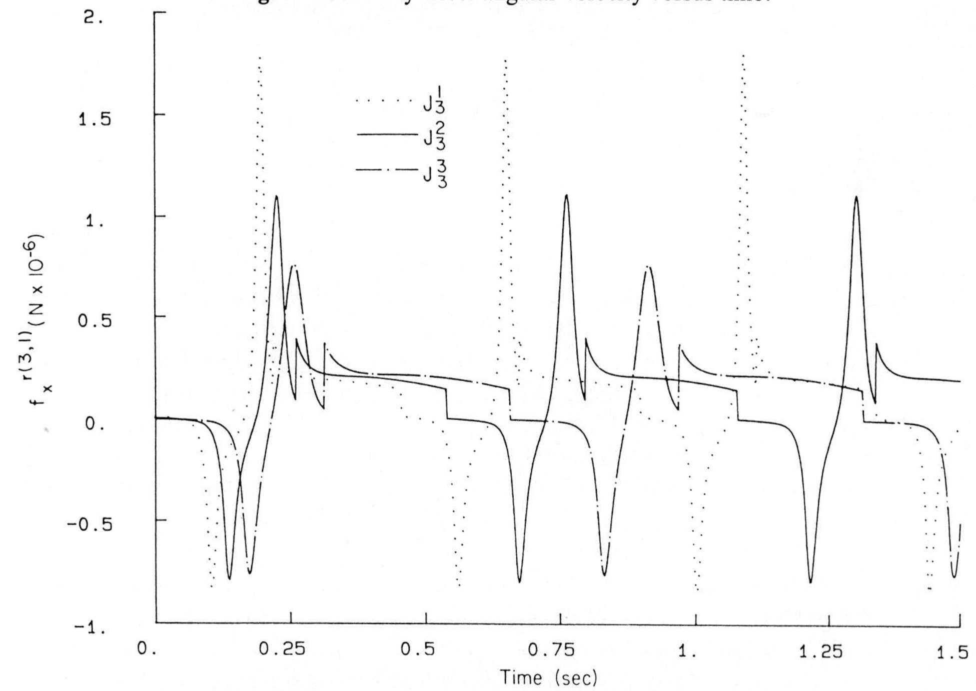

Figure 8.3.5 x-Component of flywheel bearing reaction force versus time.

inertia leads to the lowest cyclic speed, in contrast to the behavior of the compressor of Section 8.2.

Finally, the x component of reaction force between the flywheel and ground, at the revolute joint between bodies 3 and 1, is plotted versus time in Fig. 8.3.5. The greatest variation in this example occurs with the smallest flywheel inertia.

8.4 DYNAMIC ANALYSIS OF A COIL SPRING

In many applications, a relatively long coil spring is used to arrest the motion of a moving body and to return it to its original position. While a coil spring is a continuum of mass distributed along the length of the spring, a reasonable model is formed by discretizing the mass and stiffness of the spring. The lumped mass model shown in Fig. 8.4.1 is constructed by dividing the spring into equal lengths and approximating the mass effects of each component as a lumped mass m. The masses are connected by springs with stiffness k.

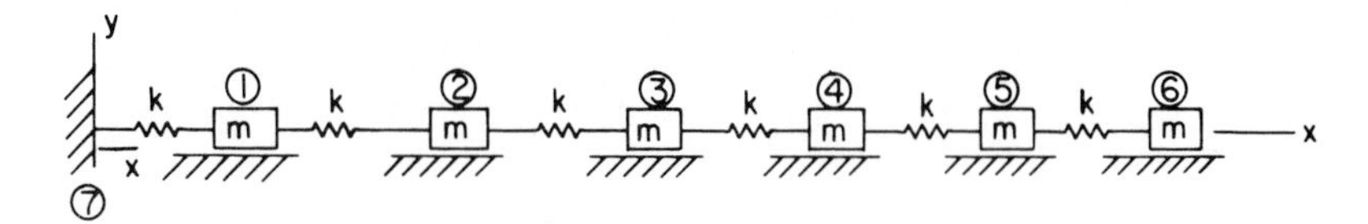

Figure 8.4.1 Model of coil spring.

8.4.1 Surge Waves

To study surge wave behavior, the motion of the system is initiated by giving body 6 an initial velocity in the negative x direction, which might be imparted by impact with another body. The kinematics of the system is defined by prescribing a translational joint between each of the masses and ground, which is body 7. Each spring is attached at the centers of mass of the bodies it connects. The mass of each body is $m = 2.5$ kg, and each spring has the same spring rate $k = 200$ N/m and free length $\ell_0 = 0.2$ m. Therefore, the equilibrium position of body i is $x_i = i \times \ell_0$. Since gravitational force acts in the negative y direction, it has no influence on the motion. The initial velocities of bodies 1 through 5 are zero, and the initial velocity of body 6 is $\dot{x}_6 = -0.1$ m/s, simulating impact on the end of the spring by an external body.

The numerical results of this dynamic simulation include plots of position versus time for each body in Fig. 8.4.2. Note that with even this crude model of wave propagation in the spring, the wave nature of motion, including reflection characteristics at the left end, is accounted for. If a larger number of masses had been used in the simulation, an even closer approximation to actual wave behavior would be produced.

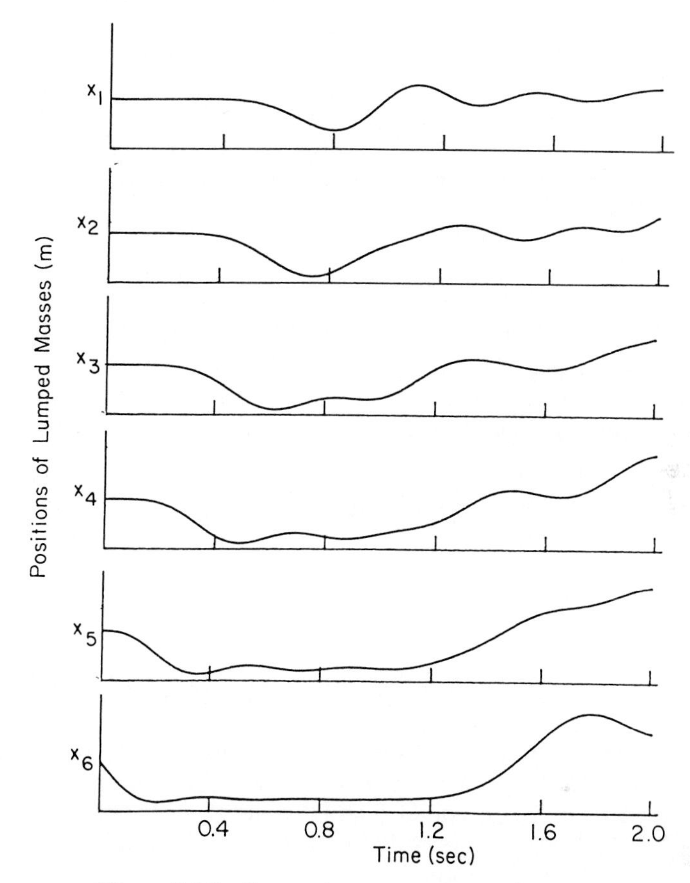

Figure 8.4.2 Lumped mass positions versus time.

8.4.2 Impact and Chattering

In the next application, a heavy mass M (body 8) moves to the left and hits the right end of the spring (body 6), as shown in Fig. 8.4.3. To model the impact between bodies 6 and 8, a unilateral spring with the force-displacement characteristics shown in Fig. 8.4.4 is used. The spring generates no force when its deflection is positive; that is, when bodies 6 and 8 are not in contact. It generates a very large compressive force when the spring deflection is negative; that is, when bodies 6 and 8 are in contact.

For this model, $M = 100$ kg, $m = 2.5$ kg, $k = 50$ N/m, $\dot{x}_8(0) = -0.1$ m/s, and all other velocities at $t = 0$ are zero. Figure 8.4.5 shows the resulting positions of the heavy mass and each of the lumped masses as functions of time. Impact occurs as the simulation begins.

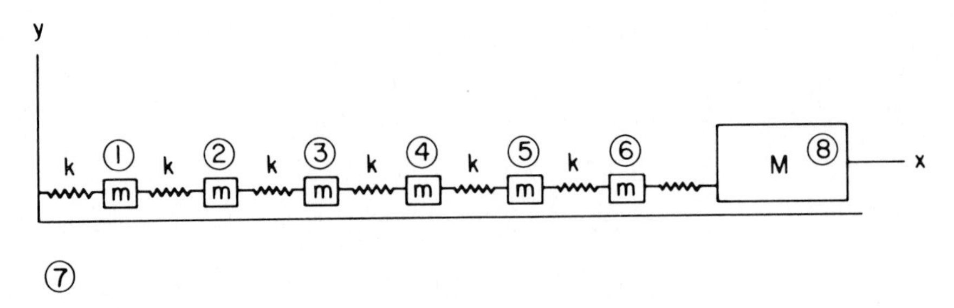

Figure 8.4.3 Model of impact to a long spring.

The maximum displacement Δx_8 of body 8 can be estimated by equating the initial kinetic energy of body 8 to the strain energy of a spring that is equivalent to the six springs in series, that is,

$$\tfrac{1}{2}M(\dot{x}_8)^2 \approx \tfrac{1}{2}k_e(\Delta x_8)^2$$

where $k_e = k/6$. Thus,

$$\tfrac{1}{2}(100)(-0.1)^2 \approx \tfrac{1}{2}(\tfrac{50}{6}(\Delta x_8)^2$$

and

$$\Delta x_8 = \sqrt{6/50} = 0.346\ \text{m}$$

The maximum deflection obtained by dynamic simulation, as shown in Fig. 8.4.5, is $\Delta x_8 = 0.327$ m. The difference is due to the kinetic energy of the lumped masses; that is, when bodies 6 and 8 come to rest, bodies 1 through 5 are still in motion. However, since body 8 is much heavier than the lumped spring masses, the approximation is good.

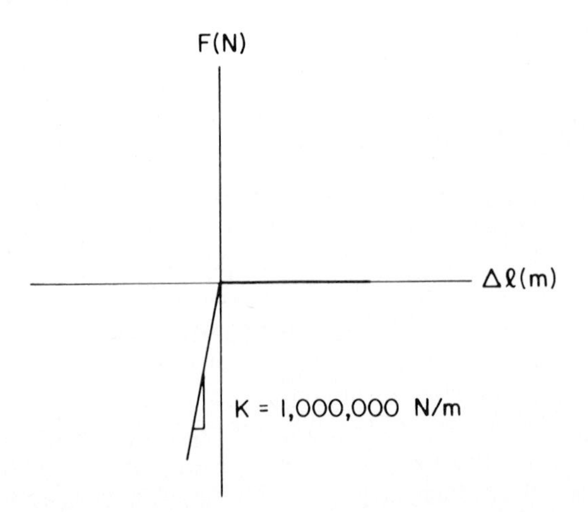

Figure 8.4.4 Unilateral spring characteristic.

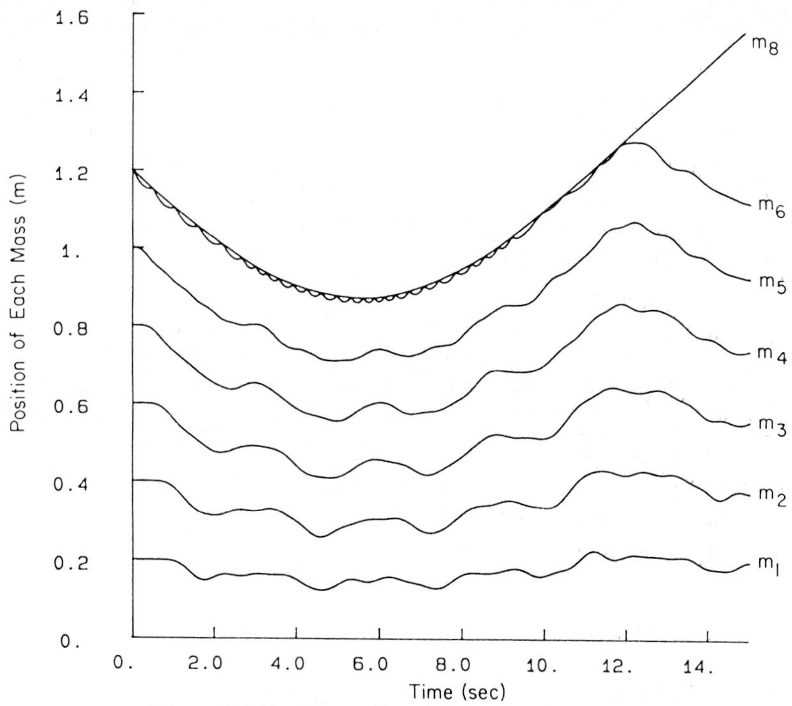

Figure 8.4.5 Chattering and surge wave, $k = 50$ N/m.

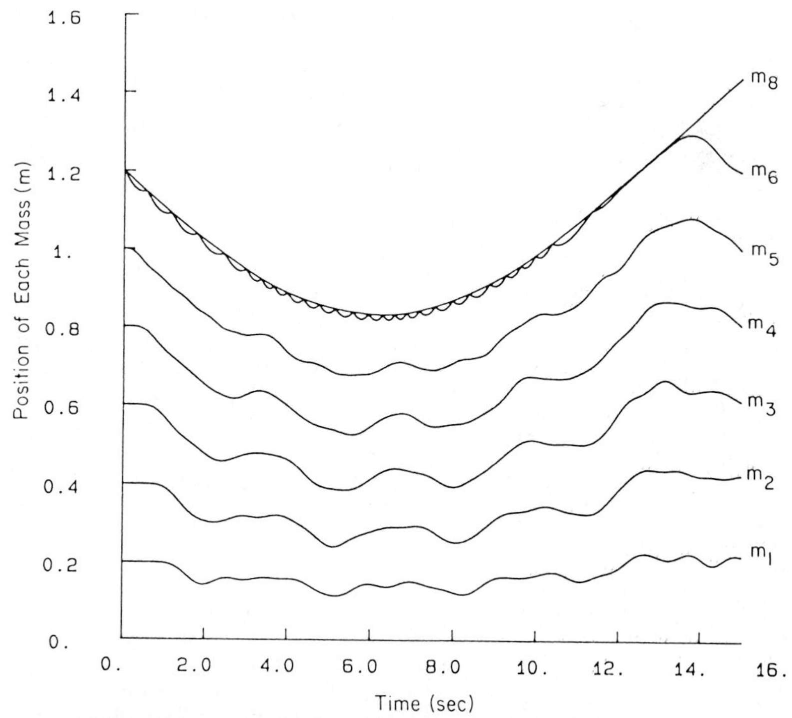

Figure 8.4.6 Chattering and surge wave, $k = 40$ N/m.

Since the spring between bodies 6 and 8 is very stiff and bilinear, chattering behavior (multiple, high-frequency contacts) occurs between bodies 6 and 8. Bodies 6 and 8 finally separate when $t \approx 11.8$ s, at which time their positions are $x_6 = x_8 \approx 1.27$ m. Since the static equilibrium position is at $x_6 = 1.2$ m, they would separate there if the spring had no mass. However, the spring has a total mass of 15 kg (6×2.5 kg). Hence, it still has kinetic energy when $x_6 = 1.2$, which carries mass 6 beyond $x_6 = 1.2$ m.

Two models, with $k = 40$ and 30 N/m, are simulated, with the results plotted in Figs. 8.4.6 and 8.4.7. The separation time is delayed to 13.4 and 15.3 s, respectively, while the separation position remains approximately the same.

The chattering motion is shown in an expanded time scale in Fig. 8.4.8, which shows the positions of bodies 6 and 8 with $k = 30$, 40, 50 N/m. As k increases, the period of chattering motion decreases, which is due to an increase in wave propagation speed in the spring.

The previous model is now modified to include a unilateral damper between bodies 6 and 8. The logic of the unilateral damper is similar to the unilateral spring in Fig. 8.4.4; that is, no damping force acts when the spring deflection is positive, while a damping force of $200\ \text{N} \cdot \text{s}/m \times \dot{\ell}$ acts when the spring deflection

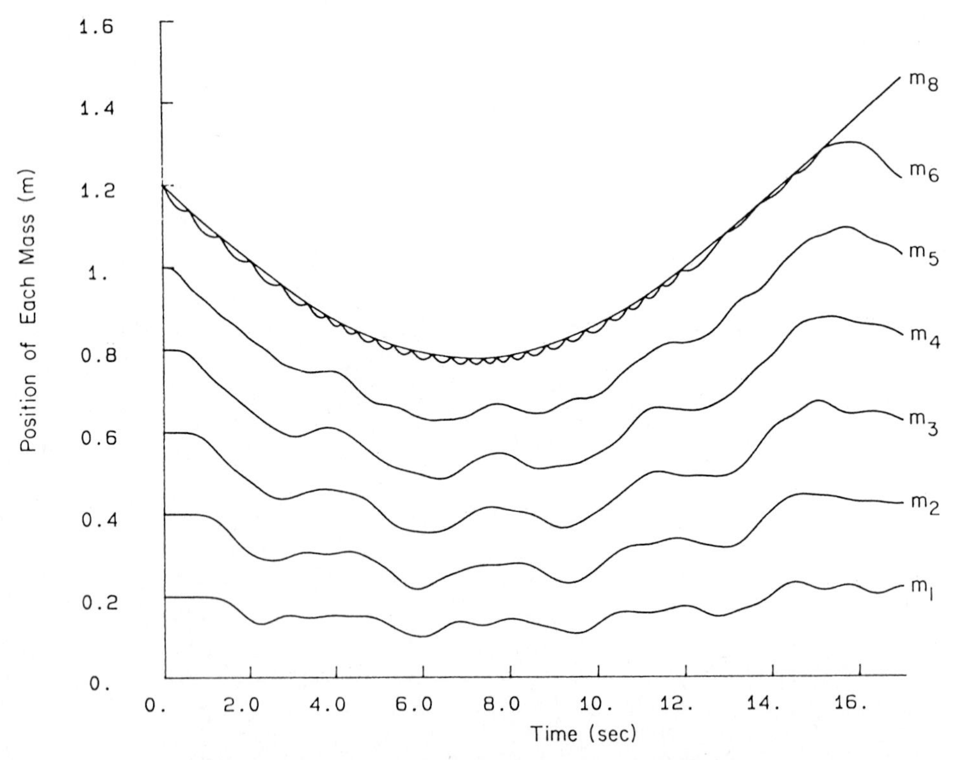

Figure 8.4.7 Chattering and single wave, $k = 30$ N/m.

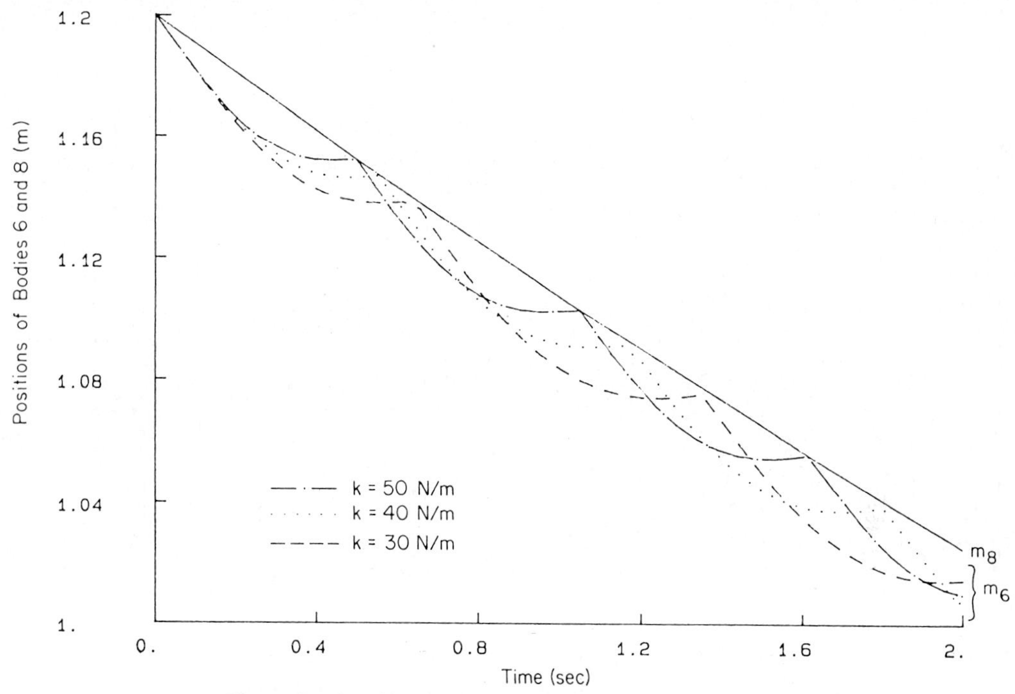

Figure 8.4.8 Chattering between bodies 6 and 8 with no damping.

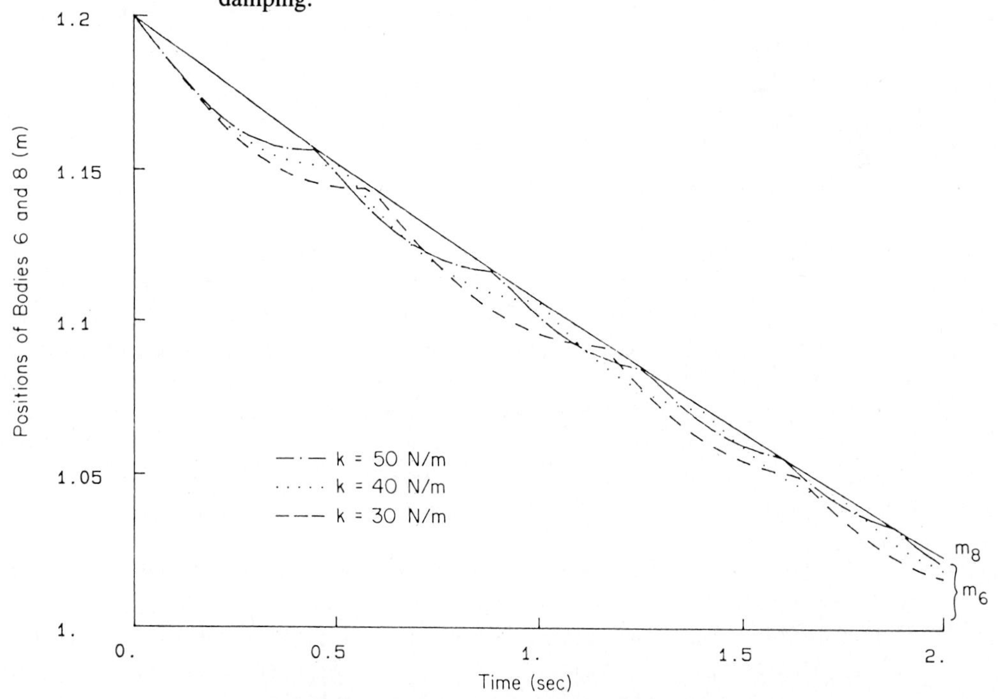

Figure 8.4.9 Chattering between bodies 6 and 8 with damping.

is negative. As shown in Fig. 8.4.9, chattering relative displacement and period are decreased, compared to the case with no damping. It was also observed that the separation position and time remain approximately the same, as in the case with no damping, although plots are not provided.

8.5 DYNAMIC ANALYSIS OF A VALVE-LIFTER MECHANISM

The cam-driven valve mechanism studied in Section 5.6 is analyzed here dynamically. The kinematic model of Section 5.6 is augmented with a valve spring, as shown in the engine of Fig. 1.1.1. Inverse dynamic analysis is carried out to assist in the design of the valve spring.

The masses and moments of inertia of the components of the mechanism of Fig. 5.6.1 are given in Table 8.5.1. The nominal angular velocity of the cam is 3000 rpm (revolutions per minute); that is, 100π rad/s.

TABLE 8.5.1 Mass properties of Valve Lifter Components

Body no.	1	2	3	4	5
Mass (g)	1.0	30.0	120.0	150.0	60.0
Moment of inertia ($g \cdot cm^2$)	1.0	15.0	2250.0	1800.0	800.0

Three runs are made, with varying values of spring constant, $k = 10$, 25, and 40 N/cm. The free length of the spring is 9 cm in each case. Plots of pushrod reaction force with the cam are given in Fig. 8.5.1.

With $k = 10$ N/cm, a negative reaction force occurs between the pushrod and the cam at about $t = 0.01$ s, when the pushrod reaches its highest point. This is due to the inertia of the parts. In an actual cam–flat-faced follower, separation would thus occur. Since the formulation of the cam–flat-faced follower joint does not allow separation, the reaction force appears to be negative. This unacceptable behavior can be corrected by increasing the spring constant of the valve spring. As seen in Fig. 8.5.1, with $k = 25$ and 40 N/cm, the negative reaction force does not appear. These designs would, therefore, be acceptable at 3000 rpm.

Three more runs are made with the nominal spring constant $k = 25$ N/cm and cam speeds of 1500, 3000, and 4500 rpm. Plots of pushrod reaction force are shown in Fig. 8.5.2. With 1500 and 3000 rpm, negative reaction force does not occur. However, at 4500 rpm the reaction force becomes negative. Hence, 4500 rpm is not an acceptable operating speed for a valve spring with $k =$ 25 N/cm.

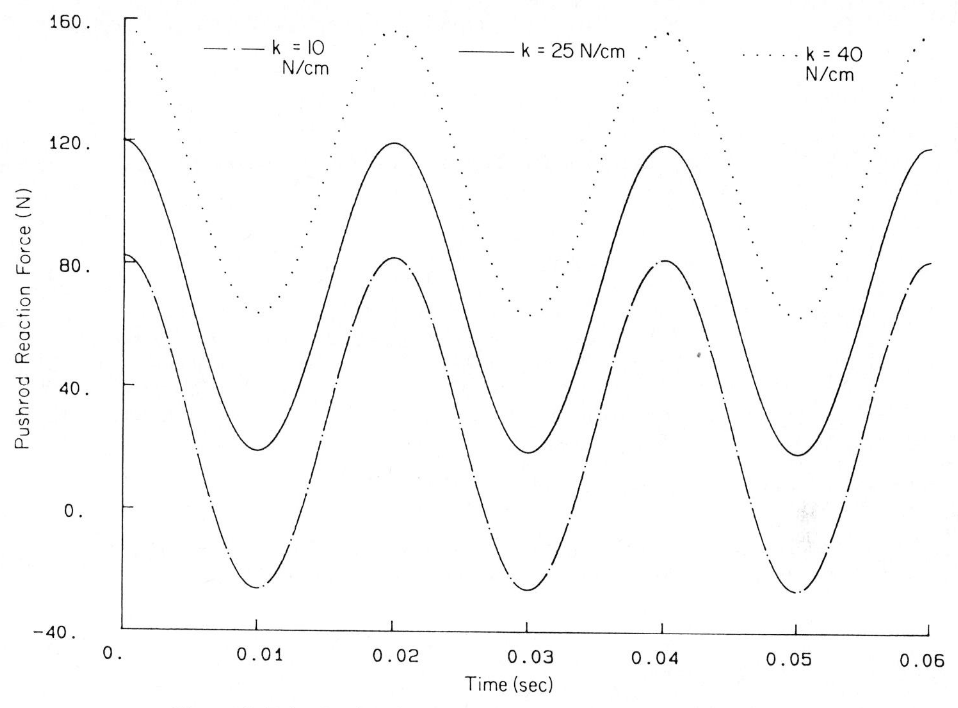

Figure 8.5.1 Pushrod reaction force versus time, 3000 rpm.

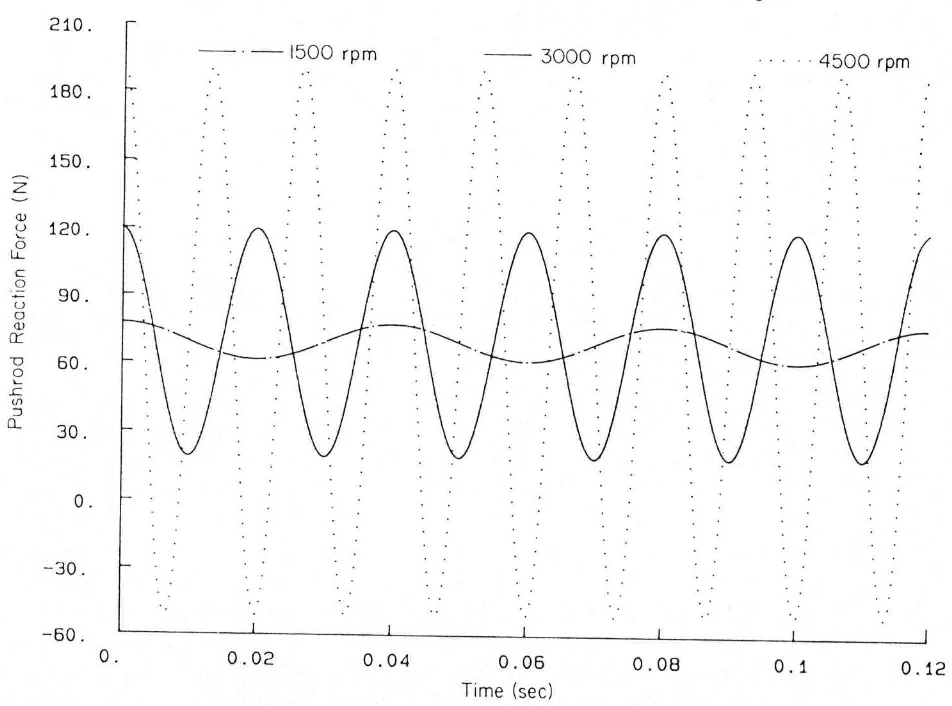

Figure 8.5.2 Pushrod reaction force versus time, $k =$ 25 N/m.

DADS PROJECTS

8.1. Perform a dynamic analysis of the film follower of DADS Project 5.1 using additional data as follows:

Inertial Properties of Bodies

Body	Mass (kg)	Polar Moment of Inertia (kg · m²)	Centroid
1	1.0	1.0	0.0
2	0.0012	2.25×10^{-8}	Midpoint of *AB*
3	0.01	9.2×10^{-8}	50 from *B*
4	0.0025	4.5×10^{-8}	Midpoint *CD*

The reader must be reminded that the body-fixed frames need to be redefined for dynamic analysis. Consider a few ways to drive the mechanism dynamically. One suggestion is to apply a torque to crank *AB*.

8.2. Perform a dynamic analysis of the web cutter of DADS Project 5.2 using additional data as follows:

Inertial Properties of Bodies

Body	Mass (kg)	Polar Moment of Inertia (kg · m²)	Centroid
1	1.0	1.0	0.0
2	0.1	1.5×10^{-5}	Midpoint of *AB*
3	1.0	0.03	*C*
4	2.0	0.04	(0.14, 0.07) with respect to x_4'-y_4' of DADS Project 5.2

For the definition of body-fixed frames and driving method, refer to DADS Project 8.1.

8.3. Perform a dynamic analysis of the rack and pinion mechanism of DADS Project 5.3. The masses of the pinion, rack, and coupler are 0.05, 0.1, and 0.05 kg, respectively. Redefine the body-fixed frames and apply torque for dynamic analysis. An alternative model can be constructed by replacing the coupler with a revolute–revolute joint.

8.4. Perform an inverse dynamic DADS analysis for the slider–crank mechanism of Section 8.2. Obtain the constraint reaction torque for the kinematic driver used. Use the resulting torque information as an applied torque to perform a dynamic DADS run and compare the results.

8.5. Perform inverse dynamic and dynamic runs for the quick-return mechanism of Section 8.3.

Part Two

SPATIAL SYSTEMS

Chapters 9 through 12 are devoted to Cartesian kinematics and dynamics of systems in which all bodies move in space. Many of the basic concepts needed have been developed in Part One. Extension of the mechanical system analysis from the planar to the spatial setting, however, requires the introduction of three-dimensional vector analysis and, of particular importance, analysis of the orientation of a body in space. The matrix, vector calculus, and numerical methods developed in Part One are adequate in almost all respects for the treatment of spatial systems.

Chapter 9 begins with an extension of vector analysis to the spatial setting and provides a self-contained treatment of the rotation and orientation of bodies in space. Angular velocity and virtual rotation are introduced and illustrated in preparation for the development of a unified theory of spatial kinematics. Euler parameter orientation coordinates are introduced and their properties are developed to support spatial kinematic and dynamic analysis. The remainder of Chapter 9 presents a spatial kinematic constraint formulation that parallels the developments presented in Chapter 3 for planar systems. Since the numerical methods for spatial kinematics are identical to those for planar kinematics, the numerical methods of Chapter 4 need not be repeated for spatial applications.

Chapter 10 is devoted to modeling and analysis methods for spatial system kinematics. Of particular importance in spatial kinematic analysis is model definition, using an extensive library of kinematic couplings that occur in spatial kinematics. Numerical examples are presented to illustrate the difficulties that can be encountered in modeling. Guidelines are presented to assist the engineer in proper model selection and analysis.

Spatial equations of motion for a single rigid body and for multibody systems are derived in Chapter 11, including the inertia properties of bodies and a variational form of the equations of motion. The basic development is identical in form to that presented in Chapter 6 for planar systems, with the exception that nonlinear terms associated with angular velocity arise in the spatial case. Equations of motion are formed in terms of both angular velocity and Euler parameter generalized coordinates. Numerical integration methods for equations of motion in Euler parameter generalized coordinates are identical to those presented in Chapter 7. Some modifications in integration methods, to exploit the angular velocity form of equations of motion, are presented. Finally, Chapter 12 presents applications of the Cartesian dynamics formulation to assist the reader in gaining experience and confidence in using the methods.

The reader is cautioned that the analytical and algebraic complexity of spatial system kinematic and dynamic analysis is substantially greater than has been experienced in the study of planar systems in Part One. To achieve a practical capability in spatial system kinematic and dynamic analysis, the engineer must master the basic concepts of three-dimensional vector analysis and the use of generalized coordinates to define the position and orientation of bodies in space. While the algebra and calculus of these methods may appear to be complicated, they are easily understood if the reader builds on the intuitive concepts of planar kinematic and dynamic analysis developed in Part One. A combined analytical and physically intuitive approach to spatial system kinematics and dynamics can aid immeasurably in understanding what might otherwise be viewed as abstract mathematical manipulations.

CHAPTER NINE

Spatial Cartesian Kinematics

The same approach developed for planar kinematic analysis in Chapter 3 is employed in this chapter for the kinematic analysis of spatial systems. Some extension to concepts of vector analysis in three-dimensional space is required, but the same basic algebraic vector approach and notation remain valid. The principal extension from planar to spatial kinematics is the complexity of defining the orientation of a body in space. Coordinate transformations are defined, as in the plane, and concepts of angular velocity and virtual rotation are introduced as vector quantities that are required for kinematic and dynamic analysis. A fundamental difference between planar and spatial kinematic analysis is the use of angular velocity and acceleration variables in spatial analysis that are not time derivatives of generalized coordinates. Euler parameter orientation generalized coordinates are defined, and their properties are developed for use in kinematic and dynamic analysis. A library of kinematic and driving constraints is derived, and their variations, in terms of both virtual rotation and Euler parameter variations, are calculated. Finally, velocity and acceleration equations are derived.

The reader is cautioned that the analytical sophistication of spatial kinematics, particularly in defining orientation of bodies in space, is more demanding than corresponding concepts in planar system kinematics. It is, nevertheless, critically important that the geometric aspects of the spatial kinematic system definition be clearly understood, in order for the engineer to effectively model and analyze the spatial kinematics of mechanical systems. This chapter has been written presuming that the reader has a clear physical understanding of the concepts of planar kinematics of mechanical systems that are developed in Part One. Even though substantial algebraic and analytical intricacy arises in the kinematics of spatial systems, geometric concepts of kinematic constraints, mechanical system modeling, and kinematic analysis are very similar to corresponding concepts in planar kinematics. The reader is, therefore, encouraged to draw parallels and analogies between spatial kinematics and kinematics of planar systems in order to exploit an intuitive mastery of planar kinematics. Adherence to physically based concepts will assist the reader in maintaining a degree of sanity while making his or her way through the following sections, which involve seemingly endless analytical and algebraic manipulations.

9.1 VECTORS IN SPACE

Vectors in space have many of the same properties as in the plane, but they have some additional properties that were not encountered in Chapter 2. The principal focus of this section is on the properties of vectors in space, building on the basic properties of vectors in the plane that were developed in Chapter 2.

Geometric Vectors The concept of a vector in space may be introduced in a geometric setting, with no requirement for identification of a reference frame. In this setting, a *geometric vector,* or simply a vector, is defined as the directed line segment from one point in space to another point in space. Vector $\vec{a}$ in Fig. 9.1.1, beginning at point A and ending at point B, is denoted by the $\vec{\ }$ notation in its geometrical sense. The *magnitude of a vector* $\vec{a}$ is its length (the distance between A and B) and is denoted by a or $|\vec{a}|$. Note that the length of a vector is positive if points A and B do not coincide and is zero only when they coincide. A vector with zero length is denoted as $\vec{0}$ and is called the *zero vector.*

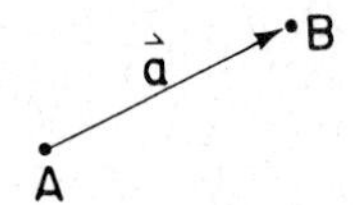

Figure 9.1.1 Vector from point A to point B.

Multiplication of a vector $\vec{a}$ by a nonnegative scalar $\alpha \geqslant 0$ is defined as a vector $\alpha\vec{a}$ in the same direction as $\vec{a}$, but having magnitude αa. *A unit vector,* having a length of 1 unit, in the direction $\vec{a} \neq \vec{0}$ is $(1/a)\vec{a}$. Multiplication of a vector $\vec{a}$ by a negative scalar $\beta < 0$ is defined as the vector with magnitude $|\beta|\, a$ and direction opposite to that of $\vec{a}$. The *negative of a vector* is obtained by multiplying the vector by -1. It is the vector with the same magnitude but opposite direction. These definitions are identical to those for the planar case.

Example 9.1.1: Let points A and B in Fig. 9.1.1 be located in an orthogonal x-y-z reference frame, as shown in Fig. 9.1.2. The distance between

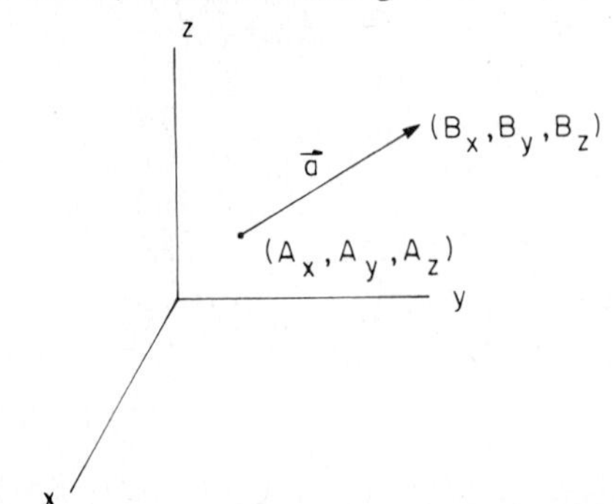

Figure 9.1.2 Vector located in orthogonal reference frame.

points A and B, with coordinates (A_x, A_y, A_z) and (B_x, B_y, B_z), respectively, is the length of $\vec{a}$; that is,

$$|\vec{a}| = [(B_x - A_x)^2 + (B_y - A_y)^2 + (B_z - A_z)^2]^{1/2}$$

Two vectors $\vec{a}$ and $\vec{b}$ are added according to the *parallelogram rule,* as shown in Fig. 9.1.3. The parallelogram used in this construction is formed in the plane that contains the intersecting vectors $\vec{a}$ and $\vec{b}$. The *vector sum* is written as

$$\vec{c} = \vec{a} + \vec{b} \tag{9.1.1}$$

Addition of vectors and multiplication of vectors by scalars obey the following rules [21]:

$$\begin{aligned} \vec{a} + \vec{b} &= \vec{b} + \vec{a} \\ (\alpha + \beta)\vec{a} &= \alpha\vec{a} + \beta\vec{a} \end{aligned} \tag{9.1.2}$$

where α and β are scalars.

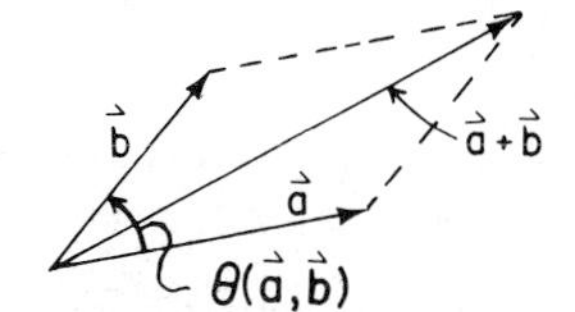

Figure 9.1.3 Addition of vectors.

Example 9.1.2: Using the definition of vector addition, the plane formed by intersecting vectors $\vec{a}$ and $\vec{b}$ in Fig. 9.1.3 may be called the x'-y' plane, with the x' axis along vector $\vec{a}$, as shown in Fig. 9.1.4. Plane trigonometry can now be used to calculate the length of vector $\vec{a} + \vec{b}$ and the angle $\theta(\vec{a}, \vec{a} + \vec{b})$ between $\vec{a}$ and $\vec{a} + \vec{b}$:

$$|\vec{a} + \vec{b}| = [(a + b \cos \theta(\vec{a}, \vec{b}))^2 + (b \sin \theta(\vec{a}, \vec{b}))^2]^{1/2}$$

$$\theta(\vec{a}, \vec{a} + \vec{b}) = \text{Arctan}\left[\frac{b \sin \theta(\vec{a}, \vec{b})}{a + b \cos \theta(\vec{a}, \vec{b})}\right]$$

These calculations characterize the vector $\vec{a} + \vec{b}$, but they are not convenient.

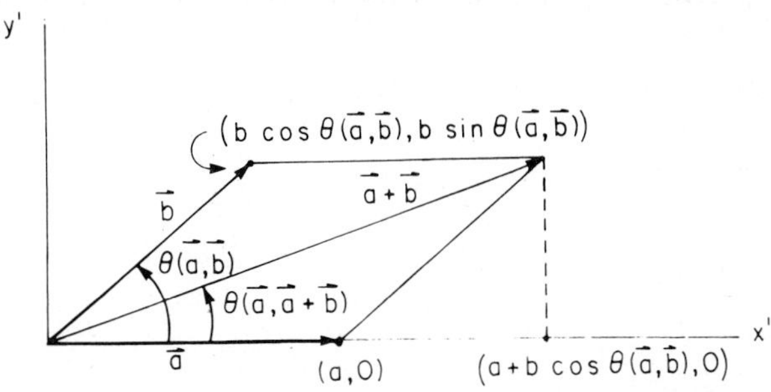

Figure 9.1.4 Sum of vectors in plane of intersection.

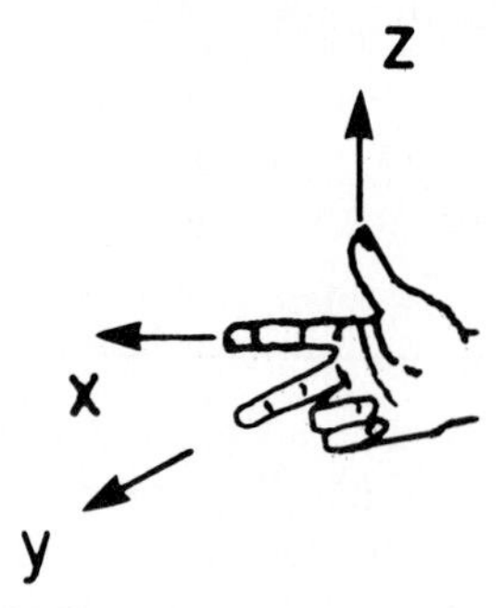

Figure 9.1.5 Right-hand orthogonal reference frame.

Orthogonal reference frames are used extensively in representing vectors. Use in this text is limited to *right-handed x-y-z* orthogonal reference frames, that is, with mutually orthogonal x, y, and z axes that are ordered by the finger structure of the right hand, as shown in Fig. 9.1.5. Such a frame is called a *Cartesian reference frame.*

A vector $\vec{a}$ can be resolved into components a_x, a_y, and a_z along the x, y, and z axes of a Cartesian reference frame, as shown in Fig. 9.1.6. These components are called the *Cartesian components of the vector.* The *unit coordinate vectors* $\vec{i}$, $\vec{j}$, and $\vec{k}$ are unit vectors that are directed along the x, y, and z axes, respectively, as shown in Fig. 9.1.6. In vector notation,

$$\vec{a} = a_x\vec{i} + a_y\vec{j} + a_z\vec{k} \tag{9.1.3}$$

Denote the angle from vector $\vec{a}$ to vector $\vec{b}$ in the plane that contains them by $\theta(\vec{a}, \vec{b})$, with counterclockwise as positive about the normal to the plane of the vectors that points toward the viewer, as shown in Fig. 9.1.3. In terms of angles between the vector $\vec{a}$ and the positive x, y, and z axes, the *components of a vector* $\vec{a}$ are

$$\begin{aligned} a_x &= a \cos \theta(\vec{i}, \vec{a}) \\ a_y &= a \cos \theta(\vec{j}, \vec{a}) \\ a_z &= a \cos \theta(\vec{k}, \vec{a}) \end{aligned} \tag{9.1.4}$$

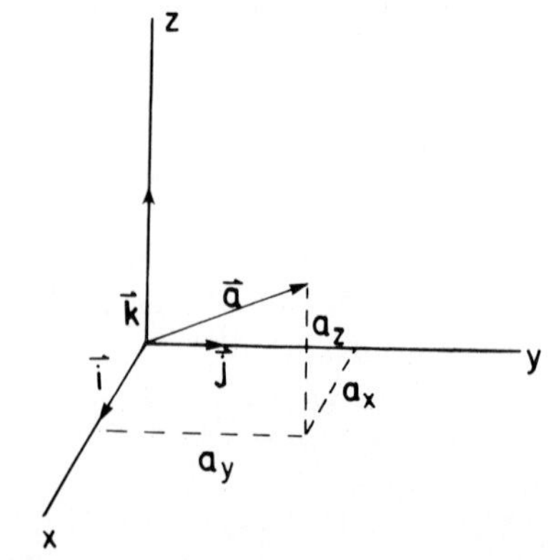

Figure 9.1.6 Components of a vector.

The quantities $\cos\theta(\vec{i}, \vec{a})$, $\cos\theta(\vec{j}, \vec{a})$, and $\cos\theta(\vec{k}, \vec{a})$ are called the *direction cosines of vector* $\vec{a}$. Note that if the viewer had been on the back side of the plane of Fig. 9.1.3 the counterclockwise angle from $\vec{a}$ and $\vec{b}$ would be $2\pi - \theta(\vec{a}, \vec{b})$. However, $\cos(2\pi - \theta) = \cos\theta$, so the viewpoint does not influence the direction cosines.

Addition of vectors $\vec{a}$ and $\vec{b}$ may be expressed in terms of their components, using Eq. 9.1.2, as

$$\begin{aligned}\vec{c} = \vec{a} + \vec{b} &= (a_x + b_x)\vec{i} + (a_y + b_y)\vec{j} + (a_z + b_z)\vec{k} \\ &\equiv c_x\vec{i} + c_y\vec{j} + c_z\vec{k}\end{aligned} \tag{9.1.5}$$

where c_x, c_y, and c_z are the Cartesian components of vector $\vec{c}$. Thus, addition of vectors occurs component by component. Using this idea, three vectors $\vec{a}$, $\vec{b}$, and $\vec{c}$ may be added to show (Prob. 9.1.1) that

$$(\vec{a} + \vec{b}) + \vec{c} = \vec{a} + (\vec{b} + \vec{c}) \tag{9.1.6}$$

The *scalar product* (sometimes called the *dot product*) of two vectors $\vec{a}$ and $\vec{b}$ is defined as the product of their magnitudes and the cosine of the angle between them; that is,

$$\vec{a} \cdot \vec{b} = ab\cos\theta(\vec{a}, \vec{b}) \tag{9.1.7}$$

This definition is purely geometric, so it is independent of the reference frame in which the vectors are represented.

Note that if two vectors $\vec{a}$ and $\vec{b}$ are nonzero (i.e., $a \neq 0$ and $b \neq 0$) then their scalar product is zero if and only if $\cos\theta(\vec{a}, \vec{b}) = 0$. Two nonzero vectors are said to be *orthogonal vectors* if their scalar product is zero.

Since $\theta(\vec{b}, \vec{a}) = 2\pi - \theta(\vec{a}, \vec{b})$ and $\cos(2\pi - \theta) = \cos\theta$, the order of terms appearing on the right side of Eq. 9.1.7 is immaterial. Thus,

$$\vec{a} \cdot \vec{b} = \vec{b} \cdot \vec{a} \tag{9.1.8}$$

The scalar product of two vectors may be interpreted as the product of the magnitude of one of the vectors times the projection of the other vector onto that vector. To see this, refer to Fig. 9.1.4, where the projection of vector $\vec{b}$ onto $\vec{a}$ has length $b\cos(\vec{a}, \vec{b})$.

Based on the definition of the scalar product, the following identities hold for the unit coordinate vectors $\vec{i}$, $\vec{j}$, and $\vec{k}$:

$$\begin{aligned}\vec{i} \cdot \vec{j} = \vec{j} \cdot \vec{k} = \vec{k} \cdot \vec{i} = 0 \\ \vec{i} \cdot \vec{i} = \vec{j} \cdot \vec{j} = \vec{k} \cdot \vec{k} = 1\end{aligned} \tag{9.1.9}$$

For any vector $\vec{a}$,

$$\vec{a} \cdot \vec{a} = aa\cos 0 = a^2$$

While not obvious on geometrical grounds, the scalar product satisfies the relation [21]

$$(\vec{a} + \vec{b}) \cdot \vec{c} = \vec{a} \cdot \vec{c} + \vec{b} \cdot \vec{c} \tag{9.1.10}$$

Using Eq. 9.1.10 and the identities of Eq. 9.1.9, a direct calculation (Prob. 9.1.2) yields

$$\vec{a} \cdot \vec{b} = a_x b_x + a_y b_y + a_z b_z \qquad \textbf{(9.1.11)}$$

Note that the concepts and properties of vectors in space discussed thus far are elementary extensions of ideas from vectors in the plane. A new concept for spatial vectors is the *vector product* (sometimes called the *cross product*) of two vectors $\vec{a}$ and $\vec{b}$, defined as the vector

$$\vec{a} \times \vec{b} = ab \sin \theta(\vec{a}, \vec{b})\vec{u} \qquad \textbf{(9.1.12)}$$

where $\vec{u}$ is the unit vector that is orthogonal (perpendicular) to the plane of intersection of vectors $\vec{a}$ and $\vec{b}$, taken in the positive right-hand coordinate direction, as shown in Fig. 9.1.7. If the viewer were behind the plane of Fig. 9.1.7, the unit normal to the plane would be $-\vec{u}$ and the counterclockwise angle from $\vec{a}$ to $\vec{b}$ would be $2\pi - \theta(\vec{a}, \vec{b})$. Then the vector product would be

$$\begin{aligned} \vec{a} \times \vec{b} &= ab \sin(2\pi - \theta(\vec{a}, \vec{b}))(-\vec{u}) \\ &= ab \sin \theta(\vec{a}, \vec{b})\vec{u} \end{aligned}$$

since $\sin(2\pi - \theta) = -\sin \theta$. This is the same result as in Eq. 9.1.12, so the viewpoint does not influence evaluation of the vector product. Since the definition of vector product is purely geometrical, the result is independent of the reference frame in which the vectors are represented.

Since reversal of the order of vectors $\vec{a}$ and $\vec{b}$ in Eq. 9.1.12 yields the opposite direction for the unit vector $\vec{u}$,

$$\vec{b} \times \vec{a} = -\vec{a} \times \vec{b} \qquad \textbf{(9.1.13)}$$

Analogous to Eq. 9.1.10, the vector product satisfies [21]

$$(\vec{a} + \vec{b}) \times \vec{c} = \vec{a} \times \vec{c} + \vec{b} \times \vec{c} \qquad \textbf{(9.1.14)}$$

Since $\theta(\vec{a}, \vec{a}) = 0$, for any vector $\vec{a}$,

$$\vec{a} \times \vec{a} = a^2 \sin 0\vec{u} = \vec{0}$$

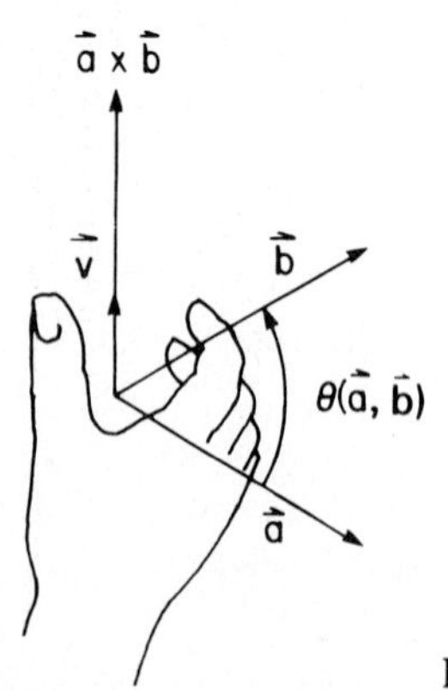

Figure 9.1.7 Vector product.

From the definition of unit coordinate vectors and vector product, the following identities are valid:

$$\begin{aligned}
\vec{i} \times \vec{i} &= \vec{j} \times \vec{j} = \vec{k} \times \vec{k} = \vec{0} \\
\vec{i} \times \vec{j} &= -\vec{j} \times \vec{i} = \vec{k} \\
\vec{j} \times \vec{k} &= -\vec{k} \times \vec{j} = \vec{i} \\
\vec{k} \times \vec{i} &= -\vec{i} \times \vec{k} = \vec{j}
\end{aligned} \tag{9.1.15}$$

Using the identities of Eq. 9.1.15 and the property of vector product of Eq. 9.1.14, the vector product of two vectors may be expanded and written in terms of their components as (Prob. 9.1.3)

$$\begin{aligned}
\vec{c} = \vec{a} \times \vec{b} &= (a_y b_z - a_z b_y)\vec{i} + (a_z b_x - a_x b_z)\vec{j} + (a_x b_y - a_y b_x)\vec{k} \\
&\equiv c_x \vec{i} + c_y \vec{j} + c_z \vec{k}
\end{aligned} \tag{9.1.16}$$

Algebraic Vectors Recall from Eq. 9.1.3 that a geometric vector $\vec{a}$ can be written in component form in a Cartesian x-y-z frame as

$$\vec{a} = a_x \vec{i} + a_y \vec{j} + a_z \vec{k}$$

The geometric vector $\vec{a}$ is thus uniquely defined by its Cartesian components, which may be written in matrix notation as

$$\mathbf{a} = \begin{bmatrix} a_x \\ a_y \\ a_z \end{bmatrix} = [a_x, a_y, a_z]^T \tag{9.1.17}$$

This is the *algebraic representation of a geometric vector.*

Note that the algebraic representation of vectors is dependent on the Cartesian reference frame selected, that is, vectors $\vec{i}$, $\vec{j}$, and $\vec{k}$. Some of the purely geometric properties of vectors are thus lost, and the properties of the reference frame that is used in defining components of vectors come into play. This apparent problem will in fact become a valuable tool in spatial kinematics and dynamics.

An *algebraic vector* is defined as a column matrix. When an algebraic vector represents a geometric vector in three-dimensional space, it has three components. Algebraic vectors with more than three components will also be employed in the kinematics and dynamics of multibody systems. In the case where $\mathbf{a} = [a_1, \ldots, a_n]^T$, the algebraic vector $\mathbf{a}$ is called an *n vector* and is said to belong to *n-dimensional real space,* denoted R^n.

If two vectors $\vec{a}$ and $\vec{b}$ are represented in algebraic form as

$$\begin{aligned}
\mathbf{a} &= [a_x, a_y, a_z]^T \\
\mathbf{b} &= [b_x, b_y, b_z]^T
\end{aligned} \tag{9.1.18}$$

then their vector sum $\vec{c} = \vec{a} + \vec{b}$ of Eq. 9.1.5 is represented in algebraic form by

(Prob. 9.1.4)

$$\mathbf{c} = \mathbf{a} + \mathbf{b} \tag{9.1.19}$$

Example 9.1.3: The algebraic representation of vectors

$$\vec{a} = \vec{i} + 2\vec{j} + 3\vec{k}$$
$$\vec{b} = -\vec{i} + \vec{j} - \vec{k}$$

is

$$\mathbf{a} = [1, 2, 3]^T$$
$$\mathbf{b} = [-1, 1, -1]^T$$

The algebraic representation of the sum $\vec{c} = \vec{a} + \vec{b}$ is

$$\mathbf{c} = \mathbf{a} + \mathbf{b} = [0, 3, 2]^T$$

Two geometric vectors are equal if and only if the Cartesian components of the vectors are equal; that is, $\mathbf{a} = \mathbf{b}$. Multiplication of a vector $\vec{a}$ by a scalar α occurs component by component, so the geometric vector $\alpha\vec{a}$ is represented by the algebraic vector $\alpha\mathbf{a}$. Since there is a one-to-one correspondence between geometric vectors and 3×1 algebraic vectors that are formed from their Cartesian components in a specified Cartesian reference frame, no distinction other than notation will be made between them in the remainder of this text.

The scalar product of two geometric vectors may be expressed in algebraic form, using the result of Eq. 9.1.11, as

$$\vec{a} \cdot \vec{b} = a_x b_x + a_y b_y + a_z b_z = \mathbf{a}^T\mathbf{b} \tag{9.1.20}$$

Example 9.1.4: The scalar product of vectors $\vec{a}$ and $\vec{b}$ (or $\mathbf{a}$ and $\mathbf{b}$) in Example 9.1.3 is

$$\vec{a} \cdot \vec{b} = \mathbf{a}^T\mathbf{b} = [1, 2, 3]\begin{bmatrix} -1 \\ 1 \\ -1 \end{bmatrix} = -2$$

From the definition of scalar product,

$$\mathbf{a}^T\mathbf{b} = ab \cos \theta(\mathbf{a}, \mathbf{b})$$

and, with $a = (\mathbf{a}^T\mathbf{a})^{1/2} = \sqrt{14}$ and $b = (\mathbf{b}^T\mathbf{b})^{1/2} = \sqrt{3}$, $\cos \theta(\mathbf{a}, \mathbf{b}) = -2/\sqrt{42}$. Thus, $\theta(\mathbf{a}, \mathbf{b}) = 1.88$ or 4.40 rad.

A skew-symmetric matrix $\tilde{\mathbf{a}}$ associated with an algebraic vector $\mathbf{a} = [a_x, a_y, a_z]^T$ is defined as

$$\tilde{\mathbf{a}} \equiv \begin{bmatrix} 0 & -a_z & a_y \\ a_z & 0 & -a_x \\ -a_y & a_x & 0 \end{bmatrix} \tag{9.1.21}$$

Note that an overhead $\tilde{}$ (pronounced tilde) indicates that the components of the vector are used to generate a skew-symmetric 3×3 matrix. Conversely, any 3×3 skew-symmetric matrix of the form

$$\mathbf{B} = \begin{bmatrix} 0 & b_{12} & b_{13} \\ -b_{12} & 0 & b_{23} \\ -b_{13} & -b_{23} & 0 \end{bmatrix}$$

can be written as $\mathbf{B} = \tilde{\mathbf{b}}$, where $\mathbf{b} = [b_1, b_2, b_3]^T$, with $b_1 = -b_{23}$, $b_2 = b_{13}$, and $b_3 = -b_{12}$; that is,

$$\mathbf{B} = \tilde{\mathbf{b}} = \begin{bmatrix} -b_{23} \\ b_{13} \\ -b_{12} \end{bmatrix}$$

The vector product $\vec{c} = \vec{a} \times \vec{b}$, which is expanded in component form in Eq. 9.1.6, can thus be written in algebraic vector form as

$$\mathbf{c} = \tilde{\mathbf{a}}\mathbf{b} = \begin{bmatrix} a_y b_z - a_z b_y \\ a_z b_x - a_x b_z \\ a_x b_y - a_y b_x \end{bmatrix} \tag{9.1.22}$$

This result is the reason the $\tilde{}$ operator is introduced. It gives a convenient and computationally practical means of evaluating the vector product of two vectors that are represented in algebraic form.

Example 9.1.5: The algebraic representation of the vector product $\vec{c} = \vec{a} \times \vec{b}$ of $\vec{a}$ and $\vec{b}$ in Example 9.1.3 is

$$\mathbf{c} = \tilde{\mathbf{a}}\mathbf{b} = \begin{bmatrix} 0 & -3 & 2 \\ 3 & 0 & -1 \\ -2 & 1 & 0 \end{bmatrix} \begin{bmatrix} -1 \\ 1 \\ -1 \end{bmatrix} = \begin{bmatrix} -5 \\ -2 \\ 3 \end{bmatrix}$$

For later use, it is helpful to develop some standard properties of the $\tilde{}$ operation. First, note that

$$\tilde{\mathbf{a}}^T = \begin{bmatrix} 0 & a_z & -a_y \\ -a_z & 0 & a_x \\ a_y & -a_x & 0 \end{bmatrix} = -\tilde{\mathbf{a}} \tag{9.1.23}$$

Also, for a scalar α,

$$\alpha\tilde{\mathbf{a}} = \begin{bmatrix} 0 & -\alpha a_z & \alpha a_y \\ \alpha a_z & 0 & -\alpha a_x \\ -\alpha a_y & \alpha a_x & 0 \end{bmatrix} = \widetilde{(\alpha\mathbf{a})} \tag{9.1.24}$$

For any vectors **a** and **b**, a direct calculation (Prob. 9.1.5) shows that

$$\tilde{\mathbf{a}}\mathbf{b} = -\tilde{\mathbf{b}}\mathbf{a} \tag{9.1.25}$$

which agrees with the vector product relation $\vec{a} \times \vec{b} = -\vec{b} \times \vec{a}$ of Eq. 9.1.13.

Direct calculation shows that

$$\tilde{\mathbf{a}}\mathbf{a} = \mathbf{0} \tag{9.1.26}$$

Hence, by Eq. 9.1.23,

$$(\tilde{\mathbf{a}}\mathbf{a})^T = \mathbf{a}^T\tilde{\mathbf{a}}^T = -\mathbf{a}^T\tilde{\mathbf{a}} = \mathbf{0} \tag{9.1.27}$$

It can be verified by direct calculation (Prob. 9.1.6) that

$$\tilde{\mathbf{a}}\tilde{\mathbf{b}} = \mathbf{b}\mathbf{a}^T - \mathbf{a}^T\mathbf{b}\mathbf{I} \tag{9.1.28}$$

where **I** is the 3×3 identity matrix. Direct calculation also shows that (Prob. 9.1.7)

$$\widetilde{(\tilde{\mathbf{a}}\mathbf{b})} = \mathbf{b}\mathbf{a}^T - \mathbf{a}\mathbf{b}^T \tag{9.1.29}$$

$$= \tilde{\mathbf{a}}\tilde{\mathbf{b}} - \tilde{\mathbf{b}}\tilde{\mathbf{a}} \tag{9.1.30}$$

From Eqs. 9.1.29 and 9.1.30,

$$\tilde{\mathbf{a}}\tilde{\mathbf{b}} + \mathbf{a}\mathbf{b}^T = \tilde{\mathbf{b}}\tilde{\mathbf{a}} + \mathbf{b}\mathbf{a}^T \tag{9.1.31}$$

Finally, it can be verified by direct calculation that

$$\widetilde{(\mathbf{a} + \mathbf{b})} = \tilde{\mathbf{a}} + \tilde{\mathbf{b}} \tag{9.1.32}$$

Matrix implementation of vector operations permits the systematic organization of calculations, which is especially valuable for computer applications.

Example 9.1.6: To define the position and orientation of a Cartesian x'-y'-z' reference frame with its origin at point P, in the x-y-z reference frame, it is first essential to locate point P with vector $\mathbf{r}^P$, as shown in Fig. 9.1.8. If a second point Q is located on the x' axis by vector $\mathbf{r}^Q$, the unit vector **f** along the x' axis is

$$\mathbf{f} = \frac{1}{|\mathbf{r}^Q - \mathbf{r}^P|}(\mathbf{r}^Q - \mathbf{r}^P)$$

If a third point R that is not on the x' axis is located in the x'-y' plane by vector $\mathbf{r}^R$, then the vector product of **f** and $\mathbf{r}^R - \mathbf{r}^P$ is a vector in the direction of the z' axis; that is,

$$\mathbf{h} = \frac{1}{|\tilde{\mathbf{f}}(\mathbf{r}^R - \mathbf{r}^P)|}\tilde{\mathbf{f}}(\mathbf{r}^R - \mathbf{r}^P)$$

Finally, the unit vector **g** along the y' axis is

$$\mathbf{g} = \tilde{\mathbf{h}}\mathbf{f}$$

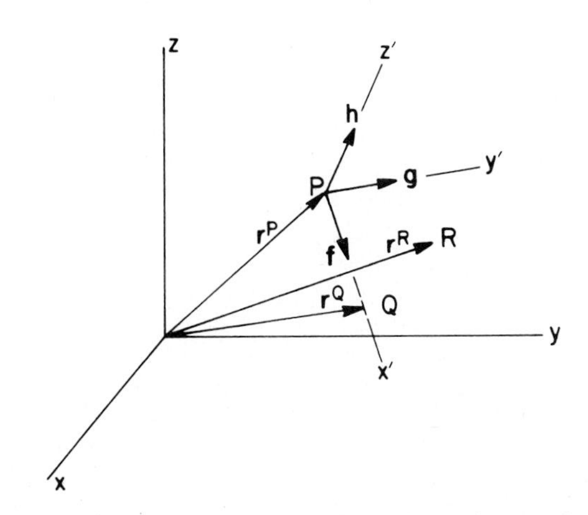

Figure 9.1.8 Points defining a Cartesian reference frame.

Thus, three distinct noncollinear points P, Q, and R in the x'-y' plane define unit vectors **f**, **g**, and **h** along the x', y', and z' axes, hence defining the position and orientation of the x'-y'-z' Cartesian reference frame.

Differentiation of Vectors In analyzing velocities and accelerations, the time derivatives of vectors that locate points must be calculated. Consider a time-dependent vector $\vec{a}(t)$ with time-dependent components $\mathbf{a} \equiv \mathbf{a}(t) = [a_x(t), a_y(t), a_z(t)]^T$ in an *x-y-z stationary Cartesian reference frame* that does not depend on time; that is, $\vec{i}$, $\vec{j}$, and $\vec{k}$ are constant. The *time derivative of a vector* $\vec{a}$ is the *velocity* of the point that is located by the vector, denoted by

$$\dot{\vec{a}} \equiv \frac{d}{dt}\vec{a}(t) = \frac{d}{dt}[a_x(t)\vec{i} + a_y(t)\vec{j} + a_z(t)\vec{k}]$$

$$= \left[\frac{d}{dt}a_x(t)\right]\vec{i} + \left[\frac{d}{dt}a_y(t)\right]\vec{j} + \left[\frac{d}{dt}a_z(t)\right]\vec{k}$$

Note that this is only valid if $\vec{i}$, $\vec{j}$, and $\vec{k}$ are not time dependent. In algebraic vector notation, this is

$$\dot{\mathbf{a}} \equiv \frac{d}{dt}\mathbf{a}(t) = \left[\frac{d}{dt}a_x(t), \frac{d}{dt}a_y(t), \frac{d}{dt}a_z(t)\right]^T \tag{9.1.33}$$

Thus, for vectors that are written in terms of their components in a stationary Cartesian reference frame, the derivative of a vector is obtained by differentiating its components.

Example 9.1.7: A particle P moves with constant speed along a circle of radius R in the x-y plane of a stationary Cartesian reference frame, as shown in

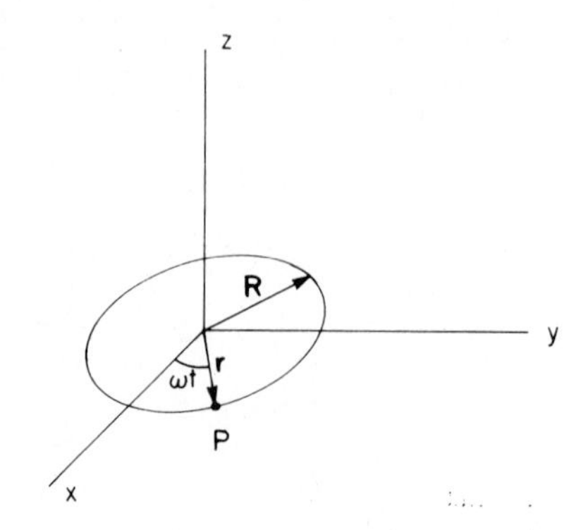

Figure 9.1.9 Particle moving on circular path with constant speed.

Fig. 9.1.9. Its position vector is

$$\mathbf{r} = \begin{bmatrix} R \cos \omega t \\ R \sin \omega t \\ 0 \end{bmatrix}$$

where R and ω are constant. Its velocity is thus

$$\dot{\mathbf{r}} = \begin{bmatrix} -R\omega \sin \omega t \\ R\omega \cos \omega t \\ 0 \end{bmatrix}$$

The derivative of the sum of two vectors is (Prob. 9.1.9)

$$\frac{d}{dt}(\mathbf{a}(t) + \mathbf{b}(t)) = \dot{a} + \dot{b} \tag{9.1.34}$$

which is analogous to the differentiation rule of calculus that the derivative of a sum is the sum of the derivatives.

The following vector forms of the *product rule of differentiation* are also valid (Prob. 9.1.10):

$$\frac{d}{dt}(\alpha\mathbf{a}) = \dot{\alpha}\mathbf{a} + \alpha\dot{\mathbf{a}} \tag{9.1.35}$$

$$\frac{d}{dt}(\mathbf{a}^T\mathbf{b}) = \dot{\mathbf{a}}^T\mathbf{b} + \mathbf{a}^T\dot{\mathbf{b}} \tag{9.1.36}$$

$$\frac{d}{dt}(\tilde{\mathbf{a}}\mathbf{b}) = \dot{\tilde{\mathbf{a}}}\mathbf{b} + \tilde{\mathbf{a}}\dot{\mathbf{b}} \tag{9.1.37}$$

where $\alpha(t)$ is a scalar function of time. Note also that

$$\dot{\tilde{\mathbf{a}}} = \tilde{\dot{\mathbf{a}}} \tag{9.1.38}$$

Example 9.1.8: If the length of a vector $\mathbf{a}(t)$ is fixed, that is, $\mathbf{a}(t)^T\mathbf{a}(t) = c$, where c is a constant, then Eq. 9.1.36 yields

$$\dot{\mathbf{a}}^T\mathbf{a} = 0 \tag{9.1.39}$$

If $\mathbf{a}$ is a position vector that locates a point, then $\dot{\mathbf{a}}$ is the velocity of that point. Hence, Eq. 9.1.39 indicates that the velocity of a point whose distance from the origin is constant is orthogonal to the position vector of the point. Note that Eq. 9.1.39 is satisfied by the velocity vector of Example 9.1.7, which is to be expected since $\mathbf{a}^T\mathbf{a} = R^2$.

The second time derivative of $\mathbf{a}(t)$ is the *acceleration* of the point that is located by the vector, denoted as

$$\ddot{\mathbf{a}} \equiv \frac{d}{dt}(\dot{\mathbf{a}}(t)) = \left[\frac{d^2}{dt^2}a_x(t), \frac{d^2}{dt^2}a_y(t), \frac{d^2}{dt^2}a_z(t)\right]^T \tag{9.1.40}$$

Thus, for vectors that are written in terms of their components in a stationary Cartesian reference frame, acceleration may be calculated in terms of the second time derivatives of the components of the vector.

Example 9.1.9: The acceleration of the particle that is located by the vector $\mathbf{r}(t)$ in Example 9.1.7 is

$$\ddot{\mathbf{r}} = \begin{bmatrix} -R\omega^2\cos\omega t \\ -R\omega^2\sin\omega t \\ 0 \end{bmatrix} = -\omega^2\mathbf{r}$$

This is the classical *centripetal acceleration* of a point that moves on a circular path, since the direction of $\ddot{\mathbf{r}}$ is opposite to the direction of $\mathbf{r}$.

9.2 KINEMATICS OF A RIGID BODY IN SPACE

A *rigid body* is defined as being made up of a continuum of particles that are constrained not to move relative to one another. While actual bodies are never perfectly rigid, deformation effects are often negligible when considering the motion of a machine that is made up multiple bodies. For this reason, most attention in the study of the dynamics of machines is devoted to modeling individual bodies in a system as being rigid. This section develops the relations needed for the kinematic and dynamic analysis of rigid bodies in space.

Substantial technical differences arise in the kinematic analysis of rigid bodies in space, as compared to rigid bodies in the plane. Most of the attention in this section is on methods of defining the orientation of rigid bodies in space.

Reference Frames for the Location and Orientation of a Body in Space Consider, for purposes of introduction, the rectangular body shown in Fig. 9.2.1. The geometry of the body, usually defined by key points on the body, may be defined in an x'-y'-z' reference frame that is fixed to the body, called a *body-fixed reference frame.* The x'-y'-z' frame is coincident with the x-y-z global reference frame in the position shown in Fig. 9.2.1. This body-fixed reference frame might be thought of as a *drafting board reference frame* in engineering practice. That is, the engineer defines points, lines, and surfaces on a machine or structural component in terms of front view, side view, and top view to create a three-dimensional definition of the component under consideration. Thus, data that define the geometry of the body are specified in the x'-y'-z' frame.

Consider next the translated and rotated position of the body shown in Fig. 9.2.2. Translation of the body may easily be defined by the vector $\mathbf{r}$ from the origin of the x-y-z reference frame to the origin of the x'-y'-z' frame. Specification of orientation is, however, more complex. Nevertheless, once the x'-y'-z' frame is oriented relative to the x-y-z reference frame, all points of interest on the body can be defined in the x-y-z reference frame. Most of the attention in this section focuses on transformations from generally oriented body-fixed frames to the x-y-z reference frame.

To see that the orientation of a body in space raises delicate technical questions, consider two rotation sequences for a body from the orientation shown in Fig. 9.2.1. Rotations of the body by angles of $\pi/2$ are to be made about its body-fixed x' and y' axes. Consider first the rotation sequence shown in Fig. 9.2.3, where the first rotation is about the body-fixed x' axis, which initially coincides with the reference frame x axis. The second rotation is about the

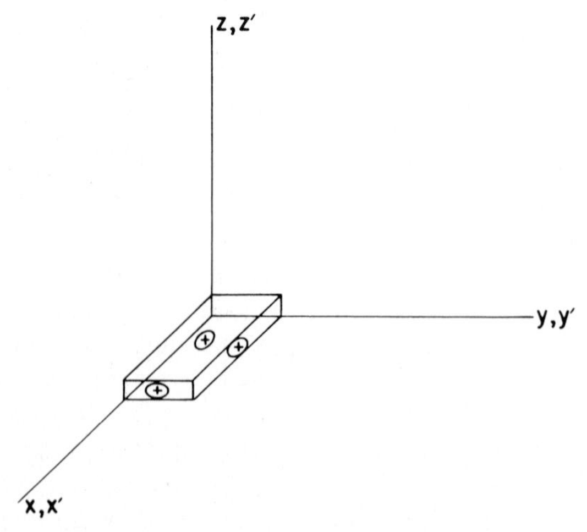

Figure 9.2.1 Body in reference position.

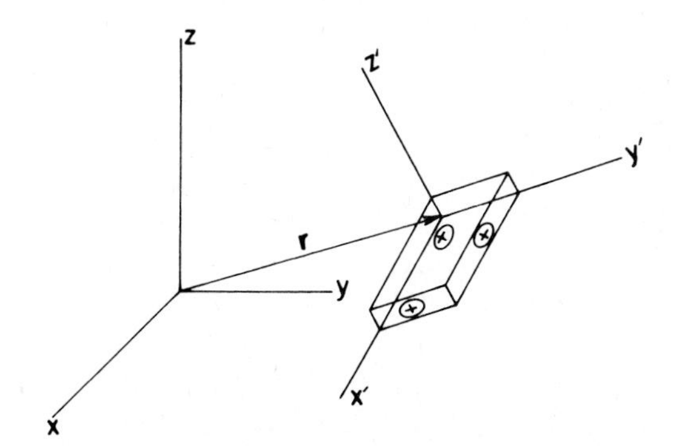

Figure 9.2.2 Body translated and rotated.

body-fixed y' axis, when it coincides with the reference frame z axis, as shown in Fig. 9.2.3a. The resulting orientation is shown in Fig. 9.2.3b.

Consider next the same rotations about body-fixed x' and y' axes, but in reverse order. First, the body is rotated $\pi/2$ about its body-fixed y' axis, which initially coincides with the reference frame y axis, as shown in Fig. 9.2.4a. The body is then rotated $\pi/2$ about its body-fixed x' axis, when it coincides with the negative reference frame z axis, as shown in Fig. 9.2.4a. The resulting orientation is shown in Fig. 9.2.4b.

Since the orientations shown in Figs. 9.2.3b and 9.2.4b are distinctly different, it is clear that the order of rotation is important in defining the orientation of a body in space. Much as in the case of matrix multiplication, in which the order of terms in the product of two matrices is critically important, the order of rotations is likewise important. This shows that large-amplitude rotation is not a vector quantity, since the order of rotations is not commutative; that is, orders of rotation cannot be inverted.

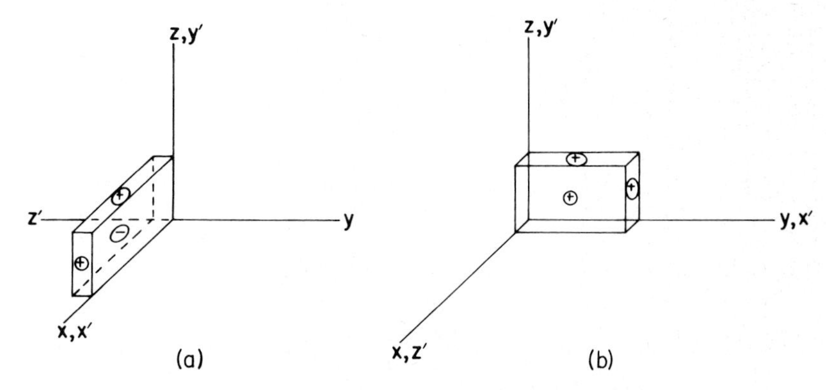

Figure 9.2.3 Rotation of body by $\pi/2$ about x' and then y' axes.

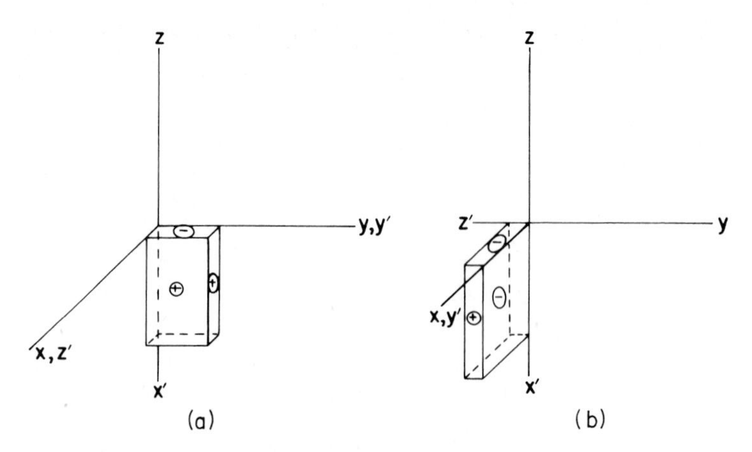

Figure 9.2.4 Rotations of body by $\pi/2$ about y' and then x' axes.

Transformation of Coordinates. It is shown in Section 9.1 that a geometric vector is uniquely represented by an algebraic vector that contains components of the geometric vector in a Cartesian reference frame. The components of a vector, however, are defined in a specific Cartesian reference frame. Consider a second Cartesian x'-y'-z' frame with the same origin as the x-y-z frame, as shown in Fig. 9.2.5. Unit x', y', and z' coordinate vectors are denoted by $\vec{f}$, $\vec{g}$, and $\vec{h}$, respectively, and unit x, y, and z coordinate vectors are denoted by $\vec{i}$, $\vec{j}$, and $\vec{k}$.

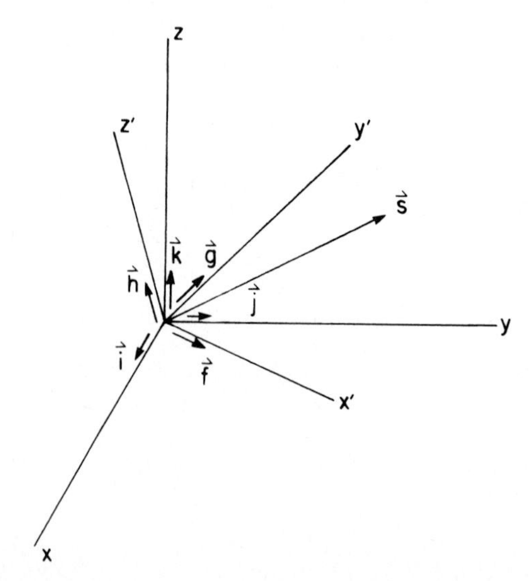

Figure 9.2.5 Two Cartesian reference frames.

A vector $\vec{s}$ in space can be represented in either of the frames as

$$\vec{s} = s_x\vec{i} + s_y\vec{j} + s_z\vec{k} \tag{9.2.1}$$

or

$$\vec{s} = s_{x'}\vec{f} + s_{y'}\vec{g} + s_{z'}\vec{h} \tag{9.2.2}$$

where

$$s_x = \vec{s}\cdot\vec{i}, \qquad s_y = \vec{s}\cdot\vec{j}, \qquad s_z = \vec{s}\cdot\vec{k} \tag{9.2.3}$$

and

$$s_{x'} = \vec{s}\cdot\vec{f}, \qquad s_{y'} = \vec{s}\cdot\vec{g}, \qquad s_{z'} = \vec{s}\cdot\vec{h} \tag{9.2.4}$$

The algebraic vectors that define $\vec{s}$ in the two frames are

$$\mathbf{s} = [s_x, s_y, s_z]^T \tag{9.2.5}$$

in the x-y-z frame and

$$\mathbf{s}' = [s_{x'}, s_{y'}, s_{z'}]^T \tag{9.2.6}$$

in the x'-y'-z' frame.

It is clear that there is a relation between $\mathbf{s}$ and $\mathbf{s}'$, since they are defined by the same geometric vector $\vec{s}$. To establish this relation, expand the $\vec{f}$, $\vec{g}$, and $\vec{h}$ unit vectors in terms of the $\vec{i}$, $\vec{j}$, and $\vec{k}$ unit vectors as

$$\begin{aligned}
\vec{f} &= a_{11}\vec{i} + a_{21}\vec{j} + a_{31}\vec{k} \\
\vec{g} &= a_{12}\vec{i} + a_{22}\vec{j} + a_{32}\vec{k} \\
\vec{h} &= a_{13}\vec{i} + a_{23}\vec{j} + a_{33}\vec{k}
\end{aligned} \tag{9.2.7}$$

where a_{ij} are the following direction cosines:

$$\begin{aligned}
a_{11} &= \vec{i}\cdot\vec{f} = \cos\theta(\vec{i}, \vec{f}) \\
a_{12} &= \vec{i}\cdot\vec{g} = \cos\theta(\vec{i}, \vec{g}) \\
a_{13} &= \vec{i}\cdot\vec{h} = \cos\theta(\vec{i}, \vec{h}) \\
a_{21} &= \vec{j}\cdot\vec{f} = \cos\theta(\vec{j}, \vec{f}) \\
a_{22} &= \vec{j}\cdot\vec{g} = \cos\theta(\vec{j}, \vec{g}) \\
a_{23} &= \vec{j}\cdot\vec{h} = \cos\theta(\vec{j}, \vec{h}) \\
a_{31} &= \vec{k}\cdot\vec{f} = \cos\theta(\vec{k}, \vec{f}) \\
a_{32} &= \vec{k}\cdot\vec{g} = \cos\theta(\vec{k}, \vec{g}) \\
a_{33} &= \vec{k}\cdot\vec{h} = \cos\theta(\vec{k}, \vec{h})
\end{aligned} \tag{9.2.8}$$

Substituting from Eq. 9.2.7 into Eq. 9.2.2 yields

$$\begin{aligned}
\vec{s} = {} & (a_{11}s_{x'} + a_{12}s_{y'} + a_{13}s_{z'})\vec{i} \\
& + (a_{21}s_{x'} + a_{22}s_{y'} + a_{23}s_{z'})\vec{j} \\
& + (a_{31}s_{x'} + a_{32}s_{y'} + a_{33}s_{z'})\vec{k}
\end{aligned}$$

Equating corresponding right sides of this representation of $\vec{s}$ and Eq. 9.2.1,

$$\begin{aligned} s_x &= a_{11}s_{x'} + a_{12}s_{y'} + a_{13}s_{z'} \\ s_y &= a_{21}s_{x'} + a_{22}s_{y'} + a_{23}s_{z'} \\ s_z &= a_{31}s_{x'} + a_{32}s_{y'} + a_{33}s_{z'} \end{aligned} \tag{9.2.9}$$

In matrix form, this is

$$\mathbf{s} = \mathbf{A}\mathbf{s}' \tag{9.2.10}$$

where **A** is called the *direction cosine matrix* or *rotation transformation matrix:*

$$\mathbf{A} = \begin{bmatrix} a_{11} & a_{12} & a_{13} \\ a_{21} & a_{22} & a_{23} \\ a_{31} & a_{32} & a_{33} \end{bmatrix} \tag{9.2.11}$$

The direction cosine matrix **A** has a very special property. If the x-y-z component forms of unit vectors $\vec{f}$, $\vec{g}$, and $\vec{h}$ are denoted by **f**, **g**, and **h** and the x-y-z components of vectors $\vec{i}$, $\vec{j}$, and $\vec{k}$ are denoted by **i**, **j**, and **k**, then

$$\begin{aligned} \mathbf{i} &= [1, 0, 0]^T \\ \mathbf{j} &= [0, 1, 0]^T \\ \mathbf{k} &= [0, 0, 1]^T \end{aligned} \tag{9.2.12}$$

and Eq. 9.2.8 shows that

$$\begin{aligned} \mathbf{f} &= [a_{11}, a_{21}, a_{31}]^T \\ \mathbf{g} &= [a_{12}, a_{22}, a_{32}]^T \\ \mathbf{h} &= [a_{13}, a_{23}, a_{33}]^T \end{aligned}$$

Therefore, the matrix **A** of Eq. 9.2.11 can be written as

$$\mathbf{A} = [\mathbf{f}, \mathbf{g}, \mathbf{h}] \tag{9.2.13}$$

Since the unit vectors **f**, **g**, and **h** are orthogonal, expansions of the matrix product shows that (Prob. 9.2.3)

$$\mathbf{A}^T\mathbf{A} = \mathbf{I} \tag{9.2.14}$$

Thus, $\mathbf{A}^T = \mathbf{A}^{-1}$ and the direction cosine matrix **A** is an *orthogonal matrix.* This special property permits an easy inversion of Eq. 9.2.10 to obtain

$$\mathbf{s}' = \mathbf{A}^T\mathbf{s} \tag{9.2.15}$$

Thus, transforming between algebraic vectors that represent the same physical vector in the x-y-z and x'-y'-z' Cartesian reference frames is a trivial matter.

Example 9.2.1: The z' axis of the x'-y'-z' Cartesian frame is fixed in the x-y plane, and the x' axis is rotated an angle ϕ above the x-y plane, as shown in

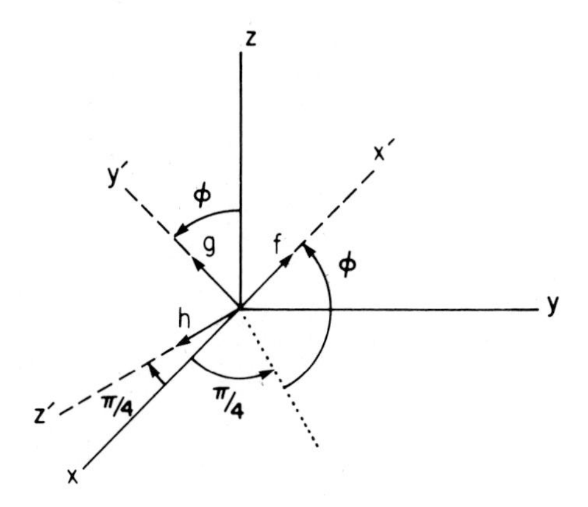

Figure 9.2.6 Orientation of a reference frame.

Fig. 9.2.6. The **f**, **g**, and **h** unit vectors are thus

$$\mathbf{f} = \left[\frac{\sqrt{2}}{2}\cos\phi, \frac{\sqrt{2}}{2}\cos\phi, \sin\phi\right]^T$$

$$\mathbf{g} = \left[-\frac{\sqrt{2}}{2}\sin\phi, -\frac{\sqrt{2}}{2}\sin\phi, \cos\phi\right]^T$$

$$\mathbf{h} = \left[\frac{\sqrt{2}}{2}, -\frac{\sqrt{2}}{2}, 0\right]^T$$

The transformation matrix from the x'-y'-z' frame to the x-y-z frame may be evaluated using Eq. 9.2.13.

When the origins of the x-y-z and x'-y'-z' frames do not coincide, the foregoing analysis is applied between the x'-y'-z' and a translated x-y-z frame, as shown in Fig. 9.2.7. If the algebraic vector $\mathbf{s}'^P$ locates point P in the x'-y'-z' frame, then in the translated x-y-z frame this vector is $\mathbf{A}\mathbf{s}'^P$. Thus,

$$\mathbf{r}^P = \mathbf{r} + \mathbf{A}\mathbf{s}'^P \tag{9.2.16}$$

where $\mathbf{r}$ is the vector from the origin of the x-y-z reference frame to the origin of the x'-y'-z' frame, as shown in Fig. 9.2.7.

The nine direction cosines in matrix $\mathbf{A}$ define the orientation of the x'-y'-z' frame relative to the x-y-z reference frame, but they are not independent. Equation 9.2.14 provides six equations among the nine direction cosines, since in component form Eq. 9.2.14 is

$$\sum_{k=1}^{3} a_{ki}a_{kj} = \begin{cases} 1, & i = j \\ 0, & i \neq j \end{cases}, \qquad i, j = 1, 2, 3$$

and interchanging $i \neq j$ yields the same equation, because $\mathbf{A}^T\mathbf{A} = \mathbf{I}$ is symmetric.

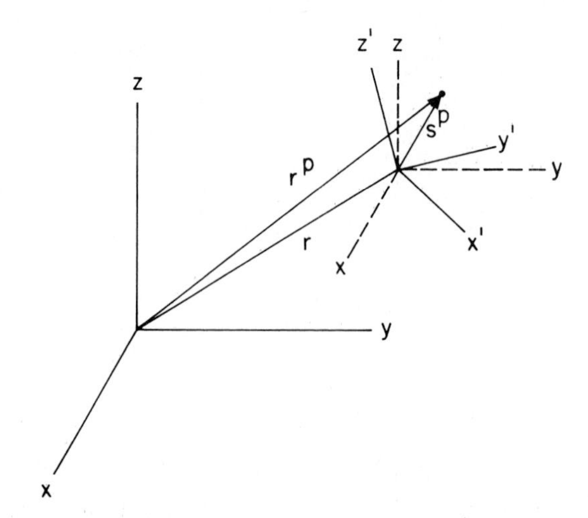

Figure 9.2.7 Translation and rotation of a reference frame.

To show that there are three rotational degrees of freedom, it must be shown that these equations impose six independent constraints on the nine components of $\mathbf{A}$.

First note that by Eq. 9.2.13, $\mathbf{A} = [\mathbf{f}, \mathbf{g}, \mathbf{h}]$, where $\mathbf{f}$, $\mathbf{g}$, and $\mathbf{h}$ are unit vectors along the x', y', and z' axes, respectively. Denote the vector of nine direction cosines in $\mathbf{A}$ as

$$\mathbf{q} = \begin{bmatrix} \mathbf{f} \\ \mathbf{g} \\ \mathbf{h} \end{bmatrix}$$

The constraints imposed on these variables by the diagonal and upper triangular parts of Eq. 9.2.14 are

$$\mathbf{\Phi}(\mathbf{q}) \equiv \begin{bmatrix} \frac{1}{2}\mathbf{f}^T\mathbf{f} - \frac{1}{2} \\ \frac{1}{2}\mathbf{g}^T\mathbf{g} - \frac{1}{2} \\ \frac{1}{2}\mathbf{h}^T\mathbf{h} - \frac{1}{2} \\ \mathbf{f}^T\mathbf{g} \\ \mathbf{f}^T\mathbf{h} \\ \mathbf{g}^T\mathbf{h} \end{bmatrix} = \mathbf{0}$$

Let $\mathbf{q}_0$ satisfy $\mathbf{\Phi}(\mathbf{q}_0) = \mathbf{0}$. The Jacobian of the constraint function $\mathbf{\Phi}(\mathbf{q})$ is

$$\mathbf{\Phi}_{\mathbf{q}} = \begin{bmatrix} \mathbf{f}^T & \mathbf{0} & \mathbf{0} \\ \mathbf{0} & \mathbf{g}^T & \mathbf{0} \\ \mathbf{0} & \mathbf{0} & \mathbf{h}^T \\ \mathbf{g}^T & \mathbf{f}^T & \mathbf{0} \\ \mathbf{h}^T & \mathbf{0} & \mathbf{f}^T \\ \mathbf{0} & \mathbf{h}^T & \mathbf{g}^T \end{bmatrix}$$

The rows of $\mathbf{\Phi_q}$ are linearly independent if and only if the columns of $\mathbf{\Phi_q}^T$ are linearly independent, that is, if and only if

$$\mathbf{\Phi_q}^T \mathbf{u} = \mathbf{0}$$

implies $\mathbf{u} = \mathbf{0}$. Expanding

$$\mathbf{\Phi_q}^T \mathbf{u} = \begin{bmatrix} \mathbf{f} & \mathbf{0} & \mathbf{0} & \mathbf{g} & \mathbf{h} & \mathbf{0} \\ \mathbf{0} & \mathbf{g} & \mathbf{0} & \mathbf{f} & \mathbf{0} & \mathbf{h} \\ \mathbf{0} & \mathbf{0} & \mathbf{h} & \mathbf{0} & \mathbf{f} & \mathbf{g} \end{bmatrix} \begin{bmatrix} u_1 \\ u_2 \\ u_3 \\ u_4 \\ u_5 \\ u_6 \end{bmatrix} = \mathbf{0}$$

and reordering variables yields

$$[\mathbf{f}, \mathbf{g}, \mathbf{h}] \begin{bmatrix} u_1 \\ u_4 \\ u_5 \end{bmatrix} = \mathbf{A} \begin{bmatrix} u_1 \\ u_4 \\ u_5 \end{bmatrix} = \mathbf{0}$$

$$[\mathbf{f}, \mathbf{g}, \mathbf{h}] \begin{bmatrix} u_4 \\ u_2 \\ u_6 \end{bmatrix} = \mathbf{A} \begin{bmatrix} u_4 \\ u_2 \\ u_6 \end{bmatrix} = \mathbf{0}$$

$$[\mathbf{f}, \mathbf{g}, \mathbf{h}] \begin{bmatrix} u_5 \\ u_6 \\ u_3 \end{bmatrix} = \mathbf{A} \begin{bmatrix} u_5 \\ u_6 \\ u_3 \end{bmatrix} = \mathbf{0}$$

Since $|\mathbf{A}^T\mathbf{A}| = |\mathbf{I}| = 1 = |\mathbf{A}^T|\,|\mathbf{A}| = |\mathbf{A}|\,|\mathbf{A}| = |\mathbf{A}|^2$, $|\mathbf{A}| = 1$ and $\mathbf{A}$ is nonsingular. Thus, the previous three equations imply that $\mathbf{u} = \mathbf{0}$. Hence, $\mathbf{\Phi_q}$ has full row rank and there is a 6×6 nonsingular submatrix of $\mathbf{\Phi_q}$ with nonzero determinant. Defining the variables in $\mathbf{q}$ that correspond to columns of the nonsingular submatrix as dependent, these six variables can be written explicitly as functions of the remaining three variables uniquely in some neighborhood of $\mathbf{q}_0$ (by the implicit function theorem). Thus, there are three degrees of rotational freedom of a rigid body in space.

While the nine direction cosines, subject to the constraints of Eq. 9.2.14, could be adopted as generalized coordinates that define the orientation of the x'-y'-z' frame, this is neither practical nor convenient. Thus, other orientation generalized coordinates are sought.

Example 9.2.2: As a special case, consider the x-y-z and x'-y'-z' frames shown in Fig. 9.2.8, with z and z' axes coincident. The angle of rotation of the x' axis relative to the x axis, with counterclockwise as positive, is denoted as ϕ.

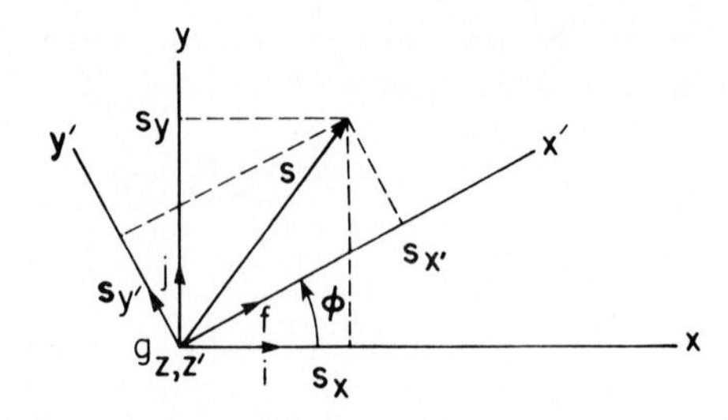

Figure 9.2.8 Reference frames with coincident z and z' axes.

From Eq. 9.2.8,

$$a_{11} = a_{22} = \cos\phi$$

$$a_{21} = \cos\left(\frac{\pi}{2} - \phi\right) = \sin\phi$$

$$a_{12} = \cos\left(\frac{\pi}{2} + \phi\right) = -\sin\phi$$

$$a_{33} = 1$$

$$a_{31} = a_{32} = a_{13} = a_{23} = 0$$

Thus the direction cosine matrix of Eq. 9.2.11 yields the transformation

$$\mathbf{s} = \begin{bmatrix} s_x \\ s_y \\ s_z \end{bmatrix} = \begin{bmatrix} \cos\phi & -\sin\phi & 0 \\ \sin\phi & \cos\phi & 0 \\ 0 & 0 & 1 \end{bmatrix} \begin{bmatrix} s_{x'} \\ s_{y'} \\ s_{z'} \end{bmatrix} = \mathbf{As}' \tag{9.2.17}$$

This contains the planar rotation relations from Chapter 2, that is,

$$\begin{bmatrix} s_x \\ s_y \end{bmatrix} = \begin{bmatrix} \cos\phi & -\sin\phi \\ \sin\phi & \cos\phi \end{bmatrix} \begin{bmatrix} s_{x'} \\ s_{y'} \end{bmatrix} \tag{9.2.18}$$

$$\begin{bmatrix} s_{x'} \\ s_{y'} \end{bmatrix} = \begin{bmatrix} \cos\phi & \sin\phi \\ -\sin\phi & \cos\phi \end{bmatrix} \begin{bmatrix} s_x \\ s_y \end{bmatrix} \tag{9.2.19}$$

Useful relations may be obtained by noting that, for any vectors $\mathbf{s}'$ and $\mathbf{v}'$ in the x'-y'-z' frame, the vector product of $\mathbf{s}$ and $\mathbf{v}$ in the x-y-z frame may be obtained by forming the vector product of $\mathbf{s}'$ and $\mathbf{v}'$ and transforming the result to the x-y-z frame; that is,

$$\tilde{\mathbf{s}}\mathbf{v} = \mathbf{A}(\tilde{\mathbf{s}}'\mathbf{v}')$$

Using $\mathbf{v} = \mathbf{A}\mathbf{v}'$,

$$(\tilde{\mathbf{s}}\mathbf{A})\mathbf{v}' = (\mathbf{A}\tilde{\mathbf{s}}')\mathbf{v}'$$

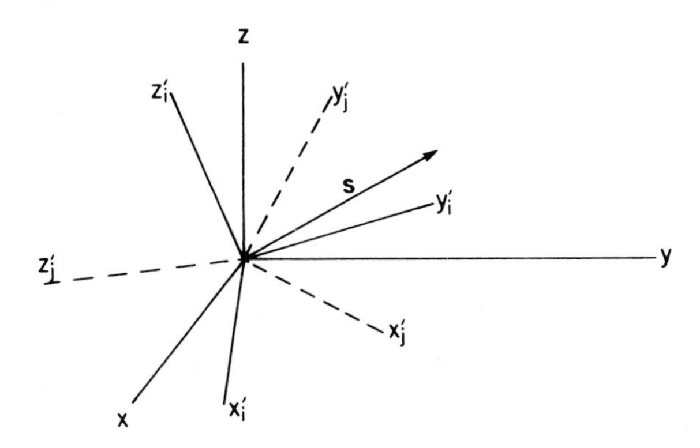

Figure 9.2.9 Reference frames with coincident origins.

Since this must hold for arbitrary $\mathbf{v}'$ (Prob. 9.2.4),

$$\tilde{\mathbf{s}}\mathbf{A} = \mathbf{A}\tilde{\mathbf{s}}' \tag{9.2.20}$$

Postmultiplying by $\mathbf{A}^T$ and using Eq. 9.2.14 yields the desired result:

$$\widetilde{(\mathbf{A}\mathbf{s}')} = \tilde{\mathbf{s}} = \mathbf{A}\tilde{\mathbf{s}}'\mathbf{A}^T \tag{9.2.21}$$

Similarly,

$$\widetilde{(\mathbf{A}^T\mathbf{s})} = \tilde{\mathbf{s}}' = \mathbf{A}^T\tilde{\mathbf{s}}\mathbf{A} \tag{9.2.22}$$

Consider the pair of Cartesian x_i'-y_i'-z_i' and x_j'-y_j'-z_j' frames shown in Fig. 9.2.9. An arbitrary vector $\mathbf{s}$ in the x-y-z frame has representations $\mathbf{s}_i'$ and $\mathbf{s}_j'$ in the x_i'-y_i'-z_i' and x_j'-y_j'-z_j' frames, respectively; that is,

$$\mathbf{s} = \mathbf{A}_i\mathbf{s}_i' = \mathbf{A}_j\mathbf{s}_j' \tag{9.2.23}$$

where $\mathbf{A}_i$ and $\mathbf{A}_j$ are transformation matrices from the x_i'-y_i'-z_i' and x_j'-y_j'-z_j' frames to the x-y-z frame, respectively.

Since $\mathbf{A}_i$ and $\mathbf{A}_j$ are orthogonal matrices, Eq. 9.2.23 yields

$$\mathbf{s}_i' = \mathbf{A}_i^T\mathbf{A}_j\mathbf{s}_j' \equiv \mathbf{A}_{ij}\mathbf{s}_j' \tag{9.2.24}$$

Since $\mathbf{s}_j'$ is an arbitrary vector,

$$\mathbf{A}_{ij} = \mathbf{A}_i^T\mathbf{A}_j \tag{9.2.25}$$

is the transformation matrix from the x_j'-y_j'-z_j' frame to the x_i'-y_i'-z_i' frame. A direct calculation shows that $\mathbf{A}_{ij}$ is an orthogonal matrix (Prob. 9.2.5).

Example 9.2.3: In numerous applications, kinematic constraints dictate that certain axes of the x_i'-y_i'-z_i' and x_j'-y_j'-z_j' frames remain parallel; for example, the vectors $\mathbf{h}_i$ and $\mathbf{h}_j$ shown in Fig. 9.2.10 are to be parallel. The angle θ of rotation, measured positive as counterclockwise from $\mathbf{f}_i$ to $\mathbf{f}_j$, is a key quantity that is to be

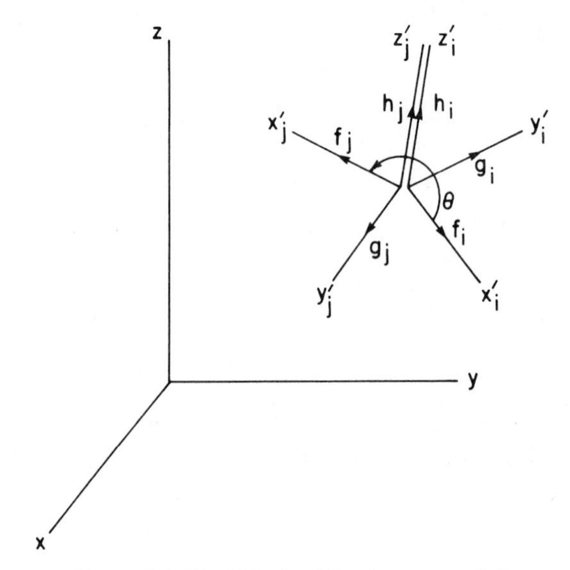

Figure 9.2.10 Triads with z' axes parallel.

calculated. From the definition of scalar product and the fact that the coordinate vectors shown in Fig. 9.2.10 are unit vectors,

$$\mathbf{f}_i^T\mathbf{f}_j = \cos\theta \tag{9.2.26}$$

Similarly, from the definition of vector product,

$$\tilde{\mathbf{f}}_i\mathbf{f}_j = \mathbf{h}_i \sin\theta$$

Taking the scalar product of both sides of this equation with $\mathbf{h}_i$ and using the fact that $\tilde{\mathbf{f}}_i\mathbf{h}_i = -\mathbf{g}_i$,

$$\sin\theta = \mathbf{h}_i^T\tilde{\mathbf{f}}_i\mathbf{f}_j = \mathbf{g}_i^T\mathbf{f}_j \tag{9.2.27}$$

Writing the unit vectors in terms of the respective reference frames in which they are fixed, and using the transformation matrices from these frames to the x-y-z frame, Eq. 9.2.26 becomes

$$\cos\theta = \mathbf{f}_i'^T\mathbf{A}_i^T\mathbf{A}_j\mathbf{f}_j' \tag{9.2.28}$$

Similarly, Eq . 9.2.27 becomes

$$\sin\theta = \mathbf{g}_i'^T\mathbf{A}_i^T\mathbf{A}_j\mathbf{f}_j' \tag{9.2.29}$$

If $\sin\theta$ and $\cos\theta$ are known, the value of θ, $0 \leq \theta < 2\pi$, may be uniquely determined by taking the Arcsin of both sides of Eq. 9.2.29 and using the algebraic sign of $\cos\theta$ from Eq. 9.2.28 to uniquely evaluate θ. To simplify notation, denote $s = \sin\theta$ from Eq. 9.2.29 and $c = \cos\theta$ from Eq. 9.2.28. Then (Prob. 9.2.6), with

$$\frac{-\pi}{2} \leq \text{Arcsin}\, s \leq \frac{\pi}{2} \tag{9.2.30}$$

θ is

$$\theta = \begin{cases} \text{Arcsin } s, & \text{if } s \geq 0 \text{ and } c \geq 0 \\ \pi - \text{Arcsin } s, & \text{if } s \geq 0 \text{ and } c < 0 \\ \pi - \text{Arcsin } s, & \text{if } s < 0 \text{ and } c < 0 \\ 2\pi + \text{Arcsin } s, & \text{if } s < 0 \text{ and } c \geq 0 \end{cases} \tag{9.2.31}$$

The angle calculated in Eq. 9.2.31 is in the range $0 \leq \theta < 2\pi$. If the number n of revolutions that have occurred is important, logic must be supplied to determine n. The angle $2n\pi$ is then added to the value of θ calculated in Eq. 9.2.31.

Velocity, Acceleration, and Angular Velocity An x'-y'-z' frame is fixed in a moving body, to define the body's position and orientation, relative to an x-y-z reference frame. Consider a point P that is fixed in the x'-y'-z' frame, as shown in Fig. 9.2.11. The vector that locates P in the x-y-z reference frame is given by Eq. 9.2.16 as

$$\mathbf{r}^P = \mathbf{r} + \mathbf{A}\mathbf{s}'^P \tag{9.2.32}$$

where $\mathbf{s}'^P$ is the constant vector of coordinates of P in the x'-y'-z' frame and $\mathbf{A}$ is the direction cosine matrix of the x'-y'-z' frame relative to the stationary x-y-z frame.

Since the x'-y'-z' frame is moving and changing its orientation with time, the vector $\mathbf{r}$ and transformation matrix $\mathbf{A}$ are functions of time. The differentiation rules of Section 2.5 can be used to obtain the time derivative of $\mathbf{r}^P$ as

$$\dot{\mathbf{r}}^P = \dot{\mathbf{r}} + \dot{\mathbf{A}}\mathbf{s}'^P \tag{9.2.33}$$

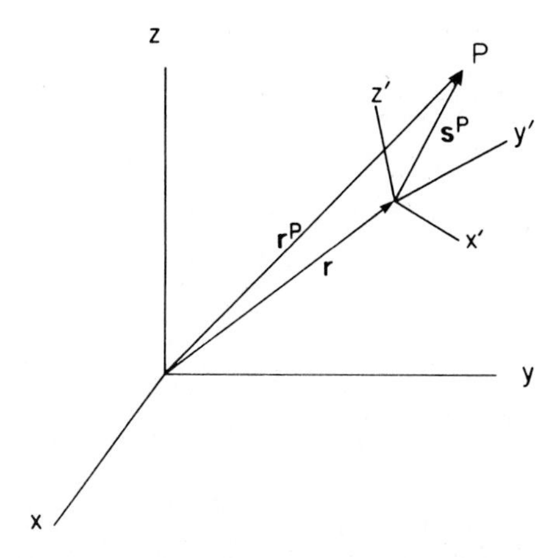

Figure 9.2.11 Point P fixed in an x'-y'-z' reference frame.

Using Eq. 9.2.15, this result may be rewritten as

$$\dot{\mathbf{r}}^P = \dot{\mathbf{r}} + \dot{\mathbf{A}}\mathbf{A}^T\mathbf{s}^P \tag{9.2.34}$$

To interpret terms in Eq. 9.2.34, it is helpful to derive an identity that involves the transformation matrix $\mathbf{A}$. Differentiating both sides of $\mathbf{A}\mathbf{A}^T = \mathbf{I}$ with respect to time,

$$\dot{\mathbf{A}}\mathbf{A}^T + \mathbf{A}\dot{\mathbf{A}}^T = \mathbf{0}$$

Thus,

$$(\dot{\mathbf{A}}\mathbf{A}^T)^T = \mathbf{A}\dot{\mathbf{A}}^T = -\dot{\mathbf{A}}\mathbf{A}^T \tag{9.2.35}$$

and $\dot{\mathbf{A}}\mathbf{A}^T$ is skew-symmetric. As noted following the definition of Eq. 9.1.21 in Section 9.1, there exists a vector $\boldsymbol{\omega}$, called the *angular velocity* of the x'-y'-z' frame, such that

$$\tilde{\boldsymbol{\omega}} = \dot{\mathbf{A}}\mathbf{A}^T \tag{9.2.36}$$

This is the definition of $\boldsymbol{\omega}$ in terms of the transformation matrix $\mathbf{A}$ and its time derivatives. Using Eq. 9.2.36 in Eq. 9.2.34 yields the *velocity equation*

$$\dot{\mathbf{r}}^P = \dot{\mathbf{r}} + \tilde{\boldsymbol{\omega}}\mathbf{s}^P \tag{9.2.37}$$

If $\mathbf{r} = \dot{\mathbf{r}} = \mathbf{0}$, this equation defines the geometry shown in Fig. 9.2.12. In geometric vector notation, this is the familiar relationship

$$\dot{\vec{r}}^P = \dot{\vec{r}} + \vec{\omega} \times \vec{s}^P$$

where $\vec{\omega}$ is the angular velocity of the x'-y'-z' frame, relative to the x-y-z reference frame. Taking the vector product in the x'-y'-z' frame and transforming to the x-y-z frame yields (Prob. 9.2.7)

$$\dot{\mathbf{r}}^P = \dot{\mathbf{r}} + \mathbf{A}\tilde{\boldsymbol{\omega}}'\mathbf{s}'^P \tag{9.2.38}$$

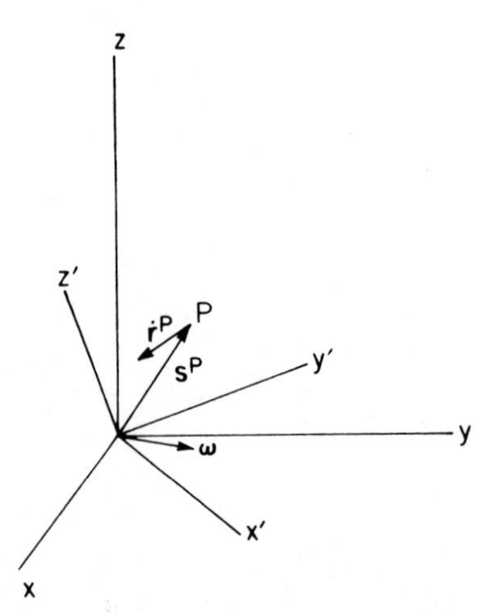

Figure 9.2.12 Angular velocity of x'-y'-z' reference frame.

Example 9.2.4: The transformation matrix $\mathbf{A}$ of Example 9.2.2 is defined in Eq. 9.2.17 as a function of the angle ϕ. For this matrix, Eq. 9.2.36 yields

$$\tilde{\boldsymbol{\omega}} = \dot{\mathbf{A}}\mathbf{A}^T = \begin{bmatrix} 0 & -1 & 0 \\ 1 & 0 & 0 \\ 0 & 0 & 0 \end{bmatrix} \dot{\phi}$$

Thus,

$$\boldsymbol{\omega} = [0, 0, \dot{\phi}]^T$$

which agrees with the physical concept of angular velocity about a fixed axis. Note also that $\boldsymbol{\omega}' = \mathbf{A}^T\boldsymbol{\omega} = \boldsymbol{\omega}$ in this case, which is expected since the z and z' axes coincide.

Multiplying both sides of Eq. 9.2.36 on the right by $\mathbf{A}$ yields the useful relationship

$$\dot{\mathbf{A}} = \tilde{\boldsymbol{\omega}}\mathbf{A} \tag{9.2.39}$$

Furthermore, using Eq. 9.2.21 with $\mathbf{s} = \boldsymbol{\omega}$, Eq. 9.2.39 yields

$$\left.\begin{aligned} \dot{\mathbf{A}} &= \mathbf{A}\tilde{\boldsymbol{\omega}}' \\ \tilde{\boldsymbol{\omega}}' &= \mathbf{A}^T\dot{\mathbf{A}} \end{aligned}\right\} \tag{9.2.40}$$

Equations 9.2.39 and 9.2.40 provide frequently used relationships between the time derivative of the transformation matrix $\mathbf{A}$ and the angular velocity $\boldsymbol{\omega}$ of the x'-y'-z' frame.

Equation 9.2.33 may be differentiated with respect to time to obtain the *acceleration equation*

$$\ddot{\mathbf{r}}^P = \ddot{\mathbf{r}} + \ddot{\mathbf{A}}\mathbf{s}'^P \tag{9.2.41}$$

Differentiating Eq. 9.2.39, the following relationships for the second time derivative of the transformation matrix $\mathbf{A}$ are obtained (Prob. 9.2.8):

$$\begin{aligned} \ddot{\mathbf{A}} &= \dot{\tilde{\boldsymbol{\omega}}}\mathbf{A} + \tilde{\boldsymbol{\omega}}\dot{\mathbf{A}} \\ &= \dot{\tilde{\boldsymbol{\omega}}}\mathbf{A} + \tilde{\boldsymbol{\omega}}\tilde{\boldsymbol{\omega}}\mathbf{A} \end{aligned} \tag{9.2.42}$$

Equation 9.2.21 may be applied to $\boldsymbol{\omega} = \mathbf{A}\boldsymbol{\omega}'$ and $\dot{\boldsymbol{\omega}} = \mathbf{A}\dot{\boldsymbol{\omega}}'$ to rewrite Eq. 9.2.42 in terms of $\boldsymbol{\omega}'$, yielding

$$\ddot{\mathbf{A}} = \mathbf{A}\dot{\tilde{\boldsymbol{\omega}}}' + \mathbf{A}\tilde{\boldsymbol{\omega}}'\tilde{\boldsymbol{\omega}}' \tag{9.2.43}$$

Example 9.2.5: Using the results of Example 9.2.4 and Eq. 9.2.38, the velocity of a point P that is defined by $\mathbf{s}'^P = [1, 1, 1]$ in the x'-y'-z' frame is

$$\dot{\mathbf{r}}^P = \mathbf{A}\tilde{\boldsymbol{\omega}}'\mathbf{s}'^P = [-\cos\phi - \sin\phi, \cos\phi - \sin\phi, 0]^T\dot{\phi}$$

where $\dot{\mathbf{r}} = \mathbf{0}$. The acceleration of point P, from Eqs. 9.2.41 and 9.2.42, is

$$\ddot{\mathbf{r}}^P = \ddot{\mathbf{A}}\mathbf{s}'^P = \begin{bmatrix} -\cos\phi - \sin\phi \\ \cos\phi - \sin\phi \\ 0 \end{bmatrix} \ddot{\phi} + \begin{bmatrix} \sin\phi - \cos\phi \\ -\cos\phi - \sin\phi \\ 0 \end{bmatrix} \dot{\phi}^2$$

The derivative relations derived here serve as the foundation for the kinematic analysis of rigid bodies in space. The reader is encouraged to become proficient in their use, in preparation for kinematic analysis in this chapter and dynamic analysis in Chapter 11 (Probs. 9.2.9 and 9.2.10). As an aid in the use of the relations derived, a summary of key formulas is provided at the end of the chapter.

Virtual Displacements and Rotations A *virtual displacement* $\delta\mathbf{r}^P$ of point P is defined as an *infinitesimal displacement* (see Section 6.1), or small displacement with time held fixed. If point P is attached to a rigid body, or equivalently if it is fixed in an x'-y'-z' frame, then a virtual displacement of point P may be represented by a virtual displacement of the origin of the x'-y'-z' frame and a variation in its direction cosines. The vector $\mathbf{r}^P$ in Eq. 9.2.32 depends on $\mathbf{r}$ and $\mathbf{A}$, so if the frame is perturbed slightly, $\mathbf{r}$ changes to $\mathbf{r} + \delta\mathbf{r}$ and $\mathbf{A}$ changes to $\mathbf{A} + \delta\mathbf{A} + \mathbf{0}(\delta\mathbf{A}^2)$.† Since the direction cosine matrix must be orthogonal, it is required that

$$(\mathbf{A} + \delta\mathbf{A} + \mathbf{0}(\delta\mathbf{A}^2))(\mathbf{A} + \delta\mathbf{A} + \mathbf{0}(\delta\mathbf{A}^2))^T$$
$$= \mathbf{A}\mathbf{A}^T + \mathbf{A}\,\delta\mathbf{A}^T + \delta\mathbf{A}\mathbf{A}^T + \delta\mathbf{A}\,\delta\mathbf{A}^T + \mathbf{0}(\delta\mathbf{A}^2) = \mathbf{I}$$

Since $\mathbf{A}$ is orthogonal, this is

$$\mathbf{A}\,\delta\mathbf{A}^T + \delta\mathbf{A}\mathbf{A}^T = \mathbf{0}(\delta\mathbf{A}^2)$$

Equating linear terms in $\delta\mathbf{A}$ on both sides (called *infinitesimals*), this is

$$\mathbf{A}\,\delta\mathbf{A}^T + \delta\mathbf{A}\mathbf{A}^T = \mathbf{0} \tag{9.2.44}$$

This result may also be obtained by taking the differential of both sides of the identity $\mathbf{A}^T\mathbf{A} = \mathbf{I}$ (Prob. 9.2.11).

Using Eq. 9.2.44, observe that

$$(\delta\mathbf{A}\mathbf{A}^T)^T = \mathbf{A}\,\delta\mathbf{A}^T = -\delta\mathbf{A}\mathbf{A}^T \tag{9.2.45}$$

so the matrix $\delta\mathbf{A}\mathbf{A}^T$ is skew-symmetric and can be represented by

$$\delta\mathbf{A}\mathbf{A}^T = \delta\tilde{\boldsymbol{\pi}} \tag{9.2.46}$$

---

† Quantities of magnitude $0(\alpha^n)$ approach zero as α^n; that is, there is a constant c such that $|0(\alpha^n)|/|\alpha^n| \leq c$. Here, they are higher-order terms in a Taylor expansion.

where the vector $\delta\boldsymbol{\pi}$ depends on both the components of $\mathbf{A}$ and $\delta\mathbf{A}$ (Prob. 9.2.12). Multiplying Eq. 9.2.46 on the right by $\mathbf{A}$,

$$\delta\mathbf{A} = \delta\tilde{\boldsymbol{\pi}}\mathbf{A} \tag{9.2.47}$$

To use the foregoing results, take the variation of Eq. 9.2.32 to obtain (Prob. 9.2.13)

$$\delta\mathbf{r}^P = \delta\mathbf{r} + \delta\mathbf{A}\mathbf{s}'^P \tag{9.2.48}$$

Using Eq. 9.2.47, this is

$$\begin{aligned}\delta\mathbf{r}^P &= \delta\mathbf{r} + \delta\tilde{\boldsymbol{\pi}}\mathbf{A}\mathbf{s}'^P \\ &= \delta\mathbf{r} + \delta\tilde{\boldsymbol{\pi}}\mathbf{s}^P\end{aligned} \tag{9.2.49}$$

Thus, $\delta\boldsymbol{\pi}$ plays the role of a rotation, called a *virtual rotation* of the x'-y'-z' frame relative to the x-y-z reference frame, with components in the latter frame. Using $\delta\boldsymbol{\pi} = \mathbf{A}\,\delta\boldsymbol{\pi}'$ and Eq. 9.2.21 in Eq. 9.2.47,

$$\begin{aligned}\delta\mathbf{A} &= \mathbf{A}\,\delta\tilde{\boldsymbol{\pi}}' \\ \delta\tilde{\boldsymbol{\pi}}' &= \mathbf{A}^T\,\delta\mathbf{A}\end{aligned} \tag{9.2.50}$$

Using these relations in Eq. 9.2.48,

$$\delta\mathbf{r}^P = \delta\mathbf{r} + \mathbf{A}\,\delta\tilde{\boldsymbol{\pi}}'\mathbf{s}'^P \tag{9.2.51}$$

which verifies that virtual rotation is a vector quantity. This fact is very important. Even though large rotation cannot be represented as a vector quantity (see Section 9.1), virtual rotation and angular velocity are vector quantities.

For a vector $\mathbf{s}'$ that is fixed in the x'-y'-z' frame, $\mathbf{s}'$ is constant and

$$\begin{aligned}\delta\mathbf{s} &= \delta\mathbf{A}\mathbf{s}' \\ &= \mathbf{A}\,\delta\tilde{\boldsymbol{\pi}}'\mathbf{s}'\end{aligned} \tag{9.2.52}$$

Using Eq. 9.1.25, this may be written as

$$\delta\mathbf{s} = -\mathbf{A}\tilde{\mathbf{s}}'\,\delta\boldsymbol{\pi}' \tag{9.2.53}$$

The δ notation associated with virtual displacements and virtual rotations may be interpreted as a *partial differential operator* of calculus. This is the case since vectors in the x'-y'-z' frame are functions of parameters that are used to define the position and orientation of the frame.

If a vector $\mathbf{h}(\mathbf{q})$ depends on a variable $\mathbf{q}$ in a differentiable way, $\delta\mathbf{h} = \mathbf{h}_{\mathbf{q}}\,\delta\mathbf{q}$. Even before the variable $\mathbf{q}$ is defined, the differential $\delta\mathbf{h}$ can be defined and used in analysis, since the linear differential operator obeys the same rules of calculus as the partial derivative operator, with time held fixed. The differential of the sum of two vectors is the sum of their differentials:

$$\delta(\mathbf{g} + \mathbf{h}) = \delta\mathbf{g} + \delta\mathbf{h} \tag{9.2.54}$$

where the differential δ operates on the quantity to its immediate right. The product rule for differentials is analogous to Eq. 2.5.13; that is,

$$\delta(\mathbf{g}^T\mathbf{h}) = \mathbf{h}^T\,\delta\mathbf{g} + \mathbf{g}^T\,\delta\mathbf{h} \tag{9.2.55}$$

Since the vector product $\tilde{}$ operator is linear (see Eq. 9.1.32), the δ and $\tilde{}$ operators may be interchanged in order of application. Thus,

$$\begin{aligned}\delta(\tilde{\mathbf{g}}\mathbf{h}) &= \delta\tilde{\mathbf{g}}\mathbf{h} + \tilde{\mathbf{g}}\,\delta\mathbf{h} \\ &= -\tilde{\mathbf{h}}\,\delta\mathbf{g} + \tilde{\mathbf{g}}\,\delta\mathbf{h}\end{aligned} \tag{9.2.56}$$

Finally, the differential of a triple product may be expanded, using the product rule differentials, to obtain

$$\begin{aligned}\delta(\mathbf{g}^T\mathbf{B}\mathbf{h}) &= \delta\mathbf{g}^T\mathbf{B}\mathbf{h} + \mathbf{g}^T\,\delta\mathbf{B}\mathbf{h} + \mathbf{g}^T\mathbf{B}\,\delta\mathbf{h} \\ &= \mathbf{h}^T\mathbf{B}^T\,\delta\mathbf{g} + \mathbf{g}^T\,\delta\mathbf{B}\mathbf{h} + \mathbf{g}^T\mathbf{B}\,\delta\mathbf{h}\end{aligned} \tag{9.2.57}$$

where $\delta\mathbf{g}^T\mathbf{B}\mathbf{h} = \mathbf{h}^T\mathbf{B}^T\,\delta\mathbf{g}$, since the transpose of a scalar is the same scalar.

These and many additional identities that involve virtual displacements and rotations, or variations in position and orientation of reference frames, can be used to advantage in the analysis of the dynamics of mechanical systems. Of particular importance in analytical kinematics and dynamics is a powerful coupling between differential calculus and vector analysis. If the abundant manipulation rules of calculus and vector analysis are used to advantage, geometrically clear results can be obtained and understood.

Example 9.2.6: To illustrate the power of combining the vector properties of virtual displacements and rotations and differential calculus, consider a commonly used constraint between two vectors $\mathbf{s}_i'$ and $\mathbf{s}_j'$ that are fixed in their respective x_i'-y_i'-z_i' and x_j'-y_j'-z_j' frames; that is, they are orthogonal:

$$\Phi = \mathbf{s}_i^T\mathbf{s}_j = 0 \tag{9.2.58}$$

Taking the differential of both sides of this equation,

$$\delta\Phi = \mathbf{s}_j^T\,\delta\mathbf{s}_i + \mathbf{s}_i^T\,\delta\mathbf{s}_j = 0 \tag{9.2.59}$$

Substituting for the differentials of the vectors from Eq. 9.2.53, the differential of Eq. 9.2.59 is obtained in terms of virtual rotations as

$$\delta\Phi = -(\mathbf{s}_j'^T\mathbf{A}_j^T\mathbf{A}_i\tilde{\mathbf{s}}_i'\,\delta\boldsymbol{\pi}_i' + \mathbf{s}_i'^T\mathbf{A}_i^T\mathbf{A}_j\tilde{\mathbf{s}}_j'\,\delta\boldsymbol{\pi}_j') = 0 \tag{9.2.60}$$

This and related easily derived differential constraint expressions, or linearized constraints, are of great value in formulating the equations of motion of constrained dynamic systems.

Example 9.2.7: As a second illustration of the power of vector properties of virtual rotations, consider the reference frames shown in Fig. 9.2.10 that are constrained to permit only relative rotation about their common $\mathbf{h}_i$ and $\mathbf{h}_j$ unit vectors. Since virtual rotations $\delta\boldsymbol{\pi}_i$ and $\delta\boldsymbol{\pi}_j$ are vector quantities, the infinitesimal relative rotation is

$$\delta\theta \mathbf{h}_i = \delta\boldsymbol{\pi}_j - \delta\boldsymbol{\pi}_i$$

Taking the scalar product of both sides with $\mathbf{h}_i$ and representing vectors in their respective reference frame components,

$$\begin{aligned} \delta\theta &= \mathbf{h}_i^T(\delta\boldsymbol{\pi}_j - \delta\boldsymbol{\pi}_i) \\ &= \mathbf{h}_i'^T(\mathbf{A}_i^T\mathbf{A}_j\,\delta\boldsymbol{\pi}_j' - \delta\boldsymbol{\pi}_i') \end{aligned} \quad \textbf{(9.2.61)}$$

Similarly, the relative angular velocity is

$$\dot{\theta} = \mathbf{h}_i'^T(\mathbf{A}_i^T\mathbf{A}_j\boldsymbol{\omega}_j' - \boldsymbol{\omega}_i') \quad \textbf{(9.2.62)}$$

Example 9.2.8: To further illustrate the power of coupling differential and vector calculus, consider a vector of generalized coordinates, $\mathbf{q}$, that defines the orientation of bodies that make up a mechanical system. Such coordinates are defined in Section 9.3, where relationships between variations in generalized coordinates and virtual rotations are established in the form

$$\begin{aligned} \delta\boldsymbol{\pi}_i' &= \mathbf{B}_i\,\delta\mathbf{q} \\ \delta\boldsymbol{\pi}_j' &= \mathbf{B}_j\,\delta\mathbf{q} \end{aligned} \quad \textbf{(9.2.63)}$$

where $\delta\mathbf{q}$ is a variation in the generalized coordinates, and $\mathbf{B}_i$ and $\mathbf{B}_j$ are matrices that depend on the generalized coordinate vector $\mathbf{q}$. Once the relationship of Eq. 9.2.63 is established, it may be substituted into Eq. 9.2.60, which was derived using vector relationships, to obtain the differential of the function Φ of Eq. 9.2.58 as

$$\delta\Phi = -(\mathbf{s}_j'^T\mathbf{A}_j^T\mathbf{A}_i\tilde{\mathbf{s}}_i'\mathbf{B}_i + \mathbf{s}_i'^T\mathbf{A}_i^T\mathbf{A}_j\tilde{\mathbf{s}}_j'\mathbf{B}_j)\,\delta\mathbf{q} \quad \textbf{(9.2.64)}$$

Since the coefficient of $\delta\mathbf{q}$ in the total differential is $\partial\Phi/\partial\mathbf{q}$,

$$\frac{\partial\Phi}{\partial\mathbf{q}} \equiv -(\mathbf{s}_j'^T\mathbf{A}_j^T\mathbf{A}_i\tilde{\mathbf{s}}_i'\mathbf{B}_i + \mathbf{s}_i'^T\mathbf{A}_i^T\mathbf{A}_j\tilde{\mathbf{s}}_j'\mathbf{B}_j) \quad \textbf{(9.2.65)}$$

9.3 EULER PARAMETER ORIENTATION GENERALIZED COORDINATES

As shown in Section 9.1, the position and orientation of a rigid body in space can be defined by locating the origin of a body-fixed x'-y'-z' frame and specifying an orthogonal direction cosine matrix that defines orientation of the x'-y'-z' frame. To characterize the orientation of the body analytically, a set of orientation generalized coordinates must be defined.

Euler's Theorem and Rotation Relations An important result that is used in defining orientation generalized coordinates is the following:

> ***Euler's Theorem:*** If the origins of two right-hand Cartesian reference frames coincide, then they may be brought into coincidence by a single rotation about some axis.

This Theorem is proved in References 4 and 9, using the orthogonality of the transformation matrix **A** between the reference frames and the eigenvalue theory of matrices.

A physical interpretation of Euler's theorem is given in Fig. 9.3.1, where the orthogonal projections of **i-f**, **j-g**, and **k-h** pairs onto the axis of rotation are equal. The axis about which rotation is to take place is called the *orientation axis* and is defined by a unit vector **u**. The angle χ of rotation, measured in the plane perpendicular to **u**, is positive counterclockwise (i.e., using the right-hand sign convention). Note that if the vector **u** had been taken in the opposite sense along the orientation axis then the angle χ shown in Fig. 9.3.1 would be replaced by $2\pi - \chi$.

Euler's theorem states that any orientation of a body can be achieved by a single rotation from a reference orientation about some axis. It is natural, therefore, to seek a representation of the direction cosine transformation matrix **A** in terms of parameters of this rotation, that is, the angle of rotation and the components of the unit vector **u** along the orientation axis.

Figure 9.3.2 shows the orientation axis, the x'-y'-z' frame, the x-y-z reference frame, an arbitrary vector **s**, and projections (or components) of **s** in

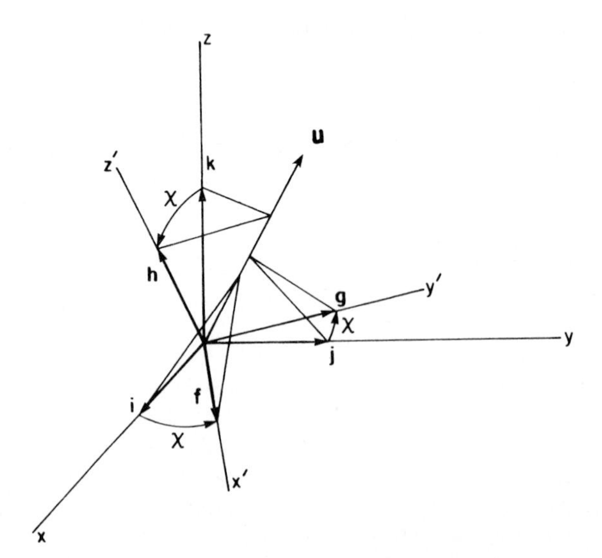

Figure 9.3.1 Euler rotation of reference frames.

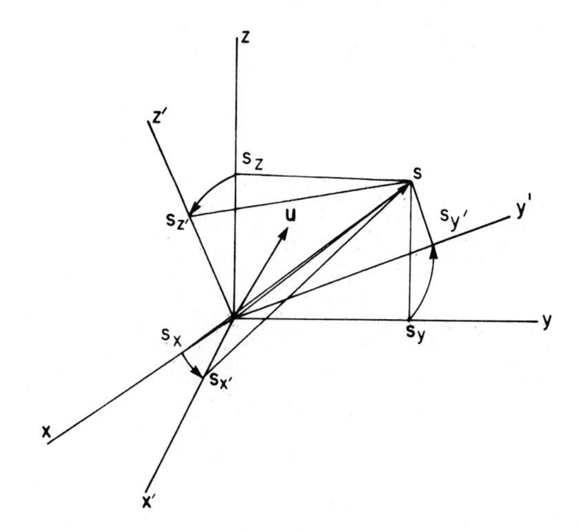

Figure 9.3.2 Vector **s** in x'-y'-z' and x-y-z frames.

each frame. The objective is to obtain relationships between components **s** and **s**′ of the vector in the x-y-z and x'-y'-z' frames, respectively. The direction cosine matrix **A** that provides the desired relation between the x'-y'-z' and x-y-z components of a vector is obtained if unit vectors **f**, **g**, and **h** along the x', y', and z' axes may be written in terms of χ and **u**; that is, Eq. 9.2.13 yields $\mathbf{A} = [\mathbf{f}, \mathbf{g}, \mathbf{h}]$.

Consider first the relation between **k** and **h**. Since **h** is obtained in the x-y-z frame by a rotation of **k** about **u**, its tip lies in the plane that is perpendicular to **u** and contains the tip of **k**, as shown in Fig. 9.3.3. The vector $\tilde{\mathbf{u}}\mathbf{k}$ lies in a parallel plane and has length equal to the radius of the circle that is formed by intersecting the plane perpendicular to **u** and the cone formed by rotating **k** about the **u** axis. The center N of the circle is located by the vector $(\mathbf{u}^T\mathbf{k})\mathbf{u}$.

A vector **a** is constructed to pass through the tip of **h** and to be perpendicular to the vector $\mathbf{k} - (\mathbf{u}^T\mathbf{k})\mathbf{u}$ from point N to the tip of **k**, as shown in Fig. 9.3.3. The vector $\mathbf{k} - (\mathbf{u}^T\mathbf{k})\mathbf{u}$ lies in the plane formed by **u** and **k**, so it is perpendicular to $\tilde{\mathbf{u}}\mathbf{k}$. Thus, vectors **a** and $\tilde{\mathbf{u}}\mathbf{k}$ are parallel. Furthermore, the lengths of vectors $\mathbf{h} - (\mathbf{u}^T\mathbf{k})\mathbf{u}$ and $\tilde{\mathbf{u}}\mathbf{k}$ are equal, since they are radii of the same circle. Thus, by simple trigonometry,

$$\mathbf{a} = \tilde{\mathbf{u}}\mathbf{k} \sin \chi$$

Similarly, the vector **b** in Fig. 9.3.3 is

$$\mathbf{b} = [\mathbf{k} - (\mathbf{u}^T\mathbf{k})\mathbf{u}] \cos \chi$$

Thus,

$$\begin{aligned} \mathbf{h} &= (\mathbf{u}^T\mathbf{k})\mathbf{u} + \mathbf{b} + \mathbf{a} \\ &= (\mathbf{u}^T\mathbf{k})\mathbf{u} + [\mathbf{k} - (\mathbf{u}^T\mathbf{k})\mathbf{u}] \cos \chi + \tilde{\mathbf{u}}\mathbf{k} \sin \chi \end{aligned}$$

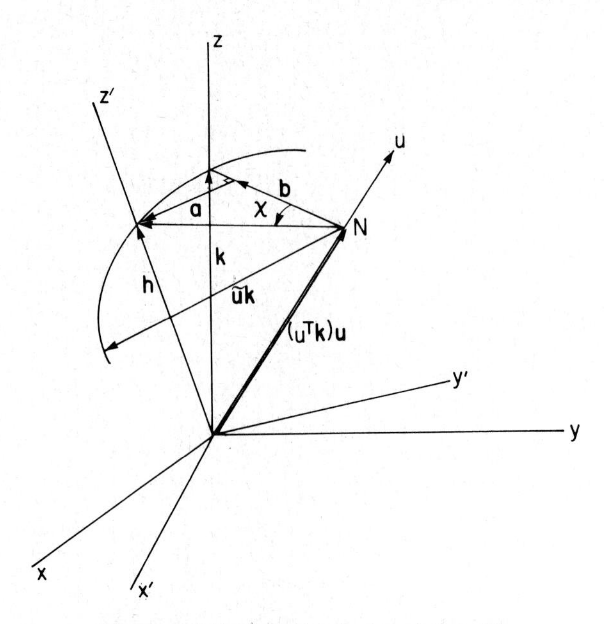

Figure 9.3.3 Relation between **k** and **h**.

A rearrangement of terms leads to

$$\mathbf{h} = \mathbf{k}\cos\chi + (\mathbf{u}^T\mathbf{k})\mathbf{u}(1-\cos\chi) + (\tilde{\mathbf{u}}\mathbf{k})\sin\chi \tag{9.3.1}$$

Euler Parameters By means of the trigonometric identities

$$1-\cos\chi = 2\sin^2\frac{\chi}{2}$$

$$\sin\chi = 2\sin\frac{\chi}{2}\cos\frac{\chi}{2}$$

$$\cos\chi = 2\cos^2\frac{\chi}{2} - 1$$

and the set of four *Euler parameters*, defined as

$$\begin{aligned} e_0 &\equiv \cos\frac{\chi}{2} \\ \mathbf{e} = \begin{bmatrix} e_1 \\ e_2 \\ e_3 \end{bmatrix} &\equiv \mathbf{u}\sin\frac{\chi}{2} \end{aligned} \tag{9.3.2}$$

Eq. 9.3.1 can be written in the form

$$\mathbf{h} = \mathbf{k}\left(2\cos^2\frac{\chi}{2} - 1\right) + 2\mathbf{u}(\mathbf{u}^T\mathbf{k})\sin^2\frac{\chi}{2} + 2\tilde{\mathbf{u}}\mathbf{k}\sin\frac{\chi}{2}\cos\frac{\chi}{2} \tag{9.3.3}$$

and manipulated to yield the desired result:

$$\mathbf{h} = [(2e_0^2 - 1)\mathbf{I} + 2(\mathbf{e}\mathbf{e}^T + e_0\tilde{\mathbf{e}})]\mathbf{k} \tag{9.3.4}$$

Since the geometric relations between **f-i** and **g-j** are identical to that between **h-k**, the relation of Eq. 9.3.4 is valid for these pairs (Prob. 9.3.1); that is,

$$\begin{aligned} \mathbf{f} &= [(2e_0^2 - 1)\mathbf{I} + 2(\mathbf{e}\mathbf{e}^T + e_0\tilde{\mathbf{e}})]\mathbf{i} \\ \mathbf{g} &= [(2e_0^2 - 1)\mathbf{I} + 2(\mathbf{e}\mathbf{e}^T + e_0\tilde{\mathbf{e}})]\mathbf{j} \end{aligned} \tag{9.3.5}$$

From Eq. 9.2.13, using Eqs. 9.3.4 and 9.3.5 and the fact that $[\mathbf{i}, \mathbf{j}, \mathbf{k}] = \mathbf{I}$,

$$\begin{aligned} \mathbf{A} &= [\mathbf{f}, \mathbf{g}, \mathbf{h}] \\ &= [(2e_0^2 - 1)\mathbf{I} + 2(\mathbf{e}\mathbf{e}^T + e_0\tilde{\mathbf{e}})][\mathbf{i}, \mathbf{j}, \mathbf{k}] \\ &= (2e_0^2 - 1)\mathbf{I} + 2(\mathbf{e}\mathbf{e}^T + e_0\tilde{\mathbf{e}}) \end{aligned} \tag{9.3.6}$$

More explicitly, the direction cosine matrix, written in terms of Euler parameters, is

$$\mathbf{A} = 2\begin{bmatrix} e_0^2 + e_1^2 - \frac{1}{2} & e_1e_2 - e_0e_3 & e_1e_3 + e_0e_2 \\ e_1e_2 + e_0e_3 & e_0^2 + e_2^2 - \frac{1}{2} & e_2e_3 - e_0e_1 \\ e_1e_3 - e_0e_2 & e_2e_3 + e_0e_1 & e_0^2 + e_3^2 - \frac{1}{2} \end{bmatrix} \tag{9.3.7}$$

Example 9.3.1: Let the x'-y'-z' frame of Fig. 9.3.1 have its origin in common with that of the x-y-z reference frame and be rotated about a fixed vector $\mathbf{u} = (1/\sqrt{3})\ [1, 1, 1]^T$ by angle ϕ. The Euler parameters for orientation of the x'-y'-z' frame are thus $e_0 = \cos(\phi/2)$ and $\mathbf{e} = \mathbf{u}\sin(\phi/2)$. The direction cosine transformation matrix, from Eq. 9.3.7, is

$$\mathbf{A} = 2\begin{bmatrix} c^2 + \dfrac{s^2}{3} - \frac{1}{2} & \dfrac{s^2}{3} - \dfrac{sc}{\sqrt{3}} & \dfrac{s^2}{3} + \dfrac{sc}{\sqrt{3}} \\ \dfrac{s^2}{3} + \dfrac{sc}{\sqrt{3}} & c^2 + \dfrac{s^2}{3} - \frac{1}{2} & \dfrac{s^2}{3} - \dfrac{sc}{\sqrt{3}} \\ \dfrac{s^2}{3} - \dfrac{sc}{\sqrt{3}} & \dfrac{s^2}{3} + \dfrac{sc}{\sqrt{3}} & c^2 + \dfrac{s^2}{3} - \frac{1}{2} \end{bmatrix}$$

where $c \equiv \cos\phi/2$ and $s \equiv \sin\phi/2$. Geometrically, a vector $\mathbf{s}'$ that is fixed in and emanates from the origin of the x'-y'-z' frame sweeps out a cone in the x-y-z frame, with axis of revolution $\mathbf{u}$ (Prob. 9.3.2).

For convenience, denote the 4×1 column vector of Euler parameters as

$$\mathbf{p} = [e_0, \mathbf{e}^T]^T = [e_0, e_1, e_2, e_3]^T \tag{9.3.8}$$

The Euler parameters are not independent, since

$$e_0^2 + \mathbf{e}^T\mathbf{e} = \cos^2\frac{\chi}{2} + \mathbf{u}^T\mathbf{u}\sin^2\frac{\chi}{2} = 1$$

That is, they must satisfy the *Euler parameter normalization constraint*

$$\mathbf{p}^T\mathbf{p} = e_0^2 + e_1^2 + e_2^2 + e_3^2 = 1 \tag{9.3.9}$$

It is possible to derive explicit formulas for the Euler parameters in terms of elements of the direction cosine matrix **A** from Eq. 9.3.7. Assume that nine direction cosines of a transformation matrix **A** are given. The *trace of* **A**, denoted by tr **A**, is defined as

$$\text{tr}\,\mathbf{A} = a_{11} + a_{22} + a_{33} \tag{9.3.10}$$

From Eq. 9.3.7,

$$\text{tr}\,\mathbf{A} = 2(3e_0^2 + e_1^2 + e_2^2 + e_3^2) - 3 = 4e_0^2 - 1 \tag{9.3.11}$$

Thus,

$$e_0^2 = \frac{\text{tr}\,\mathbf{A} + 1}{4} \tag{9.3.12}$$

Substitution of Eq. 9.3.12 into the diagonal elements of **A** in Eq. 9.3.7 results in

$$\begin{aligned} a_{11} &= 2(e_0^2 + e_1^2) - 1 \\ &= 2\left(\frac{\text{tr}\,\mathbf{A} + 1}{4} + e_1^2\right) - 1 \end{aligned}$$

Thus,

$$e_1^2 = \frac{1 + 2a_{11} - \text{tr}\,\mathbf{A}}{4} \tag{9.3.13}$$

Similarly,

$$e_2^2 = \frac{1 + 2a_{22} - \text{tr}\,\mathbf{A}}{4} \tag{9.3.14}$$

$$e_3^2 = \frac{1 + 2a_{33} - \text{tr}\,\mathbf{A}}{4} \tag{9.3.15}$$

It is interesting and computationally important to note that Eqs. 9.3.13 to 9.3.15 determine the magnitudes of the Euler parameters in terms of only the diagonal elements of the direction cosine matrix **A**. To find the algebraic signs of the Euler parameters, off-diagonal terms of **A** must be used.

Equation 9.3.9 indicates that at least one Euler parameter must be nonzero. Consider first the case $e_0 \neq 0$. The sign of e_0 may be selected as positive or negative. From Eq. 9.3.7, subtracting symmetrically placed off-diagonal terms of $\mathbf{A}$ in Eq. 9.3.7 yields

$$a_{32} - a_{23} = 4e_0e_1$$
$$a_{13} - a_{31} = 4e_0e_2$$
$$a_{21} - a_{12} = 4e_0e_3$$

or

$$\begin{aligned} e_1 &= \frac{a_{32} - a_{23}}{4e_0} \\ e_2 &= \frac{a_{13} - a_{31}}{4e_0} \\ e_3 &= \frac{a_{21} - a_{12}}{4e_0} \end{aligned} \tag{9.3.16}$$

If e_0 calculated from Eq. 9.3.12 is nonzero and its sign is selected, then Eq. 9.3.16 can be used to determine e_1, e_2, and e_3. Suppose that, for the selected sign of e_0 and the computed values of e_1, e_2, and e_3, the angle of rotation and the axis of rotation are determined to be χ and $\mathbf{u}$, respectively. If the sign of e_0 is inverted, Eq. 9.3.16 shows that the signs of e_1, e_2, and e_3 are inverted. Changing the signs of all four parameters does not influence the transformation matrix, since variable elements in the matrix are quadratic in these variables (see Eq. 9.3.7).

Either sign may be selected for $e_0 \neq 0$ and still define the same physical orientation of the body. To see this, denote by $a > 0$ the magnitude of e_0 from Eq. 9.3.12 and denote $\cos(\chi^+/2) = e_0^+ = a$ and $\cos(\chi^-/2) = e_0^- = -a$, where $0 < \chi^+/2 \leq \pi/2$ and $\pi/2 \leq \chi^-/2 < \pi$. Since the cosine function is antisymmetric about $\pi/2$, that is, $\cos(\pi/2 + \theta) = -\cos(\pi/2 - \theta)$, and $\cos(\chi^+/2) = -\cos(\chi^-/2)$, there is a θ such that $\chi^+/2 = \pi/2 + \theta$ and $\chi^-/2 = \pi/2 - \theta$. Eliminating θ,

$$\frac{\chi^+}{2} + \frac{\chi^-}{2} = \pi$$

or

$$\chi^- = 2\pi - \chi^+$$

Thus, $\sin(\chi^-/2) = \sin(\chi^+/2)$. From Eq. 9.3.16, $\mathbf{e}^- = -\mathbf{e}^+$. The unit vector $\mathbf{u}$ in Eq. 9.3.2 defines the positive sense along the orientation axis for implementation of the rotation χ by the right-hand rule. Using the above and Eq. 9.3.2, $\mathbf{u}^- = -\mathbf{u}^+$, as shown in Fig. 9.3.4. Thus, if χ^- is selected, a rotation of $2\pi - \chi^+$ is implemented counterclockwise about the $\mathbf{u}^-$ axis, or equivalently clockwise about the $\mathbf{u}^+$ axis, as shown in Fig. 9.3.4, and precisely the same orientation is achieved.

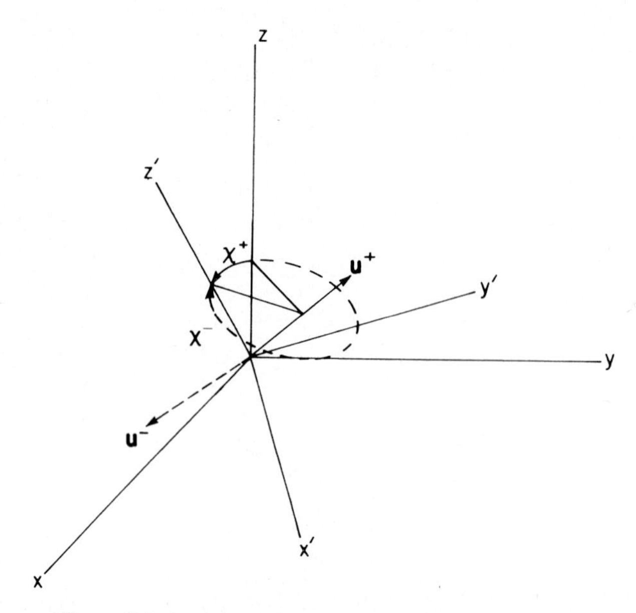

Figure 9.3.4 Alternative values of Euler parameters.

Consider next the case in which the value of e_0 calculated from Eq. 9.3.12 is zero. Then, Eq. 9.3.2 shows that $\chi = k\pi$, $k = \pm 1, \pm 3, \pm 5, \ldots$. Therefore, the sign of χ is immaterial. To find the algebraic signs of e_1, e_2, and e_3, symmetrically placed off-diagonal terms of the matrix $\mathbf{A}$ in Eq. 9.3.7 may be added to obtain

$$\begin{aligned} a_{21} + a_{12} &= 4e_1e_2 \\ a_{31} + a_{13} &= 4e_1e_3 \\ a_{32} + a_{23} &= 4e_2e_3 \end{aligned} \tag{9.3.17}$$

Since $e_0 = 0$, at least one of the three Euler parameters e_1, e_2, or e_3 from Eqs. 9.3.13 to 9.3.15 must be nonzero, and its sign may be selected as positive or negative. Then Eq. 9.3.17 can be used to determine the signs of the other two parameters.

Even though two distinct values of Euler parameters, $\mathbf{p}^+$ and $\mathbf{p}^-$ with $\mathbf{p}^- = -\mathbf{p}^+$, define the same orientation of the x'-y'-z' frame, the mapping from $\mathbf{p}$ to $\mathbf{A}(\mathbf{p})$ and hence $\mathbf{s} = \mathbf{A}(\mathbf{p})\mathbf{s}'$ is *locally one-to-one*; that is, if $|p_i^1 - p_i^2| < \delta$, $i = 1, 2, 3, 4$, for a small δ, and if $\mathbf{p}^1 \neq \mathbf{p}^2$, then $\mathbf{A}(\mathbf{p}^1) \neq \mathbf{A}(\mathbf{p}^2)$. The two alternate values of $\mathbf{p}$ are never close together, because

$$(\mathbf{p}^+ - \mathbf{p}^-)^T(\mathbf{p}^+ - \mathbf{p}^-) = 4\mathbf{p}^{+T}\mathbf{p}^+ = 4$$

Thus, for neighboring orientations that are specified by $\mathbf{A}_1$ and $\mathbf{A}_2$, the associated Euler parameters $\mathbf{p}^1$ and $\mathbf{p}^2$ can be selected to be close to one another. This observation is important in kinematic and dynamic analysis, since the objective is to solve for $\mathbf{p}(t)$ to define the time history of the orientation of a body. Thus, continuity of $\mathbf{p}(t)$ may be used to solve for a unique value of $\mathbf{p}(t)$ once the sign of $\mathbf{p}(0)$ is selected.

Properties of Euler Parameters Important relations between Euler parameters, their time derivatives, and their variations and the associated transformation matrices, angular velocities, and virtual rotations are next derived. Derivation of some of the identities is given in the text. To avoid lengthy expansions, references are given to problems at the end of this chapter where calculations are outlined and the reader is invited to fill in the details.

Two 3×4 matrices $\mathbf{E}$ and $\mathbf{G}$ are first defined as

$$\mathbf{E} \equiv [-\mathbf{e}, \tilde{\mathbf{e}} + e_0\mathbf{I}] = \begin{bmatrix} -e_1 & e_0 & -e_3 & e_2 \\ -e_2 & e_3 & e_0 & -e_1 \\ -e_3 & -e_2 & e_1 & e_0 \end{bmatrix} \tag{9.3.18}$$

$$\mathbf{G} \equiv [-\mathbf{e}, -\tilde{\mathbf{e}} + e_0\mathbf{I}] = \begin{bmatrix} -e_1 & e_0 & e_3 & -e_2 \\ -e_2 & -e_3 & e_0 & e_1 \\ -e_3 & e_2 & -e_1 & e_0 \end{bmatrix} \tag{9.3.19}$$

Observe that each row of $\mathbf{E}$ is orthogonal to $\mathbf{p}$; that is,

$$\begin{aligned} \mathbf{Ep} &= [-\mathbf{e}, \tilde{\mathbf{e}} + e_0\mathbf{I}]\begin{bmatrix} e_0 \\ \mathbf{e} \end{bmatrix} \\ &= -e_0\mathbf{e} + \tilde{\mathbf{e}}\mathbf{e} + e_0\mathbf{e} = \mathbf{0} \end{aligned} \tag{9.3.20}$$

where Eq. 9.1.26 has been used. Similarly (Prob. 9.3.4),

$$\mathbf{Gp} = \mathbf{0} \tag{9.3.21}$$

A direct calculation shows that rows of $\mathbf{E}$ (Prob. 9.3.5) and rows of $\mathbf{G}$ (Prob. 9.3.6) are orthonormal; that is,

$$\mathbf{EE}^T = \mathbf{GG}^T = \mathbf{I} \tag{9.3.22}$$

However, $\mathbf{E}^T\mathbf{E}$ is a 4×4 matrix of the form

$$\mathbf{E}^T\mathbf{E} = \begin{bmatrix} 1 - e_0^2 & -e_0\mathbf{e}^T \\ -e_0\mathbf{e} & -\mathbf{ee}^T + \mathbf{I}_3 \end{bmatrix} = \mathbf{I}_4 - \mathbf{pp}^T \tag{9.3.23}$$

where the subscript is used to emphasize that $\mathbf{I}_4$ is the 4×4 identity matrix. Similarly,

$$\mathbf{G}^T\mathbf{G} = \mathbf{I}_4 - \mathbf{pp}^T \tag{9.3.24}$$

The key relationship involving these matrices is found by evaluating the matrix product:

$$\mathbf{E}\mathbf{G}^T = [-\mathbf{e}, \tilde{\mathbf{e}} + e_0\mathbf{I}]\begin{bmatrix} -\mathbf{e}^T \\ \tilde{\mathbf{e}} + e_0\mathbf{I} \end{bmatrix}$$
$$= \mathbf{e}\mathbf{e}^T + (\tilde{\mathbf{e}} + e_0\mathbf{I})(\tilde{\mathbf{e}} + e_0\mathbf{I})$$
$$= (2e_0^2 - 1)\mathbf{I} + 2(\mathbf{e}\mathbf{e}^T + e^0\tilde{\mathbf{e}}) \tag{9.3.25}$$

Comparing Eq. 9.3.25 with the transformation matrix $\mathbf{A}$ of Eq. 9.3.6 reveals that

$$\mathbf{A} = \mathbf{E}\mathbf{G}^T \tag{9.3.26}$$

Thus the transformation matrix $\mathbf{A}$, with its quadratic terms in Euler parameters, is the product of two matrices whose terms are linear in Euler parameters.

Time Derivatives of Euler Parameters The first time derivative of Eq. 9.3.9 yields (Prob. 9.3.7)

$$\mathbf{p}^T\dot{\mathbf{p}} = \dot{\mathbf{p}}^T\mathbf{p} = 0 \tag{9.3.27}$$

Similarly, the first time derivatives of Eqs. 9.3.20 and 9.3.21 yield

$$\mathbf{E}\dot{\mathbf{p}} = -\dot{\mathbf{E}}\mathbf{p} \tag{9.3.28}$$
$$\mathbf{G}\dot{\mathbf{p}} = -\dot{\mathbf{G}}\mathbf{p} \tag{9.3.29}$$

Direct expansion, using Eqs. 9.3.18 and 9.3.19, shows (Prob. 9.3.8) that

$$\mathbf{E}\dot{\mathbf{G}}^T = \dot{\mathbf{E}}\mathbf{G}^T \tag{9.3.30}$$

Taking the time derivative of both sides of Eq. 9.3.26 and using Eq. 9.3.30,

$$\dot{\mathbf{A}} = \dot{\mathbf{E}}\mathbf{G}^T + \mathbf{E}\dot{\mathbf{G}}^T = 2\mathbf{E}\dot{\mathbf{G}}^T \tag{9.3.31}$$

The product $\mathbf{G}\dot{\mathbf{p}}$ can be expanded as

$$\mathbf{G}\dot{\mathbf{p}} = [-\mathbf{e}, -\tilde{\mathbf{e}} + e_0\mathbf{I}_3]\begin{bmatrix} \dot{e}_0 \\ \dot{\mathbf{e}} \end{bmatrix} = -\dot{e}_0\mathbf{e} - \tilde{\mathbf{e}}\dot{\mathbf{e}} + e_0\dot{\mathbf{e}}$$

Transforming both sides of this equation to skew-symmetric matrices, using the $\tilde{}$ operation on both sides, yields

$$\widetilde{(\mathbf{G}\dot{\mathbf{p}})} = -\dot{e}_0\tilde{\mathbf{e}} - \widetilde{(\tilde{\mathbf{e}}\dot{\mathbf{e}})} + e_0\tilde{\dot{\mathbf{e}}}$$
$$= -\dot{e}_0\tilde{\mathbf{e}} - \tilde{\mathbf{e}}\tilde{\dot{\mathbf{e}}} + \tilde{\dot{\mathbf{e}}}\tilde{\mathbf{e}} + e_0\tilde{\dot{\mathbf{e}}}$$
$$= -\dot{e}_0\tilde{\mathbf{e}} - \tilde{\mathbf{e}}\tilde{\dot{\mathbf{e}}} + \mathbf{e}\dot{\mathbf{e}}^T - \dot{\mathbf{e}}^T\mathbf{e}\mathbf{I} + e_0\tilde{\dot{\mathbf{e}}}$$
$$= -\dot{e}_0\tilde{\mathbf{e}} - \tilde{\mathbf{e}}\tilde{\dot{\mathbf{e}}} + \mathbf{e}\dot{\mathbf{e}}^T + e_0\dot{e}_0\mathbf{I} + e_0\tilde{\dot{\mathbf{e}}}$$
$$= [-\mathbf{e}, -\tilde{\mathbf{e}} + e_0\mathbf{I}]\begin{bmatrix} -\dot{\mathbf{e}}^T \\ \tilde{\dot{\mathbf{e}}} + \dot{e}_0\mathbf{I} \end{bmatrix}$$
$$= \mathbf{G}\dot{\mathbf{G}}^T \tag{9.3.32}$$

where Eqs. 9.1.28 and 9.1.30 and the identity $\mathbf{p}^T\dot{\mathbf{p}} = e_0\dot{e}_0 + \mathbf{e}^T\dot{\mathbf{e}} = 0$ have been used.

Relationships between Euler Parameter Derivatives and Angular Velocity Relationships between the time derivatives of Euler parameters and angular velocity vectors $\boldsymbol{\omega}$ and $\boldsymbol{\omega}'$ are needed in kinematic and dynamic analysis. Using Eqs. 9.3.26 and 9.3.31, Eq. 9.2.40 becomes

$$\tilde{\boldsymbol{\omega}}' = \mathbf{A}^T\dot{\mathbf{A}} = 2\mathbf{G}\mathbf{E}^T\mathbf{E}\dot{\mathbf{G}}^T$$

which, upon application of Eqs. 9.3.23 and 9.3.21, results in

$$\tilde{\boldsymbol{\omega}}' = 2\mathbf{G}\dot{\mathbf{G}}^T \tag{9.3.33}$$

Substituting Eq. 9.3.32 into Eq. 9.3.33 gives

$$\tilde{\boldsymbol{\omega}}' = 2\widetilde{(\mathbf{G}\dot{\mathbf{p}})}$$

Thus, the desired relationship between $\boldsymbol{\omega}'$ and $\dot{\mathbf{p}}$ is

$$\boldsymbol{\omega}' = 2\mathbf{G}\dot{\mathbf{p}} \tag{9.3.34}$$

Multiplying both sides of Eq. 9.3.34 on the left by $\mathbf{G}^T$ yields

$$\mathbf{G}^T\boldsymbol{\omega}' = 2\mathbf{G}^T\mathbf{G}\dot{\mathbf{p}}$$

which, upon application of Eqs. 9.3.24 and 9.3.27, results in the inverse transformation

$$\dot{\mathbf{p}} = \tfrac{1}{2}\mathbf{G}^T\boldsymbol{\omega}' \tag{9.3.35}$$

Using Eqs. 9.3.26 and 9.3.34, angular velocity can be written in the x-y-z reference frame as

$$\boldsymbol{\omega} = \mathbf{A}\boldsymbol{\omega}' = 2\mathbf{E}\mathbf{G}^T\mathbf{G}\dot{\mathbf{p}}$$

Application of Eqs. 9.3.24 and 9.3.27 yields

$$\boldsymbol{\omega} = 2\mathbf{E}\dot{\mathbf{p}} \tag{9.3.36}$$

The inverse transformation of Eq. 9.3.36 is (Prob. 9.3.9)

$$\dot{\mathbf{p}} = \tfrac{1}{2}\mathbf{E}^T\boldsymbol{\omega} \tag{9.3.37}$$

Relationships between Euler Parameter Variations and Virtual Rotations It was shown in Section 9.2 that the virtual displacements and rotations of a reference frame are related to variations in the generalized coordinates that are used to define position and orientation. The objective here is to develop relationships between variations in Euler parameters and virtual rotations of the x'-y'-z' frame whose orientation is defined by the Euler parameters. The principal tool in this development is the differential of calculus, which is a linear operator that satisfies the same rules of manipulation as the time derivative operator, to develop relationships between Euler parameter time derivatives and angular velocity.

To be more specific, taking the differential, or variation, of Eq. 9.3.9 yields

$$\mathbf{p}^T \delta\mathbf{p} = 0 \tag{9.3.38}$$

Since Euler parameter relations do not involve time explicitly, the differential operator satisfies the same rules as the time derivative operator, so it is not surprising that Eq. 9.3.38 is of exactly the same form as Eq. 9.3.27, with time derivative replaced by differential. Similarly, the total differential of Eq. 9.3.26 yields

$$\delta\mathbf{A} = \delta\mathbf{E}\mathbf{G}^T + \mathbf{E}\,\delta\mathbf{G}^T \tag{9.3.39}$$

which is analogous to the time derivative of the transformation matrix given in Eq. 9.3.31.

Replacing the time derivative operator by the differential operator, all Euler parameter relations that involve time derivatives can be replaced by analogous expressions in terms of differentials. Rather than repeating all these equations and identities, they will simply be used to obtain the desired relationships between variations in Euler parameters and virtual rotations.

Premultiplying both sides of Eq. 9.2.50 by $\mathbf{A}^T$,

$$\delta\tilde{\boldsymbol{\pi}}' = \mathbf{A}^T\,\delta\mathbf{A} \tag{9.3.40}$$

Substituting for $\delta\mathbf{A}$ from Eq. 9.3.39 and manipulating in exactly the same manner as in the derivation of Eq. 9.3.34 yields (Prob. 9.3.10)

$$\delta\boldsymbol{\pi}' = 2\mathbf{G}\,\delta\mathbf{p} \tag{9.3.41}$$

which provides the desired relationship between a variation in the Euler parameter vector and a virtual rotation of the x'-y'-z' frame.

Analogous to Eq. 9.3.36 (Prob. 9.3.11),

$$\delta\boldsymbol{\pi} = 2\mathbf{E}\,\delta\mathbf{p} \tag{9.3.42}$$

which relates a variation in Euler parameters to the virtual rotation vector in global coordinates.

Just as in Eq. 9.3.35, the inverse relation to Eq. 9.3.41 is (Prob. 9.3.12)

$$\delta\mathbf{p} = \tfrac{1}{2}\mathbf{G}^T\,\delta\boldsymbol{\pi}' \tag{9.3.43}$$

Finally, analogous to Eq. 9.3.37 (Prob. 9.3.13),

$$\delta\mathbf{p} = \tfrac{1}{2}\mathbf{E}^T\,\delta\boldsymbol{\pi} \tag{9.3.44}$$

These relationships provide transformations between virtual rotations and generalized coordinate variations that are written symbolically in Eq. 9.2.63. In the case of Euler parameters, the matrix $\mathbf{B}_i$ in Eq. 9.2.63 is simply $2\mathbf{G}_i$. The total differential and partial derivative calculations that are written symbolically in Eqs. 9.2.64 and 9.2.65 can now be written explicitly in terms of Euler parameters.

These results will be used extensively in formulating the Jacobian matrix of constraint equations for kinematic analysis.

Integrability of Angular Velocity and Virtual Rotation Virtual rotation $\delta\boldsymbol{\pi}'$, represented in the body-fixed frame, is related to differentials of Euler parameters in Eq. 9.3.41. This relation and the δ notation for virtual rotation raise the following question: Is $\delta\boldsymbol{\pi}'$ the differential of some function of Euler parameters? The answer is provided by investigating the right side of Eq. 9.3.41.

Consider the first component of Eq. 9.3.41, expanded using Eq. 9.3.19:

$$\delta\pi'_x = -2e_1\,\delta e_0 + 2e_0\,\delta e_1 + 2e_3\,\delta e_2 - 2e_2\,\delta e_3 \quad \mathbf{(9.3.45)}$$

According to a well-known theorem of calculus [25, 26], if functions $f_i(\mathbf{x})$, $i = 1, \ldots, n$, where $\mathbf{x} = [x_1, \ldots, x_n]^T$ are continuously differentiable, the differential form

$$\sum_i f_i(\mathbf{x})\,dx_i \quad \mathbf{(9.3.46)}$$

is *exact* (the differential of some function) if and only if

$$\frac{\partial f_i}{\partial x_j} = \frac{\partial f_j}{\partial x_i}, \qquad i, j = 1, \ldots, n \quad \mathbf{(9.3.47)}$$

Applying this test to the right side of Eq. 9.3.45, with $x_1 = e_0$, $x_2 = e_1$, $f_1 = -2e_1 = -2x_2$, and $f_2 = 2e_0 = 2x_1$,

$$\frac{\partial f_1}{\partial x_2} = \frac{\partial f_1}{\partial e_1} = -2 \neq 2 = \frac{\partial f_2}{\partial e_0} = \frac{\partial f_2}{\partial x_1}$$

Thus, the differential on the right side of Eq. 9.3.45 is not exact. This shows that *virtual rotation is not integrable*; that is, virtual rotation is not the differential of any function of Euler parameters.

Note that the coefficient of $\dot{\mathbf{p}}$ on the right side of Eq. 9.3.34 is the same as the coefficient of $\delta\mathbf{p}$ on the right side of Eq. 9.3.41. Thus, *angular velocity is not integrable*; that is, it is not the time derivative of any vector. For this reason, angular velocity is often called a *quasi-coordinate* in the dynamics literature. Apart from motivating a fancy name, this fact has important consequences in integrating equations of motion that are written in terms of angular accelerations, which are time derivatives of angular velocities. This topic will be discussed in more detail in Chapter 11.

9.4 KINEMATIC CONSTRAINTS

Kinematic constraints on the absolute position and orientation of bodies in space and on the relative position and orientation of pairs of bodies that are connected by joints are derived in this section. Conditions for parallelism and orthogonality of pairs of vectors are defined to serve as building blocks for characterizing the kinematic constraints that are encountered in applications. Constraint equations that define a substantial library of joints between bodies in space are derived to provide the foundation for spatial kinematic analysis. In anticipation of the need for velocity and acceleration analysis, first and second derivatives of constraint functions are derived.

9.4.1 Joint Definition Frames

A typical body, denoted body i, is shown in Fig. 9.4.1. Its x_i'-y_i'-z_i' body-fixed reference frame is used to position and orient the body in space. A second x_i''-y_i''-z_i'' *joint definition frame* is attached to the body, with its origin at point P_i. To orient the x_i''-y_i''-z_i'' frame, unit vectors $\mathbf{f}_i$, $\mathbf{g}_i$, and $\mathbf{h}_i$ are defined along its coordinate axes. To define $\mathbf{h}_i$, a point Q_i is defined on the z_i'' axis, a unit distance from point P_i. Next, to define $\mathbf{f}_i$, a point R_i is defined on the x_i'' axis, a unit distance from point P_i. Finally, $\mathbf{g}_i = \tilde{\mathbf{h}}_i\mathbf{f}_i$.

The foregoing procedure defines the x_i''-y_i''-z_i'' frame in terms of *joint definition points*, P_i, Q_i, and R_i, hereafter taken as defining orthogonal unit vectors $\mathbf{f}_i$, $\mathbf{g}_i$, and $\mathbf{h}_i$. In terms of the unit vectors $\mathbf{f}_i'$, $\mathbf{g}_i'$, and $\mathbf{h}_i'$, represented in the x_i'-y_i'-z_i' frame, the transformation matrix from the x_i''-y_i''-z_i'' frame to the x_i'-y_i'-z_i' frame is obtained, as in Eq. 9.2.13, as

$$\mathbf{C}_i^P = [\mathbf{f}_i', \mathbf{g}_i', \mathbf{h}_i'] \tag{9.4.1}$$

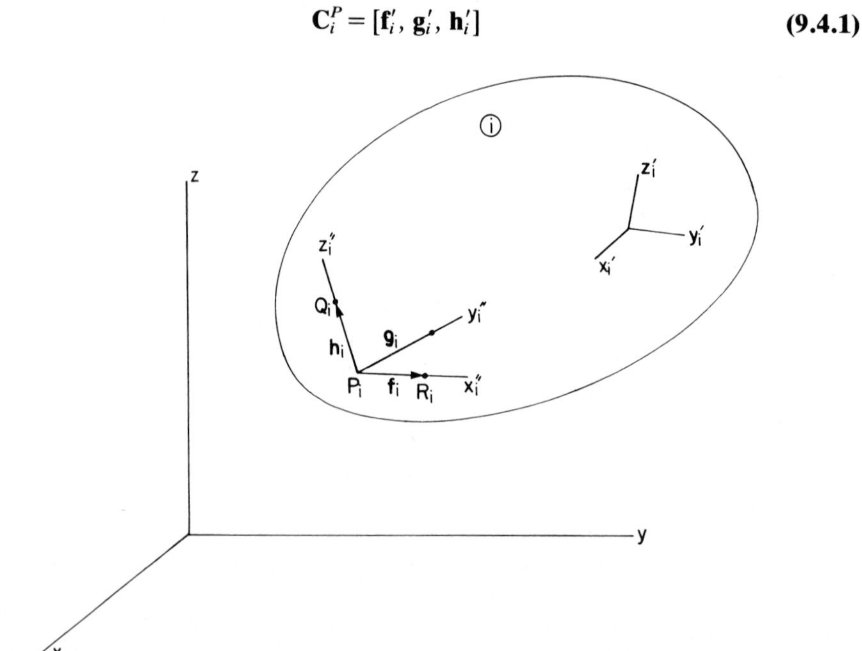

Figure 9.4.1 Construction of a joint definition frame.

Example 9.4.1: Consider the bar shown in Fig. 9.4.2, with a rotational joint at point P_2 that is pivoted about the x_2'' axis, which lies in the x_i'-y_i' plane and makes an angle of $\pi/4$ with the x_2' axis. Unit vectors along the z_2'' and x_2'' axes are, from Fig. 9.4.2,

$$\mathbf{h}_2' = [0, 0, 1]^T$$

$$\mathbf{f}_2' = \left[\frac{\sqrt{2}}{2}, \frac{-\sqrt{2}}{2}, 0\right]^T$$

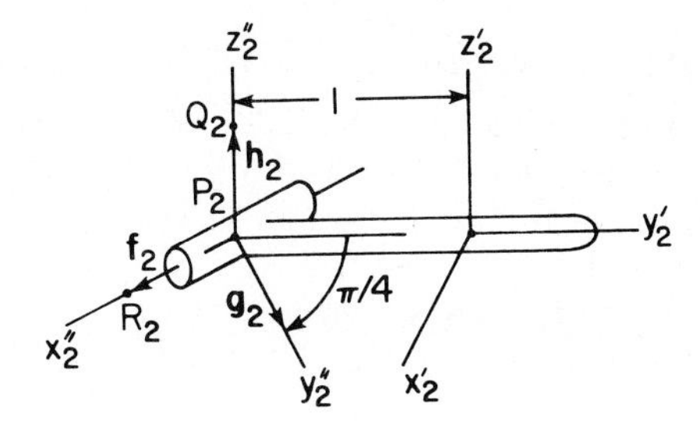

Figure 9.4.2 Oblique rotational axis.

Finally,

$$\mathbf{g}_2' = \tilde{\mathbf{h}}_2'\mathbf{f}_2' = \begin{bmatrix} 0 & -1 & 0 \\ 1 & 0 & 0 \\ 0 & 0 & 0 \end{bmatrix} \begin{bmatrix} \frac{\sqrt{2}}{2} \\ \frac{-\sqrt{2}}{2} \\ 0 \end{bmatrix} = \begin{bmatrix} \frac{\sqrt{2}}{2} \\ \frac{\sqrt{2}}{2} \\ 0 \end{bmatrix}$$

Thus, the transformation matrix $\mathbf{C}_2^P$ of Eq. 9.4.1 is

$$\mathbf{C}_2^P = [\mathbf{f}_2', \mathbf{g}_2', \mathbf{h}_2'] = \begin{bmatrix} \frac{\sqrt{2}}{2} & \frac{\sqrt{2}}{2} & 0 \\ \frac{-\sqrt{2}}{2} & \frac{\sqrt{2}}{2} & 0 \\ 0 & 0 & 1 \end{bmatrix}$$

9.4.2 Constraints on Vectors and Points

Consider the pair of rigid bodies, denoted bodies i and j, in Fig. 9.4.3. Reference points P_i and P_j and nonzero vectors $\mathbf{a}_i$ and $\mathbf{a}_j$ are fixed in bodies i and j, respectively. Kinematic constraints between pairs of bodies are often characterized by conditions of orthogonality or parallelism of pairs of such vectors. The purpose here is to derive analytical conditions with which to define a library of kinematic connections.

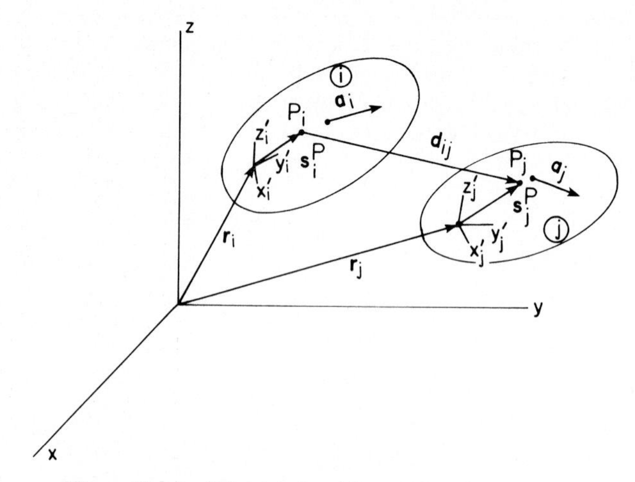

Figure 9.4.3 Vectors fixed in and between bodies.

Basic Constraints First, a necessary and sufficient condition that a pair of body-fixed nonzero vectors $\mathbf{a}_i$ and $\mathbf{a}_j$ on bodies i and j, respectively, is orthogonal is that their scalar product be zero; that is,

$$\Phi^{d1}(\mathbf{a}_i, \mathbf{a}_j) \equiv \mathbf{a}_i^T\mathbf{a}_j = 0 \tag{9.4.2}$$

where the superscript notation $d1$ indicates the first form of dot or scalar product condition. Writing the vectors $\mathbf{a}_i$ and $\mathbf{a}_j$ in terms of their respective body reference transformation matrices and body-fixed constant vectors, $\mathbf{a}_i = \mathbf{A}_i\mathbf{a}_i'$ and $\mathbf{a}_j = \mathbf{A}_j\mathbf{a}_j'$, Eq. 9.4.2 may be written in the form

$$\Phi^{d1}(\mathbf{a}_i, \mathbf{a}_j) = \mathbf{a}_i'^T\mathbf{A}_i^T\mathbf{A}_j\mathbf{a}_j' = 0 \tag{9.4.3}$$

This condition is called the *dot-1 constraint* between vectors $\mathbf{a}_i$ and $\mathbf{a}_j$. Note that it is written in terms of a pair of body-fixed constant vectors and the transformation matrices $\mathbf{A}_i$ and $\mathbf{A}_j$ that depend on the orientation generalized coordinates of bodies i and j, respectively. Thus, the dot-1 constraint restricts the relative orientation of a pair of bodies.

Example 9.4.2: The body in Example 9.4.1 is hinged relative to ground (body 1) about the y_1 axis, as shown in Fig. 9.4.4, to form an oblique pendulum. In order for the pendulum to be hinged on the y_1 axis, $\mathbf{f}_2$ must be perpendicular to $\mathbf{h}_1$ and $\mathbf{f}_1$. Hence, from Eq. 9.4.3,

$$\begin{aligned}\Phi^{d1}(\mathbf{f}_2, \mathbf{f}_1) &= \mathbf{f}_2'^T\mathbf{A}_2^T\mathbf{A}_1\mathbf{f}_1' \\ &= \left[\frac{\sqrt{2}}{2}, \frac{-\sqrt{2}}{2}, 0\right]\mathbf{A}_2^T[1, 0, 0]^T \\ &= \sqrt{2}\left(e_0^2 + e_1^2 - \tfrac{1}{2} - e_1e_2 + e_0e_3\right) = 0\end{aligned}$$

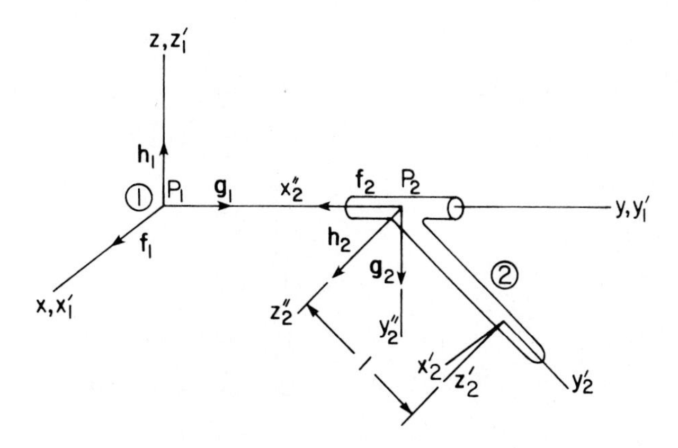

Figure 9.4.4 Oblique pendulum.

where $\mathbf{A}_1 = \mathbf{I}$ and $\mathbf{p} = [e_0, e_1, e_2, e_3]^T$ is the Euler parameter vector for body 2. Similarly,

$$\begin{aligned}\Phi^{d1}(\mathbf{f}_2, \mathbf{h}_1) &= \mathbf{f}_2'^T \mathbf{A}_2^T \mathbf{A}_1 \mathbf{h}_1' \\ &= \left[\frac{\sqrt{2}}{2}, \frac{-\sqrt{2}}{2}, 0\right] \mathbf{A}_2^T [0, 0, 1]^T \\ &= \sqrt{2}(e_1 e_3 + e_0 e_2 - e_2 e_3 - e_0 e_1) = 0\end{aligned}$$

Since $\mathbf{f}_1$, $\mathbf{h}_1$, and $\mathbf{f}_2$ are unit vectors, these two equations are equivalent to the geometric condition that $\mathbf{f}_2$ and $\mathbf{g}_1$ be parallel.

The scalar product condition can also be used to prescribe orthogonality of a body-fixed vector $\mathbf{a}_i$ and a vector $\mathbf{d}_{ij}$ between bodies, shown in Fig. 9.4.3, provided $\mathbf{d}_{ij} \neq \mathbf{0}$. Analytically, this condition is defined using the scalar product as

$$\Phi^{d2}(\mathbf{a}_i, \mathbf{d}_{ij}) = \mathbf{a}_i^T \mathbf{d}_{ij} = 0 \tag{9.4.4}$$

where the superscript $d2$ indicates a second form of dot or scalar product condition. Writing the vector $\mathbf{d}_{ij}$ as

$$\mathbf{d}_{ij} = \mathbf{r}_j + \mathbf{A}_j \mathbf{s}_j'^P - \mathbf{r}_i - \mathbf{A}_i \mathbf{s}_i'^P \tag{9.4.5}$$

Eq. 9.4.4 becomes

$$\Phi^{d2}(\mathbf{a}_i, \mathbf{d}_{ij}) = \mathbf{a}_i'^T \mathbf{A}_i^T (\mathbf{r}_j + \mathbf{A}_j \mathbf{s}_j'^P - \mathbf{r}_i) - \mathbf{a}_i'^T \mathbf{s}_i'^P = 0 \tag{9.4.6}$$

This condition is called the *dot-2 constraint* between vectors $\mathbf{a}_i$ and $\mathbf{d}_{ij}$.

Note that the dot-2 constraint is not symmetric as regards bodies i and j. To require that the vector $\mathbf{a}_j$ on body j be orthogonal to $\mathbf{d}_{ij}$, the dot-2 constraint of Eq. 9.4.6 can be used by simply interchanging indexes i and j. It is important to recall that the orthogonality condition of Eq. 9.4.6 breaks down if $\mathbf{d}_{ij} = \mathbf{0}$.

Example 9.4.3: Consider the pendulum in Example 9.4.2. To constrain P_2 to lie on the y_1' axis, two dot-2 constraints may be used to cause $\mathbf{d}_{12}$ to lie along the y_1' axis; equivalently, as long as $\mathbf{d}_{12} \neq \mathbf{0}$, $\mathbf{d}_{12}$ is perpendicular to $\mathbf{f}_1$ and $\mathbf{h}_1$. From Eq. 9.4.6,

$$\Phi^{d2}(\mathbf{f}_1, \mathbf{d}_{12}) = [1, 0, 0]\left\{\begin{bmatrix} x_2 \\ y_2 \\ z_2 \end{bmatrix} + \mathbf{A}_2\begin{bmatrix} 0 \\ -1 \\ 0 \end{bmatrix} - \begin{bmatrix} 0 \\ 0 \\ 0 \end{bmatrix}\right\} - [1, 0, 0]\begin{bmatrix} 0 \\ 0 \\ 0 \end{bmatrix}$$

$$= x_2 - 2e_1e_2 + 2e_0e_3 = 0$$

$$\Phi^{d2}(\mathbf{h}_1, \mathbf{d}_{12}) = [0, 0, 1]\left\{\begin{bmatrix} x_2 \\ y_2 \\ z_2 \end{bmatrix} + \mathbf{A}_2\begin{bmatrix} 0 \\ -1 \\ 0 \end{bmatrix}\right\}$$

$$= z_2 - 2e_2e_3 - 2e_0e_1 = 0$$

Providing $\mathbf{d}_{12} \neq \mathbf{0}$, these equations imply that $\mathbf{d}_{12}$ and $\mathbf{g}_1$ are parallel. Since they have point P_1 in common, they are collinear. Finally, since P_2 lies on the vector $\mathbf{d}_{12}$, it lies on the y_1' axis. If $\mathbf{d}_{12} = \mathbf{0}$, then points P_1 and P_2 coincide. Thus, even in this case, the equations imply that P_2 lies on the y_1' axis. Thus, the pair of dot-2 constraints is equivalent to the condition that P_2 lie on the y_1' axis.

The combined set of four constraint equations used in this example and in Example 9.4.2 has been shown to imply that P_2 lies on the y_1' axis and that $\mathbf{f}_2$ is parallel to the y axis. Since P_2 is on $\mathbf{f}_2$, $\mathbf{f}_2$ is collinear with the y_1' axis, which is the definition of the geometry of the joint. Thus, the four equations derived are equivalent to the geometry of the joint, even when $\mathbf{d}_{12} = \mathbf{0}$.

Consider next the pair of constraints derived in this example and the following pair of dot-2 constraints:

$$\Phi^{d2}(\mathbf{g}_2, \mathbf{d}_{12}) = 0$$

$$\Phi^{d2}(\mathbf{h}_2, \mathbf{d}_{12}) = 0$$

If $\mathbf{d}_{12} \neq \mathbf{0}$, these equations imply that $\mathbf{d}_{12}$ is parallel to $\mathbf{f}_2$. The pair of equations derived earlier in this example implies that P_2 lies on the y_1' axis and that $\mathbf{d}_{12}$ is orthogonal to the x_1'-y_1' plane. Hence, $\mathbf{f}_2$ is parallel to the y_1' axis, which implies the geometry of the joint.

Are the four dot-2 constraints defined in this example equivalent to the geometry of the joint? To see that the answer is no, note that when $\mathbf{d}_{12} = \mathbf{0}$ all four constraints are satisfied for any orientation of $\mathbf{f}_2$. If these four equations were used in kinematic analysis, all would be well as long as $\mathbf{d}_{12} \neq \mathbf{0}$. When P_2 coincides with P_1, however, the Jacobian of the constraints fails to have full row rank, and a singular matrix would occur in kinematic analysis. The more insidious problem is that, even if P_1 and P_2 are close but do not precisely coincide, the Jacobian becomes very ill-conditioned and numerical error occurs.

This example should adequately illustrate that constraint equations that are derived from the geometry of a joint may sometimes fail to imply the geometry of the joint. This situation cannot be tolerated in computer-aided kinematics or dynamics, since the mathematical model may fail to represent the geometry of the system.

It is often required that a pair of points on two bodies coincide. A necessary and sufficient condition for points P_i and P_j shown in Fig. 9.4.3 to coincide is that $\mathbf{d}_{ij} = \mathbf{0}$; that is,

$$\mathbf{\Phi}^s(P_i, P_j) = \mathbf{r}_j + \mathbf{A}_j\mathbf{s}_j'^P - \mathbf{r}_i - \mathbf{A}_i\mathbf{s}_i'^P = \mathbf{0} \tag{9.4.7}$$

where the superscript designation s anticipates use of this equation in defining a *spherical joint* or *ball and socket joint.* Note that this vector equation consists of three scalar equations. It is called the *spherical constraint.* Note that Eq. 9.4.7 is symmetric with respect to body indexes.

Finally, it is often required that the distance between a pair of points on adjacent bodies be fixed by a physical connection that may be thought of as a link with a spherical joint on each end. A necessary and sufficient condition that the distance between points P_i and P_j in Fig. 9.4.3 be equal to $C \neq 0$ is simply

$$\Phi^{ss}(P_i, P_j, C) = \mathbf{d}_{ij}^T\mathbf{d}_{ij} - C^2 = 0 \tag{9.4.8}$$

which is called the *distance constraint* or *spherical–spherical constraint* and consists of a single scalar equation.

If $C = 0$ in Eq. 9.4.8, then this single equation implies the vector relation $\mathbf{d}_{ij} = \mathbf{0}$ (Prob. 9.4.3). While this might seem to be a practical way to reduce the number of constraint equations, it is not. To see why, denote by $\mathbf{q}$ the vector of generalized coordinates and observe that

$$\Phi_{\mathbf{q}}^{ss} = 2\mathbf{d}_{ij}^T(\mathbf{d}_{ij})_{\mathbf{q}} = \mathbf{0}$$

since $\mathbf{d}_{ij} = \mathbf{0}$ when $C = 0$. Thus, the constraint Jacobian does not have full row rank and cannot be effectively used in kinematic analysis. For this reason, use of the distance constraint is restricted to the case $C \neq 0$.

Example 9.4.4: Consider the compound pendulum shown in Fig. 9.4.5(a), where point P_2 on body 2 is constrained to point P_1 in ground with a distance

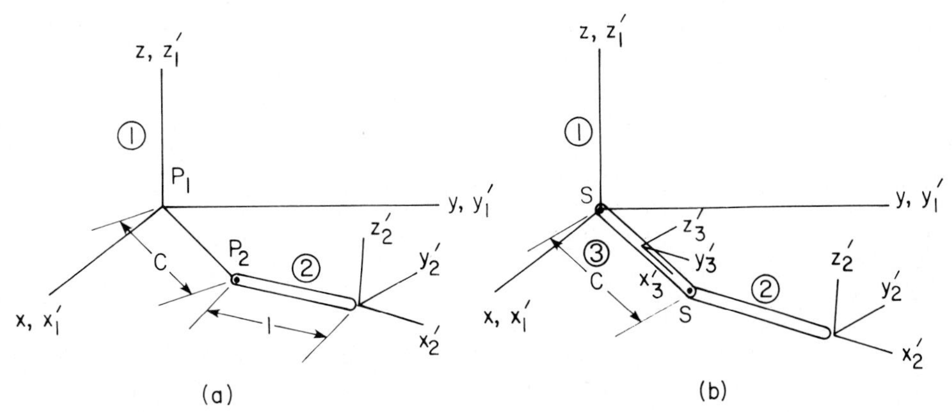

Figure 9.4.5 Compound pendulum.

constraint. From Eq. 9.4.5,

$$\begin{aligned}\mathbf{d}_{12} &= \mathbf{r}_2 + \mathbf{A}_2\mathbf{s}_2'^P - \mathbf{r}_1 - \mathbf{A}_1\mathbf{s}_1'^P \\ &= [x_2, y_2, z_2]^T + \mathbf{A}_2[-1, 0, 0]^T \\ &= [x_2 - 2e_0^2 - 2e_1^2 + 1,\ y_2 - 2e_1e_2 - 2e_0e_3,\ z_2 - 2e_1e_3 + 2e_0e_2]^T\end{aligned}$$

Substituting $\mathbf{d}_{12}$ into Eq. 9.4.8,

$$\begin{aligned}\Phi^{ss}(P_1, P_2, C) = (x_2 - 2e_0^2 - 2e_1^2 + 1)^2 + (y_2 - 2e_1e_2 - 2e_0e_3)^2 \\ + (z_2 - 2e_1e_3 + 2e_0e_2)^2 - C^2 = 0\end{aligned}$$

A second model of the compound pendulum is shown in Fig. 9.4.5(b), where the distance constraint is replaced by a third body, with spherical joints at each end. This may look kinematically equivalent to the compound pendulum in Fig. 9.4.5(a), but it is not. Since the distance constraint is a single scalar equation, the compound pendulum in Fig. 9.4.5(a) has five degrees of freedom. On the other hand, the compound pendulum in Fig. 9.4.5(b) has six degrees of freedom, because the two spherical joints at the ends of body 3 yield six constraint equations. This extra degree of freedom is accounted for as the rotational degree of freedom of body 3 about its x_3' axis. Therefore, the two models are not equivalent.

The four basic constraint equations derived thus far form the foundation for defining a library of kinematic constraints between bodies. To see that these conditions may be efficiently used in defining other geometric relationships, two parallelism conditions are derived, using the dot-1 and dot-2 constraints. Consider the pair of bodies shown in Fig. 9.4.6, with joint definition frames located at points P_i and P_j on bodies i and j, respectively.

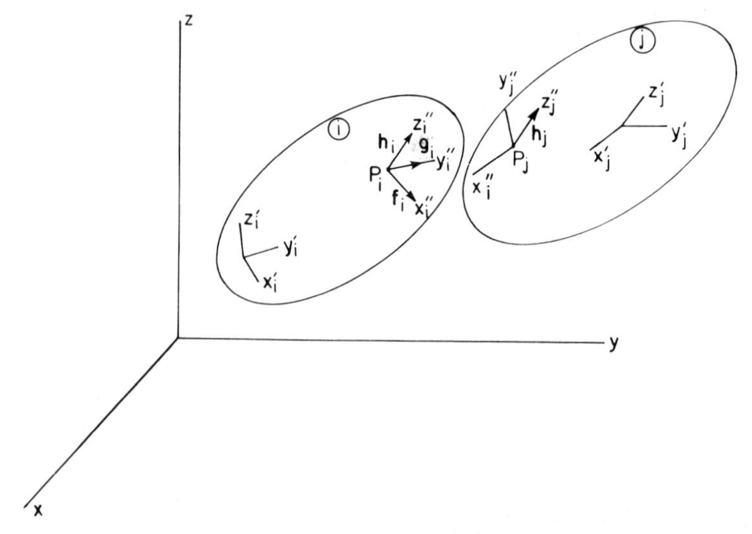

Figure 9.4.6 Parallel vectors in adjacent bodies.

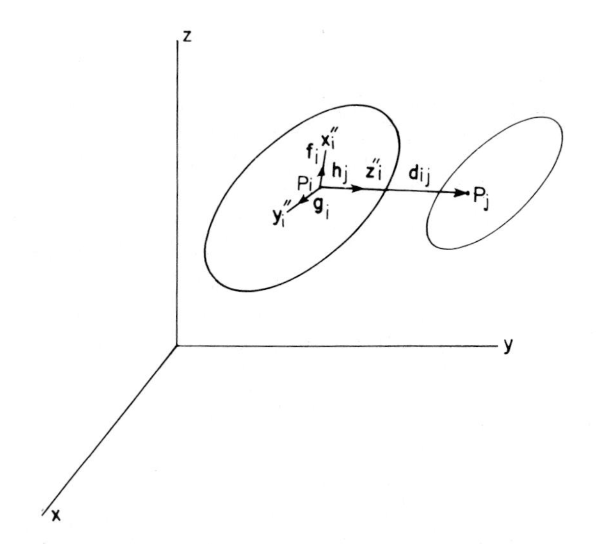

Figure 9.4.7 Parallel vectors on and between adjacent bodies.

Let the z_i'' and z_j'' axes be required to be parallel; that is, vectors $\mathbf{h}_i$ and $\mathbf{h}_j$ are to be parallel. The vector $\mathbf{h}_j$ is parallel to $\mathbf{h}_i$ if and only if it is orthogonal to $\mathbf{f}_i$ and $\mathbf{g}_i$. Thus, the condition that $\mathbf{h}_i$ and $\mathbf{h}_j$ are parallel is simply

$$\mathbf{\Phi}^{p1}(\mathbf{h}_i, \mathbf{h}_j) = \begin{bmatrix} \Phi^{d1}(\mathbf{f}_i, \mathbf{h}_j) \\ \Phi^{d1}(\mathbf{g}_i, \mathbf{h}_j) \end{bmatrix} = \mathbf{0} \tag{9.4.9}$$

where the dot-1 constraint of Eq. 9.4.3 is employed and two scalar equations are obtained. This is called the *parallel-1 constraint.*

Finally, consider the condition that the vector $\mathbf{h}_i$ along the z_i'' axis on body i is to be parallel to the vector $\mathbf{d}_{ij}$, as shown in Fig. 9.4.7. Since $\mathbf{d}_{ij} \neq \mathbf{0}$ is parallel to $\mathbf{h}_i$ if and only if it is perpendicular to $\mathbf{f}_i$ and $\mathbf{g}_i$, the second parallelism condition is

$$\mathbf{\Phi}^{p2}(\mathbf{h}_i, \mathbf{d}_{ij}) = \begin{bmatrix} \Phi^{d2}(\mathbf{f}_i, \mathbf{d}_{ij}) \\ \Phi^{d2}(\mathbf{g}_i, \mathbf{d}_{ij}) \end{bmatrix} = \mathbf{0} \tag{9.4.10}$$

which is simply an application of the dot-2 constraint of Eq. 9.4.6. It is called the *parallel-2 constraint.* Note that the parallel-2 constraint breaks down if $\mathbf{d}_{ij} = \mathbf{0}$, since the dot-2 conditions break down in this case.

Example 9.4.5: Consider again the oblique pendulum in Example 9.4.2. To keep the pivot axis of the pendulum parallel to the y_1' axis, the parallel-1 constraint may be used to cause $\mathbf{f}_2$ and $\mathbf{g}_1$ in Fig. 9.4.4 to be parallel; that is,

$$\begin{aligned}\mathbf{\Phi}^{p1}(\mathbf{f}_2, \mathbf{g}_1) &= \begin{bmatrix} \Phi^{d1}(\mathbf{h}_2, \mathbf{g}_1) \\ \Phi^{d1}(\mathbf{g}_2, \mathbf{g}_1) \end{bmatrix} \\ &= \begin{bmatrix} (\mathbf{A}_2\mathbf{h}_2')^T\mathbf{g}_1 \\ (\mathbf{A}_2\mathbf{g}_2')^T\mathbf{g}_1 \end{bmatrix} \\ &= \begin{bmatrix} [0, 0, 1]\mathbf{A}_2^T[0, 1, 0]^T \\ \left[\frac{\sqrt{2}}{2}, \frac{\sqrt{2}}{2}, 0\right]\mathbf{A}_2^T[0, 1, 0]^T \end{bmatrix} \\ &= \begin{bmatrix} 2(e_2e_3 - e_0e_1) \\ \sqrt{2}\,(e_1e_2 + e_0e_3 + e_0^2 + e_2^2 - 1/2) \end{bmatrix} = 0\end{aligned}$$

where the Euler parameters are for body 2. Constraining point P_2 to be on the y_1' axis, a complete set of constraint equations for the joint is obtained; that is, with

$$\begin{aligned}\mathbf{r}_2^P &= \mathbf{r}_2 + \mathbf{A}_2\mathbf{s}_2'^P \\ &= \begin{bmatrix} x_2 \\ y_2 \\ z_2 \end{bmatrix} + \mathbf{A}_2\begin{bmatrix} 0 \\ -1 \\ 0 \end{bmatrix} \\ &= \begin{bmatrix} x_2 - 2e_1e_2 + 2e_0e_3 \\ y_2 - 2e_0^2 - 2e_2^2 + 1 \\ z_2 - 2e_2e_3 - 2e_0e_1 \end{bmatrix}\end{aligned}$$

the remaining two constraints are

$$\begin{aligned}x_2^P &= x_2 - 2e_1e_2 + 2e_0e_3 = 0 \\ z_2^P &= z_2 - 2e_2e_3 - 2e_0e_1 = 0\end{aligned}$$

Differentials and Derivatives of Basic Constraints In anticipation that differentials and derivatives of the four basic constraint equations with respect to generalized coordinates $(\mathbf{r}_i, \mathbf{p}_i)$ and $(\mathbf{r}_j, \mathbf{p}_j)$ of bodies i and j need to be calculated, it is instructive to use the vector and differential calculus methods of Section 9.2 and identities obtained in Section 9.3 to derive the needed differential and derivative expressions. The differential of the dot-1 constraint function of Eq. 9.4.2 was derived in Example 9.2.6 of Section 9.2. From Eq. 9.2.60,

$$\delta\Phi^{d1}(\mathbf{a}_i, \mathbf{a}_j) = -(\mathbf{a}_j'^T\mathbf{A}_j^T\mathbf{A}_i\tilde{\mathbf{a}}_i'\,\delta\boldsymbol{\pi}_i' + \mathbf{a}_i'^T\mathbf{A}_i^T\mathbf{A}_j\tilde{\mathbf{a}}_j'\,\delta\boldsymbol{\pi}_j') \qquad \textbf{(9.4.11)}$$

The differential of the dot-2 constraint function of Eq. 9.4.4 is

$$\delta\Phi^{d2}(\mathbf{a}_i, \mathbf{d}_{ij}) = \mathbf{a}_i^T\,\delta\mathbf{d}_{ij} + \mathbf{d}_{ij}^T\,\delta\mathbf{a}_i$$

Using $\mathbf{d}_{ij} = \mathbf{r}_j + \mathbf{s}_j^P - \mathbf{r}_i - \mathbf{s}_i^P$ and Eq. 9.2.53, this is (Prob. 9.4.4)

$$\begin{aligned}\delta\Phi^{d2}(\mathbf{a}_i, \mathbf{d}_{ij}) &= \mathbf{a}_i'^T\mathbf{A}_i^T(\delta\mathbf{r}_j - \mathbf{A}_j\tilde{\mathbf{s}}_j'^P\,\delta\boldsymbol{\pi}_j' - \delta\mathbf{r}_i + \mathbf{A}_i\tilde{\mathbf{s}}_i'^P\,\delta\boldsymbol{\pi}_i') \\ &\quad - \mathbf{d}_{ij}^T\mathbf{A}_i\tilde{\mathbf{a}}_i'\,\delta\boldsymbol{\pi}_i' \\ &= \mathbf{a}_i'^T\mathbf{A}_i^T\,\delta\mathbf{r}_j - \mathbf{a}_i'^T\mathbf{A}_i^T\,\delta\mathbf{r}_i - \mathbf{a}_i'^T\mathbf{A}_i^T\mathbf{A}_j\tilde{\mathbf{s}}_j'^P\,\delta\boldsymbol{\pi}_j' \\ &\quad + (\mathbf{a}_i'^T\tilde{\mathbf{s}}_i'^P - \mathbf{d}_{ij}^T\mathbf{A}_i\tilde{\mathbf{a}}_i')\,\delta\boldsymbol{\pi}_i' \end{aligned} \qquad \textbf{(9.4.12)}$$

The differential of the spherical constraint of Eq. 9.4.7, using Eq. 9.2.53, is

$$\delta\boldsymbol{\Phi}^s(P_i, P_j) = \delta\mathbf{r}_j - \mathbf{A}_j\tilde{\mathbf{s}}_j'^P\,\delta\boldsymbol{\pi}_j' - \delta\mathbf{r}_i + \mathbf{A}_i\tilde{\mathbf{s}}_i'^P\,\delta\boldsymbol{\pi}_i' \qquad \textbf{(9.4.13)}$$

Finally, the differential of the distance constraint of Eq. 9.4.8 is

$$\delta\Phi^{ss}(P_i, P_j, C) = 2\mathbf{d}_{ij}^T\,\delta\mathbf{d}_{ij}$$

Using the same expansion of $\delta\mathbf{d}_{ij}$ employed in deriving Eq. 9.4.12, this is

$$\delta\Phi^{ss}(P_i, P_j, C) = 2\mathbf{d}_{ij}^T(\delta\mathbf{r}_j - \mathbf{A}_j\tilde{\mathbf{s}}_j'^P\,\delta\boldsymbol{\pi}_j' - \delta\mathbf{r}_i + \mathbf{A}_i\tilde{\mathbf{s}}_i'^P\,\delta\boldsymbol{\pi}_i') \qquad \textbf{(9.4.14)}$$

Differentials of the parallel-1 and parallel-2 constraints of Eqs. 9.4.9 and 9.4.10 are obtained through direct application of Eqs. 9.4.11 and 9.4.12.

All constraint equations encountered can be written in the form

$$\boldsymbol{\Phi}(\mathbf{r}_i, \mathbf{A}_i, \mathbf{r}_j, \mathbf{A}_j) = \mathbf{0}$$

where i and j are numbers that identify the bodies connected. Taking the differential of both sides of each equation and manipulating has led to linear equations in virtual displacements and virtual rotations. Defining the coefficients of $\delta\mathbf{r}_i$ and $\delta\mathbf{r}_j$ as $\boldsymbol{\Phi}_{\mathbf{r}_i}$ and $\boldsymbol{\Phi}_{\mathbf{r}_j}$ and the coefficients of $\delta\boldsymbol{\pi}_i'$ and $\delta\boldsymbol{\pi}_j'$ as $\boldsymbol{\Phi}_{\boldsymbol{\pi}_i'}$ and $\boldsymbol{\Phi}_{\boldsymbol{\pi}_j'}$, respectively, the linearized constraint equations are

$$\delta\boldsymbol{\Phi} = \boldsymbol{\Phi}_{r_i}\,\delta\mathbf{r}_i + \boldsymbol{\Phi}_{\boldsymbol{\pi}_i'}\,\delta\boldsymbol{\pi}_i' + \boldsymbol{\Phi}_{\mathbf{r}_j}\,\delta\mathbf{r}_j + \boldsymbol{\Phi}_{\boldsymbol{\pi}_j'}\,\delta\boldsymbol{\pi}_j' = \mathbf{0}$$

The coefficients of $\delta\mathbf{r}_i$, $\delta\mathbf{r}_j$, $\delta\boldsymbol{\pi}_i'$, and $\delta\boldsymbol{\pi}_j'$ in the differentials of Eqs. 9.4.11 to 9.4.14 are tabulated in Table 9.4.1. The reader is cautioned that since $\delta\boldsymbol{\pi}'$ is not integrable, the Jacobian matrices $\boldsymbol{\Phi}_{\boldsymbol{\pi}'}$ in Table 9.4.1 should be interpreted only as coefficients of $\delta\boldsymbol{\pi}'$ in Eqs. 9.4.11 to 9.4.14, not as derivatives of $\boldsymbol{\Phi}$ with respect to a variable $\boldsymbol{\pi}'$, since $\boldsymbol{\pi}'$ does not exist.

TABLE 9.4.1 Partial Derivaties[a] of Basic Constraint Functions

Constraint function	$\boldsymbol{\Phi}_{\mathbf{r}_i}$	$\boldsymbol{\Phi}_{\mathbf{r}_j}$	$\boldsymbol{\Phi}_{\boldsymbol{\pi}_i'}$	$\boldsymbol{\Phi}_{\boldsymbol{\pi}_j'}$
$\Phi^{d1}(\mathbf{a}_i, \mathbf{a}_j)$	$\mathbf{0}$	$\mathbf{0}$	$-\mathbf{a}_j'^T\mathbf{A}_j^T\mathbf{A}_i\tilde{\mathbf{a}}_i'$	$-\mathbf{a}_i'^T\mathbf{A}_i^T\mathbf{A}_j\tilde{\mathbf{a}}_j'$
$\Phi^{d2}(\mathbf{a}_i, \mathbf{d}_{ij})$	$-\mathbf{a}_i'^T\mathbf{A}_i^T$	$\mathbf{a}_i'^T\mathbf{A}_i^T$	$(\mathbf{a}_i'^T\tilde{\mathbf{s}}_i'^P - \mathbf{d}_{ij}^T\mathbf{A}_i\tilde{\mathbf{a}}_i')$	$-\mathbf{a}_i'^T\mathbf{A}_i^T\mathbf{A}_j\tilde{\mathbf{s}}_j'^P$
$\boldsymbol{\Phi}^s(P_i, P_j)$	$-\mathbf{I}$	$\mathbf{I}$	$\mathbf{A}_i\tilde{\mathbf{s}}_i'^P$	$-\mathbf{A}_j\tilde{\mathbf{s}}_j'^P$
$\Phi^{ss}(P_i, P_j, C)$	$-2\mathbf{d}_{ij}^T$	$2\mathbf{d}_{ij}^T$	$2\mathbf{d}_{ij}^T\mathbf{A}_i\tilde{\mathbf{s}}_i'^P$	$-2\mathbf{d}_{ij}^T\mathbf{A}_j\tilde{\mathbf{s}}_j'^P$

[a] Technically, $\boldsymbol{\Phi}_{\boldsymbol{\pi}'}$ is not a partial derivative, but $\boldsymbol{\Phi}_{\mathbf{p}} = 2\boldsymbol{\Phi}_{\boldsymbol{\pi}'}\mathbf{G}$.

Using the relationship $\delta\boldsymbol{\pi}' = 2\mathbf{G}\,\delta\mathbf{p}$ of Eq. 9.3.41, a linearized constraint equation can be written in terms of variations of Euler parameters of the bodies involved; that is,

$$\delta\boldsymbol{\Phi} = \boldsymbol{\Phi}_{\mathbf{r}_i}\,\delta\mathbf{r}_i + 2\boldsymbol{\Phi}_{\boldsymbol{\pi}_i'}\mathbf{G}_i\,\delta\mathbf{p}_i + \boldsymbol{\Phi}_{\mathbf{r}_j} + 2\boldsymbol{\Phi}_{\boldsymbol{\pi}_j'}\mathbf{G}_j\,\delta\mathbf{p}_j = \mathbf{0}$$

The coefficients of the differentials of $\delta\mathbf{r}_i$, $\delta\mathbf{r}_j$, $\delta\mathbf{p}_i$, and $\delta\mathbf{p}_j$ are thus the partial derivatives of the constraint functions with respect to these generalized coordinates. Partial derivatives with respect to $\mathbf{p}_i$ and $\mathbf{p}_j$ are thus obtained from Table 9.4.1 by multiplying $\Phi_{\pi_i'}$ and $\Phi_{\pi_j'}$ on the right by $2\mathbf{G}_i$ and $2\mathbf{G}_j$, respectively; that is,

$$\frac{\partial\boldsymbol{\Phi}}{\partial\mathbf{p}_i} \equiv \boldsymbol{\Phi}_{\mathbf{p}_i} = 2\boldsymbol{\Phi}_{\boldsymbol{\pi}_i'}\mathbf{G}_i$$

9.4.3 Absolute Constraints on a Body

Absolute constraints may be placed on the position of point P_i on body i, as shown in Fig. 9.4.8, and on the orientation of the body-fixed reference frame for body i. Six such constraint equations on individual generalized coordinates of body i may be written as

$$\begin{aligned}
\Phi^1 &= x_i^P - x_i^0 = 0 \\
\Phi^2 &= y_i^P - y_i^0 = 0 \\
\Phi^3 &= z_i^P - z_i^0 = 0 \\
\Phi^4 &= e_{1i} - e_{1i}^0 = 0 \\
\Phi^5 &= e_{2i} - e_{2i}^0 = 0 \\
\Phi^6 &= e_{3i} - e_{3i}^0 = 0
\end{aligned} \qquad \textbf{(9.4.15)}$$

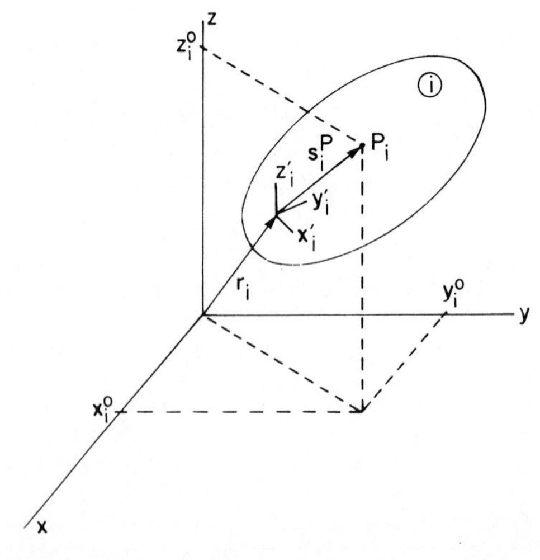

Figure 9.4.8 Absolute coordinate constraints.

Example 9.4.6: Absolute x and z constraints on points P_1 and P_2 on body 2 in Fig. 9.4.9 can be used to define a joint that permits body 2 to rotate about and slide along the global y axis (called a cylindrical joint). Writing

$$\mathbf{r}^{P_1} = \mathbf{r}_2 + \mathbf{A}_2\mathbf{s}_2^{\prime P_1}$$
$$= \begin{bmatrix} x_2 + 2e_1e_2 - 2e_0e_3 \\ y_2 + 2e_0^2 + 2e_2^2 - 1 \\ z_2 + 2e_2e_3 + 2e_0e_1 \end{bmatrix}$$
$$\mathbf{r}^{P_2} = \mathbf{r}_2 + \mathbf{A}_2\mathbf{s}_2^{\prime P_2}$$
$$= \begin{bmatrix} x_2 - 2e_1e_2 + 2e_0e_3 \\ y_2 - 2e_0^2 - 2e_2^2 + 1 \\ z_2 - 2e_2e_3 - 2e_0e_1 \end{bmatrix}$$

where Euler parameters are for body 2, absolute x and z constraints on points are

$$\begin{bmatrix} x_2 + 2e_1e_2 - 2e_0e_3 \\ z_2 + 2e_2e_3 + 2e_0e_1 \\ x_2 - 2e_1e_2 + 2e_0e_3 \\ z_2 - 2e_2e_3 - 2e_0e_1 \end{bmatrix} = \mathbf{0}$$

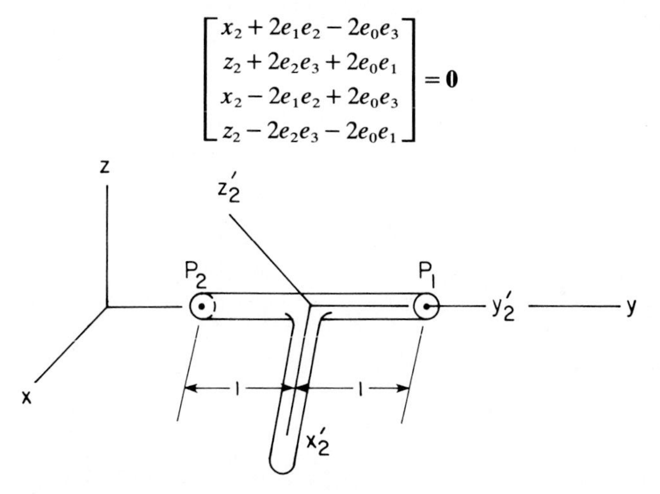

Figure 9.4.9 Cylindrical joint defined by absolute constraints on two points.

While any one of the first three constraints of Eq. 9.4.15, taken alone, is an *absolute coordinate constraint* that is physically meaningful, the last three conditions should be taken together to completely specify the orientation of body i, with the fourth Euler parameter determined by the normalization condition of Eq. 9.3.9. If all six of the constraints of Eq. 9.4.15 are employed, the position and orientation of body i relative to ground are fixed, yielding a *ground constraint.*

The vector to point P_i on body i in Fig. 9.4.8 is

$$\mathbf{r}_i^P = \mathbf{r}_i + \mathbf{A}_i\mathbf{s}_i^{\prime P}$$

The variation of this vector is, from Eq. 9.2.51,

$$\delta \mathbf{r}_i^P = \delta \mathbf{r}_i - \mathbf{A}_i \tilde{\mathbf{s}}_i'^P \, \delta \boldsymbol{\pi}_i'$$

Thus, the gradients of Φ^1, Φ^2, and Φ^3 in Eq. 9.4.15 are

$$\begin{bmatrix} \Phi^1_{\mathbf{r}_i} \\ \Phi^2_{\mathbf{r}_i} \\ \Phi^3_{\mathbf{r}_i} \end{bmatrix} = \mathbf{I}$$

$$\begin{bmatrix} \Phi^1_{\boldsymbol{\pi}_i'} \\ \Phi^2_{\boldsymbol{\pi}_i'} \\ \Phi^3_{\boldsymbol{\pi}_i'} \end{bmatrix} = -\mathbf{A}_i \tilde{\mathbf{s}}_i'^P$$

From Eqs. 9.3.19 and 9.3.43,

$$\delta \mathbf{e}_i = \frac{\partial \mathbf{e}_i}{\partial \mathbf{p}_i} \delta \mathbf{p}_i = [\mathbf{0}, \mathbf{I}] \, \delta \mathbf{p}_i = \tfrac{1}{2}(\tilde{\mathbf{e}}_i + \mathbf{e}_{0_i} \mathbf{I}) \, \delta \boldsymbol{\pi}_i'$$

Thus, the gradients of Φ^4, Φ^5, and Φ^6 in Eq. 9.4.15 are

$$\begin{bmatrix} \Phi^4_{\mathbf{r}_i} \\ \Phi^5_{\mathbf{r}_i} \\ \Phi^6_{\mathbf{r}_i} \end{bmatrix} = \mathbf{0}$$

$$\begin{bmatrix} \Phi^4_{\boldsymbol{\pi}_i'} \\ \Phi^5_{\boldsymbol{\pi}_i'} \\ \Phi^6_{\boldsymbol{\pi}_i'} \end{bmatrix} = \tfrac{1}{2}(\tilde{\mathbf{e}}_i + \mathbf{e}_{0_i} \mathbf{I})$$

An *absolute point constraint* on the position of point P_i on body i can be defined by the vector $\mathbf{r}_i^0$ in the x-y-z reference frame, as shown in Fig. 9.4.10.

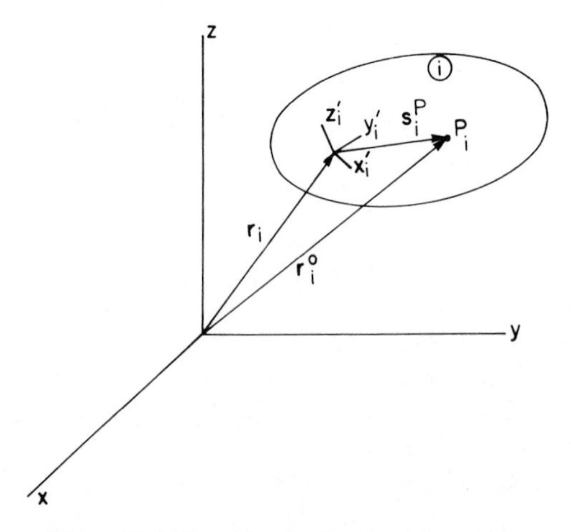

Figure 9.4.10 Absolute point constraint.

This vector condition is written in matrix form as

$$\mathbf{\Phi}^{P}(P_i) = \mathbf{r}_i + \mathbf{A}_i \mathbf{s}_i'^{P} - \mathbf{r}_i^0 = \mathbf{0} \tag{9.4.16}$$

While these three scalar constraint equations restrict the position of point P_i, they allow three rotational degrees of freedom of body i. From the expression for $\delta \mathbf{r}_i^P$ in Eq. 9.2.51,

$$\mathbf{\Phi}^{P}_{\mathbf{r}_i} = \mathbf{I}$$

$$\mathbf{\Phi}^{P}_{\boldsymbol{\pi}_i'} = -\mathbf{A}_i \tilde{\mathbf{s}}_i'^{P}$$

9.4.4 Constraints between Pairs of Bodies

A variety of *spatial joints* between pairs of bodies is employed in the construction of mechanisms and machines. Constraint equations that define a library of such joints are derived in this subsection.

The distance between points P_i and P_j on bodies i and j can be fixed and equal to $C \neq 0$, as shown in Fig. 9.4.11, by direct application of the *distance constraint* of Eq. 9.4.8; that is,

$$\Phi^{ss}(P_i, P_j, C) = 0 \tag{9.4.17}$$

This scalar constraint equation permits five relative degrees of freedom between bodies i and j.

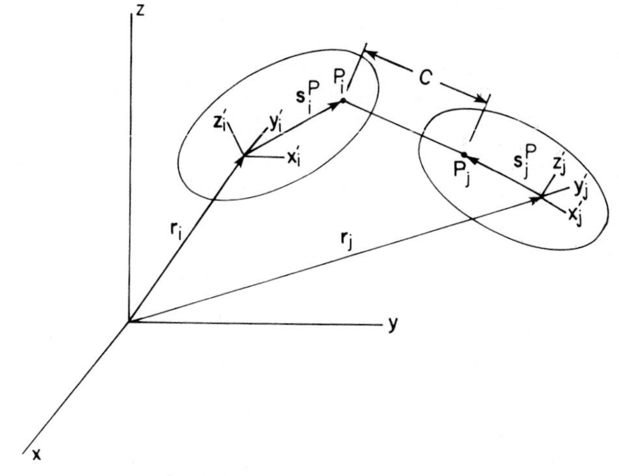

Figure 9.4.11 Distance constraint.

A *spherical joint* (or *ball and socket joint*) is defined by the condition that the center of the ball at point P_i on body i coincides with the center of the socket at P_j on body j, as shown in Fig. 9.4.12. This condition is simply the spherical constraint of Eq. 9.4.7; that is,

$$\mathbf{\Phi}^{s}(P_i, P_j) = \mathbf{0} \tag{9.4.18}$$

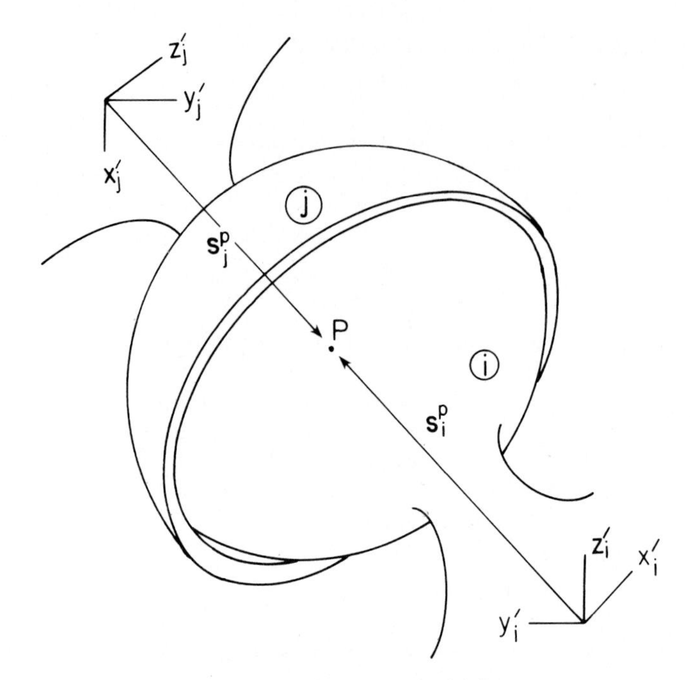

Figure 9.4.12 Spherical joint.

These three scalar constraint equations restrict only the relative position of points P_i and P_j. Three relative degrees of freedom remain.

The spherical constraint formulation does not involve a joint reference frame; so if the user does not wish to define specific frames, defaults that are parallel to the body-fixed reference frames are selected.

A *universal joint* between bodies i and j, shown in Fig. 9.4.13, is constructed with an intermediate body, or *cross*, that is pivoted in bodies i and j. The center of the cross of the universal joint is fixed in bodies i and j, defined by points P_i and P_j on the respective bodies. Points Q_i and Q_j on the arms of the cross between bodies i and j, respectively, are specified to determine the z'' axes of the joint definition frames, as shown in Fig. 9.4.13. The remaining axes of the joint definition frames are defined at the user's discretion.

The equations of constraint that characterise a universal joint are that points P_i and P_j coincide and that vectors $\mathbf{h}_i$ and $\mathbf{h}_j$ are orthogonal. These conditions are specified by the constraint equations

$$\begin{aligned} \mathbf{\Phi}^s(P_i, P_j) &= \mathbf{0} \\ \Phi^{d1}(\mathbf{h}_i, \mathbf{h}_j) &= 0 \end{aligned} \tag{9.4.19}$$

These four scalar constraint equations restrict the relative position of the bodies and rotation about the shafts of the cross. They allow two relative degrees of freedom between the bodies.

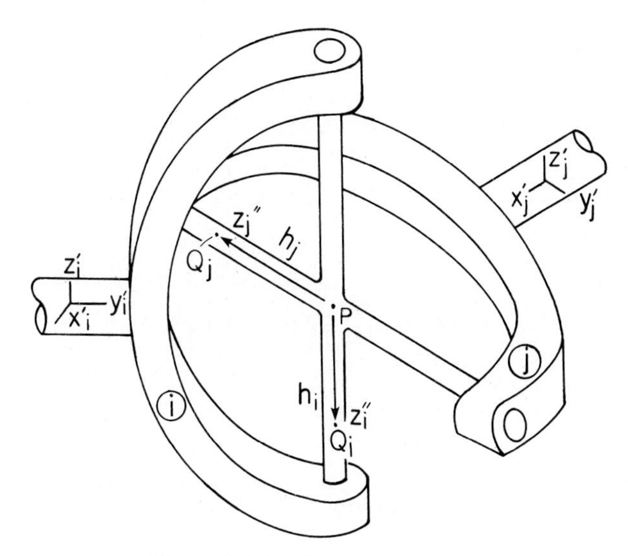

Figure 9.4.13 Universal joint.

Example 9.4.7: Consider the universal joint in Fig. 9.4.14. Let the driving angle of the first body be θ_1 and the output angle of the second body be θ_2, measured between the x-y plane and the y_2' axis. Both bodies are constrained to rotate in the x-y plane, as shown, so $\theta_2 = 0$ when $\theta_1 = 0$.

From Fig. 9.4.14,

$$\mathbf{h}_1' = [0, -1, 0]^T$$

$$\mathbf{h}_2' = [0, 0, -1]^T$$

$$\mathbf{A}_1 = \begin{bmatrix} 1 & 0 & 0 \\ 0 & \cos\theta_1 & -\sin\theta_1 \\ 0 & \sin\theta_1 & \cos\theta_1 \end{bmatrix}$$

$$\mathbf{A}_2 = \begin{bmatrix} \cos\phi & -\cos\theta_2 \sin\phi & \sin\theta_2 \sin\phi \\ \sin\phi & \cos\theta_2 \cos\phi & -\sin\theta_2 \cos\phi \\ 0 & \sin\theta_2 & \cos\theta_2 \end{bmatrix}$$

From Eq. 9.4.3,

$$\Phi^{d1}(\mathbf{h}_1, \mathbf{h}_2) = \mathbf{h}_1'^T \mathbf{A}_1^T \mathbf{A}_2 \mathbf{h}_2'$$

$$= [0, -1, 0] \begin{bmatrix} 1 & 0 & 0 \\ 0 & \cos\theta_1 & \sin\theta_1 \\ 0 & -\sin\theta_1 & \cos\theta_1 \end{bmatrix} \begin{bmatrix} \cos\phi & -\cos\theta_2 \sin\phi & \sin\theta_2 \sin\phi \\ \sin\phi & \cos\theta_2 \cos\phi & -\sin\theta_2 \cos\phi \\ 0 & \sin\theta_2 & \cos\theta_2 \end{bmatrix} \begin{bmatrix} 0 \\ 0 \\ -1 \end{bmatrix}$$

$$= -\cos\theta_1 \sin\theta_2 \cos\phi + \sin\theta_1 \cos\theta_2 = 0$$

Dividing by $\cos\theta_1 \cos\theta_2$,

$$\tan\theta_1 = \tan\theta_2 \cos\phi \tag{9.4.20}$$

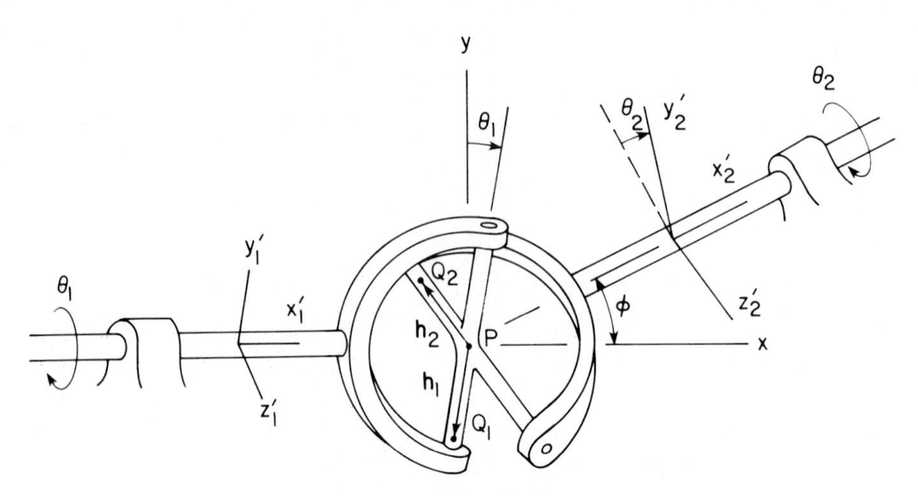

Figure 9.4.14 Universal joint example.

Hence,

$$\theta_2 = \text{Arctan}\,\frac{\tan \theta_1}{\cos \phi}$$

From Eq. 9.4.11, the variation of the dot-1 constraint is

$$-(\mathbf{h}_2'^T \mathbf{A}_2^T \mathbf{A}_1 \tilde{\mathbf{h}}_1' \, \delta\boldsymbol{\pi}_1' + \mathbf{h}_1'^T \mathbf{A}_1^T \mathbf{A}_2 \tilde{\mathbf{h}}_2' \, \delta\boldsymbol{\pi}_2') = 0$$

After substitution of $\tilde{\mathbf{h}}_i'$ and $\mathbf{A}_i$, $i = 1, 2$, with $\delta\boldsymbol{\pi}_1' = [\delta\theta_1, 0, 0]^T$ and $\delta\boldsymbol{\pi}_2' = [\delta\theta_2, 0, 0]^T$,

$$\begin{bmatrix} 0 \\ 0 \\ -1 \end{bmatrix}^T \begin{bmatrix} \cos\phi & \sin\phi & 0 \\ -\cos\theta_2 \sin\phi & \cos\theta_2 \cos\phi & \sin\theta_2 \\ \sin\theta_2 \sin\phi & -\sin\theta_2 \cos\phi & \cos\theta_2 \end{bmatrix} \begin{bmatrix} 1 & 0 & 0 \\ 0 & \cos\theta_1 & -\sin\theta_1 \\ 0 & \sin\theta_1 & \cos\theta_1 \end{bmatrix}$$

$$\times \begin{bmatrix} 0 & 0 & -1 \\ 0 & 0 & 0 \\ 1 & 0 & 0 \end{bmatrix} \begin{bmatrix} \delta\theta_1 \\ 0 \\ 0 \end{bmatrix} + \begin{bmatrix} 0 \\ -1 \\ 0 \end{bmatrix}^T \begin{bmatrix} 1 & 0 & 0 \\ 0 & \cos\theta_1 & \sin\theta_1 \\ 0 & -\sin\theta_1 & \cos\theta_1 \end{bmatrix}$$

$$\times \begin{bmatrix} \cos\phi & -\cos\theta_2 \sin\phi & \sin\theta_2 \sin\phi \\ \sin\phi & \cos\theta_2 \cos\phi & -\sin\theta_2 \cos\phi \\ 0 & \sin\theta_2 & \cos\theta_2 \end{bmatrix} \begin{bmatrix} 0 & 1 & 0 \\ -1 & 0 & 0 \\ 0 & 0 & 0 \end{bmatrix} \begin{bmatrix} \delta\theta_2 \\ 0 \\ 0 \end{bmatrix}$$

$$= -(\sin\theta_2 \cos\phi \sin\theta_1 + \cos\theta_2 \cos\theta_1)\,\delta\theta_1$$
$$+ (\cos\theta_1 \cos\theta_2 \cos\phi + \sin\theta_1 \sin\theta_2)\,\delta\theta_2 = 0 \quad \textbf{(9.4.21)}$$

Note that when $\phi = \pi/2$, hence $\cos\phi = 0$, for any $0 < \theta_2 < \pi/2$, Eq. 9.4.20 yields

$$\tan\theta_1 = 0$$

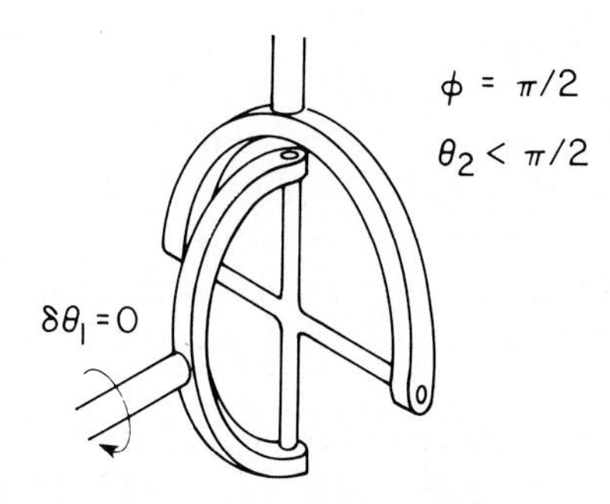

Figure 9.4.15 Singular behavior of universal joint.

which suggests that body 1 is locked. Similarly, from Eq. 9.4.21, with $\cos\phi = 0$,

$$\tan\theta_1\,\delta\theta_2 = \frac{2}{\tan\theta_2}\,\delta\theta_1$$

Since $\tan\theta_1 = 0$, this equation indicates that for any variation $\delta\theta_2$, $\delta\theta_1 = 0$. This singular behavior is interpreted as lock-up of the driving shaft, as illustrated in Fig. 9.4.15. Thus, the universal joint has a physically singular configuration, which must be avoided in applications.

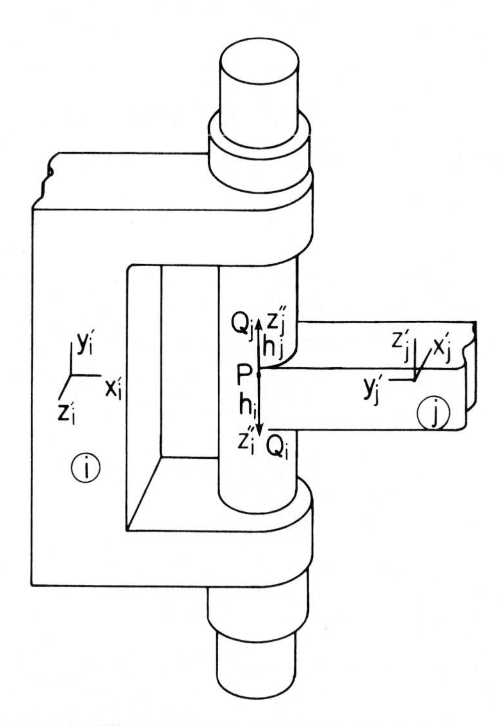

Figure 9.4.16 Revolute joint.

A *revolute joint* (or *rotational joint*) between bodies i and j is constructed with a bearing that allows relative rotation about a common axis, but precludes relative translation along this axis, as shown in Fig. 9.4.16. To define the revolute joint, the center of the joint is located on bodies i and j by points P_i and P_j. The axis of relative rotation is defined in bodies i and j by points Q_i and Q_j, and hence unit vectors $\mathbf{h}_i$ and $\mathbf{h}_j$ along the respective z'' axes of the joint definition frames. The remaining joint definition frame axes are defined at the convenience of the user.

The analytical formulation of the revolute joint is that points P_i and P_j coincide and that body-fixed vectors $\mathbf{h}_i$ and $\mathbf{h}_j$ are parallel, leading to the constraint equations

$$\begin{aligned} \mathbf{\Phi}^s(P_i, P_j) &= \mathbf{0} \\ \mathbf{\Phi}^{p1}(\mathbf{h}_i, \mathbf{h}_j) &= \mathbf{0} \end{aligned} \tag{9.4.22}$$

These five scalar constraint equations yield only one relative degree of freedom, rotation about the common axis of the bearing.

Example 9.4.8: Consider the revolute joint shown in Fig. 9.4.17. Since the orientation of body 1 does not change with respect to the x-y-z reference frame,

$$\mathbf{A}_1 = \mathbf{I}$$

From Fig. 9.4.17,

$$\begin{aligned} \mathbf{r}_1 &= \mathbf{r}_2 = \mathbf{0} \\ \mathbf{s}_1'^P &= \mathbf{s}_2'^P = \mathbf{0} \end{aligned}$$

Since body 2 rotates about the z axis, its orientation axis is

$$\mathbf{u}_2 = [0, 0, 1]^T$$

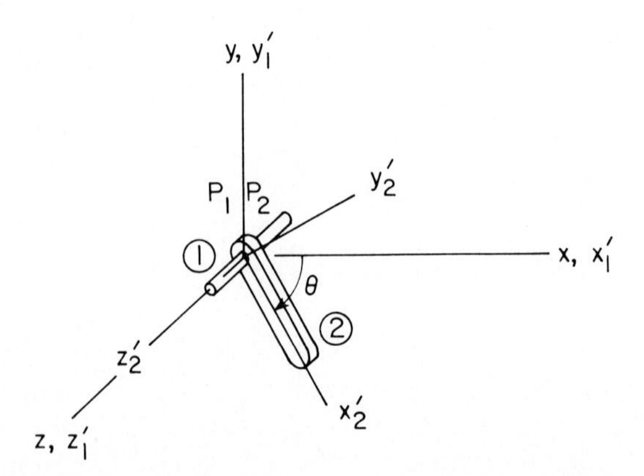

Figure 9.4.17 Revolute joint example.

Noting that rotation θ shown in Fig. 9.4.17 is clockwise, $\chi = -\theta$, and

$$e_0 = \cos\frac{\theta}{2}$$

$$\mathbf{e} = \left[0, 0, \ -\sin\frac{\theta}{2}\right]^T$$

Therefore,

$$\mathbf{A}_2 = 2\begin{bmatrix} e_0^2 - \frac{1}{2} & -e_0e_3 & 0 \\ e_0e_3 & e_0^2 - \frac{1}{2} & 0 \\ 0 & 0 & e_0^2 + e_3^2 - \frac{1}{2} \end{bmatrix}$$

$$= \begin{bmatrix} \cos\theta & \sin\theta & 0 \\ -\sin\theta & \cos\theta & 0 \\ 0 & 0 & 1 \end{bmatrix}$$

From Eq. 9.4.7,

$$\mathbf{\Phi}^s = \mathbf{r}_2 + \mathbf{A}_2\mathbf{s}_2'^P - \mathbf{r}_1 - \mathbf{A}_1\mathbf{s}_1'^P$$

$$= \begin{bmatrix} 0 \\ 0 \\ 0 \end{bmatrix} + \begin{bmatrix} \cos\theta & \sin\theta & 0 \\ -\sin\theta & \cos\theta & 0 \\ 0 & 0 & 1 \end{bmatrix}\begin{bmatrix} 0 \\ 0 \\ 0 \end{bmatrix} - \begin{bmatrix} 0 \\ 0 \\ 0 \end{bmatrix} - \mathbf{I}\begin{bmatrix} 0 \\ 0 \\ 0 \end{bmatrix} = \mathbf{0}$$

for all θ. From Eq. 9.4.9,

$$\mathbf{\Phi}^{p1}(\mathbf{h}_1, \mathbf{h}_2) = \begin{bmatrix} \Phi^{d1}(\mathbf{f}_1, \mathbf{h}_2) \\ \Phi^{d1}(\mathbf{g}_1, \mathbf{h}_2) \end{bmatrix} = \begin{bmatrix} \mathbf{f}_1'^T\mathbf{A}_1^T\mathbf{A}_2\mathbf{h}_2' \\ \mathbf{g}_1'^T\mathbf{A}_1^T\mathbf{A}_2\mathbf{h}_2' \end{bmatrix}$$

$$= \begin{bmatrix} \begin{bmatrix} 1 \\ 0 \\ 0 \end{bmatrix}^T \begin{bmatrix} \cos\theta & \sin\theta & 0 \\ -\sin\theta & \cos\theta & 0 \\ 0 & 0 & 1 \end{bmatrix}\begin{bmatrix} 0 \\ 0 \\ 1 \end{bmatrix} \\ \begin{bmatrix} 0 \\ 1 \\ 0 \end{bmatrix}^T \begin{bmatrix} \cos\theta & \sin\theta & 0 \\ -\sin\theta & \cos\theta & 0 \\ 0 & 0 & 1 \end{bmatrix}\begin{bmatrix} 0 \\ 0 \\ 1 \end{bmatrix} \end{bmatrix}$$

$$= \begin{bmatrix} [\cos\theta, \sin\theta, 0][0, 0, 1]^T \\ [-\sin\theta, \cos\theta, 0][0, 0, 1]^T \end{bmatrix} = \mathbf{0}$$

for all θ. Therefore, the revolute joint constraint equations of Eq. 9.4.22 are satisfied, for all θ.

A *cylindrical joint* is similar to a revolute joint in that it allows relative rotation about a common axis in two bodies. However, it does not preclude relative translation along this axis, as shown in Fig. 9.4.18. Joint definition points P_i and P_j are located on the common axis of rotation and additional points Q_i and Q_j on each body are defined along the axis of relative rotation, to establish the z''

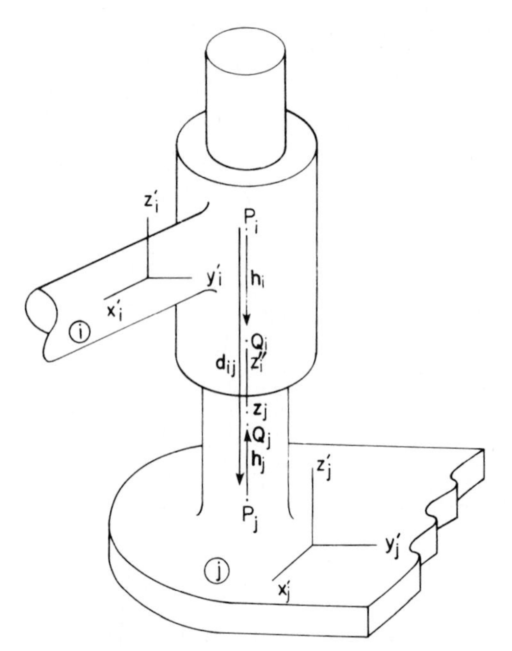

Figure 9.4.18 Cylindrical joint.

unit vectors $\mathbf{h}_i$ and $\mathbf{h}_j$ on bodies i and j, respectively. The remaining axes of the joint definition frames are defined at the convenience of the user.

The analytical definition of the cylindrical joint is that vectors $\mathbf{h}_i$ and $\mathbf{h}_j$ are collinear. Since these vectors and $\mathbf{d}_{ij}$ have points in common, collinearity is guaranteed by the conditions that $\mathbf{h}_i$ is parallel to both $\mathbf{h}_j$ and $\mathbf{d}_{ij}$, if $\mathbf{d}_{ij} \neq \mathbf{0}$; that is,

$$\begin{aligned} \mathbf{\Phi}^{p1}(\mathbf{h}_i, \mathbf{h}_j) &= \mathbf{0} \\ \mathbf{\Phi}^{p2}(\mathbf{h}_i, \mathbf{d}_{ij}) &= \mathbf{0} \end{aligned} \tag{9.4.23}$$

Note that even when $\mathbf{d}_{ij} = \mathbf{0}$, P_i and P_j coincide and vectors $\mathbf{h}_i$ and $\mathbf{h}_j$ have a point in common. Thus, the first parallelism condition of Eq. 9.4.23 implies collinearity, so Eq. 9.4.23 implies the geometry of the joint.

The cylindrical joint consists of four constraint equations; hence it allows two relative degrees of freedom between the bodies connected, relative rotation about the $\mathbf{h}_i = \mathbf{h}_j$ axis, and relative translation along this axis.

Example 9.4.9: Consider the cylindrical joint shown in Fig. 9.4.19. The only difference between this cylindrical joint and the revolute joint in Example 9.4.8 is

$$\mathbf{r}_2 = [0, 0, s]^T = \mathbf{d}_{12}$$

where s is a variable. It is shown in Example 9.4.8 that the parallel-2 constraint

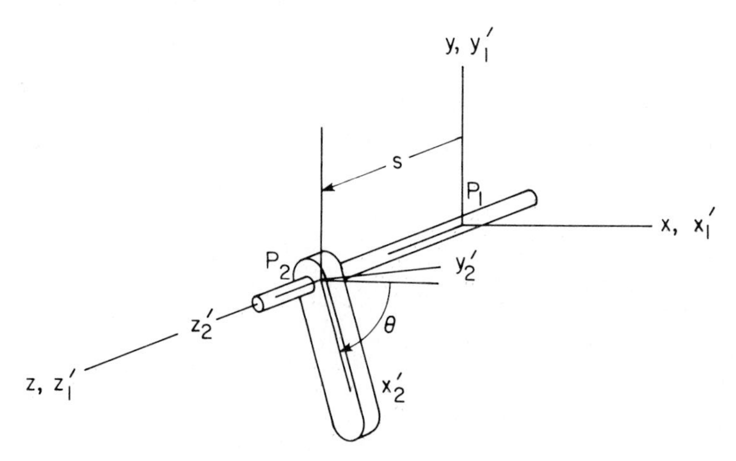

Figure 9.4.19 Cylindrical joint example.

of Eq. 9.4.23 is identically satisfied. From Eq. 9.4.10,

$$\Phi^{p2}(\mathbf{h}_1, \mathbf{d}_{12}) = \begin{bmatrix} \Phi^{d2}(\mathbf{f}_1, \mathbf{d}_{12}) \\ \Phi^{d2}(\mathbf{g}_1, \mathbf{d}_{12}) \end{bmatrix}$$

$$= \begin{bmatrix} \mathbf{f}_1'^T \mathbf{A}_1^T \mathbf{d}_{12} \\ \mathbf{g}_1'^T \mathbf{A}_1^T \mathbf{d}_{12} \end{bmatrix}$$

$$= \begin{bmatrix} \begin{bmatrix} 1 \\ 0 \\ 0 \end{bmatrix}^T \begin{bmatrix} \cos\theta & \sin\theta & 0 \\ -\sin\theta & \cos\theta & 0 \\ 0 & 0 & 1 \end{bmatrix} \begin{bmatrix} 0 \\ 0 \\ s \end{bmatrix} \\ \begin{bmatrix} 0 \\ 1 \\ 0 \end{bmatrix}^T \begin{bmatrix} \cos\theta & \sin\theta & 0 \\ -\sin\theta & \cos\theta & 0 \\ 0 & 0 & 1 \end{bmatrix} \begin{bmatrix} 0 \\ 0 \\ s \end{bmatrix} \end{bmatrix}$$

$$= \begin{bmatrix} [\cos\theta, \sin\theta, 0][0, 0, s]^T \\ [-\sin\theta, \cos\theta, 0][0, 0, s]^T \end{bmatrix} = \mathbf{0}$$

for all θ and s. Therefore, the cylindrical joint constraint of Eq. 9.4.23 is satisfied, for all θ and s.

The *translational joint* shown in Fig. 9.4.20 allows relative translation along a common axis between a pair of bodies, but precludes relative rotation about this axis. It is defined, as in the case of the cylindrical joint, by joint definition points P_i, Q_i, P_j, and Q_j along the axis of translation. The x'' axes of the joint definition frames on bodies i and j are selected so that they are perpendicular, defined by vectors $\mathbf{f}_i$ and $\mathbf{f}_j$, as shown in Fig. 9.4.20.

The analytical definition of a translational joint may be written by using the constraint equations of the cylindrical joint of Eq. 9.4.23 and adding the dot-1

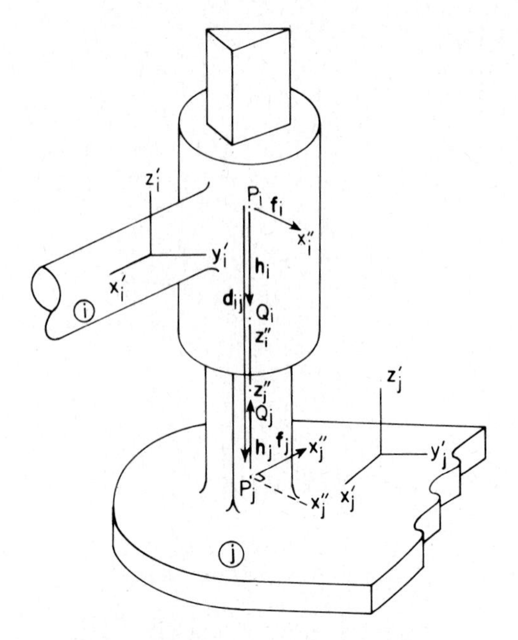

Figure 9.4.20 Translational joint.

condition that vectors $\mathbf{f}_i$ and $\mathbf{f}_j$ are orthogonal; that is,

$$\begin{aligned} \mathbf{\Phi}^{p1}(\mathbf{h}_i, \mathbf{h}_j) &= \mathbf{0} \\ \mathbf{\Phi}^{p2}(\mathbf{h}_i, \mathbf{d}_{ij}) &= \mathbf{0} \\ \Phi^{d1}(\mathbf{f}_i, \mathbf{f}_j) &= 0 \end{aligned} \tag{9.4.24}$$

Since this joint comprises five constraint equations, only one relative translational degree of freedom exists between the bodies.

Example 9.4.10: Consider the translational joint shown in Fig. 9.4.21, where

$$\mathbf{A}_1 = \mathbf{I}$$

$$\mathbf{A}_2 = \begin{bmatrix} 0 & -1 & 0 \\ 1 & 0 & 0 \\ 0 & 0 & 1 \end{bmatrix}$$

$$\mathbf{r}_1 = 0$$

$$\mathbf{r}_2 = [0, 0, s]^T$$

where s is the location of P_2 along the z axis. From Fig. 9.4.21,

$$\mathbf{s}_1'^P = \mathbf{s}_2'^P = \mathbf{0}$$

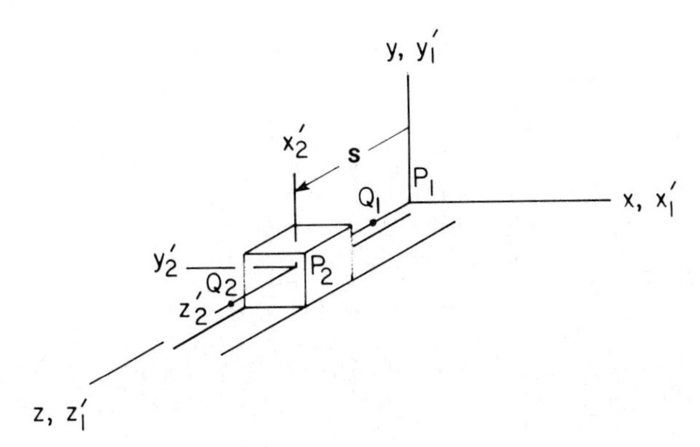

Figure 9.4.21 Translational joint example.

From Eq. 9.4.9,

$$\mathbf{\Phi}^{p1}(\mathbf{h}_1, \mathbf{h}_2) = \begin{bmatrix} \Phi^{d1}(\mathbf{f}_1, \mathbf{h}_2) \\ \Phi^{d1}(\mathbf{g}_1, \mathbf{h}_2) \end{bmatrix}$$

$$= \begin{bmatrix} [1, 0, 0]\begin{bmatrix} 0 & -1 & 0 \\ 1 & 0 & 0 \\ 0 & 0 & 1 \end{bmatrix}\begin{bmatrix} 0 \\ 0 \\ 1 \end{bmatrix} \\ [0, 1, 0]\begin{bmatrix} 0 & -1 & 0 \\ 1 & 0 & 0 \\ 0 & 0 & 1 \end{bmatrix}\begin{bmatrix} 0 \\ 0 \\ 1 \end{bmatrix} \end{bmatrix}$$

$$= \begin{bmatrix} [0, -1, 0][0, 0, 1]^T \\ [1, 0, 0][0, 0, 1]^T \end{bmatrix} = \mathbf{0}$$

From Eq. 9.4.10,

$$\mathbf{\Phi}^{p2}(\mathbf{h}_1, \mathbf{d}_{12}) = \begin{bmatrix} \Phi^{d2}(\mathbf{f}_1, \mathbf{d}_{12}) \\ \Phi^{d2}(\mathbf{g}_1, \mathbf{d}_{12}) \end{bmatrix}$$

$$= \begin{bmatrix} [1, 0, 0][0, 0, s]^T \\ [0, 1, 0][0, 0, s]^T \end{bmatrix} = \mathbf{0}$$

for all s. Finally, from Eq. 9.4.2,

$$\Phi^{d1}(\mathbf{f}_1, \mathbf{f}_2) = [1, 0, 0]\begin{bmatrix} 0 & -1 & 0 \\ 1 & 0 & 0 \\ 0 & 0 & 1 \end{bmatrix}\begin{bmatrix} 1 \\ 0 \\ 0 \end{bmatrix}$$

$$= [0, -1, 0][1, 0, 0]^T = 0$$

Therefore, the translational joint constraint equations of Eq. 9.4.24 are satisfied for all s.

The *screw joint* shown in Fig. 9.4.22 is a cylindrical joint between bodies i and j, with the additional condition that relative translation along the common axis of rotation is specified by a *screw pitch* α times the relative angle of rotation between the bodies. The relative angle θ of rotation is defined as the angle between the body-fixed x_i'' and x_j'' axes, counterclockwise taken as positive, including the cumulative angle of rotation. In terms of vectors shown in Fig. 9.4.22, the advance condition for the screw is

$$\mathbf{h}_i^T\mathbf{d}_{ij} = \alpha(\theta + 2n\pi - \theta_0) \tag{9.4.25}$$

where θ_0 is the angle between the body-fixed x'' axes when $P_i = P_j$ and n is the cumulative number of revolutions; that is, $0 \leq (1/\alpha)\mathbf{h}_i^T\mathbf{d}_{ij} - 2n\pi + \theta_0 \leq 2\pi$.

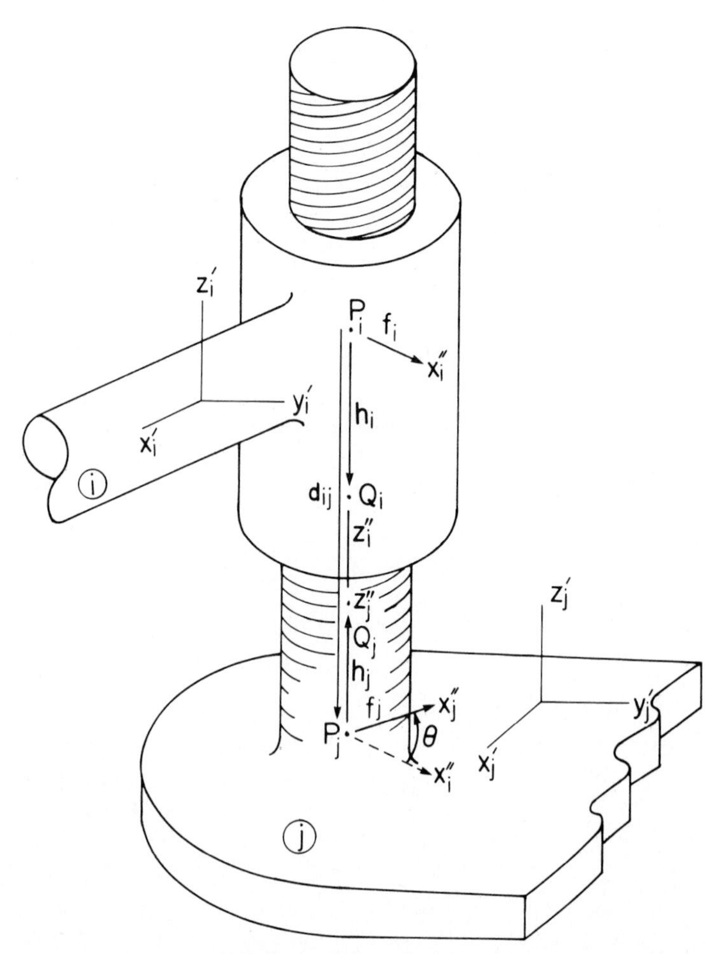

Figure 9.4.22 Screw joint.

The screw joint constraint equations are thus the cylindrical joint equations of Eq. 9.4.23 and

$$\Phi^{scr}(\mathbf{h}_i, \mathbf{d}_{ij}, \alpha, \theta_0) = \Phi^{d2}(\mathbf{h}_i, \mathbf{d}_{ij}) - \alpha(\theta + 2n\pi - \theta_0) = 0 \qquad \textbf{(9.4.26)}$$

The differential of Eq. 9.4.26 is calculated using the differential of the dot-2 constraint in Eq. 9.4.12 and the differential of θ in Eq. 9.2.61. Similarly, the partial derivatives of Φ^{scr} are obtained from the differentials of the dot-2 constraint in Table 9.4.1 and differentials of θ from Eq. 9.2.61, using $\delta\boldsymbol{\pi}' = 2\mathbf{G}\,\delta\mathbf{p}$ from Eq. 9.3.41, to obtain

$$\begin{aligned}
&\frac{\partial\theta}{\partial\mathbf{r}_i} = \frac{\partial\theta}{\partial\mathbf{r}_i} = \mathbf{0}\\
&\theta_{\pi_i'} = -\mathbf{h}_i'^T\\
&\frac{\partial\theta}{\partial\mathbf{p}_i} = -2\mathbf{h}_i'^T\mathbf{G}_i\\
&\theta_{\pi_j'} = \mathbf{h}_i'^T\mathbf{A}_i^T\mathbf{A}_j\\
&\frac{\partial\theta}{\partial\mathbf{p}_j} = 2\mathbf{h}_i'^T\mathbf{A}_i^T\mathbf{A}_j\mathbf{G}_j
\end{aligned} \qquad \textbf{(9.4.27)}$$

Example 9.4.11: Consider the cylindrical joint in Example 9.4.9 and Eq. 9.4.25, that is,

$$\begin{aligned}
\mathbf{h}_1^T\mathbf{d}_{12} &= \mathbf{h}_1'^T\mathbf{A}_1^T\mathbf{d}_{12}\\
&= [0, 0, 1]\mathbf{I}[0, 0, s]^T\\
&= [0, 0, 1][0, 0, s]^T\\
&= s
\end{aligned}$$

Therefore, Eq. 9.4.25 is satisfied if

$$s = \alpha(\theta + 2n\pi - \theta_0)$$

9.4.5 Composite Constraints between Pairs of Bodies

Quite often in applications, a pair of bodies is connected by an intermediate body (or *coupler*) that serves only to define the kinematic constraints between bodies that are connected, called a *composite joint.* In such cases, it is convenient and computationally efficient to derive equivalent kinematic constraints between the pair of bodies connected, without introducing the coupler as a separate body, with its associated Cartesian generalized coordinates. Several such composite joints are derived in this section.

The *spherical–spherical composite joint* shown in Fig. 9.4.23 consists of bodies i and j and a coupler that contains spherical joints at each end, attached to

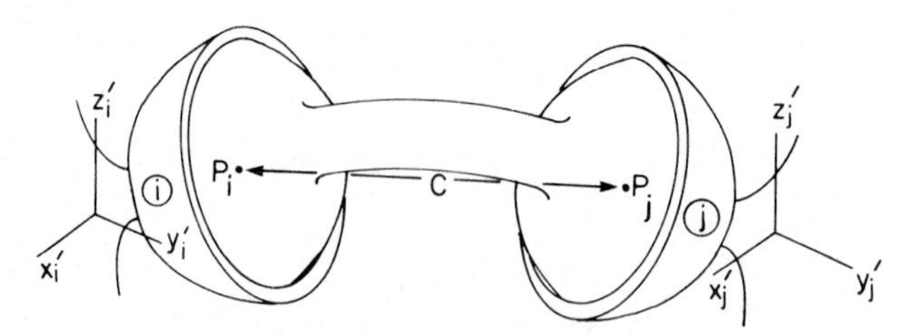

Figure 9.4.23 Spherical–spherical composite joint.

bodies i and j. The joint is defined by locating joint definition points P_i and P_j in bodies i and j, respectively, and by specifying the distance $C \neq 0$ between spherical joints on the coupler, as shown in Fig. 9.4.23. The orientation of joint definition frames is arbitrary, with defaults taken as parallel to the body-fixed reference frames.

The analytical definition of the spherical–spherical joint is simply that the distance between points P_i and P_j be equal to $C \neq 0$; that is,

$$\Phi^{ss}(P_i, P_j, C) = 0 \tag{9.4.28}$$

Since this joint is characterized by only one constraint equation, it allows five relative degrees of freedom between the bodies connected.

The *revolute–spherical composite joint* shown in Fig. 9.4.24 is comprised of a coupler with a spherical joint connected to body j and a revolute joint connected to body i. Joint definition points P_j and P_i define the centers of the respective joints on the pair of bodies. The axis of rotation in body i is defined by the z_i'' joint definition axis, and hence the unit vector $\mathbf{h}_i$. The remaining axes of the joint definition frames are arbitrary. It is required that the vector $\mathbf{d}_{ij} \neq \mathbf{0}$

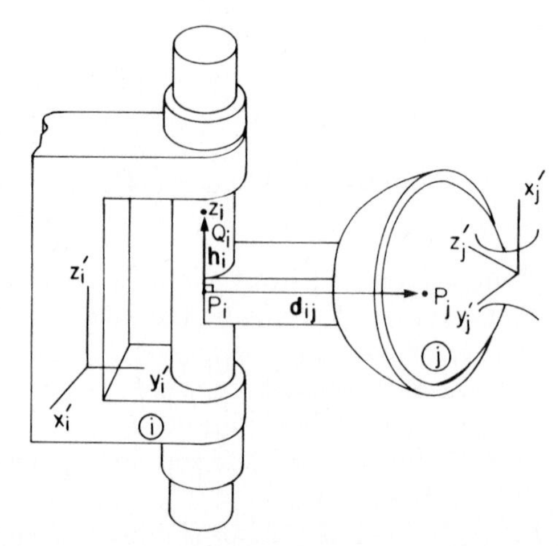

Figure 9.4.24 Revolute–spherical composite joint.

between points P_i and P_j be orthogonal to the axis of rotation of the revolute joint through P_i in the coupler.

The analytical definition of the revolute–spherical joint is that the distance between points P_i and P_j be equal to $C \neq 0$ and that vectors $\mathbf{h}_i$ and $\mathbf{d}_{ij}$ be orthogonal; that is,

$$\begin{aligned} \Phi^{ss}(P_i, P_j, C) &= 0 \\ \Phi^{d2}(\mathbf{h}_i, \mathbf{d}_{ij}) &= 0 \end{aligned} \tag{9.4.29}$$

Since two constraint equations characterize this joint, it allows four relative degrees of freedom between the bodies connected.

The *revolute–revolute composite joint with parallel axes* is defined by a coupler with parallel rotational joints that are at a fixed distance $C \neq 0$ apart, as shown in Fig. 9.4.25. Joint definition frames are fixed in bodies i and j, with their z'' axes along parallel axes of rotation, characterized by vectors $\mathbf{h}_i$ and $\mathbf{h}_j$. It is required that the vector $\mathbf{d}_{ij} \neq \mathbf{0}$ from P_i to P_j be orthogonal to each of the rotational axes. The remaining axes of the joint definition frames are arbitrary.

The analytical formulation of this composite joint is obtained by requiring that vectors $\mathbf{h}_i$ and $\mathbf{h}_j$ be parallel, that $\mathbf{h}_i$ and $\mathbf{d}_{ij}$ be orthogonal, and that the distance between points P_i and P_j be $C \neq 0$; that is,

$$\begin{aligned} \boldsymbol{\Phi}^{p1}(\mathbf{h}_i, \mathbf{h}_j) &= \mathbf{0} \\ \Phi^{d2}(\mathbf{h}_i, \mathbf{d}_{ij}) &= 0 \\ \Phi^{ss}(P_i, P_j, C) &= 0 \end{aligned} \tag{9.4.30}$$

Since this constraint is characterized by four scalar equations, it allows two relative degrees of freedom between the bodies connected.

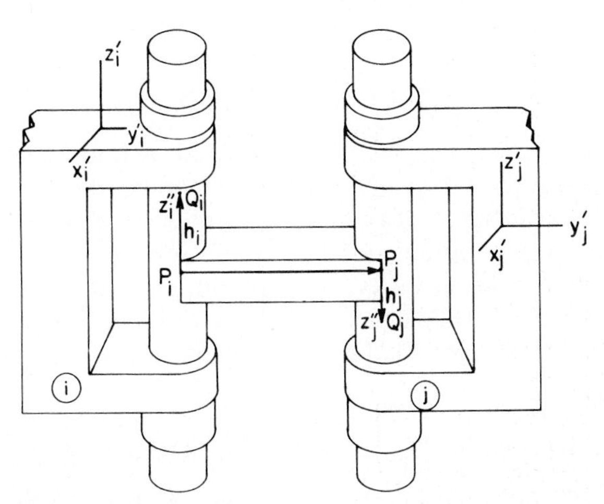

Figure 9.4.25 Revolute–revolute composite joint with parallel axes.

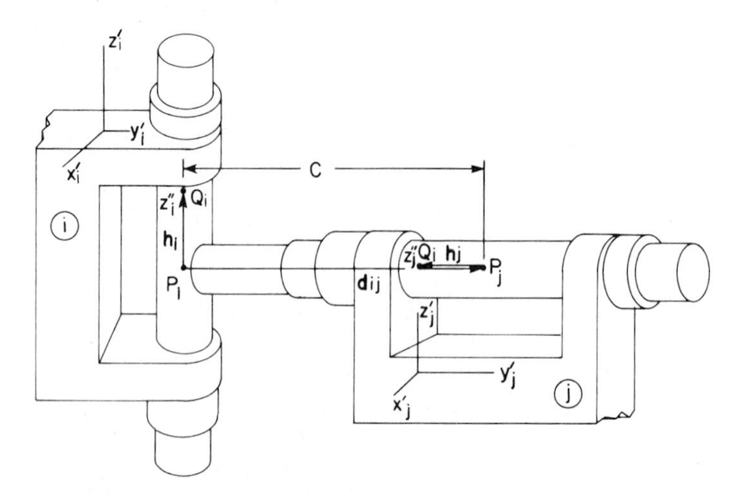

Figure 9.4.26 Revolute–revolute composite joint with orthogonal intersecting axes.

The *revolute–revolute composite joint with orthogonal intersecting axes* shown in Fig. 9.4.26 consists of a coupler with revolute joints connected to bodies i and j. In this case, the axes of the revolute joints, z_i'' and z_j'', intersect and are orthogonal. The z'' axes in bodies i and j are characterized by unit vectors $\mathbf{h}_i$ and $\mathbf{h}_j$, respectively. The remaining axes of the joint definition frames are arbitrary.

The analytical definition of this joint is obtained by requiring that $\mathbf{h}_i$ be perpendicular to $\mathbf{d}_{ij} \neq \mathbf{0}$, $\boldsymbol{h}_j$ be parallel to $\mathbf{d}_{ij}$, and the distance between points P_i and P_j be $C \neq 0$; that is,

$$
\begin{aligned}
\Phi^{d2}(\mathbf{h}_i, \mathbf{d}_{ij}) &= 0 \\
\boldsymbol{\Phi}^{p2}(\mathbf{h}_j, \mathbf{d}_{ji}) &= \mathbf{0} \\
\Phi^{ss}(P_i, P_j, C) &= 0
\end{aligned}
\tag{9.4.31}
$$

Since this composite joint is characterized by four scalar equations, it allows two relative degrees of freedom between the bodies connected.

The *revolute–cylindrical composite joint* shown in Fig. 9.4.27 consists of a coupler that is constrained to body i by a revolute joint about the $\mathbf{h}_i$ axis on body i and to body j through a cylindrical joint about the $\mathbf{h}_j$ axis. Vectors $\mathbf{h}_i$ and $\mathbf{h}_j$ are required to be orthogonal. The geometrical conditions that define the revolute–cylindrical joint are that $\mathbf{h}_j$ be perpendicular to $\mathbf{h}_i$ and that $\mathbf{h}_j$ pass through point P_i.

The analytical definition of this joint is obtained by requiring that $\mathbf{h}_i$ and $\mathbf{h}_j$ be perpendicular and that, providing $\mathbf{d}_{ij} \neq \mathbf{0}$, $\mathbf{h}_j$ be parallel to $\mathbf{d}_{ij}$; that is,

$$
\begin{aligned}
\Phi^{d1}(\mathbf{h}_i, \mathbf{h}_j) &= 0 \\
\boldsymbol{\Phi}^{p2}(\mathbf{h}_j, \mathbf{d}_{ji}) &= \mathbf{0}
\end{aligned}
\tag{9.4.32}
$$

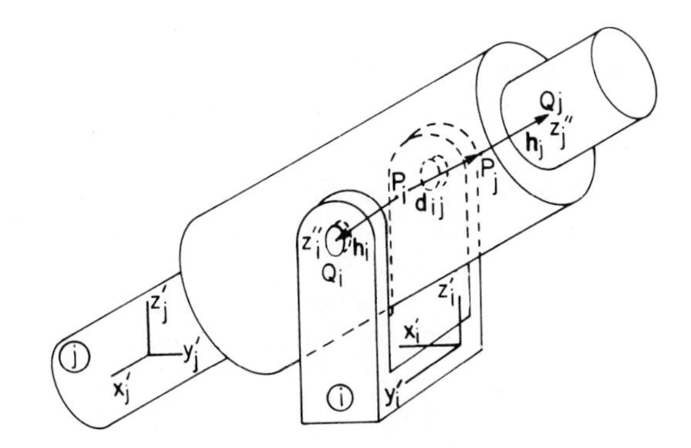

Figure 9.4.27 Revolute–cylindrical composite joint.

Note that if $\mathbf{d}_{ij} = \mathbf{0}$ then $P_i = P_j$ and the geometric conditions of the joint are satisfied. Thus, Eq. 9.4.32 is equivalent to the geometric definition of the joint. Since three scalar constraint equations comprise the definition of the revolute–cylindrical joint, there are three relative degrees of freedom between bodies i and j.

The *revolute–translational composite joint* shown in Fig. 9.4.28 is comprised of bodies i and j and a coupler between the bodies that is pivoted about $\mathbf{h}_i$ in body i. Body j can translate along vector $\mathbf{h}_j$, but cannot rotate about this axis relative to the coupler. The geometric conditions that define the revolute–translational composite joint include the conditions of the revolute–cylindrical joint, but in addition require that $\mathbf{f}_j$ and $\mathbf{h}_i$ remain parallel.

Analytical conditions that define the revolute–translational joint are that vectors $\mathbf{h}_i$ and $\mathbf{f}_j$ be parallel and that, if $\mathbf{d}_{ij} \neq \mathbf{0}$, it must be parallel to vector $\mathbf{h}_j$: that is,

$$
\begin{aligned}
\mathbf{\Phi}^{p1}(\mathbf{h}_i, \mathbf{f}_j) &= \mathbf{0} \\
\mathbf{\Phi}^{p2}(\mathbf{h}_j, \mathbf{d}_{ij}) &= \mathbf{0}
\end{aligned}
\qquad \textbf{(9.4.33)}
$$

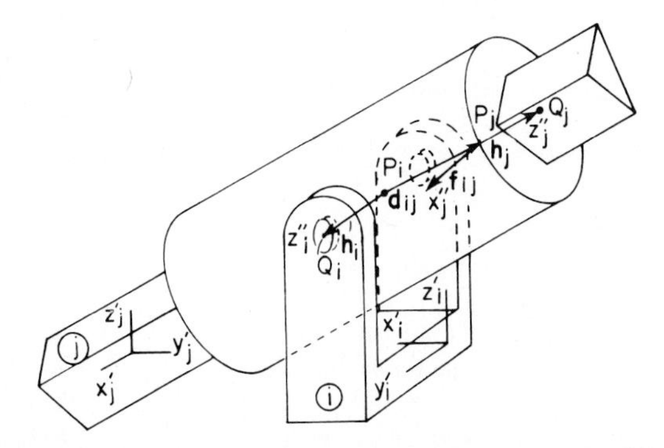

Figure 9.4.28 Revolute–translational composite joint.

In the case where $\mathbf{d}_{ij} = \mathbf{0}$, the second of Eqs. 9.4.33 reduces to the condition that $P_i = P_j$; hence $\mathbf{h}_j$ passes through point P_i. The four scalar constraint equations of Eq. 9.4.33 allow two relative degrees of freedom between bodies i and j.

Example 9.4.12: Consider the slider–crank mechanism shown in Fig. 9.4.29, where the slider (body 2) is to translate along the global x axis. Several kinematically equivalent models can be created using composite joints as couplers between bodies 1 and 2 and different combinations of absolute constraints on the slider. Each model has two bodies, suppressing ground, yielding 14 generalized coordinates.

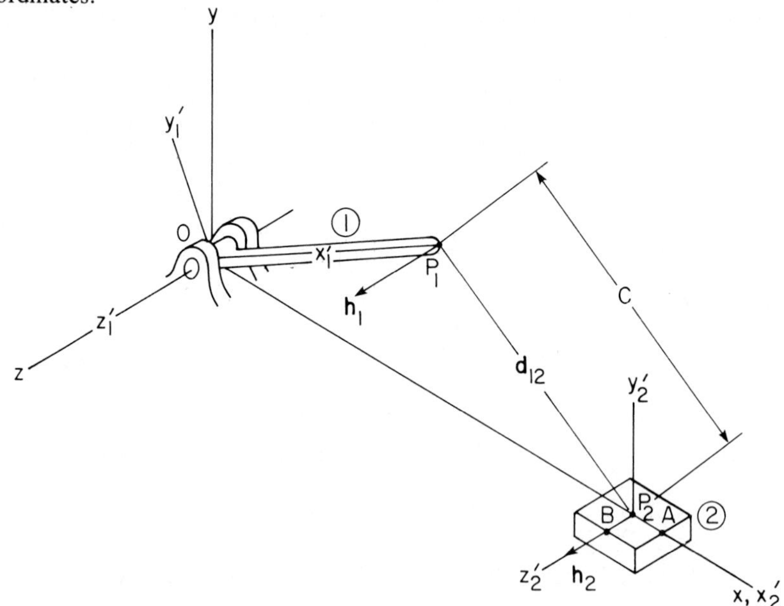

Figure 9.4.29 Slider–crank with composite joint.

(1) Model 1: Spherical-Spherical Joint	
Body 2: absolute constraints	
$y_2'^P = 0$	1
$z_2'^P = 0$	1
$e_1 = 0$	1
$e_2 = 0$	1
$e_3 = 0$	1
Spherical–spherical joint (coupler)	1
Revolute joint at O	5
Euler parameter normalization constraints	2
	13
DOF = 14 − 13 = 1.	

(2) Model 2: Revolute–Spherical Joint	
Body 2: absolute constraints	
$y_2^P = 0$	1
$e_1 = 0$	1
$e_2 = 0$	1
$e_3 = 0$	1
Revolute–spherical joint (coupler)	2
Revolute joint at O	5
Euler parameter normalization constraints	2
	13
DOF = 14 − 13 = 1.	

(3) Model 3: Revolute–Revolute Joint (Parallel Axes)

Vectors $\mathbf{h}_1$ and $\mathbf{h}_2$ shown in Fig. 9.4.29 define the revolute–revolute joint; that is, it is required that $\mathbf{h}_1$ and $\mathbf{h}_2$ be parallel.

Body 2: absolute constraints	
$y_2^P = 0$	1
$y_2^A = 0$	1
Revolute–revolute joint (coupler)	4
Revolute joint at O	5
Euler parameter normalization constraints	2
	13
DOF = 14 − 13 = 1.	

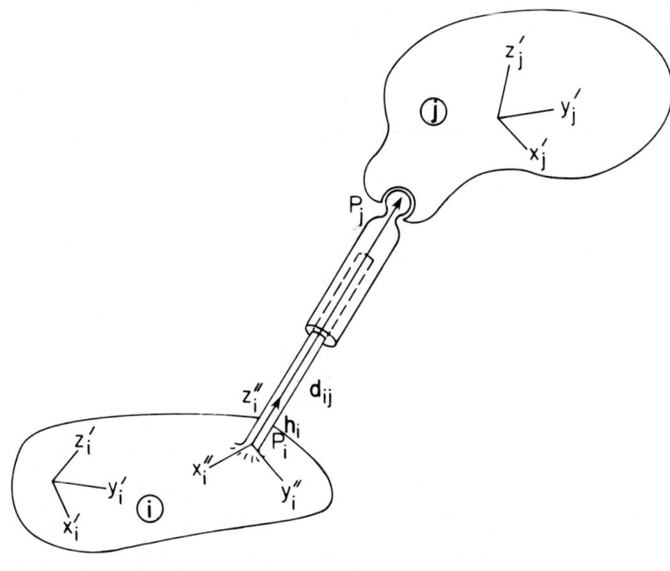

Figure 9.4.30 Strut composite joint.

The *strut composite joint* shown in Fig. 9.4.30 consists of a coupler that has a cylindrical joint about the fixed vector $\mathbf{h}_i$ in body i and a spherical joint at point P_j in body j. Point P_j on the strut lies on the axis of the coupler's cylindrical joint. The geometrical definition of this composite joint is that vector $\mathbf{h}_i$ passes through point P_j.

Providing $\mathbf{d}_{ij} \neq \mathbf{0}$, the geometry of the joint is guaranteed by the condition that $\mathbf{h}_i$ and $\mathbf{d}_{ij}$ be parallel; that is,

$$\mathbf{\Phi}^{p2}(\mathbf{h}_i, \mathbf{d}_{ij}) = \mathbf{0} \tag{9.4.34}$$

Note that when $\mathbf{d}_{ij} = \mathbf{0}$ points P_j and P_i coincide, so $\mathbf{h}_i$ passes through P_j. Thus, Eq. 9.4.34 defines the geometry of the strut, even when $\mathbf{d}_{ij} = \mathbf{0}$.

9.5 DRIVING CONSTRAINTS

Actuators may be employed to specify the time history of the position or orientation of one body relative to another or relative to ground in a mechanical system. Such time-dependent constraints are called *driving constraints*. A library of driving constraints that may be used for the kinematic analysis of mechanical systems is presented in this section.

In the case where constraints placed on the x, y, or z coordinates of point P_i on body i, as shown in Fig. 9.4.8, are dependent on time, the first three constraints of Eq. 9.4.15 may be written as time-dependent *absolute drivers*; that is,

$$\begin{aligned} \Phi^{1d} &= x_i^P - C_1(t) = 0 \\ \Phi^{2d} &= y_i^P - C_2(t) = 0 \\ \Phi^{3d} &= z_i^P - C_3(t) = 0 \end{aligned} \tag{9.5.1}$$

Each of these drivers restricts one degree of freedom of the motion of body i, relative to the x-y-z global reference frame.

In the case where the distance $C \neq 0$ defined in Fig. 9.4.11 and Eq. 9.4.17 is specified as a function of time (e.g., by a hydraulic or electrical actuator), a time-dependent *distance driver* is defined of the form

$$\Phi^{ssd} = \Phi^{ss}(P_i, P_j, 0) - (C(t))^2 \tag{9.5.2}$$

where $C(t) \neq 0$ is the distance between points P_i and P_j that are connected by the actuator. This scalar constraint equation eliminates one relative degree of freedom between bodies i and j.

If a pair of bodies is connected by a translational, cylindrical, screw, or strut joint, there is an axis of relative translation defined by the collinear body-fixed vector $\mathbf{h}_i$ and vector $\mathbf{d}_{ij}$, as shown in Fig. 9.5.1. The directed distance from point P_i to P_j is $\mathbf{h}_i^T\mathbf{d}_{ij}$, which is constrained to be equal to a specified function $C(t)$. In the notation of the dot-2 constraint function of Eq. 9.4.6, the *relative translational*

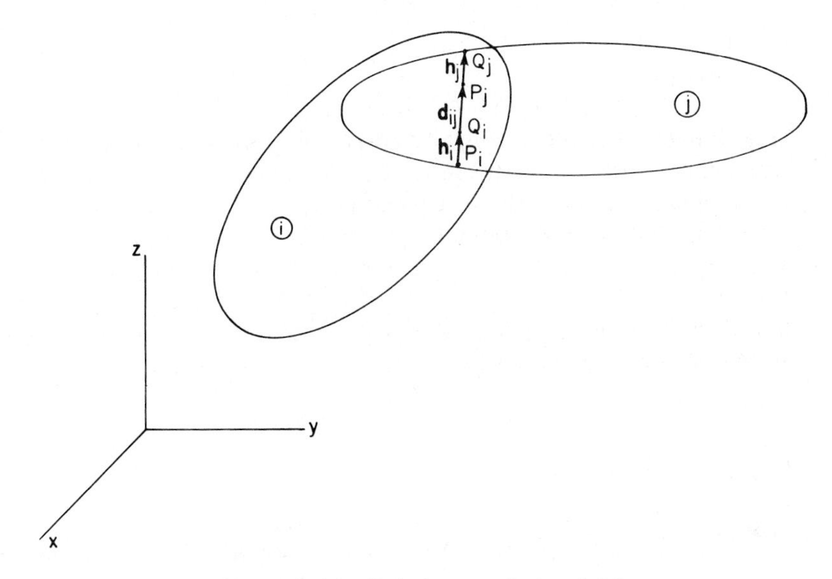

Figure 9.5.1 Relative translational driver.

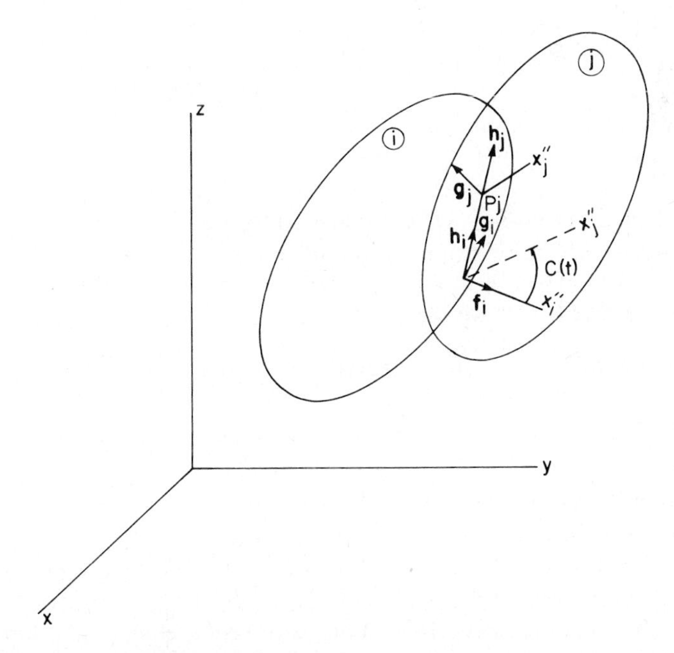

Figure 9.5.2 Relative rotational driver.

driver in translational, cylindrical, and strut screw joints is

$$\Phi^{td} = \Phi^{d2}(\mathbf{h}_i, \mathbf{d}_{ij}) - C(t) = 0 \tag{9.5.3}$$

Note that $C(t)$ can be zero or negative. This driving constraint is added to the kinematic constraint equations associated with a translational, cylindrical, screw, or strut joint to impose the driving condition.

The angle from the body-fixed x_i'' axis to the x_j'' axis, measured counterclockwise as positive, in a revolute, cylindrical, or screw joint is specified as $C(t)$, as shown in Fig. 9.5.2. The analytical definition of the *relative rotational driver* in revolute, cylindrical, and screw joints is obtained using the relative angle of rotation of Eq. 9.2.31; that is,

$$\Phi^{rotd} \equiv \theta + 2n\pi - C(t) = 0 \tag{9.5.4}$$

where n is the number of revolutions that have occurred; that is, $0 \leq C(t) - 2n\pi < 2\pi$.

Variations of each of the foregoing driving constraints are identical to the previously derived variations of corresponding kinematic constraints, since time is suppressed in calculating the variations with respect to position and orientation.

9.6 POSITION, VELOCITY, AND ACCELERATION ANALYSIS

A subtle distinction between the kinematic analysis of spatial and planar systems involves the use of angular velocity and angular acceleration variables in spatial analysis that are not time derivatives of generalized coordinates. Since position analysis requires direct computation with generalized coordinates, a different set of variables is employed for position and for velocity and acceleration analysis. The formulation of the constraint equations and their variations from Sections 9.4 and 9.5, together with the basic relations obtained in Sections 9.1 through 9.3, are used here to derive expressions that contribute to the equations for position, velocity, and acceleration of spatial systems.

9.6.1 Position Analysis

The generalized coordinate vector for body i in a system is

$$\mathbf{q}_i = \begin{bmatrix} \mathbf{r}_i \\ \mathbf{p}_i \end{bmatrix} \tag{9.6.1}$$

The composite set of generalized coordinates for the entire system is

$$\mathbf{q} = [\mathbf{q}_1^T, \mathbf{q}_2^T, \ldots, \mathbf{q}_{nb}^T]^T \tag{9.6.2}$$

where nb is the number of bodies in the system.

In addition to the kinematic and driving constraints derived in Sections 9.4 and 9.5, the Euler parameter generalized coordinates of each body must satisfy the normalization constraint of Eq. 9.3.9:

$$\Phi_i^{\mathbf{p}} = \mathbf{p}_i^T \mathbf{p}_i - 1 = 0, \qquad i = 1, \ldots, nb \tag{9.6.3}$$

The vector of *Euler parameter normalization constraint* equations is thus

$$\mathbf{\Phi}^{\mathbf{p}} = [\Phi_1^{\mathbf{p}}, \ldots, \Phi_{nb}^{\mathbf{p}}]^T = \mathbf{0} \tag{9.6.4}$$

The combined system of kinematic, driving, and Euler parameter normalization constraint equations that determines the position and orientation of the system is

$$\mathbf{\Phi}(\mathbf{q}, t) \equiv \begin{bmatrix} \mathbf{\Phi}^K(\mathbf{q}) \\ \mathbf{\Phi}^D(\mathbf{q}, t) \\ \mathbf{\Phi}^{\mathbf{p}}(\mathbf{q}) \end{bmatrix} = \mathbf{0} \tag{9.6.5}$$

It is presumed, for the purpose of kinematic analysis, that an adequate number of independent driving constraints has been specified so that Eq. 9.6.5 comprises $7nb$ equations in $7nb$ generalized coordinates.

To solve the nonlinear position equations in Eq. 9.6.5, the Jacobian matrix of the system must be calculated. First, the gradient of the kinematic constraint equations with respect to the generalized coordinates of body i is

$$\mathbf{\Phi}_{\mathbf{q}_i}^K = [\mathbf{\Phi}_{\mathbf{r}_i}^K, \mathbf{\Phi}_{\mathbf{p}_i}^K] = [\mathbf{\Phi}_{\mathbf{r}_i}^K, 2\mathbf{\Phi}_{\boldsymbol{\pi}_i'}^K \mathbf{G}_i] \tag{9.6.6}$$

where the matrices $\mathbf{\Phi}_{\mathbf{r}_i}^K$ and $\mathbf{\Phi}_{\boldsymbol{\pi}_i'}^k$ are derived for each of the kinematic constraints in Section 9.4. Similarly, the gradient of driving constraints with respect to the generalized coordinates of body i is

$$\mathbf{\Phi}_{\mathbf{q}_i}^D = [\mathbf{\Phi}_{\mathbf{r}_i}^D, \mathbf{\Phi}_{\mathbf{p}_i}^D] = [\mathbf{\Phi}_{\mathbf{r}_i}^D, 2\mathbf{\Phi}_{\boldsymbol{\pi}_i'}^D \mathbf{G}_i] \tag{9.6.7}$$

where the matrices $\mathbf{\Phi}_{\mathbf{r}_i}^D$ and $\mathbf{\Phi}_{\boldsymbol{\pi}_i'}^D$ are calculated in Section 9.5 for each of the driving constraints. Finally, a direct calculation of the ith Euler parameter normalization constraint of Eq. 9.6.3 yields

$$\Phi_{i_{\mathbf{q}_i}}^{\mathbf{p}} = [\mathbf{0}, 2\mathbf{p}_i^T] \tag{9.6.8}$$

For purposes of velocity and acceleration analysis, the variation of the Euler parameter normalization constraint is obtained, using the differential of Eqs. 9.6.3 and 9.3.43, as

$$\delta\Phi_i^{\mathbf{p}} = 2\mathbf{p}_i^T\,\delta\mathbf{p}_i = \mathbf{p}_i^T\mathbf{G}_i^T\,\delta\boldsymbol{\pi}_i' = 0$$

for arbitrary $\delta\boldsymbol{\pi}_i'$. Thus, using Eq. 9.3.21,

$$\Phi_{i_{\boldsymbol{\pi}_i'}}^{\mathbf{p}} = \mathbf{p}_i^T\mathbf{G}_i^T = \mathbf{0} \tag{9.6.9}$$

The derivatives of kinematic, driving, and Euler parameter normalization constraints of Eq. 9.6.6 to 9.6.8 may be combined to form the *Jacobian* of the constraint equation of Eq. 9.6.5 as

$$\mathbf{\Phi_q} = \begin{bmatrix} \mathbf{\Phi_q^K} \\ \mathbf{\Phi_q^D} \\ \mathbf{\Phi_q^P} \end{bmatrix} \tag{9.6.10}$$

If the kinematic, driving, and Euler parameter normalization constraints are independent and if DOF drivers have been specified, the Jacobian is nonsingular. Thus, provided that the system can be assembled at a nominal position, the implicit function theorem guarantees the existence of a unique solution for the position and orientation of the system in a neighborhood of the assembled configuration.

Even more pronounced than for planar kinematics in Chapter 3, the kinematic constraint equations for spatial systems are highly nonlinear and complicated, yielding little potential for an explicit solution. Therefore, an iterative technique is adopted for the solution of Eq. 9.6.5. The *Newton-Raphson method* is most commonly used. It is based on iterative computations, according to the algorithm of Section 4.5; that is,

$$\begin{aligned} \mathbf{\Phi_q}\, \Delta\mathbf{q}^{(j)} &= -\mathbf{\Phi}(\mathbf{q}^{(j)}, t) \\ \mathbf{q}^{(j+1)} &= \mathbf{q}^{(j)} + \Delta\mathbf{q}^{(j)} \end{aligned} \tag{9.6.11}$$

where $\mathbf{q}^{(0)}$ is the initial estimate of the assembled configuration and improved estimates are obtained by solving the sequence of equations in Eq. 9.6.11, until convergence is obtained. Numerical methods for solving these equations are presented in Chapter 4.

9.6.2 Velocity Analysis

Note, from Eq. 9.6.9, that the velocity equation in $\boldsymbol{\omega}'$ variables associated with the Euler parameter normalization constraint is identically satisfied and can, therefore, be ignored. Since the Euler parameter normalization velocity equation in $\boldsymbol{\omega}'$ is identically satisfied, differentiation of this equation yields an identically satisfied acceleration equation in $\dot{\boldsymbol{\omega}}'$. Therefore, for purposes of velocity and acceleration analysis, in terms of angular velocities and angular accelerations, the Euler parameter normalization constraint is not required.

Taking time derivatives of the kinematic and driving constraint equations yields the *velocity equations*

$$\sum_{i=1}^{nb} \left\{ \begin{bmatrix} \mathbf{\Phi}_{\mathbf{r}_i}^K \\ \mathbf{\Phi}_{\mathbf{r}_i}^D \end{bmatrix} \dot{\mathbf{r}}_i + \begin{bmatrix} \mathbf{\Phi}_{\boldsymbol{\pi}_i'}^K \\ \mathbf{\Phi}_{\boldsymbol{\pi}_i'}^D \end{bmatrix} \boldsymbol{\omega}_i' \right\} = \begin{bmatrix} -\mathbf{\Phi}_t^K \\ -\mathbf{\Phi}_t^D \end{bmatrix} \equiv \begin{bmatrix} \mathbf{v}^K \\ \mathbf{v}^D \end{bmatrix} \tag{9.6.12}$$

Since time does not arise explicitly in the kinematic constraint equations,

$$-\mathbf{\Phi}_t^K = \mathbf{0} \equiv \mathbf{v}^K \tag{9.6.13}$$

Time does, however, arise explicitly in the driving constraints, requiring that $\mathbf{v}^D$

be calculated for each of the driving constraints presented in Section 9.5. From Eq. 9.5.1,

$$\begin{bmatrix} v^{1d} \\ v^{2d} \\ v^{3d} \end{bmatrix} = -\begin{bmatrix} \Phi_t^{1d} \\ \Phi_t^{2d} \\ \Phi_t^{3d} \end{bmatrix} = \begin{bmatrix} \dot{C}_1(t) \\ \dot{C}_2(t) \\ \dot{C}_3(t) \end{bmatrix} \tag{9.6.14}$$

From Eq. 9.5.2,

$$v^{ssd} = -\Phi_t^{ssd} = 2C(t)\dot{C}(t) \tag{9.6.15}$$

From Eq. 9.5.3,

$$v^{td} = -\Phi_t^{td} = \dot{C}(t) \tag{9.6.16}$$

Finally, from Eq. 9.5.4,

$$v^{rotd} = -\Phi_t^{rotd} = \dot{C}(t) \tag{9.6.17}$$

These vectors and Eq. 9.6.13 form the right side of the velocity equation in Eq. 9.6.12, for $\dot{\mathbf{r}}_i$ and $\boldsymbol{\omega}_i'$. If $\dot{\mathbf{p}}_i$ is desired, it can be obtained from Eq. 9.3.35.

It is interesting to note that the coefficient matrix in the velocity equations of Eq. 9.6.12 is different from the Jacobian matrix in position analysis. Therefore, when velocity analysis is carried out using angular velocities, the velocity coefficient matrix must be employed and care must be taken that both the Jacobian for position analysis and the coefficient matrix in velocity analysis be nonsingular.

9.6.3 Acceleration Analysis

Kinematic acceleration equations are obtained by differentiating the velocity equations of Eq. 9.6.12. Since the terms arising in Eq. 9.6.12 involve the rotation matrix $\mathbf{A}$ for each body in the system, it is helpful to recall the identity of Eq. 9.2.40:

$$\dot{\mathbf{A}} = \mathbf{A}\tilde{\boldsymbol{\omega}}' \tag{9.6.18}$$

Note that explicit time dependence in constraint equations arises only in the driving constraints of Section 9.5 and that, in each of these constraints, the functions involved are sums of functions that depend on only generalized coordinates or time. Thus, $\boldsymbol{\Phi}_{\mathbf{q}t} = \mathbf{0}$ and differentiation of Eq. 9.6.12 yields the *acceleration equation*

$$\sum_{i=1}^{nb} \left\{ \begin{bmatrix} \boldsymbol{\Phi}_{\mathbf{r}_i}^K \\ \boldsymbol{\Phi}_{\mathbf{r}_i}^D \end{bmatrix} \ddot{\mathbf{r}}_i + \begin{bmatrix} \boldsymbol{\Phi}_{\boldsymbol{\pi}_i}^K \\ \boldsymbol{\Phi}_{\boldsymbol{\pi}_i'}^D \end{bmatrix} \dot{\boldsymbol{\omega}}_i' \right\} = -\begin{bmatrix} \boldsymbol{\Phi}_{tt}^K \\ \boldsymbol{\Phi}_{tt}^D \end{bmatrix} - \sum_{i=1}^{nb} \left\{ \begin{bmatrix} \dot{\boldsymbol{\Phi}}_{\mathbf{r}_i}^K \\ \dot{\boldsymbol{\Phi}}_{\mathbf{r}_i}^D \end{bmatrix} \dot{r}_i + \begin{bmatrix} \dot{\boldsymbol{\Phi}}_{\boldsymbol{\pi}_i'}^K \\ \dot{\boldsymbol{\Phi}}_{\boldsymbol{\pi}_i'}^D \end{bmatrix} \boldsymbol{\omega}_i' \right\} \equiv \begin{bmatrix} \gamma^K \\ \gamma^D \end{bmatrix} \tag{9.6.19}$$

To evaluate the terms in Eq. 9.6.19, the matrices presented in Table 9.4.1 and in related constraint variation identities derived in Sections 9.4 and 9.5 may be differentiated to evaluate terms on the right of Eq. 9.6.19. For the dot-1 constraint, differentiation of terms arising in Table 9.4.1, using Eq. 9.6.18, yields (Prob. 9.6.1)

$$\begin{aligned}\gamma^{d1} &= \mathbf{a}_j'^T[\mathbf{A}_j^T\mathbf{A}_i\tilde{\omega}_i' - \tilde{\omega}_j'\mathbf{A}_j^T\mathbf{A}_i]\tilde{\mathbf{a}}_i'\omega_i' \\ &\quad + \mathbf{a}_i'^T[\mathbf{A}_i^T\mathbf{A}_j\tilde{\omega}_j' - \tilde{\omega}_i'\mathbf{A}_i^T\mathbf{A}_j]\tilde{\mathbf{a}}_j'\omega_j' \\ &= -\mathbf{a}_j'^T[\mathbf{A}_j^T\mathbf{A}_i\tilde{\omega}_i'\tilde{\omega}_i' + \tilde{\omega}_j'\tilde{\omega}_j'\mathbf{A}_j^T\mathbf{A}_i]\mathbf{a}_i \\ &\quad + 2\omega_j'^T\tilde{\mathbf{a}}_j'\mathbf{A}_j^T\mathbf{A}_i\tilde{\mathbf{a}}_i'\omega_i'\end{aligned} \tag{9.6.20}$$

Similarly, for the dot-2 constraint (Prob. 9.6.2),

$$\begin{aligned}\gamma^{d2} &= -\mathbf{a}_i'^T\tilde{\omega}_i'\mathbf{A}_i^T(\dot{\mathbf{r}}_i - \dot{\mathbf{r}}_j) \\ &\quad + [(\dot{\mathbf{r}}_j + \mathbf{A}_j\tilde{\omega}_j'\mathbf{s}_j'^P - \dot{\mathbf{r}}_i - \mathbf{A}_i\tilde{\omega}_i'\mathbf{s}_i'^P)^T\mathbf{A}_i\tilde{\mathbf{a}}_i' + \mathbf{d}_{ij}^T\mathbf{A}_i\tilde{\omega}_i'\tilde{\mathbf{a}}_i']\omega_i' \\ &\quad + \mathbf{a}_i'^T[\mathbf{A}_i^T\mathbf{A}_j\tilde{\omega}_j' - \tilde{\omega}_i'\mathbf{A}_i^T\mathbf{A}_j]\tilde{\mathbf{s}}_j'^P\omega_j' \\ &= 2\omega_i'^T\tilde{\mathbf{a}}_i'\mathbf{A}_i^T(\dot{\mathbf{r}}_i - \dot{\mathbf{r}}_j) + 2\mathbf{s}_j'^{PT}\tilde{\omega}_j'\mathbf{A}_j^T\mathbf{A}_i\tilde{\omega}_i'\mathbf{a}_i' \\ &\quad -\mathbf{s}_i'^{PT}\tilde{\omega}_i'\tilde{\omega}_i'\mathbf{a}_i' - \mathbf{s}_j'^{PT}\tilde{\omega}_j'\tilde{\omega}_j'\mathbf{A}_j^T\mathbf{A}_i\mathbf{a}_i' - \mathbf{d}_{ij}^T\mathbf{A}_i\tilde{\omega}_i'\tilde{\omega}_i'\mathbf{a}_i'\end{aligned} \tag{9.6.21}$$

For the spherical constraint (Prob. 9.6.3),

$$\begin{aligned}\gamma^s &= -\mathbf{A}_i\tilde{\omega}_i'\tilde{\mathbf{s}}_i'^P\omega_i' + \mathbf{A}_j\tilde{\omega}_j'\tilde{\mathbf{s}}_j'^P\omega_j \\ &= \mathbf{A}_i\tilde{\omega}_i'\tilde{\omega}_i'\mathbf{s}_i'^P - \mathbf{A}_j\tilde{\omega}_j'\tilde{\omega}_j'\mathbf{s}_j'^P\end{aligned} \tag{9.6.22}$$

Finally, for the spherical–spherical constraint (Prob. 9.6.4),

$$\begin{aligned}\gamma^{ss} &= 2[\dot{\mathbf{r}}_j + \mathbf{A}_j\tilde{\omega}_j'\mathbf{s}_j'^P - \dot{\mathbf{r}}_i - \mathbf{A}_i\tilde{\omega}_i'\mathbf{s}_i'^P]^T \\ &\quad \times [(\dot{\mathbf{r}}_i - \dot{\mathbf{r}}_j) - \mathbf{A}_i\tilde{\mathbf{s}}_i'^P\omega_i' + \mathbf{A}_j\tilde{\mathbf{s}}_j'^P\omega_j'] \\ &\quad - 2\mathbf{d}_{ij}^T[\mathbf{A}_i\tilde{\omega}_i'\tilde{\mathbf{s}}_i'^P\omega_i - \mathbf{A}_j\tilde{\omega}_j'\tilde{\mathbf{s}}_j'^P\omega_j'] \\ &= -2(\dot{\mathbf{r}}_j - \dot{\mathbf{r}}_i)^T(\dot{\mathbf{r}}_j - \dot{\mathbf{r}}_i) + 2\mathbf{s}_j'^{PT}\tilde{\omega}_j'\tilde{\omega}_j'\mathbf{s}_j'^P \\ &\quad + 2\mathbf{s}_i'^{PT}\tilde{\omega}_i'\tilde{\omega}_i'\mathbf{s}_i'^P - 4\mathbf{s}_j'^{PT}\tilde{\omega}_j'\mathbf{A}_j^T\mathbf{A}_i\tilde{\omega}_i'\mathbf{s}_i'^P \\ &\quad + 4(\dot{\mathbf{r}}_j - \dot{\mathbf{r}}_i)(\mathbf{A}_j\tilde{\mathbf{s}}_j'^P\omega_j' - \mathbf{A}_i\tilde{\mathbf{s}}_i'^P\omega_i') \\ &\quad -2\mathbf{d}_{ij}^T[\mathbf{A}_i\tilde{\omega}_i'\tilde{\mathbf{s}}_i'^P\omega_i' - \mathbf{A}_j\tilde{\omega}_j'\tilde{\mathbf{s}}_j'^P\omega_j']\end{aligned} \tag{9.6.23}$$

Using differentials of the absolute constraints of Eq. 9.4.15, their velocity equations may be differentiated to obtain (Prob. 9.6.5)

$$\begin{aligned}\begin{bmatrix}\gamma^1\\ \gamma^2\\ \gamma^3\end{bmatrix} &= \mathbf{A}_i\tilde{\omega}_i'\tilde{\mathbf{s}}_i'^P\omega_i' \\ &= -\mathbf{A}_i\tilde{\omega}_i'\tilde{\omega}_i'\mathbf{s}_i'^P\end{aligned} \tag{9.6.24}$$

$$\begin{bmatrix}\gamma^4\\ \gamma^5\\ \gamma^6\end{bmatrix} = -\tfrac{1}{2}(\dot{\tilde{\mathbf{e}}}_i + \dot{e}_{0,i}\mathbf{I})\omega_i' \tag{9.6.25}$$

For the point constraint of Eq. 9.4.16 (Prob. 9.6.6),

$$\begin{aligned}\boldsymbol{\gamma}^P &= \mathbf{A}_i\tilde{\boldsymbol{\omega}}_i'\tilde{\mathbf{s}}_i'^P\boldsymbol{\omega}_i' \\ &= -\mathbf{A}_i\tilde{\boldsymbol{\omega}}_i'\tilde{\boldsymbol{\omega}}_i'\mathbf{s}_i'^P\end{aligned} \tag{9.6.26}$$

Finally, for the angle θ in the screw joint of Eq. 9.4.27 (Prob. 9.6.7),

$$\gamma^\theta = -\mathbf{h}_i'^T(\mathbf{A}_i^T\mathbf{A}_j\tilde{\boldsymbol{\omega}}_j' - \tilde{\boldsymbol{\omega}}_i'\mathbf{A}_i^T\mathbf{A}_j)\boldsymbol{\omega}_j' \tag{9.6.27}$$

Thus, for the screw joint of Eq. 9.4.26, Eqs. 9.6.21 and 9.6.27 yield

$$\gamma^{scr} = \gamma^{d2} - \alpha\gamma^\theta \tag{9.6.28}$$

For the driving constraints of Eq. 9.5.1, which depend explicitly on time, the foregoing derivative results and direct differentiation of time-dependent terms yield

$$\begin{bmatrix}\gamma^{1d}\\ \gamma^{2d}\\ \gamma^{3d}\end{bmatrix} = \begin{bmatrix}\gamma^1\\ \gamma^2\\ \gamma^3\end{bmatrix} + \begin{bmatrix}\ddot{C}_1(t)\\ \ddot{C}_2(t)\\ \ddot{C}_3(t)\end{bmatrix} \tag{9.6.29}$$

For the spherical–spherical driving constraint of Eq. 9.5.2,

$$\gamma^{ssd} = \gamma^{ss} + 2C(t)\ddot{C}(t) + 2\dot{C}(t)\dot{C}(t) \tag{9.6.30}$$

For the translational driving constraint of Eq. 9.5.3,

$$\gamma^{td} = \gamma^{d2} + \ddot{C}(t) \tag{9.6.31}$$

Finally, for the rotational driving constraint of Eq. 9.5.4,

$$\gamma^{\text{rotd}} = \gamma^\theta + \ddot{C}(t) \tag{9.6.32}$$

The right sides of acceleration equations presented in Eqs. 9.6.20 to 9.6.32 may now be substituted into Eq. 9.6.19 to obtain a set of equations to determine accelerations $\ddot{\mathbf{r}}_i$ and $\dot{\boldsymbol{\omega}}_j'$. Note that the coefficient matrix in these acceleration equations is identical to the coefficient matrix in the velocity equations of Eq. 9.6.12, yielding efficiency in calculating the solution of the acceleration equations.

PROBLEMS

Section 9.1

9.1.1. Use the Cartesian component forms of vectors $\vec{a}$, $\vec{b}$, and $\vec{c}$ to verify that Eq. 9.1.6 is valid.

9.1.2. Carry out the expansion indicated to verify that Eq. 9.1.11 is valid.

9.1.3. Carry out the expansion indicated to verify that Eq. 9.1.16 is valid.

9.1.4. Use the component representation of physical vectors and the associated properties of vector addition (Eq. 9.1.5) to verify that Eq. 9.1.19 is valid.

9.1.5. Expand both sides of Eq. 9.1.25 and verify that it is valid.

9.1.6. Verify Eq. 9.1.28 by expanding both sides and show that they are equal.

9.1.7. Verify Eqs. 9.1.29 and 9.1.30 by expanding both sides and show that they are equal.

9.1.8. Given points P with $\mathbf{r}^P = [1, 1, 1]^T$, Q with $\mathbf{r}^Q = [2, 1, 0]^T$, and R with $\mathbf{r}^R = [2, 2, 0]^T$, construct unit vectors $\mathbf{f}$, $\mathbf{g}$, and $\mathbf{h}$ that define an x'-y'-z' Cartesian reference frame with origin at point P, using the method of Example 9.1.6.

9.1.9. Use the definition of matrix addition and properties of differentiation to verify that Eq. 9.1.34 is valid.

9.1.10. Verify that Eqs. 9.1.35 to 9.1.37 are valid.

Section 9.2

9.2.1. Find the direction cosine representations of unit vectors in the same direction as the vectors $\mathbf{a} = [1, 1, 1]^T$ and $\mathbf{b} = [1, 1, 0]^T$.

9.2.2. If the vector $\mathbf{a}$ in Prob. 9.2.1 is along the x' axis of a Cartesian x'-y'-z' reference frame and if $\mathbf{b}$ is in the first quadrant of its x'-y' plane, find the unit coordinate vectors $\mathbf{f}$, $\mathbf{g}$, and $\mathbf{h}$. *Hint:* Since $\mathbf{a}$ and $\mathbf{b}$ lie in the x'-y' plane, $\tilde{\mathbf{a}}\mathbf{b}$ must be along the z' axis. Form the transformation matrix from the x'-y'-z' frame to the x-y-z frame.

9.2.3. Expand the product $\mathbf{A}^T\mathbf{A}$ using Eq. 9.2.13 to show that Eq. 9.2.14 is correct.

9.2.4. If $(\mathbf{B} - \mathbf{C})\mathbf{a} = \mathbf{0}$ for arbitrary $\mathbf{a}$, show that $\mathbf{B} = \mathbf{C}$, where $\mathbf{B}$ and $\mathbf{C}$ are $m \times n$ matrices and $\mathbf{a}$ is in R^n.

9.2.5. If $\mathbf{A}_i$ and $\mathbf{A}_j$ are orthogonal matrices, show that $\mathbf{A}_{ij}$ of Eq. 9.2.25 is orthogonal.

9.2.6. Verify that Eq. 9.2.31 uniquely defines the angle θ, $0 \leq \theta < 2\pi$, in Fig. 9.2.10.

9.2.7. Show that Eq. 9.2.38 is correct.

9.2.8. Show that Eqs. 9.2.42 and 9.2.43 are correct.

9.2.9. Use Eqs. 9.2.41 to 9.2.43 to show that

$$\begin{aligned}\ddot{\mathbf{r}}^P &= \ddot{\mathbf{r}} - \tilde{\mathbf{s}}^P\dot{\boldsymbol{\omega}} - \tilde{\boldsymbol{\omega}}\tilde{\mathbf{s}}^P\boldsymbol{\omega} \\ &= \ddot{\mathbf{r}} - \mathbf{A}\tilde{\mathbf{s}}'^P\dot{\boldsymbol{\omega}}' - \mathbf{A}\tilde{\boldsymbol{\omega}}'\tilde{\mathbf{s}}'^P\boldsymbol{\omega}'\end{aligned}$$

9.2.10. Write the vector $\mathbf{r}^P$ for the point $\mathbf{s}'^P$ of Example 9.2.4, with $\mathbf{r} = \mathbf{0}$, explicitly as a function of ϕ. Differentiate the result to obtain $\dot{\mathbf{r}}^P$ and $\ddot{\mathbf{r}}^P$ and compare with the results presented in Examples 9.2.4 and 9.2.5.

9.2.11 If matrices $\mathbf{A}_{m\times p}$ and $\mathbf{B}_{p\times n}$ depend on a variable $\mathbf{q}$, then

$$\delta\mathbf{A} = \left[\frac{\partial a_{ij}}{\partial \mathbf{q}}\delta\mathbf{q}\right] \quad \text{and} \quad \delta\mathbf{B} = \left[\frac{\partial b_{ij}}{\partial \mathbf{q}}\delta\mathbf{q}\right]$$

are linear in $\delta\mathbf{q}$. Use this result to show that

$$\delta(\mathbf{AB}) = (\delta\mathbf{A})\mathbf{B} + \mathbf{A}(\delta\mathbf{B})$$

In particular, show that this result and $\mathbf{A}^T\mathbf{A} = \mathbf{I}$ lead to Eq. 9.2.44.

9.2.12. Use Eq. 9.2.46 to show that the virtual rotation in Example 9.2.2 is $\delta\boldsymbol{\pi} = \delta\boldsymbol{\pi}' = [0, 0, \delta\phi]^T$. *Hint:* Follow the calculation carried out in Example 9.2.4.

9.2.13 Use the results of Prob. 9.2.11 and the fact that $\mathbf{s}'^P$ is constant to obtain Eq. 9.2.48.

Section 9.3

9.3.1. Repeat the derivation of Eq. 9.3.4 for the **f-i** and **g-j** relations to show that Eq. 9.3.5 is correct.

9.3.2. Show that as ϕ is incremented by 120° in Example 9.3.1, beginning from $\phi = 0$, that the vector $\mathbf{s}' = [1, 0, 0]^T$ becomes $\mathbf{s}(0) = \mathbf{s}'$, $\mathbf{s}(120°) = [0, 0, 1]^T$, $\mathbf{s}(240°) = [0, 0, 1]^T$, and $\mathbf{s}(360°) = \mathbf{s}'$. Interpret these results physically in terms of the cone that is swept out by $\mathbf{s}$ as ϕ varies.

9.3.3. Use the relations of Eqs. 9.3.11 to 9.3.15 with $\mathbf{A}$ of Eq. 9.2.17 to recover the geometry of the rotation of Example 9.2.2.

9.3.4. Show that each row of $\mathbf{G}$ is orthogonal to $\mathbf{p}$ (Eq. 9.3.20).

9.3.5. Show that the rows of $\mathbf{E}$ are orthogonal unit vectors (Eq. 9.3.22) (a) using the expanded form of matrix $\mathbf{E}$ in Eq. 9.3.18, and (b) using the compact form of matrix $\mathbf{E}$ in Eq. 9.3.18 and evaluating the matrix product

$$\mathbf{E}\mathbf{E}^T = [-\mathbf{e}, \tilde{\mathbf{e}} + e_0\mathbf{I}]\begin{bmatrix} -\mathbf{e}^T \\ -\tilde{\mathbf{e}} + e_0 I \end{bmatrix}$$

9.3.6. Repeat Prob. 9.3.5 for matrix $\mathbf{G}$.

9.3.7. Take the time derivative of the identity $\mathbf{p}^T\mathbf{p} = 1$ to show that $\mathbf{p}^T\dot{\mathbf{p}} = 0$ and $\dot{\mathbf{p}}^T\mathbf{p} = 0$.

9.3.8. Use the definitions of Eqs. 9.3.18 and 9.3.19 to expand the products $\dot{\mathbf{E}}\mathbf{G}^T$ and $\dot{\mathbf{E}}\mathbf{G}^T$ and show that they are equal.

9.3.9. Show that the inverse transformation of Eq. 9.3.37 is valid.

9.3.10. Carry out the detailed derivation of Eq. 9.3.41, following the procedure outlined in the text.

9.3.11. Carry out the detailed derivation of Eq. 9.3.42, following the procedure outlined in the text.

9.3.12. Carry out the detailed derivation of Eq. 9.3.43, following the procedure outlined in the text.

9.3.13. Carry out the detailed derivation of Eq. 9.3.44, following the procedure outlined in the text.

Section 9.4

9.4.1 For Fig. P9.4.1, write three constraint equations that require that the bottom plane (x_2'-y_2') of body 2 slides on the x-y plane (x_1'-y_1'). *Hint:* The origin of the body 2 reference frame must be in the x-y plane, and z_2' is parallel to the z axis if and only if the x_2' and y_2' axes are perpendicular to the z axis.

9.4.2. Body 2 is constrained to translate along the y axis, as shown in Fig. P9.4.2. Show that the constraint equations of Problem 9.4.1 plus $x_2 = 0$ and $\Phi^{d2}(\mathbf{d}_{12}, \mathbf{f}_2) = 0$ define the constraint, provided $\mathbf{d}_{12} \neq \mathbf{0}$. Show an orientation of body 2 that satisfies these constraint equations when $\mathbf{d}_{12} = \mathbf{0}$, but violates the geometry of the constraint.

9.4.3. Show that if $C = 0$ in Eq. 9.4.8 then $\mathbf{d}_{ij} = \mathbf{0}$.

9.4.4. Carry out manipulations to show that Eq. 9.4.12 is valid.

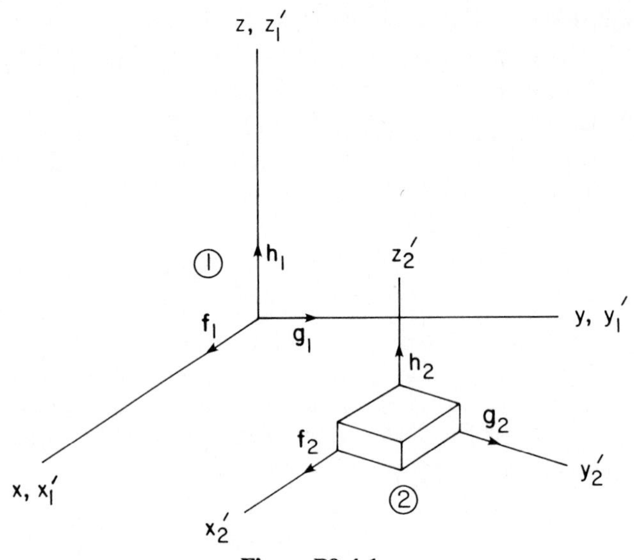

Figure P9.4.1

Section 9.6

9.6.1. Verify that Eq. 9.6.20 is valid.

9.6.2. Verify that Eq. 9.6.21 is valid.

9.6.3. Verify that Eq. 9.6.22 is valid.

9.6.4. Verify that Eq. 9.6.23 is valid.

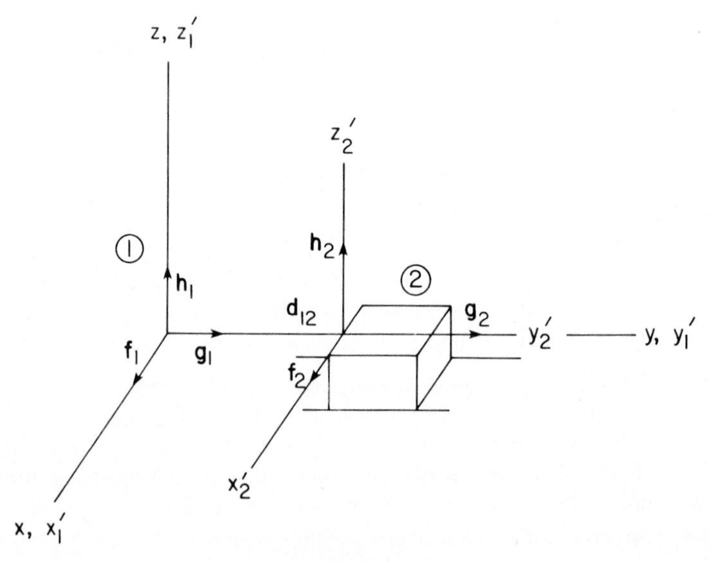

Figure P9.4.2

9.6.5. Verify that Eqs. 9.6.24 and 9.6.25 are valid.

9.6.6. Verify that Eq. 9.6.26 is valid.

9.6.7. Verify that Eq. 9.6.27 is valid.

SUMMARY OF KEY FORMULAS

Algebraic Vectors

$$\tilde{\mathbf{a}} = \begin{bmatrix} 0 & -a_z & a_y \\ a_z & 0 & -a_x \\ -a_y & a_x & 0 \end{bmatrix} \quad \mathbf{(9.1.21)}$$

$$\tilde{\mathbf{a}}^T = -\tilde{\mathbf{a}}, \qquad \tilde{\mathbf{a}}\mathbf{b} = -\tilde{\mathbf{b}}\mathbf{a}, \qquad \tilde{\mathbf{a}}\mathbf{a} = \mathbf{0} \quad \mathbf{(9.1.23, 25, 26)}$$

$$\tilde{\mathbf{a}}\tilde{\mathbf{b}} = \mathbf{b}\mathbf{a}^T - \mathbf{a}^T\mathbf{b}\mathbf{I} \quad \mathbf{(9.1.28)}$$

$$\widetilde{(\tilde{\mathbf{a}}\mathbf{b})} = \mathbf{b}\mathbf{a}^T - \mathbf{a}\mathbf{b}^T = \tilde{\mathbf{a}}\tilde{\mathbf{b}} - \tilde{\mathbf{b}}\tilde{\mathbf{a}} \quad \mathbf{(9.1.29, 30)}$$

$$\tilde{\mathbf{a}}\tilde{\mathbf{b}} + \mathbf{a}\mathbf{b}^T = \tilde{\mathbf{b}}\tilde{\mathbf{a}} + \mathbf{b}\mathbf{a}^T, \qquad \widetilde{(\mathbf{a} + \mathbf{b})} = \tilde{\mathbf{a}} + \tilde{\mathbf{b}} \quad \mathbf{(9.1.31, 32)}$$

Transformation of Coordinates

$$\mathbf{s} = \mathbf{A}\mathbf{s}', \quad \mathbf{s}' = \mathbf{A}^T\mathbf{s}, \quad \mathbf{A}^T\mathbf{A} = \mathbf{A}\mathbf{A}^T = \mathbf{I} \quad \mathbf{(9.2.10, 14, 15)}$$

$$\mathbf{A} = [\mathbf{f}, \mathbf{g}, \mathbf{h}], \quad \mathbf{r}^P = \mathbf{r} + \mathbf{s}^P = \mathbf{r} + \mathbf{A}\mathbf{s}'^P \quad \mathbf{(9.2.13, 16)}$$

$$\tilde{\mathbf{s}} = \mathbf{A}\tilde{\mathbf{s}}'\mathbf{A}^T, \quad \tilde{\mathbf{s}}' = \mathbf{A}^T\tilde{\mathbf{s}}\mathbf{A}, \quad \mathbf{A}_{ij} = \mathbf{A}_i^T\mathbf{A}_j \quad \mathbf{(9.2.21, 22, 25)}$$

Velocity, Acceleration, and Angular Velocity

$$\dot{\mathbf{r}}^P = \dot{\mathbf{r}} + \dot{\mathbf{A}}\mathbf{s}'^P = \dot{\mathbf{r}} + \tilde{\boldsymbol{\omega}}\mathbf{s}^P = \dot{\mathbf{r}} + \mathbf{A}\tilde{\boldsymbol{\omega}}'\mathbf{s}'^P \quad \mathbf{(9.2.33, 37, 38)}$$

$$\dot{\mathbf{A}} = \tilde{\boldsymbol{\omega}}\mathbf{A} = \mathbf{A}\tilde{\boldsymbol{\omega}}', \quad \tilde{\boldsymbol{\omega}}' = \mathbf{A}^T\dot{\mathbf{A}} \quad \mathbf{(9.2.39, 40)}$$

$$\ddot{\mathbf{r}}^P = \ddot{\mathbf{r}} + \ddot{\mathbf{A}}\mathbf{s}'^P \quad \mathbf{(9.2.41)}$$

$$\ddot{\mathbf{A}} = \tilde{\dot{\boldsymbol{\omega}}}\mathbf{A} + \tilde{\boldsymbol{\omega}}\tilde{\boldsymbol{\omega}}\mathbf{A} = \mathbf{A}\tilde{\dot{\boldsymbol{\omega}}}' + \mathbf{A}\tilde{\boldsymbol{\omega}}'\tilde{\boldsymbol{\omega}}' \quad \mathbf{(9.2.42, 43)}$$

Virtual Displacements and Rotations

$$\delta\mathbf{A} = \delta\tilde{\boldsymbol{\pi}}\mathbf{A} = \mathbf{A}\,\delta\tilde{\boldsymbol{\pi}}', \qquad \delta\tilde{\boldsymbol{\pi}}' = \mathbf{A}^T\,\delta\mathbf{A} \quad \mathbf{(9.2.47, 50)}$$

$$\delta\mathbf{r}^P = \delta\mathbf{r} + \delta\mathbf{A}\mathbf{s}'^P = \delta\mathbf{r} + \delta\tilde{\boldsymbol{\pi}}\mathbf{s}^P = \delta\mathbf{r} + \mathbf{A}\,\delta\tilde{\boldsymbol{\pi}}'\mathbf{s}'^P \quad \mathbf{(9.2.48, 49, 51)}$$

$$\delta\mathbf{s} = \delta\mathbf{A}\mathbf{s}' = \mathbf{A}\,\delta\tilde{\boldsymbol{\pi}}'\mathbf{s}' = -\mathbf{A}\tilde{\mathbf{s}}'\,\delta\boldsymbol{\pi}' \quad \mathbf{(9.2.52, 53)}$$

$$\delta(\mathbf{g}^T\mathbf{h}) = \mathbf{h}^T\,\delta\mathbf{g} + \mathbf{g}^T\,\delta\mathbf{h}, \qquad \delta(\tilde{\mathbf{g}}\mathbf{h}) = \tilde{\mathbf{g}}\,\delta\mathbf{h} - \tilde{\mathbf{h}}\,\delta\mathbf{g} \quad \mathbf{(9.2.55, 56)}$$

Euler Parameter Definitions

$$e_0 = \cos(\chi/2), \qquad \mathbf{e} = \mathbf{u}\sin(\chi/2) \quad \mathbf{(9.3.2)}$$

$$\mathbf{A} = (2e_0^2 - 1)\mathbf{I} + (\mathbf{e}\mathbf{e}^T + e_0\tilde{\mathbf{e}}) \quad \mathbf{(9.3.6)}$$

$$\mathbf{p} = [e_0, \mathbf{e}^T]^T = [e_0, e_1, e_2, e_3]^T, \quad \mathbf{p}^T\mathbf{p} = 1 \qquad \textbf{(9.3.8, 9)}$$

Euler Parameter Relations

$$\mathbf{E} = [-\mathbf{e}, \tilde{\mathbf{e}} + e_0\mathbf{I}]_{3\times 4}, \qquad \mathbf{G} = [-\mathbf{e}, -\tilde{\mathbf{e}} + e_0\mathbf{I}]_{3\times 4} \qquad \textbf{(9.3.18, 19)}$$

$$\mathbf{E}\mathbf{E}^T = \mathbf{G}\mathbf{G}^T = \mathbf{I}, \qquad \mathbf{E}^T\mathbf{E} = \mathbf{G}^T\mathbf{G} = \mathbf{I}_4 - \mathbf{p}\mathbf{p}^T \qquad \textbf{(9.3.22, 23, 24)}$$

$$\mathbf{A} = \mathbf{E}\mathbf{G}^T \qquad \textbf{(9.3.26)}$$

$$\boldsymbol{\omega}' = 2\mathbf{G}\dot{\mathbf{p}}, \qquad \dot{\mathbf{p}} = \tfrac{1}{2}\mathbf{G}^T\boldsymbol{\omega}' \qquad \textbf{(9.3.34, 35)}$$

$$\boldsymbol{\omega} = 2\mathbf{E}\dot{\mathbf{p}}, \qquad \dot{\mathbf{p}} = \tfrac{1}{2}\mathbf{E}^T\boldsymbol{\omega} \qquad \textbf{(9.3.36, 37)}$$

$$\delta\boldsymbol{\pi}' = 2\mathbf{G}\,\delta\mathbf{p}, \qquad \delta\mathbf{p} = \tfrac{1}{2}\mathbf{G}^T\,\delta\boldsymbol{\pi}' \qquad \textbf{(9.3.41, 43)}$$

$$\delta\boldsymbol{\pi} = 2\mathbf{E}\,\delta\mathbf{p}, \qquad \delta\mathbf{p} = \tfrac{1}{2}\mathbf{E}^T\,\delta\boldsymbol{\pi} \qquad \textbf{(9.3.42, 44)}$$

CHAPTER TEN

Spatial Kinematic Modeling and Analysis

10.1 MODELING AND ANALYSIS TECHNIQUES

The basic techniques for kinematic modeling of spatial mechanisms and machines are identical to those discussed in Section 5.1 for planar systems. The principal and very important difference between modeling planar and spatial kinematic systems concerns the redundancy of constraints. In spatial systems, it is easy to implement what appear to be proper constraints, only to find that redundancies exist.

Example 10.1.1: Consider modeling the slider–crank mechanism of Fig. 10.1.1 with three revolute joints and one translational joint. If Euler parameters are used for orientation, each body has seven generalized coordinates. There are, therefore, 28 generalized coordinates for the four bodies. The revolute and translational joints each have five constraint equations, yielding 20 constraint equations. In addition, the six constraints for fixing ground and four Euler parameter normalization constraints yield a total of 30 constraint equations. Since the actual mechanism is intended to have one degree of freedom, there must be three redundant constraints. To understand why this is the case, note that if the revolute joint axes in bodies 4, 1, and 2 at points A, B, and C in Fig. 10.1.1 are all parallel then the revolute axes in bodies 2 and 3 at point C are automatically parallel, yielding two redundant constraint equations. Finally, the revolute joints at points A and B require bodies 1 and 2 to move in the x-y plane, and the translational joint between body 3 and ground also requires that the origin of the body-fixed reference frame in body 3 lie in the x-y plane. Therefore, the constraint on the z coordinate of the revolute joint at point C is redundant. These conditions define the three degrees of redundancy in this overly constrained model.

As an alternative check on redundancy, the *manufacturing imperfection test* outlined in step 1(b) of the procedure suggested in Section 5.1 can be applied to this mechanism. The imperfect configuration of the slider–crank shown in Fig.

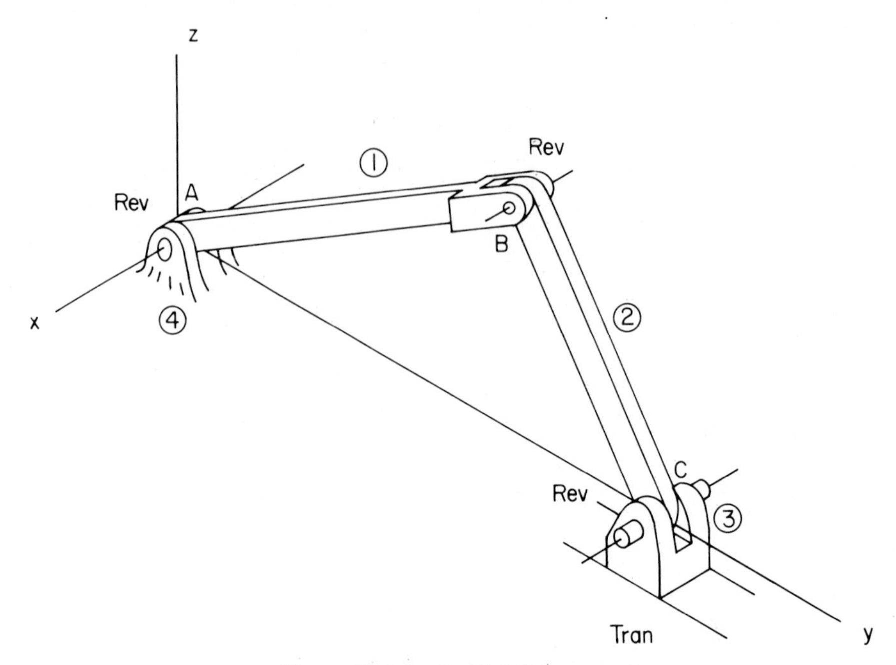

Figure 10.1.1 Spatial slider–crank.

10.1.2 has a misaligned revolute joint axis in ground at point A. If all other revolute joint axes are parallel, then a misalignment of the revolute joint axes between bodies 2 and 3 exists at point C, preventing assembly. This implies two degrees of redundancy, since two parameters must be adjusted to precisely align the axes of the revolute joint at point C. A third degree of redundancy exists, since the center point of the revolute joint on body 2 does not lie in the y-z plane; hence there is an offset, which defines one additional degree of redundancy.

An alternative model might be considered in which the translational joint between bodies 3 and 4 is replaced by absolute x and z constraints on point C in body 3. This reduces the number of constraint equations by 3 and creates a situation in which the number of generalized coordinates minus the number of constraints is one, which is desired. While the counting check of step 1(a) of Section 5.1 is thus satisfied, this is still not a good model. To see why, consider again the misalignment shown in Fig. 10.1.2. The offset inconsistency in the revolute joint between bodies 2 and 3 remains, so the system cannot be assembled. This apparent difficulty could be overcome by replacing the revolute joint between bodies 2 and 3 by a cylindrical joint, which has one fewer constraint equations. As a result, the mechanism can be assembled, as shown in Fig. 10.1.3. While the mechanism can be assembled, an extra degree of freedom has inadvertently been created that permits body 3 to rotate about the cylindrical

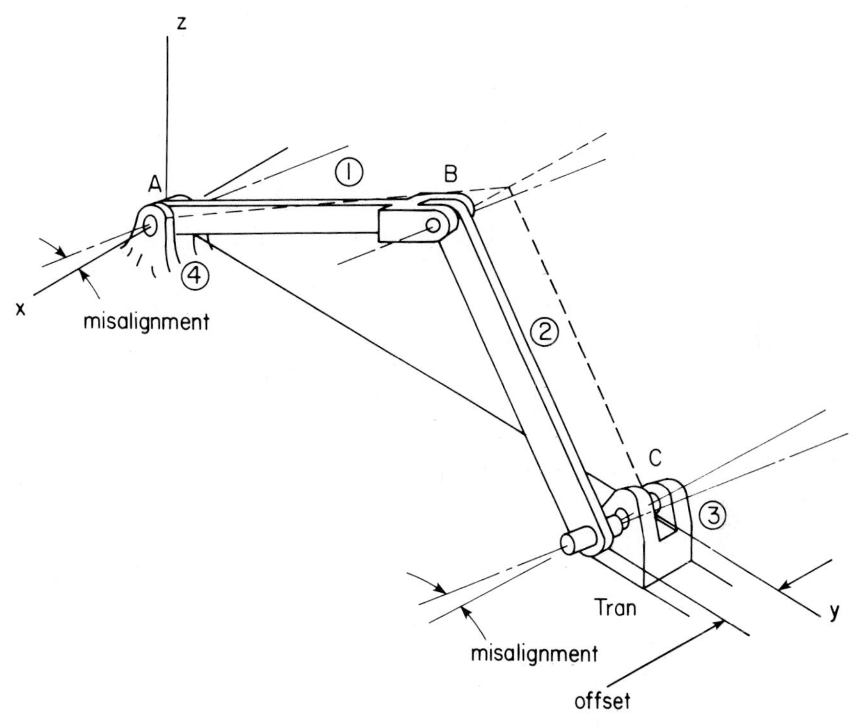

Figure 10.1.2 Imperfect spatial slider–crank.

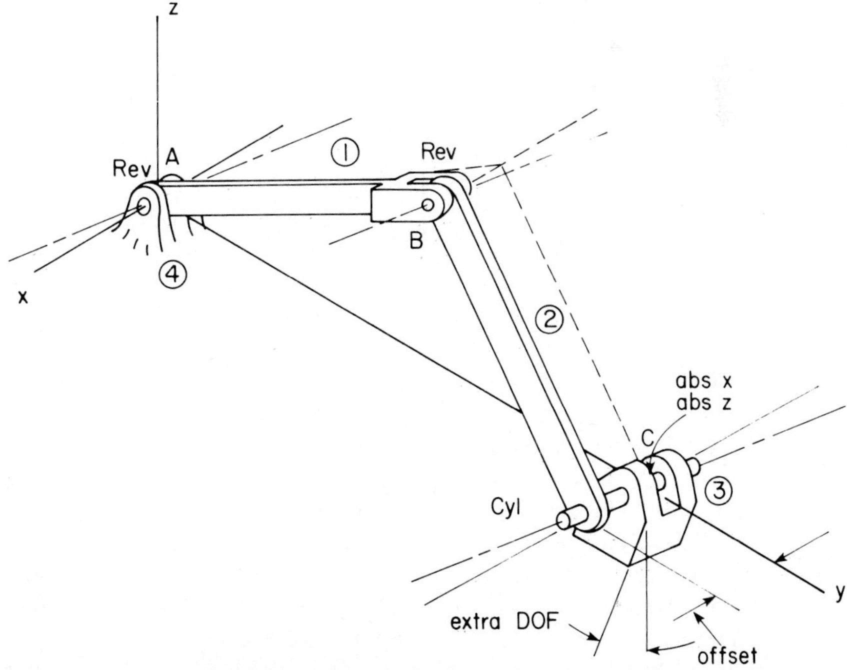

Figure 10.1.3 Modified spatial slider–crank.

joint axis, which is not desired in the actual mechanism. Thus, careless adjustment in constraint definition, based only on counting constraint equations, can lead to either or both redundant constraints and unwanted extra degrees of freedom.

Apart from the need for careful definition of independent kinematic constraints, the kinematic analysis of spatial systems is a direct extension of the kinematic analysis of planar systems. Kinematic analyses of realistic mechanisms are carried out in the remaining sections of this chapter using the DADS computer code [27]. The effects of design variations are studied to illustrate the use of the methods of Chapter 9 in support of the design of mechanical systems.

10.2 KINEMATIC ANALYSIS OF A SPATIAL SLIDER–CRANK MECHANISM

10.2.1 Model

The spatial slider–crank mechanism shown in Fig. 10.2.1 is modeled using four bodies. The model is defined as follows:

Model	
Bodies	
Four bodies	$nc = 28$
Constraints	
Revolute joint (crank, ground)	5
Spherical joint (crank, connecting rod)	3
Revolute–cylindrical joint (connecting rod, slider)	3
Translational joint (slider, ground)	5
Distance constraint (connecting rod, slider)	1
Ground constraint	6
Euler parameter normalization constraint	4
DOF $= 28 - 27 = 1$.	$nh = 27$

The motion of the system can be defined by requiring that the orientation of the crank (body 1) be some function of time. This is equivalent to imposing a driving constraint so that the remaining degrees of freedom are determined.

To define a kinematic joint, six points (three points on each body) are chosen, depending on the type of joint that is intended. These points, P_i, Q_i, R_i, P_j, Q_j, and R_j, defined in their respective centroidal body-fixed reference frames on bodies i and j, form joint reference triads.

For the revolute joint at point A in Fig. 10.2.1, the common point in the joint is defined by points P_1 and P_4 given in Table 10.2.1. Points Q_1 and Q_4 in Table 10.2.1 are chosen to define the axis of rotation in the bodies. Vectors P_1Q_1 and P_4Q_4 define the z'' axes of the joint reference triads on each body. Points R_1

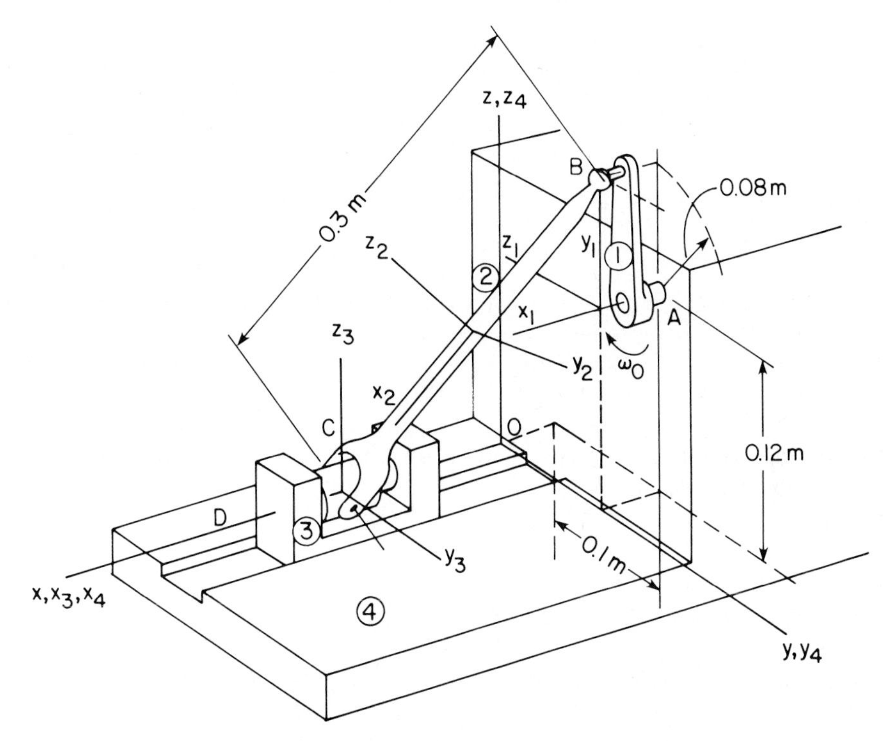

Figure 10.2.1 Spatial slider–crank.

and R_4 in Table 10.2.1 define the joint x'' axes. These six points define the revolute joint at point A in the model.

TABLE 10.2.1 Data for Revolute Joint

Body \ Point	P x'	P y'	P z'	Q x'	Q y'	Q z'	R x'	R y'	R z'
Crank ①	0.0	0.0	0.0	1.0	0.0	0.0	0.0	0.0	1.0
Ground ④	0.0	0.1	0.12	1.0	0.1	0.12	0.0	0.1	1.12

For the spherical joint at point B in Fig. 10.2.1, points P_1 and P_2 define the common point in the joint. Points Q_1, Q_2, R_1, and R_2 can be arbitrarily chosen to define the joint reference triads. These six points for the spherical joint at point B in the model are defined in Table 10.2.2.

For the revolute–cylindrical joint at point C in Fig. 10.2.1, points P_2, P_3, Q_2, and Q_3 are chosen to define the axis of rotation of the joint, and hence the z'' axes of joint reference triads. Points R_2 and R_3 then define the x'' axes of the

TABLE 10.2.2 Data for Spherical Joint

Body \ Point	P x'	P y'	P z'	Q x'	Q y'	Q z'	R x'	R y'	R z'
Crank ①	0.0	0.08	0.0	0.0	0.08	1.0	1.0	0.08	0.0
Connecting rod ②	−0.15	0.0	0.0	−0.15	0.0	1.0	1.15	0.0	0.0

TABLE 10.2.3 Data for Revolute–Cylindrical Joint

Body \ Point	P x'	P y'	P z'	Q x'	Q y'	Q z'	R x'	R y'	R z'
Connecting rod ②	0.15	0.0	0.0	0.15	1.0	0.0	1.15	0.0	0.0
Slider ③	0.2	0.0	0.0	1.2	0.0	0.0	0.2	1.0	0.0

joint reference triads. These six points for the revolute–cylindrical joint in the model are defined in Table 10.2.3.

For the translational joint at point D in Fig. 10.2.1, points P_3, P_4, Q_3, and Q_4 are chosen to define the common lines of translation of the joint, which are the z'' axes of the joint reference triads. Points R_3 and R_4 then define the x'' axes of the joint reference triads. These six points for the translational joint are defined in Table 10.2.4.

TABLE 10.2.4 Data for Translational Joint

Body \ Point	P x'	P y'	P z'	Q x'	Q y'	Q z'	R x'	R y'	R z'
Slider ③	0.0	0.0	0.0	1.0	0.0	0.0	0.0	1.0	0.0
Ground ④	0.2	0.0	0.0	1.2	0.0	0.0	0.2	1.0	0.0

For the distance constraint between the connecting rod and slider, P_2 and P_3 are points between which the distance is fixed. Points Q_2, R_2, Q_3, and R_3 can be arbitrarily chosen to define the joint reference triads. These six points for the distance constraint are defined in Table 10.2.5.

TABLE 10.2.5 Data for Distance Constraint

Body \ Point	P x'	P y'	P z'	Q x'	Q y'	Q z'	R x'	R y'	R z'	Distance
Connecting rod ②	0.15	0.0	0.0	0.15	1.0	0.0	0.15	0.0	1.0	1.0
Slider ③	1.0	0.0	0.0	1.0	1.0	0.0	1.0	0.0	1.0	

10.2.2 Assembly

The position and orientation of each body reference frame in the global reference frame is estimated for initial assembly analysis. Table 10.2.6 provides estimates of generalized coordinates (abbreviated GC) for the model. Euler parameters are used to specify the orientation of each body. Table 10.2.7 shows the resulting assembled configuration.

TABLE 10.2.6 Position and Orientation Estimates

Body \ GC	x	y	z	e_1	e_2	e_3
Crank ①	0.0	0.1	0.12	0.71	0.0	0.0
Connecting rod ②	0.1	0.05	0.1	−0.21	0.40	−0.1
Slider ③	0.2	0.0	0.0	0.0	0.0	0.0
Ground ④	0.0	0.0	0.0	0.0	0.0	0.0

TABLE 10.2.7 Assembled Configuration

Body \ GC	x	y	z	e_0	e_1	e_2	e_3
Crank ①	0.00002	0.09982	0.12005	0.72090	0.69306	0.00004	−0.00004
Connecting rod ②	0.09993	0.05183	0.09998	0.88723	−0.21202	0.39833	−0.09569
Slider ③	0.19959	0.00057	−0.00008	1.0	0.0	−0.00012	−0.00025
Ground ④	0.0	0.0	0.0	1.0	0.0	0.0	0.0

10.2.3 Driver Specification

The crank can only rotate in the revolute joint. Taking the relative angle θ in the joint as the driving coordinate, it is specified that the crank rotate at $\omega_0 = 2\pi$ rad/s. The driver is thus specified by the condition

$$\theta = 2\pi t$$

10.2.4 Analysis

Three runs are made with varying connecting rod lengths of 0.3, 0.27, and 0.24 m. Lock-up occurs when the length of the connecting rod is less than 0.2362 m, as shown in Fig. 10.2.2. The position, velocity, and acceleration histories of point C on the slider are shown in Figs. 10.2.3, 10.2.4, and 10.2.5. Notice the high peak of acceleration in Fig. 10.2.5 when $\ell = 0.24$ m, which is a near singular design.

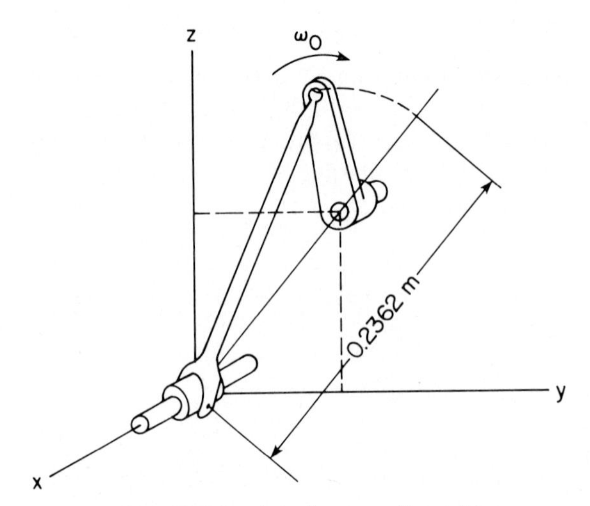

Figure 10.2.2 A lock-up configuration.

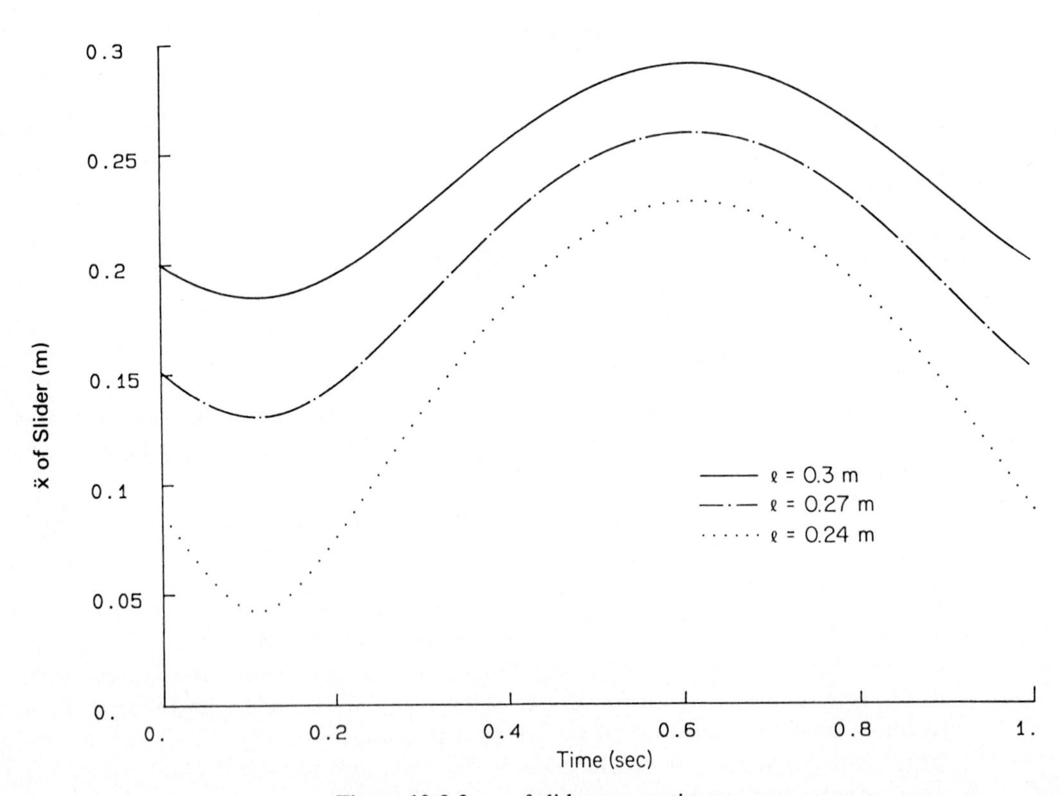

Figure 10.2.3 x of slider versus time.

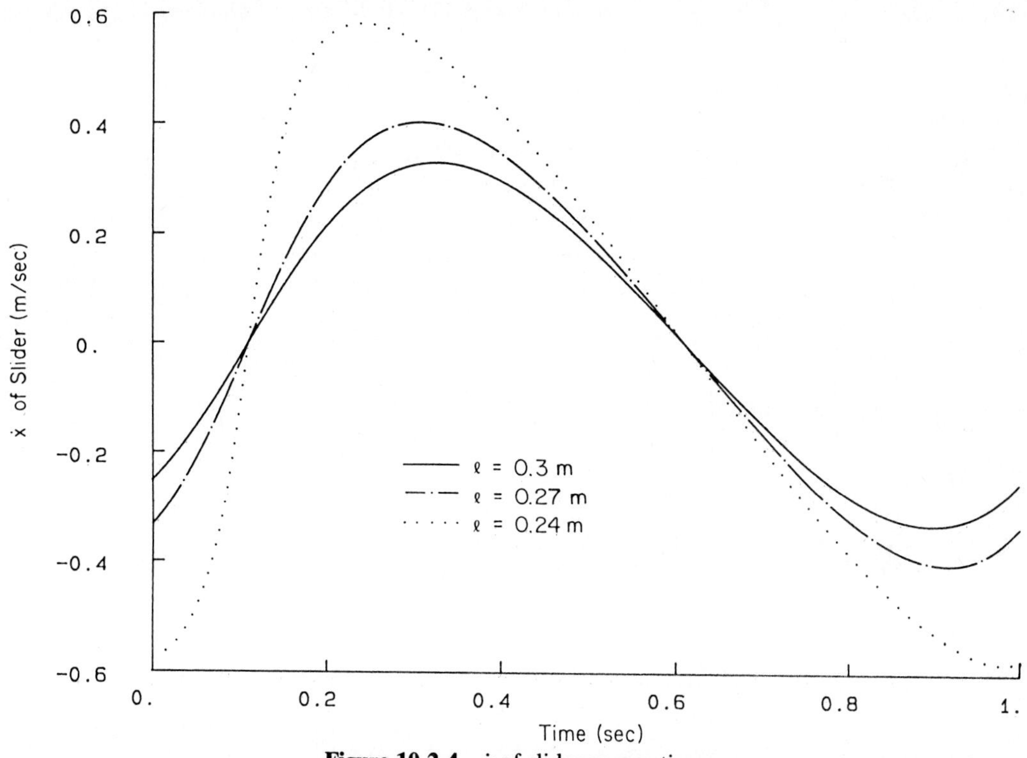

Figure 10.2.4 $\dot{x}$ of slider versus time.

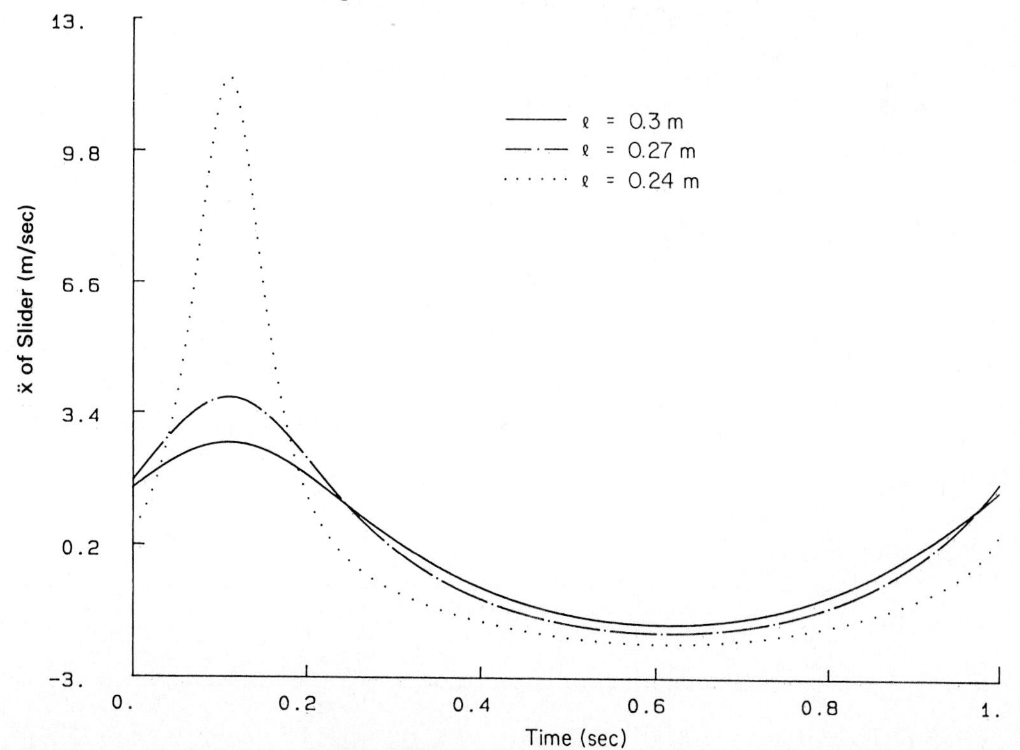

Figure 10.2.5 $\ddot{x}$ of slider versus time.

10.3 KINEMATIC ANALYSIS OF A SPATIAL FOUR-BAR MECHANISM

10.3.1 Alternative Models

A spatial four-bar mechanism can be modeled in many different ways. Two kinematically equivalent models are developed and analyzed here. In model 1 (Fig. 10.3.1), each link in the mechanism and ground is modeled as a body. Revolute, spherical, and universal joints are used to model this mechanism, as follows:

Model 1	
Bodies	
Four bodies	$nc = 28$
Constraints	
Revolute joints (links 1 and 3, ground): A	5
D	5
Universal joint (link 1, link 2): B	4
Spherical joint (link 2, link 3): C	3
Ground constraints (link 4):	6
Euler parameter normalization constraints	4
	$nh = 27$
DOF = 28 − 27 = 1.	

For kinematic analysis, the one remaining degree of freedom is eliminated by imposing a driving constraint.

To define each kinematic joint, six points are chosen according to the type of joint that is intended. For a revolute joint, points P_i and P_j are chosen to locate a common point in the joint. Points Q_i and Q_j are chosen to define the axis of rotation. Vectors P_iQ_i and P_jQ_j define the z'' axes of the joint reference triad on each body. Points R_i and R_j define the joint x'' axes. Revolute joint definition data for model 1 are given in Table 10.3.1.

TABLE 10.3.1 Revolute Joint Data, Model 1

Joint *A*

Body \ Point	*P* x'	*P* y'	*P* z'	*Q* x'	*Q* y'	*Q* z'	*R* x'	*R* y'	*R* z'
Ground ④	0.0	0.0	0.0	1.0	0.0	0.0	0.0	0.0	1.0
Link ①	0.0	0.0	0.0	1.0	0.0	0.0	0.0	0.0	1.0

Joint *D*

Body \ Point	*P* x'	*P* y'	*P* z'	*Q* x'	*Q* y'	*Q* z'	*R* x'	*R* y'	*R* z'
Link ③	0.0	3.7	0.0	1.0	3.7	0.0	0.0	2.7	0.0
Ground ④	−4.0	−8.5	0.0	−4.0	−9.5	0.0	−3.0	−8.5	0.0

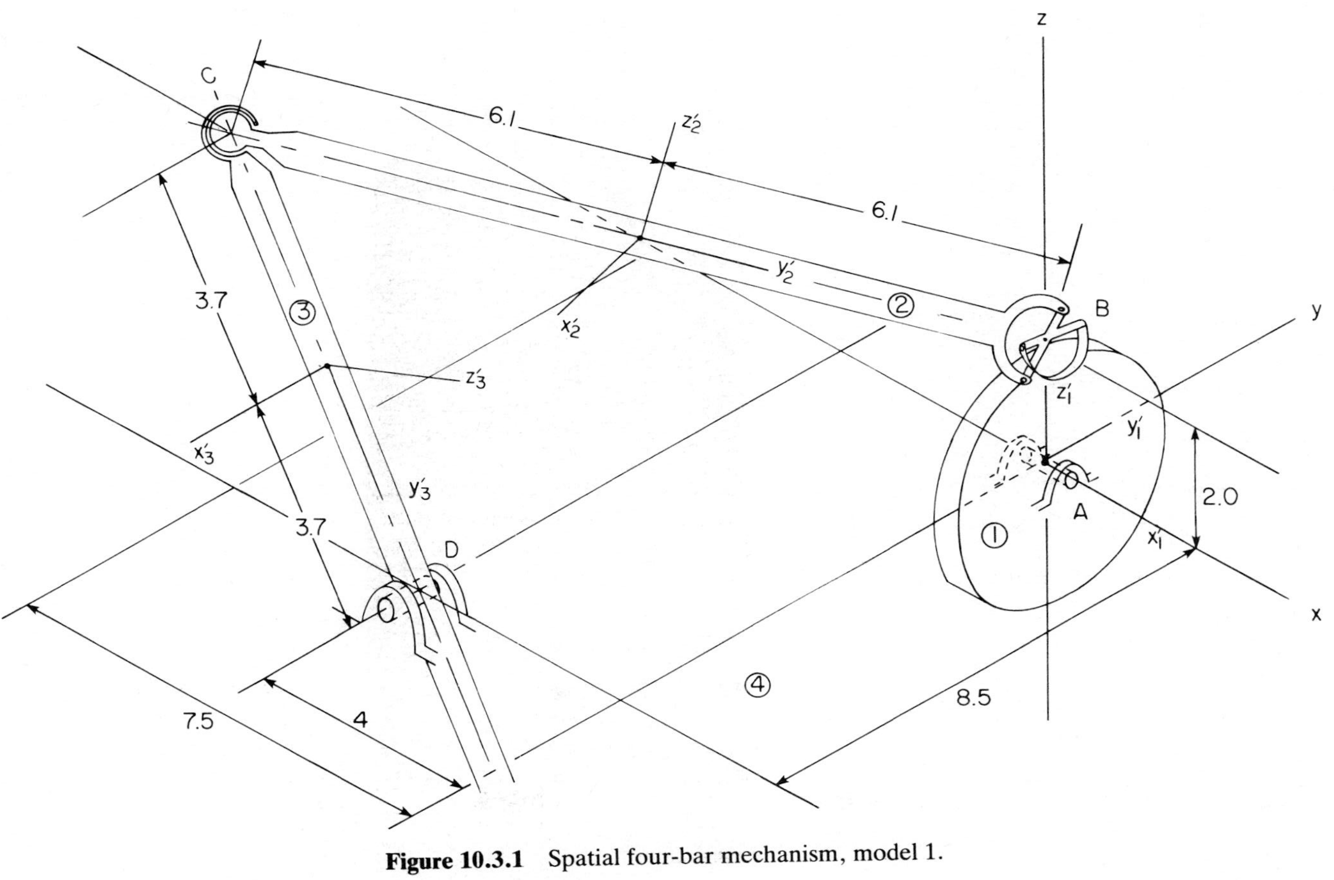

Figure 10.3.1 Spatial four-bar mechanism, model 1.

TABLE 10.3.2 Universal Joint Data, Model 1

Body \ Point	P x'	P y'	P z'	Q x'	Q y'	Q z'	R x'	R y'	R z'
Link ①	0.0	0.0	2.0	0.750	−0.662	2.0	0.244	0.277	2.929
Link ②	0.0	6.1	0.0	0.0	6.1	1.0	0.0	6.1	0.0

For a universal joint, the common points in the joint are P_i and P_j. Vectors P_iQ_i and P_jQ_j define the axes about which bodies are allowed to rotate. In addition, vectors P_iQ_i and P_jQ_j should be orthogonal to each other. Points R_i and R_j define the x'' axes of the joint reference triads. Universal joint definition data for joint B of model 1 are given in Table 10.3.2.

For spherical joints, P_i and P_j define the common point in the joint. Points Q_i, Q_j, R_i, and R_j are chosen to define the joint reference triads. These six points for the spherical joint at point C in model 1 are defined in Table 10.3.3.

TABLE 10.3.3 Spherical Joint Data, Model 1

Body \ Point	P x'	P y'	P z'	Q x'	Q y'	Q z'	R x'	R y'	R z'
Link ②	0.0	−6.1	0.0	0.0	−6.1	1.0	0.0	−5.1	0.0
Link ③	0.0	−3.7	0.0	0.0	−3.7	1.0	0.0	−2.7	0.0

In model 2 (Fig. 10.3.2), link BC is modeled as the coupler in a spherical–spherical composite joint. The other joints are the same as in model 1. This mechanism model is defined as follows:

Model 2

Bodies	
Three bodies	$nc = 21$
Constraints	
Revolute joints 2 (links 1 and 2, ground): A	5
D	5
Spherical–spherical joint (link 1, link 2): BC	1
Ground constraints (link 3)	6
Euler parameter normalization constraints	3
	$nh = 20$

DOF = 21 − 20 = 1.

To define the spherical–spherical joint between bodies 1 and 2, points P_1 and P_2 are used to locate the spherical joint on each body. Points Q_1, Q_2, R_1, and R_2

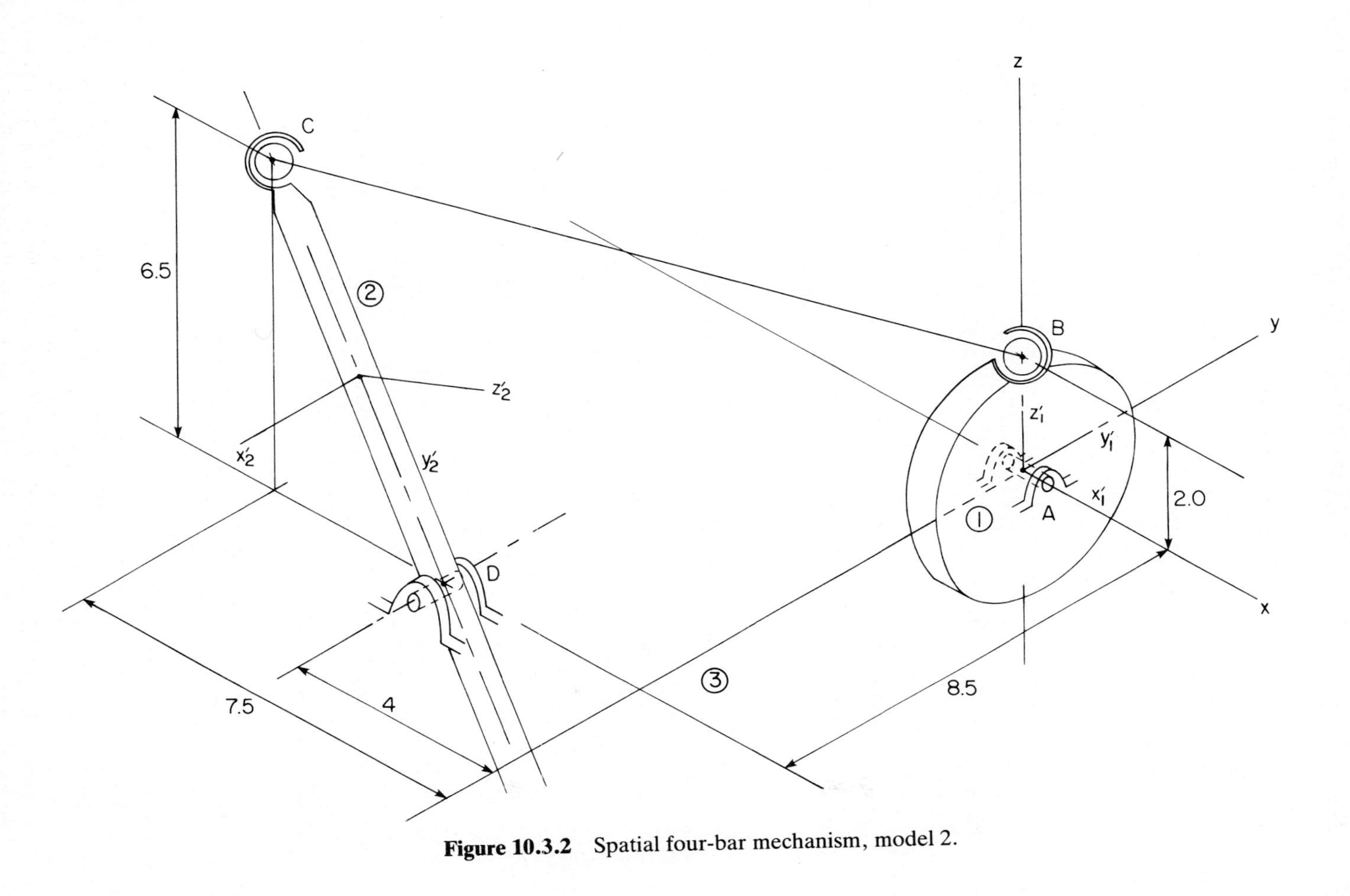

Figure 10.3.2 Spatial four-bar mechanism, model 2.

TABLE 10.3.4 Spherical–Spherical Joint Data, Model 2

Body \ Point	P x'	P y'	P z'	Q x'	Q y'	Q z'	R x'	R y'	R z'
Link ①	0.0	0.0	2.0	0.0	1.0	2.0	1.0	0.0	2.0
Link ②	0.0	−3.7	0.0	1.0	−3.7	0.0	0.0	−2.7	0.0

are chosen to define the z'' and x'' axes of the joint reference triads. The length of the link is the distance between points P_1 and P_2, 12.19 m in this model. Spherical–spherical joint point definition data for model 2 are given in Table 10.3.4.

10.3.2 Assembly Analysis

Initial estimates for the position and orientation generalized coordinates (abbreviated GC) of each body reference frame with respect to the global reference frame are given in Table 10.3.5. Euler parameters are used to specify the

TABLE 10.3.5 Position and Orientation Estimates, Model 1

Body \ GC	x	y	z	e_1	e_2	e_3
Link ①	0.0	0.0	0.0	0.0	0.0	0.0
Link ②	−3.75	−4.25	4.25	−0.29	−0.27	−0.26
Link ③	−5.75	−8.5	3.25	−0.36	0.36	−0.61
Ground ④	0.0	0.0	0.0	0.0	0.0	0.0

orientation of each body. Table 10.3.6 shows the resulting assembled configuration.

For model 2, the position and orientation of link 2 are not required, because it is not modeled as a body. The remaining data in Tables 10.3.5 and 10.3.6 are valid for model 2.

TABLE 10.3.6 Assembled Configuration, Model 1

Body \ GC	x	y	z	e_0	e_1	e_2	e_3
Link ①	0.00016	0.00005	−0.00027	1.0	0.00001	−0.00004	0.00005
Link ②	−3.75300	−4.25020	4.25530	0.87944	−0.29098	−0.27410	−0.25910
Link ③	−5.75300	−8.50010	3.25530	0.60687	−0.36245	0.36247	−0.60684
Ground ④	0.0	0.0	0.0	1.0	0.0	0.0	0.0

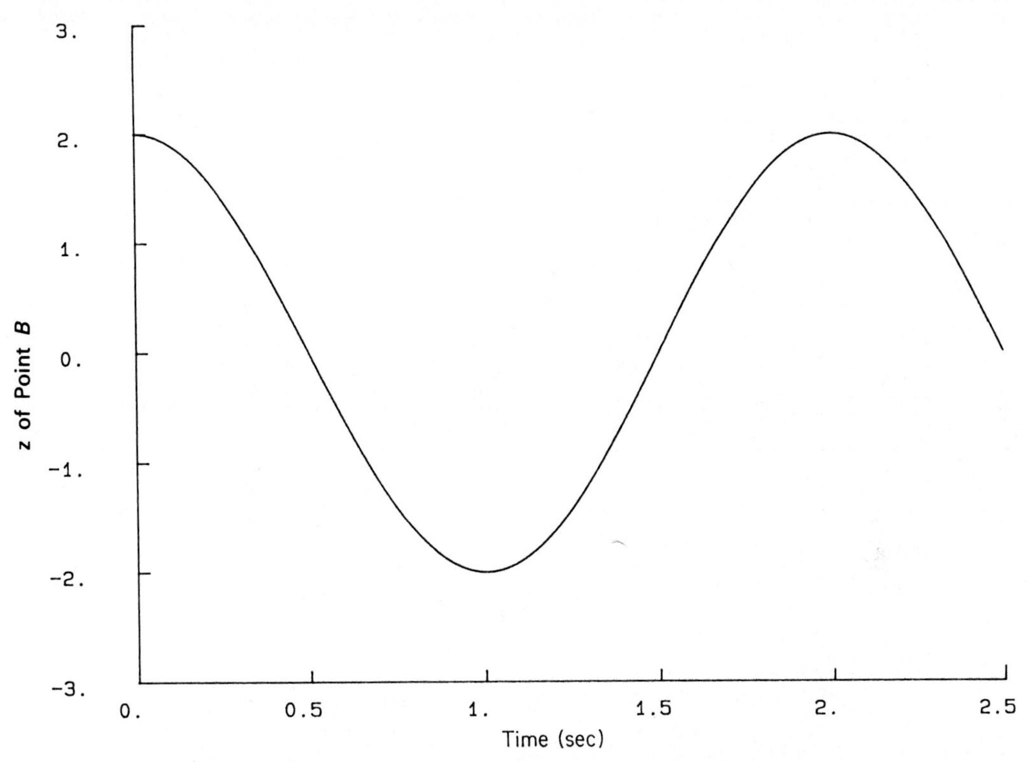

Figure 10.3.3 z coordinate of point B versus time, model 1.

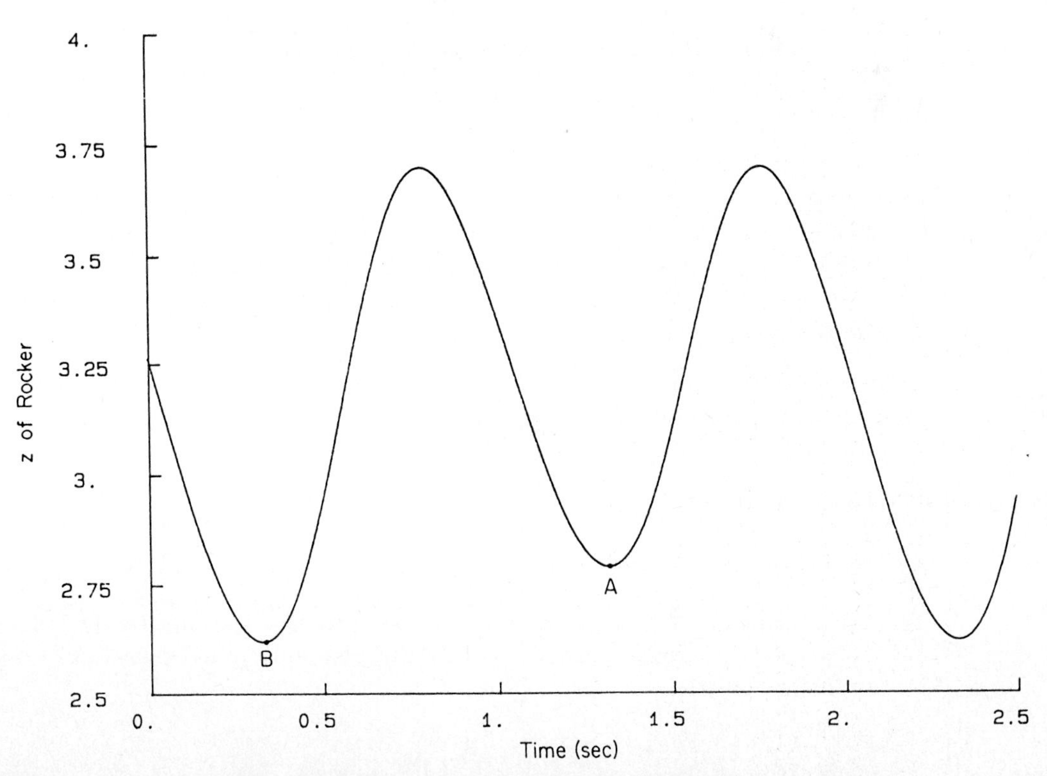

Figure 10.3.4 z coordinate of body 3 versus time, model 1.

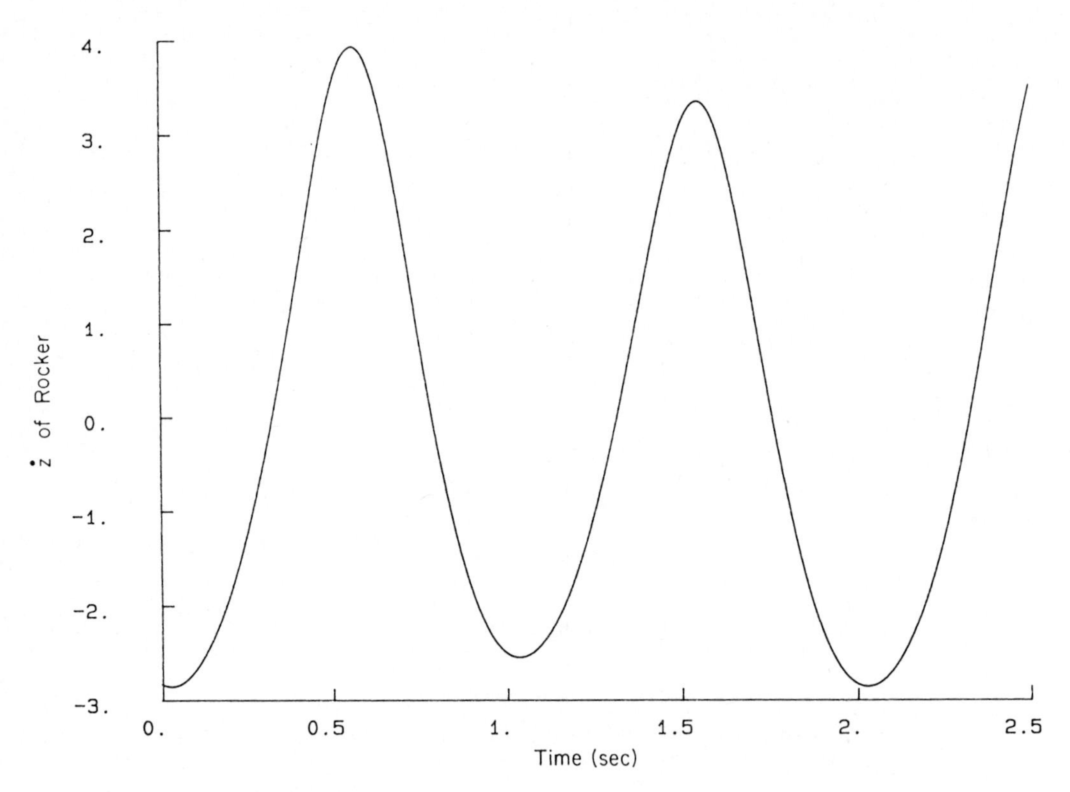

Figure 10.3.5 z velocity of body 3 versus time, model 1.

10.3.3 Driver Specification

Since each model has one kinematic degree of freedom, one driver is specified to rotate link AB about the global x axis. Taking the relative angle θ as the driven coordinate, it is specified that link AB rotate at π rad/s. The driver is thus

$$\theta = \pi t$$

10.3.4 Analysis

Some typical plots of the results for model 1 are shown in Figs. 10.3.3 to 10.3.5. Identical results are obtained for model 2.

10.4 KINEMATIC ANALYSIS OF AN AIR COMPRESSOR

10.4.1 Model

The air compressor shown in Fig. 10.4.1 has six pistons. The disk (body 3) has one end of each of the six connecting rods attached and evenly spaced (60°) on

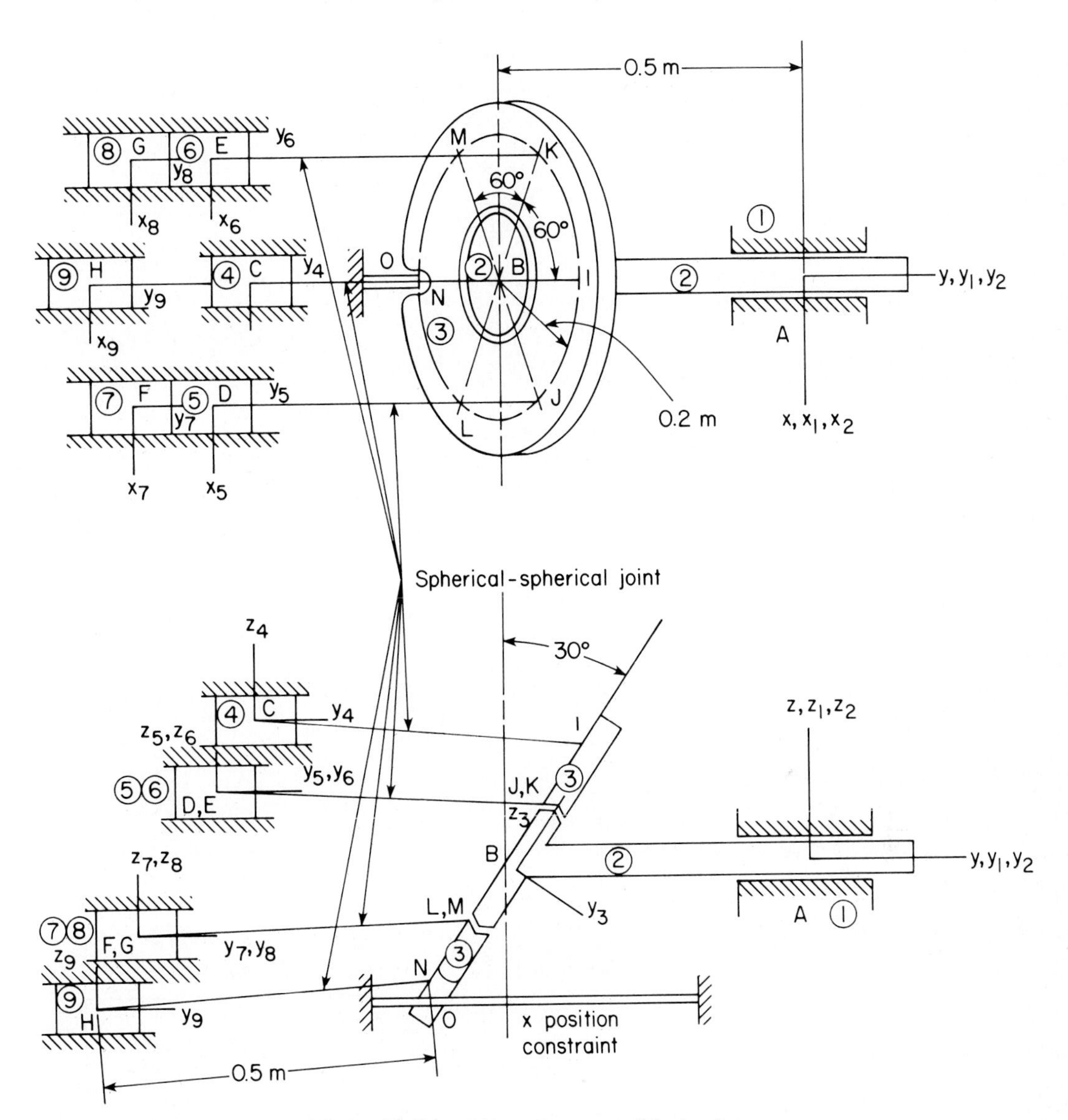

Figure 10.4.1 Air compressor with six pistons.

the circumference of a circle of radius 0.2 m. The disk is connected to the rotor (body 2) by a revolute joint whose axis of rotation is perpendicular to the disk and 30° from the axis of the rotor. As the rotor turns, the disk is prevented from rotating by an x position absolute constraint at point O, which models a slot in the disk through which a bar parallel to the global y axis passes. Canting of the disk generates reciprocating motion of the pistons as the rotor turns. Unlike the connecting rods of an automobile engine, the connecting rods of the compressor have spherical joints at each end. They are modeled here as spherical–spherical joints between the disk and pistons. The compressor model is defined as follows:

Model	
Bodies	
Nine bodies	$nc = 63$
Constraints	
Revolute joints (bodies 1 and 3, 2): A	5
B	5
Translational joints (bodies 4, 5, 6, 7, 8, and 9, 1)	
C, D, E, F, G, and H	6×5
Distance constraints (bodies 4, 5, 6, 7, 8, and 9, 3)	
C–I, D–J, E–K, F–L, G–M, and H–N	6×1
Position constraint (body 3): O	1
Ground constraints	6
Euler parameter normalization constraints	9
	$nh = 62$
$\text{DOF} = 63 - 62 = 1$.	

Revolute joint data, translational joint data, and spherical–spherical joint data are summarized in Tables 10.4.1 to 10.4.3.

TABLE 10.4.1 Revolute Joint Data

Joint	Body \ Point		P x'	P y'	P z'	Q x'	Q y'	Q z'	R x'	R y'	R z'
A	Ground	①	0.0	0.0	0.0	0.0	1.0	0.0	1.0	0.0	0.0
	Rotor	②	0.0	0.0	0.0	0.0	1.0	0.0	1.0	0.0	0.0
B	Rotor	②	0.0	−0.5	0.0	0.0	0.3660	−0.5	1.0	−0.5	0.0
	Disk	③	0.0	0.0	0.0	0.0	1.0	0.0	1.0	0.0	0.0

10.4.2 Assembly

The position and orientation of each body reference frame in the global frame is estimated for initial assembly analysis. Table 10.4.4 provides position and orientation estimates for the generalized coordinates (abbreviated GC), using Euler parameters to specify the orientation of each body. Table 10.4.5 shows the resulting assembled configuration.

TABLE 10.4.2 Translational Joint Data

Joint	Body \ Point		P x′	P y′	P z′	Q x′	Q y′	Q z′	R x′	R y′	R z′
C	Ground	①	0.0	−1.0	0.2	0.0	0.0	0.2	1.0	−1.0	0.2
	Piston 1	④	0.0	0.0	0.0	0.0	1.0	0.0	1.0	0.0	0.0
D	Ground	①	0.1732	−1.0	0.1	0.1732	0.0	0.1	1.1732	−1.0	0.1
	Piston 2	⑤	0.0	0.0	0.0	0.0	1.0	0.0	1.0	0.0	0.0
E	Ground	①	−0.1732	−1.0	0.1	−0.1732	0.0	0.1	0.8268	−1.0	0.1
	Piston 3	⑥	0.0	0.0	0.0	0.0	1.0	0.0	1.0	0.0	0.0
F	Ground	①	0.1732	−1.0	−0.1	0.1732	0.0	−0.1	1.1732	−1.0	−0.1
	Piston 4	⑦	0.0	0.0	0.0	0.0	1.0	0.0	1.0	0.0	0.0
G	Ground	①	−0.1732	−1.0	−0.1	−0.1732	0.0	−0.1	0.8268	−1.0	−0.1
	Piston 5	⑧	0.0	0.0	0.0	0.0	1.0	0.0	1.0	0.0	0.0
H	Ground	①	0.0	−1.0	−0.2	0.0	0.0	−0.2	1.0	−1.0	−0.2
	Piston 6	⑨	0.0	0.0	0.0	0.0	1.0	0.0	1.0	0.0	0.0

TABLE 10.4.3 Spherical–Spherical Joint Data

Joint	Body \ Point		P x′	P y′	P z′	Q x′	Q y′	Q z′	R x′	R y′	R z′	Distance
C-I	Disk	③	0.0	0.0	0.2	0.0	1.0	0.2	1.0	0.0	0.2	0.5
	Piston 1	④	0.0	0.0	0.0	0.0	1.0	0.0	1.0	0.0	0.0	
D-J	Disk	③	0.1732	0.0	0.1	0.1732	1.0	0.1	1.1732	0.0	0.1	0.5
	Piston 2	⑤	0.0	0.0	0.0	0.0	1.0	0.0	1.0	0.0	0.0	
E-K	Disk	③	−0.1732	0.0	0.1	−0.1732	1.0	0.1	0.8268	0.0	0.1	0.5
	Piston 3	⑥	0.0	0.0	0.0	0.0	1.0	0.0	1.0	0.0	0.0	
F-L	Disk	③	0.1732	0.0	−0.1	0.1732	1.0	−0.1	1.1732	0.0	−0.1	0.5
	Piston 4	⑦	0.0	0.0	0.0	0.0	1.0	0.0	1.0	0.0	0.0	
G-M	Disk	③	−0.1732	0.0	−0.1	−0.1732	1.0	−0.1	0.8268	0.0	−0.1	0.5
	Piston 5	⑧	0.0	0.0	0.0	0.0	1.0	0.0	1.0	0.0	0.0	
H-N	Disk	③	0.0	0.0	−0.2	0.0	1.0	−0.2	1.0	0.0	−0.2	0.5
	Piston 6	⑨	0.0	0.0	0.0	0.0	1.0	0.0	1.0	0.0	0.0	

TABLE 10.4.4 Position and Orientation Estimates

Body \ GC		x	y	z	e_1	e_2	e_3
Ground	①	0.0	0.0	0.0	0.0	0.0	0.0
Rotor	②	0.0	0.0	0.0	0.0	0.0	0.0
Disk	③	0.0	−0.5	0.0	−0.26	0.0	0.0
Piston 1	④	0.0	−1.0	0.2	0.0	0.0	0.0
Piston 2	⑤	0.17	−1.0	0.1	0.0	0.0	0.0
Piston 3	⑥	−0.17	−1.0	0.1	0.0	0.0	0.0
Piston 4	⑦	0.17	−1.0	−0.1	0.0	0.0	0.0
Piston 5	⑧	−0.17	−1.0	−0.1	0.0	0.0	0.0
Piston 6	⑨	0.0	−1.0	−0.2	0.0	0.0	0.0

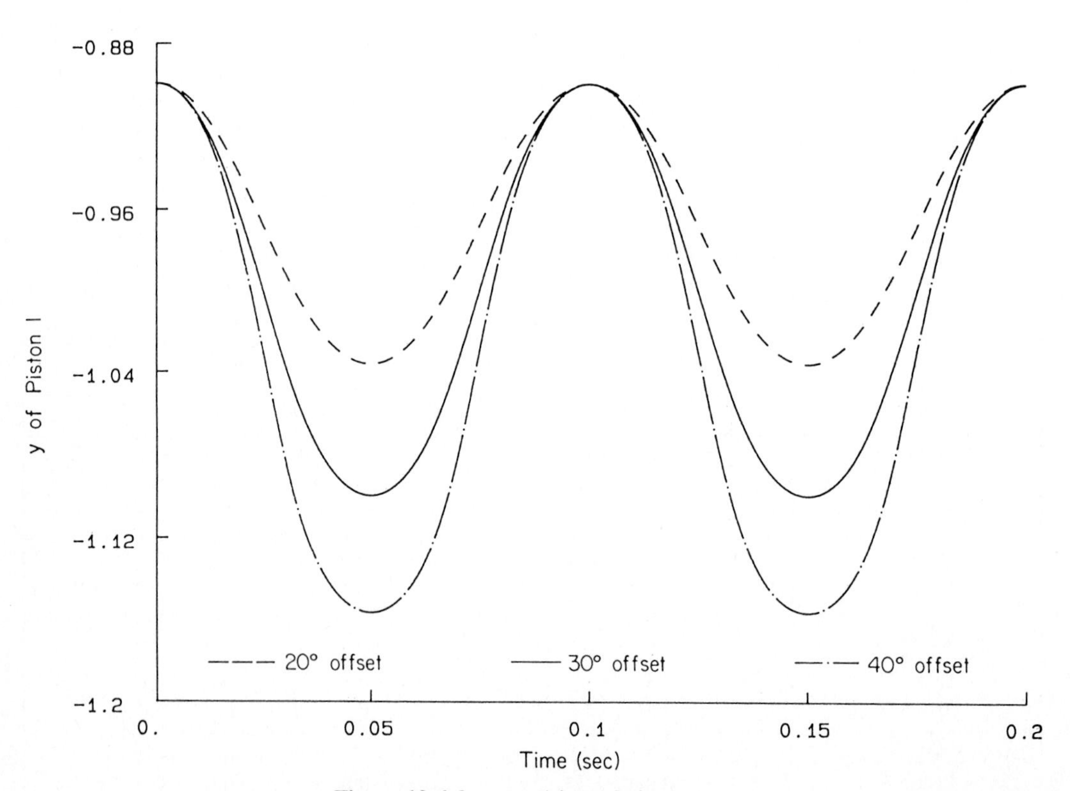

Figure 10.4.2 y position of piston 1.

TABLE 10.4.5 Assembled Configuration

Body \ GC	x	y	z	e_0	e_1	e_2	e_3
Ground ①	0.0	0.0	0.0	1.0	0.0	0.0	0.0
Rotor ②	0.0	−0.00090	0.00026	1.00020	0.00019	0.0	0.0
Disk ③	0.0	−0.50203	0.00020	1.96597	−0.25844	0.0	0.0
Piston 1 ④	0.0	−0.90244	0.19992	1.0	0.0	0.0	0.0
Piston 2 ⑤	0.17317	−0.95243	0.09999	1.0	0.0	0.0	0.0
Piston 3 ⑥	−0.17317	−0.95243	0.09999	1.0	0.0	0.0	0.0
Piston 4 ⑦	0.17317	−1.05130	−0.10002	1.0	0.0	0.0	0.0
Piston 5 ⑧	−0.17317	−1.05130	−0.10002	1.0	0.0	0.0	0.0
Piston 6 ⑨	0.0	−1.10110	−0.20000	1.0	0.0	0.0	0.0

10.4.3 Driver Specification

The rotor speed is chosen as 600 rpm (revolutions per minute), so the relative angle driven in the revolute joint between ground and the rotor is

$$\theta = 62.832t$$

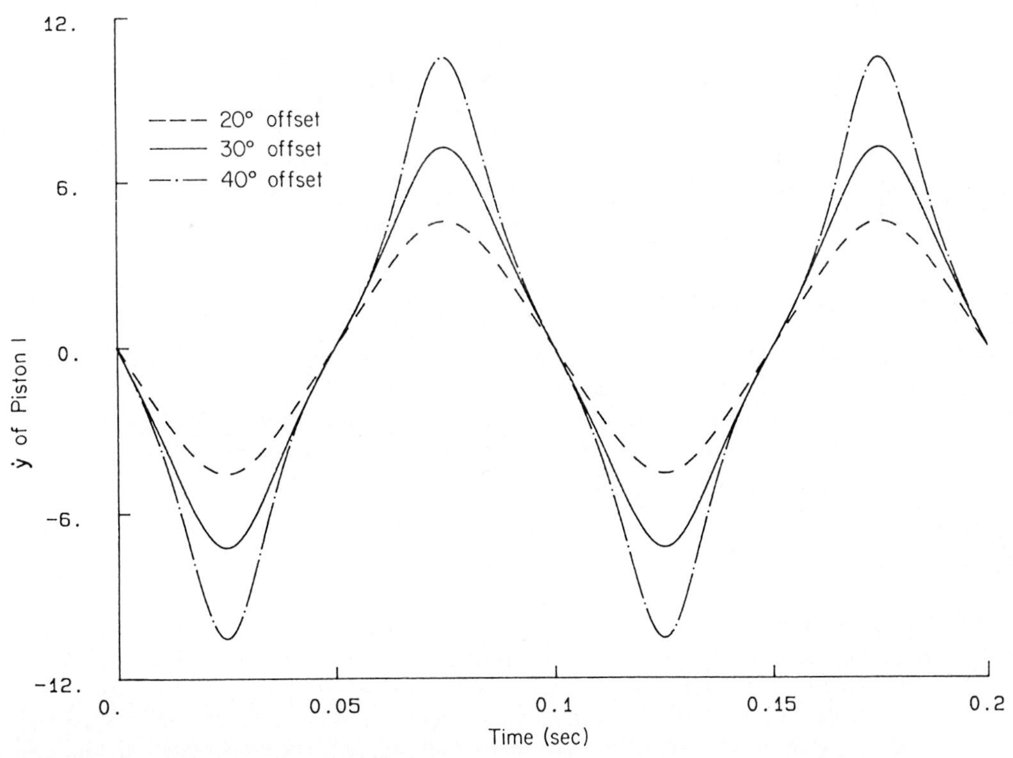

Figure 10.4.3 y velocity of piston 1.

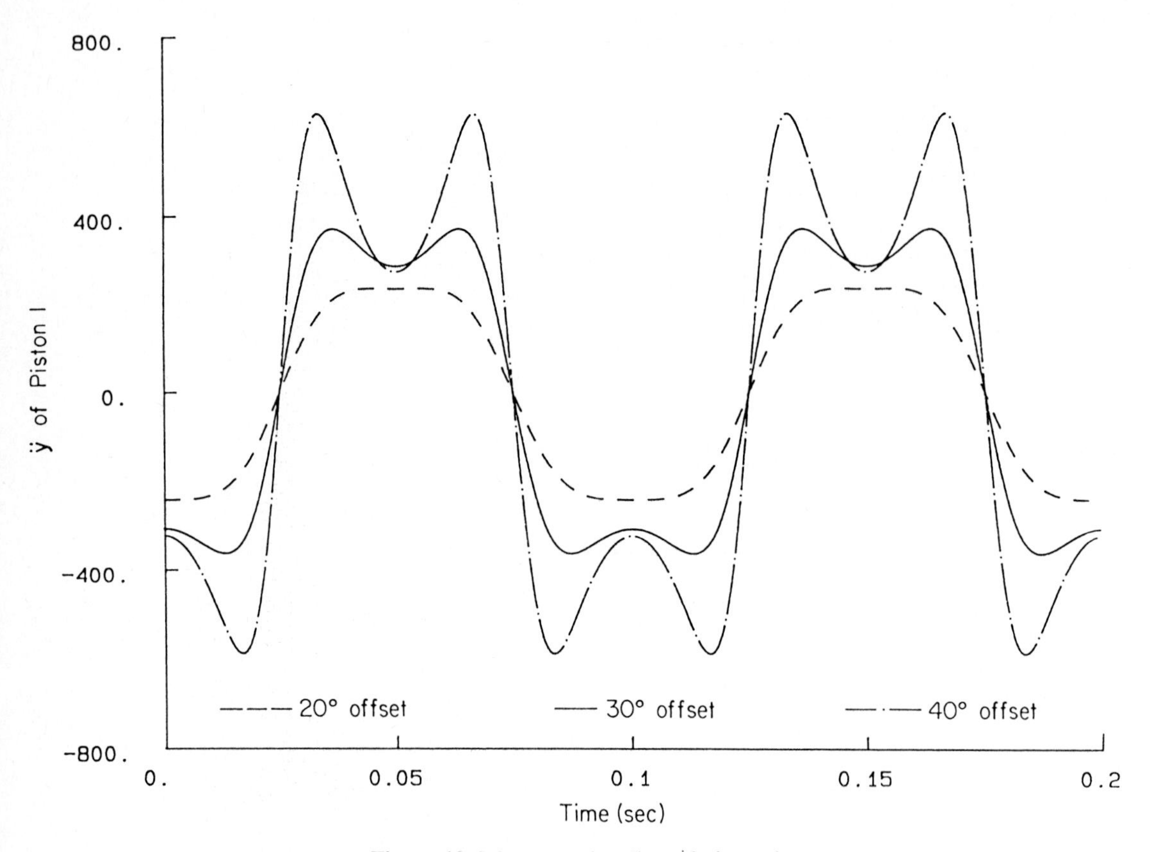

Figure 10.4.4 y acceleration of piston 1.

10.4.4 Analysis

Three different runs are made with the offset angle between the axis of the rotor and the axis of revolution equal to 20°, 30°, and 40°.

As shown in Fig. 10.4.2, the model with 40° offset angle generates the longest stroke of the piston. It also has the most extreme variations in velocity and acceleration of the pistons, as shown in Figs. 10.4.3 and 10.4.4.

PROBLEMS

DADS Projects

10.1. Set up a three-body DADS model of the spatial slider–crank mechanism in Fig. 10.2.1. The bodies are the crank, slider, and ground. Use revolute and translational joints between the crank and ground and the slider and ground, respectively, as in Section 10.2. Model the connecting rod as a spherical–spherical constraint between the crank and slider.

Using the estimates given in Section 10.2.2, carry out assembly and show that identical results are obtained with the new model. Use the driver of Section 10.2.3 and repeat the analysis carried out in Section 10.2.4. Verify that identical results are obtained.

10.2. Set up a fifteen-body DADS model of the air compressor in Fig. 10.4.1. Replace the spherical–spherical joints that model the connecting rods with six bodies. Use a spherical joint to connect each connecting rod to the disk (body 3) and a universal joint to connect it to the associated piston. The purpose of the universal joint is to control rotation of the connecting rod about its own axis.

Repeat the analysis of Sections 10.4.2 to 10.4.4 and show that identical results are obtained.

CHAPTER ELEVEN

Dynamics of Spatial Systems

Many of the basic ideas used in Chapter 6 for planar dynamics are applicable for spatial dynamics. The main difference is that angular velocity effects in spatial dynamics are more intricate. They lead to nonlinear terms in the equations of motion of even a single body. A self-contained derivation of the equations of motion of spatial rigid body systems is presented here, including a definition of the inertia matrix and development of its basic properties. The numerical aspects of spatial dynamic analysis are presented in this chapter only as regards differences from corresponding methods for the dynamics of planar systems in Chapter 7.

Greater emphasis is placed in this chapter on the formulation of spatial equations of motion for multibody systems and less on manipulation with analytical examples than was the case in the study of planar system dynamics in Chapter 6. The reason for this distinction is the algebraic complexity of representing the orientation of bodies in space and the more extensively nonlinear character of spatial equations of motion than was the case in the study of planar system dynamics. In spite of the analytical complexity of spatial dynamic equations, the mathematical theory and numerical methods developed and illustrated for planar systems in Part I are directly applicable and provide a firm foundation for computer implementation of solution methods. It is suggested that the reader concentrate on the similarity in form of the equations of spatial multibody dynamics and the simpler and more intuitively appealing equations of planar dynamics. Similarity in the form of the equations is exploited in creating reliable computer implementations.

From the viewpoint of the user of established computer software, such as the DADS program [27], essentially all the algebraic and numerical complexity is hidden in the implementation of computer intensive calculations. This happy circumstance is an illustration of a situation in which equations of spatial dynamics that are virtually intractable from the point of view of the human analyst are practically implementable on modern high-speed digital computers, whose ever-increasing power makes spatial kinematic and dynamic analysis practical. It is the responsibility of the engineer to fully understand the basic formulation, however, in order to avoid analytical and numerical difficulties. As is shown in this chapter, if physically meaningful models are generated and care is taken to define the data that characterize spatial systems, reliable numerical results may be obtained and understood.

11.1 EQUATIONS OF MOTION OF A RIGID BODY

Equations of Motion from Newton's Laws Consider the rigid body shown in Fig. 11.1.1, which is located in space by a vector $\mathbf{r}$ and a set of generalized coordinates that defines the orientation of the x'-y'-z' body-fixed frame, relative to an inertial x-y-z reference frame. A differential mass $dm(P)$ at a typical point P is located on the body vector $\mathbf{s}^P$; that is, $\mathbf{s}'^P$ is constant. As a model of a rigid body, let a distance constraint act between each differential element in the body.

Forces that act on a differential element of mass at point P include the external force $\mathbf{F}(P)$ per unit of mass at point P and the internal force $\mathbf{f}(P, R)$ per unit of masses at points P and R, as shown in Fig. 11.1.1. Internal forces in this model of a rigid body are due only to gravitational interaction and distance constraints, so they are collinear (see the historical footnote on this subject in Section 6.1).

Newton's equation of motion for differential mass $dm(P)$ is

$$\ddot{\mathbf{r}}^P\,dm(P) - \mathbf{F}(P)\,dm(P) - \int_m \mathbf{f}(P, R)\,dm(R)\,dm(P) = \mathbf{0} \qquad \textbf{(11.1.1)}$$

where integration of the internal force $\mathbf{f}(P, R)$ is taken over the entire body.

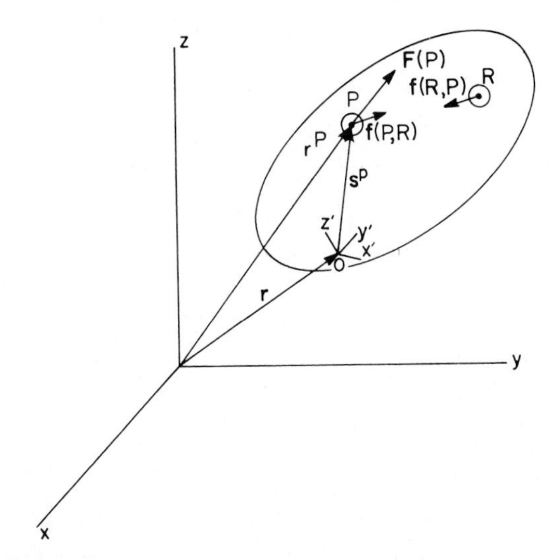

Figure 11.1.1 Forces acting on a rigid body in space.

Let $\delta\mathbf{r}^P$ denote a *virtual displacement* of point P; that is, an infinitesimal variation in the location of point P that is consistent with the allowed motion of point P. Premultiplying both sides of Eq. 11.1.1 by $\delta\mathbf{r}^{PT}$ and integrating over the total mass of the body yields

$$\int_m \delta\mathbf{r}^{PT}\ddot{\mathbf{r}}^P\,dm(p) - \int_m \delta\mathbf{r}^{PT}\mathbf{F}(P)\,dm(P) - \int_m\int_m \delta\mathbf{r}^{PT}\mathbf{f}(P, R)\,dm(R)\,dm(P) = 0 \tag{11.1.2}$$

Manipulation of the double integral that appears in Eq. 11.1.2 shows that (Prob. 11.1.1)

$$\begin{aligned}\int_m\int_m &\delta\mathbf{r}^{PT}\mathbf{f}(P, R)\,dm(R)\,dm(P)\\ &= \frac{1}{2}\int_m\int_m \delta\mathbf{r}^{PT}\mathbf{f}(P, R)\,dm(R)\,dm(P)\\ &\quad + \frac{1}{2}\int_m\int_m \delta\mathbf{r}^{RT}\mathbf{f}(R, P)\,dm(P)\,dm(R)\\ &= \frac{1}{2}\int_m\int_m (\delta\mathbf{r}^P - \delta\mathbf{r}^R)^T\mathbf{f}(P, R)\,dm(P)\,dm(R)\end{aligned} \tag{11.1.3}$$

where the first equality simply rewrites the expression and interchanges dummy variables P and R of integration in the second integral. The second equality follows from interchange of the order of integration and uses Newton's third law of action and reaction; that is, $\mathbf{f}(P, R)\,dm(P)\,dm(R) = -\mathbf{f}(R, P)\,dm(R)\,dm(P)$.

For a rigid body, with distance constraints between all points,

$$(\mathbf{r}^P - \mathbf{r}^R)^T(\mathbf{r}^P - \mathbf{r}^R) = C$$

Taking the differential of both sides yields

$$(\delta\mathbf{r}^P - \delta\mathbf{r}^R)^T(\mathbf{r}^P - \mathbf{r}^R) = 0 \tag{11.1.4}$$

Since the internal force $\mathbf{f}(P, R)$ in the present model of a rigid body acts between points P and R; that is,

$$\mathbf{f}(P, R) = k(\mathbf{r}^P - \mathbf{r}^R) \tag{11.1.5}$$

where k is a constant, from Eq. 11.1.4, the right side of Eq. 11.1.3 is zero.

Collinearity of $\mathbf{f}(P, R)$ and $\mathbf{f}(R, P)$ (i.e., Eq. 11.1.5) is not implied by Newton's third law for general force fields. Examples of noncollinearity of $\mathbf{f}(P, R)$ and $\mathbf{f}(R, P)$ include charged particles in an electric field, dipoles in a magnetic field, and particles in a nonuniform gravitational field. In the presence of such effects, the integral on the right of Eq. 11.1.3 may not be zero and such forces must be accounted for as externally applied forces. Note that the right side of Eq. 11.1.3 is the virtual work of the internal forces in a rigid body. Attention in this text is focused on rigid bodies, for which the internal forces are workless.

Substituting Eqs. 11.1.4 and 11.1.5 into Eq. 11.1.3,

$$\int_m \int_m \delta\mathbf{r}^{PT}\mathbf{f}(P, R)\, dm(P)\, dm(R) = 0 \qquad \textbf{(11.1.6)}$$

Using this result, Eq. 11.1.2 simplifies to

$$\int_m \delta\mathbf{r}^{PT}\ddot{\mathbf{r}}^P\, dm(P) - \int_m \delta\mathbf{r}^{PT}\mathbf{F}(P)\, dm(P) = 0 \qquad \textbf{(11.1.7)}$$

which must hold for all virtual displacements $\delta\mathbf{r}^P$ that are consistent with constraints on the body.

To take full advantage of Eq. 11.1.7, the virtual displacement of point P may be written in terms of a virtual displacement of the origin of the x'-y'-z' frame and a virtual rotation of the body. From Eq. 9.2.49, a virtual displacement of point P is

$$\delta\mathbf{r}^P = \delta\mathbf{r} - \mathbf{A}\tilde{\mathbf{s}}'^P\, \delta\boldsymbol{\pi}' \qquad \textbf{(11.1.8)}$$

Similarly, from Eqs. 9.2.41 and 9.2.43, the acceleration of point P may be written as

$$\begin{aligned}\ddot{\mathbf{r}}^P &= \ddot{\mathbf{r}} + \ddot{\mathbf{A}}\mathbf{s}'^P \\ &= \ddot{\mathbf{r}} + \mathbf{A}\dot{\tilde{\boldsymbol{\omega}}}'\mathbf{s}'^P + \mathbf{A}\tilde{\boldsymbol{\omega}}'\tilde{\boldsymbol{\omega}}'\mathbf{s}'^P \end{aligned} \qquad \textbf{(11.1.9)}$$

Substituting Eqs. 11.1.8 and 11.1.9 into the variational equation of Eq. 11.1.7 yields

$$\int_m (\delta\mathbf{r}^T + \delta\boldsymbol{\pi}'^T\tilde{\mathbf{s}}'^P\mathbf{A}^T)(\ddot{\mathbf{r}} + \mathbf{A}\dot{\tilde{\boldsymbol{\omega}}}'\mathbf{s}'^P + \mathbf{A}\tilde{\boldsymbol{\omega}}'\tilde{\boldsymbol{\omega}}'\mathbf{s}'^P)\, dm(P)$$

$$- \int_m (\delta\mathbf{r}^T + \delta\boldsymbol{\pi}'^T\tilde{\mathbf{s}}'^P\mathbf{A}^T)\mathbf{F}(P)\, dm(P) = 0 \qquad \textbf{(11.1.10)}$$

for all $\delta\mathbf{r}$ and $\delta\boldsymbol{\pi}'$ that are consistent with constraints that act on the body. Expanding the integrals of Eq. 11.1.10 yields (Prob. 11.1.2)

$$\begin{aligned}\delta\mathbf{r}^T\ddot{\mathbf{r}}\int_m dm(P) &+ \delta\mathbf{r}^T(\mathbf{A}\dot{\tilde{\boldsymbol{\omega}}}' + \mathbf{A}\tilde{\boldsymbol{\omega}}'\tilde{\boldsymbol{\omega}}')\int_m \mathbf{s}'^P\, dm(P) \\ &+ \delta\boldsymbol{\pi}'^T\int_m \tilde{\mathbf{s}}'^P\, dm(P)\mathbf{A}^T\ddot{\mathbf{r}} + \delta\boldsymbol{\pi}'^T\int_m \tilde{\mathbf{s}}'^P\dot{\tilde{\boldsymbol{\omega}}}'\mathbf{s}'^P\, dm(P) \\ &+ \delta\boldsymbol{\pi}'^T\int_m \tilde{\mathbf{s}}'^P\, \tilde{\boldsymbol{\omega}}'\tilde{\boldsymbol{\omega}}'\mathbf{s}'^P\, dm(P) - \delta\mathbf{r}^T\int_m \mathbf{F}(P)\, dm(P) \\ &- \delta\boldsymbol{\pi}'^T\int_m \tilde{\mathbf{s}}'^P\mathbf{F}'(P)\, dm(P) = 0 \end{aligned} \qquad \textbf{(11.1.11)}$$

for all $\delta\mathbf{r}$ and $\delta\boldsymbol{\pi}'$ that are consistent with constraints.

Equations of Motion with Centroidal Body-Fixed Reference Frame To simplify the form of Eq. 11.1.11, a body-fixed x'-y'-z' reference frame is selected with its origin at the *center of mass* (or *centroid*) of the body. It is thus called a *centroidal body reference frame.* The total mass is

$$m \equiv \int_m dm(P) \tag{11.1.12}$$

and, by definition of the centroid,

$$\int_m \mathbf{s}'^P \, dm(P) = \mathbf{0} \tag{11.1.13}$$

The total external force acting on the body is simply

$$\mathbf{F} \equiv \int_m \mathbf{F}(P) \, dm(P) \tag{11.1.14}$$

and the *moment* (or *torque*) of the external forces with respect to the origin of the x'-y'-z' frame is

$$\mathbf{n}' \equiv \int_m \tilde{\mathbf{s}}'^P \mathbf{F}'(P) \, dm(P) \tag{11.1.15}$$

The fourth integral in Eq. 11.1.11 may be written as

$$\begin{aligned} \int_m \tilde{\mathbf{s}}'^P \tilde{\dot{\boldsymbol{\omega}}}' \mathbf{s}'^P \, dm(P) &= -\left(\int_m \tilde{\mathbf{s}}'^P \tilde{\mathbf{s}}'^P \, dm(P) \right) \dot{\boldsymbol{\omega}}' \\ &\equiv \mathbf{J}' \dot{\boldsymbol{\omega}}' \end{aligned} \tag{11.1.16}$$

where the constant *inertia matrix* (or *inertia tensor*) $\mathbf{J}'$ with respect to the x'-y'-z' centroidal frame is defined as

$$\begin{aligned} \mathbf{J}' &\equiv -\int_m \tilde{\mathbf{s}}'^P \tilde{\mathbf{s}}'^P \, dm(P) \\ &= \int_m \begin{bmatrix} (y'^P)^2 + (z'^P)^2 & -x'^P y'^P & -x'^P z'^P \\ -x'^P y'^P & (x'^P)^2 + (z'^P)^2 & -y'^P z'^P \\ -x'^P z'^P & -y'^P z'^P & (x'^P)^2 + (y'^P)^2 \end{bmatrix} dm(P) \end{aligned} \tag{11.1.17}$$

where each term in the matrix on the right is integrated separately. The diagonal elements in the matrix $\mathbf{J}'$ are called *moments of inertia* and the off-diagonal terms are called *products of inertia.* Since symmetrically placed off-diagonal elements are equal, $\mathbf{J}'$ is symmetric; that is, $\mathbf{J}' = \mathbf{J}'^T$.

The integrand of the fifth integral in Eq. 11.1.11 may be expanded, using Eqs. 9.1.25 and 9.1.31, as

$$\begin{aligned} \tilde{\mathbf{s}}'^P \tilde{\boldsymbol{\omega}}' \tilde{\boldsymbol{\omega}}' \mathbf{s}'^P &= -\tilde{\mathbf{s}}'^P \tilde{\boldsymbol{\omega}}' \tilde{\mathbf{s}}'^P \boldsymbol{\omega}' \\ &= -\tilde{\boldsymbol{\omega}}' \tilde{\mathbf{s}}'^P \tilde{\mathbf{s}}'^P \boldsymbol{\omega}' - \boldsymbol{\omega}' \mathbf{s}'^{PT} \tilde{\mathbf{s}}'^P \boldsymbol{\omega}' - \mathbf{s}'^P \boldsymbol{\omega}'^T \tilde{\boldsymbol{\omega}}' \mathbf{s}'^P \\ &= -\tilde{\boldsymbol{\omega}}' \tilde{\mathbf{s}}'^P \tilde{\mathbf{s}}'^P \boldsymbol{\omega}' \end{aligned}$$

Integrating both sides over the mass of the body yields

$$\int_m \tilde{\mathbf{s}}'^P \tilde{\boldsymbol{\omega}}' \tilde{\boldsymbol{\omega}}' \mathbf{s}'^P \, dm(P) = \tilde{\boldsymbol{\omega}}' \left(- \int_m \tilde{\mathbf{s}}'^P \tilde{\mathbf{s}}'^P \, dm(P) \right) \boldsymbol{\omega}' = \tilde{\boldsymbol{\omega}}' \mathbf{J}' \boldsymbol{\omega}' \qquad \mathbf{(11.1.18)}$$

Substituting Eqs. 11.1.12 to 11.1.18 into Eq. 11.1.11 yields the *variational Newton–Euler equations of motion* for a rigid body with a centroidal body-fixed reference frame,

$$\delta \mathbf{r}^T [m\ddot{\mathbf{r}} - \mathbf{F}] + \delta \boldsymbol{\pi}'^T [\mathbf{J}' \dot{\boldsymbol{\omega}}' + \tilde{\boldsymbol{\omega}}' \mathbf{J}' \boldsymbol{\omega}' - \mathbf{n}'] = 0 \qquad \mathbf{(11.1.19)}$$

which must hold for all virtual displacements $\delta \mathbf{r}$ and virtual rotations $\delta \boldsymbol{\pi}'$ of the centroidal x'-y'-z' frame that are consistent with constraints that act on the body.

Example 11.1.1: A body shown in Fig. 11.1.2 is constrained so that point P on its z' axis moves on the surface of a unit sphere. This constraint is defined by

$$\Phi = (\mathbf{r} + \mathbf{A}\mathbf{h}')^T (\mathbf{r} + \mathbf{A}\mathbf{h}') - 1 = 0$$

where $\mathbf{h}'$ is the z' axis unit vector, $\mathbf{h}' = [0, 0, 1]^T$. Taking the variation of both sides of this equation yields

$$\begin{aligned} \delta\Phi &= 2(\mathbf{r} + \mathbf{A}\mathbf{h}')^T (\delta\mathbf{r} + \delta\mathbf{A}\mathbf{h}') \\ &= 2(\mathbf{r} + \mathbf{A}\mathbf{h}')^T \delta\mathbf{r} + 2(\mathbf{r} + \mathbf{A}\mathbf{h}')^T \mathbf{A}\, \delta\tilde{\boldsymbol{\pi}}'\mathbf{h}' \\ &= 2(\mathbf{r} + \mathbf{A}\mathbf{h}')^T \delta\mathbf{r} - (2\mathbf{r}^T \mathbf{A}\tilde{\mathbf{h}}' + 2\mathbf{h}'^T \tilde{\mathbf{h}}')\, \delta\boldsymbol{\pi}' \\ &= 2(\mathbf{r} + \mathbf{A}\mathbf{h}')^T \delta\mathbf{r} - (2\mathbf{r}^T \mathbf{A}\tilde{\mathbf{h}}')\, \delta\boldsymbol{\pi}' = 0 \end{aligned}$$

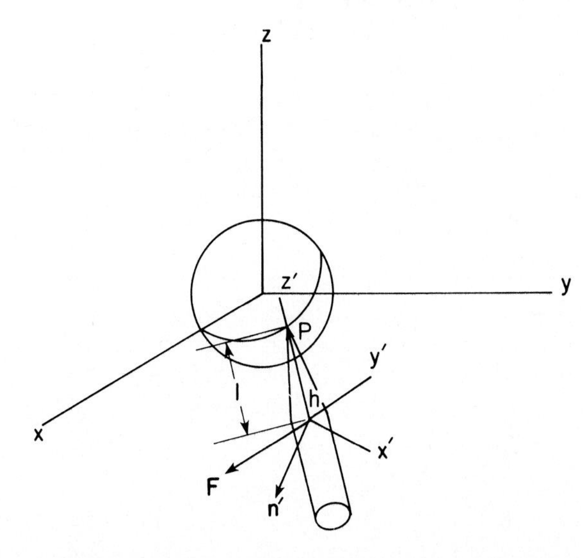

Figure 11.1.2 Pendulum with point on unit sphere.

Since Eq. 11.1.19 must hold for all $\delta\mathbf{r}$ and $\delta\boldsymbol{\pi}'$ that satisfy the above equation, the Lagrange multiplier theorem guarantees the existence of a multiplier λ such that

$$\delta\mathbf{r}^T[m\ddot{\mathbf{r}} - \mathbf{F} + 2\lambda(\mathbf{r} + \mathbf{A}\mathbf{h}')] + \delta\boldsymbol{\pi}'^T[\mathbf{J}'\dot{\boldsymbol{\omega}}' + \tilde{\boldsymbol{\omega}}'\mathbf{J}'\boldsymbol{\omega}' - \mathbf{n}' + 2\lambda\tilde{\mathbf{h}}'\mathbf{A}^T\mathbf{r}] = 0$$

which must hold for all $\delta\mathbf{r}$ and $\delta\boldsymbol{\pi}'$. Thus, the equations of motion are

$$m\ddot{\mathbf{r}} + 2(\mathbf{r} + \mathbf{A}\mathbf{h}')\lambda = \mathbf{F}$$

$$\mathbf{J}'\dot{\boldsymbol{\omega}}' + 2\tilde{\mathbf{h}}'\mathbf{A}^T\mathbf{r}\lambda = \mathbf{n}' - \tilde{\boldsymbol{\omega}}'\mathbf{J}'\boldsymbol{\omega}'$$

and the constraint equation.

If no constraints act on a body, (i.e., it is free to move in space), then $\delta\mathbf{r}$ and $\delta\boldsymbol{\pi}'$ are arbitrary and their coefficients in Eq. 11.1.19 must be zero. This yields the *Newton–Euler equations of motion* for an unconstrained body:

$$\begin{aligned} m\ddot{\mathbf{r}} &= \mathbf{F} \\ \mathbf{J}'\dot{\boldsymbol{\omega}}' &= \mathbf{n}' - \tilde{\boldsymbol{\omega}}'\mathbf{J}'\boldsymbol{\omega}' \end{aligned} \tag{11.1.20}$$

The Newton–Euler equations of Eq. 11.1.20 may be thought of as differential equations in $\mathbf{r}$ and $\boldsymbol{\omega}'$. In some cases, they may be integrated for these variables. Integration for orientation of the body is possible, however, only after a set of orientation generalized coordinates, such as Euler parameters, is introduced. This is required, since $\boldsymbol{\omega}'$ cannot be integrated directly (Prob. 11.1.3).

Equations of Motion for a Body with One Point Fixed in Space Consider the special case of a body, such as the top shown in Fig. 11.1.3, with a point O'' fixed in the inertial reference frame. Since point O'' will not generally be the center of mass of the body, to avoid confusion, define an x''-y''-z'' body-fixed reference frame with origin at point O''. Since point O'' is

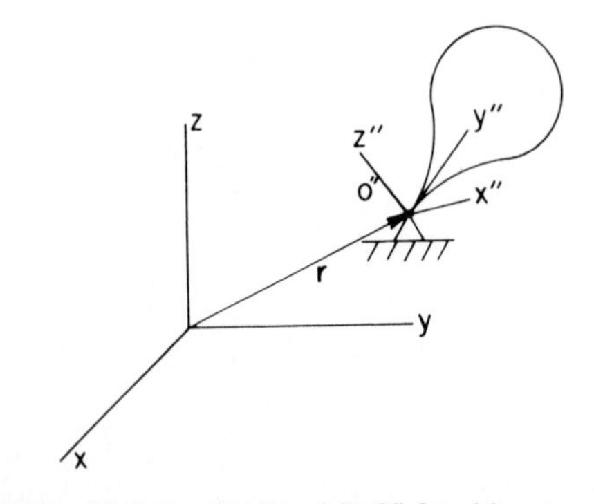

Figure 11.1.3 Body with O'' fixed in space.

fixed in space, **r** is constant, its velocity and acceleration are zero, and kinematically admissible virtual displacements of point O'' are zero; that is,

$$\begin{aligned}\mathbf{r} &= \mathbf{c}\\ \dot{\mathbf{r}} &= \ddot{\mathbf{r}} = \mathbf{0}\\ \delta\mathbf{r} &= \mathbf{0}\end{aligned} \tag{11.1.21}$$

The general variational form of the equations of motion in a noncentroidal reference frame of Eq. 11.1.11 can be written in terms of the x''-y''-z'' frame, using the relations of Eq. 11.1.21. Using Eqs. 11.1.15 to 11.1.18, all of which are valid for the noncentroidal x''-y''-z'' reference frame, Eq. 11.1.11 reduces to

$$\delta\boldsymbol{\pi}''^{T}[\mathbf{J}''\dot{\boldsymbol{\omega}}'' + \tilde{\boldsymbol{\omega}}''\mathbf{J}''\boldsymbol{\omega}'' - \mathbf{n}''] = 0 \tag{11.1.22}$$

for all virtual rotations $\delta\boldsymbol{\pi}''$ that are consistent with any additional constraints that may be imposed on motion of the body.

If no orientation constraints are imposed on the body; for example, if the body of Fig. 11.1.3 is free to move as a pendulum or a spinning top with a spherical joint at point O'', then $\delta\boldsymbol{\pi}''$ of Eq. 11.1.22 is arbitrary and its coefficient must be zero, leading to the *Euler equations of motion*:

$$\mathbf{J}''\dot{\boldsymbol{\omega}}'' = \mathbf{n}'' - \tilde{\boldsymbol{\omega}}''\mathbf{J}''\boldsymbol{\omega}'' \tag{11.1.23}$$

Since these equations are nonlinear in angular velocities that appear on the right side, closed-form solution is generally not possible.

Example 11.1.2: As a special case of a body with a point fixed in space, consider a pendulum that is constrained to rotate about the global y axis, as shown in Fig. 11.1.4. The body-fixed x''-y''-z'' frame is selected so that the y'' axis coincides with the global y axis and is the axis of rotation. By construction of the

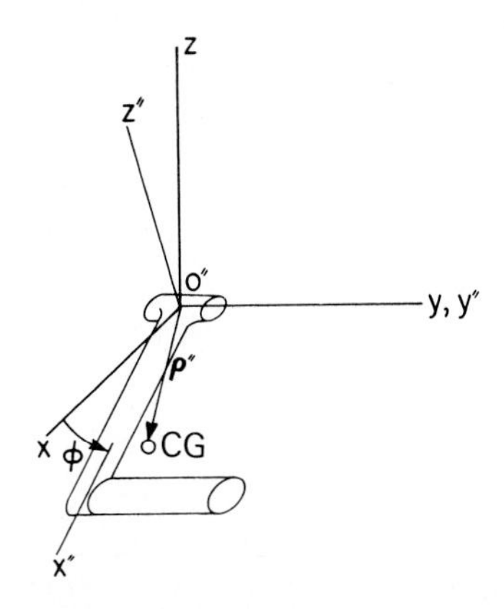

Figure 11.1.4 Pendulum rotating about y axis.

body-fixed reference frame and definition of the angle ϕ of rotation defined in Fig. 11.1.4, the transformation from the x''-y''-z'' frame to the inertial x-y-z frame is

$$\mathbf{s} \equiv \mathbf{A}\mathbf{s}'' = \begin{bmatrix} \cos\phi & 0 & \sin\phi \\ 0 & 1 & 0 \\ -\sin\phi & 0 & \cos\phi \end{bmatrix} \mathbf{s}''$$

A direct calculation indicates that angular velocity, acceleration, and virtual rotation will be about the y'' axis and are related to the angle ϕ by

$$\begin{aligned} \boldsymbol{\omega}'' &= [0, \dot{\phi}, 0]^T \\ \dot{\boldsymbol{\omega}}'' &= [0, \ddot{\phi}, 0]^T \\ \delta\boldsymbol{\pi}'' &= [0, \delta\phi, 0]^T \end{aligned} \tag{11.1.24}$$

Substituting these relations into Eq. 11.1.22 yields the variational form of the equations of motion of the pendulum as

$$[0, \delta\phi, 0]\left\{\begin{bmatrix} J_{x''y''}\ddot{\phi} \\ J_{y''y''}\ddot{\phi} \\ J_{z''y''}\ddot{\phi} \end{bmatrix} + \begin{bmatrix} J_{z''y''}\dot{\phi}^2 \\ 0 \\ -J_{x''y''}\dot{\phi}^2 \end{bmatrix} - \begin{bmatrix} n_{x''} \\ n_{y''} \\ n_{z''} \end{bmatrix}\right\} = 0 \tag{11.1.25}$$

The gravitational force acts at the center of mass and is $\mathbf{F} = -mg\mathbf{k}$, in the x-y-z frame. In the x''-y''-z'' frame, this is

$$\mathbf{F}'' = \mathbf{A}^T\mathbf{F} = mg\begin{bmatrix} \sin\phi \\ 0 \\ -\cos\phi \end{bmatrix}$$

The torque $\mathbf{n}''$ is thus, from Eq. 11.1.15,

$$\mathbf{n}'' = \tilde{\boldsymbol{\rho}}''\mathbf{F}'' = mg\begin{bmatrix} -\rho_{y''}\cos\phi \\ \rho_{z''}\sin\phi + \rho_{x''}\cos\phi \\ -\rho_{y''}\sin\phi \end{bmatrix}$$

Since $\delta\phi$ in Eq. 11.1.25 is arbitrary, its coefficient must be zero, yielding the differential equation of motion of the pendulum as

$$J_{y''y''}\ddot{\phi} = n_{y''} = mg(\rho_{z''}\sin\phi + \rho_{x''}\cos\phi) \tag{11.1.26}$$

Note that only the moment of inertia about the y'' axis of the body appears in the equation of motion.

To determine reaction torques that act at the rotational joint located at point O'' in Fig. 11.1.4, the general form of reaction torque may be written as

$$\mathbf{T}'' = [T_{x''}, 0, T_{z''}]^T \tag{11.1.27}$$

The y'' component of the reaction torque is zero, since the rotational joint permits free rotation of the body about that axis. Adding the reaction torque of Eq. 11.1.27 to Eq. 1.1.22 yields a variational equation of motion that accounts for all torques acting on the body. Since motion of the body, with point O'' fixed in space, under the action of the reaction torques is equivalent to imposing the rotational joint constraint, the virtual rotation $\delta\boldsymbol{\pi}''$, with $\mathbf{T}''$ included, is arbitrary, and the coefficient of $\delta\boldsymbol{\pi}''$ in Eq. 11.1.22 must be zero, leading to Eq. 11.1.26 and

$$\begin{aligned} T_{x''} &= J_{z''y''}\dot{\phi}^2 + J_{x''y''}\ddot{\phi} - n_{x''} \\ T_{z''} &= -J_{x''y''}\dot{\phi}^2 + J_{z''y''}\ddot{\phi} - n_{z''} \end{aligned} \tag{11.1.28}$$

Thus, if Eq. 11.1.26 is solved for ϕ, $\dot{\phi}$, and $\ddot{\phi}$, the reaction torques due to asymmetry of the pendulum are determined by Eq. 11.1.28. This result shows that the products of inertia have a substantial influence on the forces that act in a system, if asymmetry is present.

11.2 PROPERTIES OF CENTROID AND MOMENTS AND PRODUCTS OF INERTIA

Centroid To find the centroid of a body, let a noncentroidal body-fixed x''-y''-z'' reference frame with origin at O'' be given. Let the centroid be designated as point O', and let the x'-y'-z' reference frame have its origin at O', as shown in Fig. 11.2.1.

The vector $\boldsymbol{\rho}''$ in the x''-y''-z'' frame that locates the centroid O' may be determined by writing $\mathbf{s}'^P = \mathbf{C}(\mathbf{s}''^P - \boldsymbol{\rho}'')$, where $\mathbf{C}$ is the constant orthogonal transformation matrix from the x''-y''-z'' frame to the x'-y'-z' frame; that is,

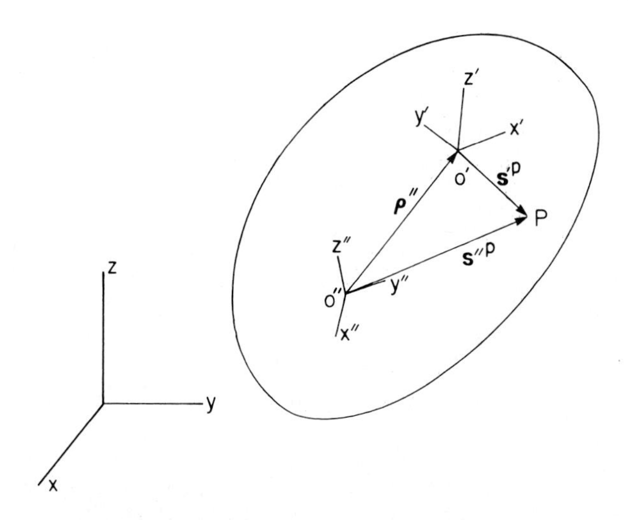

Figure 11.2.1 Location of centroid.

$\mathbf{s}' = \mathbf{C}\mathbf{s}''$. Using Eq. 11.1.13,

$$\mathbf{0} = \int_m \mathbf{s}'^P \, dm(P) = \int_m \mathbf{C}(\mathbf{s}''^P - \boldsymbol{\rho}'') \, dm(P)$$

$$= \mathbf{C} \int_m \mathbf{s}''^P \, dm(P) - \mathbf{C}\boldsymbol{\rho}'' \int_m dm(P)$$

$$= \mathbf{C} \int_m \mathbf{s}''^P \, dm(P) - m \, \mathbf{C}\boldsymbol{\rho}''$$

Premultiplying both sides of this equation by $\mathbf{C}^T = \mathbf{C}^{-1}$,

$$\boldsymbol{\rho}'' = \frac{1}{m} \int_m \mathbf{s}''^P \, dm(P) \tag{11.2.1}$$

Example 11.2.1: Consider the homogeneous body shown in Fig. 11.2.2. By Eq. 11.2.1, its centroid in the x''-y''-z'' frame is

$$\boldsymbol{\rho}'' = \frac{1}{m} \int_0^{10} \int_0^{10} \int_{9-0.2y''}^{11+0.2y''} \gamma [x'', y'', z'']^T \, dz'' \, dy'' \, dx''$$

$$= \frac{\gamma}{m} \int_0^{10} \int_0^{10} [x''(2 + 0.4y''), y''(2 + 0.4y''), 20 + 4y'']^T \, dy'' \, dx''$$

$$= \frac{\gamma}{m} \int_0^{10} \left[40x'', \frac{700}{3}, 400 \right]^T dx''$$

$$= \frac{\gamma}{m} \left[2000, \frac{7000}{3}, 4000 \right]^T$$

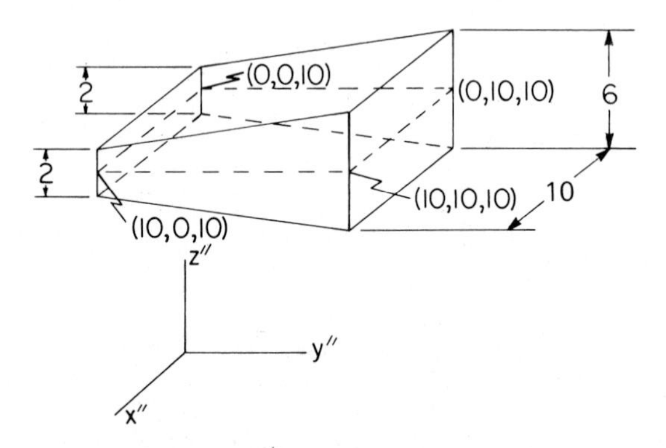

Figure 11.2.2 Homogeneous body.

where γ is mass density of the material. The mass of the body is $m = \gamma 400$. Thus, the centroid is located by

$$\boldsymbol{\rho}'' = [5, \tfrac{35}{6}, 10]^T$$

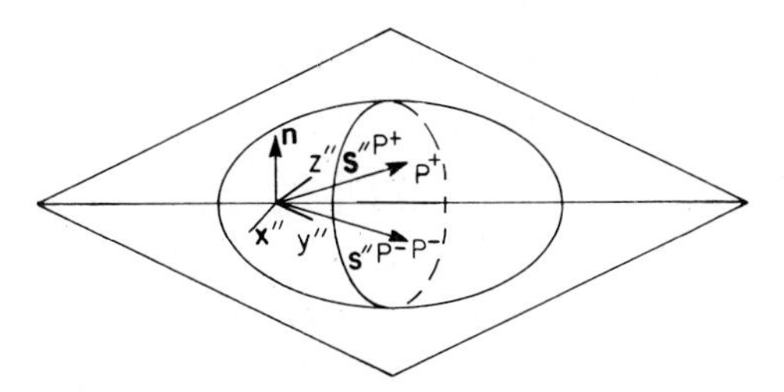

Figure 11.2.3 Body with plane of symmetry.

Consider the body shown in Fig. 11.2.3, for which the location of points and mass distribution are symmetric about a plane with normal $\mathbf{n}$. Select a body x''-y''-z'' frame so that its origin is on the plane of symmetry. Thus, for points P^+ and P^- and associated differential masses that are symmetrically placed with respect to the *plane of symmetry*,

$$\begin{gathered} \mathbf{n}^T \mathbf{s}''^{P^-} = -\mathbf{n}^T \mathbf{s}''^{P^+} \\ dm\,(P^-) = dm\,(P^+) \end{gathered} \tag{11.2.2}$$

Taking the scalar product of both sides of Eq. 11.2.1 with $\mathbf{n}$ and using Eq. 11.2.2 to change variables of integration,

$$\begin{aligned} \mathbf{n}^T \boldsymbol{\rho}'' &= \frac{1}{m} \int_m \mathbf{n}^T \mathbf{s}''^P \, dm(P) \\ &= \frac{1}{m} \int_{m^+} \mathbf{n}^T \mathbf{s}''^{P^+} \, dm(P^+) + \frac{1}{m} \int_{m^-} \mathbf{n}^T \mathbf{s}''^{P^-} \, dm(P^-) \\ &= \frac{1}{m} \int_{m^+} \mathbf{n}^T \mathbf{s}''^{P^+} \, dm(P^+) + \frac{1}{m} \int_{m^+} \mathbf{n}^T (-\mathbf{s}''^{P^+}) \, dm(P^+) \\ &= 0 \end{aligned} \tag{11.2.3}$$

Equation 11.2.3 shows that if a body has a plane of symmetry, then the centroid lies on that plane. This agrees with the result of Example 11.2.1, in which the planes $x'' = 5$ and $z'' = 10$ are planes of symmetry. If a body has an *axis of symmetry*; that is, every plane that contains the axis of symmetry is a plane of symmetry, then the centroid lies on the axis of symmetry. Note that if a body is geometrically symmetric, but is asymmetrically nonhomogeneous, these helpful geometric results are no longer valid.

Transformation of Inertia Matrices The inertia matrix with respect to the x''-y''-z'' frame in Fig. 11.2.1 is defined as

$$\mathbf{J}'' = -\int_m \tilde{\mathbf{s}}''^P \tilde{\mathbf{s}}''^P \, dm(P) \tag{11.2.4}$$

To obtain a relationship between $\mathbf{J}''$ and the inertia matrix $\mathbf{J}'$ with respect to the x'-y'-z' centroidal reference frame, substitute $\mathbf{s}''^P = \boldsymbol{\rho}'' + \mathbf{C}^T\mathbf{s}'^P$ into Eq. 11.2.4 to obtain

$$\mathbf{J}'' = -\int_m (\tilde{\boldsymbol{\rho}}'' + \widetilde{\mathbf{C}^T\mathbf{s}'^P})(\tilde{\boldsymbol{\rho}}'' + \widetilde{\mathbf{C}^T\mathbf{s}'^P})\, dm(P)$$

Using Eq. 9.2.21 and expanding terms, this becomes

$$\mathbf{J}'' = -m\tilde{\boldsymbol{\rho}}''\tilde{\boldsymbol{\rho}}'' - \tilde{\boldsymbol{\rho}}''\mathbf{C}^T \int_m \tilde{\mathbf{s}}'^P\, dm(P)\mathbf{C} - \mathbf{C}^T \int_m \tilde{\mathbf{s}}'^P\, dm(P)\mathbf{C}\tilde{\boldsymbol{\rho}}'' - \mathbf{C}^T \int_m \tilde{\mathbf{s}}'^P\tilde{\mathbf{s}}'^P\, dm(P)\mathbf{C} \tag{11.2.5}$$

Since the x'-y'-z' frame is centroidal,

$$\int_m \tilde{\mathbf{s}}'^P\, dm(P) = \widetilde{\left(\int_m \mathbf{s}'^P\, dm(P)\right)} = \mathbf{0}$$

Using this result and Eq. 11.1.17, Eq. 11.2.5 becomes

$$\mathbf{J}'' = \mathbf{C}^T\mathbf{J}'\mathbf{C} - m\tilde{\boldsymbol{\rho}}''\tilde{\boldsymbol{\rho}}'' \tag{11.2.6}$$

where $\mathbf{C}$ is the orthogonal transformation matrix from the x''-y''-z'' frame to the centroidal x'-y'-z' frame; that is, $\mathbf{s}' = \mathbf{C}\mathbf{s}''$. Using Eq. 9.1.28, Eq. 11.2.6 may be written in the more conventional form [4, 7, 9, 35]

$$\mathbf{J}'' = \mathbf{C}^T\mathbf{J}'\mathbf{C} + m(\boldsymbol{\rho}''^T\boldsymbol{\rho}''\mathbf{I} - \boldsymbol{\rho}''\boldsymbol{\rho}''^T) \tag{11.2.7}$$

As a special case, if $\mathbf{C} = \mathbf{I}$ (i.e., if the x'-y'-z' and x''-y''-z'' frames are parallel) then moments of inertia on the diagonal of Eq. 11.2.7 are related by

$$\begin{aligned} J_{x''x''} &= J_{x'x'} + m(\rho_{y''}^2 + \rho_{z''}^2) \\ J_{y''y''} &= J_{y'y'} + m(\rho_{x''}^2 + \rho_{z''}^2) \\ J_{z''z''} &= J_{z'z'} + m(\rho_{x''}^2 + \rho_{y''}^2) \end{aligned} \tag{11.2.8}$$

Thus, moments of inertia with respect to a noncentroidal x''-y''-z'' frame are those with respect to the parallel centroidal x'-y'-z' frame plus the mass of the body times the square of the distances between the respective parallel axes. This is the so-called *parallel axis theorem.* Products of inertia for parallel axes are obtained from off-diagonal terms in Eq. 11.2.7 as

$$\begin{aligned} J_{x''y''} &= J_{x'y'} - m\rho_{x''}\rho_{y''} \\ J_{x''z''} &= J_{x'z'} - m\rho_{x''}\rho_{z''} \\ J_{y''z''} &= J_{y'z'} - m\rho_{y''}\rho_{z''} \end{aligned} \tag{11.2.9}$$

Consider a special case in which one of the planes of an x''-y''-z'' frame is a plane of symmetry (e.g., the x''-y'' plane). Then, using the symmetry argument employed in deriving Eq. 11.2.3 (Prob. 11.2.2),

$$J_{z''x''} = J_{z''y''} = 0 \tag{11.2.10}$$

Similarly, if the x''-z'' plane is a plane of symmetry,

$$J_{x''y''} = J_{z''y''} = 0 \tag{11.2.11}$$

and if the y''-z'' plane is a plane of symmetry,

$$J_{x''y''} = J_{x''z''} = 0 \tag{11.2.12}$$

If any two coordinate planes of the x''-y''-z'' frame are planes of symmetry (i.e., have both geometric and mass distribution symmetry), then, from Eqs. 11.2.10 to 11.2.12, all products of inertia are zero (Prob. 11.2.3). This is a very helpful property, since bodies in mechanical systems often have two planes of symmetry. A common case of pairs of planes of symmetry is homogeneous bodies of revolution about one of the axes of a reference frame.

Example 11.2.2: The moments and products of inertia of the body of Fig. 11.2.2 with respect to a centroidal x'-y'-z' frame that is parallel to the x''-y''-z'' frame can be calculated using the parallel axis theorem. Using the results of Example 11.2.1 in Eq. 11.2.8,

$$J_{x'x'} = J_{x''x''} - 400\gamma\left[\left(\frac{35}{6}\right)^2 + 100\right]$$

$$J_{y'y'} = J_{y''y''} - 400\gamma[25 + 100]$$

$$J_{z'z'} = J_{z''z''} - 400\gamma\left[25 + \left(\frac{35}{6}\right)^2\right]$$

which may be evaluated using the results of Prob. 11.2.3. Since the x'-y' and y'-z' planes are planes of symmetry,

$$J_{x'y'} = J_{x'z'} = J_{y'z'} = 0$$

Principal Axes Even if a body is not symmetric, it is possible to find a centroidal reference frame with respect to which the products of inertia are all zero. Let the x''-y''-z'' and x'-y'-z' frames have their origins at the centroid of a body. Then, $\boldsymbol{\rho}'' = \mathbf{0}$. Presume $\mathbf{J}''$ is known, in general, as a full symmetric matrix. Multiplying Eq. 11.2.7 on the left by $\mathbf{C}$ and multiplying it on the right by $\mathbf{C}^T$,

$$\mathbf{J}' = \mathbf{C}\mathbf{J}''\mathbf{C}^T \tag{11.2.13}$$

The task is now to find an orthogonal transformation matrix $\mathbf{C}$ from the x''-y''-z'' frame to the x'-y'-z' frame for which $\mathbf{J}'$ is diagonal.

From the definition of $\mathbf{J}''$ in Eq. 11.2.4, for any vector $\mathbf{a}$ in R^3,

$$\mathbf{a}^T\mathbf{J}''\mathbf{a} = -\int_m \mathbf{a}^T\tilde{\mathbf{s}}''\tilde{\mathbf{s}}''\mathbf{a}\,dm$$

$$= \int_m \mathbf{a}^T\tilde{\mathbf{s}}''^T\tilde{\mathbf{s}}''\mathbf{a}\,dm$$

$$= \int_m (\tilde{\mathbf{s}}''\mathbf{a})^T(\tilde{\mathbf{s}}''\mathbf{a})\,dm \geq 0 \tag{11.2.14}$$

Thus, $\mathbf{J}''$ is *positive semidefinite.* In fact, providing that mass density is nowhere zero in the body, $\mathbf{a}^T\mathbf{J}''\mathbf{a} = \mathbf{0}$ implies that $\tilde{\mathbf{s}}''\mathbf{a} = \mathbf{0} = -\tilde{\mathbf{a}}\mathbf{s}''$, for all $\mathbf{s}''$ in the body. If there are three linearly independent vectors $\mathbf{s}_i''$, $i = 1, 2, 3$, to points in the body, then

$$\tilde{\mathbf{a}}[\mathbf{s}_1'', \mathbf{s}_2'', \mathbf{s}_3''] = \mathbf{0}$$

and since $[\mathbf{s}_1'', \mathbf{s}_2'', \mathbf{s}_3'']$ is nonsingular, $\tilde{\mathbf{a}} = \mathbf{0}$ and $\mathbf{a} = \mathbf{0}$. Thus, in this case, $\mathbf{J}''$ is *positive definite.* This is the case for any homogeneous body whose volume is not zero.

The 3×3 positive semidefinite matrix $\mathbf{J}''$ has three orthonormal *eigenvectors,* denoted $\mathbf{f}''$, $\mathbf{g}''$, and $\mathbf{h}''$ in the x''-y''-z'' frame, with corresponding nonnegative eigenvalues [22], ζ_i, $i = 1, 2, 3$, ordered such that $\zeta_1 \geq \zeta_2 \geq \zeta_3 \geq 0$, where

$$\begin{aligned} \mathbf{J}''\mathbf{f}'' &= \zeta_1\mathbf{f}'' \\ \mathbf{J}''\mathbf{g}'' &= \zeta_2\mathbf{g}'' \\ \mathbf{J}''\mathbf{h}'' &= \zeta_3\mathbf{h}'' \end{aligned} \tag{11.2.15}$$

Since $-\mathbf{g}''$ and $-\mathbf{h}''$ are also eigenvectors, their signs can be arranged so that $\mathbf{f}''$, $\mathbf{g}''$, and $\mathbf{h}''$ are unit coordinate vectors for a right-hand x'-y'-z' Cartesian frame. The transformation matrix $\mathbf{C}^T$ from the x'-y'-z' frame to the x''-y''-z'' frame is, by Eq. 9.2.13,

$$\mathbf{C}^T = [\mathbf{f}'', \mathbf{g}'', \mathbf{h}''] \tag{11.2.16}$$

Substituting $\mathbf{C}^T$ from Eq. 11.2.16 into Eq. 11.2.13,

$$\begin{aligned} \mathbf{J}' &= \mathbf{C}\mathbf{J}''\mathbf{C}^T \\ &= [\mathbf{f}'', \mathbf{g}'', \mathbf{h}'']^T\mathbf{J}''[\mathbf{f}'', \mathbf{g}'', \mathbf{h}''] \\ &= [\mathbf{f}'', \mathbf{g}'', \mathbf{h}'']^T[\mathbf{J}''\mathbf{f}'', \mathbf{J}''\mathbf{g}'', \mathbf{J}''\mathbf{h}''] \\ &= [\mathbf{f}'', \mathbf{g}'', \mathbf{h}'']^T[\zeta_1\mathbf{f}'', \zeta_2\mathbf{g}'', \zeta_3\mathbf{h}''] \\ &= [\mathbf{f}'', \mathbf{g}'', \mathbf{h}'']^T[\mathbf{f}'', \mathbf{g}'', \mathbf{h}'']\begin{bmatrix} \zeta_1 & 0 & 0 \\ 0 & \zeta_2 & 0 \\ 0 & 0 & \zeta_3 \end{bmatrix} \\ &= \begin{bmatrix} \zeta_1 & 0 & 0 \\ 0 & \zeta_2 & 0 \\ 0 & 0 & \zeta_3 \end{bmatrix} \end{aligned} \tag{11.2.17}$$

Thus, in the x'-y'-z' frame, $J_{x'x'} = \zeta_1$, $J_{y'y'} = \zeta_2$, $J_{z'z'} = \zeta_3$, and $J_{x'y'} = J_{x'y'} = J_{y'z'} = 0$. The axes of such a centroidal frame are called *principal axes* and the associated moments of inertia are called *principal moments of inertia.* By construction, the principal products of inertia are zero.

Example 11.2.3: The triangular plate shown in Fig. 11.2.4 has negligible thickness, so the mass is considered concentrated in the x-y plane, with mass density γ per unit area. Evaluating the integrals in Eq. 11.1.17 in the x''-y''-z'' frame,

$$
\begin{aligned}
J''_{x} &= \gamma \int_0^1 \int_0^{1-x} y^2\, dy\, dx \\
&= \gamma \int_0^1 \frac{(1-x)^3}{3}\, dx \\
&= \frac{\gamma}{12}
\end{aligned}
$$

$$
J''_{yy} = J''_{xx} = \frac{\gamma}{12}
$$

$$
\begin{aligned}
J''_{zz} &= \gamma \int_0^1 \int_0^{1-x} (x^2 + y^2)\, dy\, dx \\
&= \gamma \int_0^1 \left[x^2(1-x) + \frac{(1-x)^3}{3} \right] dx \\
&= \frac{\gamma}{6}
\end{aligned}
$$

$$
\begin{aligned}
J''_{xy} &= -\gamma \int_0^1 \int_0^{1-x} xy\, dy\, dx \\
&= -\gamma \int_0^1 \frac{x(1-x)^2}{2}\, dx \\
&= -\frac{\gamma}{24}
\end{aligned}
$$

$$
J''_{xz} = J''_{yz} = 0
$$

Thus, the inertia matrix in the x''-y''-z'' frame is

$$
\mathbf{J}'' = \frac{\gamma}{24} \begin{bmatrix} 2 & -1 & 0 \\ -1 & 2 & 0 \\ 0 & 0 & 4 \end{bmatrix}
$$

The centroid is located by Eq. 11.2.1, with $m = \gamma/2$:

$$
\begin{aligned}
\boldsymbol{\rho}'' &= \frac{2}{\gamma} \int_0^1 \int_0^{1-x} [x, y, 0]^T \gamma\, dy\, dx \\
&= [\tfrac{1}{3}, \tfrac{1}{3}, 0]^T
\end{aligned}
$$

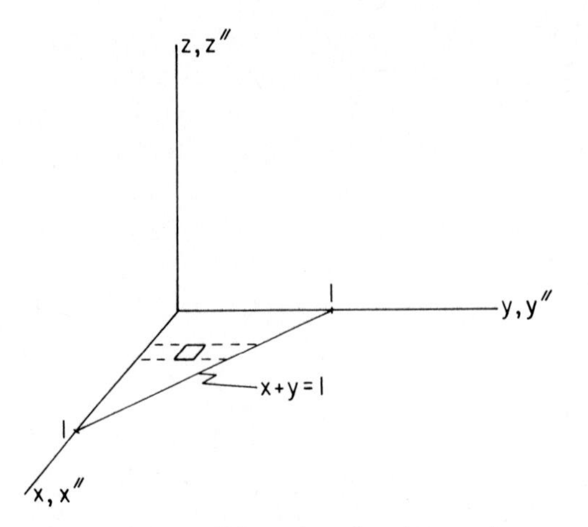

Figure 11.2.4 Triangular plate in x-y plane.

For a centroidal x'-y'-z' frame parallel to the x''-y''-z'' noncentroidal frame shown in Fig. 11.2.5, the parallel axis theorem applies and, from Eqs. 11.2.8 and 11.2.9,

$$\mathbf{J}' = \mathbf{J}'' - m\begin{bmatrix} (\rho_{y''}^2 + \rho_{z''}^2) & -\rho_{x''}\rho_{y''} & -\rho_{x''}\rho_{z''} \\ -\rho_{x''}\rho_{y''} & (\rho_{x''}^2 + \rho_{z''}^2) & -\rho_{y''}\rho_{z''} \\ -\rho_{x''}\rho_{z''} & -\rho_{y''}\rho_{z''} & (\rho_{x''}^2 + \rho_{y''}^2) \end{bmatrix}$$

$$= \frac{\gamma}{72}\begin{bmatrix} 2 & 1 & 0 \\ 1 & 2 & 0 \\ 0 & 0 & 4 \end{bmatrix}$$

The eigenvalues of $\mathbf{J}'$ are obtained by solving the *characteristic equation* [22]:

$$|\mathbf{J}' - \zeta\mathbf{I}| = \begin{bmatrix} \frac{\gamma}{36} - \zeta & \frac{\gamma}{72} & 0 \\ \frac{\gamma}{72} & \frac{\gamma}{36} - \zeta & 0 \\ 0 & 0 & \frac{\gamma}{18} - \zeta \end{bmatrix}$$

$$= \left(\frac{\gamma}{36} - \zeta\right)^2\left(\frac{\gamma}{18} - \zeta\right) - \left(\frac{\gamma}{18} - \zeta\right)\left(\frac{\gamma}{72}\right)^2$$

$$= \left[\left(\frac{\gamma}{36} - \zeta\right)^2 - \left(\frac{\gamma}{72}\right)^2\right]\left(\frac{\gamma}{18} - \zeta\right) = 0$$

whose solutions are

$$\zeta_1 = \frac{\gamma}{18}, \qquad \zeta_2 = \frac{\gamma}{24}, \qquad \zeta_3 = \frac{\gamma}{72}$$

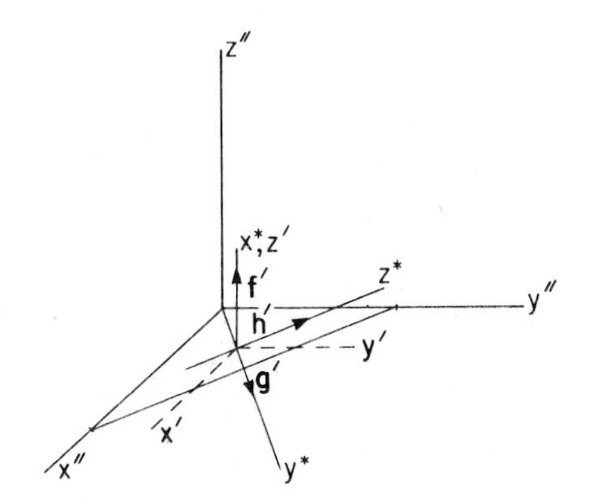

Figure 11.2.5 Principal axes for triangular plate.

Substituting these values of ζ into Eq. 11.2.15 yields the eigenvectors

$$\mathbf{f}' = [0, 0, 1]^T$$

$$\mathbf{g}' = \left[\frac{1}{\sqrt{2}}, \frac{1}{\sqrt{2}}, 0\right]^T$$

$$\mathbf{h}' = \left[-\frac{1}{\sqrt{2}}, \frac{1}{\sqrt{2}}, 0\right]^T$$

Principal axes are shown in Fig. 11.2.5, in which the inertia matrix is

$$\mathbf{J}^* = \begin{bmatrix} \frac{\gamma}{18} & 0 & 0 \\ 0 & \frac{\gamma}{24} & 0 \\ 0 & 0 & \frac{\gamma}{72} \end{bmatrix}$$

Inertia Properties of Complex Bodies Components of machines are often made up of combinations of subcomponents that have standard shapes (e.g., circles, disks, cylinders, spheres, and rectangular solids). A typical example of such a component is shown in Fig. 11.2.6, in which each subcomponent and void are of some standard shape, typical of those resulting from standard manufacturing processes. The objective of this subsection is to develop expressions for the inertia properties of the composite body using easily calculated properties of individual subcomponents.

Let an x''-y''-z'' frame be fixed in the composite body in a convenient location and orientation (e.g., as shown in Fig. 11.2.6). Using this frame, the centroid of the composite body may be obtained using the definition of Eq. 11.2.1, employing the property that an integral over the entire mass may be

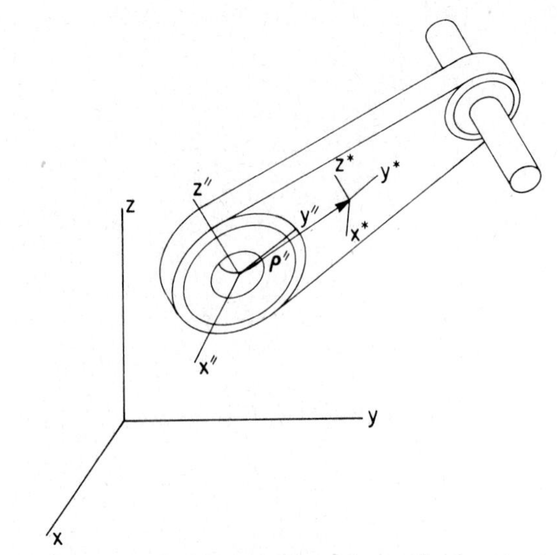

Figure 11.2.6 Composite body made up of subcomponents.

written as the sum of integrals over subcomponents m_i, to obtain

$$\begin{aligned}\boldsymbol{\rho}'' &= \frac{1}{m}\sum_{i=1}^{k}\int_{m_i} \mathbf{s}''^P\, dm(P) \\ &= \frac{1}{m}\sum_{i=1}^{k} m_i \boldsymbol{\rho}_i''\end{aligned} \tag{11.2.18}$$

where $m = \sum_{i=1}^{k} m_i$. To use this result, the centroid $\boldsymbol{\rho}_i''$ is first located in the x''-y''-z'' frame, using mass and centroid location information from Table 11.2.1 or from direct numerical calculation. Equation 11.2.18 is then used to locate the centroid of the composite body in the x''-y''-z'' frame.

The inertia matrix $\mathbf{J}^*$ with respect to the x^*-y^*-z^* centroidal reference frame of the composite body (e.g., as shown in Fig. 11.2.6) is now to be calculated. The notation x^*-y^*-z^* is selected here to avoid confusion with the centroidal x_i'-y_i'-y_i' frame of subcomponent i. The defining equation of Eq. 11.1.17 may be written and its integral evaluated as a sum of integrals over subcomponents that make up the composite body to obtain

$$\begin{aligned}\mathbf{J}^* &= -\int_m \tilde{\mathbf{s}}^{*P}\tilde{\mathbf{s}}^{*P}\, dm(P) \\ &= \sum_{i=1}^{k} -\int_{m_i} \tilde{\mathbf{s}}^{*P}\tilde{\mathbf{s}}^{*P}\, dm(P) \\ &= \sum_{i=1}^{k} \mathbf{J}_i^*\end{aligned} \tag{11.2.19}$$

To use Eq. 11.2.19, the inertia matrix $\mathbf{J}_i^*$ of subcomponent i in the x^*-y^*-z^* frame must be calculated. To do this, a convenient x_i'-y_i'-z_i' centroidal frame is defined for subcomponent i in the composite body. Denote by $\mathbf{J}_i'$ the inertia matrix of subcomponent i, with respect to its selected centroidal x'-y'-z' frame and by $\boldsymbol{\rho}_i'$ the vector that locates the centroid of the subcomponent in the x^*-y^*-z^* frame. Since the x^*-y^*-z^* frame is not centroidal for the individual

TABLE 11.2.1 Inertia Properties of Some Homogeneous Bodies

Body	Mass and Moment of Inertia (γ = mass density) (A = cross-sectional area)
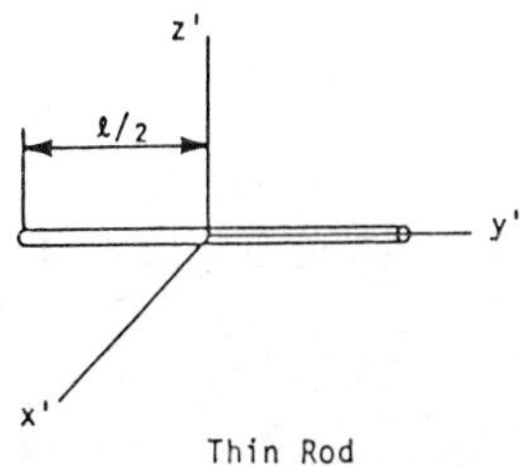 Thin Rod	$m = \gamma \ell A$ $J_{x'x'} = J_{z'z'} = \frac{m}{12}\ell^2$ $J_{y'y'} = 0$
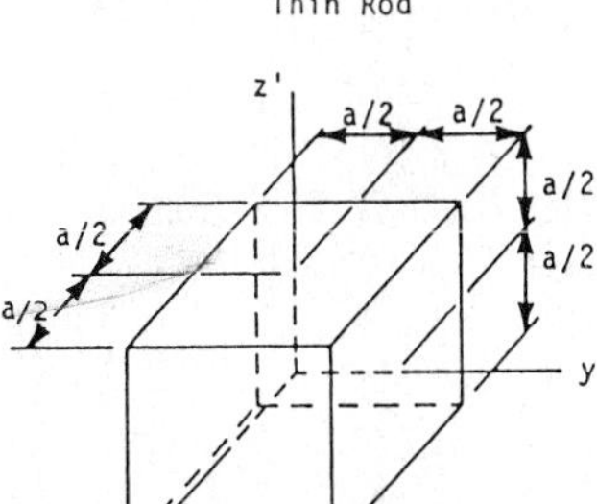 Cube	$m = \gamma a^3$ $J_{x'x'} = J_{y'y'} = J_{z'z'} = \frac{1}{6}ma^2$
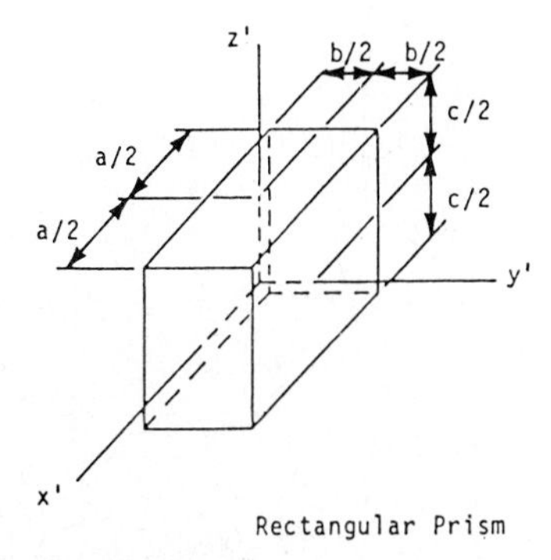 Rectangular Prism	$m = \gamma abc$ $J_{x'x'} = \frac{1}{12}m(b^2 + c^2)$ $J_{y'y'} = \frac{1}{12}m(a^2 + c^2)$ $J_{z'z} = \frac{1}{12}m(a^2 + b^2)$

TABLE 11.2.1 Continued

Body	Mass and Moment of Inertia (γ = mass density) (A = cross-sectional area)
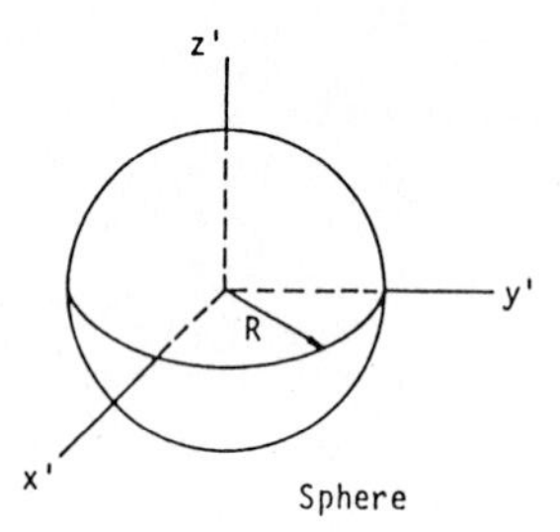 Sphere	$m = \frac{4}{3}\pi\gamma R^2$ $J_{x'x'} = J_{y'y'} = J_{z'z'} = \frac{2}{5}mR^2$
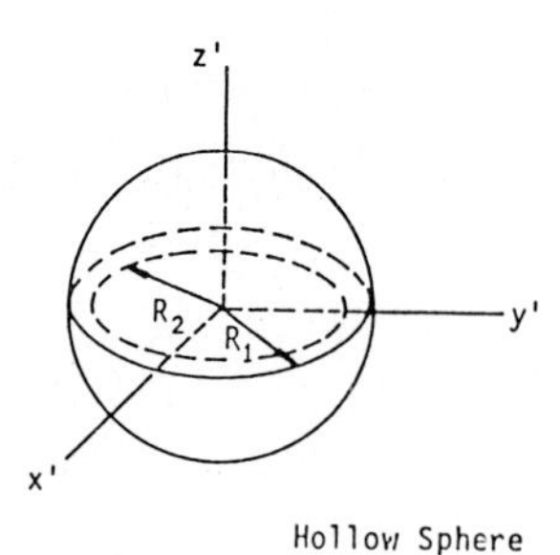 Hollow Sphere	$m = \frac{4}{3}\pi\gamma(R_1^2 - R_2^2)$ $J_{x'x'} = J_{y'y'} = J_{z'z'} = \frac{2}{5}\, m \frac{R_1^5 - R_2^5}{R_1^3 R_2^3}$
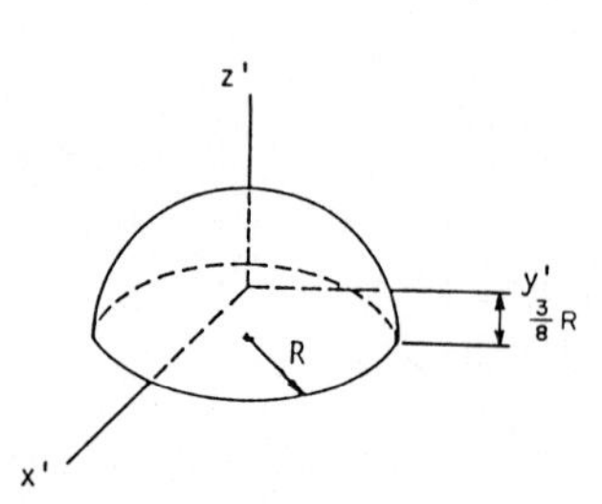 Hemisphere	$m = \frac{2}{3}\pi\gamma R^3$ $J_{x'x'} = J_{y'y'} = \frac{83}{320}mR^2$ $J_{z'z'} = \frac{2}{5}mR^2$
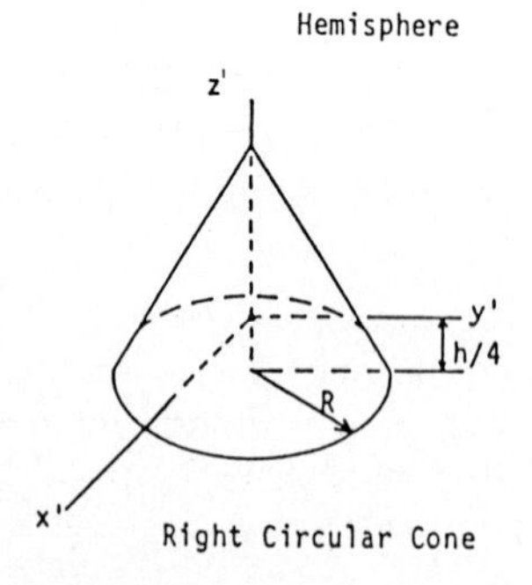 Right Circular Cone	$m = \frac{1}{3}\pi\gamma R^2 h$ $J_{x'x'} = J_{y'y'} = \frac{3}{80}m(4R^2 + h^2)$ $J_{z'z'} = \frac{3}{10}mR^2$

TABLE 11.2.1 Continued

Body	Mass and Moment of Inertia (γ = mass density) (A = cross-sectional area)
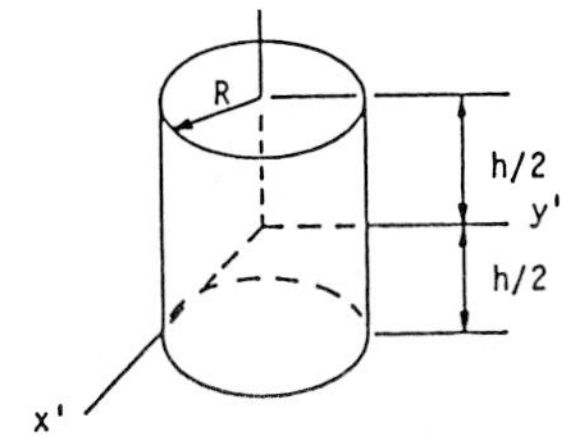 Right Circular Cylinder	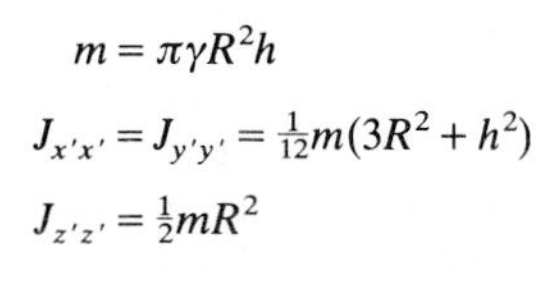 $m = \pi\gamma R^2 h$ $J_{x'x'} = J_{y'y'} = \frac{1}{12}m(3R^2 + h^2)$ $J_{z'z'} = \frac{1}{2}mR^2$
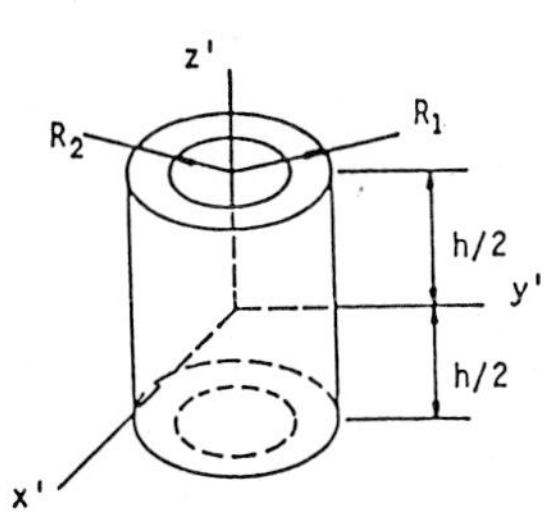 Hollow Right Circular Cylinder	$m = \pi\gamma h(R_1^2 - R_1^2)$ $J_{x'x'} = J_{y'y'} = \frac{1}{2}m(3R_1^2 + 3R_2^2 + h^2)$ $J_{z'z'} = \frac{1}{2}m(R_1^2 + R_2^2)$

subcomponents, the inertia transformation of Eq. 11.2.7 must be used to calculate the inertia matrix of subcomponent i with respect to the composite body centroidal x^*-y^*-z^* frame. Denoting by $\mathbf{C}_i'$ the transformation matrix from the x^*-y^*-z^* frame to the x_i'-y_i'-z_i' frame, Eq. 11.2.7 yields

$$\mathbf{J}_i^* = \mathbf{C}_i'^T\mathbf{J}_i'\mathbf{C}_i' + m_i(\boldsymbol{\rho}_i'^T\boldsymbol{\rho}_i'\mathbf{I} - \boldsymbol{\rho}_i'\boldsymbol{\rho}_i'^T) \tag{11.2.20}$$

This result may be substituted into Eq. 11.2.19 to obtain the desired inertia matrix for the composite body. The inertia properties of a few homogeneous bodies for use in such calculations are provided in Table 11.2.1.

If there are voids in a composite body, Eq. 11.2.19 may be used by assigning $-\mathbf{J}_i^*$ as the inertia matrix of the void, where $\mathbf{J}_i^*$ is the inertia matrix of a body that would occupy the void, with the same material density as the body from which it is removed.

Example 11.2.4: The body of Fig. 11.2.7 is made up of a cylinder with hemispherical ends, all made of the same homogeneous material of density γ. Since all three coordinate planes are planes of symmetry, the centroid is at the origin of the x-y-z frame and all products of inertia with respect to this frame are

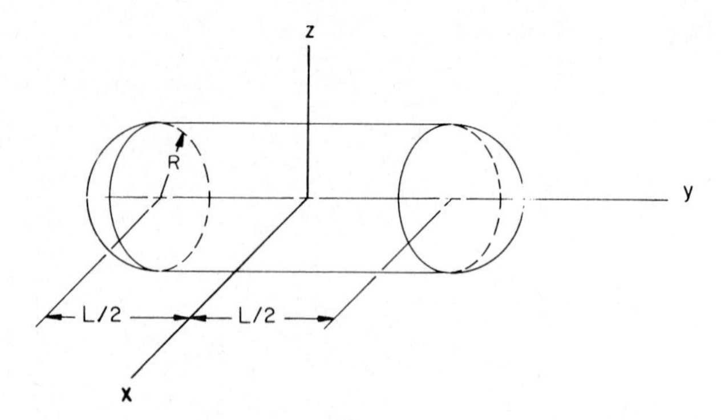

Figure 11.2.7 Cylinder with spherical ends.

zero. Using Eqs. 11.2.19 and 11.2.20, with $\mathbf{C}_i' = \mathbf{I}$, and Table 11.2.1, the moments of inertia of the composite body are

$$J_{xx} = J_{zz} = \frac{\pi\gamma R^2 L}{12}(3R^2 + L^2) + 2\left[\frac{83\pi\gamma R^5}{480} + \left(\frac{2\pi\gamma R^3}{3}\right)\left(\frac{4L + 3R}{8}\right)^2\right]$$

$$J_{yy} = \frac{\pi\gamma R^4 L}{2} + 2\left[\frac{4\pi\gamma R^5}{15}\right]$$

Example 11.2.5: The body of Fig. 11.2.8 is formed by a rectangular bar with a hole of radius $R = c/2$ about the x axis. Since all three coordinate planes are planes of symmetry, the centroid is at the origin of the x-y-z frame and the products of inertia with respect to this frame are zero.

From Eq. 11.2.20, with $\mathbf{J}_1$ as the inertia matrix of the bar without a hole

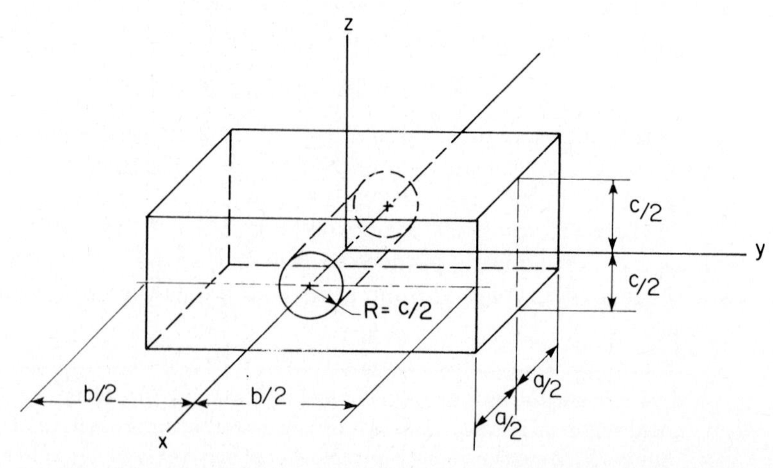

Figure 11.2.8 Rectangular solid with hole.

and $\mathbf{J}_2$ as the inertia matrix of a cylinder that would occupy the hole,

$$\mathbf{J} = \mathbf{J}_1 - \mathbf{J}_2$$

Using Table 11.2.1,

$$J_{xx} = \frac{\gamma abc}{12}(b^2 + c^2) - \frac{\pi\gamma c^4 a}{32}$$

$$J_{yy} = \frac{\gamma abc}{12}(a^2 + c^2) - \frac{\pi\gamma c^2 a}{48}\left(\frac{3c^2}{4} + a^2\right)$$

$$J_{zz} = \frac{\gamma abc}{12}(a^2 + b^2) - \frac{\pi\gamma c^2 a}{48}\left(\frac{3c^2}{4} + a^2\right)$$

The relatively elementary examples studied in this section show that the transformation relations for inertial properties and the theory of calculating centroids and inertia properties of complex bodies can be used systematically. The reader may, however, despair over the extent of such computations for complex composite bodies, such as automobile bodies, aircraft landing gear assemblies, vending machine components, and robot arms. Indeed, manual application of the methods used in this section would lead to extensive algebraic manipulations that are extremely time consuming and prone to error when carried out analytically by the engineer.

The availability of large-scale computer codes for system dynamic simulation creates new opportunities for large-scale applications, but requires detailed inertia data that have not normally been calculated in engineering practice. This dilemma may be alleviated by emerging computer-aided design and computer-aided engineering systems that contain extensive geometric modeling software that describes the geometry of complex mechanical assemblies. The methods developed and illustrated in this section form the foundation for computer implementation with geometric modelers to determine the locations of centroids and moments and products of inertia with respect to the reference frames specified by the engineer. The technical challenge in implementing such methods is computer generation of the geometric definition of complex bodies, which is the domain of solid modeling methods that are emerging in the form of practical computational tools in modern computer-aided design and computer-aided engineering systems.

Once geometric and material property information is defined and entered into a computer data base, implementation of the analytical methods presented in this section for calculating the locations of centroids and moments and products of inertia is a relatively simple matter. The viewpoint taken for the remainder of this text is that such modern computer-aided engineering tools are available to the engineer for the computation of the inertial properties of machine components, or that the engineer is willing to manually carry out the tedious computations illustrated in this section to obtain the inertial properties of bodies in mechanical systems.

11.3 EQUATIONS OF MOTION OF CONSTRAINED SPATIAL SYSTEMS

Consider now nb bodies that form a constrained multibody system. The composite set of generalized coordinates for the system is

$$\begin{aligned} \mathbf{r} &= [\mathbf{r}_1^T, \mathbf{r}_2^T, \ldots, \mathbf{r}_{nb}^T]^T \\ \mathbf{p} &= [\mathbf{p}_1^T, \mathbf{p}_2^T, \ldots, \mathbf{p}_{nb}^T]^T \end{aligned} \tag{11.3.1}$$

The kinematic and driving constraints that act on the system are derived in Chapter 9 in the form

$$\mathbf{\Phi}(\mathbf{r}, \mathbf{p}, t) = \mathbf{0} \tag{11.3.2}$$

In addition, the Euler parameter normalization constraints

$$\mathbf{\Phi}^{\mathbf{p}} \equiv \begin{bmatrix} \mathbf{p}_1^T\mathbf{p}_1 - 1 \\ \vdots \\ \mathbf{p}_{nb}^T\mathbf{p}_{nb} - 1 \end{bmatrix} = \mathbf{0} \tag{11.3.3}$$

must hold.

Constrained equations of motion are derived in this section using the results of Section 11.1. The equations of motion are first written in terms of angular accelerations and then in terms of second derivatives of Euler parameters.

11.3.1 Newton–Euler Form of Constrained Equations of Motion

To implement the variational Newton–Euler equations, Eq. 11.1.19 is evaluated for each body in the system and the resulting equations are added to obtain a Newton–Euler variational equation of motion for the system. To simplify notation, define

$$\begin{aligned} \delta\mathbf{r} &= [\delta\mathbf{r}_1^T, \delta\mathbf{r}_2^T, \ldots, \delta\mathbf{r}_{nb}^T]^T \\ \mathbf{M} &\equiv \operatorname{diag}(m_1\mathbf{I}_3, m_2\mathbf{I}_3, \ldots, m_{nb}\mathbf{I}_3) \\ \delta\boldsymbol{\pi}' &= [\delta\boldsymbol{\pi}_1'^T, \delta\boldsymbol{\pi}_2'^T, \ldots, \delta\boldsymbol{\pi}_{nb}'^T]^T \\ \mathbf{F} &\equiv [\mathbf{F}_1^T, \mathbf{F}_2^T, \ldots, \mathbf{F}_{nb}^T]^T \\ \mathbf{J}' &\equiv \operatorname{diag}(\mathbf{J}_1', \mathbf{J}_2', \ldots, \mathbf{J}_{nb}') \\ \boldsymbol{\omega}' &\equiv [\boldsymbol{\omega}_1'^T, \boldsymbol{\omega}_2'^T, \ldots, \boldsymbol{\omega}_{nb}'^T]^T \\ \mathbf{n}' &\equiv [\mathbf{n}_1'^T, \mathbf{n}_2'^T, \ldots, \mathbf{n}_{nb}'^T]^T \\ \tilde{\boldsymbol{\omega}}' &\equiv \operatorname{diag}(\tilde{\boldsymbol{\omega}}_1', \tilde{\boldsymbol{\omega}}_2', \ldots, \tilde{\boldsymbol{\omega}}_{nb}') \end{aligned} \tag{11.3.4}$$

Using this notation, the sum of Eqs. 11.1.19 over all bodies in the system may be written as (Prob. 11.3.1)

$$\delta\mathbf{r}^T[\mathbf{M}\ddot{\mathbf{r}} - \mathbf{F}] + \delta\boldsymbol{\pi}'^T[\mathbf{J}'\dot{\boldsymbol{\omega}}' + \tilde{\boldsymbol{\omega}}'\mathbf{J}'\boldsymbol{\omega}' - \mathbf{n}'] = 0$$

which must hold for all virtual displacements $\delta\mathbf{r}$ and virtual rotations $\delta\boldsymbol{\pi}'$ that are consistent with constraints. Forces and torques that act on the system may be partitioned into *applied forces and torques* $\mathbf{F}^A$ and $\mathbf{n}'^A$ and *constraint forces and torques* $\mathbf{F}^C$ and $\mathbf{n}'^C$, respectively. For all constraints treated in this text, forces of constraint do no work as long as virtual displacements and rotations are consistent with constraints; that is,

$$\delta\mathbf{r}^T\mathbf{F}^C + \delta\boldsymbol{\pi}'^T\mathbf{n}'^C = 0$$

Thus, with $\mathbf{F} = \mathbf{F}^A + \mathbf{F}^C$, $\mathbf{n}' = \mathbf{n}'^A + \mathbf{n}'^C$, and this relation, the *variational equation of motion for a constrained system* is

$$\delta\mathbf{r}^T[\mathbf{M}\ddot{\mathbf{r}} - \mathbf{F}^A] + \delta\boldsymbol{\pi}'^T[\mathbf{J}'\dot{\boldsymbol{\omega}}' + \tilde{\boldsymbol{\omega}}'\mathbf{J}'\boldsymbol{\omega}' - \mathbf{n}'^A] = 0 \tag{11.3.5}$$

which must hold for all kinematically admissible virtual displacements and rotations.

Virtual displacements $\delta\mathbf{r}$ and virtual rotations $\delta\boldsymbol{\pi}'$ are kinematically admissible for constraints of Eq. 11.3.2 if

$$\boldsymbol{\Phi}_{\mathbf{r}}\,\delta\mathbf{r} + \boldsymbol{\Phi}_{\pi'}\,\delta\boldsymbol{\pi}' = \mathbf{0} \tag{11.3.6}$$

where $\boldsymbol{\Phi}_{\mathbf{r}}$ and $\boldsymbol{\Phi}_{\pi'}$ are assembled using the results of Chapter 9. Euler parameter normalization constraints are excluded, since they are automatically satisfied by $\delta\boldsymbol{\pi}'$. Since Eq. 11.3.5 must hold for all $\delta\mathbf{r}$ and $\delta\boldsymbol{\pi}'$ that satisfy Eq. 11.3.6, by the Lagrange multiplier theorem of Section 6.3, there exists a Lagrange multiplier vector $\boldsymbol{\lambda}$ such that

$$\delta\mathbf{r}^T[\mathbf{M}\ddot{\mathbf{r}} - \mathbf{F}^A + \boldsymbol{\Phi}_{\mathbf{r}}^T\boldsymbol{\lambda}] + \delta\boldsymbol{\pi}'^T[\mathbf{J}'\dot{\boldsymbol{\omega}}' + \tilde{\boldsymbol{\omega}}'\mathbf{J}'\boldsymbol{\omega}' - \mathbf{n}'^A + \boldsymbol{\Phi}_{\pi'}^T\boldsymbol{\lambda}] = 0 \tag{11.3.7}$$

for arbitrary $\delta\mathbf{r}$ and $\delta\boldsymbol{\pi}'$. The coefficients of these arbitrary variations must be zero, yielding the *constrained Newton–Euler equations* of motion

$$\begin{aligned} \mathbf{M}\ddot{\mathbf{r}} + \boldsymbol{\Phi}_{\mathbf{r}}^T\boldsymbol{\lambda} &= \mathbf{F}^A \\ \mathbf{J}'\dot{\boldsymbol{\omega}}' + \boldsymbol{\Phi}_{\pi'}^T\boldsymbol{\lambda} &= \mathbf{n}'^A - \tilde{\boldsymbol{\omega}}'\mathbf{J}'\boldsymbol{\omega}' \end{aligned} \tag{11.3.8}$$

To complete the equations of motion, acceleration equations associated with the kinematic constraints of Eq. 11.3.2 must be obtained. The *velocity equation* obtained in Section 9.6.2, by taking the time derivative of Eq. 11.3.2, is

$$\boldsymbol{\Phi}_{\mathbf{r}}\dot{\mathbf{r}} + \boldsymbol{\Phi}_{\pi'}\boldsymbol{\omega}' = -\boldsymbol{\Phi}_t \equiv \mathbf{v} \tag{11.3.9}$$

The time derivative of this equation, calculated in Section 9.6.3, yields the *acceleration equation*

$$\boldsymbol{\Phi}_{\mathbf{r}}\ddot{\mathbf{r}} + \boldsymbol{\Phi}_{\pi'}\dot{\boldsymbol{\omega}}' = \boldsymbol{\gamma} \tag{11.3.10}$$

where the vector $\boldsymbol{\gamma}$ is defined in Chapter 9 for each constraint equation.

Combining Eqs. 11.3.8 and 11.3.10, the *system acceleration equations* are

$$\begin{bmatrix} \mathbf{M} & \mathbf{0} & \mathbf{\Phi}_{\mathbf{r}}^T \\ \mathbf{0} & \mathbf{J}' & \mathbf{\Phi}_{\pi'}^T \\ \mathbf{\Phi}_{\mathbf{r}} & \mathbf{\Phi}_{\pi'} & \mathbf{0} \end{bmatrix} \begin{bmatrix} \ddot{\mathbf{r}} \\ \dot{\boldsymbol{\omega}}' \\ \boldsymbol{\lambda} \end{bmatrix} = \begin{bmatrix} \mathbf{F}^A \\ \mathbf{n}'^A - \tilde{\boldsymbol{\omega}}'\mathbf{J}'\boldsymbol{\omega}' \\ \boldsymbol{\gamma} \end{bmatrix} \tag{11.3.11}$$

These equations of motion, taken with the kinematic constraint equations of Eq. 11.3.2 and the velocity equations of Eq. 11.3.9, yield a mixed system of *differential–algebraic equations of motion* for the system.

Technically, Eq. 11.3.11 is a system of mixed first-order differential–algebraic equations for velocity variables $\dot{\mathbf{r}}$ and $\boldsymbol{\omega}'$ and the algebraic variables $\boldsymbol{\lambda}$. It is not a second-order differential–algebraic system, since angular velocity $\boldsymbol{\omega}'$ is not integrable. Equation 11.3.11 must be augmented by first-order kinematic velocity equations in terms of Euler parameter time derivatives and the kinematic constraint equations of Eq. 11.3.2. The details of formulating such a reduced first-order system of differential–algebraic equations of motion are discussed in more detail in Section 11.6.

Just as shown in Section 6.3 for planar systems, if $\delta\dot{\mathbf{r}}^T\mathbf{M}\,\delta\dot{\mathbf{r}} + \delta\boldsymbol{\omega}'^T\mathbf{J}'\,\delta\boldsymbol{\omega}' > 0$ for all nonzero $\delta\dot{\mathbf{r}}$ and $\delta\boldsymbol{\omega}'$ such that $\mathbf{\Phi}_{\mathbf{r}}\,\delta\dot{\mathbf{r}} + \mathbf{\Phi}_{\pi'}\,\delta\boldsymbol{\omega}' = \mathbf{0}$, and if $[\mathbf{\Phi}_{\mathbf{r}}, \mathbf{\Phi}_{\pi'}]$ has full row rank, then the coefficient matrix in Eq. 11.3.11 is nonsingular (Prob. 11.3.2).

Initial conditions on position and orientation and on velocity must be provided to define the dynamics of a system. Since the orientation of a body is specified by Euler parameters, initial conditions on the position and orientation must be specified in the form

$$\mathbf{\Phi}^I(\mathbf{r}, \mathbf{p}, t_0) = \mathbf{0} \tag{11.3.12}$$

where $\mathbf{r}$ and $\mathbf{p}$ must satisfy Eq. 11.3.2 at t_0 and the Euler parameter normalization constraints of Eq. 11.3.3. Equations 11.3.2, 11.3.3, and 11.3.12 should uniquely determine $\mathbf{r}(t_0)$ and $\mathbf{p}(t_0)$.

If angular velocity is used in formulating the equations of motion, the velocity initial conditions should be given in the form

$$\mathbf{B}_{\mathbf{r}}^I\dot{\mathbf{r}} + \mathbf{B}_{\omega'}^I\boldsymbol{\omega}' = \mathbf{v}^I \tag{11.3.13}$$

at $t = t_0$, where $\mathbf{B}_{\mathbf{r}}^I$, $\mathbf{B}_{\omega'}^I$, and $\mathbf{v}^I$ may depend on $\mathbf{r}$ and $\mathbf{p}$ at t_0. Furthermore, Eqs. 11.3.9 and 11.3.13 should uniquely determine $\dot{\mathbf{r}}(t_0)$ and $\boldsymbol{\omega}'(t_0)$.

11.3.2 Euler Parameter Form of Constrained Equations of Motion

The equations of motion can also be written in terms of the Euler parameters, with Lagrange multipliers to account for both the kinematic and Euler parameter normalization constraints. To write the velocity equation of Eq. 11.3.9 in terms of the Euler parameter derivatives, Eq. 9.3.34 may be used to obtain

$$\mathbf{\Phi}_{\mathbf{r}}\dot{\mathbf{r}} + \mathbf{\Phi}_{\mathbf{p}}\dot{\mathbf{p}} = -\mathbf{\Phi}_t \equiv \mathbf{v} \tag{11.3.14}$$

where

$$\mathbf{\Phi}_{\mathbf{p}_i} = 2\mathbf{\Phi}_{\boldsymbol{\pi}'_i}\mathbf{G}_i \tag{11.3.15}$$

To write the constraint acceleration equation of Eq. 11.3.10 in terms of Euler parameters, a relation between $\dot{\boldsymbol{\omega}}'$ and $\ddot{\mathbf{p}}$ is needed. Differentiating Eq. 9.3.34,

$$\dot{\boldsymbol{\omega}}' = 2\mathbf{G}\ddot{\mathbf{p}} + 2\dot{\mathbf{G}}\dot{\mathbf{p}}$$

From Eq. 9.3.19,

$$\begin{aligned}\dot{\mathbf{G}}\dot{\mathbf{p}} &= [-\dot{\mathbf{e}}, \dot{\tilde{\mathbf{e}}} + \dot{e}_0\mathbf{I}]\begin{bmatrix} \dot{e}_0 \\ \dot{\mathbf{e}} \end{bmatrix} \\ &= -\dot{\mathbf{e}}\dot{e}_0 - \dot{\tilde{\mathbf{e}}}\dot{\mathbf{e}} + \dot{e}_0\dot{\mathbf{e}} \\ &= \mathbf{0}\end{aligned} \tag{11.3.16}$$

Thus,

$$\dot{\boldsymbol{\omega}}' = 2\mathbf{G}\ddot{\mathbf{p}} \tag{11.3.17}$$

To obtain the inverse relation to Eq. 11.3.17, differentiate Eq. 9.3.35 to obtain

$$\ddot{\mathbf{p}} = \tfrac{1}{2}(\dot{\mathbf{G}}^T\boldsymbol{\omega}' + \mathbf{G}^T\dot{\boldsymbol{\omega}}') \tag{11.3.18}$$

Using Eq. 9.3.34,

$$\dot{\mathbf{G}}^T\boldsymbol{\omega}' = 2\dot{\mathbf{G}}^T\mathbf{G}\dot{\mathbf{p}} \tag{11.3.19}$$

Differentiating both sides of Eq. 9.3.24,

$$\dot{\mathbf{G}}^T\mathbf{G} = -[\mathbf{G}^T\dot{\mathbf{G}} + \dot{\mathbf{p}}\mathbf{p}^T + \mathbf{p}\dot{\mathbf{p}}^T]$$

Substituting this result into Eq. 11.3.19 and using Eqs. 9.3.27 and 11.3.16,

$$\dot{\mathbf{G}}^T\boldsymbol{\omega}' = -2\dot{\mathbf{p}}^T\dot{\mathbf{p}}\mathbf{p} \tag{11.3.20}$$

From Eqs. 9.3.35 and 9.3.22,

$$\dot{\mathbf{p}}^T\dot{\mathbf{p}} = \tfrac{1}{4}\boldsymbol{\omega}'^T\mathbf{G}\mathbf{G}^T\boldsymbol{\omega}' = \tfrac{1}{4}\boldsymbol{\omega}'^T\boldsymbol{\omega}'$$

Substituting this result and Eq. 11.3.20 into Eq. 11.3.18 yields the desired result:

$$\ddot{\mathbf{p}} = \tfrac{1}{2}\mathbf{G}^T\dot{\boldsymbol{\omega}}' - \tfrac{1}{4}\boldsymbol{\omega}'^T\boldsymbol{\omega}'\mathbf{p} \tag{11.3.21}$$

Note the quadratic term in $\boldsymbol{\omega}'$ in this relation, which is more complicated than Eq. 9.3.35 for $\dot{\mathbf{p}}$ in terms of $\boldsymbol{\omega}'$.

Substituting Eq. 11.3.17 into Eq. 11.3.10 and using Eq. 11.3.15, the kinematic acceleration equations are obtained, in terms of Euler parameters, as

$$\mathbf{\Phi}_{\mathbf{r}}\ddot{\mathbf{r}} + \mathbf{\Phi}_{\mathbf{p}}\ddot{\mathbf{p}} = \boldsymbol{\gamma} \tag{11.3.22}$$

where $\boldsymbol{\gamma}$ is defined in Section 9.6.3.

Differentiating the Euler parameter constraint equation of Eq. 11.3.3 yields the Euler parameter velocity equation

$$\mathbf{\Phi_p^p \dot{p}} = \mathbf{0} \tag{11.3.23}$$

where

$$\mathbf{\Phi_p^p} = 2\begin{bmatrix} \mathbf{p}_1^T & \mathbf{0} & \mathbf{0} \dots \mathbf{0} \\ \mathbf{0} & \mathbf{p}_2^T & \mathbf{0} \dots \mathbf{0} \\ \vdots & & \\ \mathbf{0} & \mathbf{0} & \mathbf{0} \dots \mathbf{p}_{nb}^T \end{bmatrix}$$

Differentiating Eq. 11.3.23 yields the Euler parameter acceleration equation

$$\mathbf{\Phi_p^p \ddot{p}} = -2\begin{bmatrix} \dot{\mathbf{p}}_1^T \dot{\mathbf{p}}_1 \\ \vdots \\ \dot{\mathbf{p}}_{nb}^T \dot{\mathbf{p}}_{nb} \end{bmatrix} \equiv \boldsymbol{\gamma}^{\mathbf{p}} \tag{11.3.24}$$

Similarly, the total differential of Eq. 11.3.2 may be calculated to obtain

$$\mathbf{\Phi_r}\,\delta\mathbf{r} + \mathbf{\Phi_p}\,\delta\mathbf{p} = \mathbf{0} \tag{11.3.25}$$

which defines kinematically admissible virtual displacements of the system. Likewise, variations in Euler parameters must satisfy

$$\mathbf{\Phi_p^p}\,\delta\mathbf{p} = \mathbf{0} \tag{11.3.26}$$

which is obtained by taking the differential of Eq. 11.3.3.

Substituting for $\delta\boldsymbol{\pi}'$ from Eq. 9.3.41, for $\boldsymbol{\omega}'$ from Eq. 9.3.34, for $\dot{\boldsymbol{\omega}}'$ from Eq. 11.3.17, and using Eqs. 9.3.24, 9.3.29, 9.3.32, Eq. 11.3.5 may be rewritten as

$$\delta\mathbf{r}^T[\mathbf{M\ddot{r}} - \mathbf{F}^A] + \delta\mathbf{p}^T[4\mathbf{G}^T\mathbf{J}'\mathbf{G\ddot{p}} - 8\dot{\mathbf{G}}^T\mathbf{J}'\dot{\mathbf{G}}\mathbf{p} - 2\mathbf{G}^T\mathbf{n}'^A] = 0 \tag{11.3.27}$$

where

$$\mathbf{G} = \mathrm{diag}(\mathbf{G}_1, \mathbf{G}_2, \dots, \mathbf{G}_{nb})$$

Equation 11.3.27 must hold for all $\delta\mathbf{p}$ and $\delta\mathbf{r}$ that satisfy Eqs. 11.3.25 and 11.3.26.

Applying the Lagrange multiplier theorem to append the virtual displacement and Euler parameter variation constraints of Eqs. 11.3.25 and 11.3.26, there exist associated Lagrange multipliers $\boldsymbol{\lambda}$ and $\boldsymbol{\lambda}^{\mathbf{p}}$ such that

$$\delta\mathbf{r}^T[\mathbf{M\ddot{r}} - \mathbf{F}^A + \mathbf{\Phi_r}^T\boldsymbol{\lambda}] + \delta\mathbf{p}^T[4\mathbf{G}^T\mathbf{J}'\mathbf{G\ddot{p}} - 8\dot{\mathbf{G}}^T\mathbf{J}'\dot{\mathbf{G}}\mathbf{p} - 2\mathbf{G}^T\mathbf{n}'^A + \mathbf{\Phi_p}^T\boldsymbol{\lambda} + \mathbf{\Phi_p^p}^T\boldsymbol{\lambda}^{\mathbf{p}}] = 0 \tag{11.3.28}$$

which must hold for arbitrary $\delta\mathbf{r}$ and $\delta\mathbf{p}$. Since coefficients of these arbitrary variations must be zero, and appending the acceleration equations of Eqs. 11.3.22 and 11.3.24, the *Euler parameter system acceleration equation* is obtained as

$$\begin{bmatrix} \mathbf{M} & \mathbf{0} & \mathbf{\Phi_r}^T & \mathbf{0} \\ \mathbf{0} & 4\mathbf{G}^T\mathbf{J}'\mathbf{G} & \mathbf{\Phi_p}^T & \mathbf{\Phi_p}^{PT} \\ \mathbf{\Phi_r} & \mathbf{\Phi_p} & \mathbf{0} & \mathbf{0} \\ \mathbf{0} & \mathbf{\Phi_p^p} & \mathbf{0} & \mathbf{0} \end{bmatrix} \begin{bmatrix} \ddot{\mathbf{r}} \\ \ddot{\mathbf{p}} \\ \boldsymbol{\lambda} \\ \boldsymbol{\lambda}^{\mathbf{p}} \end{bmatrix} = \begin{bmatrix} \mathbf{F}^A \\ 2\mathbf{G}^T\mathbf{n}'^A + 8\dot{\mathbf{G}}^T\mathbf{J}'\dot{\mathbf{G}}\mathbf{p} \\ \boldsymbol{\gamma} \\ \boldsymbol{\gamma}^{\mathbf{p}} \end{bmatrix} \tag{11.3.29}$$

This system of equations, taken with the kinematic and Euler parameter normalization constraints of Eqs. 11.3.2 and 11.3.3 and the associated velocity equations of Eqs. 11.3.14 and 11.3.23, comprise the mixed system of *differential–algebraic equations of motion* of the system in terms of Euler parameters. A direct extension of the argument outlined in Prob. 11.3.2 may be used to show that the coefficient matrix in Eq. 11.3.29 is nonsingular (Prob. 11.3.3).

Initial conditions must be given on $\mathbf{r}$ and $\mathbf{p}$ at t_0, as in Eq. 11.3.12. For velocities, however, initial conditions should be given on $\dot{\mathbf{r}}$ and $\dot{\mathbf{p}}$. Since initial conditions are most naturally given on angular velocities, rather than on $\dot{\mathbf{p}}$, it is desirable to retain initial conditions in the form of Eq. 11.3.13. This can be done by substituting for $\boldsymbol{\omega}'$ from Eq. 9.3.34 into Eq. 11.3.13 to obtain

$$\mathbf{B}_r^I\dot{\mathbf{r}} + 2\mathbf{B}_{\omega'}\mathbf{G}\dot{\mathbf{p}} = \mathbf{v}^I \qquad \textbf{(11.3.30)}$$

which, with Eqs. 11.3.14 and 11.3.23, determine $\dot{\mathbf{r}}(t_0)$ and $\dot{\mathbf{p}}(t_0)$.

Given a complete set of initial conditions on $\mathbf{r}$, $\mathbf{p}$, $\dot{\mathbf{r}}$, and $\dot{\mathbf{p}}$ that satisfy Eqs. 11.3.2, 11.3.3, 11.3.14, and 11.3.23, at t_0, Eq. 11.3.29, and Eqs. 11.3.2, 11.3.3, 11.3.14, and 11.3.23 for $t > t_0$ constitute a mixed system of second-order differential–algebraic equations of motion for a constrained mechanical system. This contrasts with the first-order character of Eq. 11.3.11, as noted earlier. The expanded system of mixed differential–algebraic equations in $\mathbf{r}$ and $\mathbf{p}$ may thus be integrated, using any of the techniques developed in Chapter 7 for solving differential–algebraic equations of motion for planar systems. Only increased dimensionality and the detailed form of the equations of motion change.

A note of some importance regarding the difference in forms of Eqs. 11.3.11 and 11.3.29 is in order. The upper-left terms in the coefficient matrix on the left of Eq. 11.3.11 are constant matrices $\mathbf{M}$ and $\mathbf{J}'$, whereas the generalized coordinate dependent matrix $\mathbf{G}^T\mathbf{J}'\mathbf{G}$ appears on the left of Eq. 11.3.29. Even though both coefficient matrices in Eqs. 11.3.11 and 11.3.29 are nonsingular for meaningful physical systems, the appearance of generalized coordinate dependent terms in the matrix of Eq. 11.3.29 may influence the performance of matrix factorization algorithms and numerical integration methods. These considerations are discussed in more detail in Section 11.6.

11.4 INTERNAL FORCES

As in Chapter 6 for planar systems, the internal forces associated with springs, dampers, and actuators can be accounted for using the principle of virtual work. The virtual work approach is even more attractive for spatial systems than for planar systems, since relationships for virtual rotations are available.

11.4.1 Translational Spring–Damper–Actuator

Consider first the *translational spring–damper–actuator* (or TSDA) shown in Fig. 11.4.1, which connects points P_i and P_j on bodies i and j, respectively. The vector

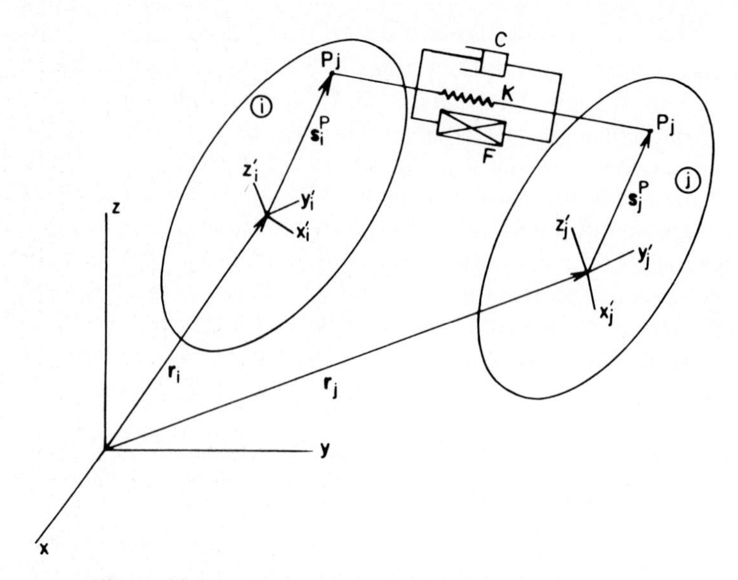

Figure 11.4.1 Translational spring–damper–actuator.

from P_i to P_j is

$$\mathbf{d}_{ij} = \mathbf{r}_j + \mathbf{A}_j\mathbf{s}_j'^P - \mathbf{r}_i - \mathbf{A}_i\mathbf{s}_i'^P \tag{11.4.1}$$

Thus, the length ℓ of the spring–damper–actuator is given by

$$\ell^2 = \mathbf{d}_{ij}^T\mathbf{d}_{ij} \tag{11.4.2}$$

Differentiating with respect to time yields

$$2\ell\dot{\ell} = 2\mathbf{d}_{ij}^T\dot{\mathbf{d}}_{ij} \tag{11.4.3}$$

which may be solved for the time rate of change of length as

$$\begin{aligned}\dot{\ell} &= \left(\frac{\mathbf{d}_{ij}}{\ell}\right)^T(\dot{\mathbf{r}}_j + \dot{\mathbf{A}}_j\mathbf{s}_j'^P - \dot{\mathbf{r}}_i - \dot{\mathbf{A}}_i\mathbf{s}_i'^P) \\ &= \left(\frac{\mathbf{d}_{ij}}{\ell}\right)^T(\dot{\mathbf{r}}_j - \mathbf{A}_j\tilde{\mathbf{s}}_j'^P\boldsymbol{\omega}_j' - \dot{\mathbf{r}}_i + \mathbf{A}_i\tilde{\mathbf{s}}_i'^P\boldsymbol{\omega}_i')\end{aligned} \tag{11.4.4}$$

where Eq. 9.2.40 has been used to write the time derivatives of transformation matrices in terms of angular velocities.

Note that if ℓ approaches zero an indeterminate fraction occurs in the first term of Eq. 11.4.4. In this case, L'Hospital's rule [25, 26] may be used to obtain

$$\lim_{\ell\to 0}\frac{\mathbf{d}_{ij}}{\ell} = \dot{\mathbf{d}}_{ij}/\dot{\ell}\,|_{\ell=0} \tag{11.4.5}$$

which is used only when the length of the spring–damper–actuator approaches zero.

The magnitude of the force that acts in the spring–damper–actuator, with tension taken as positive, is

$$f = k(\ell - \ell_0) + c\dot{\ell} + F(\ell, \dot{\ell}) \quad \textbf{(11.4.6)}$$

where k is the spring coefficient, c is the damping coefficient, and $F(\ell, \dot{\ell})$ is a general actuator force. The virtual work of this force is simply

$$\delta W = -f\,\delta\ell \quad \textbf{(11.4.7)}$$

where the variation $\delta\ell$ in length is obtained by taking the differential of Eq. 11.4.2 and dividing by ℓ, as in the derivation of Eq. 11.4.4, to obtain

$$\delta\ell = \left(\frac{\mathbf{d}_{ij}}{\ell}\right)^T (\delta\mathbf{r}_j - \mathbf{A}_j\tilde{\mathbf{s}}_j'^P\,\delta\boldsymbol{\pi}_j' - \delta\mathbf{r}_i + \mathbf{A}_i\tilde{\mathbf{s}}_i'^P\,\delta\boldsymbol{\pi}_i') \quad \textbf{(11.4.8)}$$

where Eq. 9.2.50 has been used to expand the differential of the transformation matrix. Substituting this result into Eq. 11.4.7 yields

$$\delta W = -\frac{f}{\ell}\mathbf{d}_{ij}^T(\delta\mathbf{r}_j - \mathbf{A}_j\tilde{\mathbf{s}}_j'^P\,\delta\boldsymbol{\pi}_j' - \delta\mathbf{r}_i + \mathbf{A}_i\tilde{\mathbf{s}}_i'^P\,\delta\boldsymbol{\pi}_i') \quad \textbf{(11.4.9)}$$

The coefficients of virtual displacements and virtual rotations in Eq. 11.4.9 yield the angular orientation form of the TSDA generalized force as

$$\begin{aligned}\mathbf{Q}_i &= f/\ell\begin{bmatrix}\mathbf{d}_{ij}\\ \tilde{\mathbf{s}}_i'^P\mathbf{A}_i^T\mathbf{d}_{ij}\end{bmatrix}\\ \mathbf{Q}_j &= -f/\ell\begin{bmatrix}\mathbf{d}_{ij}\\ \tilde{\mathbf{s}}_j'^P\mathbf{A}_j^T\mathbf{d}_{ij}\end{bmatrix}\end{aligned} \quad \textbf{(11.4.10)}$$

In the case where the generalized forces associated with Euler parameter orientation coordinates are desired, the virtual rotations in Eq. 11.4.9 may be written in terms of the variations in Euler parameters, using Eq. 9.3.41. Making this substitution and identifying the coefficients of virtual displacements and variations in Euler parameters, the Euler parameter form of the TSDA generalized forces is obtained as

$$\begin{aligned}\mathbf{Q}_i &= f/\ell\begin{bmatrix}\mathbf{d}_{ij}\\ 2\mathbf{G}_i^T\tilde{\mathbf{s}}_i'^P\mathbf{A}_i^T\mathbf{d}_{ij}\end{bmatrix}\\ \mathbf{Q}_j &= -f/\ell\begin{bmatrix}\mathbf{d}_{ij}\\ 2\mathbf{G}_j^T\tilde{\mathbf{s}}_j'^P\mathbf{A}_j^T\mathbf{d}_{ij}\end{bmatrix}\end{aligned} \quad \textbf{(11.4.11)}$$

The effect of the TSDA force on the equations of motion is accounted for by inserting the angular orientation generalized forces of Eq. 11.4.10 into the right side of Eq. 11.3.11 to obtain the angular acceleration equations of motion.

Similarly, the Euler parameter generalized forces of Eq. 11.4.11 may be substituted into the right side of Eq. 11.3.29 to obtain the Euler parameter acceleration equations.

11.4.2 Rotational Spring–Damper–Actuator

A *rotational spring–damper–actuator* (or RSDA) can act about the common axis of a revolute, cylindrical, or screw joint between bodies i and j, as shown in Fig. 11.4.2. The cumulative relative angle of rotation is $\theta + 2n\pi$, where n is the number of revolutions from the free angle of the spring and θ is measured from the joint definition x_i'' axis in body i to the joint definition x_j'' axis in body j. The angle θ is determined by Eq. 9.2.31. Its differential $\delta\theta$ and time derivative $\dot{\theta}$ are determined by Eqs. 9.2.61 and 9.2.62.

The magnitude of the torque exerted by the RSDA, with a counterclockwise torque exerted on the x_j''-y_j''-z_j'' frame relative to the x_i''-y_i''-z_i'' frame as positive, is

$$n = k_\theta(\theta + 2n_{\text{rev}}\pi) + c_\theta\dot{\theta} + N(\theta + 2n_{\text{rev}}\pi,\ \dot{\theta}) \tag{11.4.12}$$

where k_θ is the spring coefficient, spring torque is zero when x_i'' and x_j'' initially coincide and $n_{\text{rev}} = 0$, c_θ is the damping coefficient, and $N(\theta + 2n\pi,\ \dot{\theta})$ is a general actuator torque. Since n is positive counterclockwise and $\delta\theta$ is positive counterclockwise, the virtual work of this torque is

$$\delta W = -n\,\delta\theta \tag{11.4.13}$$

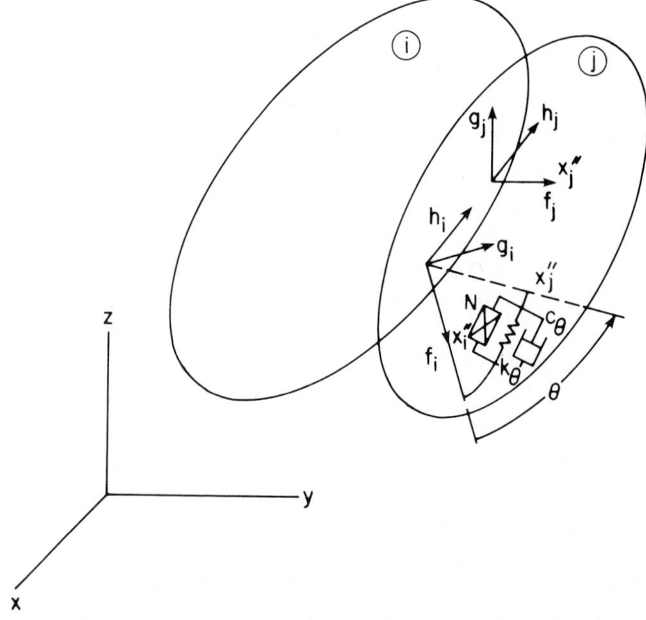

Figure 11.4.2 Rotational spring–damper–actuator.

Equation 9.2.62 may be used to write $\dot{\theta}$ in terms of angular velocities; that is,

$$\dot{\theta} = -\mathbf{h}_i'^T(\mathbf{A}_i^T\mathbf{A}_j\boldsymbol{\omega}_j' - \boldsymbol{\omega}_i') \qquad \textbf{(11.4.14)}$$

Similarly, substituting from Eq. 9.2.61 into Eq. 11.4.13,

$$\delta W = -n\mathbf{h}_i'^T(\mathbf{A}_i^T\mathbf{A}_j\,\delta\boldsymbol{\pi}_j' - \delta\boldsymbol{\pi}_i') \qquad \textbf{(11.4.15)}$$

Thus, the generalized forces due to this element on bodies i and j are

$$\mathbf{Q}_i = \begin{bmatrix} \mathbf{0} \\ n\mathbf{h}_i' \end{bmatrix}$$

$$\mathbf{Q}_j = -\begin{bmatrix} \mathbf{0} \\ n\mathbf{A}_j^T\mathbf{A}_i\mathbf{h}_i' \end{bmatrix} \qquad \textbf{(11.4.16)}$$

If generalized forces associated with Euler parameter generalized coordinates are desired, then $\mathbf{Q}_i$ and $\mathbf{Q}_j$ in Eq. 11.4.15 must be multiplied on the left by $2\mathbf{G}_i$ and $2\mathbf{G}_j$, respectively.

11.5 INVERSE DYNAMICS, EQUILIBRIUM ANALYSIS, AND REACTION FORCES IN JOINTS

The inverse dynamics of kinematically driven spatial systems follow exactly the same approach as for planar systems presented in Section 6.4. The kinematics problem is first solved for position, velocity, and acceleration using the formulation of Chapter 9. The equations of motion are then solved algebraically for Lagrange multipliers associated with the constraints. Finally, using the results presented later in this section, the reaction forces and torques associated with both the kinematic and driving constraints are calculated. While the details of the terms in the equations of constraint and equations of spatial motion differ, the matrix form of these equations is identical to that encountered for planar systems in Section 6.4, so precisely the same analysis sequence and matrix subroutines may be used for spatial inverse dynamic analysis.

Equilibrium analysis for spatial systems follows exactly the same pattern as in the case of planar systems in Section 6.5. While detailed expressions in the equations of equilibrium are different in the case of spatial systems, precisely the same matrix form is encountered and the same basic principles may be employed as in Section 6.5. Either dynamic settling to an equilibrium position, direct solution of the equilibrium equations, or minimum total potential energy for conservative systems may be used to determine equilibrium positions. The numerical considerations in spatial system equilibrium analysis are identical to those in planar systems presented in Section 6.5.

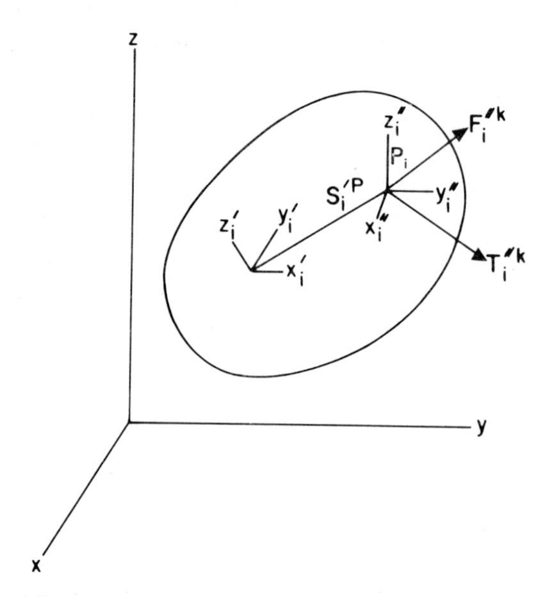

Figure 11.5.1 Reaction force and torque on body i at point P.

Consider a typical joint k, with joint definition point P (see Fig. 11.5.1), constraint equations $\mathbf{\Phi}^k = \mathbf{0}$, and associated Lagrange multipliers $\boldsymbol{\lambda}^k$. In terms of the x_i''-y_i''-z_i'' joint definition frame, joint reaction forces are $\mathbf{F}_i''^k$ and reaction torques are $\mathbf{T}_i''^k$, as shown in Fig. 11.5.1. Equating the virtual work of these reaction forces and torques to the negative of the variational terms in the equations of motion associated with constraint k, as in Eqs. 6.6.1 and 6.6.2,

$$-(\delta \mathbf{r}_i^T \mathbf{\Phi}_{\mathbf{r}_i}^{kT} \boldsymbol{\lambda}^k + \delta \boldsymbol{\pi}_i'^T \mathbf{\Phi}_{\boldsymbol{\pi}_i'}^{kT} \boldsymbol{\lambda}^k) = \delta \mathbf{r}_i''^{PT} \mathbf{F}_i''^k + \delta \boldsymbol{\pi}_i''^T \mathbf{T}_i''^k \tag{11.5.1}$$

which must hold for arbitrary virtual displacements.

To take advantage of Eq. 11.5.1, all virtual displacements must be represented in the joint definition frame. From Eq. 9.2.51,

$$\delta \mathbf{r}_i^P = \delta \mathbf{r}_i - \mathbf{A}_i \tilde{\mathbf{s}}_i'^P \, \delta \boldsymbol{\pi}_i' \tag{11.5.2}$$

In terms of virtual displacement and rotation, represented in joint definition frame coordinates,

$$\begin{aligned} \delta \mathbf{r}_i^P &= \mathbf{A}_i \mathbf{C}_i \, \delta \mathbf{r}_i''^P \\ \delta \boldsymbol{\pi}_i' &= \mathbf{C}_i \, \delta \boldsymbol{\pi}_i'' \end{aligned} \tag{11.5.3}$$

where $\mathbf{C}_i$ is the direction cosine transformation matrix from the x_i''-y_i''-z_i'' joint definition frame to the x_i'-y_i'-z_i' centroidal reference frame. From Eqs. 11.5.2 and 11.5.3,

$$\delta \mathbf{r}_i = \mathbf{A}_i \mathbf{C}_i \, \delta \mathbf{r}_i''^P + \mathbf{A}_i \tilde{\mathbf{s}}_i'^P \mathbf{C}_i \, \delta \boldsymbol{\pi}_i'' \tag{11.5.4}$$

Substituting the virtual rotation and displacement from Eqs. 11.5.3 and 11.5.4 into Eq. 11.5.1,

$$\delta \mathbf{r}_i^{nPT}(\mathbf{F}_i^{\prime\prime k} + \mathbf{C}_i^T \mathbf{A}_i^T \mathbf{\Phi}_{\mathbf{r}_i}^{kT} \boldsymbol{\lambda}^k) + \delta \boldsymbol{\pi}_i^{\prime\prime T}(\mathbf{T}_i^{\prime\prime k} + \mathbf{C}_i^T \mathbf{\Phi}_{\boldsymbol{\pi}_i}^{kT} \boldsymbol{\lambda}^k + \mathbf{C}_i^T \tilde{\mathbf{s}}_i^{\prime PT} \mathbf{A}_i^T \mathbf{\Phi}_{\mathbf{r}_i}^{kT} \boldsymbol{\lambda}^k) = 0 \quad \textbf{(11.5.5)}$$

which must hold for arbitrary $\delta \mathbf{r}_i^{\prime\prime P}$ and $\delta \boldsymbol{\pi}_i^{\prime\prime}$. Therefore,

$$\begin{aligned} \mathbf{F}_i^{\prime\prime k} &= -\mathbf{C}_i^T \mathbf{A}_i^T \mathbf{\Phi}_{\mathbf{r}_i}^{kT} \boldsymbol{\lambda}^k \\ \mathbf{T}_i^{\prime\prime k} &= -\mathbf{C}_i^T (\mathbf{\Phi}_{\boldsymbol{\pi}_i}^{kT} - \tilde{\mathbf{s}}_i^{\prime P} \mathbf{A}_i^T \mathbf{\Phi}_{\mathbf{r}_i}^{kT}) \boldsymbol{\lambda}^k \end{aligned} \quad \textbf{(11.5.6)}$$

which are the desired expressions for joint reaction forces and torques on body i at joint k in the joint reference frame.

11.6 NUMERICAL CONSIDERATIONS IN SOLVING SPATIAL DIFFERENTIAL–ALGEBRAIC EQUATIONS OF MOTION

If the Euler parameter form of the equations of motion in Eq. 11.3.29 is employed, with the kinematic constraints of Eq. 11.3.2 and Euler parameter normalization constraints of Eq. 11.3.3, then any of the methods presented in Chapter 7 for integrating mixed differential–algebraic equations of motion may be directly applied. This approach has been used successfully and continues to serve as a reliable method for dynamic analysis.

If, on the other hand, the Newton–Euler form of the acceleration equations of Eq. 11.3.11, written in terms of angular accelerations, is employed, some modification of the methods presented in Chapter 7 is required. One alternative is to solve Eq. 11.3.11 for $\ddot{\mathbf{r}}$ and $\dot{\boldsymbol{\omega}}'$ and then to apply Eq. 11.3.21 to calculate $\ddot{\mathbf{p}}$ and integrate for Euler parameter generalized coordinates, just as would have been done if Eq. 11.3.29 had been used to determine $\ddot{\mathbf{p}}$. Even this slight modification in the basic algorithm is attractive, since the coefficient submatrix in the (2, 2) position in Eq. 11.3.11 is constant, whereas the corresponding matrix in Eq. 11.3.29 depends on generalized coordinates and must be reevaluated each time the acceleration equations are evaluated.

Both of the foregoing alternatives work with second time derivatives of generalized coordinates. Both, therefore, permit direct implementation of all four of the numerical integration algorithms of Section 7.3. Since the derivatives of kinematic and Euler parameter constraint equations with respect to all generalized coordinates are readily computed, independent generalized coordinates can be identified and the generalized coordinate partitioning or hybrid algorithm of Section 7.3 can be directly implemented. At the time of this writing, the hybrid algorithm is implemented with Euler parameter second time derivatives in the DADS software system [27] for the dynamic analysis of constrained mechanical systems. This algorithm is both theoretically sound and has been demonstrated to be computationally reliable for broad classes of applications.

Improved methods for the numerical solution of mixed differential–algebraic systems, however, will likely be developed and implemented in the near future.

Another alternative for formulating mixed differential–algebraic equations for constrained mechanical systems is to define an intermediate variable $\mathbf{s} \equiv \dot{\mathbf{r}}$ and to form a first-order system of differential–algebraic equations, using Eqs. 9.3.35 and 11.3.11, as

$$\begin{bmatrix} \mathbf{M} & \mathbf{0} & \mathbf{\Phi}_{\mathbf{r}}^T \\ \mathbf{0} & \mathbf{J}' & \mathbf{\Phi}_{\pi'}^T \\ \mathbf{\Phi}_{\mathbf{r}} & \mathbf{\Phi}_{\pi'} & \mathbf{0} \end{bmatrix} \begin{bmatrix} \dot{\mathbf{s}} \\ \dot{\boldsymbol{\omega}}' \\ \boldsymbol{\lambda} \end{bmatrix} = \begin{bmatrix} \mathbf{F} \\ \mathbf{n}' - \tilde{\boldsymbol{\omega}}'\mathbf{J}'\boldsymbol{\omega}' \\ \boldsymbol{\gamma} \end{bmatrix}$$

$$\dot{\mathbf{r}} = \mathbf{s} \tag{11.5.7}$$

$$\dot{\mathbf{p}} = \tfrac{1}{2}\mathbf{G}^T(\mathbf{p})\boldsymbol{\omega}'$$

In addition, the kinematic and Euler parameter normalization constraints of Eqs. 11.3.2 and 11.3.3 and the constraint velocity equations of Eq. 11.3.9 must hold; that is,

$$\mathbf{\Phi}(\mathbf{r}, \mathbf{p}, t) = \mathbf{0}$$

$$\mathbf{\Phi}^{\mathbf{p}}(\mathbf{p}) = \mathbf{0} \tag{11.5.8}$$

$$\mathbf{\Phi}_{\mathbf{r}}\dot{\mathbf{r}} + \mathbf{\Phi}_{\pi'}\boldsymbol{\omega}' = \boldsymbol{\nu}$$

The first-order system of differential–algebraic equations of Eqs. 11.5.7 and 11.5.8, taken with initial conditions that satisfy Eq. 11.5.8 at the initial time, determines the motion of the constrained mechanical system. It is shown in Reference 48 that this system may be numerically solved using the direct numerical integration method of Section 7.3.

PROBLEMS

Section 11.1

11.1.1. Carry out the manipulations in detail to derive Eq. 11.1.3.

11.1.2. Carry out the manipulations in detail to derive Eq. 11.1.11.

11.1.3. Show that $\boldsymbol{\omega}'$ is not a total differential, because it cannot be integrated. (*Hint*: Use Eq. 9.3.34 to express $\boldsymbol{\omega}'$ in terms of Euler parameter differentials and show that each component of $\boldsymbol{\omega}'$ is not an exact differential.)

11.1.4. Let the pendulum of Example 11.1.2 be permitted to rotate about and slide along the y axis of an inertial reference frame, as shown in Fig. P11.1.4. Write the equations of motion in terms of ϕ and s.

Section 11.2

11.2.1. Calculate the moments of inertia of the body of Example 11.2.1 (Fig. 11.2.2) with respect to the x''-y''-z'' frame. Use these results, $\boldsymbol{\rho}''$ from Example 11.2.1, and the

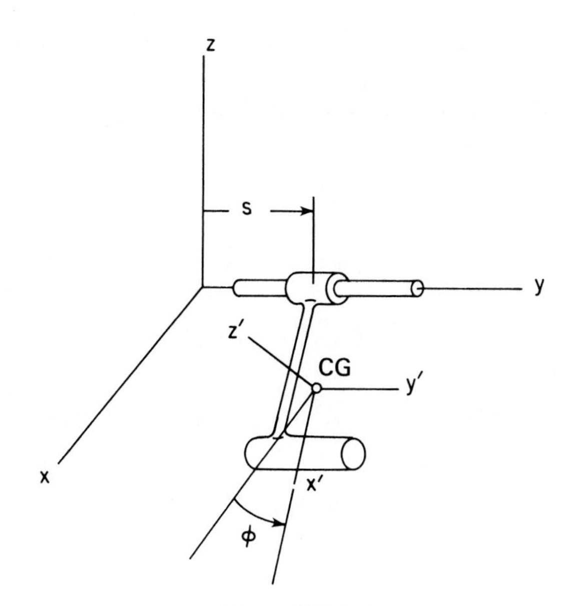

Figure P11.1.4

parallel axis transformation of Example 11.2.2 to evaluate the moments of inertia of the body with respect to the centroidal x'-y'-z' frame.

11.2.2. Verify that Eqs. 11.2.10 to 11.2.12 are valid under the stated conditions.

11.2.3. Show that, if any pair of coordinate planes are planes of symmetry for a body, then all products of inertia with respect to this body-fixed reference frame are zero.

Section 11.3

11.3.1. Expand Eq. 11.3.5, using the definitions of Eq. 11.3.4, and show that it is just

$$\sum_{i=1}^{nb} \{\delta\mathbf{r}_i^T[m_i\ddot{\mathbf{r}}_i - \mathbf{F}_i^A] + \delta\boldsymbol{\pi}_i'^T[\mathbf{J}_i'\dot{\boldsymbol{\omega}}_i' + \tilde{\boldsymbol{\omega}}_i'\mathbf{J}_i'\boldsymbol{\omega}_i' - \mathbf{n}_i'^A]\} = 0$$

11.3.2. Use the argument presented in Section 6.3 to show that the coefficient matrix in Eq. 11.3.11 is nonsingular if kinetic energy is positive for all nonzero virtual velocities that satisfy the homogeneous velocity equations of Eq. 11.3.9 and if the constraint Jacobian has full row rank.

11.3.3. Repeat the argument of Prob. 11.3.2 to show that, under the same hypotheses, the coefficient matrix in Eq. 11.3.29 is nonsingular.

SUMMARY OF KEY FORMULAS

Newton–Euler Equations of Motion of a Body

$$\delta\mathbf{r}^T[m\ddot{\mathbf{r}} - \mathbf{F}] + \delta\boldsymbol{\pi}'^T[\mathbf{J}'\dot{\boldsymbol{\omega}}' + \tilde{\boldsymbol{\omega}}'\mathbf{J}'\boldsymbol{\omega}' - \mathbf{n}'] = 0 \qquad \textbf{(11.1.19)}$$

$$\delta\mathbf{r}^T[m\ddot{\mathbf{r}} - \mathbf{F}] + \delta\mathbf{p}^T[4\mathbf{G}^T\mathbf{J}'\mathbf{G}\ddot{\mathbf{p}} - 8\dot{\mathbf{G}}^T\mathbf{J}'\dot{\mathbf{G}}\mathbf{p} - 2\mathbf{G}^T\mathbf{n}'] = 0 \qquad \textbf{(11.3.27)}$$

Inertia Properties

$$\int_m \mathbf{s}'^P \, dm(P) = \mathbf{0}, \qquad \boldsymbol{\rho}'' = \frac{1}{m}\int_m \mathbf{s}''^P \, dm(P) \qquad \textbf{(11.1.13, 11.2.1)}$$

$$\mathbf{J}' = -\int_m \tilde{\mathbf{s}}'^P \tilde{\mathbf{s}}'^P \, dm(P), \qquad \mathbf{J}'' = -\int_m \tilde{\mathbf{s}}''^P \tilde{\mathbf{s}}''^P \, dm(P) \qquad \textbf{(11.1.17, 11.2.4)}$$

$$\mathbf{J}'' = \mathbf{C}^T\mathbf{J}'\mathbf{C} + m(\boldsymbol{\rho}''^T\boldsymbol{\rho}''\mathbf{I} - \boldsymbol{\rho}''\boldsymbol{\rho}''^T), \qquad \mathbf{s}' = \mathbf{C}\mathbf{s}'' \qquad \textbf{(11.2.7)}$$

$$\boldsymbol{\rho}'' = \frac{1}{m}\sum_{i=1}^{k} m_i\boldsymbol{\rho}_i'', \qquad \mathbf{J}' = \sum_{i=1}^{k} \mathbf{J}_i' \qquad \textbf{(11.2.18, 19)}$$

Lagrange Multiplier Constrained Equations of Motion

$$\begin{bmatrix} \mathbf{M} & \mathbf{0} & \boldsymbol{\Phi}_{\mathbf{r}}^T \\ \mathbf{0} & \mathbf{J}' & \boldsymbol{\Phi}_{\boldsymbol{\pi}'}^T \\ \boldsymbol{\Phi}_{\mathbf{r}} & \boldsymbol{\Phi}_{\boldsymbol{\pi}'} & \mathbf{0} \end{bmatrix} \begin{bmatrix} \ddot{\mathbf{r}} \\ \dot{\boldsymbol{\omega}}' \\ \boldsymbol{\lambda} \end{bmatrix} = \begin{bmatrix} \mathbf{F}^A \\ \mathbf{n}'^A - \tilde{\boldsymbol{\omega}}'\mathbf{J}'\boldsymbol{\omega}' \\ \boldsymbol{\gamma} \end{bmatrix} \qquad \textbf{(11.3.11)}$$

$$\begin{bmatrix} \mathbf{M} & \mathbf{0} & \boldsymbol{\Phi}_{\mathbf{r}}^T & \mathbf{0} \\ \mathbf{0} & 4\mathbf{G}^T\mathbf{J}'\mathbf{G} & \boldsymbol{\Phi}_{\mathbf{p}}^T & \mathbf{0} \\ \boldsymbol{\Phi}_{\mathbf{r}} & \boldsymbol{\Phi}_{\mathbf{p}} & \mathbf{0} & \mathbf{0} \\ \mathbf{0} & \boldsymbol{\Phi}_{\mathbf{p}}^{\mathbf{p}} & \mathbf{0} & \mathbf{0} \end{bmatrix} \begin{bmatrix} \ddot{\mathbf{r}} \\ \ddot{\mathbf{p}} \\ \boldsymbol{\lambda} \\ \boldsymbol{\lambda}^{\mathbf{p}} \end{bmatrix} = \begin{bmatrix} \mathbf{F}^A \\ 2\mathbf{G}^T\mathbf{n}'^A + 8\dot{\mathbf{G}}^T\mathbf{J}'\dot{\mathbf{G}}\mathbf{p} \\ \boldsymbol{\gamma} \\ \boldsymbol{\gamma}^{\mathbf{p}} \end{bmatrix} \qquad \textbf{(11.3.29)}$$

Joint Reaction Forces

$$\mathbf{F}_i''^k = -\mathbf{C}_i^T\mathbf{A}_i^T\boldsymbol{\Phi}_{\mathbf{r}_i}^{kT}\boldsymbol{\lambda}^k, \qquad \mathbf{T}_i''^k = -\mathbf{C}_i^T(\boldsymbol{\Phi}_{\boldsymbol{\pi}_i}^{kT} + \tilde{\mathbf{s}}_i'^{PT}\mathbf{A}_i^T\boldsymbol{\Phi}_{\mathbf{r}_i}^{kT})\boldsymbol{\lambda}^k \qquad \textbf{(11.5.6)}$$

CHAPTER TWELVE

Spatial Dynamic Modeling and Analysis

12.1 MODELING AND ANALYSIS TECHNIQUES

Once a sound kinematic model of a spatial system is defined, the remaining ingredients for dynamic analysis include the specification of masses and moments of inertia, initial conditions on position and velocity, and forces that act on the system. A typical sequence in the dynamic analysis of a mechanical system includes the definition of an equilibrium position, inverse dynamic analysis with a kinematically driven system to determine the required driving and reaction forces, and transient dynamic analysis under the influence of applied forces. Examples are analyzed dynamically in this chapter, using the DADS computer code [27], to illustrate the use of the formulation presented in Chapter 11 to support the design of mechanical systems.

12.2 DYNAMIC ANALYSIS OF A SPATIAL SLIDER–CRANK MECHANISM

12.2.1 Model

The spatial slider–crank mechanism shown in Fig. 10.2.1 is used here to demonstrate dynamic analysis. The system configuration and definition of each joint are as presented in the kinematic analysis of Section 10.2. The only additional data required here are the inertia properties of each body, the forces that act, and the initial conditions of motion. The masses and moments of inertia are presented in Table 12.2.1.

TABLE 12.2.1 Inertia Properties of Slider–Crank Components

Body		Mass	$I_{x'x'}$	$I_{y'y'}$	$I_{z'z'}$	$I_{x'y'}$	$I_{y'z'}$	$I_{z'x'}$
Crank	①	0.12	0.0001	0.00001	0.0001	0.0	0.0	0.0
Connecting rod	②	0.5	0.004	0.0004	0.004	0.0	0.0	0.0
Slider	③	2.0	0.0001	0.0001	0.0001	0.0	0.0	0.0
Ground	④	1.0	1.0	1.0	1.0	0.0	0.0	0.0

12.2.2 Inverse Dynamic Analysis

The crank is first required to rotate at a constant angular velocity of 2π rad/s (60 rpm). Torques required at the crank to achieve motion of the system that is dictated by this driver are analyzed. Figure 12.2.1 shows three driving torque curves for connecting rod lengths of $\ell = 0.3$, 0.27, and 0.24 m. At about $t = 0.1$ s, when the slider velocity changes direction, sudden changes in driving torque occur. Note that the variation in torque is most severe for the near singular case

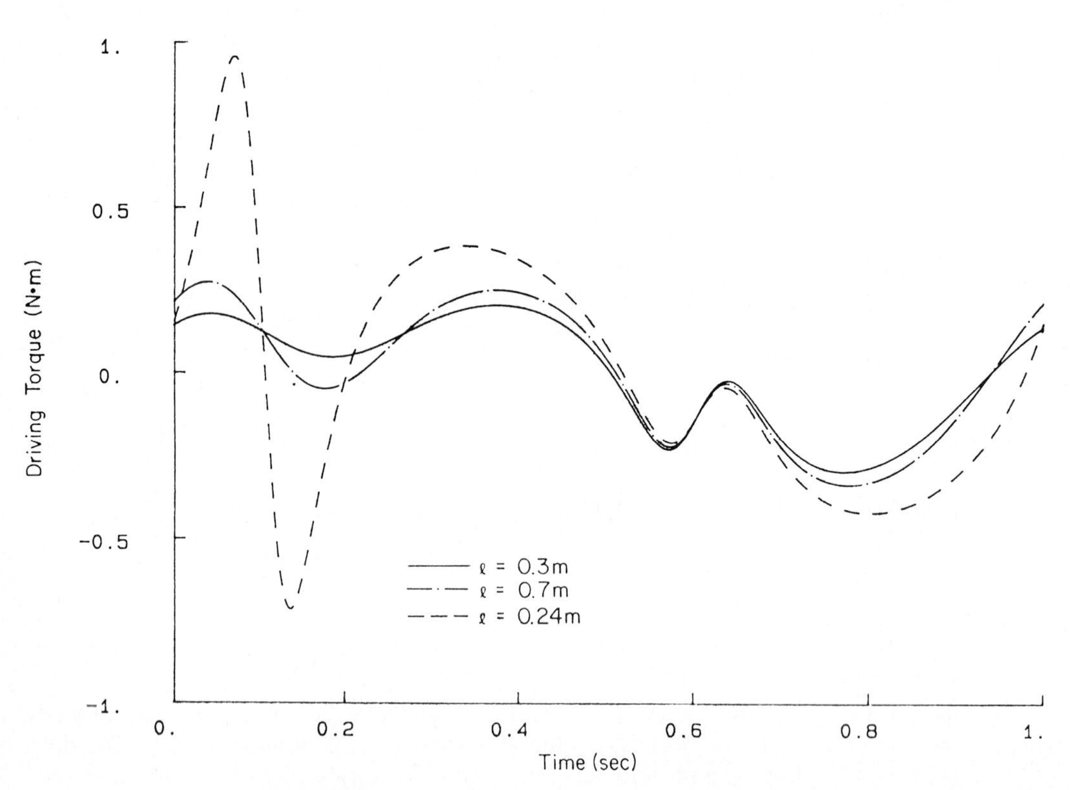

Figure 12.2.1 Driving torque for a spatial slider–crank mechanism.

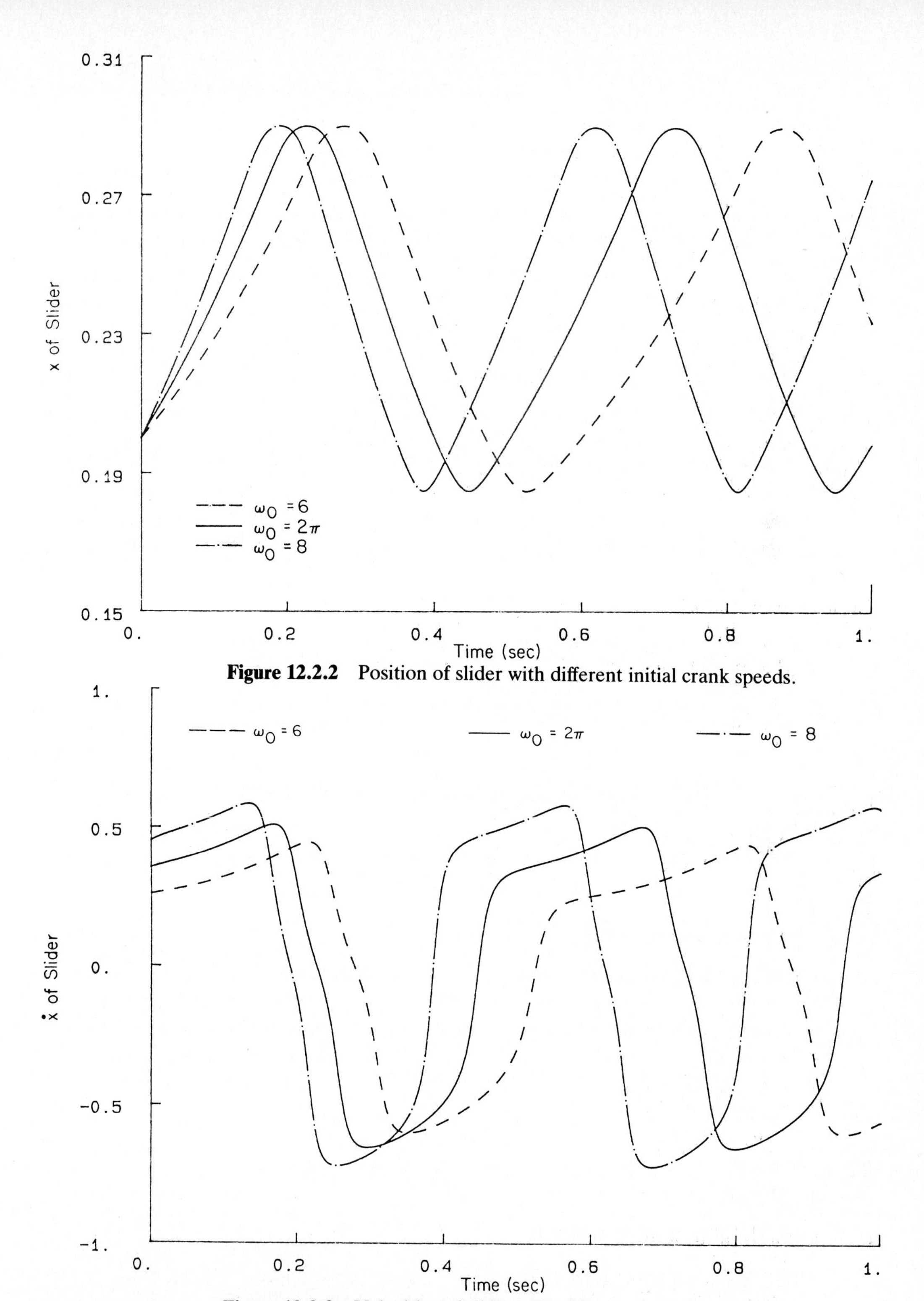

Figure 12.2.2 Position of slider with different initial crank speeds.

Figure 12.2.3 Velocities of slider with different initial crank speeds.

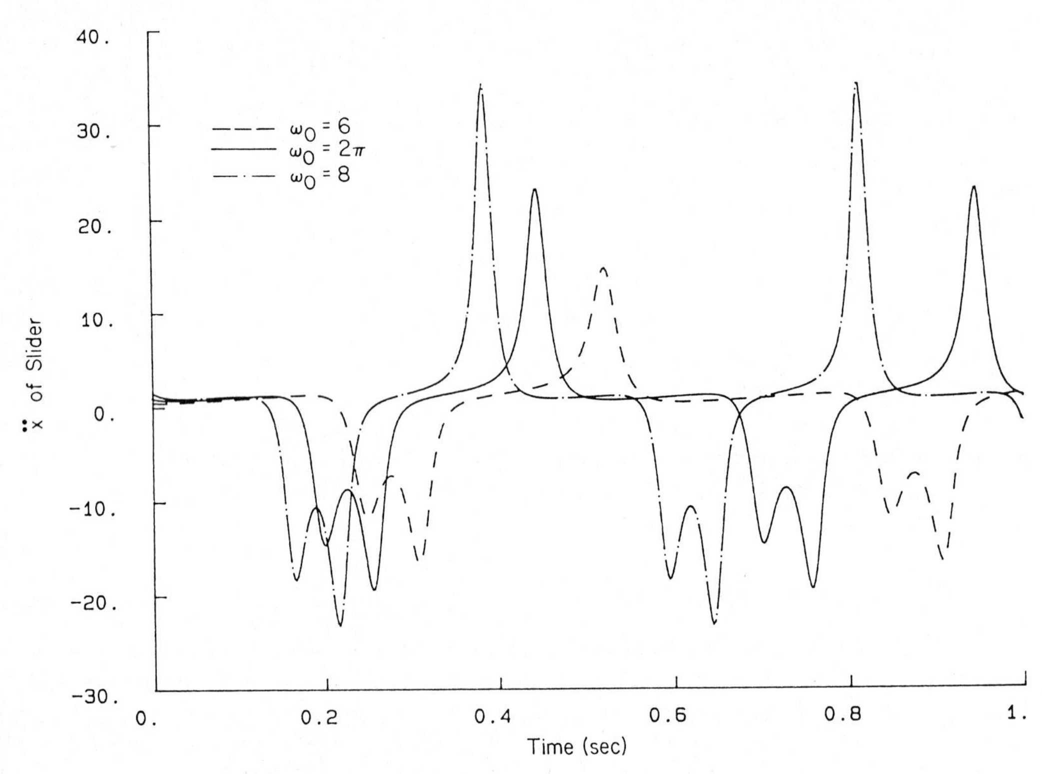

Figure 12.2.4 Accelerations of slider with different initial crank speeds.

$\ell = 0.24$ m, when the slider changes its direction near the plane in which the crank rotates (i.e., near $t = 0.1$ s).

12.2.3 Dynamic Analysis

Three dynamic runs are executed for the nominal design ($\ell = 0.3$ m), with three different initial angular velocities of the crank, $\omega_0 = 6$ rad/s (43.6 rpm), 2π rad/s (60 rpm), and 8 rad/s (76.4 rpm). Plots of positions, velocities, and accelerations of the slider are shown in Figs. 12.2.2 to 12.2.4. Since the system is conservative, plots are periodic, with different periods according to the initial angular velocities.

12.3 DYNAMIC ANALYSIS OF A SPATIAL FOUR-BAR MECHANISM

12.3.1 Alternative Models

The kinematic models of the four-bar mechanism used here are the same as those used in Section 10.3.1 for kinematic analysis. Model 1 in Fig. 10.3.1 consists of

TABLE 12.3.1 Inertia Properties of Four-Bar Mechanism, Model 1

Body	Mass	$I_{x'x'}$	$I_{y'y'}$	$I_{z'z'}$	$I_{x'y'}$	$I_{y'z'}$	$I_{z'x'}$
Link ①	2.0	4.0	2.0	0.0	0.0	0.0	0.0
Link ②	1.0	12.4	0.01	0.0	0.0	0.0	0.0
Link ③	1.0	4.54	0.01	0.0	0.0	0.0	0.0
Ground	1.0	1.0	1.0	1.0	0.0	0.0	0.0

four bodies (including ground) and uses universal and spherical joints. Model 2 in Fig. 10.3.2 consists of three bodies and uses a spherical–spherical composite joint.

The masses and moments of inertia of the components of model 1 are presented in Table 12.3.1. Properties for the bodies that make up model 2 are obtained by simply ignoring link 2 in Table 12.3.1.

12.3.2 Equilibrium Analysis

Gravity is taken as acting in the negative z direction. If gravitational force is the only external force acting on the system, it is expected that the equilibrium position of the system will have link 3 rotated to reach lower z coordinates of centroids than the position shown in Fig. 10.3.1 and defined quantitatively in Table 10.3.5. The resulting equilibrium configuration is presented in Table 12.3.2. Note that the z coordinates of links 2 and 3 are near their local minima at point A in Fig. 10.3.4. Since the heights of their centers of mass dominate the total potential energy of the system, the relative minimum point A of link 3 in Fig. 10.3.4, which corresponds to the configuration of Table 12.3.2, is a position of relative minimum total potential energy; that is, it is a stable equilibrium configuration.

Note that the z coordinate of link 3 could reach the global minimum point B of Fig. 10.3.4 if the initial estimate used to start the minimization algorithm had been nearer this configuration. A different equilibrium position would thus have

TABLE 12.3.2 Equilibrium Position of Four-Bar Mechanism, Model 1, under Gravitational Load

Body	x	y	z	e_0	e_1	e_2	e_3
Link ①	0.0	0.0	0.0	0.4722	−0.8815	0.0	0.0
Link ②	0.4281	−3.418	2.238	0.3080	−0.1269	0.9065	−0.2595
Link ③	−1.572	−8.5	2.792	0.2932	−0.6435	0.6435	−0.2932
Ground	0.0	0.0	0.0	1.0	0.0	0.0	0.0

TABLE 12.3.3 Equilibrium Position of Four-Bar Mechanism, Model 2, under Gravitational Load

Body	x	y	z	e_0	e_1	e_2	e_3
Link ①	0.0	0.0	0.0	0.8665	0.4992	0.0	0.0
Link ②	−6.610	−8.5	2.622	0.6530	−0.2713	0.2713	−0.6530
Ground	0.0	0.0	0.0	1.0	0.0	0.0	0.0
Point A	0.0	−1.730	1.003				

been achieved. This is another illustration of the effect of nonlinearities in the kinematics and dynamics of machines, leading to multiple solutions.

For model 2, the resulting equilibrium configuration is given in Table 12.3.3.

12.3.3 Inverse Dynamic Analysis

Inverse dynamic analysis is carried out by imposing a constant angular velocity of link 1, as in Section 10.3.3. Plots of the required driving torque and the x component of the reaction force at revolute joint A are presented in Fig. 12.3.1.

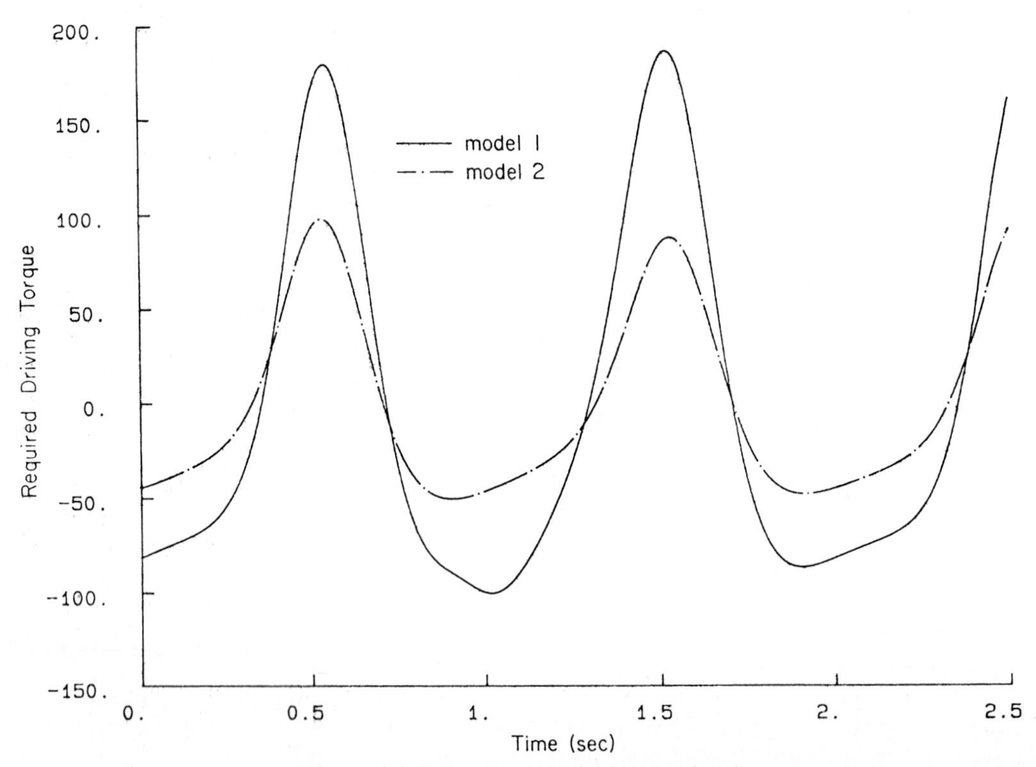

Figure 12.3.1(a) Driving torque at point A.

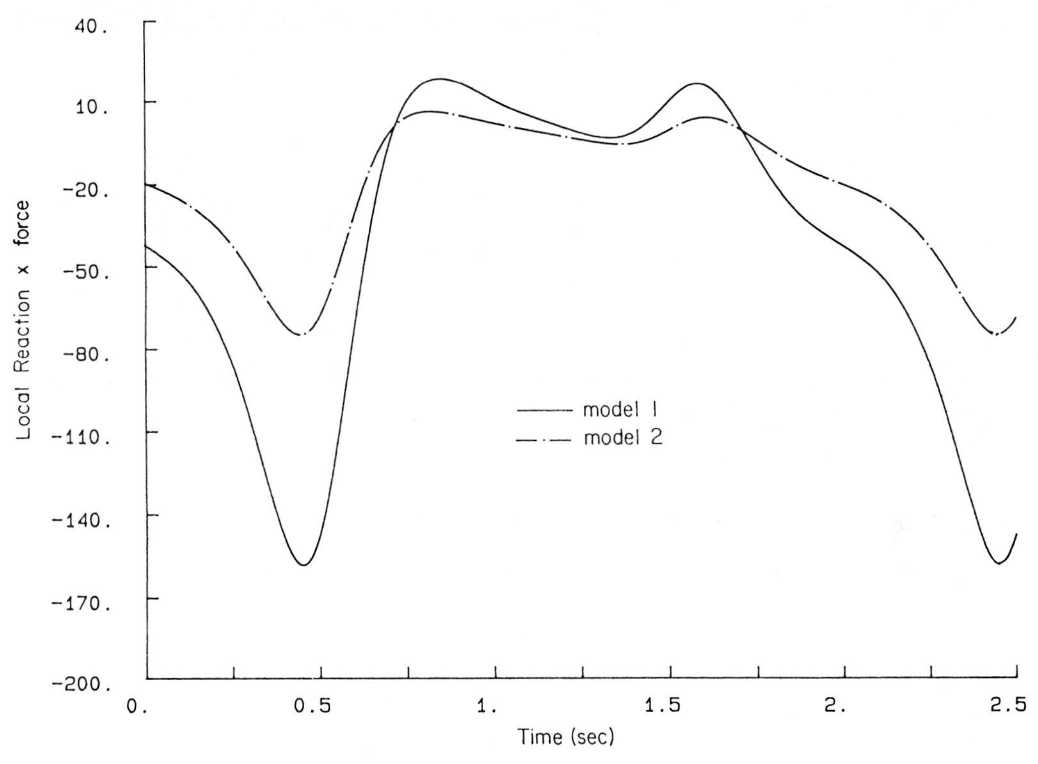

Figure 12.3.1(b) Reaction *x* force at point *A*.

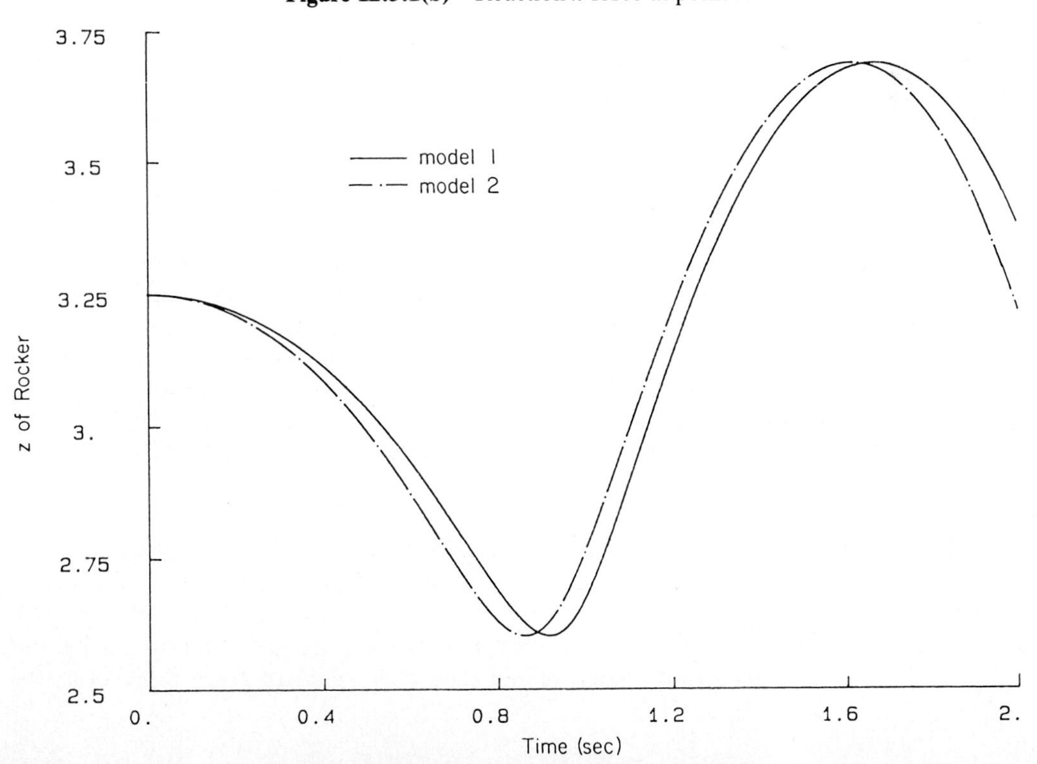

Figure 12.3.2(a) Position of centroid of link *CD*.

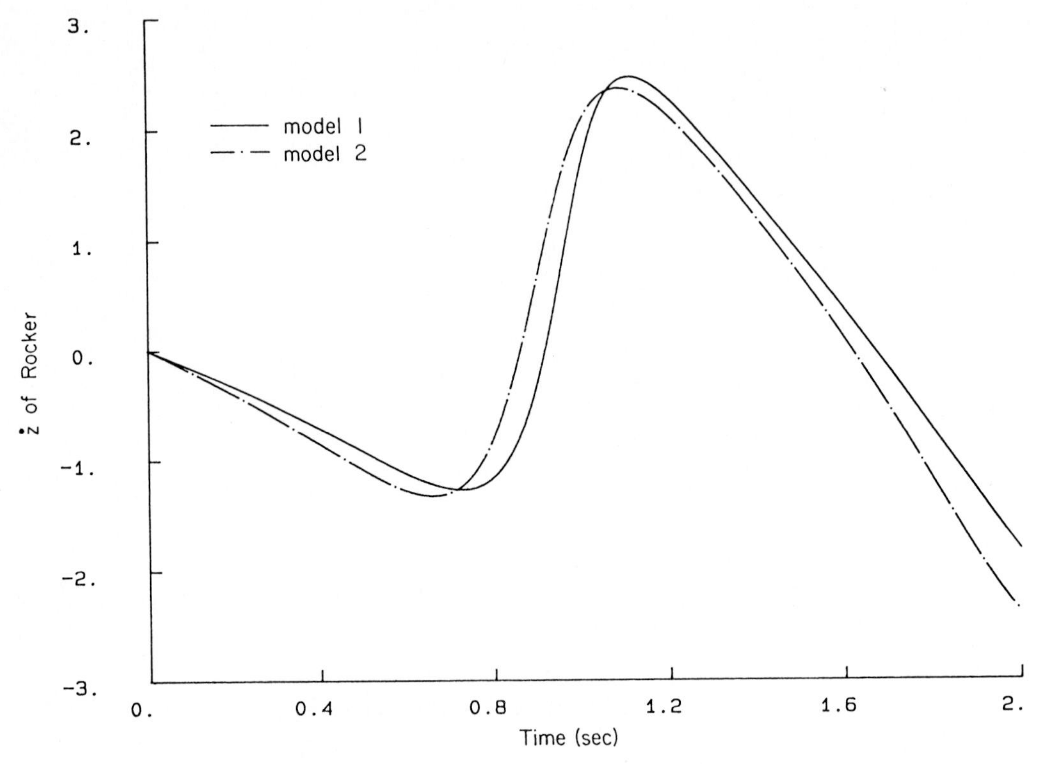

Figure 12.3.2(b) Velocity of centroid of link *CD*.

Note that, due to different inertia properties, the required driving torques and reaction forces in models 1 and 2 are different.

12.3.4 Dynamic Analysis

Dynamic response of the spatial four-bar mechanism is analyzed by imposing a counterclockwise torque $n = 10\,\mathrm{N \cdot m}$ on link 1. With the same initial position shown in Fig. 10.3.1 (Table 10.3.5), the system is started from rest. Plots of the position and velocity of the centroid of link *CD* (link 3 in model 1 and link 2 in model 2) are given in Fig. 12.3.2. Because the inertia properties of link *BC* are ignored in model 2, different dynamic response is obtained. Note that the basic character of motion is the same, but the response of model 2 is faster due to its lower inertia.

12.4 DYNAMIC ANALYSIS OF AN AIR COMPRESSOR

12.4.1 Model

Dynamic analysis of the air compressor shown in Fig. 10.4.1 is now carried out. The system configuration and definition of each joint are given in Section 10.4.

TABLE 12.4.1 Inertia Properties of Air Compressor

Body		Mass	$I_{x'x'}$	$I_{y'y'}$	$I_{z'z'}$	$I_{x'y'}$	$I_{y'z'}$	$I_{z'x'}$
Ground	①	1.0	1.0	1.0	1.0	0.0	0.0	0.0
Rotor	②	10.0	0.1	0.01	0.1	0.0	0.0	0.0
Disk	③	2.0	0.04	0.08	0.04	0.0	0.0	0.0
Piston 1	④	1.5	0.002	0.002	0.002	0.0	0.0	0.0
Piston 2	⑤	1.5	0.002	0.002	0.002	0.0	0.0	0.0
Piston 3	⑥	1.5	0.002	0.002	0.002	0.0	0.0	0.0
Piston 4	⑦	1.5	0.002	0.002	0.002	0.0	0.0	0.0
Piston 5	⑧	1.5	0.002	0.002	0.002	0.0	0.0	0.0
Piston 6	⑨	1.5	0.002	0.002	0.002	0.0	0.0	0.0

The only additional data required are the inertia properties of each body, the initial conditions of motion for the system, and the forces that act. The masses and moments of inertia are presented in Table 12.4.1.

To simulate the air pressure that acts on each piston, a force curve characteristic is chosen, as shown in Fig. 12.4.1. The distance *A*-*B* is the full stroke of the piston. The valve opens when the pressure is at its maximum, at point *C*. Dimensions *A*, *B*, and *C* are chosen such that the compression ratio is

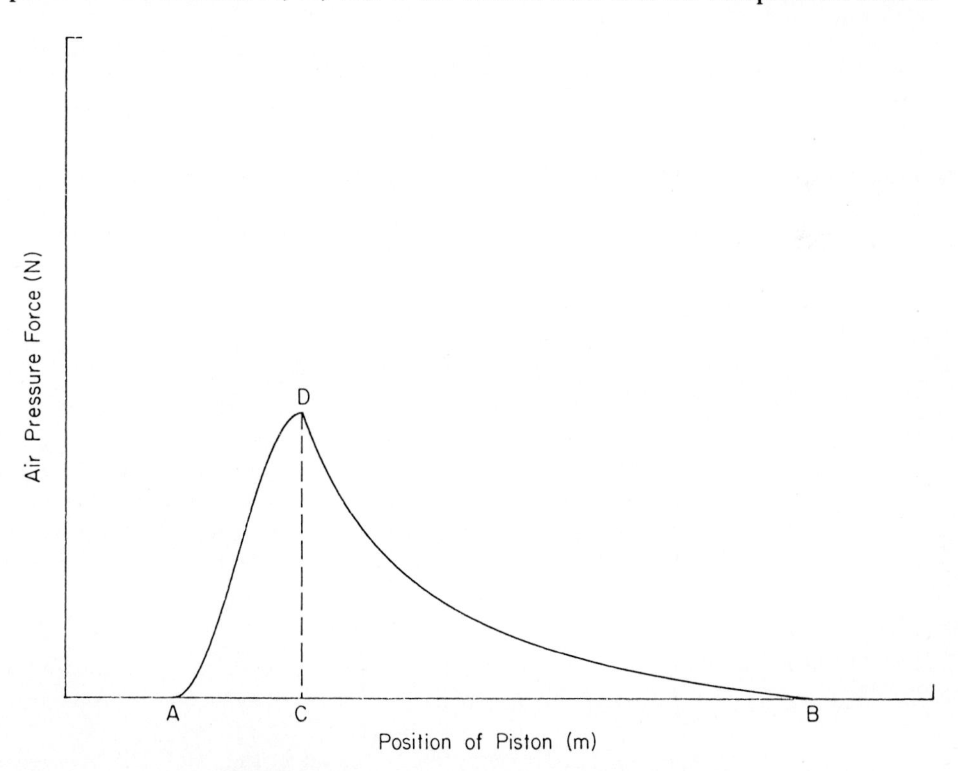

Figure 12.4.1 Air pressure force characterisitic.

5:1. For the nominal tilt angle of 30° for the plate, the stroke AB is 0.2 m. Hence, $AC = 0.04$ m. Force curve DB in Fig. 12.4.1 is

$$F_{DB} = \frac{F_{max}}{y - y_A} - F_{min}$$

where y is piston position, y_A is piston position at A (left end of stroke), F_{max} is the area of piston (m^2) × maximum pressure (N/m^2), and F_{min} is the force due to air pressure at the right end of the stroke B, which is subtracted to make the gauge pressure zero at B. For this model, $y_A = -1.1$ m, $y_B = -0.9$ m, and the piston bore diameter is 0.1 m.

Force curve AD in Fig. 12.4.1 is

$$F_{AD} = -\frac{F_C}{2}\cos\frac{\pi}{y_C - y_A}(y - y_A) + \frac{F_C}{2}$$

where F_C is force due to air pressure at C and y_C is piston position at C.

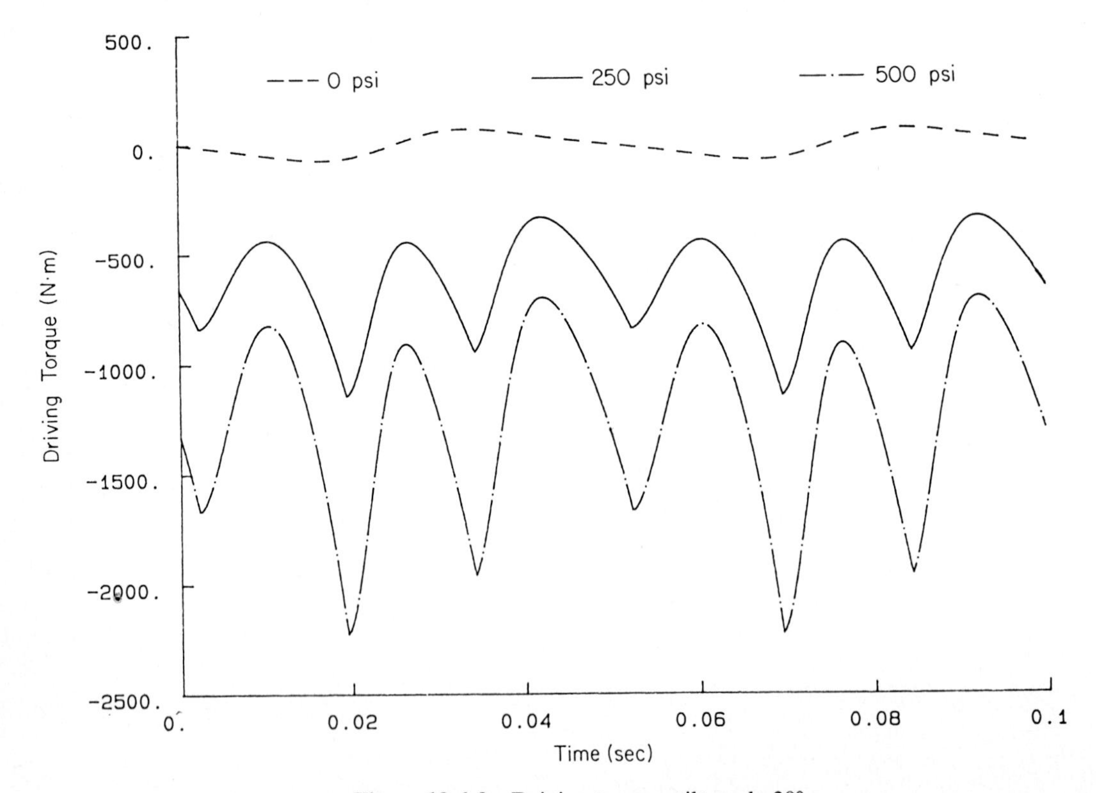

Figure 12.4.2 Driving torque, tilt angle 30°.

12.4.2 Inverse Dynamic Analysis

The rotor is required to rotate at a constant angular velocity of 20π rad/s (600 rpm). To achieve the motion of the system that is dictated by this driver, the torques required at the rotor are analyzed. Figure 12.4.2 shows three driving torque curves, with maximum air pressures of 500, 250, and 0 psi. The x position constraint force shown in Fig. 12.4.3 is essential information for the design of the ball–socket–slider mechanism that prevents the disk from rotating.

Similar analyses are carried out and plotted in Figs. 12.4.4 and 12.4.5, with a tilt angle of the disk of 40°.

12.4.3 Dynamic Analysis

Dynamic simulations are carried out with a constant torque applied to the rotor of the compressor, simulating an external power source such as an electric motor. The constant torque is calculated by integrating the force curve in Fig. 12.4.1, dividing by 2π (one full revolution), and multiplying by 6. Runs are made with three different initial rotor angular velocities of 60 rad/s (573 rpm), 70 rad/s (668 rpm), and 80 rad/s (764 rpm).

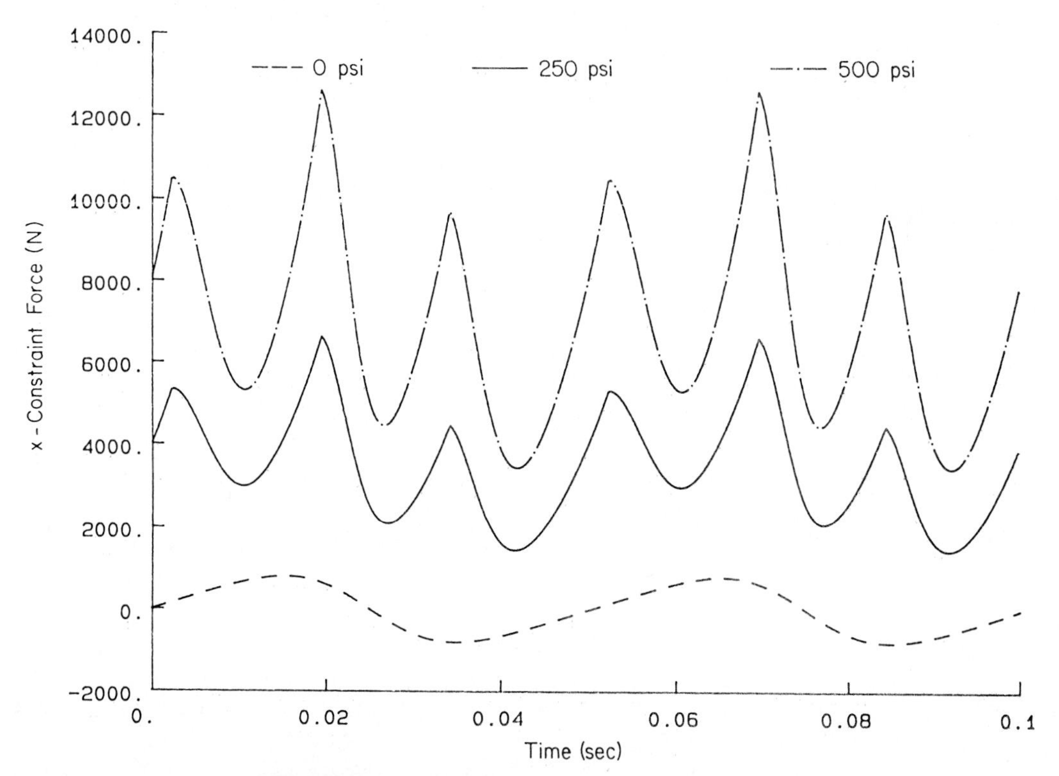

Figure 12.4.3 x position constraint force, tilt angle 30°.

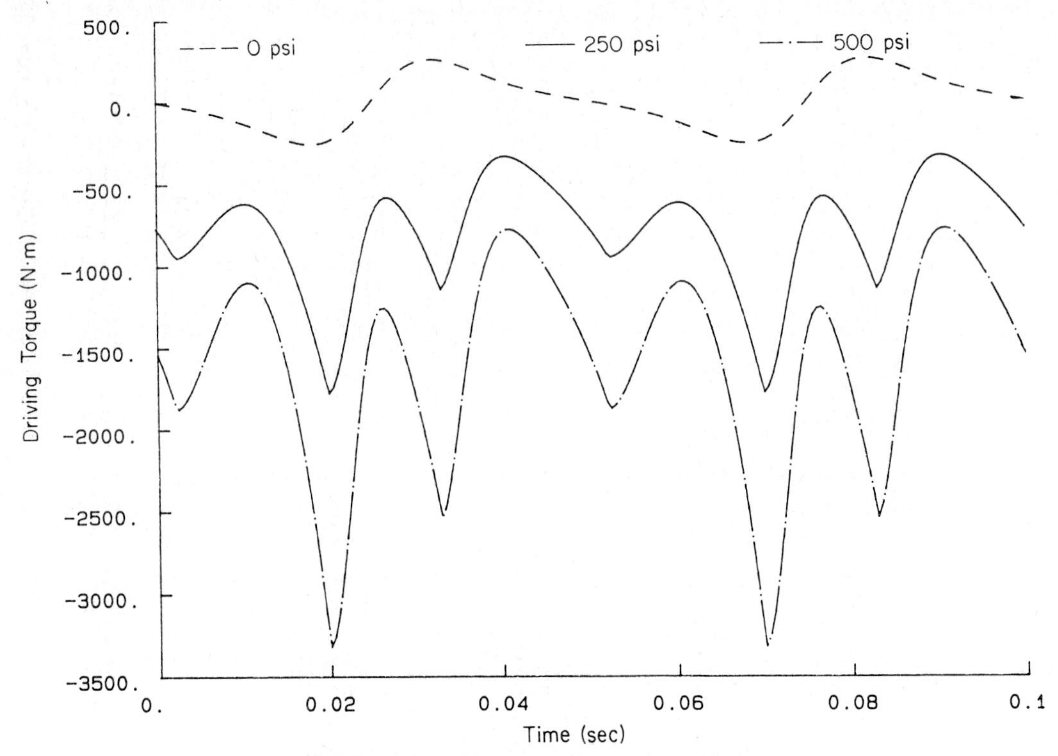

Figure 12.4.4 Driving torque, tilt angle 40°.

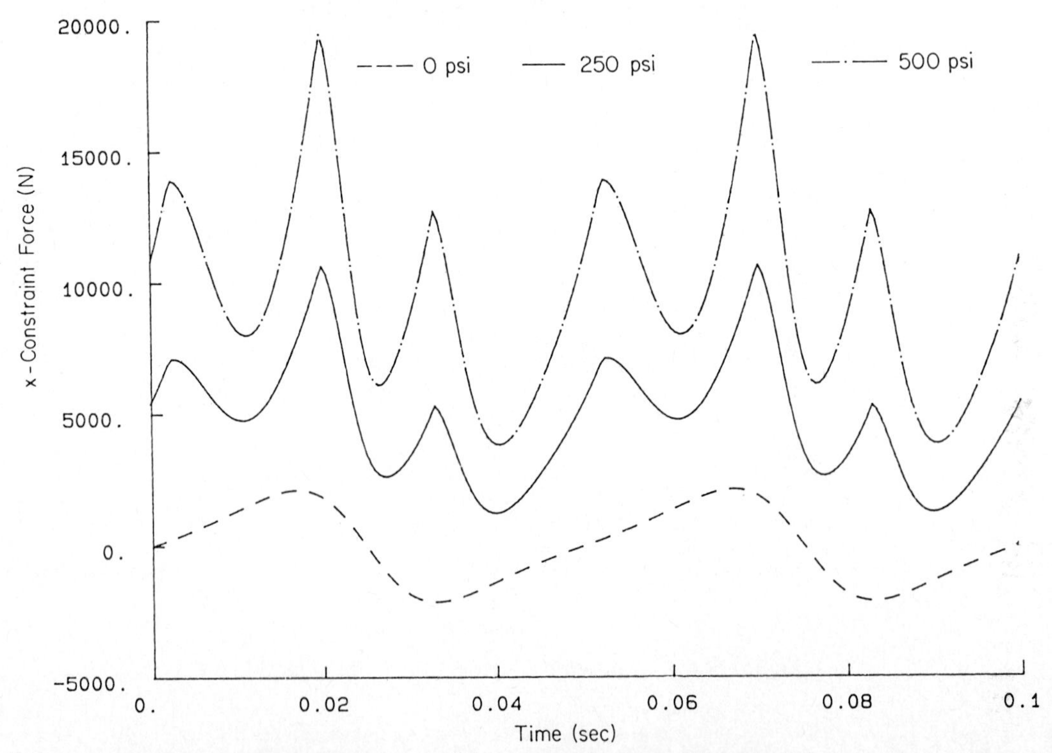

Figure 12.4.5 *x* position constraint force, tilt angle 40°.

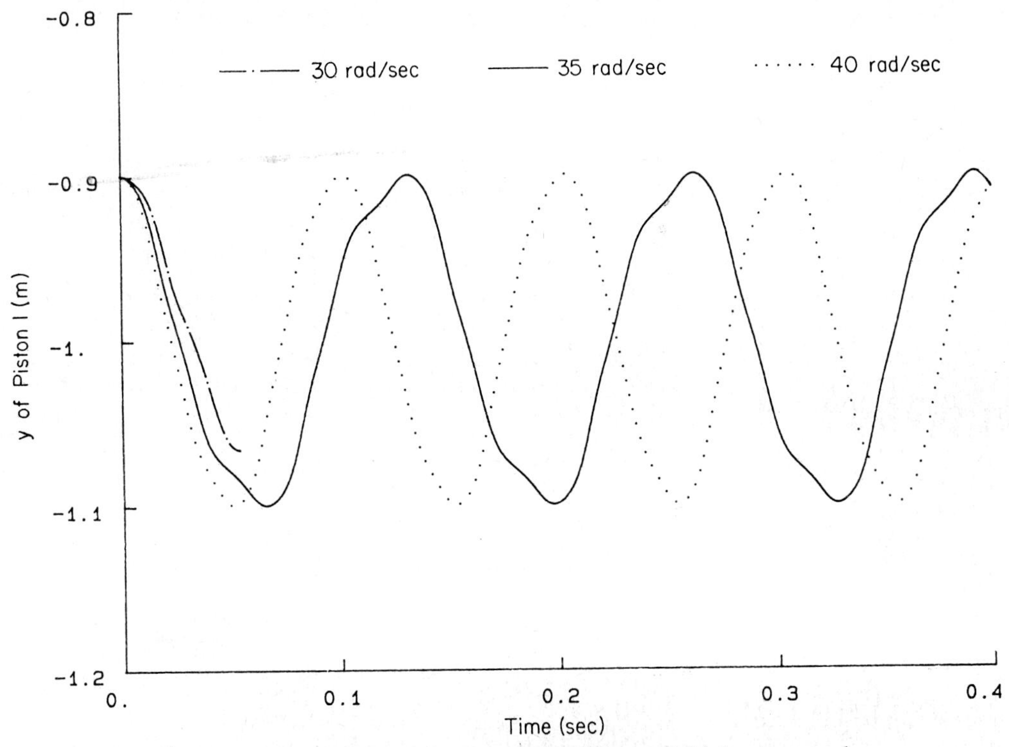

Figure 12.4.6 Position of piston 1 with different initial rotor speeds.

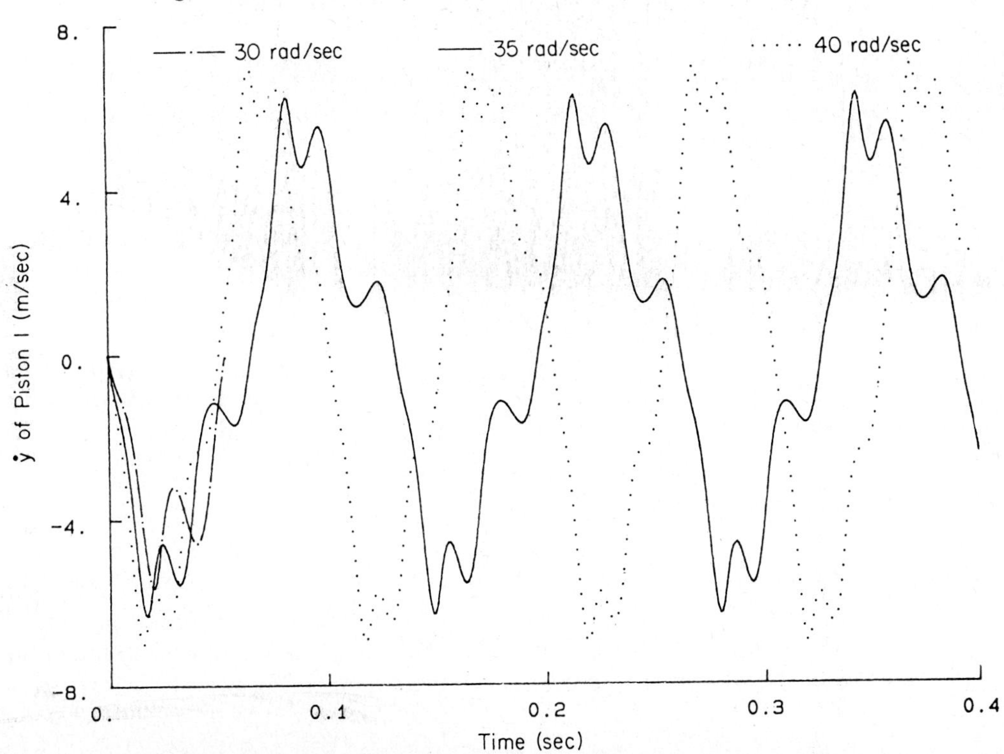

Figure 12.4.7 Velocities of piston 1 with different initial rotor speeds.

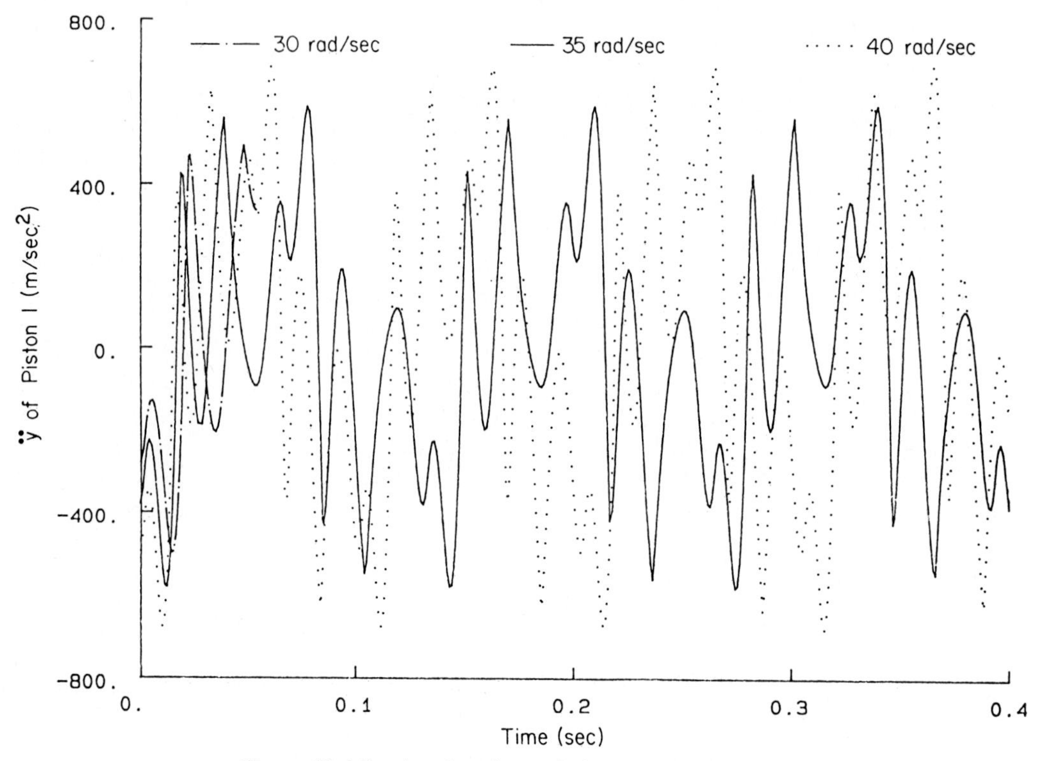

Figure 12.4.8 Accelerations of piston 1 with different initial rotor speeds.

The constant torque is selected to overcome the energy loss in one cycle of the rotor due to the external air pressure force. Therefore, if the initial angular velocity ω_0, and hence the kinetic energy of the system, is not sufficient, motion may stop before one full cycle is completed, as shown for $\omega_0 = 30$ rad/s in Fig. 12.4.6. If the initial angular velocity ω_0, and hence the kinetic energy, is sufficient, then the applied torque yields periodic motion, as shown in Fig. 12.4.6. Plots of the velocities and accelerations of piston 1 are shown in Figs. 12.4.7 and 12.4.8, respectively.

12.5 DYNAMICS OF A VEHICLE

One of the most significant applications of the modeling and analysis of dynamic systems is vehicle system analysis. This section studies the modeling and analysis of the passenger car of Fig. 1.1.10. The objective of this analysis is to predict the motion of the vehicle in response to steering input.

12.5.1 Alternative Models

Two dynamic models of the automobile are created for analysis. Model 1 is composed of bodies 1 through 6 in Fig. 12.5.1, with the front suspension of Fig. 12.5.2. Body 1 is the chassis, bodies 2 to 5 are front and rear wheel assemblies, and body 6 is a steering rack. Model 2 adds a pair of roll-stabilizing bars (bodies 7 and 8) to model 1. The torsional compliance of these bars is modeled by translational springs between their ends and the suspension lower control arms. Masses and moments of inertia of the eight bodies are defined in Table 12.5.1. Products of inertia are all zero.

Each front wheel assembly includes a McPherson strut, whose kinematic connections with the chassis and steering rack are shown in Figs. 12.5.1 and 12.5.2. Each wheel assembly is connected to the chassis by a strut joint at the top (B-C in Fig. 12.5.2) to represent the McPherson strut and a revolute–spherical joint at the bottom (A-B in Fig. 12.5.2) to represent the lower control arm. Each rear wheel assembly includes a trailing arm that is connected to the chassis by a revolute joint. The lateral position of the steering rack is controlled by a pinion on the steering column. It is connected to the chassis by a translational joint. The tie rods that connect the rack and wheel assemblies are modeled as distance constraints. Roll-stabilizing bars (bodies 7 and 8) are connected to the chassis by revolute joints at points J and K, respectively.

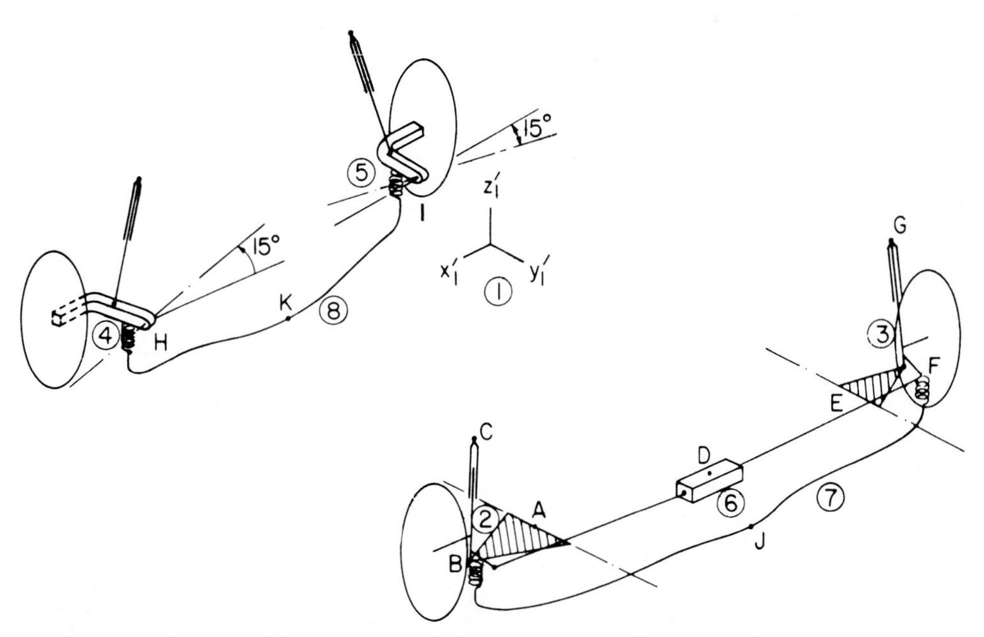

Figure 12.5.1 Automobile suspension schematic.

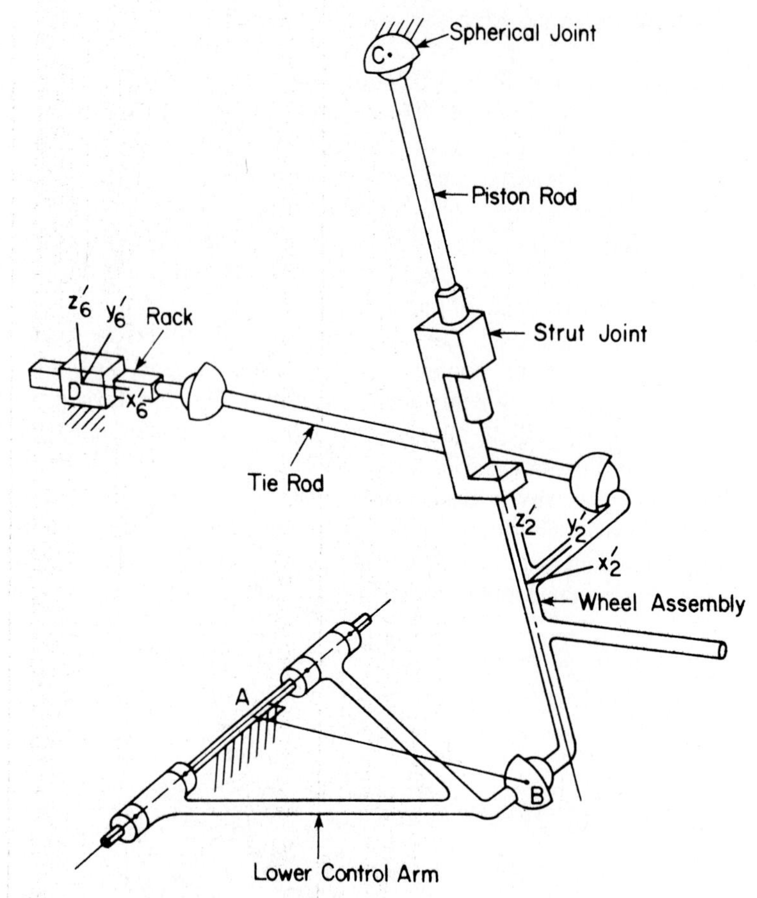

Figure 12.5.2 McPherson strut front suspension.

TABLE 12.5.1 Inertia Properties of the Vehicle

Body		Mass	$I_{x'x'}$	$I_{y'y'}$	$I_{z'z'}$
Chassis	①	1247.5	1240.1	209.6	139.9
Wheel assembly (FR)	②	37.6	1.0	0.5	1.0
Wheel assembly (FL)	③	37.6	1.0	0.5	1.0
Wheel assembly (RR)	④	43.0	0.67	1.5	1.2
Wheel assembly (RL)	⑤	43.0	0.67	1.5	1.2
Rack	⑥	0.1	0.01	0.01	0.01
Stabilizing bar (F)	⑦	1.0	0.1	0.01	0.01
Stabilizing bar (R)	⑧	1.0	0.1	0.01	0.01

The kinematic model of model 2 is described as follows:

Bodies	
Eight bodies	$nc = 56$
Constraints	
Revolute–spherical joint: A-B	2
Strut joint: B-C	2
Translational joint: D	5
Revolute–spherical joint: E-F	2
Strut joint: F-G	2
Revolute joint: H	5
Revolute joint: I	5
Revolute joint: J	5
Revolute joint: K	5
Distance constraint: ②-⑥	1
Distance constraint: ③-⑥	1
Driving constraint: D	1
Euler parameter normalization constraint	8
	$nh = 44$
$\text{DOF} = 56 - 44 = 12$.	

Tables 12.5.2 to 12.5.6 define the kinematic joint definition data for model 2. Eight internal TSDA force elements are used to model the vehicle suspension system, including the roll-stabilizing bars. Table 12.5.7 defines the attachment points and physical characteristics of these elements.

TABLE 12.5.2 Revolute–Spherical Joint Data

Joint A-B

Body \ Point	P x'	P y'	P z'	Q x'	Q y'	Q z'	R x'	R y'	R z'
Chassis ①	0.33	1.1735	−0.3923	0.322	0.8675	−0.3893	0.33	1.1735	0.0
Wheel assembly ②	0.0507	0.0321	−0.2044	0.0507	0.0321	1.0	0.0507	1.0	−0.2044

Distance = 0.203

Joint E-F

Body \ Point	P x'	P y'	P z'	Q x'	Q y'	Q z'	R x'	R y'	R z'
Chassis ①	−0.33	1.1735	−0.3923	−0.322	0.8675	−0.3893	−0.33	1.1175	0.0
Wheel assembly ③	−0.0507	0.0321	−0.2044	−0.0507	0.0321	1.0	−0.0507	1.0	−0.2044

Distance = 0.203

TABLE 12.5.3 Strut Joint Data

Joint *B-C*

Body \ Point	P x'	P y'	P z'	Q x'	Q y'	Q z'	R x'	R y'	R z'
Chassis ①	0.5011	1.1162	0.1998	0.5011	1.1162	1.0	0.5011	2.0	0.1998
Wheel assembly ②	0.0	0.0	0.0	1.0	0.0	0.0	0.0	1.0	0.0

Joint *F-G*

Body \ Point	P x'	P y'	P z'	Q x'	Q y'	Q z'	R x'	R y'	R z'
Chassis ①	−0.5011	1.1162	0.1998	−0.5011	1.1162	1.0	−0.5011	2.0	0.1998
Wheel assembly ③	0.0	0.0	0.0	1.0	0.0	0.0	0.0	1.0	0.0

TABLE 12.5.4 Translational Joint Data

Body \ Point	P x'	P y'	P z'	Q x'	Q y'	Q z'	R x'	R y'	R z'
Chassis ①	0.0	1.2675	−0.3748	1.0	1.2675	−0.3748	0.0	2.0	−0.3748
Rack and pinion ⑥	0.0	0.0	0.0	1.0	0.0	0.0	0.0	1.0	0.0

TABLE 12.5.5 Revolute Joint Data

Joint *H*

Body \ Point	P x'	P y'	P z'	Q x'	Q y'	Q z'	R x'	R y'	R z'
Chassis ①	0.2026	−1.0095	−0.285	−0.7633	−1.2683	−0.285	0.2026	−1.0095	0.0
Wheel assembly ④	−0.3774	0.278	0.0093	−1.343	0.0192	0.0093	−0.3774	0.278	0.0

Joint *I*

Body \ Point	P x'	P y'	P z'	Q x'	Q y'	Q z'	R x'	R y'	R z'
Chassis ①	−0.2026	−1.0095	−0.285	0.7633	−1.2683	−0.285	−0.2026	−1.0095	0.0
Wheel assembly ⑤	0.3774	0.278	0.0093	1.343	0.0192	0.0093	0.774	0.278	1.0

Joint *J*

Body \ Point	P x'	P y'	P z'	Q x'	Q y'	Q z'	R x'	R y'	R z'
Chassis ①	0.0	1.061	−0.3263	1.0	1.061	−0.3263	0.0	2.0	−0.3263
Stabilizing bar ⑦	0.0	0.0	0.0	1.0	0.0	0.0	0.0	1.0	0.0

TABLE 12.5.5 continued

Joint K

Body \ Point		P x'	y'	z'	Q x'	y'	z'	R x'	y'	z'
Chassis	①	0.0	−1.4075	−0.1813	1.0	−1.4075	−0.1813	0.0	2.0	−0.1813
Stabilizing bar	⑧	0.0	0.0	0.0	1.0	0.0	0.0	0.0	1.0	0.0

TABLE 12.5.6 Distance Constraint Data

Distance ②–⑥

Body \ Point		P x'	y'	z'	Q x'	y'	z'	R x'	y'	z'
Wheel assembly	②	0.07	0.155	−0.186	0.07	0.155	1.0	0.07	1.0	−0.186
Rack and pinion	⑥	0.305	0.0	0.0	0.305	0.0	1.0	0.305	1.0	0.0

Distance = 0.3742

Distance ③–⑥

Body \ Point		P x'	y'	z'	Q x'	y'	z'	R x'	y'	z'
Wheel assembly	③	−0.07	0.155	−0.186	−0.07	0.155	1.0	−0.07	1.0	−0.186
Rack and pinion	⑥	−0.305	0.0	0.0	−0.305	0.0	1.0	−0.305	1.0	−0.0

Distance = 0.3742.

Tire forces must be considered to complete the vehicle model. Tire vertical force F_{ver} is calculated as

$$F_{ver} = k_t \times d$$

where k_t is the tire spring constant and d is tire vertical deformation, which is

TABLE 12.5.7 TSDA Data

TSDA	P_i	P_j	k	C	L_0
①-②	(0.5011, 1.1162, 0.1998)	(0.0, 0.0, 0.0)	38,600	150	0.4992
①-③	(−0.5011, 1.1162, 0.1998)	(0.0, 0.0, 0.0)	38,600	150	0.4992
①-④	(0.58, −1.2875, 0.2)	(0.0, 0.0, 0.0)	38,600	200	0.5641
①-⑤	(−0.58, −1.2875, 0.2)	(0.0, 0.0, 0.0)	38,600	200	0.5641
⑦-②	(0.509, 0.1405, 0.02)	(−0.0897, 0.0168, −0.2119)	83,400	0	0.1011
⑦-③	(−0.509, 0.1405, 0.02)	(0.0897,0.0168, −0.2119)	83,400	0	0.1011
⑧-④	(0.392, 0.168, −0.0474)	(−0.1832, 0.048, −0.0423)	9,330	0	0.0943
⑧-⑤	(−0.392, 0.168, −0.0474)	(0.1832, 0.048, −0.0423)	9,330	0	0.0943

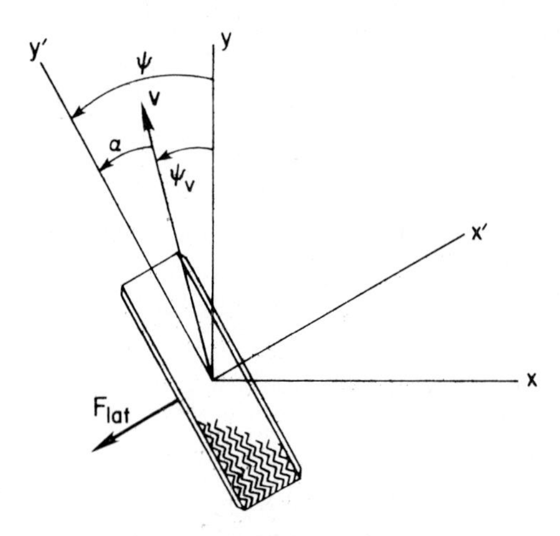

Figure 12.5.3 Tire schematic.

defined by the height of the spindle above the road surface. Tire lateral force F_{lat} is generated by lateral slip of the tire on the road surface and is approximated as

$$F_{\text{lat}} = C_\alpha \times \alpha$$

where C_α is tire cornering stiffness and α is the tire slip angle. This lateral force cannot exceed μF_{ver}, where $\mu < 1$ is the coefficient of friction between the tire and the road surface.

Figure 12.5.3 schematically defines the tire slip angle α, which is the angle between the tire heading and the tire velocity vector $\mathbf{v}$. The tire is assumed to be vertical and the road surface is the global x-y plane. A local x'-y' reference frame is fixed in the spindle to which the wheel is attached. Angles ψ and ψ_v are called the tire yaw (or heading) and velocity angles, respectively, as shown in Fig. 12.5.3. The tire slip angle is thus

$$\alpha = \psi - \psi_v \tag{12.5.1}$$

Angle ψ is calculated from the definition of the Euler parameters of the spindle, presuming the tire is vertical; that is,

$$\cos\frac{\psi}{2} = e_0$$

or

$$\psi = 2\cos^{-1} e_0, \qquad 0 \leq \psi < 2\pi \tag{12.5.2}$$

The tire velocity angle ψ_v is

$$\psi_v = -\tan^{-1}\frac{v_x}{v_y} \tag{12.5.3}$$

TABLE 12.5.8 Position and Orientation Estimates

Body \ GC		x	y	z	e_1	e_2	e_3
Chassis	①	0.0	0.0	0.55	0.0	0.0	0.0
Wheel assembly	②	0.6	1.17	0.34	0.0	0.0	0.0
Wheel assembly	③	−0.6	1.17	0.34	0.0	0.0	0.0
Wheel assembly	④	0.58	−1.29	0.26	0.0	0.0	0.0
Wheel assembly	⑤	0.58	−1.29	0.26	0.0	0.0	0.0
Rack	⑥	0.0	1.26	0.16	0.0	0.0	0.0
Stabilizing bar	⑦	0.0	1.05	0.21	0.0	0.0	0.0
Stabilizing bar	⑧	0.0	−1.40	0.38	0.0	0.0	0.0

12.5.2 Equilibrium Analysis

Static equilibrium analysis of model 1 is performed using dynamic analysis of the vehicle. Estimates of the position and orientation of the vehicle are tabulated in Table 12.5.8. Figures 12.5.4 and 12.5.5 show settling of the vehicle to an equilibrium position due to damping in the suspension system. Table 12.5.9 defines the static equilibrium configuration obtained.

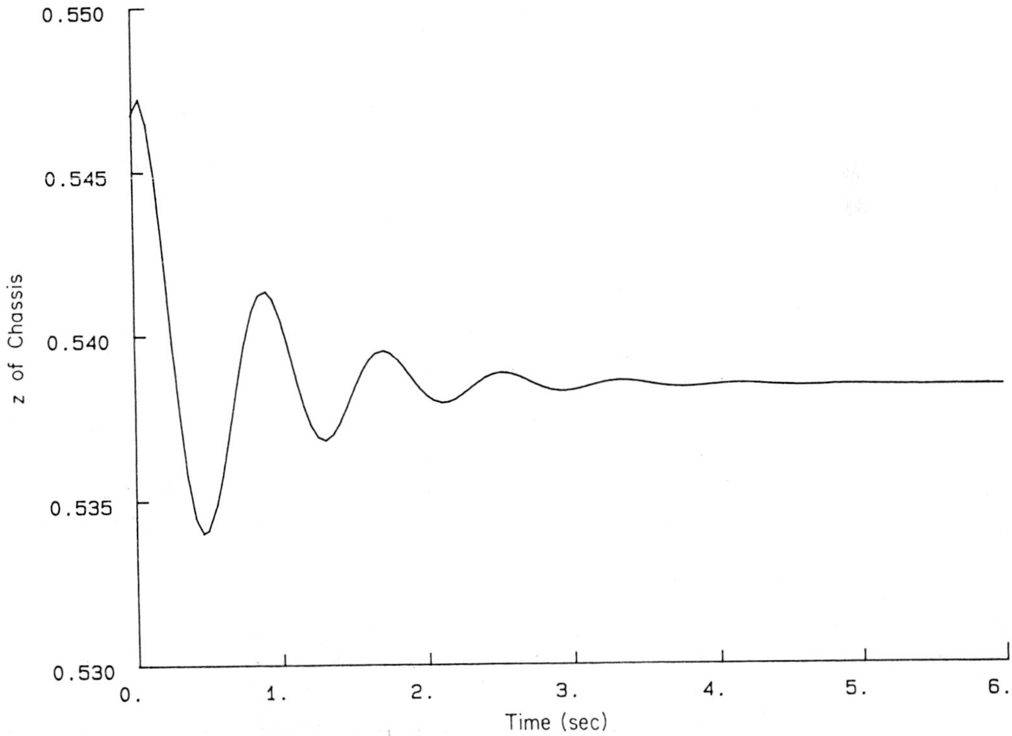

Figure 12.5.4 Vehicle equilibrium height (model 1).

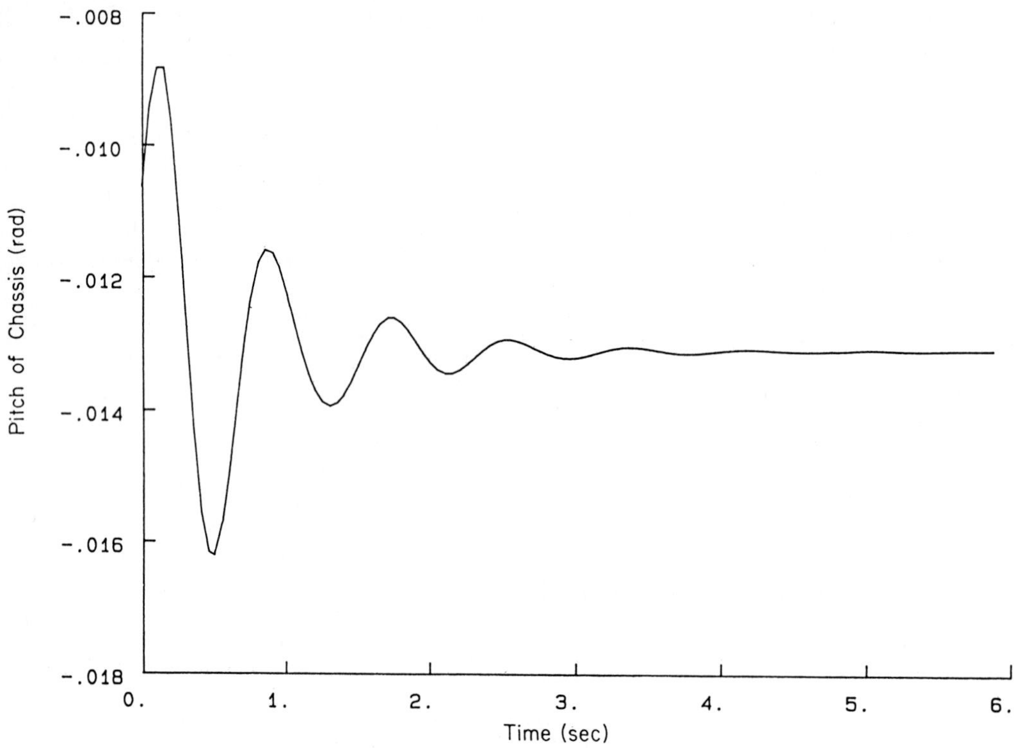

Figure 12.5.5 Vehicle equilibrium pitch (model 1).

TABLE 12.5.9 Equilibrium Configuration

Body \ GC	x	y	z	e_1	e_2	e_3
Chassis ①	0.0	0.0	0.5385	−0.0065	0.0	−0.0005
Wheel assembly ②	0.6014	1.1761	0.3449	−0.0022	−0.0065	−0.001
Wheel assembly ③	−0.5992	1.1772	0.3449	−0.0023	0.0065	−0.0002
Wheel assembly ④	0.5788	−1.2918	0.2617	−0.0075	−0.0003	−0.0005
Wheel assembly ⑤	−0.5812	−1.2907	0.2617	−0.0075	0.0003	−0.0005
Rack ⑥	0.0012	1.2625	0.1471	−0.0065	0.0	−0.0005
Stabilizing bar ⑦	0.001	1.0566	0.1983	0.0675	0.0	−0.0005
Stabilizing bar ⑧	−0.0013	−1.4097	0.3756	−0.0508	0.0	−0.0002

12.5.3 Steering Specification

In both models 1 and 2, a driving constraint is specified for translation of the rack, relative to the chassis, to simulate steering. Figure 12.5.6 shows three curves of rack position as functions of time. With curves 1 and 2, lane change maneuvers of the vehicle are expected. A circular path of the vehicle is expected with curve 3.

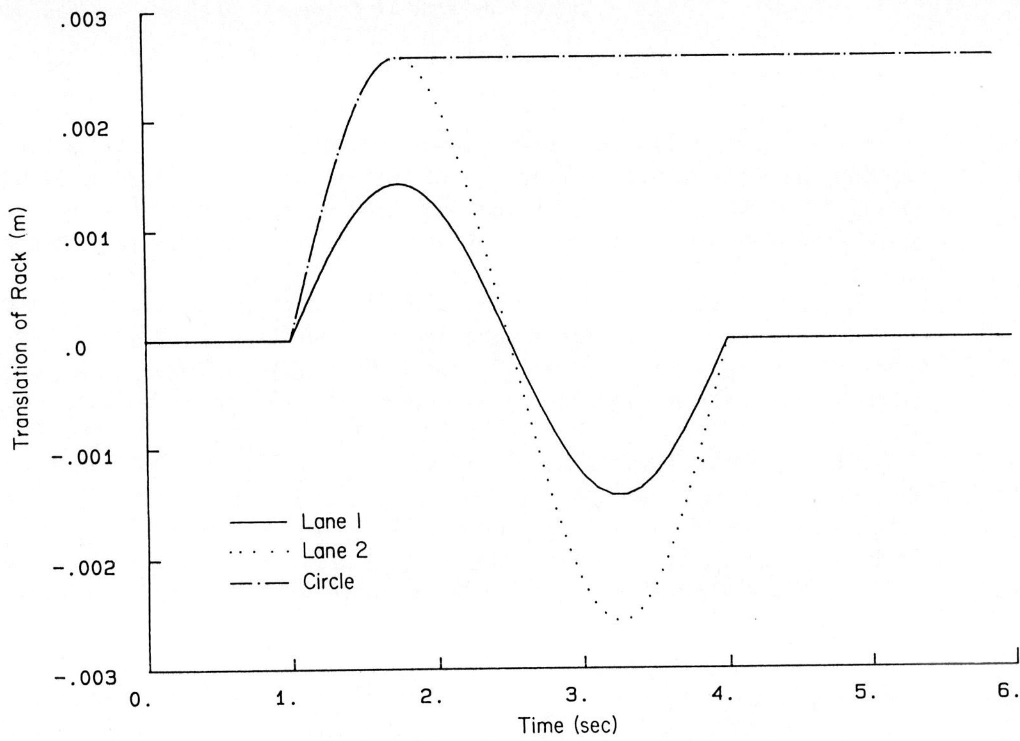

Figure 12.5.6 Rack driver for steering.

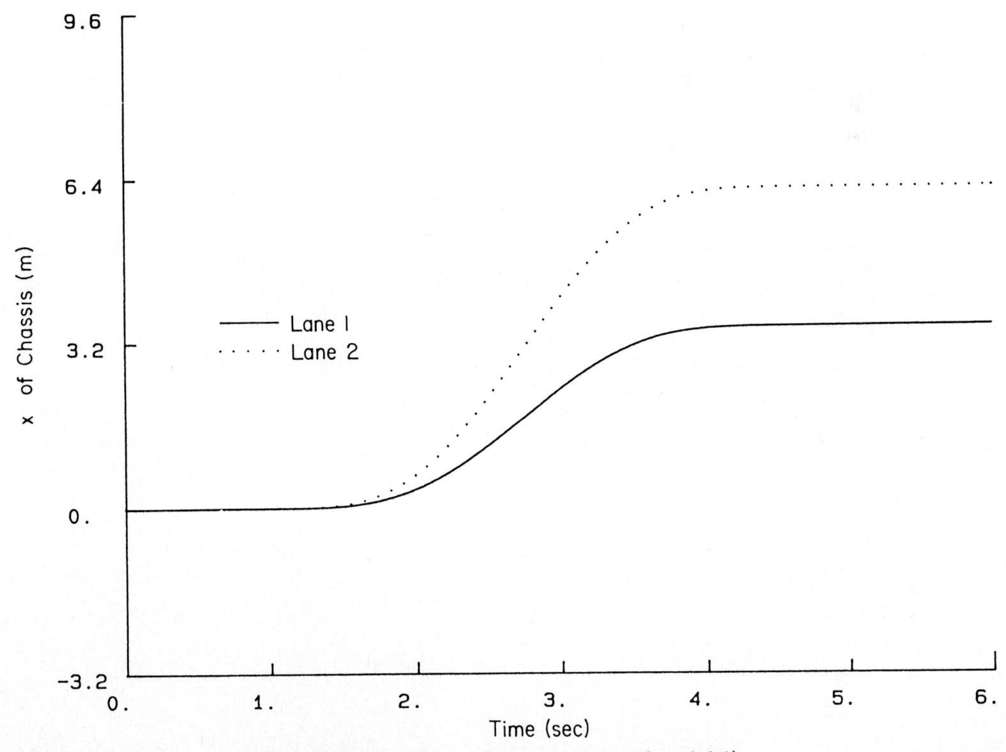

Figure 12.5.7 Lane–change position (model 1).

12.5.4 Dynamic Analysis

Lane Changes The dynamic response of the vehicle is analyzed by imposing an initial forward velocity $\dot{y}_0 = 55$ mph of the vehicle. Figures 12.5.7 and 12.5.8 show lateral positions and accelerations of model 1 for the two lane-change maneuvers. The pitch, roll, and yaw of the vehicle in lane change 2 are plotted in Fig. 12.5.9.

To see the stabilizing effect of roll-stabilizing bars, the dynamic response of model 2 in the lane-change maneuvers is obtained. Roll motions of the vehicle are compared in Fig. 12.5.10 for both models and with both lane-change drivers. The roll-stabilizer effect in model 2 is clearly observed.

Circular Path Three different initial speeds, $\dot{y}_0 = 20$, 40, and 55 mph, of the vehicle are used with steering input 3 of Fig. 12.5.6. The paths of the vehicle from these simulations are plotted in Fig. 12.5.11. For $\dot{y}_0 = 55$ mph, the maximum tire slip force occurs; that is, $F_{lat} = \mu F_{ver}$, until the speed of the vehicle substantially decreases due to large lateral slip. Figure 12.5.12 shows

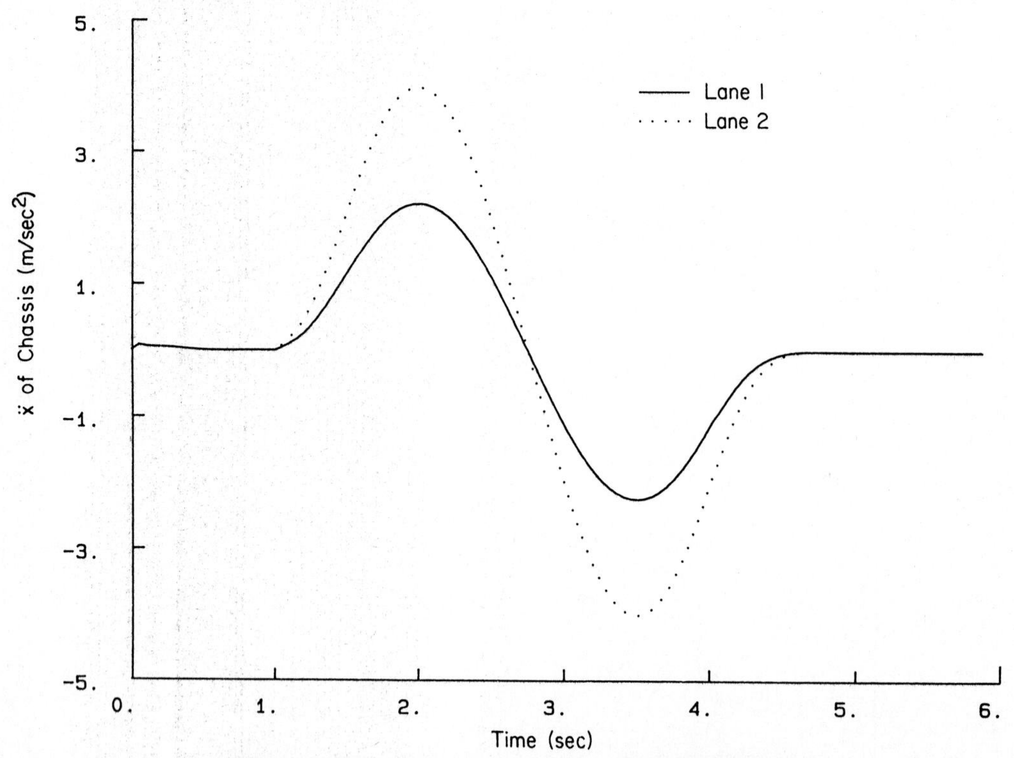

Figure 12.5.8 Lane–change acceleration (model 1).

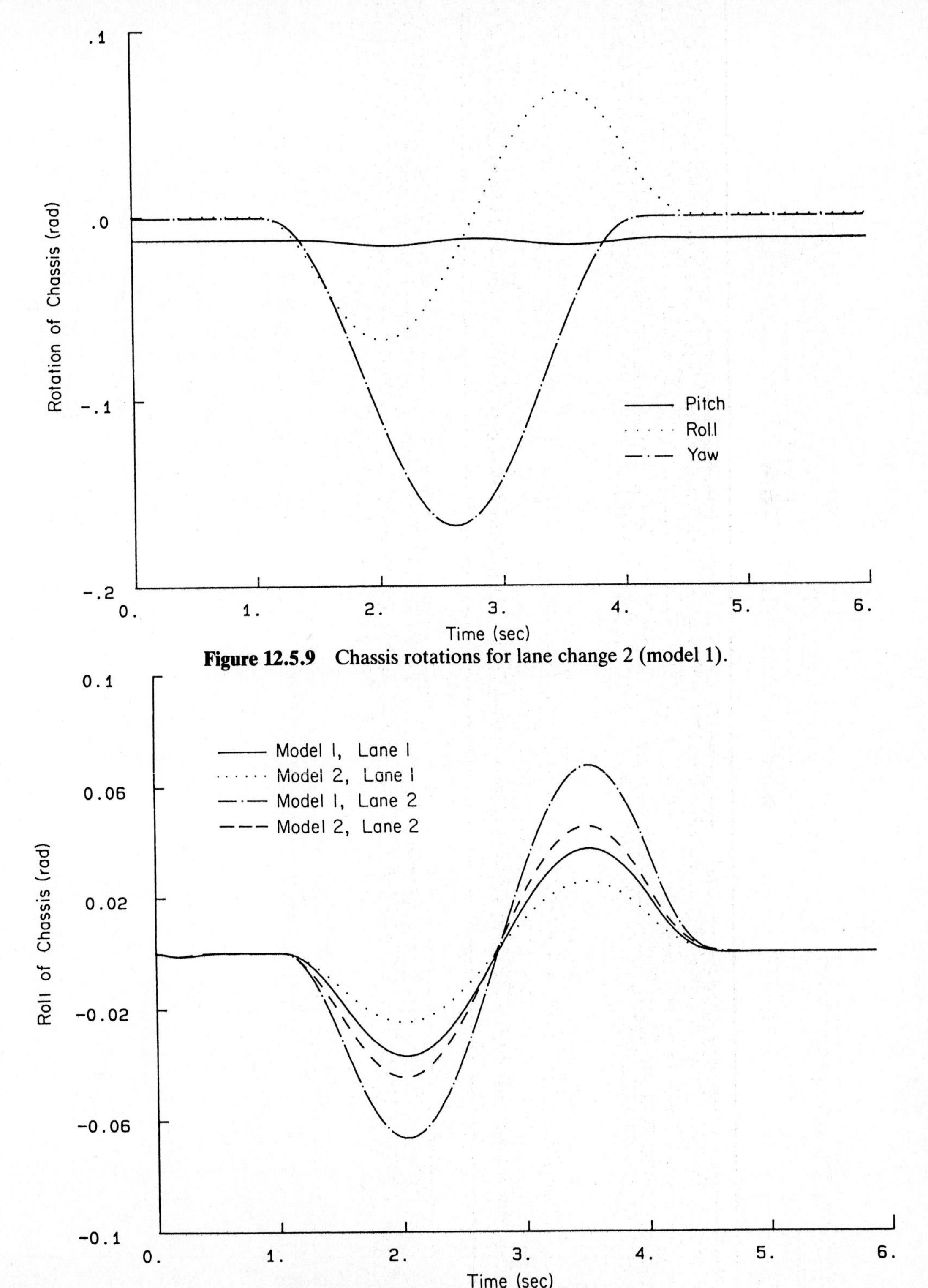

Figure 12.5.9 Chassis rotations for lane change 2 (model 1).

Figure 12.5.10 Chassis roll in lane changes.

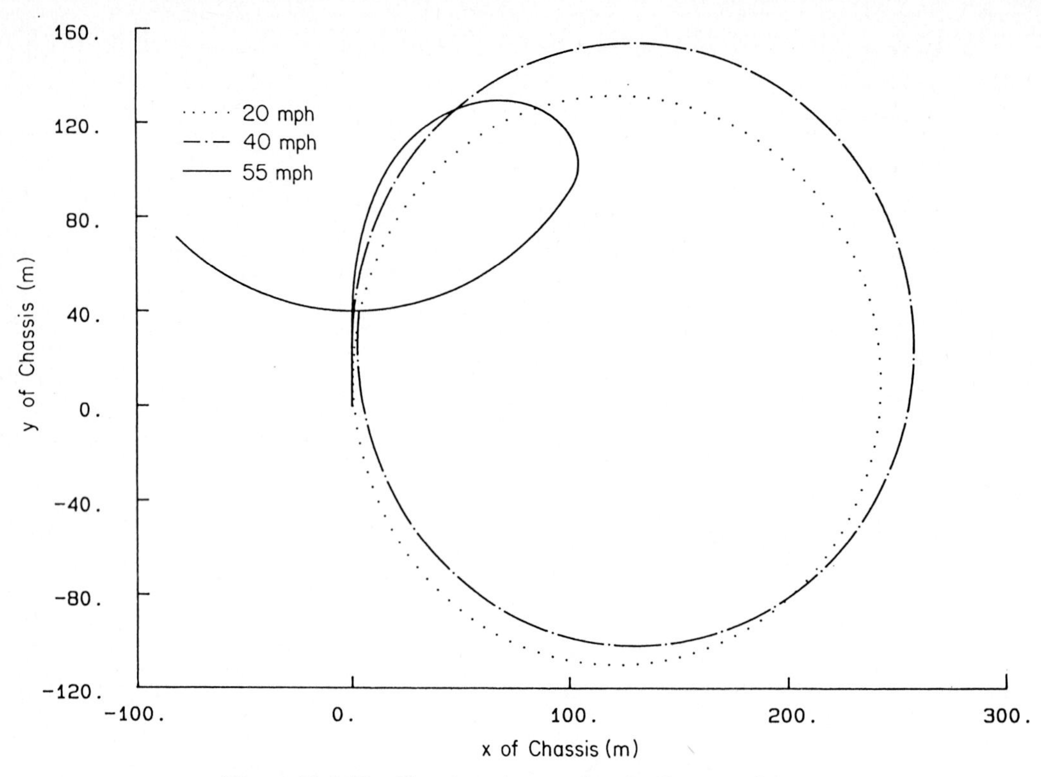

Figure 12.5.11 Chassis trajectory for circular steer input.

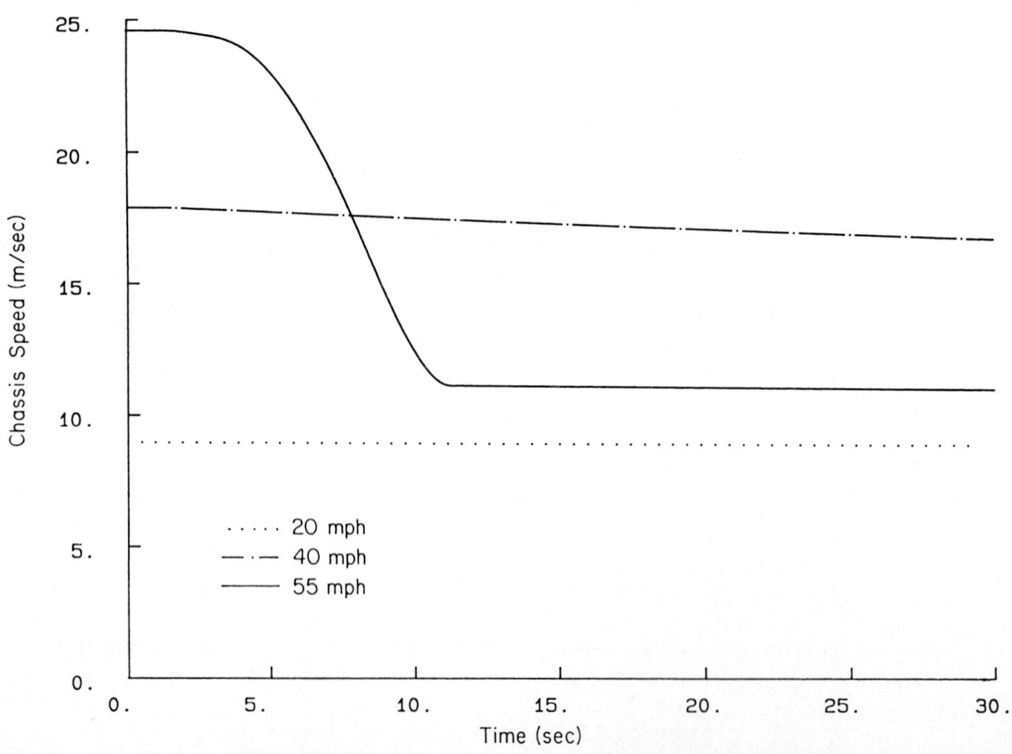

Figure 12.5.12 Chassis speed for cicular steer input.

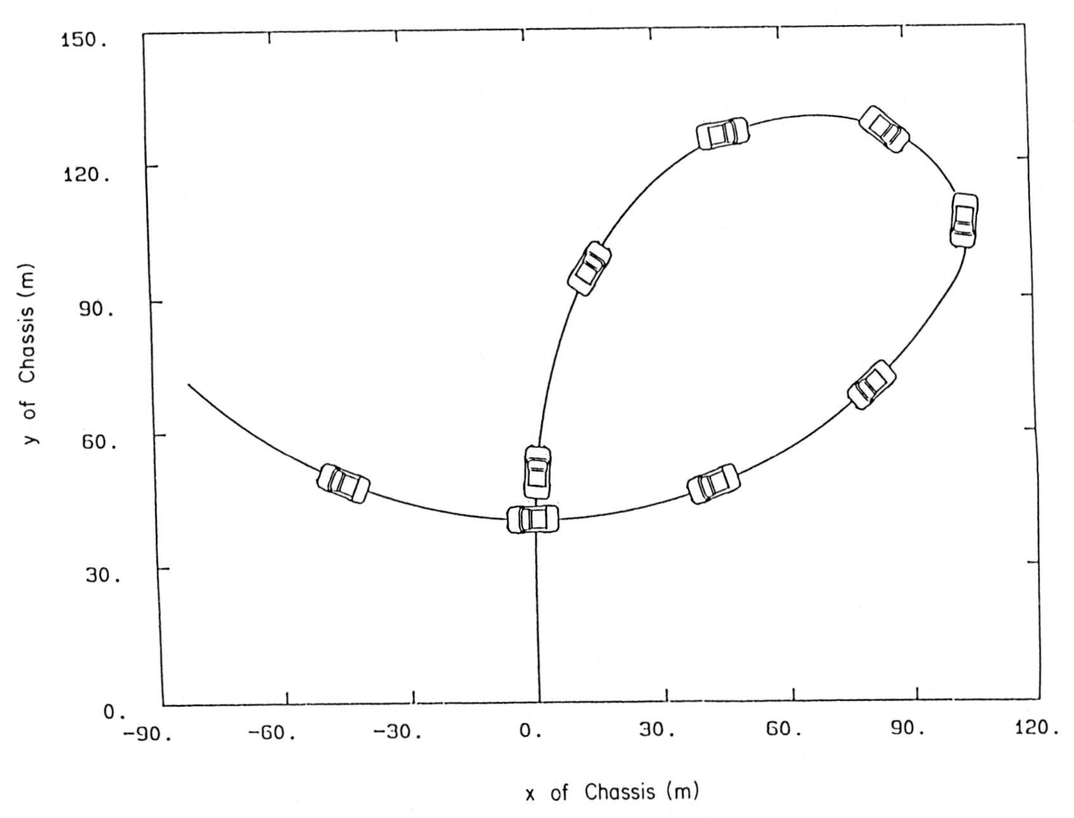

Figure 12.5.13 Vehicle orientation in circular steer maneuver (55 mph).

the speeds of the vehicle in the three different simulations. As expected, considerable speed loss (reduction in kinetic energy) occurs in the 55 mph case, since dissipative friction forces act while the vehicle is slipping laterally.

Figure 12.5.13 shows that the vehicle is not tangent to the path during the maximum slip period in the $\dot{y}_0 = 55$ mph simulation.

12.6 DYNAMIC ANALYSIS OF A GOVERNOR MECHANISM

12.6.1 Alternative Models

The governor mechanism shown in Fig. 12.6.1 is composed of five bodies, including ground (body 1). Bodies 2 to 5 are shaft, ball 1, ball 2, and collar, respectively. The arms attached to the balls are connected to the shaft by revolute joints whose axes are parallel, both perpendicular to the shaft. The

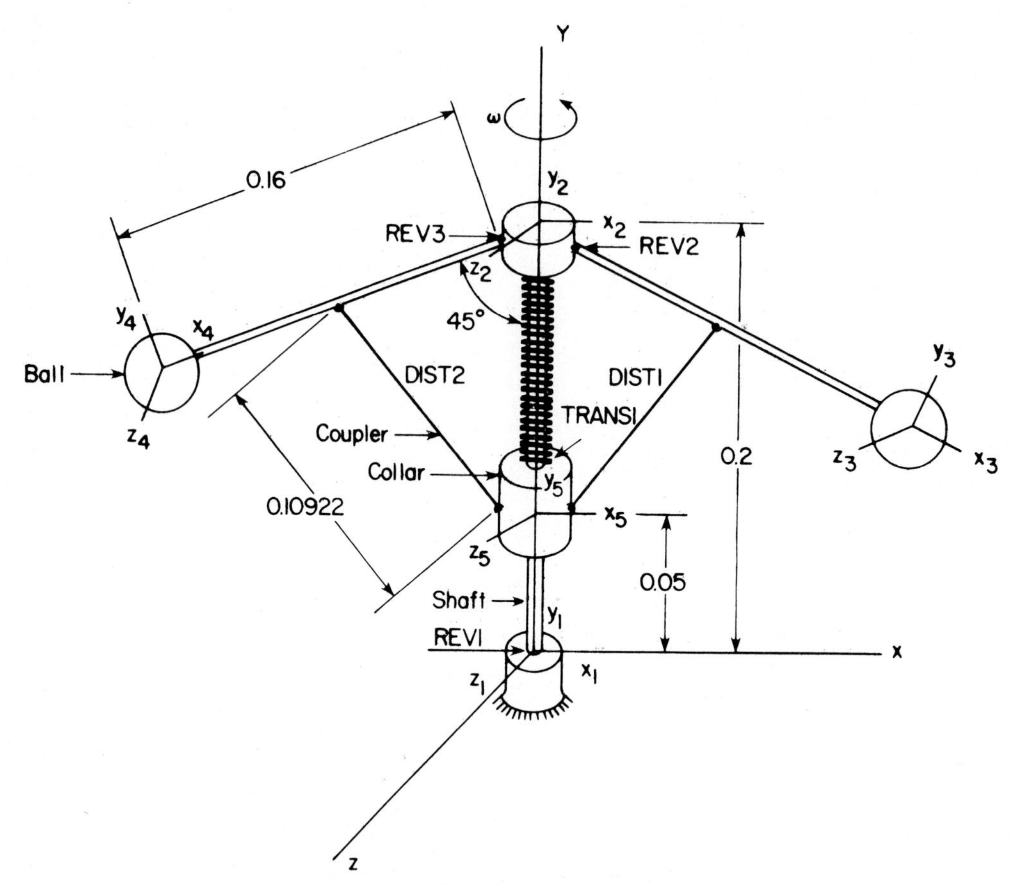

Figure 12.6.1 Governor mechanism.

collar is connected to the shaft by a translational joint. Couplers that connect the ball arms and collar are modeled as distance constraints. A TSDA element is attached between the shaft and collar. The inertia properties of the components of the system are defined in Table 12.6.1.

The intended function of the governor is to maintain a nearly constant

TABLE 12.6.1 Inertia Properties of Governor Mechanism

Body		Mass	$I_{x'}$	$I_{y'}$	$I_{z'}$	$I_{x'y'}$	$I_{y'z'}$	$I_{x'z'}$
Ground	①	1.0	1.0	1.0	1.0	0.0	0.0	0.0
Spindle	②	200.0	25.0	50.0	25.0	0.0	0.0	0.0
Ball 1	③	1.0	0.1	0.1	0.1	0.0	0.0	0.0
Ball 2	④	1.0	0.1	0.1	0.1	0.0	0.0	0.0
Collar	⑤	1.0	0.15	0.125	0.15	0.0	0.0	0.0

angular speed ω of the shaft under a varying resisting torque from the machine being driven. If the angular velocity of the shaft decreases, the balls drop and hence lower the collar. A linkage attached to the collar then opens the fuel feed to the engine, which generates an increased torque and leads to a speed-up of the shaft. As a result, the balls rise toward their nominal position as the shaft angular velocity approaches the desired value.

The elements of the kinematic model are as follows:

Bodies	
Five bodies	$nc = 35$
Constraints	
Distance constraint: DIST1	1
Distance constraint: DIST2	1
Revolute joint: REV1	5
Revolute joint: REV2	5
Revolute joint: REV3	5
Translational joint: TRANS1	5
Ground constraint	6
Normalization constraint	5
Euler parameter	
	$nh = 33$

$\text{DOF} = 35 - 33 = 2.$

Data for the distance constraints, revolute joints, and translational joint are tabulated in Tables 12.6.2 to 12.6.4. Data for the TSDA element are given in Table 12.6.5. Three different models (1, 2, and 3) are distinguished, with spring constants $k = 1000$, 2000, and 3000 N/m, respectively.

TABLE 12.6.2 Data for Distance Constraints

DIST1

Body \ Point	*P* x'	*P* y'	*P* z'	*Q* x'	*Q* y'	*Q* z'	*R* x'	*R* y'	*R* z'
Ball 1 ③	−0.08	0.0	0.0	−0.08	0.0	1.0	1.08	0.0	0.0
Collar ⑤	0.0	0.0	0.0	0.0	0.0	1.0	1.0	0.0	0.0

Distance = 0.10922

DIST2

Body \ Point	*P* x'	*P* y'	*P* z'	*Q* x'	*Q* y'	*Q* z'	*R* x'	*R* y'	*R* z
Ball 2 ④	0.08	0.0	0.0	0.08	0.0	1.0	1.08	0.0	0.0
Collar ⑤	0.0	0.0	0.0	0.0	0.0	1.0	1.0	0.0	0.0

Distance = 0.10922

TABLE 12.6.3 Data for Revolute Joints

REV1

Body \ Point	P x'	P y'	P z'	Q x'	Q y'	Q z'	R x'	R y'	R z'
Ground ①	0.0	0.0	0.0	0.0	1.0	0.0	0.0	0.0	1.0
Spindle ②	0.0	−0.2	0.0	0.0	−1.2	0.0	0.0	−0.2	1.0

REV2

Body \ Point	P x'	P y'	P z'	Q x'	Q y'	Q z'	R x'	R y'	R z'
Spindle ②	0.0	0.0	0.0	0.0	0.0	1.0	1.0	0.0	0.0
Ball 1 ③	−0.16	0.0	0.0	−0.16	0.0	1.0	1.0	0.0	0.0

REV3

Body \ Point	P x'	P y'	P z'	Q x'	Q y'	Q z'	R x'	R y'	R z'
Spindle ②	0.0	0.0	0.0	0.0	0.0	1.0	1.0	0.0	0.0
Ball 2 ④	0.16	0.0	0.0	0.16	0.0	1.0	1.0	0.0	0.0

TABLE 12.6.4 Data for Translational Joint

Body \ Point	P x'	P y'	P z'	Q x'	Q y'	Q z'	R x'	R y'	R z'
Spindle ②	0.0	0.0	0.0	0.0	1.0	0.0	0.0	0.0	1.0
Collar ⑤	0.0	0.0	0.0	0.0	1.0	0.0	0.0	0.0	1.0

TABLE 12.6.5 Data for Translational Spring–Damper–Actuator

Body \ Point	P x'	P y'	P z'	Q x'	Q y'	Q z'	R x'	R y'	R z'
Spindle ②	0.0	0.0	0.0	0.0	0.0	1.0	1.0	0.0	0.0
Collar ⑤	0.0	0.0	0.0	0.0	0.0	1.0	1.0	0.0	0.0

Spring constant, $k = 1000,\ 2000,\ 3000$ N/m
Damping rate, $c = 30$ kg/s
Free length of spring, $L_0 = 0.15$ m

12.6.2 Steady-State Analysis

Based on the configuration of Fig. 12.6.1, steady-state motion is determined by selecting the desired angular speed ω of the spindle. Using dynamic force balancing and $\omega = 11.0174$ rad/s, at a slope of 45° of the ball arms, the collar is to be stationary. Table 12.6.6 defines the steady-state configuration of the system.

TABLE 12.6.6 Data for Steady-State Configuration

Body	x	y	z	e_0	e_1	e_2	e_3
Ground ①	0.0	0.0	0.0	1.0	0.0	0.0	0.0
Spindle ②	0.0	0.2	0.0	1.0	0.0	0.0	0.0
Ball 1 ③	0.11314	0.08686	0.0	0.9239	0.0	0.0	0.3827
Ball 2 ④	−0.11314	0.08686	0.0	0.9239	0.0	0.0	0.3827
Collar ⑤	0.0	0.05	0.0	1.0	0.0	0.0	0.0

12.6.3 External Torque

An external torque T_e due to the load driven by the shaft is applied to the spindle, as shown in Fig. 12.6.2. To compensate for this torque, which tends to reduce shaft speed, the torque T_s applied to the spindle by the engine as a result of fuel fed by a collar height variation $\Delta\ell$ is modeled as

$$T_s = C \times \Delta\ell \tag{12.6.1}$$

where C is the torque generated due to fuel fed to the engine by a unit $\Delta\ell$, which is spring deformation (vertical movement of collar). Three different values $C = 7500$, 12,500, and 17,500, corresponding to increasing engine power, are used to study the system dynamic response.

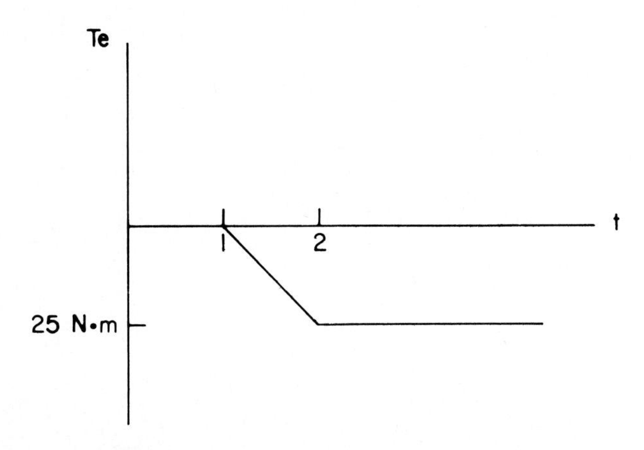

Figure 12.6.2 External torque on spindle.

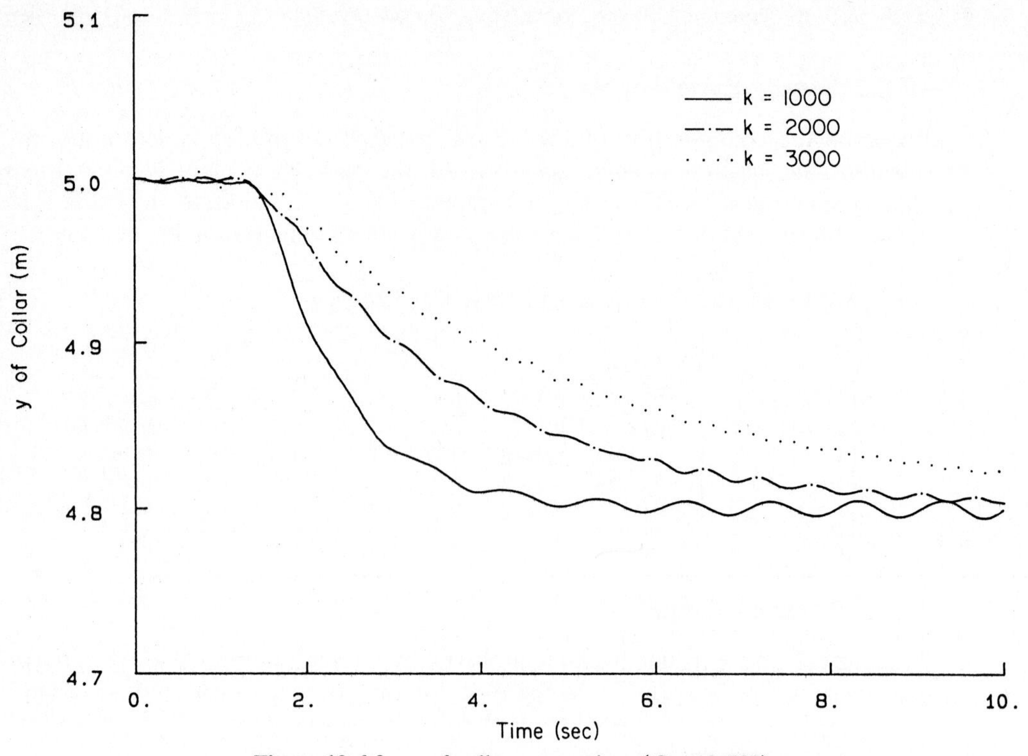

Figure 12.6.3 y of collar versus time ($C = 12{,}500$).

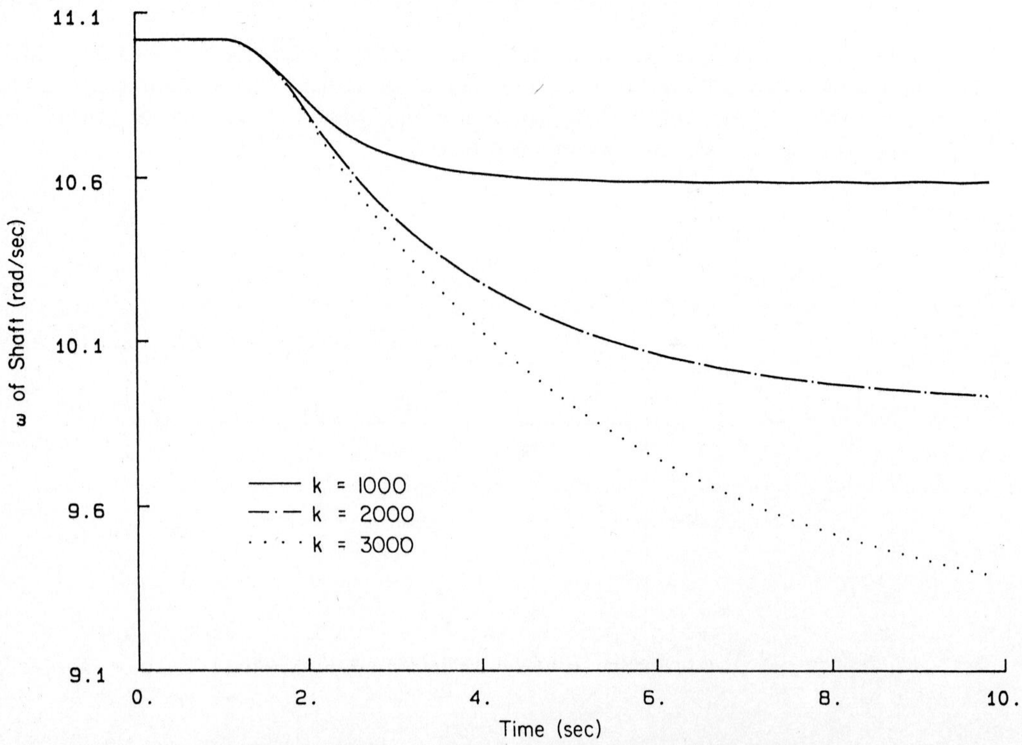

Figure 12.6.4 ω of shaft versus time ($C = 12{,}500$).

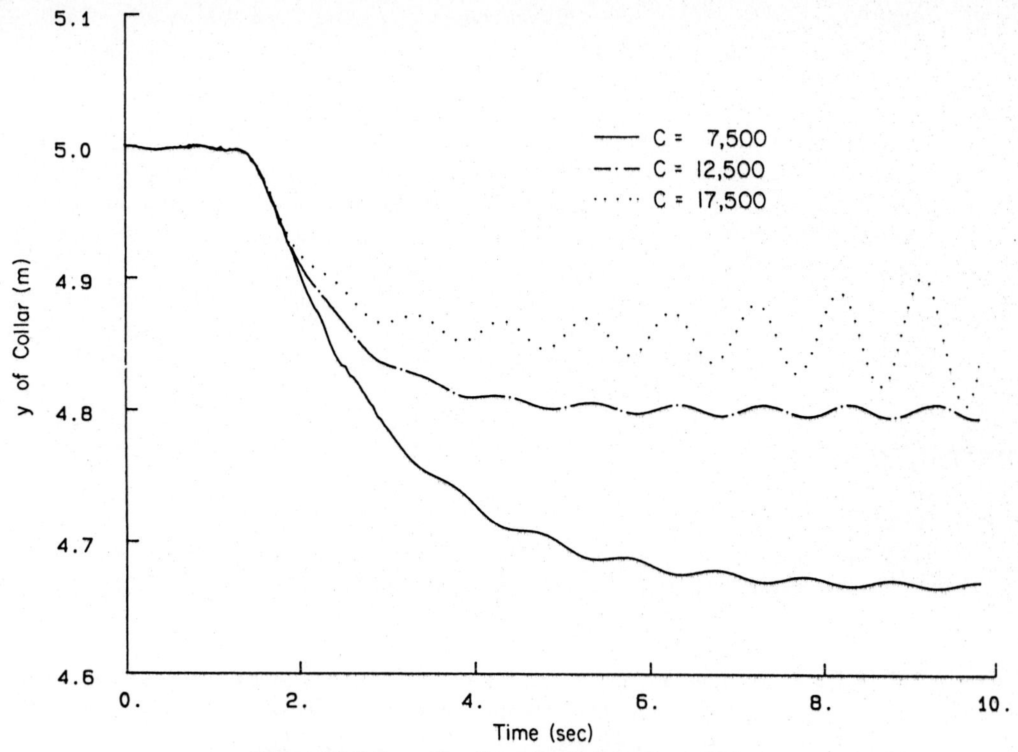

Figure 12.6.5 y of collar versus time ($k = 1000$).

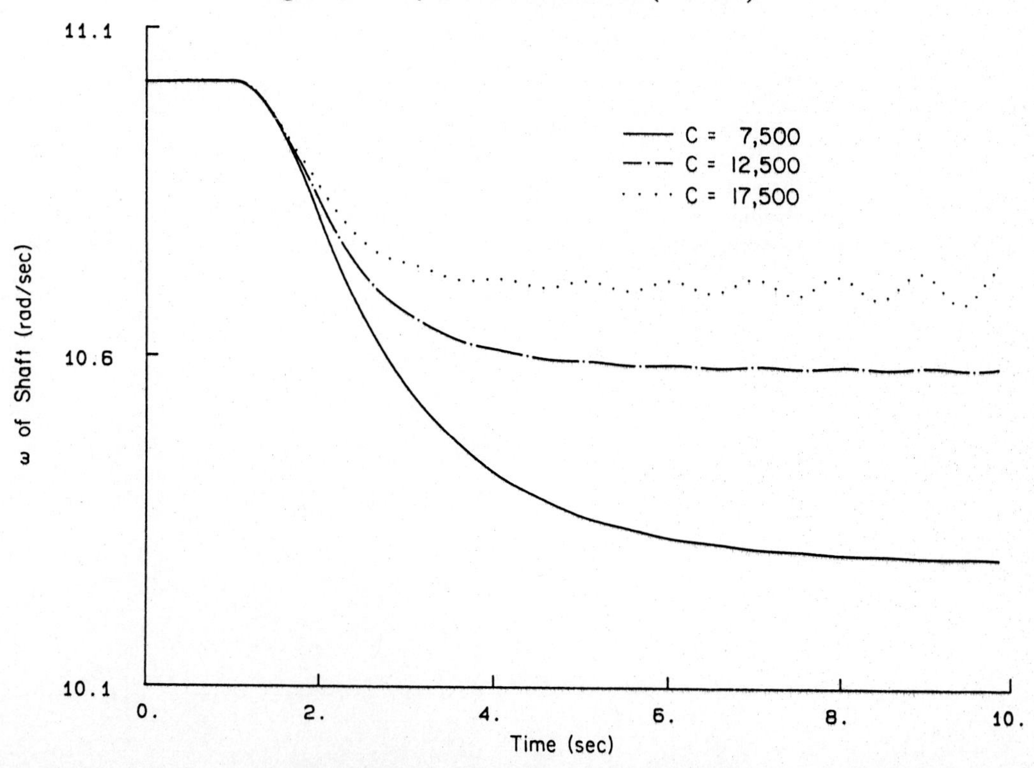

Figure 12.6.6 ω of shaft versus time ($k = 1000$).

12.6.4 Dynamic Analysis

Starting with the given steady-state configuration and $C = 12{,}500$, dynamic simulations are performed for three different values of the spring rate k. Figures 12.6.3 and 12.6.4 show variations in the vertical position of the collar and angular speed of the shaft. Note that the stiffer the spring, the longer the transition time from the initial steady state to another steady state.

Dynamic responses are plotted in Figs. 12.6.5 and 12.6.6 for different torque feedback rates, with $k = 1000$. As the feedback rate becomes larger (more powerful engine), the system may become unstable.

PROBLEMS

DADS Projects

12.1 Using the three body model of DADS Project 10.1, repeat the dynamic analysis of the spatial slider–crank mechanism presented in Section 12.2. Analyze the differences in required driving torque and dynamic response from those presented in Sections 12.2.2 and 12.2.3, due to the elimination of connecting rod inertias.

12.2 Using the data in Section 12.4 and the following inertial properties of each connecting rod, carry out a dynamic analysis of the fifteen-body air compressor model of DADS Project 10.2

Mass	$I_{x'x'}$	$I_{y'y'}$	$I_{z'z'}$	$I_{x'y'}$	$I_{y'z'}$	$I_{z'x'}$
0.5	0.04	0.04	0.001	0.0	0.0	0.0

where the center of mass of the connecting rod is at its midpoint and its z' axis is along the rod.

Evaluate the differences in driving torque, reaction forces, and dynamic response from those presented in Section 12.4, due to the inertias of the connecting rods.

References

1. Beyer, R., *The Kinematic Synthesis of Mechanisms,* McGraw-Hill, New York, 1963.
2. Hirschhorn, J., *Kinematics and Dynamics of Plane Mechanisms,* McGraw-Hill, New York, 1962.
3. Paul, B., *Kinematics and Dynamics of Planar Machinery,* Prentice-Hall, Englewood Cliffs, N.J., 1979.
4. Wittenburg, J., *Dynamics of Systems of Rigid Bodies,* Teubner, Stuttgart, 1977.
5. Soni, A. H., *Mechanism Synthesis and Analysis,* McGraw-Hill, New York, 1974.
6. Suh, C. H., and Radcliffe, C. W., *Kinematics and Mechanisms Design,* Wiley, New York, 1978.
7. Greenwood, D. T., *Principles of Dynamics,* 2nd ed., Prentice-Hall, Englewood Cliffs, N.J., 1988.
8. Kane, T. R., and Levinson, D. A., *Dynamics: Theory and Applications,* McGraw-Hill, New York, 1985.
9. Goldstein, H., *Classical Mechanics,* 2nd ed., Addison-Wesley, Reading, Mass., 1980.
10. Noble, B., and Hussain, M. A., "Applications of MACSYMA to Calculations in Dynamics," *Computer Aided Analysis and Optimization of Mechanical System Dynamics* (ed. E. J. Haug), Springer-Verlag, Heidelberg, 1984.
11. Zienkiewiez, O., *The Finite Element Method,* McGraw-Hill, New York, 1977.
12. Gallagher, R. H., *Finite Element Analysis: Fundamentals,* Prentice-Hall, Englewood Cliffs, N.J., 1975.
13. Chua, L. O., and Lin, P.-M., *Computer Aided Analysis of Electronic Circuits,* Prentice-Hall, Englewood Cliffs, N.J., 1975.
14. Calahan, D. A., *Computer Aided Network Design,* McGraw-Hill, New York, 1972.
15. Paul, B., and Krajcinovic, D., "Computer Analysis of Machines with Planar Motion—Part I: Kinematics; Part II: Dynamics," *Journal of Applied Mechanics,* Vol. 37, pp. 697–712, 1970.
16. Chace, M. A., and Smith, D. A., "DAMN—A Digital Computer Program for the Dynamic Analysis of Generalized Mechanical Systems," SAE paper 710244, January 1971.
17. Sheth, P. N., and Uicker, J. J., Jr., "IMP (Integrated Mechanisms Program), A Computer Aided Design Analysis System for Mechanisms and Linkages," *Journal of Engineering for Industry,* Vol. 94, pp. 454–464, 1972.
18. Orlandea, N., Chace, M. A., and Calahan, D. A., "A Sparsity-Oriented Approach to the Dynamic Analysis and Design of Mechanical Systems, Parts I and II," *Journal of Engineering for Industry,* Vol. 99, pp. 773–784, 1977.

19. Wehage, R. A., and Haug, E. J., "Generalized Coordinate Partitioning for Dimension Reduction in Analysis of Constrained Dynamic Systems," *Journal of Mechanical Design,* Vol. 104, No. 1, pp. 247–255, 1982.
20. Chung, I. S., and others, *Dynamic Analysis of Three Dimensional Constrained Mechanical Systems Using Euler Parameters,* Technical Report No. 81-11, Center for Computer Aided Design, College of Engineering, University of Iowa, Iowa City, Iowa, October, 1981.
21. Davis, H. F., *Introduction to Vector Analysis,* 4th ed., Allyn & Bacon, Newton, Mass., 1979.
22. Strang, G., *Linear Algebra and Its Applications,* 2nd ed., Academic Press, New York 1980.
23. Erdman, A. G., and Sandor, G. N., *Advanced Mechanism Design: Analysis and Synthesis,* Vols. I and II, Prentice-Hall, Englewood Cliffs, N.J., 1984.
24. de Boor, C., *A Practical Guide to Splines,* Springer-Verlag, New York, 1978.
25. Goffman, C., *Calculus of Several Variables,* Harper & Row, New York, 1965.
26. Corwin, L. J., and Szczarba, R. H., *Multivariable Calculus,* Marcel Dekker, New York, 1982.
27. *DADS User's Manual,* Computer Aided Design Software Incorporated, P.O. Box 203, Oakdale, Iowa, 52319.
28. Haug, E. J., and Arora, J. S., *Applied Optimal Design,* Wiley-Interscience, New York, 1979.
29. Fletcher, R., and Powell, M. J. D., "A Rapidly Convergent Descent Method for Minimization," *Computer Journal,* Vol. 6, pp. 163–180, 1963.
30. Hildebrand, F. B., *Introduction to Numerical Analysis,* McGraw-Hill, New York, 1956.
31. Atkinson, K. E., *An Introduction to Numerical Analysis,* Wiley, New York, 1978.
32. Truesdell, C. A., *Essays in the History of Mechanics,* Springer-Verlag, New York, 1968.
33. Fiacco, A. V., and McCormick, G. P., *Nonlinear Programming: Sequential Unconstrained Minimization Techniques,* Wiley, New York, 1968.
34. Langhaar, H. L., *Energy Methods in Applied Mechanics,* Wiley, New York, 1962.
35. Haug, E. J., *Intermediate Dynamics,* Allyn & Bacon, Newton, Mass., 1989.
36. Shampine, L. F., and Gordon, M. K., *Computer Solution of Ordinary Differential Equations: The Initial Value Problem,* Freeman, San Francisco, 1975.
37. Gear, C. W., *Numerical Initial Value Problems in Ordinary Differential Equations,* Prentice-Hall, Englewood Cliffs, N.J., 1971.
38. Enright, W. H., "Numerical Methods for Systems of Initial Value Problems—The State of the Art," *Computer Aided Analysis and Optimization of Mechanical System Dynamics* (ed. E. J. Haug), Springer-Verlag, Heidelberg, 1984, pp. 309–322.
39. Petzold, L. D., "Differential/Algebraic Equations Are Not ODE's," *SIAM Journal of Scientific and Statistical Computing,* Vol. 3, No. 3, pp. 367–384, 1982.
40. Baumgarte, J., "Stabilization of Constraints and Integrals of Motion," *Computer Methods in Applied Mechanics and Engineering,* Vol. 1, pp. 1–16, 1972.
41. Mani, N. K., and Haug, E. J., "Singular Value Decomposition for Dynamic System Design Sensitivity Analysis," *Engineering with Computers,* Vol. 1, pp. 103–109, 1985.
42. Park, T., "A Hybrid Constraint Stabilization—Generalized Coordinate Partitioning Method for Machine Dynamics," *Journal of Mechanisms, Transmissions, and Automation in Design,* Vol. 108, No. 2, 1986, pp. 211–216.

43. Lotstedt, P., and Petzold, R., "Numerical Solution of Nonlinear Differential Equations with Algebraic Constraints," Sandia Report SAND 83-8877, Sandia National Laboratories, Albuquerque, N.M. 87185, 1983.
44. Coddington, E. A., and Levinson, N., *Theory of Ordinary Differential Equations,* McGraw-Hill, New York, 1955.
45. Park, T., Haug, E. J., and Yim H. J., "Automated Mechanism and Machine Theory, Kinematic Feasibility Evaluation and Analysis of Mechanical Systems," *Mechanism and Machine Theory,* 1988.
46. Duff, I. S., Erisman, A. M., and Reid, J. K., *Direct Methods for Sparse Matrices,* Oxford University Press, New York, 1986.
47. Duff, I. S., *MA28-A Set of Fortran Subroutines for Sparse Unsymmetric Linear Equations,* AIRE Harwell, Didcot, Oxon, U.K., 1977.
48. Kwon, O. K., and Haug, E. J., *An Index One Formulation of Differential–Algebraic Equations for Mechanical System Dynamics,* Center for Computer Aided Design, Technical Report 87-9, University of Iowa, Iowa City, Iowa, 1987.

Symbol Index

Subject Index